Limits

Properties of Limits as $x \to a$

Suppose that c is any constant, n is a positive integ [illegible] L_1, and $\lim_{x \to a} g(x) = L_2$, where L_1 and L_2 are real numbers.

1. $\lim_{x \to a} c = c$

2. $\lim_{x \to a} x = a$

3. $\lim_{x \to a} x^n = a^n$

4. $\lim_{x \to a} \left[f(x) \pm g(x) \right] = \lim_{x \to a} f(x) \pm \lim_{x \to a} g(x) = L_1 \pm L_2$

5. $\lim_{x \to a} \left[f(x) \cdot g(x) \right] = \left[\lim_{x \to a} f(x) \right] \cdot \left[\lim_{x \to a} g(x) \right] = L_1 \cdot L_2$

6. $\lim_{x \to a} \left[\frac{f(x)}{g(x)} \right] = \frac{\lim_{x \to a} f(x)}{\lim_{x \to a} g(x)} = \frac{L_1}{L_2}$ where $L_2 \neq 0$

7. $\lim_{x \to a} \left[c \cdot f(x) \right] = c \cdot \lim_{x \to a} f(x) = c \cdot L_1$

8. $\lim_{x \to a} \sqrt[n]{f(x)} = \sqrt[n]{\lim_{x \to a} f(x)} = \sqrt[n]{L_1} = (L_1)^{1/n}$ where $(L_1)^{1/n}$ is defined.

9. If p is a polynomial function, then $\lim_{x \to a} p(x) = p(a)$.

Summary of Limits for Rational Functions as $x \to \pm\infty$

Consider the function $\frac{a_n x^n + a_{n-1} x^{n-1} + \ldots + a_0}{b_m x^m + b_{m-1} x^{m-1} + \ldots + b_0}$, where $a_n \neq 0$ and $b_m \neq 0$.

Case 1: For $m = n$, $\lim_{x \to \pm\infty} f(x) = \frac{a_n}{b_m}$.

Case 2: For $m > n$, $\lim_{x \to \pm\infty} f(x) = 0$.

Case 3: For $m < n$, $\lim_{x \to +\infty} f(x) = +\infty$.

(Or $-\infty$ depending on the signs of a_n and b_m.)

Essential Calculus
with Applications

D. Franklin Wright
Spencer P. Hurd
Bill D. New

Editor: Phillip Bushkar
Developmental Editor: Marcel Prevuznak
Preproduction Editor: Jennifer Knowles Butler
Production Editors: Harding Brumby, Cynthia Ellison, Mandy Glover, Larry Wadsworth, Jr.
Editorial Assistants: Kimberly Cumbie, Kelly Epperson, Ben Hamner, D. Kanthi, Bethany Loftis, Nina Miller, B. Syam Prasad, Susan Rackley, Eric Wilder

Layout: QSI (Pvt.) Ltd.; U. Nagesh, E. Jeevan Kumar
Art: Ayvin Samonte

A division of Quant Systems, Inc.

Library of Congress Control Number: 2005937101

Printed in the United States of America

ISBN:

Student Hardcover:	Student Bundle
978-0-918091-95-6	978-0-918091-93-2

CONTENTS

PREFACE TO THE SECOND EDITION

Purpose and Style

Essential Calculus with Applications (second edition) provides a solid introductory base for the mastery of basic calculus. In particular, business and social science majors will be well prepared for the analysis and evaluation of function-based analytical materials in their fields. MBA and economics students will find a wealth of examples which will serve to introduce topics studied in more advanced courses.

With feedback from helpful reviewers and three years of classroom experience, this edition has many special features which will be of use to faculty in forming a base course in elementary differential and integral calculus.

1. First, the superb layout from Hawkes Publishing eliminates the clutter which dominates the pages of many math books and which intimidates most students.
2. The algebra necessary to solve typical problems is fully illustrated in the examples in each section, but an instructor will easily be able to establish parameters for his or her students with respect to calculator usage. Indeed, the text is written with the assumption that a student will have a calculator or graphing utility available. Of course, NCTM standards for high school algebra assume a graphical utility will be used, and NCTM/NCATE standards for (math) teacher education require expertise in technology for teacher candidates.
3. The use of the Hawkes Publishing software which accompanies the text supports the learning of mathematics in ways measurably superior to courses without such assistance, and the publisher will be happy to provide details of studies which support this assertion.
4. Most topics are presented numerically or geometrically and only then algebraically. For example, (one-sided and two-sided) limits are presented informally, numerically, and without epsilon-delta arguments. Also, the derivative of $y = \ln x$ in Section 5.3 is obtained numerically first and only then is an algebra derivation shown.
5. It will be easy for an instructor to emphasize certain topics because each topic has a great wealth of homework exercises.
6. There is sufficient material for a second semester of calculus which is often required for MBA students, officer candidates (ROTC), advanced placement students, and science or social science majors.

The style is informal but we have worked to keep mathematical accuracy, and the frequent text boxes which emphasize selected formulas and statements will be of great use to the student for (self) study and review.

Applications of the Derivative

CHAPTER 4

Did You Know?

In the late 1500's two remarkable astronomers came to the forefront of the scientific community, Tycho Brahe (1546 – 1601) and Galileo Galilei (1564 – 1642). Brahe was the first astronomer to take detailed records of the movements of planetary objects using his invention which was the predecessor of the sextant (his successor, Kepler, used these meticulous records to formulate his laws of planetary motion). Galileo was the first to study far-reaching parts of the galaxy with his improvements upon the telescope (he claimed to have seen mountains on the moon, and to have proved that the Milky Way was made up of tiny stars). These men not only made vital contributions to the scientific community, revolutionizing astronomical measurement and observation instruments, but were also two of history's most colorful scientists, passing many unique and intriguing stories on to posterity.

The popular story associated with Brahe has to do with his nose. Brahe lost his nose in what is said to have been a duel with a fellow student over who was the better mathematician! From then on he wore a prosthetic nose made of a gold and silver alloy.

Galileo, on learning of an instrument called the telescope, set out to build his own. After successfully building a replica of the reported telescope, he went on to make great improvements to the instrument in future revisions, including increasing the known magnification (which was 4 times) to around 9 times! He then sold the exclusive rights of manufacture of this powerful device (difficult to guarantee even today) to the Venetian senate, effectively pirating another man's invention and swindling the government in one business deal. The senate later froze his salary, finally realizing what Galileo had done.

The lenses in a telescope are called convex lenses. Convex lenses are thicker at the center than at the edges, while concave lenses are thicker at the edges than at the center. Both convex and concave lenses have concavity in the mathematical sense of the word. If one were to draw out the "graph" of these lenses, one would see that they both form a parabola.

These two famous astronomers and their colleague Johannes Kepler (1571-1630) were mathematicians first and used ideas usually thought of as part of calculus in their calculations. Formal trigonometry (Chapter 9) and the new logarithms (Chapter 5) increased the value of mathematics to the newly emerging sciences of the Renaissance period. When the calculus of Newton appeared (1668-1700), it was exactly what was needed to accelerate their developments.

Did You Know?:

A feature at the beginning of every chapter presents some interesting math history related to the chapter at hand.

Introduction:

Presented before the first section of every chapter, this feature provides an introduction to the subject of the chapter and its purpose.

In this chapter, we point out that f' is a function which has its own derivative called the second derivative of f. This second derivative is used to describe and understand $f(x)$ itself. We also develop the "Second Derivative Test" and show how both the first derivative and second derivative can be used in curve sketching. Higher order derivatives are discussed but are not applied here. The ideas in this chapter on sketching are mainly applied to polynomial functions; however, the student should be aware that the ideas presented apply to sketching the graphs of most types of functions.

While we have discussed applications throughout the first three chapters, the emphasis has been on developing rules and techniques for differentiation. Now we focus on the real-world applications of these techniques and show how calculus can be used to maximize profit, minimize inventory cost, reduce time or distance of travel, or, in general, to optimize a function.

The last section on differentials shows how the notation dy and dx can be used to indicate small changes in the variables y and x, respectively. Although the applications of differentials are somewhat limited, understanding the idea of differentials and being able to use corresponding notation will prove valuable when integration is described in Chapters 6 and 7.

4.1 Higher-Order Derivatives: Concavity and the Second Derivative Test

Objectives

After completing this section, you will be able to:

1. *Find the second derivative of a function.*
2. *Determine the concavity of a function.*
3. *Locate the points of inflection of a function by hand and by using a graphing calculator.*

Objectives:

The objectives provide the students with a clear and concise list of skills presented in each section.

Higher-Order Derivatives

Given a function f, we have defined and discussed the derivative f', called the **first derivative**. We have seen that there may be x-values for which f is defined but for which the derivative does not exist. However, in general, the functions defined and applied in this book have slope defined at most points for which the function is defined. This means that f' is itself a function which has its own derivative defined at most points in its domain, and it too can be differentiated. The **second derivative**, f'', if it exists, is the derivative of the first derivative. The **third derivative**, f''' (or $f^{(3)}$), if it exists, is the derivative of the second derivative, and so on.

SECTION 6.2 Integration by Substitution

The technique used in Example 2, multiplying by 1 in the form $\frac{1}{5}\cdot 5$ and then factoring out $\frac{1}{5}$ from the integral, is valid because $\frac{1}{5}$ is a constant. The same technique is not valid with variables. For example, inserting $\frac{1}{x}\cdot x$ is valid, but then factoring out $\frac{1}{x}$ from the integral is not valid.

Example 3: Integration by Substitution with e

Find $\int 7xe^{x^2+3}dx$.

Solution: Let $u = x^2 + 3$. Then $du = 2x\,dx$.
We adjust for the 2 by introducing $\frac{1}{2}\cdot 2$ into the integrand.

$\int 7xe^{x^2+3}dx = 7\int e^{x^2+3}\cdot\frac{1}{2}\cdot 2x\,dx$ — Use the Constant Multiple Rule with 7 and $\frac{1}{2}\cdot 2 = 1$ since we need $2x\,dx$ for du.

$= \frac{7}{2}\int e^u du$ — Substitute and use the Constant Multiple Rule with $\frac{7}{2}$.

$= \frac{7}{2}e^u + C$ — By Formula IV.

$= \frac{7}{2}e^{x^2+3} + C$

Example 4: Integration by Substitution

Find $\int \frac{x^2}{(x^3-1)^5}dx$.

Solution: Choose $u = x^3 - 1$. Then $du = 3x^2\,dx$.

$\int \frac{x^2}{(x^3-1)^5}dx = \int\left(\frac{\frac{1}{3}\cdot 3x^2 dx}{(x^3-1)^5}\right)$ Mult get 3

Notes:

Notes highlight common mistakes and lend additional clarity to more subtle details.

Definition Boxes:

Definitions are presented in highly visible boxes for easy reference.

Examples:

Examples are denoted with titled headers indicating the problem solving skill being presented. Each section contains many carefully explained examples complete with tables, diagrams, and graphs. Examples are presented in an easy to understand step-by-step fashion and annotated for additional clarity.

SECTION 1.6 Functions

Function, Domain, and Range

*Let D and R be two sets of real numbers. A **function** f is a rule that matches each number x in D with exactly one number y (or f(x)) in R. D is called the **domain** of f, and R is called the **range** of f.*

Note: In a function, there is only one y-value (or f(x)-value) for each value of x.

Intuitively, a function can be thought of as an input-output machine (or a magic box). Values of x are input (such as numbers of books), and values of y [or $f(x)$] are output (such as dollars representing costs). For our purposes, we can think of the machine as an algebraic expression, such as $5x + 1000$. (See Figure 1.6.1.)

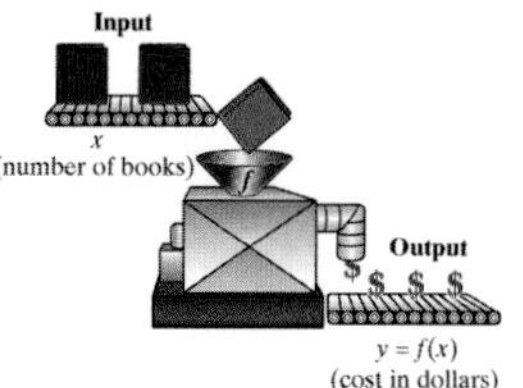

Figure 1.6.1 "A Function Machine"

We say that x is the **independent variable** and that y [or $f(x)$] is the **dependent variable** because we choose the value of x but the value of y depends on our independent choice.

Example 1: Function Evaluation

For the familiar function $f(x) = x^2$, evaluate the following:

a. $f(3)$ — Read "f of 3."
Solution: $f(3) = 3^2 = 9$ — 3 is substituted for x.

b. $f(5)$
Solution: $f(5) = 5^2 = 25$ — 5 is substituted for x.

c. $f(-2)$
Solution: $f(-2) = (-2)^2 = 4$ — -2 is substituted for x.

41

SECTION 2.2 Rates of Change and Derivatives

press TRACE and then type the first x-value and press ENTER. Note that the values $x = 1$ and $y = 1$ are listed at the bottom of the screen. Type the next x-value and press ENTER.

Continue to type the x-values in the first column, one after the other, without having to retype the TRACE command. Without pressing TRACE again, the corresponding y-values can be obtained and written down.

Next, with the graph displayed on the screen, type 2ND TRACE and select item six (type 6, and ENTER). Now type the desired x-value (start with −1). `dy/dx = -2` appears at the bottom of the screen (see Figure 2.2.7). The symbol dy/dx is another notation for *derivative.*

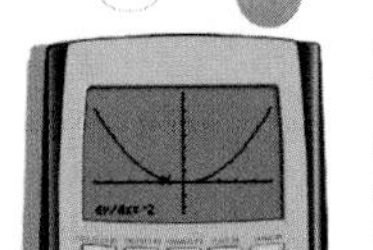
Figure 2.2.7

Retype 2ND TRACE, select item 6 again, and input the next x-value. Continue until all of the x-values in the table have been used, meanwhile writing down the appropriate slopes.

The calculator has the ability to draw a tangent line and calculate the line's equation, just as easily performing the previous calculation. With the graph of $y = x^2$ on the screen, press 2ND PRGM (this will activate the DRAW function) and select `5: Tangent(` (see Figure 2.2.8).

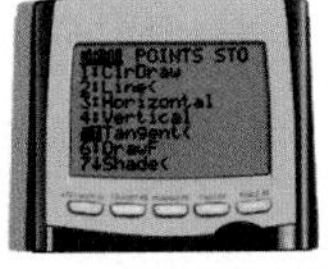

Next, type in the desired x-value, for instance $x = 4$, and press ENTER. The calculator draws a tangent line to th

graph at $x = 4$ and displays pertinent information at th bottom of the screen as shown in Figure 2.2.9. The i is:

`x=4`
`y=8x+-16.`

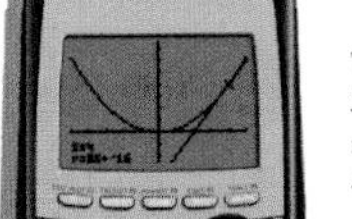
Figure 2.2.9

The equation of the tangent line is $y =$ y-intercept of −16. To solve for the re repeat the previous commands 2ND, P x-value.

When the table is complete, you should b slope of the function $y = x^2$. Table 2.2.2 completed for the integers between −1 an

Calculator Instruction:

Step-by-step instructions are presented to introduce students to basic graphing skills with a TI-84 Plus calculator along with actual screen shots of a TI-84 Plus for visual reference.

CHAPTER 4 Applications of the Derivative

4.3 Exercises

For each of the rational functions in Exercises 1 – 12, find **a.** *any vertical asymptotes,* **b.** *any horizontal asymptotes, and* **c.** *any oblique asymptotes.*

1. $f(x)=\frac{1}{x-4}$ **2.** $f(x)=-\frac{3}{x+6}$ **3.** $f(x)=\frac{2x}{x+8}$

4. $f(x)=\frac{5x}{2x+1}$ **5.** $f(x)=\frac{x+2}{x^2+1}$ **6.** $f(x)=\frac{x-7}{x^2+3}$

7. $f(x)=\frac{5x^2}{3x^2-2x-1}$ **8.** $f(x)=\frac{2x^2}{x^2+3x}$ **9.** $f(x)=\frac{x^2-4}{x}$

10. $f(x)=\frac{3x^2+2}{x}$ **11.** $f(x)=\frac{x^2+1}{x+1}$ **12.** $f(x)=\frac{x^2-5}{x-2}$

In Exercises 13 – 24, sketch the graph of each rational function. Show any related asymptotes on each graph.

13. $f(x)=-\frac{2}{x+5}$ **14.** $f(x)=\frac{4}{x-3}$ **15.** $f(x)=\frac{2x}{x+1}$

16. $f(x)=\frac{3x}{x-2}$ **17.** $f(x)=\frac{x-2}{x-1}$ **18.** $f(x)=\frac{x+4}{2x+1}$

19. $f(x)=2x+\frac{2}{x}$ **20.** $f(x)=3x+\frac{12}{x}$ **21.** $f(x)=\frac{3x^2+6}{x}$

22. $f(x)=\frac{2x^2+1}{3x}$ **23.** $f(x)=\frac{x^3-3x^2+1}{(x-10)(x-3)}$

24. $f(x)=\frac{x^3+x^2-13x+100}{(x-3)(x-10)}$

25. Junker Renovation completely overhauls junked or abandoned cars. Data shows their 1970's models hold their value quite well. The value $F(x)$ of one of these cars is given by $F(x)=70-\frac{15x}{x+1}$ where x is the number of years since repurchase and F is in hundreds of dollars.

a. What is the initial resale price of a car?
b. Find all asymptotes.
c. Sketch the function.
d. What is the long term value of one of these cars?

344

Exercises:

Each section concludes with a selection of exercises designed to allow the student to practice skills and master concepts. The skill level is graduated and successful completion of the the exercises should produce mastery of the material.

Chapter 1 Index of Key Ideas and Terms

Section 1.1 Real Numbers and Number Lines

Types of Numbers pages 2 – 4

- **Natural numbers:** $\mathbb{N} = \{1, 2, 3, \ldots\}$
- **Whole numbers:** $\mathbb{W} = \{0, 1, 2, 3, \ldots\}$
- **Integers:** $\mathbb{Z} = \{\ldots, -3, -2, -1, 0, 1, 2, 3, \ldots\}$
- **Rational numbers (Fractions):** $\left\{\begin{array}{l}\mathbb{Q} = \left\{\frac{a}{b} \middle| a \text{ and } b \text{ are integers with } b \neq 0\right\} \\ \mathbb{Q} = \{\text{repeating infinite decimal numbers}\}\end{array}\right.$
- **Irrational numbers:** $\mathbb{I} = \{\text{nonrepeating infinite decimal numbers}\}$
- **Real numbers:** $\mathbb{R} = \{\text{all rational and irrational numbers}\}$

Properties of Addition page 5

Assume that a, b, and c represent real numbers.

Commutative Property of Addition
$a + b = b + a$

Associative Property of Addition
$a + (b + c) = (a + b) + c$

Additive Identity
$a + 0 = a$

Additive Inverse
$a + (-a) = 0$

Properties of Multiplication page 5

Assume that a, b, and c represent real numbers.

Commutative Property of Multiplication
$a \cdot b = b \cdot a$

Associative Property of Multiplication
$a \cdot (b \cdot c) = (a \cdot b) \cdot c$

Multiplicative Identity
$a \cdot 1 = a$

Multiplicative Inverse
$a \cdot \frac{1}{a} = 1$

Distributive Property (of Multiplication over Addition)

Assume that a, b, and c represent real numbers.

$a(b + c) = a \cdot b + a \cdot c$
$(b + c)a = b \cdot a + c \cdot a$

Index of Key Ideas and Terms:

Each chapter contains an index highlighting the main concepts and skills presented in the chapter along with full definitions and page numbers for easy reference.

Real-World Applications:

Many examples directly apply concepts introduced in the section to the solution of real-world problems in many diverse fields.

55. Population. The population of a community is growing at a rate given by $\frac{dP}{dt} = 120 - 15t^{\frac{1}{2}}$ people per year. Find a function to describe the population t years from now if the present population is 8600 people.

56. Rodent control. Animal control officers have implemented a program to eliminate rats in a community. They estimate that the population of rats is changing at a rate of $\frac{dP}{dt} = 24t^{\frac{1}{2}} - 40t$ rats per month. Find a function for the rat population t months from now if the current population is estimated to be 6300 rats.

57. Air quality. The air quality control office estimates that for a population of x thousand people, the level of pollution in the air is increasing at a rate of $\frac{dL}{dx} = 0.2 + 0.002x$ parts per million per thousand people. Find a function to estimate the level of the pollutants if the level is 5.4 parts per million when the population is 20,000 people.

58. Ecology. Biologists are treating a stream contaminated with bacteria. The level of contamination is changing at a rate of $\frac{dN}{dt} = -\frac{960}{t^2} - 240$ bacteria per cubic centimeter per day, where t is the number of days since the treatment began. Find a function $N(t)$ to estimate the level of contamination if the level after 1 day was about 5720 bacteria per cubic centimeter.

59. Height. An object is projected vertically so that the velocity after t seconds is given by $v(t) = 96 - 32t$ feet per second.
a. Find the height function $s(t)$ if $s(0) = 18$ feet.
b. What will be the height after 3 seconds?

60. Distance. A vehicle travels in a straight line for t minutes with a velocity $v(t) = 72t - 6t^2$ feet per minute, for $0 \le t \le 10$.
a. Find the distance function $s(t)$ if $s(0) = 0$.
b. How far will the vehicle travel in 5 minutes?
c. How far will the vehicle travel in 10 minutes?

Hawkes Learning Systems: Essential Calculus

The Indefinite Integral

Hawkes Learning Systems:

Each section's exercises are followed by a feature which highlights corresponding lessons in *HLS: Essential Calculus with Applications* software for ease of professor-assigned or student-motivated assignments, review, and instruction.

Chapter 1 Test

In Exercises 1 – 4, simplify each expression so that it contains only positive exponen Assume that all variables represent nonzero real numbers.

1. $3x^3\cdot x^{-4}$ **2.** $6x^3\cdot 2x^{-1}$ **3.** $\frac{x^7}{x^3}$ **4.** $\frac{2x^2}{5x^4}$

In Exercises 5 – 8, simplify each expression. Negative exponents may remain in the answer if they are in the numerator. Assume that all variables represent nonzero real numbers.

5. $\frac{x^2\cdot x}{x^3\cdot x^{-4}}$ **6.** $(4x^2y^{-2})(-5x^4y^{-3})$ **7.** $\frac{(x^2y^{-2})^3}{(x^3y)^2}$ **8.** $\left(\frac{x^{-2}y^{-1}}{2x^2y^3}\right)^{-2}$

In Exercises 9 – 14, simplify each expression. Assume that all variables represent positive real numbers.

9. $9^{\frac{3}{2}}$ **10.** $16^{-\frac{3}{4}}$ **11.** $3x^{\frac{1}{2}}\cdot 8x^{-\frac{1}{3}}$ **12.** $\frac{x^3}{x^{\frac{5}{4}}}$

13. $\left(\frac{27x^3}{y^6}\right)^{\frac{1}{3}}$ **14.** $\frac{\left(x^{\frac{1}{2}}y\right)^{\frac{2}{3}}}{x^{\frac{2}{3}}y^{-1}}$

In Exercises 15 – 22, change each expression to an equivalent expression in radical notation or exponential notation. Assume that each variable represents a positive real number.

15. $x^{\frac{8}{3}}$ **16.** $7x^{\frac{2}{5}}y^{\frac{3}{5}}$ **17.** $(3x+1)^{-\frac{1}{2}}$ **18.** $(x+6)^{\frac{2}{3}}$

19. $3\sqrt[4]{x^3}$ **20.** $-2\sqrt[3]{x^2y}$ **21.** $\sqrt[5]{4x-3}$ **22.** $\frac{6}{5\sqrt{9-x^2}}$

In Exercises 23 – 28, perform the indicated operations

23. $(7x^2+2x-1)+(8x-4)$ **24.** $(4x^2+$

25. $(6x^2+x-10)-(x^3+x^2+x-4)$ **26.** $(x^3+4$

27. $\left[(x^2-4x)-(2x^2+6x+1)\right]+\left[(3x^2-2x+7)-(x^2\right.$

28. $\left[(2x^2-x)+(3x+2)\right]-\left[(x-3)-(x^2+1)\right]$

In Exercises 29 – 32, find the indicated products.

29. $(7x+6)(7x-6)$ **30.** $(2x$

31. $(2x+5y)(4x^2-10xy+25y^2)$ **32.** $x^2($

Chapter Test:

Chapter tests follow each chapter and are designed to serve as a practice test for actual examinations given in class or the lab.

Cumulative Review

In Exercises 1 – 4, use the graph of y = f(x) to determine the limits.

1. $\lim_{x\to -2^-}(f(x))$ **2.** $\lim_{x\to -2^+}(f(x))$

3. $\lim_{x\to 0^-}(f(x))$ **4.** $\lim_{x\to 0^+}(f(x))$

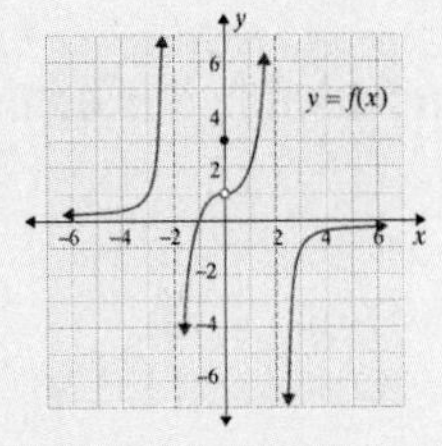

In Exercises 5 – 8, determine the indicated limits (if they exist). In each case, make a table with 5 (x, y) pairs to justify your answer.

5. $\lim_{x\to 4^-}\left(\frac{2x^3-8x^2+6x-24}{x-4}\right)$ **6.** $\lim_{x\to -1^+}\left(\frac{2x^2+3\sqrt{x+1}}{\sqrt{x+1}}\right)$

7. $\lim_{x\to 3^-}\left(\frac{x^2-9}{x-3}\right)$ **8.** $\lim_{x\to 4^+}\left(\frac{9x-54}{x-6}\right)$

9. Give examples of functions whose left- and right-hand limits (as x approaches 4):
- **a.** Are not equal.
- **b.** Are equal.
- **c.** Whose left-hand limit is 2 and whose right-hand limit is 3.

10. If y represents the number of 5-gallon drums of heating oil consumed by a utility company and x represents the number of weeks since January 1st, what does the derivative represent?

11. Suppose (2.0, 3) and (2.10, 3.3) are two points on the graph of a function $y-f(x)$. Estimate the slope of the tangent line to $f(x)$ at $x = 2$.

In Exercises 12 – 14, determine the formula for the derivative of the given function.

12. $f(x)=6x^2$ **13.** $f(x)=6x^2(x-1)$ **14.** $f(x)=14\sqrt{x}$

Cumulative Review:

A cumulative review spanning all previous chapters is given following the Chapter Test in Chapters 3 through 10. This serves to improve retention of material and build student confidence. These are also an excellent study tool for midterms, finals, and other multi-chapter examinations.

Additional Features

Calculator Problems: Each problem is designed to highlight the usefulness of a calculator in solving certain complex problems but maintain the necessity of understanding the concepts behind the problem.

Chapter Test: Provides an opportunity for the students to practice the skills presented in the chapter in a test format.

Cumulative Review: As new concepts build on previous concepts, the cumulative review provides the student with an opportunity to continually reinforce existing skills while practicing newer skills.

Answers: Answers are provided for odd numbered section exercises and for all even and odd numbered exercises in the Chapter Tests and Cumulative Reviews.

Teachers' Edition:

Answers: Answers to all the exercises are conveniently located in the margins next to the problems.

Teaching Notes: Suggestions for more in-depth classroom discussions and alternate methods and techniques are located in the margins.

Also included in this edition:

- New reader friendly layout
- Arrangement of chapters for better flow, continuity, and progression
- Calculator Instructions (New emphasis on the graphing calculator)
- Calculator Problems

Content

The TI-84 plus graphing calculator is used throughout for motivation and exposition. Each instructor using the text will have to decide standards and outcomes expected of students. It is possible to use the text and a calculator and greatly minimize algebra expectations, but most will easily find a middle ground. We mention that engineering majors usually are required to use a more powerful calculator which can differentiate and integrate algebra expressions. Experienced instructors will have no trouble determining a syllabus based on usual expectations and the sophistication of the entering students, but we give a brief overview of each chapter and indicate a minimal set of sections recommended for every user.

Chapter 1 - This chapter contains a review of several algebra topics critical to the development of calculus. Recommended sections:
1.2 Integer Exponents
1.3 Fractional Exponents and Radicals
1.5 Lines and their Graphs
1.6 Functions
1.7 Functions and Their Graphs: A Calculator Section

Chapter 2 - This chapter introduces limits, slopes, and the derivative. Recommended sections:
2.1 Limits
2.2 Rates of Change and Derivatives
2.3 Slope and Rate of Change Considered Algebraically
2.4 Applications: Marginal Analysis

Chapter 3 - This chapter deals with the three basic rules for differentiation, the Product Rule, the Quotient Rule, and the Chain Rule. Recommended sections:
3.1 Product and Quotient Rules
3.2 The Chain Rule and the General Power Rule
3.3 Implicit Differentiation and Related Rates
3.4 Local Extrema and the First Derivative Test

Chapter 4 - This chapter contains applications of the derivative. Recommended sections:
4.1 Higher-Order Derivatives: Concavity and the Second Derivative Test
4.2 Curve Sketching: Polynomial Functions
And at least one from
4.3 Curve Sketching: Rational Functions
4.4 Business Applications

Chapter 5 - This chapter introduces the exponential function and the natural logarithm function. Recommended sections:
5.1 Exponential Functions
5.2 The Algebra of the Natural Logarithm Function
5.3 Differentiation of Logarithmic Functions
5.4 Differentiation of Exponential Functions
5.5 Applications of Exponential Functions

Chapter 6 - This chapter introduces the indefinite and definite integrals. Recommended sections:
6.1 The Indefinite Integral
6.3 The Fundamental Theorem of Calculus and the Definite Integral
6.4 Area (with Applications)
And at least one of
6.5 Area Between Two Curves (with Applications)
6.6 Differential Equations

End of First Semester

A second semester can be offered using the text. It is recommended that one review all of Chapter 6 carefully then cover the indicated sections in Chapters 7 through 10.

Chapter 7 - This chapter contains a selection of topics which apply integration. Recommended sections:
7.1 Integration by Parts
7.2 Annuities and Income Streams
7.3 Tables of Integrals
7.4 Improper Integrals
7.5 Probability
7.6 Volume

Chapter 8 - This chapter takes the student into three dimensions with multivariate calculus. Recommended sections:
8.1 Functions of Several Variables
8.2 Partial Derivatives
8.3 Local Extrema for Functions of Two Variables
8.6 Double Integrals

Chapter 9 - This chapter extends differentiation and integration to the trigonometric functions. The basics of triangle trigonometry are minimally presented so that the calculus can be emphasized. Recommended sections:
9.1 The Trigonometric Functions
9.2 Derivatives of the Trigonometric Functions
9.3 Integration of the Trigonometric Functions
9.4 Inverse Trigonometric Functions

Chapter 10 - This chapter provides a pared down introduction to sequences, Taylor polynomials, and power series. Recommended sections:
10.1 Infinite Sequences
10.2 Taylor Polynomials
10.3 Taylor Series, Infinite Expressions, and Their Applications.

Acknowledgements

I would like to thank Developmental Editor Marcel Prevuznak for managing the tasks of the many assistants who worked on the second edition, and for keeping my nose to the grindstone. My congratulations go to the layout staff who have typeset such a lovely textbook.

Many thanks go to the manuscript reviewers who had many terrific ideas which improved the presentations in each chapter.

Marlow Dorrough, *University of Mississippi*
Stanley Matsuda, *Crafton Hills College*
Ron Palcic, *Johnson County Community College*
Stan Perrine, *Charleston Southern University*
Harriet Roadman, *New River Community College*
Rodger Hammons, *Morehead State University*

A special thanks goes to James Hawkes for beginning the process, for inviting me to participate in one of the most challenging and satisfying professional tasks I have ever tried, and for sharing with me so many of his key ideas which in the end have improved the excellent first edition.

A special thanks goes to my co-authors, D. Franklin Wright and Bill New, who worked so hard to make the first edition a success.

Spencer P. Hurd

TO THE STUDENT

The goal of this text and of your instructor is for you to succeed in calculus. Certainly, you should make this your goal as well. What follows is a brief discussion about developing good work habits and using the features of this text to your best advantage. For you to achieve the greatest return on your investment of time and energy, you should practice the following three rules of learning.

1. Reserve a block of time for study every day.
2. Study what you don't know.
3. Don't be afraid to make mistakes.

How to use this book

The following six-step guide will not only make using this book a more worthwhile and efficient task, but it will also help you benefit more from classroom lectures or the assistance that you receive in a math lab.

1. Try to look over the assigned section(s) before attending class or lab. In this way, new ideas may not sound so foreign when you hear them mentioned again. This will also help you see where you need to ask questions about material that seems difficult to you.

2. Read examples carefully. They have been chosen and written to show you all of the problem-solving steps that you need to be familiar with. You might even try to solve example problems on your own before studying the solutions that are given.

3. Work the section exercises faithfully as they are assigned. Problem-solving practice is the single most important element in achieving success in any math class, and there is no good substitute for actually doing this work yourself. Demonstrating that you can think independently through each step of each type of problem will also give you confidence in your ability to answer questions on quizzes and exams. Check the Answer Key periodically while working section exercises to be sure that you have the right ideas and are proceeding in the right manner.

4. Use the Chapter Index of Key Ideas and Terms as a recap when you begin to prepare for a Chapter Test. It will reference all the major ideas that you should be familiar with from that chapter and indicate where you can turn if review is needed. You can also use the Chapter Index as a final checklist once you feel you have completed your review and are prepared for the Chapter Test.

5. Chapter Tests are provided so that you can practice for the tests that are actually given in class or lab. To simulate a test situation, block out a one-hour, uninterrupted period in a quiet place where your only focus is on accurately completing the Chapter Test. Use the Answer Key at the back of the book as a self-check only after you have completed all of the questions on the test.

6. Cumulative Reviews will help you retain the skills that you acquired in studying earlier chapters. They appear after every chapter beginning with Chapter 3. Approach them in much the same manner as you would the Chapter Tests in order to keep all of your skills sharp throughout the entire course.

How to Prepare for an Exam

Gaining Skill and Confidence

The stress that many students feel while trying to succeed in mathematics is what you have probably heard called "math anxiety." It is a real-life phenomenon, and many students experience such a high level of anxiety during mathematics exams in particular that they simply cannot perform to the best of their abilities. It is possible to overcome this stress simply by building your confidence in your ability to do mathematics and by minimizing your fears of making mistakes.

No matter how much it may seem that in mathematics you must either be right or wrong, with no middle ground, you should realize that you can be learning just as much from the times that you make mistakes as you can from the times that your work is correct. Success will come. Don't think that making mistakes at first means that you'll never be any good at mathematics. Learning mathematics requires lots of practice. Most importantly, it requires a true confidence in yourself and in the fact that with practice and persistence the mistakes will become fewer, the successes will become greater, and you will be able to say, "I can do this."

Showing What You Know

If you have attended class or lab regularly, taken good notes, read your textbook, kept up with homework exercises, and asked for help when it was needed, then you have already made significant progress in preparing for an exam and conquering any anxiety. Here are a few other suggestions to maximize your preparedness and minimize your stress.

1. Give yourself enough time to review. You will generally have several days advance notice before an exam. Set aside a block of time each day with the goal of reviewing a manageable portion of the material that the test will cover. Don't cram!
2. Work lots of problems to refresh your memory and sharpen your skills. Go back to redo selected exercises from all of your homework assignments.
3. Reread your text and your notes, and use the Chapter Index of Key Ideas and Terms and the Chapter Test to recap major ideas and do a self-evaluated test simulation.
4. Be sure that you are well-rested so that you can be alert and focused during the exam.
5. Don't study up to the last minute. Give yourself some time to wind down before the exam. This will help you to organize your thoughts and feel more calm as the test begins.
6. As you take the test, realize that its purpose is not to trick you, but to give you and your instructor an accurate idea of what you have learned. Good study habits, a positive attitude, and confidence in your own ability will be reflected in your performance on any exam.
7. Finally, you should realize that your responsibility does not end with taking the exam. When your instructor returns your corrected exam, you should review your instructor's comments and any mistakes that you might have made. Take the opportunity to learn from this important feedback about what you have accomplished, where you could work harder, and how you can best prepare for future exams.

HAWKES LEARNING SYSTEMS: Essential Calculus with Applications

Overview

This multimedia courseware allows students to become better problem-solvers by creating a mastery level of learning in the classroom. The software includes "overview," "instruct," "practice," "tutor," and "certify" modes in each lesson, allowing students to learn through step-by-step interactions with the software. The automated homework system's tutorial and assessment modes extend instructional influence beyond the classroom. Intelligence is what makes the tutorials so unique. By offering intelligent tutoring and mastery level testing to measure what has been learned, the software extends the instructor's ability to influence students to solve problems. This courseware can be ordered either separately or bundled together with this text.

Minimum Requirements

In order to run *HLS: Essential Calculus with Applications*, you will need:

1 GHz or faster processort
Windows® 2000 or later
128 MB RAM (256 MB recommended)
200 MB hard drive space
800x600 resolution (1024x768 recommended)
Internet Explorer 6.0 or later
CD-ROM drive

Getting Started

Before you can run *HLS: Essential Calculus with Applications*, you will need an access code. This 30 character code is <u>your</u> personal access code. To obtain an access code, go to **http://www.hawkeslearning.com** and follow the links to the access code request page (unless directed otherwise by your instructor.)

Installation

Insert the *HLS: Essential Calculus with Applications* CD#1 into the CD-ROM drive. Next, double-click on the **My Computer** icon and double-click on the CD-ROM drive. Then double-click on Setup.exe.

The complete installation will install the entire product except the multimedia files on your hard drive.

After selecting the desired installation option, follow the on-screen instructions to complete your installation of *HLS: Essential Calculus with Applications*.

Starting the Courseware

After you have installed *HLS: Essential Calculus with Applications* on your computer, to run the courseware, select Start/Programs/Hawkes Learning Systems/Essential Calculus with Applications.

You will be prompted to enter your access code with a message box similar to the following:

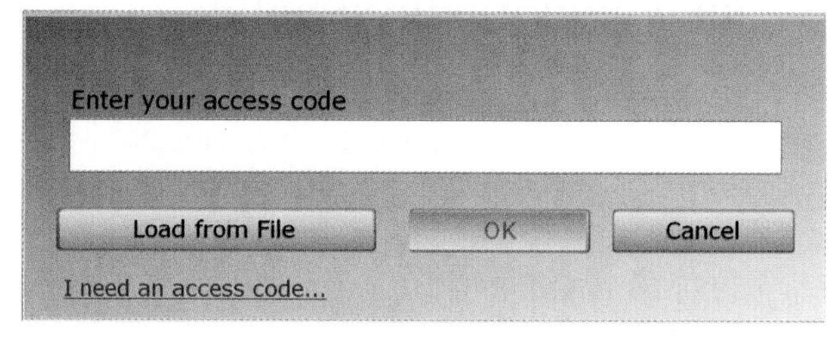

Type your access code into the box provided. When you are finished, press OK.

If you typed in your access code correctly, you will be prompted to save the code to disk. If you choose to save your code to disk, typing in the access code each time you run *HLS: Essential Calculus with Applications* will not be necessary. Instead, select the Load from File button when prompted to enter your access code and choose the path to your saved access code.

Now that you have entered your access code and saved it to disk, you are ready to run a lesson. From the table of contents screen, choose the appropriate chapter and choose the lesson you wish to run.

Features

Each lesson in *HLS: Essential Calculus with Applications* has five modes: Overview, Instruct, Practice, Tutor, and Certify.

Overview: Overview provides you with a brief overview of the lesson. It presents an example of the type of question you will see in Practice and Certify, shows you how to input an answer, lists some of the specific features of the lesson, and tells you how many correct answers are needed to pass Certify.

Instruct: Instruct provides an expository on the material covered in the lesson in a multimedia environment. This same instruct mode can be accessed via the tutor mode.

Practice: Practice allows you to hone your problem-solving skills. It provides an unlimited number of randomly generated problems. Practice also provides access to the Tutor mode by selecting the Tutor button located by the Submit button.

Tutor: Tutor mode is broken up into several parts: Instruct, Explain Error, Step-by-Step, and Solution.

1. Instruct, which can also be selected directly from Practice mode, contains a multimedia lecture of the material covered in a lesson.
2. Explain Error is active whenever a problem is incorrectly answered. It will attempt to explain the error that caused you to incorrectly answer the problem.
3. Step-by-Step is an interactive "step through" of the problem. It breaks each problem into several steps, explains to you each step in solving the problem, and asks you a question about the step. After you answer the last step correctly, you have solved the problem.
4. Solution will provide you with a detailed, "worked-out" solution to the problem.

Throughout the Tutor, you will see words or phrases colored green with a dashed underline. These are called Hot Words. Clicking on a Hot Word will provide you with more information on the word or phrase.

Certify: Certify is the testing mode. You are given a finite number of problems and a certain number of strikes (problems you can get wrong). If you answer the required number of questions, you will receive a certification code and a certificate. Write down your certification code and/or print out your certificate. The certification code will be used by your instructor to update your records. Note that the Tutor is not available in Certify.

Integration of Courseware and Textbook

Although both are excellent when used alone, the *Essential Calculus with Applications* textbook and *HLS: Essential Calculus with Applications* courseware were designed to be used simultaneously. To facilitate this task, great care has been taken to integrate the text and courseware. Each lesson in the text is concluded by the icon below.

This icon indicates which *HLS: Essential Calculus with Applications* lessons you should run in order to test yourself on the subject material and to review the contents of a chapter.

Support

If you have questions about *HLS: Essential Calculus with Applications* or are having technical difficulties, we can be contacted at the following:

Phone: (843) 571-2825
Email: techsupport@hawkeslearning.com
Web: www.hawkeslearning.com

Our support hours are 8:30 am to 5:30 pm, Eastern Time, Monday through Friday.

Functions, Models, and Graphs

CHAPTER 1

Did You Know?

Euler

The mathematician who invented the notation for functions, $f(x)$, which you will study in this chapter, was Leonhard Euler (1707 – 1783) of Switzerland. Euler was one of the most prolific mathematical researchers of all time, and he lived during a period in which mathematics was making great progress. He studied mathematics, theology, medicine, astronomy, physics, and oriental languages before he began a career as a court philosopher-mathematician. His professional life was spent at St. Petersburg Academy by invitation of Catherine I of Russia, at the Berlin Academy under Frederick the Great of Prussia, and again at the St. Petersburg Academy under Catherine the Great. The collected works of Euler fill 80 volumes, and for almost 50 years after Euler's death, the publications of the St. Petersburg Academy continued to include articles by him.

Euler was blind the last 17 years of his life, but he continued his mathematical research by writing on a large slate and dictating to a secretary. He was responsible for the conventionalization of many mathematical symbols such as $f(x)$ for function notation, i for $\sqrt{-1}$, e for the base of the natural logarithms, π for the ratio of circumference to diameter of a circle, and Σ for the summation symbol.

From the age of 20 to his death, Euler was busy adding to knowledge in every branch of mathematics. He wrote with modern symbolism, and his work in calculus was particularly outstanding. Euler had a rich family life, having had 13 children, and he not only contributed to mathematics but reformed the Russian system of weights and measures, supervised the government pension system in Prussia and the government geographic office in Russia, designed canals, and worked in many areas of physics, including acoustics and optics. It was said of Euler, by the French academician François Arago, that he could calculate without apparent effort "just as men breathe and eagles sustain themselves in the air."

An interesting story is told about Euler's meeting with the French philosopher Diderot at the Russian court. Diderot had angered the Czarina by his antireligious views, and Euler was called to the court to debate Diderot. Diderot was told that the great mathematician Euler had an algebraic proof that God existed. Euler walked in towards Diderot and said, "Monsieur, $\dfrac{a+b^n}{n} = x$, therefore God exists, respond." Diderot, who had no understanding of algebra, was unable to respond.

Much of calculus is related to the interplay between the algebra of functions and the geometry of their graphs. This geometry is concerned with how the curve varies with its peaks, valleys, and steepness. We measure this variation and change at a point with a number called the slope at the point. This slope is exactly the rate of change of the function at the point, which is discussed in Chapter 2. This rate of change is also the slope of a certain line, which is called the tangent line at the point. The tangent line at some point on a general curve is exactly analogous to the tangent to a circle at a point.

Even though straight lines, their equations, slopes, and graphs, are studied thoroughly in both beginning algebra and intermediate algebra, these topics and others are reviewed in detail in Chapter 1, and their importance in a calculus course cannot be overstated.

1.1 Real Numbers and Number Lines

Objectives

After completing this section, you will be able to:

1. *Identify types of numbers.*
2. *Determine if given numbers are greater than, less than, or equal to other given numbers.*
3. *Interpret set-builder notation.*
4. *Determine absolute values.*

One of the most important steps in learning the subject matter in any course is to become familiar with the basic vocabulary. In this course, we will be working with the types of numbers listed here. These types of numbers are considered fundamental knowledge for any course in mathematics, and you should be familiar with their names and what numbers are represented in each category.

Natural numbers: $\mathbb{N} = \{1, 2, 3, \ldots\}$

Whole numbers: $\mathbb{W} = \{0, 1, 2, 3, \ldots\}$

Integers: $\mathbb{Z} = \{\ldots, -3, -2, -1, 0, 1, 2, 3, \ldots\}$

Rational numbers (Fractions): $\begin{cases} \mathbb{Q} = \left\{\dfrac{a}{b}, \text{where } a \text{ and } b \text{ are integers with } b \neq 0\right\} \\ \mathbb{Q} = \{\text{repeating infinite decimal numbers}\} \end{cases}$

Irrational numbers: $\mathbb{I} = \{\text{nonrepeating infinite decimal numbers}\}$

Real numbers: $\mathbb{R} = \{\text{all rational and irrational numbers}\}$

The diagram shown in Figure 1.1.1 illustrates the interrelationships among these various types of numbers.

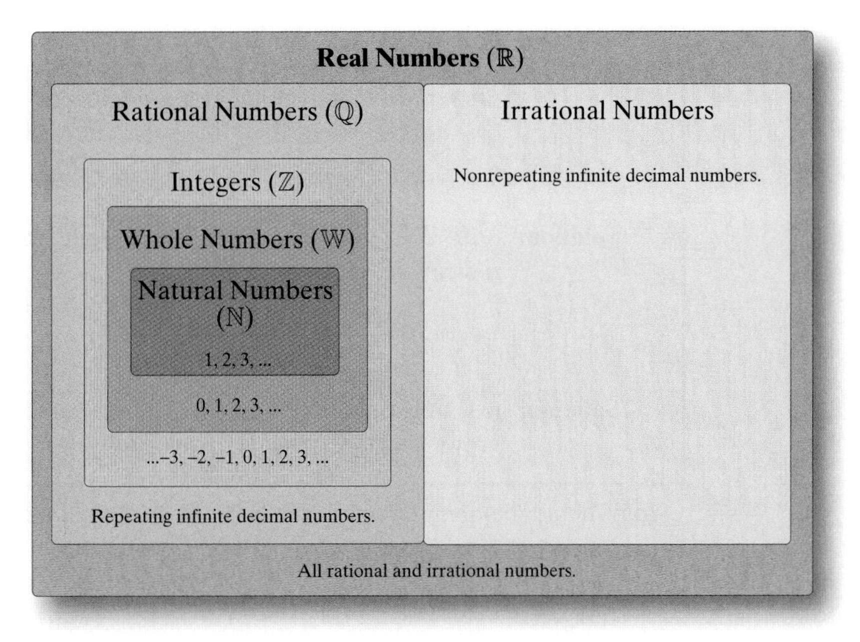

Figure 1.1.1

Example 1: Types of Numbers

a. $\pi = 3.14159265358979...$

Solution: π is an irrational number. Written as a decimal number, it is an infinite nonrepeating decimal number. (It is customary to write "..." to indicate the decimal is not terminating. Since there is *no* block of digits which is shown as repeating, you infer there is no repeating block, and thus infer that π is irrational.)

b. $\frac{1}{7} = 0.142857142857...$

Solution: $\frac{1}{7}$ is a rational number. We can also write $\frac{1}{7}$ as $0.\overline{142857}$, where the bar is used to indicate the repeating pattern of digits. (Here the "..." occurs with a block of six digits which appear to repeat.)

c. $\frac{3}{4} = 0.75$

Solution: $\frac{3}{4}$ is a rational number. We can treat 0.75 as an infinite repeating decimal in the form of $0.75\overline{0}$, where the repeating pattern is all 0's. Such decimal numbers are also called **terminating decimals**.

continued on next page ...

d. $-\frac{2}{3} = -0.66666\ldots = -0.\overline{6}$

Solution: $-\frac{2}{3}$ is a rational number. We also can write $-\frac{2}{3} = \frac{-2}{3} = \frac{2}{-3}$, where the numerator and denominator are integers.

e. $\sqrt{25} = 5$

Solution: $\sqrt{25}$ is a rational number, integer, and whole number. We will discuss radicals, such as square roots and cube roots, in detail in Section 1.3.

f. $\alpha = 1.020030004000050000006\ldots.$

Solution: α is not a rational number. The size of the blocks of consecutive zeros grows and never repeats.

You should be familiar with the following characteristics of the various types of numbers.

Integers:

1. Integers consist of the numbers in the set $\{\ldots, -3, -2, -1, 0, 1, 2, 3, \ldots\}$.
2. Integers fall into three categories:
   a. Positive integers $(1, 2, 3, \ldots)$
   b. Zero (0)
   c. Negative integers $(\ldots, -3, -2, -1)$
3. Every integer is also a rational number because it can be written with a denominator of 1. For example:

$$0 = \frac{0}{1}, \quad 1 = \frac{1}{1}, \quad 2 = \frac{2}{1}, \quad 3 = \frac{3}{1}, \quad \text{and so on.}$$

$$-1 = \frac{-1}{1}, \quad -2 = \frac{-2}{1}, \quad -3 = \frac{-3}{1}, \quad \text{and so on.}$$

Rational Numbers:

If a fraction has a numerator and nonzero denominator that are integers, it can be written as an infinite repeating decimal. (**Note:** This includes terminating decimals.) For example,

$$\frac{1}{3} = 0.3333\ldots = 0.\overline{3} \quad \text{and} \quad \frac{5}{8} = 0.625\overline{0} = 0.625.$$

Irrational Numbers:

There are some radicals, such as $\sqrt{2} = 1.414\ldots$, $\sqrt{3} = 1.732\ldots$, and $\sqrt{5} = 2.236\ldots$, that are irrational numbers. Accuracy to seven or eight decimal places can be found with a calculator. (Radicals will be reviewed in Section 1.3.)

The following properties of addition and multiplication with real numbers will be used throughout the text and are listed here for reference.

Properties of Addition and Multiplication

Assume that a, b, and c represent real numbers.

For Addition	Name of Property	For Multiplication
$a+b=b+a$	***Commutative***	$a \cdot b = b \cdot a$
$a+(b+c)=(a+b)+c$	***Associative***	$a \cdot (b \cdot c) = (a \cdot b) \cdot c$
$a+0=a$	***Identity***	$a \cdot 1 = a$
$a+(-a)=0$	***Inverse***	$a \cdot \frac{1}{a} = 1,\ a \neq 0$
Distributive *(for multiplication over addition)* $a(b+c)=a \cdot b + a \cdot c$		

Real numbers can be represented graphically with number lines. Mathematicians have shown that there is a one-to-one (1–1) correspondence between the real numbers and the points on a line. Thus a number line is called a **real number line**, and the points on the line are considered to be the graphs of the real numbers. The terms **number** and **point** are used interchangeably.

The concept of order with real numbers can also be illustrated with number lines. On a horizontal number line, **smaller numbers are always to the left of the larger numbers**.

Example 2: Number Line

Graph the numbers in the given set A on a number line. Use heavy dots and estimate the placement of points that are not integers.

$$A = \left\{ -\sqrt{25},\ -3,\ -1.5,\ 0,\ \frac{3}{4},\ \sqrt{3},\ \pi,\ 4 \right\}$$

Solution:

$-\sqrt{25}$ -3 -1.5 0 $\frac{3}{4}$ $\sqrt{3}$ π 4

The symbols used to indicate order (or inequalities) are given in the following definition.

Symbols for Order

$<$	***is less than***	$>$	***is greater than***
$\leq$	***is less than or equal to***	$\geq$	***is greater than or equal to***

Example 3: Symbols for Order

Arrange the following numbers in order from smallest to largest using the most appropriate of the four symbols for order: 17, 1.7, −1.7, −17, 0, $\frac{20}{3}$, $\frac{40}{6}$, 20, and −20.

Solution: $-20 < -17 < -1.7 < 0 < 1.7 < \frac{20}{3} \leq \frac{40}{6} < 17 < 20$

Set-builder notation is convenient for representing sets of real numbers with specified properties (such as being less than or greater than some specified value).

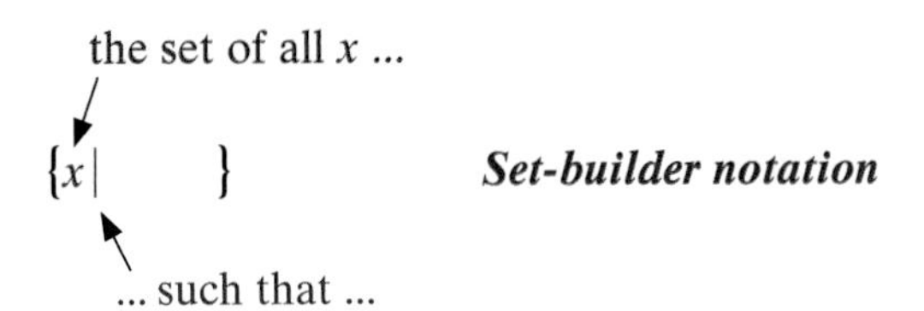

For example, $\{x \mid x < 5\}$ is read "the set of all x such that x is less than 5." The graphs of such sets of numbers are shaded portions of real numbers lines, as shown in the following examples.

Example 4: Set-Builder Notation

a. Graph the set of real numbers $\{x \mid x < 3\}$.

Solution:

3

The open circle indicates that 3 **is not** included in the set.

b. Graph the set of real numbers $\{x \mid x \geq -1.5\}$.

Solution:

-1.5

The closed circle indicates that -1.5 **is** included in the set.

An **interval** on a number line (or on the x- or y-axis in a coordinate system) is a set of numbers and is indicated using set builder notation or is described with parentheses and brackets.

The closed interval $[a, b]$ includes the endpoints. $[a, b] = \{x \mid a \leq x \leq b\}$.

The open interval (a, b) does not include the endpoints. $(a, b) = \{x \mid a < x < b\}$.

An interval which includes only one endpoint may be referred to as half-open or half-closed.

$[a, b) = \{x \mid a \leq x < b\}$. $(a, b] = \{a < x \leq b\}$.

The ∞ symbol will be used to indicate that an interval is not bounded on one or both ends; for example,

$$(-\infty, b) = \{x \mid x < b\}$$
$$(-\infty, b] = \{x \mid x \leq b\}$$
$$(-\infty, \infty) = \{x \mid x \text{ is a real number}\} = \mathbb{R}$$
$$(a, \infty) = \{x \mid x > a\}$$
$$[a, \infty) = \{x \mid x \geq a\}$$

Now we consider the concept of **absolute value**, symbolized with vertical bars, $|\ |$. Graphically, the absolute value of a number is its distance from 0. Thus the two numbers $+6$ and -6 have the same absolute value because they are the same distance from 0 (Figure 1.1.2). Since distance is nonnegative, the absolute value of any number is nonnegative.

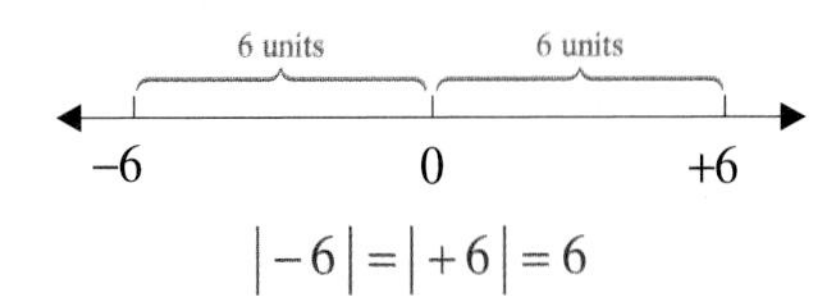

$|-6| = |+6| = 6$

Figure 1.1.2

The following definition of absolute value is often useful when we are working with numbers as algebraic expressions.

Absolute Value

For any real number a,

$$|a| = \begin{cases} a & \text{if } a \geq 0 \\ -a & \text{if } a < 0. \end{cases}$$

Note that in the second part of the definition $-a$ is positive if a is negative. For example, if $a = -5$, then $-a = -(-5) = +5$. Thus

$$|a| = |-5| = -(-5) = +5.$$

Example 5: Absolute Value

Find the absolute value of each of the numbers **a.** -10, **b.** -2.3, **c.** 4.5, and **d.** 0.

Solutions: **a.** $|-10| = -(-10) = 10$

b. $|-2.3| = -(-2.3) = 2.3$

c. $|4.5| = 4.5$

d. $|0| = 0$

Example 6: Absolute Value Equations

a. What real numbers satisfy the equation $|x| = 4$?

Solution: Since $|4| = 4$ and $|-4| = 4$, x can be either 4 or -4.

b. What real numbers satisfy the equation $|x| = -1.7$?

Solution: This equation has no solution ($\varnothing$) because the absolute value of any real number is nonnegative.

1.1 Exercises

In Exercises 1 – 6, list the numbers from the set $A = \left\{-6, -\sqrt{7}, -\sqrt{4}, -\frac{5}{4}, -1.2, 0, \frac{3}{8}, \sqrt{3}, \sqrt{10}, \sqrt{16}, 5.2\right\}$ *that belong to each of the sets described. Then graph each set of answers on a number line. (*Hint: using a calculator to convert numbers in set A to decimal form may be helpful.)

1. $\{x \mid x$ is a whole number$\}$

2. $\{x \mid x$ is an integer$\}$

3. $\{x \mid x$ is a positive integer$\}$

4. $\{x \mid x$ is a irrational number$\}$

5. $\{x \mid x$ is a rational number$\}$

6. $\{x \mid x$ is a real number$\}$

In Exercises 7 – 12, choose the word that correctly completes each statement.

7. If x is a rational number, then x is (never, sometimes, always) a real number.
8. If x is a rational number, then x is (never, sometimes, always) an irrational number.
9. If x is an integer, then x is (never, sometimes, always) a whole number.
10. If x is a real number, then x is (never, sometimes, always) a rational number.
11. If x is a rational number, then x is (never, sometimes, always) an integer.
12. If x is a positive integer, then x is (never, sometimes, always) a whole number.

In Exercises 13 – 22, graph each of the sets of real numbers; then write each as an interval using the symbol $-\infty$ or ∞.

13. $\{x|x \le 7\}$ **14.** $\{x|x < -9\}$ **15.** $\{x|x > -3.4\}$

16. $\{x|x \ge 4.9\}$ **17.** $\{x|x \le 0\}$ **18.** $\{x|x > 0\}$

19. $\left\{x \middle| x < \frac{16}{3}\right\}$ **20.** $\left\{x \middle| x < -\frac{8}{3}\right\}$ **21.** $\left\{x \middle| x \ge -\frac{7}{2}\right\}$

22. $\left\{x \middle| x \ge -\frac{11}{4}\right\}$

In Exercises 23 – 32, find the value of each expression.

23. $|1.4|$ **24.** $|-5.3|$ **25.** $\left|-\frac{3}{8}\right|$ **26.** $\left|\frac{4}{3}\right|$

27. $\left|-\sqrt{2}\right|$ **28.** $|0|$ **29.** $|-4|+|8|$ **30.** $|2|+|-8.4|$

31. $|-3.6|-|-4.2|$ **32.** $|-9.2|-|6.6|$

In Exercises 33 – 42, find the real numbers that satisfy the given equation.

33. $|x| = 8$ **34.** $|x| = 10.3$ **35.** $|x| = \frac{7}{16}$ **36.** $|x| = \frac{9}{5}$

37. $|x| = -2.5$ **38.** $|x| = \sqrt{3}$ **39.** $|x| = 3\sqrt{2}$ **40.** $|x| = -0.8$

41. $|x| = x$ **42.** $|x| = -x$

43. Let $x = 0.999\overline{9}$. Is $x = 1$ or is $x < 1$? Is x a rational number?

Hawkes Learning Systems: Essential Calculus

Real Numbers and Number Lines

1.2 Integer Exponents

Objectives

After completing this section, you will be able to:

Simplify algebraic expressions with constant or single-variable bases using the properties of integer exponents.

The three categories of integer exponents are defined as follows.

Zero Exponent

$a^0 = 1$ *for any nonzero number or nonzero algebraic expression* a. *(Note:* 0^0 *is undefined.)*

Positive Integer Exponents

Suppose a is any nonzero number or nonzero algebraic expression. Then

$$a^1 = a$$

$$a^2 = a(a^1); \text{ and, when } a^{n-1} \text{ is defined,}$$

$$a^n = a(a^{n-1}).$$

More simply, $a \cdot a = a^2$, $a \cdot a \cdot a = a^3$, and so on.

Negative Integer Exponents

If $a \neq 0$, then $a^{-n} = \dfrac{1}{a^n}$ *for any positive integer n.*

In particular, $a^{-1} = \dfrac{1}{a}$ and $a^{-2} = \dfrac{1}{a^2}$, etc.

Zero is a special number which demands different rules for exponents

The Number Zero

0^0 *is not defined.*

$0^n = 0$ *for any positive integer n.*

Example 1: Writing Exponents

Write the following numbers without exponents:

a. 3^4 **b.** 2^{-3} **c.** $\frac{4}{5^{-1}}$ **d.** $(-4)^4$ **e.** -4^4

Solutions: **a.** $3^4 = 3 \cdot 3 \cdot 3 \cdot 3 = 81$ **b.** $2^{-3} = \frac{1}{2^3} = \frac{1}{2 \cdot 2 \cdot 2} = \frac{1}{8}$

c. $\frac{4}{5^{-1}} = \frac{4}{1/5} = 4\left(\frac{5}{1}\right) = 20$ **d.** $(-4)^4 = (-4) \cdot (-4) \cdot (-4) \cdot (-4) = 256$

e. $-4^4 = -(4^4) = -(4) \cdot (4) \cdot (4) \cdot (4) = -256$

The properties of integer exponents listed below can be used to simplify algebraic expressions.

Properties of Exponents

If a and b are nonzero real numbers and m and n are integers, then

1. $a^m \cdot a^n = a^{m+n}$
2. $a^0 = 1$
3. $a^{-1} = \frac{1}{a}$
4. $a^{-n} = \frac{1}{a^n}$
5. $\frac{a^m}{a^n} = a^{m-n}$
6. $\sqrt[n]{a} = a^{\frac{1}{n}}$ $(n \neq 0)$
7. $(a^m)^n = a^{mn}$
8. $(ab)^n = a^n b^n$
9. $\left(\frac{a}{b}\right)^n = \frac{a^n}{b^n}$

Example 2: Positive Exponents

Using the properties of exponents, simplify the following expressions so that they contain only positive exponents. Assume that all variables represent nonzero real numbers.

a. $x^3 \cdot x^5$

Solution: $x^3 \cdot x^5 = x^{3+5} = x^8$ Using property 1.

b. $\frac{(y^{-3})^2}{y^{-8}}$

Solution: $\frac{(y^{-3})^2}{y^{-8}} = \frac{y^{-6}}{y^{-8}} = y^{-6-(-8)} = y^{-6+8} = y^2$ Using property 7 and then 5.

continued on next page ...

c. $\left(\frac{y^{-2}}{y^{-6}}\right)^{-2}$ Example c. demonstrates two different ways to simplify the same expression.

Solution: $\left(\frac{y^{-2}}{y^{-6}}\right)^{-2}=\frac{\left(y^{-2}\right)^{-2}}{\left(y^{-6}\right)^{-2}}=\frac{y^4}{y^{12}}=y^{4-12}=y^{-8}=\frac{1}{y^8}$ Using properties 9, 7, 5, and 4 (in that order)

Or, by using a different sequence of steps:

$\left(\frac{y^{-2}}{y^{-6}}\right)^{-2}=\left(y^{-2-(-6)}\right)^{-2}=\left(y^4\right)^{-2}=y^{-8}=\frac{1}{y^8}$ Using properties 5, 7, and 4 (in that order)

Example 3: Negative Exponents

Consolidate terms and simplify the following expressions. Do not use negative exponents. Assume that all variables represent nonzero real numbers.

a. $\frac{2x^3\cdot 6y}{4xy^{-1}}$

Solution: $\frac{2x^3\cdot 6y}{4xy^{-1}}=\frac{12x^3y}{4xy^{-1}}=3x^{3-1}y^{1-(-1)}=3x^2y^2$ Simplify the constants first, then apply property 5.

b. $\frac{8xy^{-2}}{2x^2y^{-2}}$

Solution: $\frac{8xy^{-2}}{2x^2y^{-2}}=4x^{1-2}y^{-2-(-2)}=4x^{-1}=\frac{4}{x}$ Simplify the constants first, then apply property 5 and property 3.

Example 4: Negative Exponents

c. Given the expression $\left(\frac{3ab^{-3}}{a^{-2}b^3}\right)^{-3}$, remove parentheses and express with all variables in the numerator. Express constants without exponents.

Solution: $\left(\frac{3ab^{-3}}{a^{-2}b^3}\right)^{-3}=\left(3a^3b^{-6}\right)^{-3}=3^{-3}a^{-9}b^{18}=\frac{a^{-9}b^{18}}{27}$ Using properties 5, 8, and 4 (in that order).

1.2 Exercises

In Exercises 1 – 8, evaluate each expression.

1. 7^3 **2.** 3^4 **3.** 6^{-2} **4.** 5^{-3}

5. -2^4 **6.** $(-4)^3$ **7.** $-2\cdot 5^2$ **8.** $3\cdot 2^{-4}$

In Exercises 9 – 26, simplify each expression so that it contains only positive exponents. Assume that all variables represent nonzero real numbers.

9. x^{-2} **10.** t^{-3} **11.** $5y^{-1}$ **12.** $7x^{-2}$

13. $x^2 \cdot x^5$ **14.** $y^4 \cdot y$ **15.** $x^3 \cdot x^{-2}$ **16.** $b^2 \cdot b^{-4}$

17. $x^{-5} \cdot x^{-2}$ **18.** $x^4 \cdot x^{-5}$ **19.** $a^0 \cdot a^{-3}$ **20.** $x^3 \cdot x^{-1}$

21. $\frac{y^5}{y^3}$ **22.** $\frac{s^7}{s^4}$ **23.** $\frac{x^8}{x^3}$ **24.** $\frac{x^8}{x^2}$

25. $\frac{x^3}{x^6}$ **26.** $\frac{a^0}{a^4}$

In Exercises 27 – 50 simplify each expression. Negative exponents may remain in the answer if they are in the numerator. Assume that all variables represent nonzero real numbers. Answers may vary.

27. $\frac{x^2}{x^{-1}}$ **28.** $\frac{t^4}{t^{-4}}$ **29.** $\frac{b^2}{b^{-2}}$ **30.** $\frac{x}{x^{-3}}$

31. $\frac{x^5x^{-2}}{x^{-3}}$ **32.** $\frac{y^2y^{-3}}{y^4}$ **33.** $\frac{a^6a^0}{a^{-2}}$ **34.** $\frac{x^6x^{-2}}{x^4}$

35. $\frac{y^8y^{-3}}{\left(y^2\right)^3}$ **36.** $\frac{\left(s^{-2}\right)^3}{ss^{-3}}$ **37.** $\frac{\left(x^{-2}\right)^4}{x^{-2}x^{-1}}$ **38.** $\frac{t^3t^{-5}}{\left(t^2\right)^4}$

39. $\left(\frac{x^2x^{-1}}{x^5x}\right)^{-2}$ **40.** $\left(\frac{x^{-3}x^0}{x^2x}\right)^{-3}$ **41.** $\frac{a^{-1}b^2}{3ab^{-5}}$ **42.** $\frac{5x^3y^{-1}}{6x^{-2}y^{-3}}$

43. $\frac{4x^5y^3}{3x^{-1}y^5}$ **44.** $\frac{9s^{-4}t^2}{2st^{-3}}$ **45.** $\left(\frac{2x^2y^{-1}}{3xy^2}\right)^{-3}$ **46.** $\left(\frac{-3x^{-2}y^4}{x^0y^3}\right)^{-1}$

47. $\left(\frac{6st}{s^2t^{-3}}\right)^{-2}$ **48.** $\left(\frac{2xy^4}{3x^2y^2}\right)^{-3}$ **49.** $\left(\frac{5^{-1}x^3y^{-2}}{xy^{-1}}\right)^2$ **50.** $\left(\frac{a^2b^{-3}}{3^{-1}a^{-1}b}\right)^3$

Hawkes Learning Systems: Essential Calculus

Integer Exponents

1.3 Fractional Exponents and Radicals

Objectives

After completing this section, you will be able to:

1. *Understand the meaning of nth root.*
2. *Evaluate expressions of the form $a^{\frac{m}{n}}$.*
3. *Simplify expressions using the properties of fractional exponents.*

Fractional Exponents

To understand the meaning of a fractional exponent, we begin with the exponent $\frac{1}{n}$, where n is a positive integer. For positive a, $a^{\frac{1}{n}}$ is the number which, raised to the nth power, gives a.

$$\left(a^{\frac{1}{n}}\right)^n = a^{\frac{1}{n}\cdot n} = a^1 = a$$

For example,

$$25^{\frac{1}{2}} = 5 \quad \text{since} \quad 5^2 = 25$$

and

$$8^{\frac{1}{3}} = 2 \quad \text{since} \quad 2^3 = 8.$$

More generally, for positive a, positive b, and positive n,

$$a^{\frac{1}{n}} = b \quad \text{if and only if} \quad a = b^n.$$

Later on we will discuss 0, negative values for a, and the radical symbol $\sqrt{}$.

Principal n^{th} Root

For any positive real number a and any positive integer n, the number denoted $a^{\frac{1}{n}}$ is a positive real number called the principal n^{th} root of a, and

$$\left(\boldsymbol{a^{\frac{1}{n}}}\right)^{\boldsymbol{n}} = \boldsymbol{a}.$$

Example 1: Evaluating Square Roots

Simplify $81^{\frac{1}{2}}$.

Solution: $81^{\frac{1}{2}} = 9$. If the exponent is $\frac{1}{2}$, the root is called the **square root**, with the understanding that it is nonnegative. Thus 9 is the square root (or **positive square root** or **principal square root**) of 81. Since $(-9)^2 = 81$, 81 has two square roots, and –9 is called the **negative square root** of 81.

NOTES

Remember that the notation $81^{\frac{1}{2}}$ indicates the positive square root.

Example 2: Evaluating Cube Roots

Simplify $125^{\frac{1}{3}}$.

Solution: $125^{\frac{1}{3}} = 5$. If the exponent is $\frac{1}{3}$, the root is called the **cube root**. Thus 5 is the cube root of 125.

Special considerations must be made for negative values of a in the expression $a^{\frac{1}{n}}$. If a is negative and n is even, then $a^{\frac{1}{n}}$ is **not a real number** because even powers of real numbers are always nonnegative. For example, $(-4)^{\frac{1}{2}}$ is not a real number because there is no real number that, when squared, gives –4. Numbers such as $(-4)^{\frac{1}{2}}$ and $(-10)^{\frac{1}{4}}$ are a part of the complex number system and are not a part of this course. If n is odd, then $a^{\frac{1}{n}}$ is a real number regardless of whether a is positive or negative. The following note summarizes the basic rules of exponents.

NOTES

1. If $a > 0$, then $a^{\frac{1}{n}}$ is a positive real number and $\left(a^{\frac{1}{n}}\right)^n = a$.

2. If $a < 0$ and n is even, $a^{\frac{1}{n}}$ is not a real number.

3. If $a < 0$ and n is odd, $a^{\frac{1}{n}}$ is a negative real number and $\left(a^{\frac{1}{n}}\right)^n = a$.

4. If $a = 0, a^{\frac{1}{n}} = 0^{\frac{1}{n}} = 0$.

Next, we consider expressions such as $8^{\frac{2}{3}}$ and $81^{\frac{3}{4}}$ in which the exponent is a fraction in the form $\frac{m}{n}$.

The General Form $a^{\frac{m}{n}}$

If the following three expressions are defined, then they are equal to each other:

$$a^{\frac{m}{n}} = \left(a^{\frac{1}{n}}\right)^m = \left(a^m\right)^{\frac{1}{n}}.$$

Example 3: Simplifying Expressions with Rational Exponents

Simplify each expression.

a. $8^{\frac{2}{3}}$

Solution: $8^{\frac{2}{3}} = \left(8^{\frac{1}{3}}\right)^2 = (2)^2 = 4$

or

$8^{\frac{2}{3}} = \left(8^2\right)^{\frac{1}{3}} = (64)^{\frac{1}{3}} = 4.$

The definition allows us to evaluate the expression two ways.

Generally, the first approach, taking a root first and then raising this root to a power, is easier because the numbers are smaller. Part **b.** illustrates this fact.

b. $81^{\frac{3}{4}}$

Solution: $81^{\frac{3}{4}} = \left(81^{\frac{1}{4}}\right)^3 = (3)^3 = 27$

or

$81^{\frac{3}{4}} = \left(81^3\right)^{\frac{1}{4}} = (531{,}441)^{\frac{1}{4}} = 27.$

Most people would need a calculator to evaluate the last expression.

All the previous properties stated in Section 1.2 for integer exponents also apply for fractional exponents. With this knowledge, we can simplify expressions such as

$$x^{\frac{1}{3}} \cdot x^{\frac{1}{6}} \text{ and } \left(x^{-\frac{3}{4}} \cdot x^{\frac{1}{3}}\right)^2.$$

Example 4: Simplifying Expressions with Rational Exponents

Simplify each expression. Assume that x represents a positive real number.

a. $x^{\frac{1}{3}} \cdot x^{\frac{1}{6}}$

Solution: $x^{\frac{1}{3}} \cdot x^{\frac{1}{6}} = x^{\frac{1}{3}+\frac{1}{6}} = x^{\frac{2}{6}+\frac{1}{6}} = x^{\frac{3}{6}} = x^{\frac{1}{2}}$

b. $\left(x^{-\frac{3}{4}} \cdot x^{\frac{1}{3}}\right)^2$

Solution: $\left(x^{-\frac{3}{4}} \cdot x^{\frac{1}{3}}\right)^2 = \left(x^{-\frac{9}{12}+\frac{4}{12}}\right)^2 = \left(x^{-\frac{5}{12}}\right)^2 = x^{-\frac{10}{12}} = x^{-\frac{5}{6}}$ or $= \frac{1}{x^{\frac{5}{6}}}$

Radical Notation

Another notation, closely related to fractional exponents, is **radical notation**: $\sqrt[n]{a}$. The symbol $\sqrt{}$ is called a **radical sign**, n is called the **index**, and a is called the **radicand**.

NOTES If no index is given, the index is then understood to be 2.

The following definition shows the relationship between fractional exponents and radicals.

Radical Notation

If a is a real number, n is a positive integer, and $a^{\frac{1}{n}}$ is a real number, then

$$\boldsymbol{a^{\frac{1}{n}} = \sqrt[n]{a}.}$$

Thus we have

$$a^{\frac{1}{2}} = \sqrt{a}, \quad a^{\frac{1}{3}} = \sqrt[3]{a}, \quad a^{\frac{1}{4}} = \sqrt[4]{a}, \quad \text{and so on.}$$

Also, adopting previous definitions, we have

$$a^{\frac{m}{n}} = \left(\sqrt[n]{a}\right)^m = \sqrt[n]{a^m}.$$

Example 5: Conversion to Radical Notation

Change each expression to an equivalent expression in radical notation. All variable expressions represent positive real numbers.

a. $x^{\frac{2}{3}}$

Solution: $x^{\frac{2}{3}} = \sqrt[3]{x^2}$

b. $\left(x^2+1\right)^{\frac{1}{2}}$

Solution: $\left(x^2+1\right)^{\frac{1}{2}} = \sqrt{x^2+1}$

c. $\frac{1}{2}(x+2)^{-\frac{1}{2}}$

Solution: $\frac{1}{2}(x+2)^{-\frac{1}{2}} = \frac{1}{2}\cdot\frac{1}{(x+2)^{\frac{1}{2}}} = \frac{1}{2(x+2)^{\frac{1}{2}}} = \frac{1}{2\sqrt{x+2}}$

Example 6: Conversion to Exponential Notation

Change each expression to an equivalent expression in exponential notation. All variable expressions represent positive real numbers.

a. $\sqrt{x^2-4}$

Solution: $\sqrt{x^2-4} = \left(x^2-4\right)^{\frac{1}{2}}$

Remember that because no index is given, the index is understood to be 2.

b. $8\sqrt[3]{y^2}$

Solution: $8\sqrt[3]{y^2} = 8y^{\frac{2}{3}}$

c. $\dfrac{1}{\sqrt[3]{z^3+1}}$

Solution: $\dfrac{1}{\sqrt[3]{z^3+1}} = \dfrac{1}{\left(z^3+1\right)^{\frac{1}{3}}}$ or $\left(z^3+1\right)^{-\frac{1}{3}}$

To simplify radical expressions that involve square roots of positive numbers, we need the following two properties.

Properties of Square Roots

If a and b are positive real numbers:

1. $\sqrt{ab} = \sqrt{a} \cdot \sqrt{b}$ **2.** $\sqrt{\dfrac{a}{b}} = \dfrac{\sqrt{a}}{\sqrt{b}}$

The squares of the positive integers are called **perfect square numbers**. Thus the perfect squares are

$$1, 4, 9, 16, 25, 36, 49, 64, 81, 100, 121, 144, \text{ and so on.}$$

Expressions involving square roots are considered simplified if the radical contains no square factors. Thus $\sqrt{24}$ is not simplified because 24 has the square factor 4. We can simplify $\sqrt{24}$ as

$$\sqrt{24} = \sqrt{4 \cdot 6} = \sqrt{4} \cdot \sqrt{6} = 2\sqrt{6}.$$

NOTES

CAUTION! Students should note the following are common errors. In fact, these equations are almost always false. The student should find numbers a and b which demonstrate that each of the following equations is **false in general**.

a. $\sqrt{a+b} = \sqrt{a} + \sqrt{b}$

Let $a = 64$ and $b = 36$. Then $\left.\begin{array}{l}\sqrt{a+b} = \sqrt{64+36} = \sqrt{100} = 10 \\ \sqrt{a}+\sqrt{b} = 8+6 = 14\end{array}\right\}$ are not equal.

b. $\sqrt{a^2+b^2} = a+b$

Let $a = 8$ and $b = 6$. Then $\left.\begin{array}{l}\sqrt{a^2+b^2} = \sqrt{64+36} = \sqrt{100} = 10 \\ a+b = 8+6 = 14\end{array}\right\}$ are not equal.

c. $(a+b)^2 = a^2+b^2$.

Let $a = 6$ and $b = 8$. Then $\left.\begin{array}{l}(a+b)^2 = (6+8)^2 = 14^2 = 196 \\ a^2+b^2 = 36+64 = 100\end{array}\right\}$ are not equal.

1.3 Exercises

In Exercises 1 – 32, simplify each expression. Assume that all variables represent positive real numbers.

1. $64^{\frac{1}{3}}$ **2.** $144^{\frac{1}{2}}$ **3.** $16^{\frac{3}{4}}$ **4.** $(-8)^{\frac{2}{3}}$

5. $25^{-\frac{3}{2}}$ **6.** $4^{-\frac{5}{2}}$ **7.** $-81^{\frac{1}{4}}$ **8.** $-49^{\frac{3}{2}}$

9. $32^{\frac{2}{5}}$ **10.** $64^{-\frac{2}{3}}$ **11.** $x^2 \cdot x^{\frac{1}{2}}$ **12.** $x^3 \cdot x^{\frac{2}{3}}$

13. $x^{-\frac{1}{2}} \cdot x^{\frac{1}{3}}$ **14.** $x^{\frac{3}{4}} \cdot x^{\frac{1}{2}}$ **15.** $a^2 \cdot a^{-\frac{2}{5}}$ **16.** $x^{-3} \cdot x^{\frac{4}{3}}$

17. $\dfrac{x}{x^{\frac{2}{3}}}$ **18.** $\dfrac{t^2}{t^{\frac{1}{2}}}$ **19.** $\dfrac{x^{\frac{1}{4}}}{x^2}$ **20.** $\dfrac{y^{\frac{3}{4}}}{y^4}$

21. $\dfrac{b^{\frac{4}{5}}}{b^{\frac{1}{2}}}$ **22.** $\dfrac{x^{\frac{3}{2}}}{x^{-\frac{1}{3}}}$ **23.** $\dfrac{s^{\frac{3}{4}}}{s^{-\frac{1}{8}}}$ **24.** $\dfrac{x^{\frac{1}{2}}}{x^{\frac{4}{3}}}$

25. $\dfrac{x^{\frac{1}{2}} \cdot x^{-\frac{3}{4}}}{x^{-\frac{1}{2}}}$ **26.** $\dfrac{y^{\frac{2}{3}} \cdot y^{\frac{4}{3}}}{y^{-2}}$ **27.** $\dfrac{x^{\frac{3}{2}} \cdot x^{\frac{4}{5}}}{x^{-\frac{1}{2}} \cdot x^2}$ **28.** $\dfrac{x^{-\frac{2}{3}} \cdot x^{\frac{1}{2}}}{x \cdot x^{-\frac{3}{2}}}$

29. $\left(\dfrac{27x^3y^6}{x^{-3}}\right)^{\frac{1}{3}}$ **30.** $\left(\dfrac{16a^{-4}b^3}{ab^{-1}}\right)^{\frac{1}{4}}$ **31.** $\dfrac{\left(16s^2t\right)^{\frac{1}{4}}}{\left(5s^{\frac{1}{3}}t^{-\frac{1}{2}}\right)^2}$ **32.** $\dfrac{\left(25x^2y^{-1}\right)^{\frac{1}{2}}}{\left(5x^{\frac{1}{5}}y^{\frac{3}{5}}\right)^3}$

In Exercises 33 – 42, change each expression to an equivalent expression in radical notation. Assume that each variable expression represents a positive real number.

33. $y^{\frac{1}{2}}$ **34.** $x^{\frac{2}{5}}$ **35.** $x^{-\frac{1}{4}}$ **36.** $x^{-\frac{1}{2}}$

37. $(x-2)^{\frac{4}{5}}$ **38.** $(5a+3)^{\frac{2}{3}}$ **39.** $2\left(y^2+4\right)^{-\frac{1}{2}}$ **40.** $7\left(x^2+8\right)^{-\frac{1}{3}}$

41. $\dfrac{-2}{7}\left(t^2-4\right)^{-\frac{1}{4}}$ **42.** $\dfrac{5}{8}\left(x^2+6\right)^{-\frac{1}{2}}$

In Exercises 43 – 52, change each expression to an equivalent expression in exponential notation. Use negative exponents and eliminate all fractions. Assume that each variable expression represents a positive real number.

43. $\sqrt[4]{x}$ **44.** $\sqrt[3]{b}$ **45.** $\frac{1}{\sqrt[4]{s}}$ **46.** $\frac{1}{\sqrt[5]{x^2}}$

47. $\sqrt[3]{(y+10)^2}$ **48.** $\sqrt[4]{(x+1)^3}$ **49.** $\frac{1}{\sqrt[3]{a^2+2}}$ **50.** $\frac{2}{5\sqrt[3]{x^2-6}}$

51. $\frac{4x}{9\sqrt[3]{(x+5)^2}}$ **52.** $\frac{3y}{5\sqrt[3]{y^3+8}}$

In Exercises 53 – 60, simplify each of the radical expressions. Assume that each variable represents a positive real number.

53. $\sqrt{48}$ **54.** $\sqrt{175}$ **55.** $\sqrt{\frac{9}{16}}$ **56.** $\sqrt{\frac{49}{225}}$

57. $\sqrt{150x}$ **58.** $\sqrt{112a}$ **59.** $\sqrt{108t^3}$ **60.** $\sqrt{216x^3}$

Hawkes Learning Systems: Essential Calculus

Fractional Exponents and Radicals

1.4 Polynomials

Objectives

After completing this section, you will be able to:

1. *Define a polynomial.*
2. *Classify a polynomial as a monomial, binomial, trinomial, or a polynomial with more than three terms.*
3. *Add, subtract, multiply, and factor polynomials.*

Polynomials

Constants are fixed numbers and **variables** are letters that represent unspecified numbers. A **monomial** is a single term with whole number exponents on its variables and no variable in the denominator. The general form of a **monomial in *x*** is

$$kx^n,$$

where n is a whole number and k is a constant. The constant k is called the **coefficient** and n is called the **degree**. Any monomial or algebraic sum of monomials is called a **polynomial**.

Classification of Polynomials

		Example
Monomial	*(polynomial with one term)*	$5x$
Binomial	*(polynomial with two terms)*	$7x-8$
Trinomial	*(polynomial with three terms)*	x^3+9x-4
Polynomial	*(any sum or difference of a set of monomials)*	$7x-8,\ x^3+9x-4$

The **degree of a polynomial** is the largest of the degrees of all its terms. Although the emphasis in this review will be on polynomials in one variable, polynomials may contain more than one variable. The degree of a term in more than one variable is the sum of the exponents on its variables. Thus $-4x^2y$ is a third-degree term in x and y. The expression $-4x^2y+5xy-6y^2$ is a third-degree polynomial in x and y, and $5x^3+8x^2$ is a third-degree polynomial in x.

Addition and Subtraction

Like terms (or **similar terms**) are terms that contain the same variable factors with the same exponents. Constants are like terms. We combine like terms by using the distributive property and adding (or subtracting) the coefficients.

Example 1: Simplify Polynomials

Simplify the following polynomials by combining like terms.

a. $5x^2 + 7x^2$

Solution: $5x^2 + 7x^2 = (5+7)x^2 = 12x^2$

b. $3x + 4x - 13x + 1$

Solution: $3x + 4x - 13x + 1 = (3+4-13)x + 1 = -6x + 1$

c. $4x^3 - x^3 - 5 + 3 - 3x^3$

Solution: $4x^3 - x^3 - 5 + 3 - 3x^3 = (4-1-3)x^3 - 5 + 3 = 0 \cdot x^3 - 2 = -2$

The **sum** of two or more polynomials is found by combining like terms. The **difference** of two polynomials is found by adding the opposite of each term being subtracted.

Example 2: Addition with Polynomials

Find the sum $(2x^2 + 5x - 7) + (8x^2 - 6x + 1)$.

Solution:
$$\begin{aligned}(2x^2 + 5x - 7) + (8x^2 - 6x + 1) &= 2x^2 + 8x^2 + 5x - 6x - 7 + 1 \\ &= (2+8)x^2 + (5-6)x + (-7+1) \\ &= 10x^2 - x - 6\end{aligned}$$

Example 3: Subtraction with Polynomials

Find the difference $(9x^3-4x^2+3x)-(8x^3+4x^2-5)$.

Solution:
$$\begin{aligned}(9x^3-4x^2+3x)-(8x^3+4x^2-5) &= 9x^3-4x^2+3x-8x^3-4x^2+5\\ &= 9x^3-8x^3-4x^2-4x^2+3x+5\\ &= x^3-8x^2+3x+5\end{aligned}$$

Multiplication

The product of a monomial with a polynomial of two or more terms can be found by using the distributive property and the rules of exponents.

Example 4: Multiplication with Polynomials

Use the distributive property to find the products.

a. $5x^2(2x^3+3)$

Solution:
$$\begin{aligned}5x^2(2x^3+3) &= 5x^2\cdot 2x^3+5x^2\cdot 3\\ &= 10x^5+15x^2\end{aligned}$$

b. $3x(x^2-6x+2)$

Solution:
$$\begin{aligned}3x(x^2-6x+2) &= 3x\cdot x^2+3x\cdot(-6x)+3x\cdot 2\\ &= 3x^3-18x^2+6x\end{aligned}$$

We can find the product of two binomials by using the **FOIL method**. For example,

$$(3x-2)(x+4) \xrightarrow{\text{FOIL}} (3x-2)(x+4) = 3x\cdot x+3x\cdot 4+(-2)\cdot x+(-2)\cdot 4$$
$$= 3x^2+12x-2x-8$$
$$= 3x^2+10x-8$$

Example 5: Using the FOIL Method

Use the FOIL method to find the products.

a. $(x+5)(x+6)$

Solution:

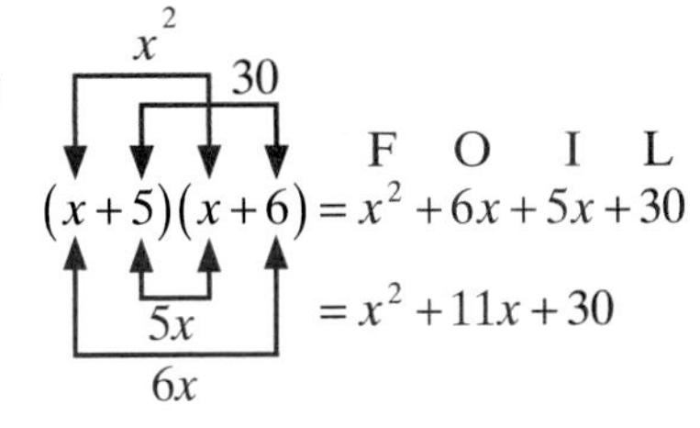

b. $(3x-2)(7x+5)$

Solution:

$21x^2$ -10

F O I L

$$(3x-2)(7x+5)=21x^2+15x-14x-10$$

$$=21x^2+x-10$$

$-14x$

$15x$

Certain products occur so frequently that they are stated as general formulas and are given names:

	Special Products	Names
I	$X^2-A^2=(X+A)(X-A)$	**Difference of two squares**
II	$(X+A)^2=X^2+2AX+A^2$	**Perfect square trinomial**
III	$(X-A)^2=X^2-2AX+A^2$	**Perfect square trinomial**
IV	$X^3+A^3=(X+A)(X^2-AX+A^2)$	**Sum of two cubes**
V	$(X-A)(X^2+AX+A^2)=X^3-A^3$	**Difference of two cubes**

NOTES

CAUTION!
Be careful to note that the two trinomials X^2-AX+A^2 and X^2+AX+A^2 in Formulas **IV** and **V** are not perfect square trinomials.

Example 6: Using Special Products

Find and name each of the following products.

a. $9x^2 - 16$

Solution: $9x^2 - 16 = (3x)^2 - (4)^2$

$= (3x+4)(3x-4)$ Difference of two squares

b. $(x+5)^2$

Solution: $(x+5)^2 = x^2 + 2 \cdot 5 \cdot x + 5^2$

$= x^2 + 10x + 25$ Perfect square trinomial

c. $(x+3)(x^2 - 3x + 9)$

Solution: $(x+3)(x^2 - 3x + 9) = x^3 + 3^3$

$= x^3 + 27$ Sum of two cubes

This product can also be found by using the distributive property and combining like terms as follows:

$$(x+3)(x^2 - 3x + 9) = (x+3)(x^2) + (x+3)(-3x) + (x+3)(9)$$

$$= x^3 + 3x^2 - 3x^2 - 9x + 9x + 27$$

$$= x^3 + 27$$

Factoring

Since factoring polynomials requires evidence of multiplication, our previous work with multiplication gives clues to factoring successfully. The following sequence of steps is recommended for factoring polynomials.

Steps for Factoring Polynomials

1. *Look for any common factors (usually monomials or binomials).*
2. *See if the product fits any of the special forms just listed.*
3. *Try the reverse of the FOIL method if the product is a trinomial.*

NOTES Factoring as a technique can be applied, especially to solve equations, even if all the coefficients are not integers. For example, since $3=\left(\sqrt{3}\right)^2$, we can write $x^2-3=\left(x-\sqrt{3}\right)\left(x+\sqrt{3}\right)$, and since $5=\left(\sqrt[3]{5}\right)^3$, we can write $x^3-5=\left(x-\sqrt[3]{5}\right)\left(x^2+\sqrt[3]{5}x+\sqrt[3]{5^2}\right)$. For the purpose of algebra review, we emphasize only integer factorizations in this chapter.

Example 7: Factoring Polynomials

Factor out any common factors in each of the following polynomials.

a. $3x^3+15x^2+21x$

Solution: $3x^3+15x^2+21x=3x\cdot x^3+3x\cdot 5x+3x\cdot 7$

$$=3x\left(x^2+5x+7\right)$$

The largest common factor is $3x$.

b. $x^2\left(x^2+2\right)+5x\left(x^2+2\right)+3\left(x^2+2\right)$

Solution: $x^2\left(x^2+2\right)+5x\left(x^2+2\right)+3\left(x^2+2\right)=\left(x^2+2\right)\left(x^2+5x+3\right)$

The binomial $\left(x^2+2\right)$ is a common factor.

Example 8: Using the FOIL Method

Use the FOIL method to factor the following polynomials.

a. $x^2+8x+12$

Solution: For $x^2+8x+12$, we have $\mathrm{F}=x^2$ and L = 12. Thus, in the following diagram, we want to find the two missing numbers such that L = 12 and $\mathrm{O}+\mathrm{I}=8x$.

$\mathrm{F}=x^2$ $\quad$ $\mathrm{L}=12$

$$x^2+8x+12=(x+\ ?\ )(x+\ ?\ )$$

I $\quad$ O

Since $6\cdot 2=12$ and $6+2=8$, we have

$$x^2+8x+12=(x+6)(x+2).$$

continued on next page ...

b. $2x^2 - 13x - 7$

Solution: For $2x^2 - 13x - 7$, we have $F = 2x^2$ and $L = -7$. In the following diagram we need factors of -7 such that $O + I = -13x$.

$F = 2x^2$ $L = -7$

$$2x^2 - 13x - 7 = (2x + \ ?\ )(x + \ ?\ )$$

I

O

We want to fill in the question marks with +1 and −7 in that order. Doing so will result in $O = 2x(-7) = -14x$, $I = 1x$, and $O + I = -14x + 1x = -13x$. Thus

$$2x^2 - 13x - 7 = (2x + 1)(x - 7).$$

Example 9: Using Special Products

Factor each of the following polynomials by using the list of special products.

a. $4x^2 - 25$

Solution: $4x^2 - 25$ is the difference of two squares:

$$4x^2 - 25 = (2x + 5)(2x - 5).$$

b. $x^2 + 12x + 36$

Solution: $x^2 + 12x + 36$ is a perfect square trinomial with $A = \frac{1}{2}(12) = 6$ and $A^2 = 6^2 = 36$:

$$x^2 + 12x + 36 = (x + 6)^2.$$

c. $x^3 - 125$

Solution: $x^3 - 125$ is the difference of two cubes with $A^3 = 5^3$:

$$x^3 - 125 = (x - 5)(x^2 + 5x + 25).$$

1.4 Exercises

In Exercises 1 – 14, perform the indicated operation and simplify the expressions.

1. $(x+3x^2)+(5-x^2)$

2. $(x^2+2x-4)+(x^2-4)$

3. $(8a^2+5a+2)+(-3a^2+9a-4)$

4. $(3x^2+5x-4)+(2x^2-2x+4)$

5. $(2x^2+3x+8)-(x^2-5x+6)$

6. $(4x^3-7x^2+3x)-(-2x^3+5x-1)$

7. $(8x^2+9)-(4x^2-3x-2)$

8. $(y^3+4y^2-7)-(3y^3+y^2+2y+1)$

9. $(a^2-3ab+b^2)+(2a^2-5ab-b^2)$

10. $(7x^2-2xy+3y^2)+(-3x^2-2xy+5y^2)$

11. $(-3x^2-2xy+5y^2)-(4x^2+3xy)$

12. $(5x^2-3xy+7y^2)-(6x^2-9xy+8y^2)$

13. $2x^2(3x^2+5x-1)$

14. $-4y^2(2y^2+5y-4)$

Fill in the missing expressions in Exercises 15 – 17.

15. $(2x+3y)^2=4x^2+\_\_\_\_+9y^2$

16. $(9x-5y)^3=729x^3-3(\_\_\_\_)x^2y+3(\_\_\_\_)xy^2-125y^3$

17. $(3x^2+8y)^2=9(\_\_\_\_)+48(\_\_\_\_)+64(\_\_\_\_)$

In Exercises 18 – 33, find the products.

18. $(3x-8)(x-5)$

19. $(7x+6)(2x-3)$

20. $(5x+11)(3x-4)$

21. $(3x-4)(4x-3)$

22. $(3x+1)^2$

23. $(4x-3)^2$

24. $(7x-4y)^2$

25. $(3x+2y)^2$

26. $(4x-5)(4x+5)$

27. $(6x+y)(6x-y)$

28. $3x^2(1+3x)$

29. $2x(x^2+3x-4)$

30. $(x+2)(x^2-2x+4)$

31. $(x+3)(x^2-3x+9)$

32. $(y-5)(y^2+5y+25)$

33. $(x+2y)(x^2-2xy+4y^2)$

In Exercises 31 – 36, simplify each expression.

34. $5a+2(a-3)-(3a+7)$

35. $11+[3x-2(1+5x)]$

36. $3y-[5-7(y+2)-6y]$

37. $10t-[8-5(3-2t)-7t]$

38. $x(x-5)+[6x-x(4-x)]$

39. $x(2x+1)-[5x-x(2x+3)]$

Fill in the missing expressions in Exercises 39 and 40.

40. $11x^2-99y^2=11(x-\_\_\_\_)(x+\_\_\_\_)$

41. $16x^3+54y^3=2(2x+\_\_\_\_)(\_\_\_\_x^2-\_\_\_\_xy+9y^2)$

In Exercises 37 – 55, factor each expression completely. (Each factor should have integer coefficients.)

42. $x^2+6x-27$

43. $s^2-5s-14$

44. $x^2+27x+50$

45. $x^2+11x-26$

46. $2x^2-98$

47. $4b^3-64b$

48. $9y^3-16y$

49. $27a^2-12$

50. $x^2+6xy+9y^2$

51. x^6-1

52. x^5-x^3

53. $25x^8-16$

54. $125y^6-27z^3$

55. $2t^3+16y^3$

56. $1600x^2+880xy+121y^2$

57. s^4-1

58. $3a^2+12ab+12b^2$

59. $100xy^2+200xy+100x$

60. $x^{21}-x^{19}$

61. If you were teaching Algebra I to ninth grade students, how would you explain the difference between a variable and a constant in an algebra expression?

Hawkes Learning Systems: Essential Calculus

Polynomials

1.5 Lines and Their Graphs

Objectives

After completing this section, you will be able to:

1. *Graph ordered pairs.*
2. *Graph linear equations.*
3. *Find the slope of a line.*
4. *Find the equation of a line given two characteristics of the line.*

Cartesian Coordinate System

In the **Cartesian coordinate system**, two perpendicular number lines, one horizontal and one vertical, are used to separate a plane into four **quadrants**. The horizontal number line is called the ***x*-axis**, and the vertical number line is called the ***y*-axis**. The point where the two axes intersect is called the **origin**.

Each point in the plane can be represented by an **ordered pair** of real numbers. Just as there is a 1–1 correspondence between the set of real numbers and the points on a number line, there is a 1–1 correspondence between the set of ordered pairs of real numbers and the points in a plane. In an ordered pair (x, y), the value of x is called the **first coordinate** (or **first component** or ***x*-coordinate**), and it represents the directed distance of the point from the y-axis. The value of y is called the **second coordinate** (or **second component** or ***y*-coordinate**), and it represents the directed distance of the point from the x-axis. Several points, including the origin, which is designated by the ordered pair (0, 0), have been **plotted** (or **graphed**) in Figure 1.5.1.

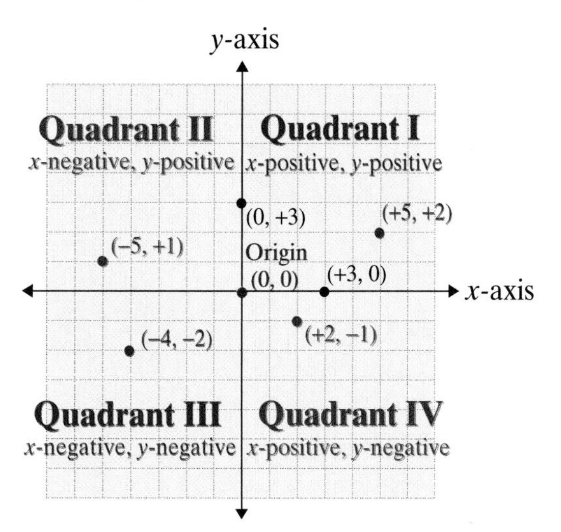

Figure 1.5.1

Straight Lines

A first degree equation in x and y is called a **linear equation**, and the graph of all the points that satisfy such an equation is a straight line. Straight lines and their properties are particularly useful in the study of calculus and its applications.

Over short ranges of the variable x, a straight line can be a very useful approximation to a graph. Straight lines are simpler to work with algebraically and numerically.

All of these ideas will be discussed in detail in later chapters.

The **slope** of a line is a measure of the steepness of the line. We will see that the slope can be **positive** (for lines that slant upward to the right), **negative** (for lines that slant downward to the right), **0** (for horizontal lines), or **undefined** (for vertical lines). Given any two points (x_1, y_1) and (x_2, y_2) on a line, we can calculate the slope of the line with the following formula. (See Figure 1.5.2.)

Slope of a Line

$$\textit{Slope} = m = \frac{\textit{rise}}{\textit{run}} = \frac{y_2 - y_1}{x_2 - x_1} = \frac{y_1 - y_2}{x_1 - x_2},$$

where $x_1 \neq x_2$.

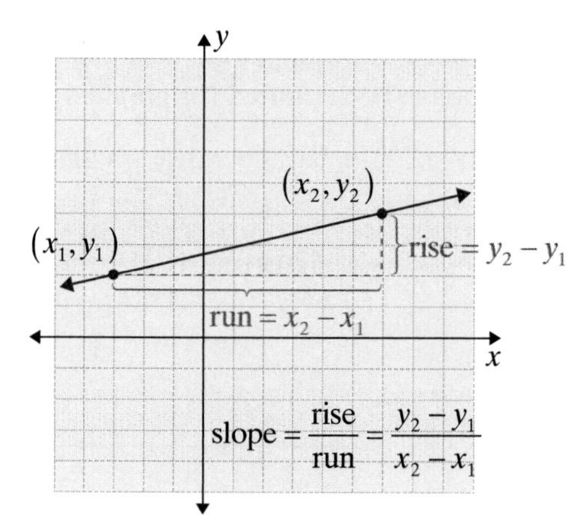

Figure 1.5.2

Example 1: Slope of a Line

Find the slope of the line that contains the two points (–2, 3) and (5, –1) and then graph the line.

Solution: Using $(x_1, y_1) = (-2, 3)$ and $(x_2, y_2) = (5, -1)$,

$$m = \frac{y_2 - y_1}{x_2 - x_1} = \frac{-1-3}{5-(-2)} = \frac{-4}{7} = -\frac{4}{7}.$$

Or, using $(x_1, y_1) = (-2, 3)$ and $(x_2, y_2) = (5, -1)$,

$$m = \frac{y_1 - y_2}{x_1 - x_2} = \frac{3-(-1)}{-2-5} = \frac{4}{-7} = -\frac{4}{7}.$$

The slope $m = -\frac{4}{7}$ indicates that y changes –4 units for every change of 7 units in x.

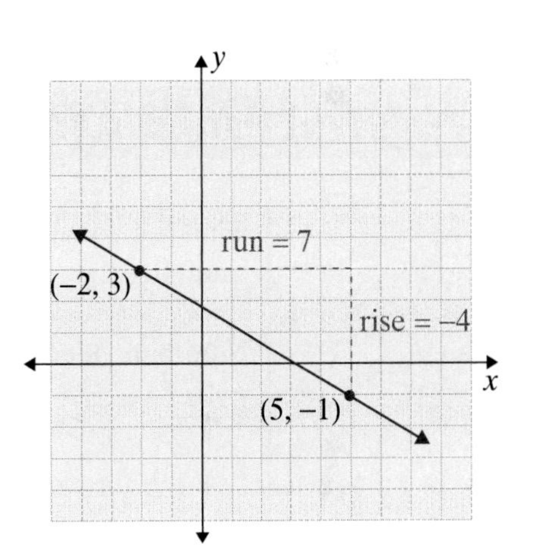

The three basic forms for the equation of a line are given in the list on the following page. Each form can be algebraically manipulated to appear in either of the other forms, but the slope-intercept form is usually preferred in calculus applications.

Forms of Linear Equations

1. ***Standard form:***

   $Ax + By = C$ *where A and B are not both* 0.

2. ***Slope-intercept form:***

   $y = mx + b$

   The slope of the line is m. The y-intercept is the point (0, *b*). *The coordinate b is also called the y-intercept.*

3. ***Point-slope form:***

   $y - y_1 = m(x - x_1)$

   The slope of the line is m, and (x_1, y_1) *is a point on the line.*

Example 2: Equation of a Line

Find the equation of the line in slope-intercept form containing the two points (1, –5) and (10, 22).

Solution: First find the slope:

$$m = \frac{y_2 - y_1}{x_2 - x_1} = \frac{22-(-5)}{10-1} = \frac{27}{9} = 3.$$

Because $y = mx + b$ and $m = 3$, we know that $y = 3x + b$. Now we only need to determine the value of b. To do so, either point can be used. Choosing the point (10, 22), we have

$$y = 3x + b \qquad \text{Slope-intercept form with } m = 3.$$

$$22 = 3(10) + b \qquad \text{Using the fact that } (x, y) = (10, 22).$$

$$22 = 30 + b$$

$$-8 = b.$$

Therefore, $y = 3x - 8$.

Check: One can check that the other point (1, –5) satisfies this equation:

$$y = 3x - 8$$

$$-5 = 3(1) - 8$$

$$-5 = -5.$$

Example 3: Equation of a Line

Find the equation of the line with slope $\frac{3}{2}$ and passing through the point (0, 1). Graph the line.

Solution: The point (0, 1) is the y-intercept. Thus we can use the slope-intercept form.

$$y = \frac{3}{2}x + 1$$ Slope-intercept form.

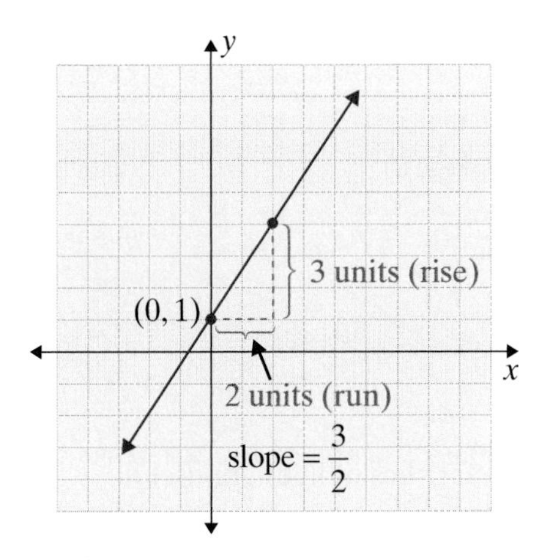

If we start at (0, 1) and move 2 units to the right (run) and 3 units up (rise), we will arrive at another point on the line. Since two points determine a line, the line through these two points is the graph of the given equation.

Example 4: Equation of a Line

Find the equation in slope-intercept form of $y - 5 = 6(x - 1)$. This is the point-slope form of a line with slope 6 which passes through (1, 5).

Solution:

$$y - 5 = 6(x - 6)$$

$$y - 5 = 6x - 6$$ distributive property

$$y = 6x - 1$$ add 5 to both sides

Now we consider the special cases of horizontal and vertical lines. The equation $y = 0 \cdot x + 3$ (or simply $y = 3$) describes a horizontal line with **slope 0**. All points with a second coordinate of 3 are on this line. The equation $0 \cdot y + x = -2$ (or simply $x = -2$) describes a vertical line with **slope undefined**. All points with first coordinate –2 are on this line. (See Figure 1.5.3 on the next page.)

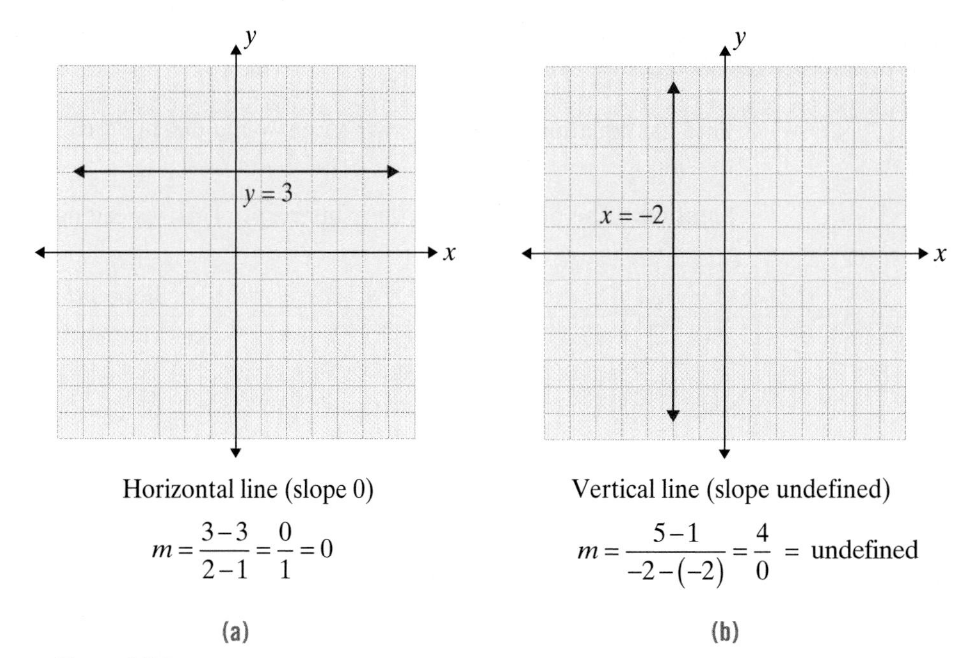

Figure 1.5.3

The following two characteristics of lines are useful:

1. **Parallel lines have the same slope**. The lines $y = 2x + 1$ and $y = 2x - 3$ are parallel because they have the same slope, 2.

2. **Perpendicular lines have slopes that are negative reciprocals of each other**. The lines $y = \frac{2}{3}x + 1$ and $y = -\frac{3}{2}x + 5$ are perpendicular because their slopes are negative reciprocals. Note that the product of their slopes is $\frac{2}{3}\left(-\frac{3}{2}\right) = -1$.

NOTES

It is helpful to remember that if a line passes through the origin, which is the point (0, 0), then the y-intercept $b = 0$; and if $b = 0$, the line goes through the origin.

1.5 Exercises

In Exercises 1 – 5, determine the slope of the line through the two points given.

1. $(3,8),(-9,4)$

2. $(8,14),(10,64)$

3. $(-3,-1),(2,14)$

4. $(1.5,6),(-2.5,-2)$

5. $(6.5,-3.1),(-2,3.3)$

In Exercises 6 – 10, determine if the line given contains the specified point.

6. $2x+3y=17;$ $(2,3)$

7. $y=6x-3;$ $(10,44)$

8. $y=-6x-4;$ $(8,-52)$

9. $y=23x-3;$ $(2,63)$

10. $7x-5y=20;$ $(-30,-46)$

For each line described in Exercises 11 – 15, write an equation in the form specified.

11. Passing through $(-1,0)$ with slope $\frac{7}{2}$; standard form

12. Passing through $(4,-1)$ and $(-2,3)$; standard form.

13. Passing through $(0,3)$ and $(-3,1)$; slope-intercept form.

14. Passing through $(5,2)$ and $(-3,6)$; point-slope form.

15. Passing through $(-3,4)$ and $(2,2)$; standard form.

In Exercises 16 – 21, write each equation in slope-intercept form, if possible. Find the slope and y-intercept; then draw the graph.

16. $x-2y=6$

17. $x+5y=10$

18. $4x-7=0$

19. $4x+2y=11$

20. $3x+4y=8$

21. $5x-6y=12$

In Exercises 22 – 26, determine if lines $\overleftrightarrow{AB}$ and $\overleftrightarrow{BC}$ are the same line.

22. $A=(0,0),\ B=(3,8),\ C=(6,16)$

23. $A=(1,-1),\ B=(3,2),\ C=(6,16)$

24. $A=(1,1),\ B=(2,4),\ C=(6,16)$

25. $A=(-6,2),\ B=(0,2),\ C=(5,8)$

26. $A=(1,3),\ B=(-1,-1),\ C=(0,2)$

In Exercises 27 – 36, determine if the two lines given are parallel, perpendicular, or neither. Explain why.

27. $y = 4x - 1$ and $y = 0.25x + 6$

28. $y = 11x + 4$ and $y = 11x - 4$

29. $2x + 3y = 17$ and $2x + 3y = 0$

30. $4x + 6y = 4$ and $3x - 2y = 4$

31. $45x + 5y = 20$ and $y = \frac{1}{9}x + 3$

32. $y = 5x + 7$ and $10x - 2y = 9$

33. $3x - y = 12$ and $-x + 3y = 36$

34. $y = \frac{1}{2}x + 8$ and $2x - 4y = 3$

35. $y = x - 15$ and $3x + 3y = 15$

36. $y = 6x - 6$ and $4x + 24y = 4$

In Exercises 37 – 42, find an equation for the line satisfying the given conditions.

37. Parallel to $x + 3y = 4$ and passing through $(3, 1)$.
38. Parallel to $2x + 3y = 5$ and passing through $(-2, 4)$.
39. Parallel to $7x - 5y = 2$ and passing through $(-1, 2)$.
40. Perpendicular to $5x + 2y = 3$ and passing through $(-4, -1)$.
41. Perpendicular to $3x - 4y = 4$ and passing through $(3, 3)$.
42. Perpendicular to $x - 6y = 4$ and passing through $(-2, 7)$.

43. Sketch $y = 3x - 4$ and $y = -3x + 2$. Draw a horizontal line through their common point. What can be said about this horizontal line?

44. Every line that does not pass through the origin determines a right triangle with the right angle at the origin. The three sides of the triangle lie on part of the line and the two axes (see Figure 1.5.4). Write a formula for the area of this triangle if the line is given by $y = mx + b$.

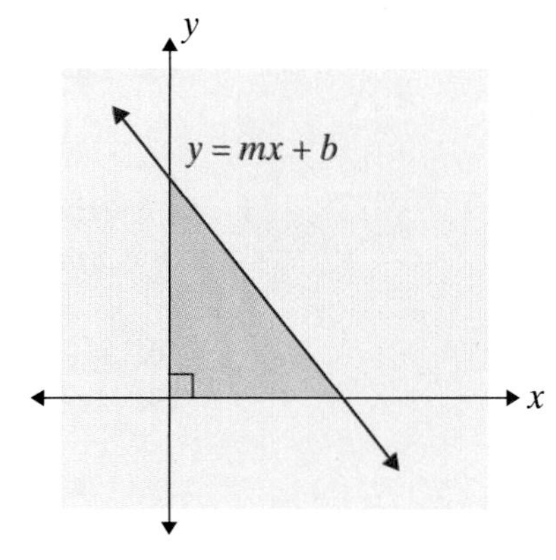

Figure 1.5.4

45. Suppose (c_1, d_1) and (c_2, d_2) are points on the graph of $y = mx + b$. Then $m = \frac{d_2 - d_1}{c_2 - c_1}$. Find an expression for b in terms of the c's and d's.

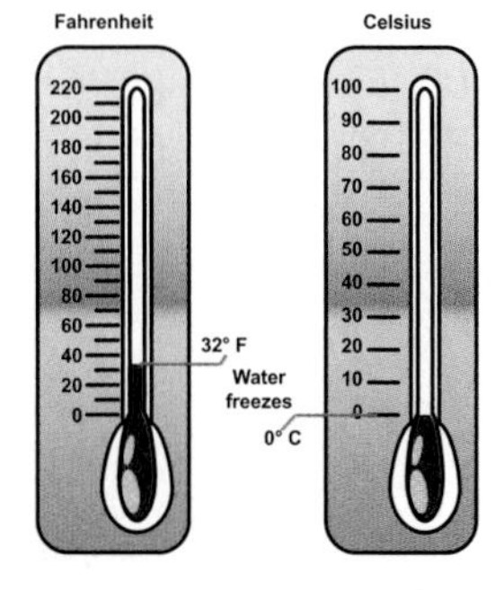

46. Water freezes at 32° Fahrenheit and at 0° Celsius. Water boils at 212° Fahrenheit and at 100° Celsius. Determine a linear relation (equation) for a temperature F (for Fahrenheit) in terms of the temperature C (for Celsius).

47. Given that the lines $y_1 = 2x_1 + b_1$ and $y_2 = -3x_2 + b_2$ intersect at $(12, 24)$, find b_1 and b_2 and sketch both of the lines.

48. A certain high school cafeteria has costs (in dollars) given by $C(x) = 0.5x + 300$, where x is the number of students served. The school charges \$0.75 per student for lunch.

a. If the school usually serves 1000 students, how much is the average daily loss in operations?

b. How many students are needed in order to break even?

49. If a vendor of hot dogs at school ball games profits \$100 on the sale of 300 hot dogs and \$130 on the sale of 360 hot dogs, how many must be sold in order to break even? Assume a linear relationship between profit and the quantity x of hot dogs.

50. Suppose a roadside stand will sell 250 pounds of shrimp for \$562.50. At the same rate, what would 100 pounds cost?

51. At one time gold sold for a price of \$325 an ounce. Six months later it sold for \$355 an ounce. What was the average monthly rate of change of the price?

52. The Castle Tee Shirt Company sells twelve dozen t-shirts with a college logo to college bookstores for \$308, and 15 dozen for \$380. Assuming a linear relationship between cost and quantity, how much would 20 dozen t-shirts costs?

Hawkes Learning Systems: Essential Calculus

Lines and Their Graphs

1.6 Functions

Objectives

After completing this section, you will be able to:

1. *Evaluate a function.*
2. *State the domain of a function.*
3. *Find the decomposition of a function.*
4. *Find the composition of a function.*
5. *Use the vertical line test to determine whether or not a graph represents a function.*
6. *Recognize and apply the difference quotient.*

Definition of a Function

Suppose that the labor costs to produce a book are \$5 per book and there are fixed costs of \$1000 (rent, light, heat, etc.) regardless of whether or not any books are produced. Then, for x books, the costs in dollars can be expressed as

$$C = 5x + 1000.$$

We say that the cost is a **function** of the number of books produced. In function notation, we write

$$C(x) = 5x + 1000$$ $C(x)$ is read "C of x."

and $$C(2000) = 5 \cdot 2000 + 1000 = 11,000.$$ 2000 is the x-value input and 11,000 is the y-value output.

Thus "C of 2000 equals 11,000." Or the cost of producing 2000 books is \$11,000.

For example, when we write that $f(4) = 2$, the point $(x, y) = (4, 2)$ is a point on the graph of $f(x)$. It may be helpful to read $f(4) = 2$ as "the value of f at $x = 4$ is 2."

The following formal definition of a **function** includes two important terms, **domain** and **range**. In this text, the domain and range will always be sets of real numbers.

Function, Domain, and Range

*Let D and R be two sets of real numbers. A **function** f is a rule that matches each number x in D with exactly one number y (or f(x)) in R. D is called the **domain** of f, and R is called the **range** of f.*

Note: In a function, there is only one y-value (or f(x)-value) for each value of x.

Intuitively, a function can be thought of as an input-output machine (or a magic box). Values of x are input (such as numbers of books), and values of y [or $f(x)$] are output (such as dollars representing costs). For our purposes, we can think of the machine as an algebraic expression, such as $5x + 1000$. (See Figure 1.6.1.)

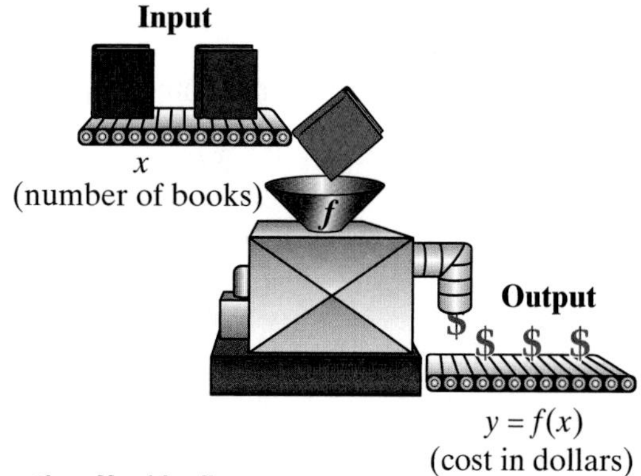

Figure 1.6.1 "A Function Machine"

We say that x is the **independent variable** and that y [or $f(x)$] is the **dependent variable** because we choose the value of x but the value of y depends on our independent choice.

Example 1: Function Evaluation

For the familiar function $f(x) = x^2$, evaluate the following:

a. $f(3)$ — Read "f of 3."
Solution: $f(3) = 3^2 = 9$ — 3 is substituted for x.

b. $f(5)$
Solution: $f(5) = 5^2 = 25$ — 5 is substituted for x.

c. $f(-2)$
Solution: $f(-2) = (-2)^2 = 4$ — -2 is substituted for x.

Example 2: Function Evaluation

For $f(x) = x^3 - 2x^2 + 3x + 100$, evaluate the following:

a. $f(2)$

Solution: $f(2) = 2^3 - 2 \cdot 2^2 + 3 \cdot 2 + 100$ Substitute 2 for x.

$= 8 - 8 + 6 + 100 = 106$

b. $f(a)$

Solution: $f(a) = a^3 - 2a^2 + 3a + 100$ Substitute a for x.

c. $f(a+1)$

Solution: $f(a+1) = (a+1)^3 - 2(a+1)^2 + 3(a+1) + 100$ Substitute $a + 1$ for x.

Expanding and combining like terms, we have:

$$f(a+1) = a^3 + 3a^2 + 3a + 1 - 2(a^2 + 2a + 1) + 3a + 3 + 100$$

$$= a^3 + 3a^2 + 3a + 1 - 2a^2 - 4a - 2 + 3a + 3 + 100$$

$$= a^3 + a^2 + 2a + 102.$$

d. $f(a) + 1$

Solution: $f(a) + 1 = a^3 - 2a^2 + 3a + 100 + 1$ Add 1 to $f(a)$.

$= a^3 - 2a^2 + 3a + 101$

NOTES

CAUTION!

As illustrated in Examples 2c and 2d, $f(a+1) \neq f(a) + 1$.

Example 3: Function Evaluation

For $F(x) = x^2 - 8x$, evaluate the following:

a. $F(5) - F(2)$

Solution: We want the difference between the two functional values $F(5)$ and $F(2)$.

$F(5) = 5^2 - 8 \cdot 5 = 25 - 40 = -15$ — Substitute 5 for x.

$F(2) = 2^2 - 8 \cdot 2 = 4 - 16 = -12$ — Substitute 2 for x.

$F(5) - F(2) = -15 - (-12) = -15 + 12 = -3$ — Find the difference.

b. $F(x + h)$

Solution: Substitute $x + h$ for x.

$$F(x+h) = (x+h)^2 - 8(x+h)$$ Substitute $x + h$ for x.

$$= x^2 + 2xh + h^2 - 8x - 8h$$

c. $F(x + h) - F(x)$

Solution: $F(x+h) - F(x) = \left[(x+h)^2 - 8(x+h)\right] - \left[x^2 - 8x\right]$ — Find the difference between $F(x + h)$ and $F(x)$.

$$= x^2 + 2xh + h^2 - 8x - 8h - x^2 + 8x$$

$$= 2xh + h^2 - 8h$$

The following example shows how a single function can be represented by different algebraic expressions, each to be used for different parts of the domain. We say that the function is **defined in pieces** or **defined piecewise.**

Example 4: Piecewise Function

A real estate broker charges a commission of 6% on sales valued up to \$300,000. For sales valued at more than \$300,000, the commission is \$6000 plus 4% of the sale price.

a. Represent the commission earned as a function R.

Solution: $R(x) = \begin{cases} 0.06x & \text{for } 0 \le x \le 300{,}000 \\ 0.04x + 6000 & \text{for } x > 300{,}000 \end{cases}$

continued on next page ...

b. Find $R(200{,}000)$.

Solution: $R(200{,}000) = 0.06(200{,}000) = \$12{,}000$

Use $R(x) = 0.06x$ since $200{,}000 \leq 300{,}000$.

c. Find $R(500{,}000)$.

Solution: $R(500{,}000) = 0.04(500{,}000) + 6000 = 20{,}000 + 6000 = \$26{,}000.$

Use $R(x) = 0.04x + 6000$ since $500{,}000 > 300{,}000$.

The letters f, g, and h (as well as F, G, and H) are commonly used to represent functions. In some cases where specific meanings are intended, we use other letters, such as C (for cost), R (for revenue), and P (for profit).

More on the Domain of a Function

The **domain** of a function f is the set of all possible values for x. The domain may be stated to be a specific set of real numbers; or, if not stated explicitly, the domain is understood to be the set of real numbers such that the corresponding functional values (or y-values) are real numbers. For polynomial functions, the domain is the set of all real numbers (the whole x-axis) unless specifically limited in a particular application.

Example 5: Domain

Find the domain of the following functions.

a. $f(x) = \dfrac{1}{x-2}$

Solution: Since no denominator can be 0, $x - 2 \neq 0$. Thus the domain consists of all real numbers but 2. We indicate this by writing $x \neq 2$.

b. $h(x) = \sqrt{x-2}$

Solution: Since we are only interested in real numbers, $x - 2$ must be nonnegative. Thus the domain is indicated by $x - 2 \geq 0$ or $x \geq 2$.

The Vertical Line Test

Many familiar formulas or equations in mathematics have useful or informative graphs. But not every such graph is the graph of a function. Remember that a function can be viewed as a rule, which, when one number is the input, then only one number is the output. If a graph shows that two or more y-values correspond to a single x-input, then the graph is not that of a function.

We can test (usually visually) whether or not a graph represents a function by using the **vertical line test**.

Vertical Line Test

*If **any** vertical line intersects a graph at more than one point, then the graph **does not** represent a function.*

As an example, the graph of $y^2 = x$ is shown in Figure 1.6.3. Since there is a vertical line that intersects the graph in more than one point, the graph does not represent a function. For the values of $x = 4$, there are two corresponding y-values, $y = 2$ and $y = -2$.

The vertical line $x = 4$ intersects the graph in two points. Therefore, the graph is **not** a function.

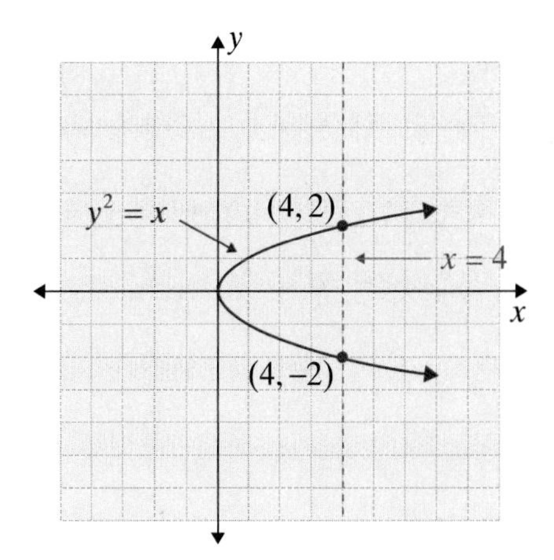

Figure 1.6.3

Example 6: Vertical Line Test

Use the vertical line test to determine whether or not each of the following graphs is a function.

a.

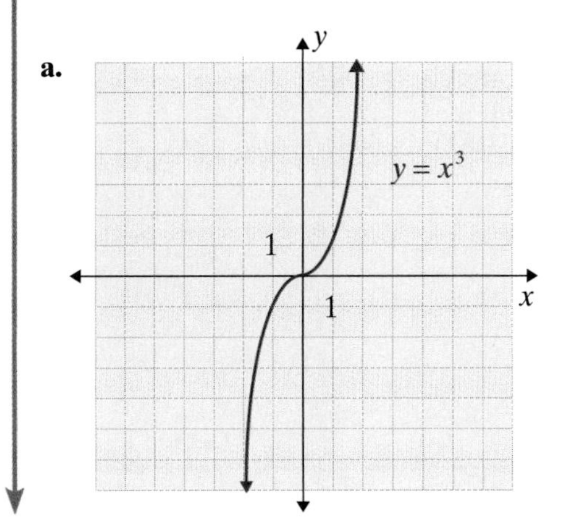

continued on next page...

Solution: The equation $y = x^3$ represents a function (which can also be written as $f(x) = x^3$). No vertical line can intersect the graph in more than one point.

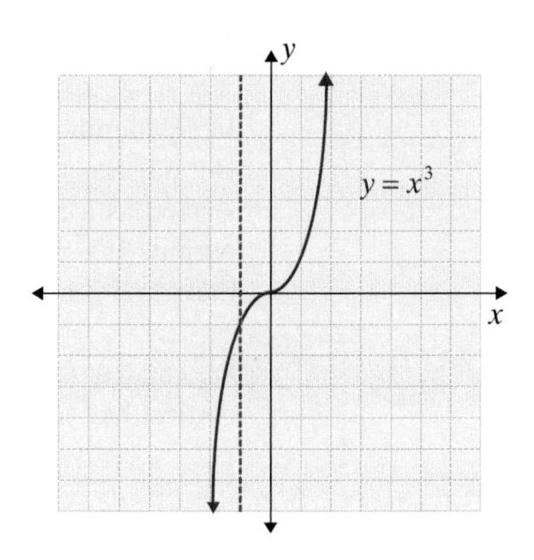

b.

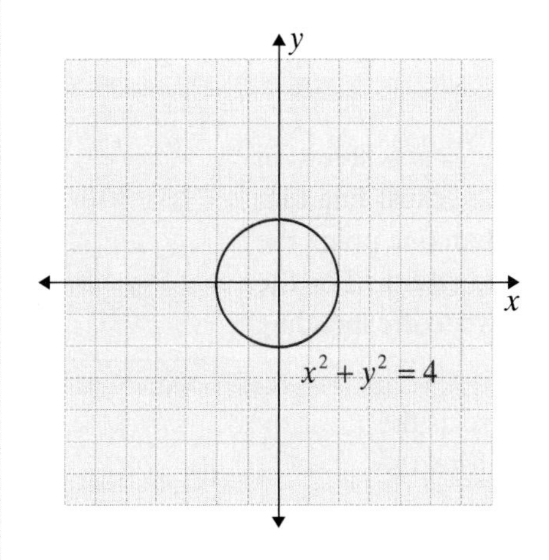

Solution: The graph of $x^2 + y^2 = 4$ is a circle. The graph shows that the equation does not represent a function. Vertical lines can be drawn that intersect the graph in more than one point.

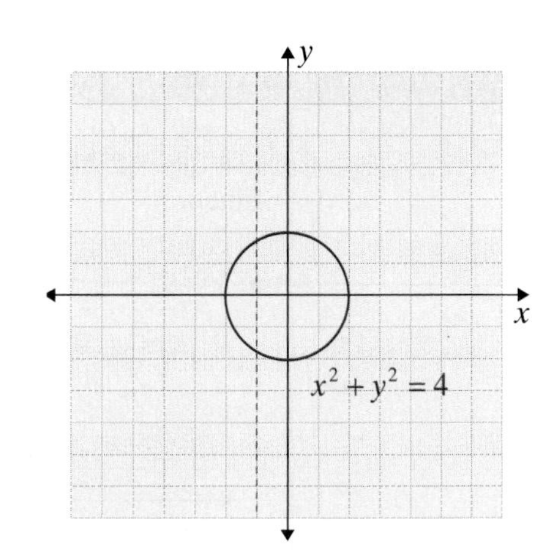

Operations with Functions

In many calculus applications we can recognize that parts of a function seem put together from other functions. For example, suppose $f(x)=\frac{3x+1}{5x-6}$. We recognize that $3x+1$ and $5x-6$ are familiar linear functions. Thus $f(x)$ is a rule which says, in effect, first find $3x+1$ and then divide by $5x-6$.

Example 7: Recognizing the Parts of a Function

Given $f(x)=x^2+6x+8$, rewrite $f(x)$ as a sum of two functions, one of which is linear.

Solution: Let $h(x)=x^2$ and $g(x)=6x+8$. Then $f(x)=h(x)+g(x)$. The function $g(x)$ is called the "linear part" of $f(x)$ and $h(x)$ is called the "quadratic part" of $f(x)$.

New Functions from Old

It is often useful, and simpler, to write $f=h+g$ instead of $f(x)=h(x)+g(x)$. In some texts, one sees $f(x)=(h+g)(x)$, but we will avoid this notation. The notation $h+g$ denotes the new function formed by addition. It is not an addition of numbers. We sometimes write

$$h+g,\quad h\cdot g,\quad \frac{h}{g},\quad h-g$$

for a new function formed from the functions h and g.

Example 8: Decomposition of a Function

Find a decomposition of $f(x)$ into two simpler functions $g(x)$ and $h(x)$. (That is, define new functions g and h and express f in terms of g and h. More than one answer is possible.)

a. $f(x)=1-3x+16\sqrt{x}$

Solution: Let $h(x)=1-3x$ and $g(x)=16\sqrt{x}$. Then $f(x)=g(x)+h(x)$.

continued on next page ...

b. $f(x)=\dfrac{3x+20}{2+13x+6x^2}$

Solution: Let $h(x)=3x+20$ and $g(x)=2+13x+6x^2$. Then $f(x)=\dfrac{h(x)}{g(x)}$.

c. $f(x)=(2x-1)\sqrt{x}$

Solution: Let $h(x)=2x-1$ and $g(x)=\sqrt{x}$. Then $f(x)=h(x)\cdot g(x)$.

Example 9: Operations with Functions

Given $f(x)=x^2+6x+8$ and $g(x)=x^2-4$, express $h(x)$ in terms of x and simplify if possible.

a. $h(x)=f(x)+g(x)$

Solution: $h(x)=f(x)+g(x)=(x^2+6x+8)+(x^2-4)=2x^2+6x+4$

b. $h(x)=f(x)-g(x)$

Solution: $h(x)=f(x)-g(x)=(x^2+6x+8)-(x^2-4)=6x+12$

c. $h(x)=f(x)\cdot g(x)$

Solution: $h(x)=f(x)\cdot g(x)=(x^2+6x+8)(x^2-4)=x^4+6x^3+4x^2-24x-32$

d. $h(x)=\dfrac{f(x)}{g(x)}$

Solution: $h(x)=\dfrac{f(x)}{g(x)}=\dfrac{x^2+6x+8}{x^2-4}=\dfrac{(x+4)(x+2)}{(x-2)(x+2)}=\dfrac{x+4}{x-2}$

If we consider the example $f(x)=\sqrt{3x+1}$, we see that none of the previous four types of combinations are suitable for expressing the nature of $f(x)=\sqrt{g(x)}$. The final, and possibly the most important, type of operation with functions, is that of **function composition**.

For example if

$$f(x) = x^2 + 3x + 2$$

and $$g(x) = x - 1,$$

then the **composite function** $(f \circ g)(x) = f(g(x))$ (read "f of g of x") is formed by substituting the entire expression for $g(x)$ in place of x in the expression for $f(x)$.

$$\begin{aligned} f(g(x)) &= (g(x))^2 + 3 \cdot g(x) + 2 \\ &= (x-1)^2 + 3(x-1) + 2 \\ &= x^2 - 2x + 1 + 3x - 3 + 2 \\ &= x^2 + x \end{aligned}$$

Composite Functions

If f and g are two functions, then the ***composite function*** *f of g is defined to be*

$$(f \circ g)(x) = f(g(x)).$$

The domain of $(f \circ g)$ is that part of the domain of g for which $(f \circ g)$ is defined.

The concept of composite function is illustrated with the function machine idea in Figure 1.6.2. The diagram helps in understanding that the domain of $f(g(x))$ is all x in the domain of g for which $g(x)$ is in the domain of f.

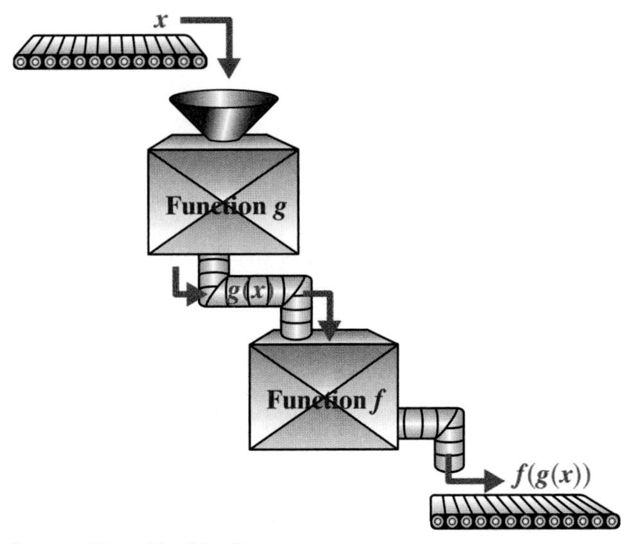

Figure 1.6.2 "A Composition Machine"

Example 10: Composite Functions

For the functions $f(x)=\sqrt{x-1}$ and $g(x)=x^2+2$, form the following composite functions.

a. $f(g(x))$

Solution: Substitute $g(x)$ for x in the expression for $f(x)$.

$$\begin{aligned} f(g(x)) &= \sqrt{g(x)-1} \\ &= \sqrt{x^2+2-1} \\ &= \sqrt{x^2+1} \end{aligned}$$

b. $g(f(x))$

Solution: Substitute $f(x)$ for x in the expression for $g(x)$.

$$\begin{aligned} g(f(x)) &= (f(x))^2+2 \\ &= \left(\sqrt{x-1}\right)^2+2 \\ &= x-1+2 = x+1 \end{aligned}$$

NOTES

Example 10 illustrates that, as a general rule, $f \circ g$ is not equal to $g \circ f$.

Example 11: Composite Functions

For the functions $F(x)=2x^2+x$ and $H(x)=3x$, find the following composite functions.

a. $F(H(x))$

Solution: $$\begin{aligned} F(H(x)) &= 2\cdot(H(x))^2+H(x) \\ &= 2(3x)^2+3x \\ &= 18x^2+3x \end{aligned}$$

b. $H(F(x))$

Solution: $H(F(x)) = 3 \cdot F(x)$

$= 3(2x^2 + x)$

$= 6x^2 + 3x$

Example 12: Composite Functions

In each case identify two functions $h(x)$ and $g(x)$ so that the given function $f(x)$ can be expressed as $f(x) = g(h(x))$.

a. $f(x) = \sqrt{3x+1}$

Solution: Let $h(x) = 3x + 1$ and $g(x) = \sqrt{x}$.

Then $f(x) = g(h(x)) = g(3x+1) = \sqrt{3x+1}$.

b. $f(x) = 3(x-5)^5$

Solution: Let $h(x) = x - 5$ and $g(x) = 3x^5$.

Then $f(x) = g(h(x)) = g(x-5) = 3(x-5)^5$.

c. $f(x) = 10^{3x+1}$

Solution: Let $h(x) = 3x + 1$ and $g(x) = 10^x$.

Then $f(x) = g(h(x)) = g(3x+1) = 10^{3x+1}$.

As a matter of fact, recognizing a natural decomposition of a function into a composition of two other functions is a simple idea that can be amazingly productive.

Example 13: Evaluating a Function

For each part of Example 12, calculate $f(1)$ by first calculating $h(1)$ and then $g(h(1))$.

a. $f(x) = \sqrt{3x+1}$

Solution: $h(1) = 3(1) + 1 = 4$

So $f(1) = g(h(1)) = g(4) = \sqrt{4} = 2$.

continued on next page ...

b. $f(x)=3(x-5)^5$

Solution: $h(1)=(1)-5=-4$

So $f(1)=g(h(1))=g(-4)=3(-4)^5=-3072.$

c. $f(x)=10^{3x+1}$

Solution: $h(1)=3(1)+1=4$

So $g(h(1))=g(4)=10^4=10{,}000.$

The Difference Quotient

Given that (x_1, y_1) and (x_2, y_2) are two points on the graph of some function $y=f(x)$, we note that $y_2=f(x_2)$ and $y_1=f(x_1)$. We will use the following notation

$$\Delta x = x_2 - x_1 = \text{change in } x \qquad \Delta x \text{ is read "delta-}x\text{."}$$
$$\Delta y = y_2 - y_1 = \text{change in } y$$

Consequently, the slope m of the segment (or line), which contains these two points, is

$$m=\frac{\Delta y}{\Delta x}=\frac{y_2-y_1}{x_2-x_1}=\frac{f(x_2)-f(x_1)}{x_2-x_1}.$$

The quotient expressed above will occur repeatedly in later sections, and we wish to emphasize that such quotients are slopes.

Example 14: The Difference Quotient

For $f(x)=8-x^2$ calculate, simplify, and interpret the following expressions.

a. $\dfrac{f(5)-f(3)}{5-3}$

Solution: $\dfrac{f(5)-f(3)}{5-3}=\dfrac{(8-5^2)-(8-3^2)}{2}=\dfrac{-17-(-1)}{2}=\dfrac{-17+1}{2}=\dfrac{-16}{2}=-8$

This quotient is the slope of the line joining $(5,-17)$ to $(3,-1)$, two points on the graph of f.

b. Calculate Δy and Δx if $(a, f(a))$ and $(a+h, f(a+h))$ are two points on the graph of $f(x)$.

Solution: $\Delta y = y_2 - y_1 = f(a+h) - f(a) = \left[8-(a+h)^2\right] - (8-a^2)$

$$= 8 - (a^2 + 2ah + h^2) - 8 + a^2$$

$$= 8 - a^2 - 2ah - h^2 - 8 + a^2$$

$$= -2ah - h^2$$

$$\Delta x = x_2 - x_1 = (a+h) - a = h$$

The difference is Δy for the two points $(a, f(a))$ and $(a + h, f(a + h))$.

c. Calculate $m = \dfrac{\Delta y}{\Delta x} = \dfrac{f(a+h)-f(a)}{h}$ for the two points in part (b).

Solution: $m = \dfrac{\Delta y}{\Delta x} = \dfrac{f(a+h)-f(a)}{h} = \dfrac{-2ah-h^2}{h} = \dfrac{h(-2a-h)}{h} = -2a - h$

The quotient is the slope of the line containing the two points in part b since $h = \Delta x$. **Note:** $x_1 = a,\ x_2 = a + h$.

1.6 Exercises

In Exercises 1 – 9, evaluate the given function for parts a. - d.

1. $f(x) = 2x - 7$.
a. $f(5)$
b. $f(-2)$
c. $f(a + 1)$
d. $f(a) + 1$

2. $f(x) = 3x + 5$.
a. $f(2)$
b. $f(-1)$
c. $f(a + 1)$
d. $f(a) + 1$

3. $f(x) = x^2 - 2x + 1$.
a. $f(-2)$
b. $f(3)$
c. $f(a + 1)$
d. $f(a) + 1$

4. $f(x) = 3x^2 - x + 2$.
a. $f(-3)$
b. $f(2)$
c. $f(a + 1)$
d. $f(a) + 1$.

5. $f(x) = x^3 + x^2 - 3x + 1$.
a. $f(-1)$
b. $f(-3)$
c. $f(a + 1)$
d. $f(a) + 1$

6. $f(x) = 2x^3 - 4x^2 + x - 6$.
a. $f(-2)$
b. $f(4)$
c. $f(a + 1)$
d. $f(a) + 1$

7. $f(x) = 4x^2 - 1$.
a. $f(3)$
b. $f(a + 2)$
c. $f(x + h)$
d. $f(-2) - f(-1)$

8. $f(x) = 2 - 3x^2$.
a. $f(5)$
b. $f(a - 3)$
c. $f(x + h)$
d. $f(3) - f(2)$

9. $f(x) = \sqrt{x+5}$. where $a \geq -7$,
a. $f(-1)$
b. $f(a + 2)$
c. $f(x + h)$
d. $f(4) - f(1)$.

10. Let $f(x)=\sqrt{x^2+1}$. Find **a.** $f(\sqrt{3})$, and **b.** $f(a+1)$.

11. Let $f(x)=\dfrac{2x+1}{x^2}$. Find **a.** $f(1)$, **b.** $f(3)$, **c.** $f(3)-f(1)$, **d.** $\dfrac{f(3)-f(1)}{3-1}$, and **e.** interpret the meaning in parts **a.** – **d.**

12. Let $f(x)=\dfrac{x^2-4}{x+3}$. Find **a.** $f(0)$, **b.** $f(2)$, **c.** $f(2)-f(0)$, **d.** $\dfrac{f(2)-f(0)}{2-0}$, and **e.** interpret the result of part **c.**

13. Let $f(x)=\begin{cases} x-4 & \text{if } x\le 2 \\ x^2-6 & \text{if } x>2 \end{cases}$. Find **a.** $f(-1)$, **b.** $f(2)$, **c.** $f(2.5)$, and **d.** $f(3)$.

14. Let $f(x)=\begin{cases} x^2 & \text{if } x<0 \\ 3x-2 & \text{if } x\ge 0 \end{cases}$. Find **a.** $f(0)$, **b.** $f(-2)$, **c.** $f(1.5)$, and **d.** $f(3)$.

Determine the domain of f(x) in Exercises 15 – 18

15. $f(x)=\dfrac{3x+1}{(x-5)(x-6)}$

16. $f(x)=\sqrt{2x+10}$

17. $f(x)=\sqrt{x^2+2}$

18. $f(x)=\dfrac{5}{\sqrt{x+10}}$

In Exercises 19 – 39, let $f(x)=\dfrac{1}{x}$, $g(x)=x^2-2x$, and $h(x)=\sqrt{x}$. *Find each of the following.*

19. $f(x)+g(x)$

20. $f(x)-g(x)$

21. $f(x)\cdot h(x)$

22. $g(x)\cdot h(x)$

23. $\dfrac{h(x)}{f(x)}$

24. $\dfrac{g(x)}{h(x)}$

25. $f(g(1))$

26. $g(f(2))$

27. $h(g(3))$

28. $g(h(2))$

29. $f(h(3))$

30. $h(f(4))$

31. $g(t+1)$

32. $f(t-3),\ t\ne 3$

33. $h(2t-1),\ t\ge \dfrac{1}{2}$

34. $f(g(t))$

35. $g(f(x))$

36. $f(h(x))$

37. $g(h(x))$

38. $h(g(u))$

39. $h(f(u))$

In Exercises 40 – 54, find $f(x+h)-f(x)$.

40. $f(x)=3x-1$

41. $f(x)=5x-2$

42. $f(x)=x^2+4$

43. $f(x)=x^2-3$

44. $f(x)=2x^2+1$

45. $f(x)=5+3x^2$

46. $f(x)=x^2-x$

47. $f(x)=x^2+2x$

48. $f(x)=3x-x^2$

49. $f(x)=4x^2-x$

50. $f(x)=2x^2-3x$

51. $f(x)=x^3$

52. $f(x)=x^3-1$

53. $f(x)=x^3+7$

54. $f(x)=x^3+5$

In Exercises 55 – 62, determine the domain of each function.

55. $f(x)=\dfrac{2x}{x-2}$

56. $f(x)=\dfrac{x-3}{x+1}$

57. $f(x)=\dfrac{4}{x^2-x-12}$

58. $f(x)=x-3$

59. $f(x)=4-3x$

60. $f(x)=\dfrac{1}{\sqrt{2x+5}}$

61. $f(x)=\begin{cases}3x+1 & \text{if } 0\le x<4\\ 5x-2 & \text{if } x\ge 4\end{cases}$

62. $f(x)=\begin{cases}2-x^2 & \text{if } x\le 2\\ x-4 & \text{if } x>2\end{cases}$

In Exercises 63 – 74, use the vertical line test to determine whether or not each graph represents a function.

63.

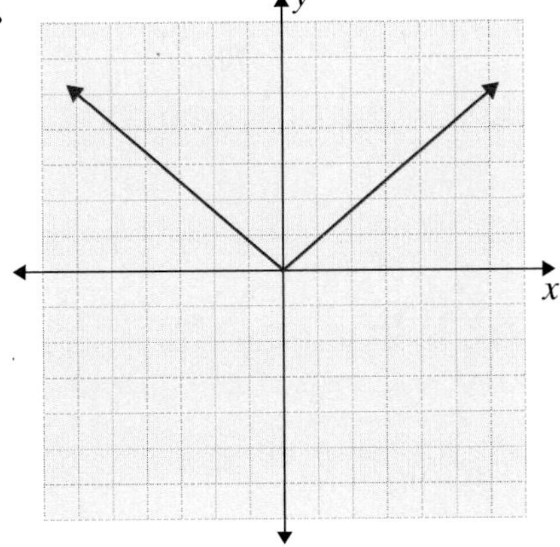

64.

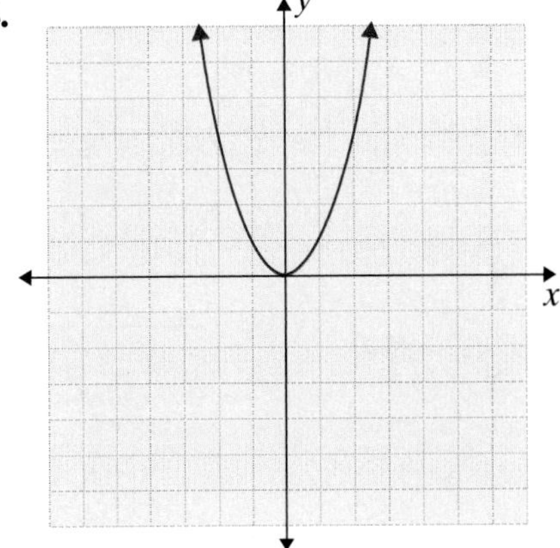

65.

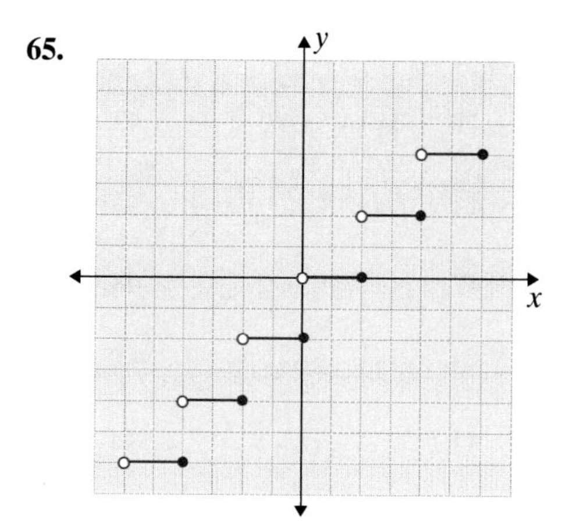

66.

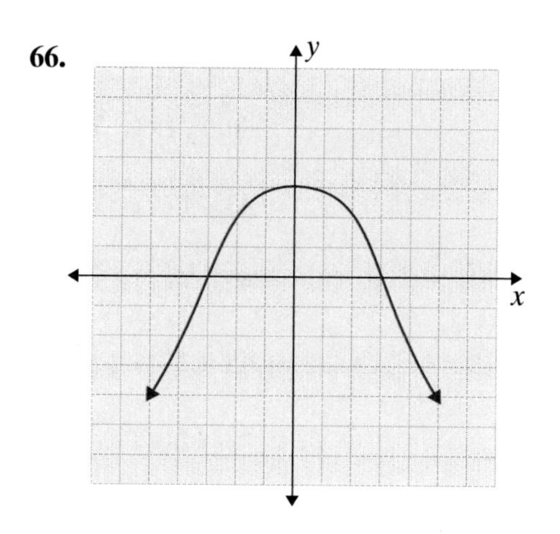

67.

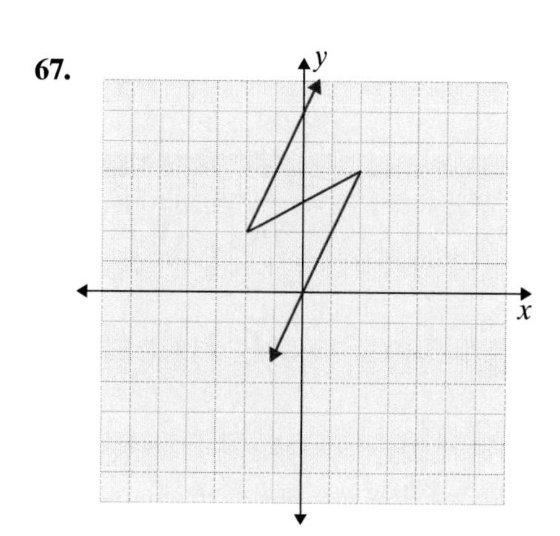

68.

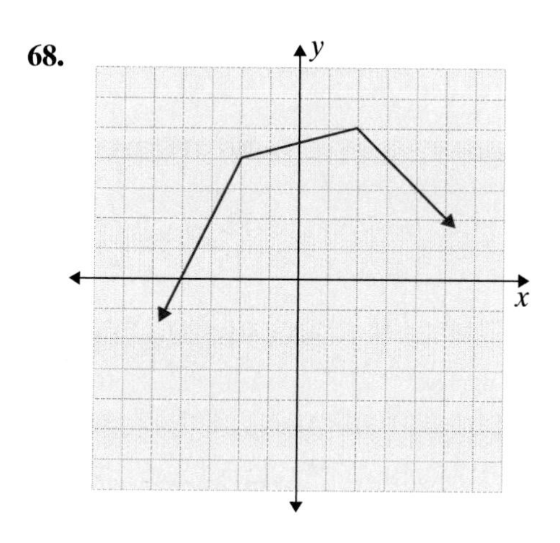

69.

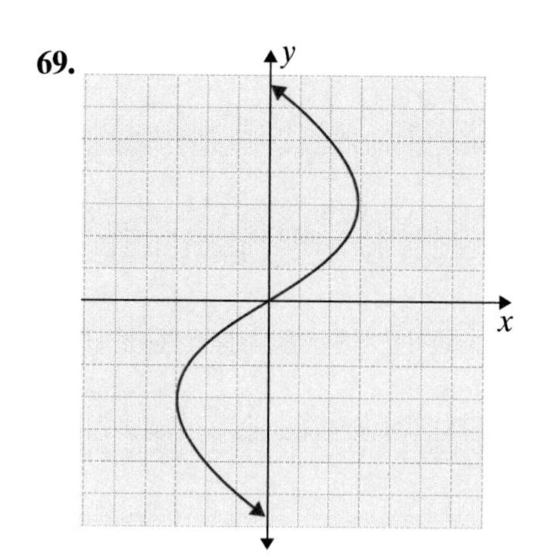

70.

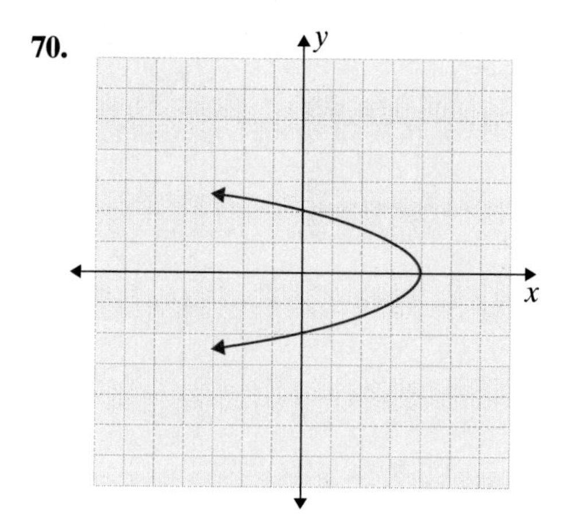

71.

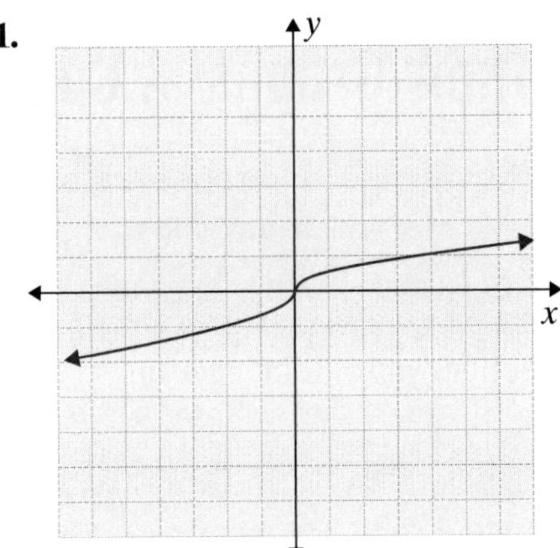

72.

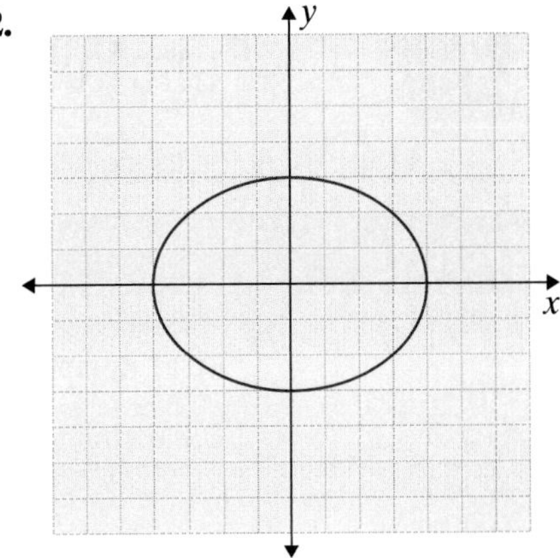

73.

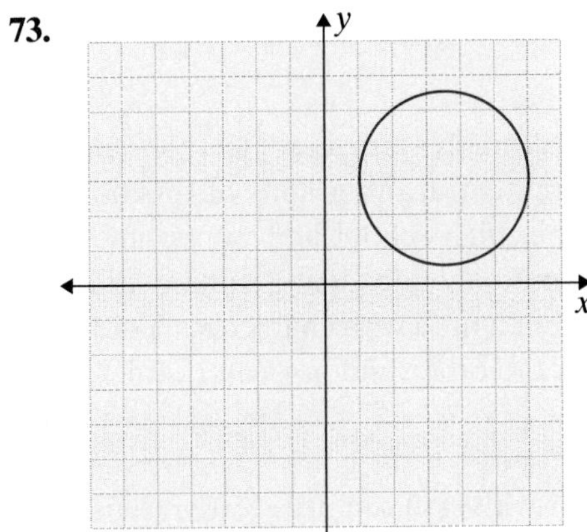

74.

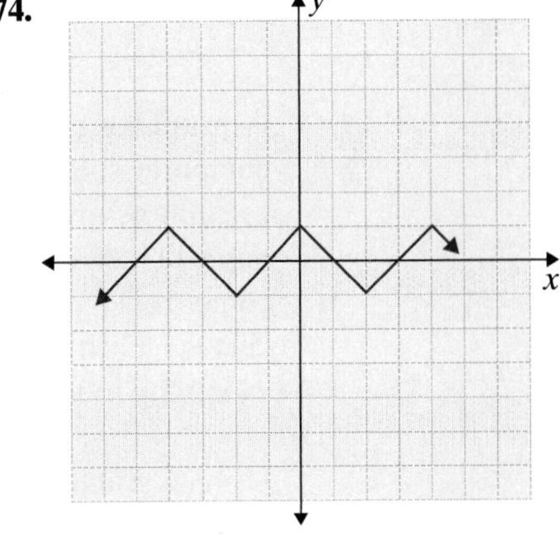

Hawkes Learning Systems: Essential Calculus

An Introduction to Functions	p. 40 - 46	Ex. 1 - 18, 55 - 74
Operations with Functions	p. 47 - 53	Ex. 19 - 54

1.7 Functions and Their Graphs: A Calculator Section

Objectives

After completing this section, you will be able to:

Use a graphing calculator to:

1. *graph a function,*
2. *make a data table of x- and y-values of a function, and*
3. *find the minimum, maximum, and zero(s) of a function.*

A **graphing utility** is the common term used to refer to a graphing calculator, or a computer software program, that draws graphs of functions. Many types of graphing calculators are introduced in high school and we assume throughout that the reader has a TI – 84 Plus calculator or an equivalent graphing utility. The insights into calculus come from patterns in algebra united with characteristics of the graphs of corresponding curves. Each supplements the other and a greater understanding is promoted by the capability to use both.

Using a graphing calculator with skill requires certain sensitivity into the input domain used and the output range of interest for a given function. For example, suppose we would like to get a reasonable graph of $y = x^2 - 20x + 100$ using the TI–84 Plus. We describe the steps specifically in the pages that follow.

For the TI–84 Plus, the steps to graph the function $y = x^2 - 20x + 100$ are:

Step 1: Turn on the calculator.

Step 2: Press the Y= key. The window you should see is shown at the right. The cursor should be blinking in the Y1 position. If there is already a formula typed into the location Y1, press the CLEAR button.

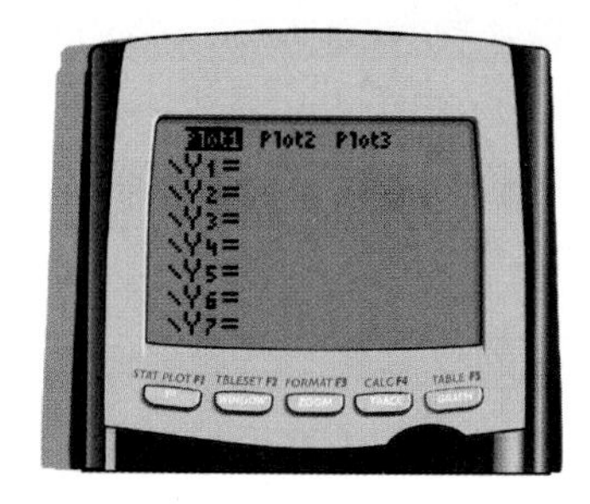

Step 3 Type X,T,θ,n x^2 − 2 0 X,T,θ,n + 1 0 0 .

This completes the process of entering the formula into the calculator's memory.

Note, the button ^ can also be used to denote exponents.

Now you must determine the set of x-values you wish to have plotted. A standard range suitable for many graphs is the x-axis interval $[-10, 10]$. Likewise, $[-10, 10]$ is often suitable for y. Let us try to create a graph using this window. Follow these steps:

Step 1: Press WINDOW.

Step 2: In the `Xmin` position type (−) `10`. This tells the calculator that $x = -10$ will be the extreme lefthand input value of x. Now press the down arrow key once and type `10` (in the `Xmax` location). This tells the calculator that $x = 10$ will be the rightmost x-value plotted and that the graph will only contain x-values in the range $-10 \leq x \leq 10$.

Step 3: Arrow down to `Xscl`, the x-scale location. This tells the calculator how far apart to place tic-marks on the x-axis. You usually want from about 10 to at most 20 tic-marks from left to right. Type 2, and down arrow once. This means that, from left to right, tic-marks will appear on the x-axis every 2 units.

Step 4: Repeat the setting for `Ymin` (the least y-value shown on the y-axis), `Ymax` (the largest y-value shown on the y-axis), and `Yscl`.

Step 5: Press GRAPH. The picture drawn is not very informative (see screen at right). The problem is that the screen cuts off at the top all y-values greater than 10. Evidently there is more to the graph, but it lies above $y = 10$. Note that when $x = 0$, y is 100. However, the picture only shows y-values up to $y = 10$. This method, checking the y-intercept, can be very helpful in determining a good range for y.

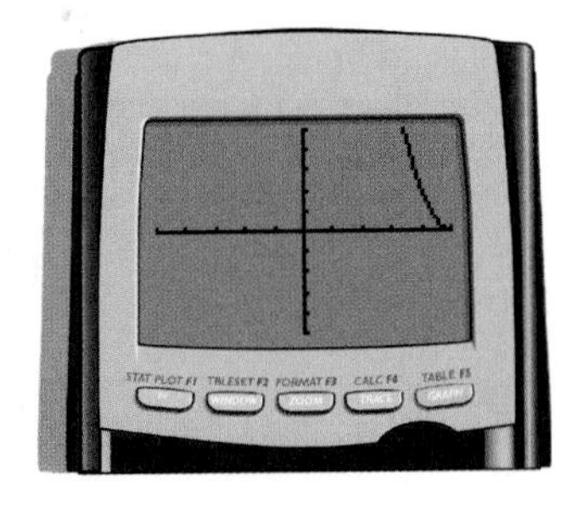

Step 6: Press WINDOW, and arrow down to `Ymax` and type `100`. Then press GRAPH. The new picture is little better. But we need to see more.

Step 7: Press WINDOW; change `Xmax` to 25, `Ymax` to 300 and `Yscl` to 25.

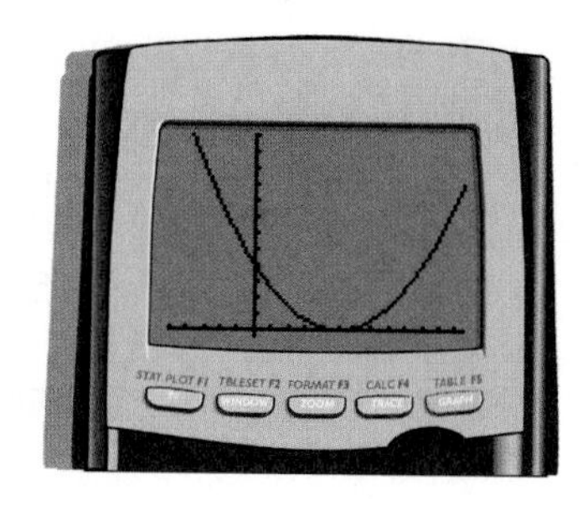

Now the characteristic parabola shape is recognizable (see screen at right). The screen is not centered on the vertex of the parabola at (10, 0) but at least the key features of the graph can be seen. The calculator has calculated about 100 y-values with uniformly separated x-values from $x = -10$ to $x = 25$. The pixels plotted have been connected with very short straight line segments. It appears that the graph just barely touches the x-axis. Let us confirm this possibility two ways, by a closer inspection visually and using algebra.

With the graph on the screen, press TRACE.

You notice that a blinking spot appears on screen and the x- and y-coordinates of the spot are displayed at the bottom. Press the right and left arrow keys to move the cursor closer to the apparent lowest point on the graph. Unfortunately, this point is very near the bottom of the screen. We need a better view.

Press ZOOM and select item `2: Zoom In` by pressing 2 and ENTER. The graph is redrawn (see screen to right) with the blinking spot centered in the screen, and the window settings have been narrowed to shrink the ranges for x and for y. Press WINDOW and note the new settings. Press GRAPH. Press TRACE.

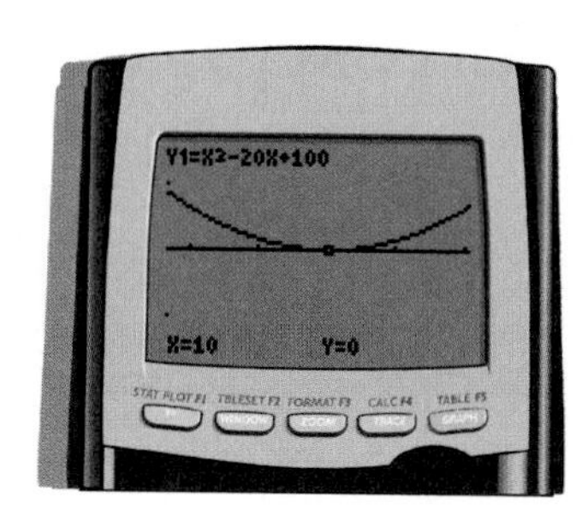

The y-value near $x = 10$ seems to be zero and we can check. Type `10` (with TRACE selected). The x-value 10 appears at the bottom. Press ENTER. The calculated value $y = f(10)$ appears, and yes, $y = 0$ (see screen at right).

Algebra confirms this since $f(x) = x^2 - 20x + 100 = (x-10)^2$. In factored form, it is now easy to see, with a glance at the formula, that $f(10) = (10 - 10)^2 = 0$. Because of the exponent 2, the factored expression reveals why all the y-values are nonnegative and why (10, 0) is the lowest point on the graph.

Example 1: Selecting Window Dimensions with a Calculator

Using a calculator as an aid, draw on a sheet of paper a sketch of $y = \frac{4x+1}{3^x}$. Determine a window of appropriate dimensions.

Solution: The steps are, press Y=, CLEAR (so to clear what might be there from previous work), (4 X,T,θ,n + 1) ÷ 3 ^ X,T,θ,n. This enters the formula for the function. Note that the parentheses **must** be used. Without parentheses, the calculator will graph $4x + \left(\frac{1}{3^x}\right)$. The character ^ is used to denote exponentiation, and the calculator computes exponents first, then division and multiplication, then addition and subtraction (remember "Please Excuse My Dear Aunt Sally"). Now we select a window. The standard window, with `Xmin` = −10, `Xmax` = 10, `Ymin` = −10, and `Ymax` = 10 is rather poor. Most of the screen is just empty space. [See Figure 1.7.1(a).]

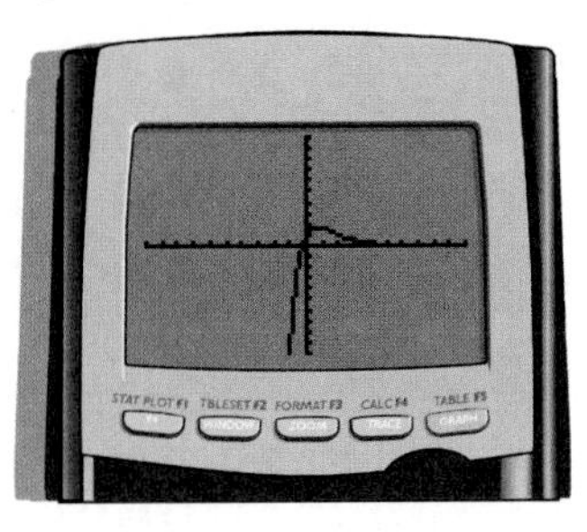

(a) First Graph

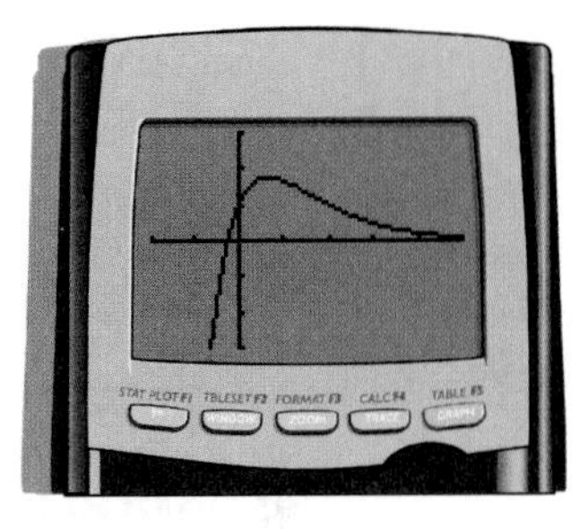

(b) Second Graph

Figure 1.7.1

Press WINDOW and edit `Xmin` = −2, `Xmax` = 5, `Xscl` = 1, `Ymin` = −3, `Ymax` = 3, and `Yscl` = 1. Press GRAPH. The second graph drawn is now very revealing and we do not need to change the window entries any further [see Figure 1.7.1(b).]

The better window entries in Example 1 can be obtained easily by trial and error. However, it is often possible (and easier) to determine a window setting by a careful analysis.

Example 2: Selecting Window Dimensions without a Calculator

Determine a sensible window for $y = \frac{6x^3 + 2x + 20}{x^2 + 1}$ without graphing. Then check by drawing the graph with a graphing utility.

Solution: For $x = 0$ we calculate (try it mentally) that the fraction reduces to 20, or that $y = 20$. So Ymax must be bigger than 20. In general, the numerator for this function will be larger than the denominator, since x^3 is bigger than x^2. If $x = 10$, y is greater than $\frac{6(10^3)}{10^2} = 60$. Lets make Ymax = 100. For large negative x, y is close to $\frac{6(x^3)}{x^2} = 6x$; which will be -60 if $x = -10$.

Lets make Ymin = -100. We will begin with the estimate that Xmin = -10 and Xmax = 10. This window turns out to be excellent (see screen at right). The key features are illustrated. If a greater range were selected for x or for y, the visual detail near the origin would be less clear. If a more narrow range were selected then the apparent large-scale behavior of the function would not be seen.

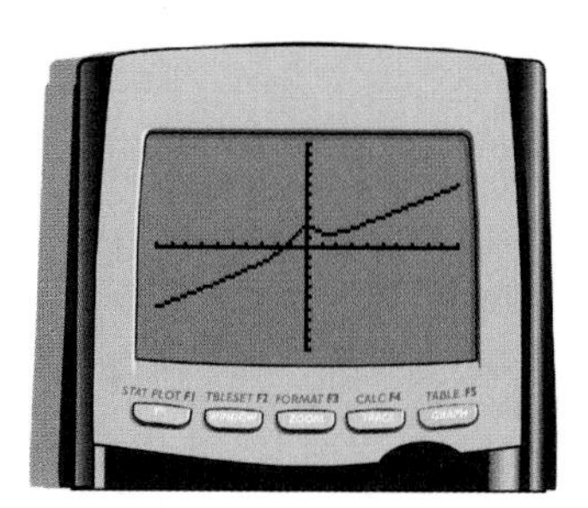

Making a Data Table Without Retyping a Formula

Suppose we want a table of x- and y-values for the function $f(x)$ in Example 2. One way is to substitute numbers into the formula and, with or without a calculator, determine the output. The TI-84 Plus provides a convenient method.

Step 1: Type the function $f(x) = \frac{(6x^3 + 2x + 20)}{(x^2 + 1)}$ into the Y_1 position.

Note: Typing the parentheses into the calculator is a critical step.

Step 2: Use the values Xmin = -10, Xmax = 10, Ymin = -100, and Ymax = 100. Press GRAPH.

Step 3: To construct the chart shown in Figure 1.7.2, press TRACE. Type (−) 3 and ENTER. The values $x = -3$ and $y = -14.8$ appear at the bottom of the screen. Write them down on a sheet of paper.

x	y
-3	-14.8
-2	-6.4
-1	6
0	20
1	14
2	14.4
3	18.8

Figure 1.7.2

Step 4: Type (−) 2 and enter. Note that $x = -2$ and $y = -6.4$ now appear on the screen; write them down.

Step 5: Continue with the remaining x-inputs.

Solving an Equation

Suppose we want to solve

$$\frac{6x^3 + 2x + 20}{x^2 + 1} = 0.$$

From the chart just above, we note that $(-2, -6.4)$ and $(-1, 6)$ are points on the graph of $f(x)$. But at some x-value between $x = -2$ and $x = -1$ there is an x-value, say a, such that $f(a) = 0$. The number a is the zero of the function f, and $(a, 0)$ is an x-intercept. That is, $x = a$ solves $f(x) = 0$. Redraw the graph of $f(x)$ with a smaller window, which must include the point $(a, 0)$, using `Xmin` = −3, `Xmax` = 3, `Ymin` = −25, and `Ymax` = 25. Press 2ND TRACE, which operates the `CALC` function. Select item 2 in the `MENU` using the down arrow key (or type 2), and ENTER. You will be prompted to reply to "`Left Bound?`" (see Figure 1.7.3). To reply, move the blinking spot using the left or right arrow keys to a position just to the left of $(a, 0)$. Do this by watching the y-values change as you move the blinking cursor. Moving leftward, the y-values will decrease from positive to negative. When this happens press ENTER. Next, you will be prompted for "`Right Bound?`" Move the cursor to the right until `Y` changes from negative to positive. Press ENTER. You will be prompted to guess the value of a. Move the cursor back once to the left and press ENTER. The result:

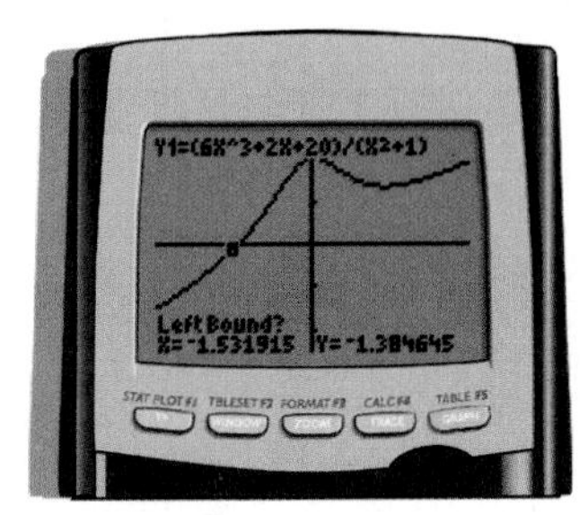

Figure 1.7.3

```
ZERO
X=-1.419485, Y=0
```

should appear on the screen. (See Figure 1.7.4.)

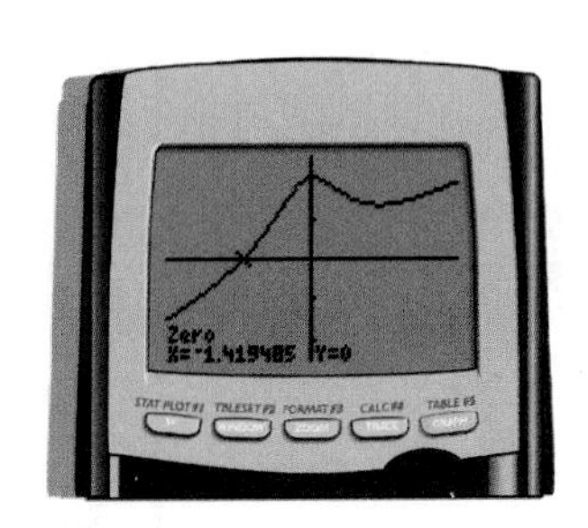

Figure 1.7.4

There is an alternate method which will locate the zero. When prompted for a left bound, just type in (–) 2 and ENTER. Then, when prompted for a right bound, type in (–) 1 and ENTER. When prompted for a guess, type (–) 1 . 5 and ENTER. The calculator determines a decimal approximation usually, rather than the exact value.

Note that in solving this equation, we created a function f and located the x-value –1.419485 so that $f(-1.419485) = 0$. An input number to a function that gives an output of zero ($y = 0$) is called a **zero of the function**. Solving an equation and finding a zero of the function are equivalent tasks.

Example 3: Rewriting Functions for Calculator Use

Rewrite each function in a form suitable for typing into a graphing utility.

a. $f(x)=\dfrac{3x}{4x-1}$

Solution: $3x \div (4x-1)$

b. $f(x)=\dfrac{x^{\frac{2}{3}}}{x+2}+3$

Solution: $x \wedge (2/3) \div (x+2)+3$

c. $f(x)=2+x+\dfrac{32+x}{\sqrt{x+1}}$

Solution: $2+x+(32+x)\div\sqrt{(x+1)}$ or

$2+x+(32+x)\div(x+1)\wedge(1/2)$

One possible method uses the square root button, followed by "$x+1)$". Instead, one could use "$(x+1)\wedge(1/2)$" or "$(x+1)\wedge.5$".

Example 4: Finding the Zero of a Function Using a Calculator

Solve $x^3-2x+3=0$.

Solution: Type the function into Y1. The number 4 is greater than any coefficient. Thus, for x more than 4 or less than –4, the y-values will be determined by the x^3 term. So, for x to the left of $x = -4$, y will be a large negative number and for x bigger than $x = 4$, y will again be a large positive number. It follows that

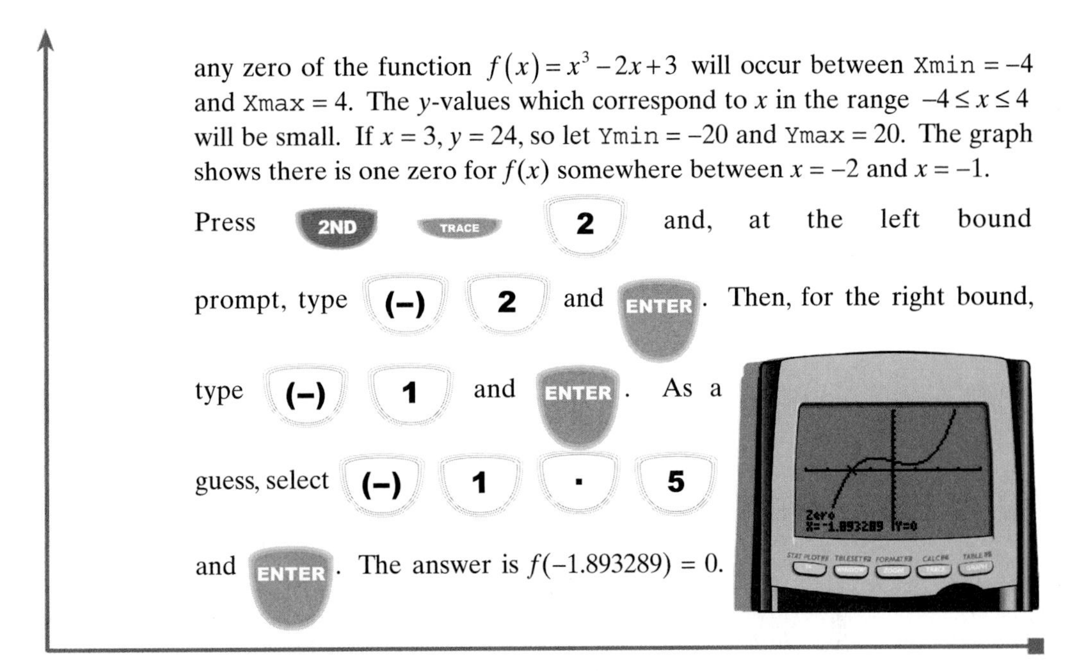

any zero of the function $f(x) = x^3 - 2x + 3$ will occur between `Xmin` $= -4$ and `Xmax` $= 4$. The y-values which correspond to x in the range $-4 \le x \le 4$ will be small. If $x = 3$, $y = 24$, so let `Ymin` $= -20$ and `Ymax` $= 20$. The graph shows there is one zero for $f(x)$ somewhere between $x = -2$ and $x = -1$.

Press 2ND TRACE 2 and, at the left bound prompt, type (–) 2 and ENTER. Then, for the right bound, type (–) 1 and ENTER. As a guess, select (–) 1 . 5 and ENTER. The answer is $f(-1.893289) = 0$.

Finding A Minimum Value

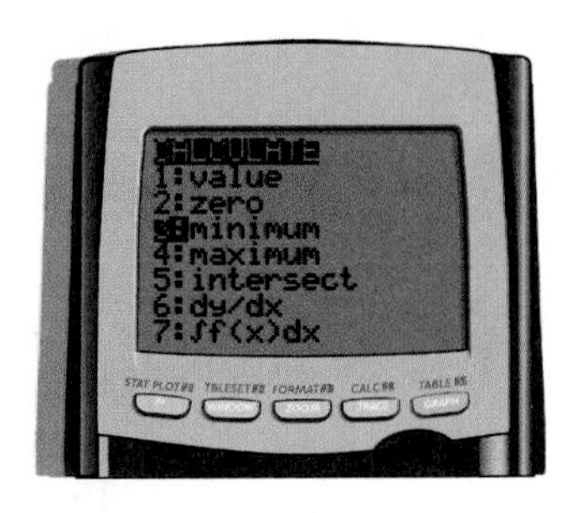

For the function just discussed, note that there is an interesting dip downward between $x = 0$ and $x = 3$. Let us determine the minimum y-value in this range. With the same function $f(x)$ typed into `Y1`, and with a window that includes the interval from $x = 0$ to $x = 3$, press GRAPH and then 2ND TRACE (for the `CALC` function) and select `3: Minimum`.

You can do this by using the down arrow and pressing ENTER, or just by typing `3`.

You will be prompted for a "`Left Bound?`" Type 1 and ENTER. You will be prompted for a "`Right Bound?`" Type 3 and ENTER. Finally, you will be prompted for a guess. Press 2 and ENTER. The value:

```
MINIMUM
X=1.394159, Y=13.264728
```

will appear on the screen. Finding a maximum value is similar.

1.7 Exercises

In Exercises 1 – 8, express each function (using parentheses, ^, ×, ÷, etc.) in a form suitable for typing into a graphing utility.

1. $f(x)=1+3x+\sqrt{x}$

2. $f(x)=3x-2+\sqrt[3]{x}$

3. $f(x)=\dfrac{3x}{\sqrt{x-1}}$

4. $f(x)=\dfrac{1+\sqrt{x}}{2-3x}$

5. $f(x)=\dfrac{(-3x+16)^4}{3x+6}$

6. $f(x)=\sqrt{2+5x+\sqrt{x}}$

7. $f(x)=\left(2x^{\frac{2}{3}}+3x^{\frac{5}{3}}\right)^5$

8. $f(x)=\left(9x^{\frac{1}{5}}+2x^{\frac{3}{5}}\right)^{10}$

In Exercises 9 – 22, graph the function given in the window $[-10, 10]$ *by* $[-10, 10]$*. Sketch the graph that appears and explain in a sentence what appears to be wrong with the picture. Then find a correct window, which reveals the significant parts of the graphs, and draw the "improved" graph.*

9. $f(x)=\dfrac{3x-25}{\sqrt{x^2+5}}$

10. $f(x)=(3x+4)^2(5x-25)^2$

11. $f(x)=(6x+30)^2(3x-15)^2$

12. $f(x)=(40+3x)\sqrt{16-x}$

13. $f(x)=35+17x-x^2-x^3$

14. $f(x)=210-80x+x^3$

15. $f(x)=35+56x-14x^2$

16. $f(x)=25+50x-10x^2$

17. $f(x)=\sqrt[3]{x^3-x^2-x-50}$

18. $f(x)=\sqrt[3]{x^4-3x^2-3x-30}$

19. $f(x)=\left(10+2x-25x^2\right)^{\frac{1}{3}}$

20. $f(x)=\left(30-11x+x^2\right)^{\frac{1}{3}}$

21. $f(x)=\left(12-6x-x^2\right)^{\frac{4}{3}}$

22. $f(x)=\left(x^3-x-100\right)^{\frac{1}{3}}$

Graph each function in the window shown in Exercises 23 – 30. Make a table of x- and y-values for the values x = –2, –1, 0, and 2. Then find all the zeros of the function in the x-interval given.

23. $f(x)=10+5x-6x^3$; $[-4, 4]$ by $[-20, 20]$

24. $f(x)=\dfrac{5^x+3}{25x^2+3x+3}-2$; $[-4, 4]$ by $[-5, 5]$

25. $f(x)=\dfrac{x\sqrt[3]{x-5}}{1+x}$; [–2, 8] by [–5, 5]

26. $f(x)=\dfrac{(2+3x-x^2)(14-x)}{\sqrt{x^2+20}}$; [–5, 20] by [–5, 5]

27. $s(t)=(t-3)(-16t^2+32t-70)$; [–3, 8] by [–200, 400]

28. $s(t)=(t^2+3)(3-40t+t^2)^{\frac{1}{2}}$; [–5, 5] by [–200, 200]

29. $u(t)=\left(3-12t+18t^{\frac{3}{2}}\right)^{\frac{1}{2}}$; [–5, 5] by [–8, 12]

30. $u(t)=\dfrac{(32-11t)^2(t+3)}{\sqrt{12+t}}$; [–5, 9] by [–1200, 1600]

For Exercises 31 – 40, graph the function in a suitable window and find the smallest y-value possible, or the smallest in the x-interval specified.

31. $f(x)=x^2-104x+2724$

32. $f(x)=\dfrac{-1-x^2-3x^3}{5^x}$

33. $f(x)=x^3-17x+5;\quad -3\le x\le 5$

34. $f(x)=\dfrac{\sqrt[3]{x}-150}{5+x^2}$

35. $f(x)=x^{1.5}-8x-15$

36. $f(x)=x^{1.8}-x-100$

37. $f(x)=2^x-50x$

38. $f(x)=(1.5)^{x+1}-(x+1)^5;\quad -5\le x\le 5$

39. $f(x)=17^x-x^{17};\quad -2\le x\le 2$

40. $f(x)=(-x)(3^{-x})$

Hawkes Learning Systems: Essential Calculus

Functions and Their Graphs

1.8 Functions and Models

Objectives

After completing this section, you will be able to:

1. *Recognize and understand various business and economic terms.*
2. *Create and evaluate mathematical models for real-life situations.*

Mathematical Modeling in Business and Economics

In this section we will introduce terms and ideas from business and economics that will be used throughout the remainder of the text. Most people are familiar with terms such as **profit** and **loss**, and many are familiar with the concepts of **supply** and **demand**. We want to represent these ideas and others with mathematical expressions. Creating a mathematical formula that describes a real-world problem is called **mathematical modeling**. The formula itself is called a **model**. Some models describe the problem accurately with a great deal of precision while others yield only approximations or merely intelligent guesses.

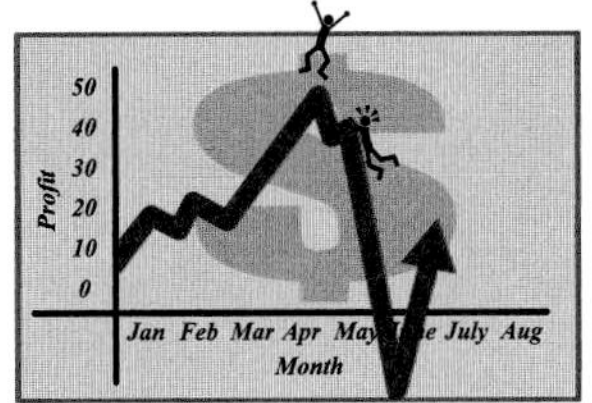

One business application that can be easily modeled with a formula is simple interest. **Simple interest** is the money paid for the use of money (called the **principal**) over a specific time.

Simple Interest

$$I = Prt$$

where P *= principal,* r *= annual rate of interest, and* t *= time in years.*

If $P = \$1000$ is invested at 8% for 2 years, then the simple interest earned is

$$\begin{aligned} I &= Prt \\ &= 1000(0.08)(2) = \$160. \end{aligned}$$

The balance (or amount A) at the end of 2 years would be the principal plus the interest,

$$A = P + I = 1000 + 160 = \$1160.$$

However, most people would prefer to have their money earning interest on interest already earned (**compound interest**).

Compound Interest

A formula (model) for the balance with annual compounding at rate r for t years is

$$\boldsymbol{A = P(1+r)^t},$$

where r is in decimal form.

Thus, using the previous data with $P = 1000$, $r = 8\% = 0.08$, and $t = 2$ years, we have

$$\begin{aligned} A &= 1000(1.08)^t \\ &= 1000(1.1664) \\ &= \$1166.40. \end{aligned}$$

Both of these models accurately describe financial situations. There are other formulas for compound interest that we will discuss later.

Unlike the formulas for simple and compound interest, models that describe general economic theories can be ambiguous and inaccurate because they are affected by uncontrollable variables such as politics, interest rates, and international crises. General theories and models such as these will be left to courses in economics.

In this text we will discuss several models related to specific situations involving particular products that we can analyze with algebra and calculus. The following terms and their interrelationships will be used.

Term	Symbol	Description
Items (or units)	x	Number of items produced.
Cost	$C(x)$	Total costs
Fixed costs	$C(0)$	Constant costs that do not depend on the number of items produced (rent, light, heat, etc.). (The y-intercept for $C(x)$.)
Variable costs	$C(x) - C(0)$	Costs that depend on the number of items produced (labor, material, etc.).
Revenue	$R(x)$	Income that depends on the number of items sold and the selling price per item.
Price	p	Selling price per item.

continued on next page ...

Term	Symbol	Description
Profit	$P(x)$	The difference between the revenue and the cost $[P(x)=R(x)-C(x)]$.
Break-even point		Point where revenue equals cost $[R(x)=C(x)$, or where $P(x)=0]$.
Supply function	$p=S(x)$	p is the price per item at which producers are willing to supply x items. (As supply x increases, the price increases.)
Demand function	$p=D(x)$	p is the price per item at which consumers are willing to buy x items. (As demand x increases, the price decreases.)
Equilibrium price	p_E	Price at which supply is equal to the demand $[S(x)=D(x)]$.
Equilibrium point	(x_E, p_E)	Point at which the supply is equal to the demand. Consumers buy (demand) all x_E items supplied when the price is p_E.
Revenue in terms of the unit price	$R(x)=x\cdot D(x)$ $=x\cdot p$	Revenue is the product of the number of items sold times the price per item. This price is set by the demand function $p=D(x)$.

A General Comment about Price

We have stated that the price per item, given in the supply and demand functions, is a function of (or dependent on) the number x of items supplied or demanded. You should be aware, however, that in the field of economics, supply and demand functions are sometimes written as

$$x=S(p) \quad \text{and} \quad x=D(p)$$

in which the number of items supplied or demanded is a function of the price. That is, price is used as the independent variable to indicate that a change in price affects the number of items supplied or demanded.

One reason we have chosen to write $p=S(x)$ and $p=D(x)$ is so that we can use these functions in conjunction with the revenue function

$$R(x)=x\cdot p=x\cdot D(x),$$

which is dependent on the number of items sold. This approach will simplify our study of the applications of calculus to economics.

Example 1: Car Rental

Suppose that the total cost of renting a car consists of a fixed cost of \$35 per day plus a variable cost of 15 cents per mile.

a. Write a cost function that represents the cost of driving x miles in one day.

b. Find the cost of driving 500 miles in one day.

Solutions: a. The formula for calculating cost is $C(x) = (\text{variable cost}) + (\text{fixed cost})$.
Let x = number of miles driven in one day,
variable cost = $\$0.15x$, and fixed cost = \$35 per day.

Inserting these into the formula gives
$C(x) = 0.15x + 35$.

b. $C(500) = 0.15(500) + 35$
$= 75 + 35 = \$110$

Find the cost by substituting $x = 500$ miles into the formula.

Note: Observe that if you do no driving the fixed costs are $C(0) = \$35$.

Example 2: Revenue

A manufacturer determines that the revenue generated by selling x units of a product is a linear function of x. If the revenue from 20 units is \$380 and the revenue from 15 units is \$285, find the revenue function.

Solution: Since the revenue function is linear, treat the given information as two points on a line: (20, 380) and (15, 285).

We use the formula for slope,

$$m = \frac{y_1 - y_2}{x_1 - x_2}$$

and the point-slope form

$$y - y_1 = m(x - x_1),$$

where x = number of units sold and y = revenue in dollars.

$$m = \frac{380 - 285}{20 - 15} = \frac{95}{5} = 19$$

Substitute the values for x_1, x_2, y_1, and y_2 into the formula for m.

continued on next page ...

$$y - 380 = 19(x - 20)$$

$$y = 19x - 380 + 380$$

$$y = 19x$$

Using the found value for m, and the known values for x_1 and y_1, solve for y in the point-slope formula.

The revenue function is $R(x) = 19x$.

This results in the revenue function, $R(x)$.

Example 3: Break-Even Point

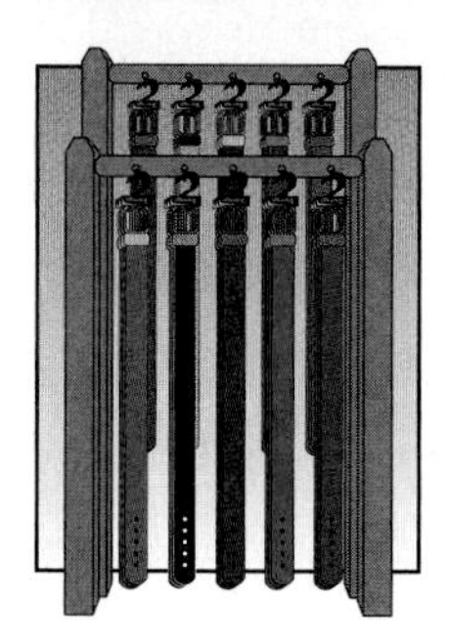

The Green-Belt Company determines that the cost of manufacturing men's belts is \$2 each plus \$300 per day in fixed costs. The company sells the belts for \$3 each. What is the break-even point?

Solution: The break-even point occurs where revenue and costs are equal. Let x = the number of belts manufactured in a day.

$$R(x) = 3x$$

$$C(x) = 2x + 300$$

For, $R(x) = C(x)$,

$$3x = 2x + 300$$

$$x = 300.$$

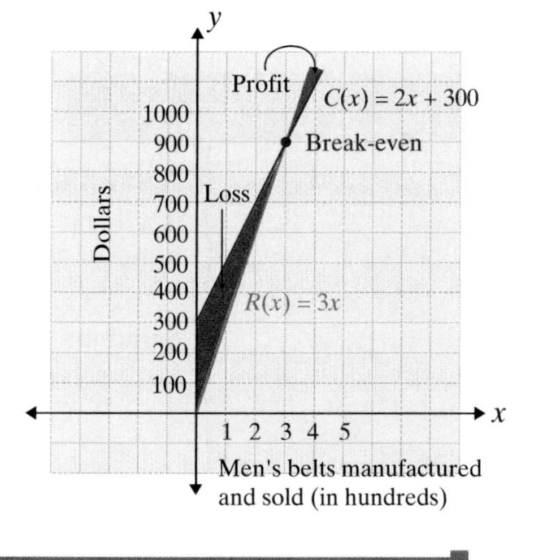

So 300 belts must be made and sold each day for the company to break even. The company must sell more than 300 belts each day to make a profit.

In Example 3 revenue and costs are linear functions and there appears to be no limit to the profit that can be made each day. To make more profit, the company need only make and sell more belts. However, in the real world, there are a limited number of customers and the company's plant physically restricts the number of belts that can be produced. Example 4 illustrates a more realistic model in which the revenue is a quadratic function, reflecting the fact that the demand function $p = D(x)$ is a decreasing function. That is, as the number of items sold increases, the price declines, and, eventually, the revenue declines.

Example 4: Modeling in Production

Suppose that a company has determined that the cost of producing x items is $500 + 140x$ and that the price it should charge for one item is $p = 200 - x$.

a. Find the cost function.
b. Find the revenue function.
c. Find the profit function.
d. Find the break-even point.

Solutions: a. The cost function is given as $C(x) = 500 + 140x$.

b. The revenue function is found by multiplying the price for one item by the number of items sold.

$$R(x) = p \cdot x$$
$$= (200 - x)x = 200x - x^2$$

c. Profit is the difference between revenue and cost.

$$P(x) = R(x) - C(x)$$
$$= (200x - x^2) - (500 + 140x)$$
$$= -x^2 + 60x - 500$$

d. To find the break-even point, set the revenue equal to the cost and solve for x.

$$R(x) = C(x)$$
$$200x - x^2 = 500 + 140x$$
$$0 = x^2 - 60x + 500$$
$$0 = (x - 10)(x - 50)$$
$$x = 10 \quad \text{or} \quad x = 50$$

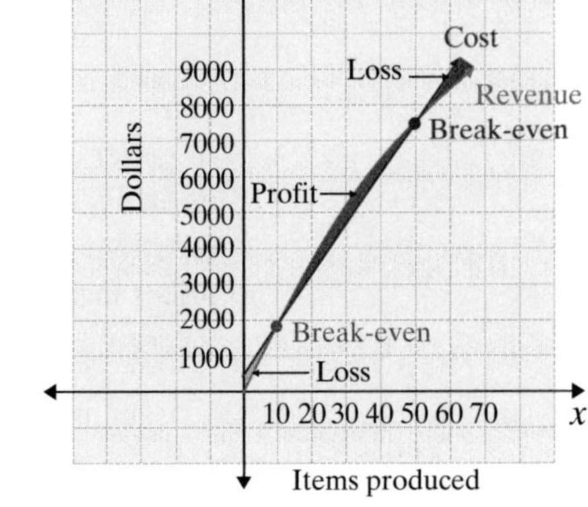

There are two break-even points.

This model shows that a profit occurs if the company produces between 10 and 50 items. We will discuss calculus techniques for maximizing profit later.

In the marketplace, price is related to both supply and demand. As we stated earlier, these basic relationships can be modeled by the following two functions.

$p = S(x)$	Supply function:	price per unit at which the supplier will supply x units of an item.
$p = D(x)$	Demand function:	price per unit at which consumers will buy x units of an item.

The equilibrium point, denoted as (x_E, p_E), is the point where the price is such that the production level (supply) is equal to the purchase level (demand). (See Figure 1.8.1.)

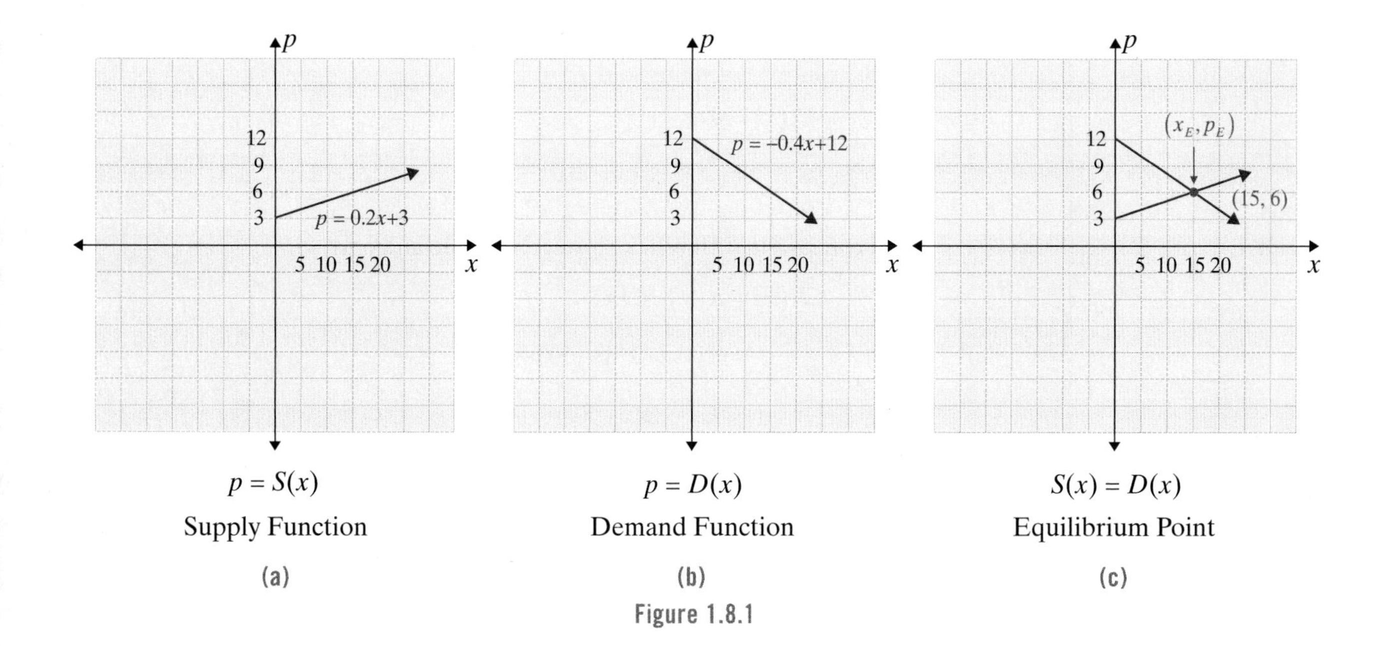

$p = S(x)$
Supply Function
(a)

$p = D(x)$
Demand Function
(b)

$S(x) = D(x)$
Equilibrium Point
(c)

Figure 1.8.1

In general, supply and demand functions are not linear functions as they are shown to be in Figure 1.8.1. However, supply functions are increasing functions, and demand functions are decreasing functions. Suppliers will happily provide more products as prices increase, and consumers will happily buy more products as prices decrease.

Example 5: Equilibrium Point

Suppose that the supply function for a particular product is

$$p = S(x) = x^2 + x + 3$$

and the demand function $p = D(x) = (x-5)^2$ where x represents thousands of units and p represents thousands of dollars. Find the equilibrium point (x_E, p_E).

Solution: We solve for x_E by setting $S(x) = D(x)$. Then we substitute this value for x into either $S(x)$ or $D(x)$ to find p_E.

$$S(x) = D(x)$$

$$x_E^2 + x_E + 3 = (x_E - 5)^2$$

$$x_E^2 + x_E + 3 = x_E^2 - 10x_E + 25$$

$$11x_E = 22$$

$$x_E = 2$$

$$p_E = 2^2 + 2 + 3 = 9$$

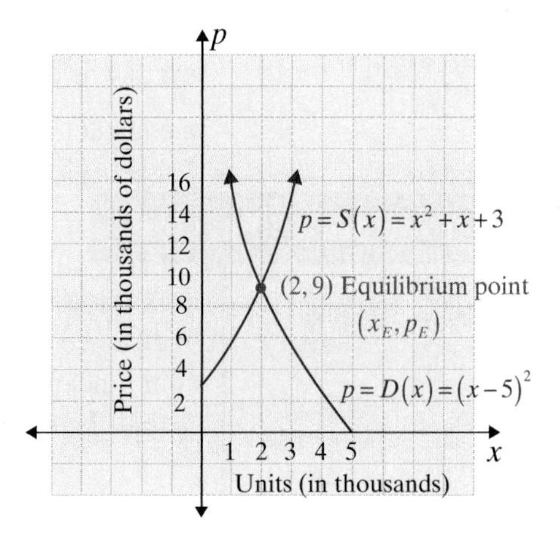

The equilibrium point occurs where $x = 2000$ units and $p = \$9000$.

Other Applications

Whenever one quantity is equal to some constant times another quantity, the first is said to **vary directly with** (or to be **directly proportional to** or just **proportional to**) the second quantity. For example, the circumference C of a circle is directly proportional to the diameter d. We have the general relationship

$$C = kd,$$

where k is called the **constant of proportionality**. For circles, we know that this constant is π. That is,

$$C = \pi d.$$

Other examples are:

$A = \pi r^2$ The area A of a circle varies directly as the square of the radius r.

$d = kw$ Hooke's Law says that the distance d that a hanging spring stretches is proportional to the weight w of the object attached to the spring. (The constant k depends on the characteristics of the spring such as its length and type of material.)

If one quantity increases as the other decreases while their product remains fixed, then the quantities are said to **vary inversely** or to be **inversely proportional**. For example, suppose $y = \frac{5}{x}$. As x increases, y is said to "decrease in proportion", and xy stays equal to 5. A famous example from physics is the following: the gravitational force between an object and the earth is inversely proportional to the square of the distance from the object to the center of the earth.

$$F = \frac{m_1 m_2}{G d^2} = \frac{k}{d^2}$$

where F is the gravitational force, d is the distance from the object to the center of the earth, k is the constant of proportionality, m_1 and m_2 are the masses of the objects, G is a constant and $k = \frac{m_1 m_2}{G}$.

Example 6: Physics

A basic law of physics states that, when an object is dropped in a near vacuum, the distance it falls varies directly with the square of the elapsed time. If an object dropped in a near vacuum falls 64 feet in 2 seconds, how far does it fall in 3 seconds? (Assume that the object does not hit the ground before 3 seconds have elapsed.)

Solution: The basic law of physics described above can be modeled by the equation

$$d = kt^2,$$

where d = distance, t = time, and k = constant of proportionality.

$$d = kt^2$$

$$64 = k(2)^2$$ First, find k by using $d = 64$ and $t = 2$.

$$64 = 4k$$

$$16 = k$$

$$d = kt^2$$

$$d = 16(3)^2 = 144 \text{ ft.}$$ Now find d by using $k = 16$ and $t = 3$.

The object will fall 144 feet in 3 seconds.

Examples 7 and 8 feature cost functions described differently from those we have discussed previously and Example 9 presents a function related to a geometric figure.

Example 7: Stock Market

Scott wants to buy a total of 500 shares in the stock market. He will buy x shares at \$4 per share and the rest at \$6 per share.

a. Write a function for the total cost of the shares.

b. What will be Scott's cost if he buys 200 shares at \$4?

Solutions: a. Since Scott will buy x shares at \$4 per share, he must buy $500 - x$ shares at \$6 per share. The total cost function is

$$C(x) = 4x + 6(500 - x) = 3000 - 2x.$$

b. If Scott buys 200 shares at \$4, then $x = 200$ and

$$C(200) = 3000 - 2 \cdot 200 = 3000 - 400 = \$2600.$$

Example 8: Phone Call Charges

Suppose that the cost of an overseas call is \$9.00 for the first 3 minutes or less plus 95 cents for each additional minute. Write a function for the cost of a call of x minutes. (Assume that a fraction of a minute over 3 minutes is charged the corresponding fraction of 95 cents.)

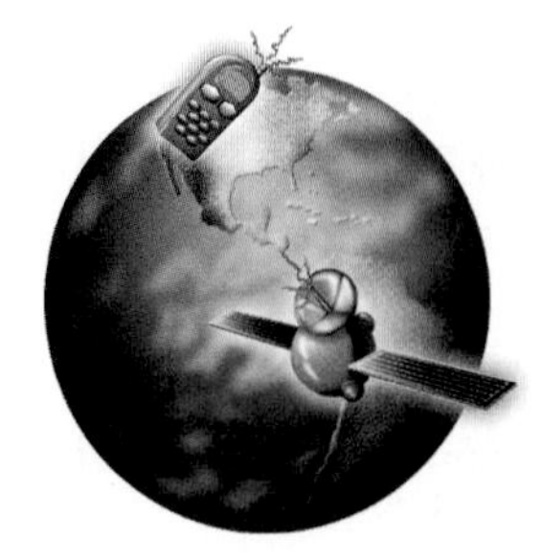

Solution: Let x = the length of the call in minutes. Then

$$C(x) = \begin{cases} 9.00 & \text{for } 0 < x \le 3 \\ 9.00 + 0.95(x - 3) & \text{for } x > 3 \end{cases}$$

The function is defined in pieces because the formula to be used depends on whether or not the call lasts longer than 3 minutes.

Example 9: Perimeter

A rectangular lot is fenced on three sides. An adjacent building forms the fourth side (see diagram). If the total length of the fencing is 48 feet, write a function that represents the area of the rectangle. Determine a good viewing rectangle and graph the function.

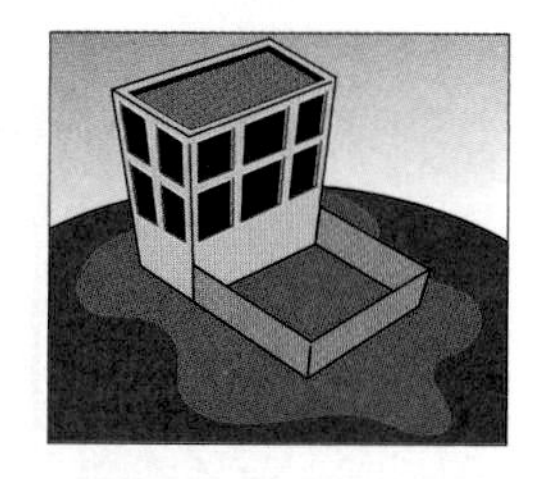

Solution: Let x = width of the rectangle. Since the total length of the fence is 48 feet and two sides are each of length x, then the length of the third side must be $48 - 2x$.

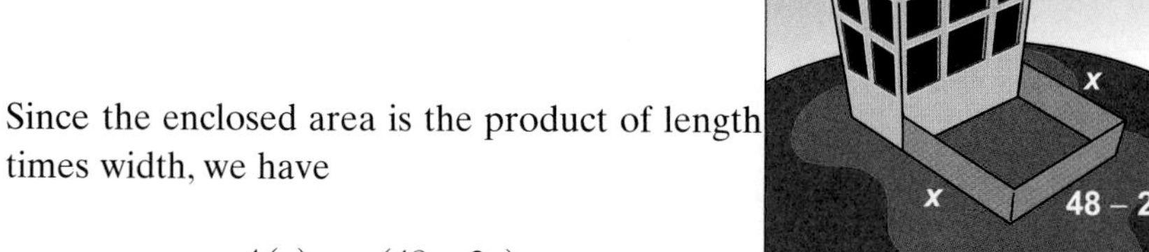

Since the enclosed area is the product of length times width, we have

$$A(x) = x\,(48 - 2x).$$

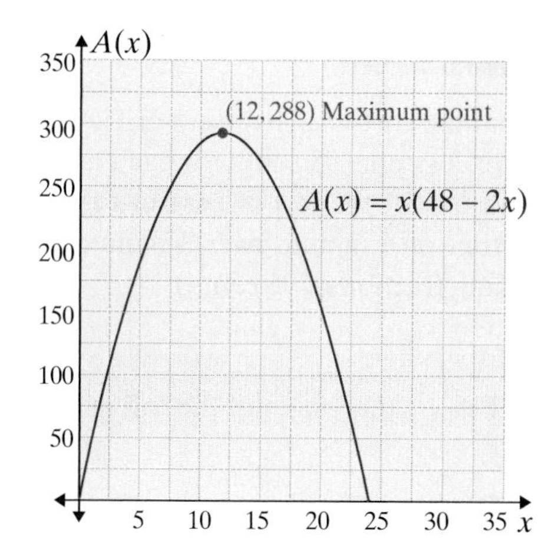

The significance of the highest point (at $x = 12$) is that when the width is 12, the maximum area is enclosed (288 ft.^{2}). Any other dimensions for the fence reduce total enclosed area.

1.8 Exercises

In Exercises 1 – 10, the quantity represented by y is related to the quantity represented by x in one of the following ways:

(a) y is directly proportional to x
(b) y is inversely proportional to x
(c) y is directly proportional to the square of x
(d) y is inversely proportional to the square of x
(e) other

Match the given formula to the best choice of (a) through (e).

1. $y = 32x$

2. $y = \dfrac{32}{x}$

3. $y = 16x^2$

4. $y = \dfrac{-32}{x^2}$

5. $y = -x$

6. $y = 2x + 3$

7. $y = \pi x$

8. $y = (x-1)^2$

9. $y = \dfrac{x}{10}$

10. $y = \dfrac{wxv}{stu}$

For Exercises 11 – 15, use the following situation:

Two students create a computer program which connects dots on a grid so that two players can play "Chase the Rabbit." They buy blank discs at a local store for \$0.50 and sell them for \$5.00. They pay another student \$27/day (5 days a week) to answer the phone, take orders down, relay customer questions, and make duplicates of the master disc.

Let x denote the number of discs made, let C(x) be the weekly total cost function (linear), and let R(x) be the revenue function.

11. **a.** Write the expression for $C(x)$.
b. Determine the weekly cost of producing 500 discs.

12. **a.** Write the revenue function.
b. How much revenue is produced by the sale of 500 discs.

13. **a.** Write the profit function.
b. Determine the profit from the sale of 500 discs.

14. How many discs must be sold in order to break even?

15. A business professor estimates that the campus craze for the game could become national, and, therefore, that the game could be marketed nationally. If 100 colleges were to become market sites, find a profit function for all 100 colleges together. Assume total profits and the number of colleges involved are directly proportional.

For Exercises 16 – 20, use the following information:

An ideal gas satisfies a law which may be stated as $\frac{PV}{T} = 0.821n$, *where P is the pressure in atmospheres (atm), V is the volume in liters, T is the temperature in degrees Kelvin (K = 273 + C, where C is Celsius), and n is the number of moles (gram molecular weights). Thus, for one mole of gas, pressure and volume are indirectly proportional for a constant temperature, pressure and temperature are directly proportional for a fixed volume, and volume and temperature are directly proportional for a fixed pressure.*

Assume, if necessary, that there is 1 mole (6.023 × 10^{23} molecules) present.

16. What volume is occupied by 1 mole of gas at a temperature of 300°K and a pressure of 2 atm?

17. A fixed volume of gas is heated from a temperature of 200°K and 0.5 atm of pressure to 300°K. What is the new pressure?

18. A gas at a pressure of 2 atm expands, at constant temperature, from 10 liters to 15 liters. What is the new pressure?

19. One mole of gas occupies a volume of 2 liters and has a temperature of 136.5°K. What is the pressure in atmospheres?

20. A gas has a volume of 2 liters, a temperature of 30°C and a pressure of 1 atm. When the gas is heated to 60°C and its volume is compressed to a volume of 1.25 liters, what is its new pressure? (Hint: For this problem, $n \neq 1$.)

Find the break-even point for the revenue and cost functions in Exercises 21 – 24.

21. $R(x) = 15x$

$C(x) = 5x + 30$

22. $R(x) = 27x$

$C(x) = 14x + 442$

23. $R(x) = 24x - 0.2x^2$

$C(x) = 8x + 300$

24. $R(x) = 9x - 0.7x^2$

$C(x) = 2x + 11.2$

Find the equilibrium point for the supply and demand functions in Exercises 25 – 28.

25. $S(x) = 2x + 3$

$D(x) = 15 - x$

26. $S(x) = 4x + 7$

$D(x) = 33 - 1.2x$

27. $S(x) = x^2 + x$

$D(x) = 35 - x$

28. $S(x) = x^2 + 3$

$D(x) = 51 - 2x$

29. **Modeling in business.** The manager of a pie shop sells his pies for \$6.50. The overhead is \$378 per day and each pie costs \$1.10 to make.
    a. Write the revenue function.
    b. Write the cost function.
    c. Write the profit function.
    d. Find the break-even point.

30. **Modeling in business.** A certain style of athletic shoe costs \$11.80 per pair to produce. The fixed costs are \$864 per week. The shoes can be sold for \$19.00 per pair.
    a. Write the revenue function.
    b. Write the cost function.
    c. Write the profit function.
    d. Find the break-even point.

31. **Modeling in manufacturing.** A manufacturer of golf clubs finds that the fixed costs are \$5780 per week and the cost of producing each set of clubs is \$73.00. Each set of clubs can be sold for \$243.00.
    a. Write the revenue function.
    b. Write the cost function.
    c. Write the profit function.
    d. Find the break-even point.

32. **Modeling in business.** A soft drink company has fixed costs of \$4000 per day. The variable costs are \$2.75 per case of soda. Each case sells for \$5.25.
    a. Write the revenue function.
    b. Write the cost function.
    c. Write the profit function.
    d. Find the break-even point.

33. **Modeling in production.** The cost of producing 200 pens is \$290. Producing 250 pens would cost \$297.50.
    a. Find the average cost per pen for additional 50 pens over 200.
    b. Assuming the total cost function is linear, write an equation for the cost of producing x pens.
    c. What are the fixed costs?

34. **Modeling in production.** The Blue Umbrella Company can produce 500 umbrellas per week at a cost of \$1800. It would cost \$1950 to produce 600 umbrellas.
    a. Find the average cost of each of the additional 100 umbrellas over 500.
    b. Assuming the total cost is a linear function, write an equation for the cost of producing x umbrellas.
    c. What are the fixed costs?

35. Revenue-profit. It has been determined that the cost of producing x units of a certain item is $11x + 500$. The demand function is given by $p = D(x) = 31 - 0.5x$.

a. Write the revenue function, and

b. Write the profit function.

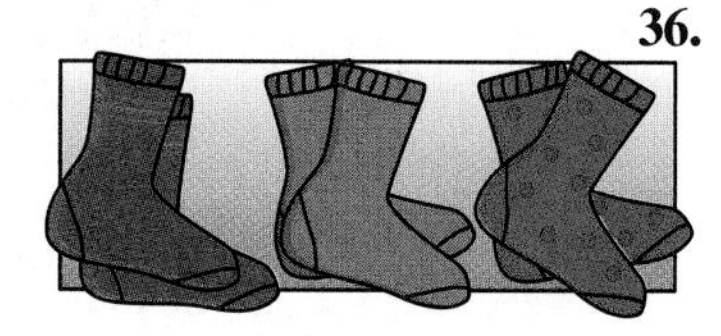

36. Modeling in sales. The manager of a men's store knows he can sell 60 pairs of a certain style of sock when the price is $1.20 per pair. If the price is $1.50, he can sell only 48 pairs of socks. The total cost function for x pairs of socks is $C(x) = 0.70x + 15$ dollars.

a. Assuming the demand function is linear, write an equation for $D(x)$.

b. Write the revenue function.

c. Write the profit function.

37. Modeling in manufacturing. A manufacturer of color televisions can sell 800 sets to his dealers at $384 each. If the price is $380, he can sell 1000 sets. The total cost of producing x television sets is $C(x) = 3600 + 250x - 0.01x^2$ dollars.

a. Assuming the demand function is linear, write an equation for $D(x)$.

b. Write the revenue function.

c. Write the profit function.

38. Revenue. Suppose the revenue R from the sale of a product is directly proportional to the number of units x of the product that are sold. Suppose also that the revenue from the sale of 65 units of the product is $1820.

a. Write a function for R in terms of x.

b. Find the revenue if 75 units are sold.

39. Interest. Suppose the annual interest I earned on an investment is directly proportional to the amount of money invested P. Suppose also that an investment of $8200 earns an annual interest of $512.50.

a. Write a function for I in terms of P.

b. Find the annual interest earned by $6000.

40. Interest What will $6000 accumulate to if it is deposited in a bank for three years and earns 5% a year with annual compounding?

41. Price. Suppose that for a certain product, the price per item p is inversely proportional to the number of items sold x. Suppose also that the price per item is $8.50 when 40 items are sold.

a. Write a function for p in terms of x.

b. Find the price if 34 items are sold.

42. Demand. Pat has decided to produce a limited number of prints from one of her paintings. She plans to issue x prints, where $0 < x \leq 50$. If she wants her revenue to be $5000, write a function for the demand $D(x)$.

43. Number of orders. The owner of a camera shop expects to sell 800 cameras of a particular style during the year. How many orders will the dealer need to place with his distributor if each order is for x cameras?

44. Salary. A salesperson's weekly salary depends on the amount of her sales. Her salary is \$250 per week plus a commission of 8% of her weekly sales in excess of \$2500. Write a function for her salary if her sales were x dollars.

45. Cost of a telephone call. For a long-distance call, the telephone company charges 65 cents for the first 3 minutes or less, plus 15 cents for each additional minute. Write a cost function for a call x minutes long.

46. Car rental. The rate for renting a car at a local agency is \$22.50 per day plus \$0.10 for each mile driven in excess of 100. If a car is rented for one day, write a function for the cost in terms of the number of miles driven.

47. Agriculture. A farmer raises strawberries. They cost him 38 cents per basket to produce. He is able to sell only 85% of those he produces. If he sells his strawberries at 75 cents per basket, find a function for his profit in terms of the number of baskets he produces.

48. Retail profit. A grocery store bought ice cream for 59 cents per quart and stored it in two freezers. During the night, one freezer "defrosted" and ruined 14 quarts. If the remaining ice cream was sold for 98 cents per quart, find a function for the profit in terms of the number of quarts bought.

49. Retail profit. It costs Liz \$12 to build a picture frame. She estimates that, if she charges x dollars per frame, she can sell $60 - x$ frames per week. Write a function for her weekly profit.

50. Retail profit. A toy retailer pays \$3 each for a particular doll. He estimates that, if he charges x dollars for each doll, he will be able to sell $300 - 20x$ dolls. Write a function for his profit.

51. Area. The perimeter of a rectangle is 276 feet. If the rectangle is x feet long, write a function for the area $A(x)$.

52. Perimeter. The area of a rectangle is 426 cm^2. If the length of the rectangle is x centimeters, write a function for the perimeter $P(x)$.

53. **Perimeter.** The area of a rectangle is 288 ft^2. If the length of the rectangle is x feet, write a function for the perimeter $P(x)$.

54. **Area.** The perimeter of a rectangle is 197 inches. If the rectangle is x inches wide, write a function for the area $A(x)$.

55. **Construction.** The maintenance department at the city zoo wants to build a pen and divide it as shown in the diagram at the right. If the department has a total of 720 feet of fencing, write a function for the area in terms of x.

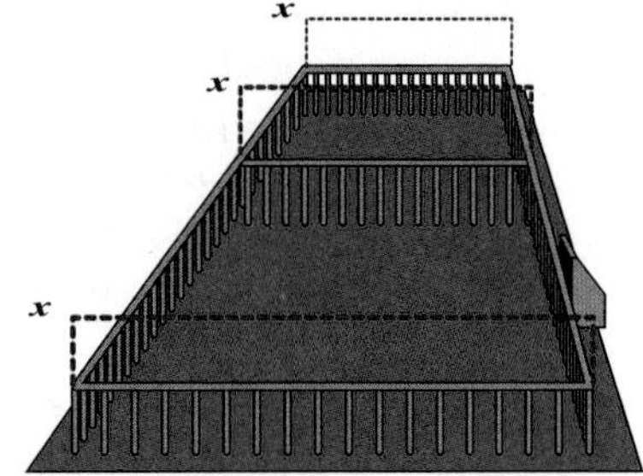

Hawkes Learning Systems: Essential Calculus

Functions and Models

Chapter 1 Index of Key Ideas and Terms

Section 1.1 Real Numbers and Number Lines

Types of Numbers pages 2 – 4

Natural numbers: $\mathbb{N} = \{1, 2, 3, \ldots\}$

Whole numbers: $\mathbb{W} = \{0, 1, 2, 3, \ldots\}$

Integers: $\mathbb{Z} = \{\ldots, -3, -2, -1, 0, 1, 2, 3, \ldots\}$

Rational numbers (Fractions): $\begin{cases} \mathbb{Q} = \left\{ \dfrac{a}{b} \mid a \text{ and } b \text{ are integers with } b \neq 0 \right\} \\ \mathbb{Q} = \{\text{repeating infinite decimal numbers}\} \end{cases}$

Irrational numbers: $\mathbb{I} = \{\text{nonrepeating infinite decimal numbers}\}$

Real numbers: $\mathbb{R} = \{\text{all rational and irrational numbers}\}$

Properties of Addition page 5

Assume that a, b, and c represent real numbers,

Commutative Property of Addition

$a + b = b + a$

Associative Property of Addition

$a + (b + c) = (a + b) + c$

Additive Identity

$a + 0 = a$

Additive Inverse

$a + (-a) = 0$

Properties of Multiplication page 5

Assume that a, b, and c represent real numbers,

Commutative Property of Multiplication

$a \cdot b = b \cdot a$

Associative Property of Multiplication

$a \cdot (b \cdot c) = (a \cdot b) \cdot c$

Multiplicative Identity

$a \cdot 1 = a$

Multiplicative Inverse

$a \cdot \frac{1}{a} = 1$

Distributive Property (of Multiplication over Addition) page 5

Assume that a, b, and c represent real numbers,

$a(b + c) = a \cdot b + a \cdot c$

$(b + c)a = b \cdot a + c \cdot a$

continued on next page ...

Section 1.1 Real Numbers and Number Lines (continued)

Symbols for Order page 6

$<$ "is less than"
$>$ "is greater than"
$\leq$ "is less than or equal to"
$\geq$ "is greater than or equal to"

Set-Builder Notation pages 6 - 7

Set-builder notation is a way of representing sets of real numbers with specified properties (such as being less than or greater than some specified value).

$\{x \mid x > 7\}$ is read "the set of all x such that x is greater than 7."

Absolute Value pages 7 - 8

The **absolute value** of a number is its distance from 0.
For any real number a,

$$|a| = \begin{cases} a & \text{if } a \geq 0 \\ -a & \text{if } a < 0. \end{cases}$$

Section 1.2 Integer Exponents

Zero Exponent page 10

$a^0 = 1$ for any nonzero number or nonzero algebraic expression a.
(Note: 0^0 is undefined.)

Properties of Exponents pages 10 - 12

For positive real numbers a and b and integers m and n:

1. $a^m \cdot a^n = a^{m+n}$
2. $a^0 = 1$
3. $a^{-1} = \dfrac{1}{a}$
4. $a^{-n} = \dfrac{1}{a^n}$
5. $\dfrac{a^m}{a^n} = a^{m-n}$
6. $\sqrt[n]{a} = a^{\frac{1}{n}}$
7. $(a^m)^n = a^{mn}$
8. $(ab)^n = a^n b^n$
9. $\left(\dfrac{a}{b}\right)^n = \dfrac{a^n}{b^n}$

Section 1.3 Fractional Exponents and Radicals

Fractional Exponents pages 14 - 15

1. If $a > 0$, then $a^{\frac{1}{n}}$ is a positive real number called the **principal** n^{th} **root** of a and $\left(a^{\frac{1}{n}}\right)^n = a$.

2. If $a < 0$ and n is even, then $a^{\frac{1}{n}}$ is not a real number.

3. If $a < 0$ and n is odd, then $a^{\frac{1}{n}}$ is a negative real number and $\left(a^{\frac{1}{n}}\right)^n = a$.

4. If $a = 0$, then $a^{\frac{1}{n}} = 0^{\frac{1}{n}} = 0$.

General Form page 16

If a is a real number, n is a positive integer, m is any nonzero integer, and $a^{\frac{1}{n}}$ is a real number, then $a^{\frac{m}{n}} = \left(a^{\frac{1}{n}}\right)^m = \left(a^m\right)^{\frac{1}{n}}$.

Radical Notation pages 17 - 19

If a is a real number, n is a positive integer, m is any nonzero integer, and $a^{\frac{1}{n}}$ is a real number, then $a^{\frac{1}{n}} = \sqrt[n]{a}$, and $a^{\frac{m}{n}} = \left(a^{\frac{1}{n}}\right)^m = \left(\sqrt[n]{a}\right)^m = \sqrt[n]{a^m}$.

Properties of Square Roots page 19

If a and b are positive real numbers:

1. $\sqrt{ab} = \sqrt{a}\sqrt{b}$ 2. $\sqrt{\dfrac{a}{b}} = \dfrac{\sqrt{a}}{\sqrt{b}}$

Section 1.4 Polynomials

Classification of Polynomials: page 22

Monomial: polynomial with one term
Binomial: polynomial with two terms
Trinomial: polynomial with three terms
Polynomial of degree *x*: any sum or difference of a set of monomials

continued on next page ...

Section 1.4 Polynomials (continued)

Like Terms (Similar Terms) page 23

Terms that contain the same variable factors with the same exponents.

Addition and Subtraction of Polynomials pages 23 - 24

Multiplication of Polynomials pages 24 - 26

FOIL Method: First, Outside, Inside, Last pages 24 - 25

Special Products pages 25 - 26

I. $(X+A)(X-A)=X^2-A^2$ Difference of Two Squares

II. $(X+A)^2=X^2+2AX+A^2$ Perfect Square Trinomial

III. $(X-A)^2=X^2-2AX+A^2$ Perfect Square Trinomial

IV. $(X+A)(X^2-AX+A^2)=X^3+A^3$ Sum of Two Cubes

V. $(X-A)(X^2+AX+A^2)=X^3-A^3$ Difference of Two Cubes

Factoring of Polynomials pages 26 - 28

1. Look for any **common factors.**
2. See if the product fits any of the **special products.**
3. Try the reverse of the **FOIL method** if the product is a trinomial.

Section 1.5 Lines and Their Graphs

Cartesian Coordinate System page 31

Quadrants

Axes

Horizontal axis (or x-axis)

Vertical axis (or y-axis)

Origin

Ordered Pairs

First coordinate (or first component)

Second coordinate (or second component)

Slope of a Line pages 32 - 33

$$m=\frac{y_2-y_1}{x_2-x_1}$$

Forms of Linear Equations pages 34 - 36

Standard Form: $Ax+By=C$, where A and B are not both 0.

Slope-Intercept Form: $y=mx+b$, b is the y-intercept.

Point-Slope Form: $y-y_1=m(x-x_1)$, (x_1, y_1) is a point on the line.

Section 1.6 Functions

Function, Domain, and Range pages 40 - 41

Let D and R be two sets of real numbers. A **function** f is a rule that matches each number x in D with exactly one number y [or $f(x)$] in R. D is called the **domain** of f, and R is called the **range** of f.

Piecewise Functions pages 43 - 44

Vertical Line Test pages 45 - 46

If **any** vertical line intersects a graph at more than one point, then the graph **does not** represent a function.

Operations with Functions pages 47 - 52

Composite Function pages 48 - 50

If f and g are two functions, then the composite function f of g is defined to be $(f \circ g)(x) = f(g(x))$.

The domain of $(f \circ g)$ is that part of domain of g for which $(f \circ g)$ is defined.

The Difference Quotient pages 52 - 53

The slope m of the segment (or line), which contains these two points, (x_1, y_1) and (x_2, y_2), is $m = \dfrac{\Delta y}{\Delta x} = \dfrac{y_2 - y_1}{x_2 - x_1} = \dfrac{f(x_2) - f(x_1)}{x_2 - x_1}$.

Section 1.7 Functions and Their Graphs: A Calculator Section

Graphing and Viewing a Function with Appropriate Window Dimensions pages 58 - 62

Making a Data Table pages 62 - 63

Solving an Equation pages 63 - 64

Finding the Zero of a Function pages 64 - 65

Finding a Minimum Value of a Function page 65

Section 1.8 Functions and Models

Simple Interest: $I = Prt$	page 68
Compound Interest: $A = P(1 + r)^t$	page 69
Other Models: Fixed costs, Price, Revenue, Break-Even Point, Supply and Demand	pages 69 - 78
Constant of Proprotionality: k	page 75

Chapter 1 Review

For a review of the topics and problems from Chapter 1, look at the following lessons from *Hawkes Learning Systems: Essential Calculus.*

Real Numbers and Number Lines
Integer Exponents
Fractional Exponents and Radicals
Polynomials
Lines and Their Graphs
An Introduction to Functions
Operations with Functions
Functions and Their Graphs: A Calculator Section
Functions and Models

Chapter 1 Test

In Exercises 1– 4, simplify each expression so that it contains only positive exponents. Assume that all variables represent nonzero real numbers.

1. $3x^3 \cdot x^{-4}$ **2.** $6x^3 \cdot 2x^{-1}$ **3.** $\frac{x^7}{x^3}$ **4.** $\frac{2x^2}{5x^4}$

In Exercises 5 – 8, simplify each expression. Negative exponents may remain in the answer if they are in the numerator. Assume that all variables represent nonzero real numbers.

5. $\frac{x^2 \cdot x}{x^3 \cdot x^{-4}}$ **6.** $(4x^2y^{-2})(-5x^4y^{-3})$ **7.** $\frac{(x^2y^{-2})^3}{(x^3y)^2}$ **8.** $\left(\frac{x^{-2}y^{-1}}{2x^2y^3}\right)^{-2}$

In Exercises 9 – 14, simplify each expression. Assume that all variables represent positive real numbers.

9. $9^{\frac{3}{2}}$ **10.** $16^{-\frac{3}{4}}$ **11.** $3x^{\frac{1}{2}} \cdot 8x^{-\frac{1}{3}}$ **12.** $\frac{x^3}{x^{\frac{5}{4}}}$

13. $\left(\frac{27x^3}{y^6}\right)^{\frac{1}{3}}$ **14.** $\frac{\left(x^{\frac{1}{2}}y\right)^{-\frac{2}{3}}}{x^{\frac{2}{3}}y^{-1}}$

In Exercises 15 – 22, change each expression to an equivalent expression in radical notation or exponential notation. Assume that each variable represents a positive real number.

15. $x^{\frac{8}{3}}$ **16.** $7x^{\frac{2}{5}}y^{\frac{3}{5}}$ **17.** $(3x+1)^{-\frac{1}{2}}$ **18.** $(x+6)^{\frac{2}{3}}$

19. $3\sqrt[4]{x^3}$ **20.** $-2\sqrt[3]{x^2y}$ **21.** $\sqrt[5]{4x-3}$ **22.** $\frac{6}{5\sqrt{9-x^2}}$

In Exercises 23 – 28, perform the indicated operations and simplify the expressions.

23. $(7x^2+2x-1)+(8x-4)$ **24.** $(4x^2+2x-7)+(5x^2+x-2)$

25. $(6x^2+x-10)-(x^3+x^2+x-4)$ **26.** $(x^3+4x^2-x)-(-2x^3+6x-3)$

27. $\left[(x^2-4x)-(2x^2+6x+1)\right]+\left[(3x^2-2x+7)-(x^2-3x+1)\right]$

28. $\left[(2x^2-x)+(3x+2)\right]-\left[(x-3)-(x^2+1)\right]$

In Exercises 29 – 32, find the indicated products.

29. $(7x+6)(7x-6)$ **30.** $(2x+9)^2$

31. $(2x+5y)(4x^2-10xy+25y^2)$ **32.** $x^2(4x+5)(2x-7)$

For each line described in Exercises 33 – 38, write the equation in the form specified.

33. Passing through (2, 5) with slope $-\frac{4}{3}$; standard form.

34. Passing through (−3, 7) with slope $\frac{5}{8}$; standard form.

35. Passing through (−8, 3) and (−4, 8); slope-intercept form.

36. Passing through (20, 9) and (30, 6); slope-intercept form.

37. Parallel to $5x - 3y = 7$ and passing through (−3, 3); slope-intercept form.

38. Perpendicular to $2x + 7y = 4$ and passing through (2, −1); standard form.

In Exercises 39 and 40, write each equation in slope-intercept form. Find the slope and y-intercept; then draw the graph.

39. $2x + 3y = 12$ **40.** $5x - 2y = 8$

In Exercises 41 – 46, find the x-intercepts, and maximum or minimum values of y. Graph the equation in a suitable window.

41. $y = 3(x-2)^2 + 5$ **42.** $y = x^2 + 6x + 5$ **43.** $y = -x^2 + 4x - 3$

44. $y = 2x^2 - 3x - 2$ **45.** $y = \frac{25 - 3x + x^4}{22 + x^2}$ **46.** $y = (3 - \sqrt{x})(1 + 2x)$

47. Let $f(x) = 7x - 4$. Find:

a. $f(2)$ **b.** $f(-3)$ **c.** $f(a+1)$ **d.** $f(a) + 1$

48. Let $f(x) = 5 - 2x^2$. Find:

a. $f(-2)$ **b.** $f(a-2)$ **c.** $f(x+h)$ **d.** $f(3) - f(2)$

49. Let $f(x) = \begin{cases} 2x+1 & \text{if } x \le -1 \\ x^2 - 2 & \text{if } x > -1 \end{cases}$. Find:

a. $f(-2)$ **b.** $f(-1)$ **c.** $f(-0.5)$ **d.** $f(0)$

In Exercises 50 – 55, let $f(x) = 5x - 2$, $g(x) = \frac{1}{x+4}$, and $h(x) = \sqrt{x+3}$. Find each of the following.

50. $f(x) + g(x)$ **51.** $h(x) \cdot g(x)$ **52.** $\frac{f(x)}{g(x)}$ **53.** $f(h(x))$

54. $g(f(x))$ **55.** $h(f(x))$

In Exercises 56 – 58, find $f(x+h) - f(x)$.

56. $f(x) = 6x + 1$ **57.** $f(x) = 2x^2 - 5x + 3$ **58.** $f(x) = x^3 - 4$

In Exercises 59 and 60, use the vertical line test to determine whether or not each of the graphs represents a function.

59.

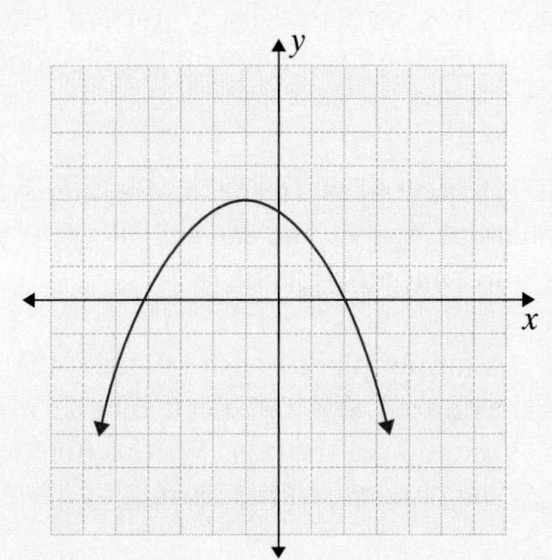

60.

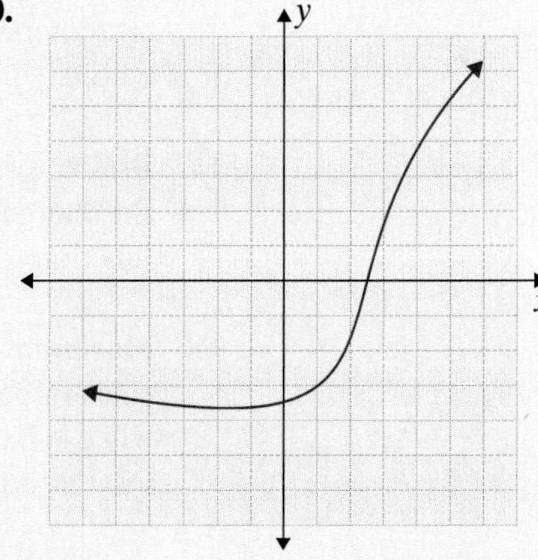

61. Break-even point. Find the break-even point if the revenue function is $R(x) = 32.50x$ and the cost function is $C(x) = 24x + 561$.

62. Equilibrium point. If the supply and demand functions for a product are given by $S(x) = x^2 + 3x$ and $D(x) = 450 - 4x$, respectively, find the equilibrium point.

63. Modeling in business. A popular bumper sticker sells for \$0.65. It costs \$0.10 to produce each sticker and the fixed costs are \$220 per week.

- **a.** Write the revenue function.
- **b.** Write the cost function.
- **c.** Write the profit function
- **d.** Find the break-even point.

64. Modeling in production. It costs \$750 to produce 12 units of a product. It has been determined that the cost of producing 15 units of the product would be \$798.

- **a.** What is the average cost of each of the additional 3 units over 12?
- **b.** Assuming the total cost is a linear function, write an equation for the cost of producing x units.
- **c.** What are the fixed costs?

65. Modeling in sales. A department store manager has determined that if the price for a necktie is \$12, the store will sell 100 neckties per month. However, only 80 neckties are sold if the price is raised to \$14.

- **a.** Write a linear function for the demand $D(x)$ where x is the number of neckties sold.
- **b.** Write the revenue function R.

66. Revenue. The management of a local amusement park found that when the admission price was \$15, the park had about 1200 customers per day. Each time the admission price was increased by \$1 (above \$15), the park lost about 200 customers. Write a function for the revenue if the admission price is increased by x dollars (above \$15).

67. Profit. It costs Lorraine \$7.50 to make a leather wallet. She estimates that if she charges x dollars per wallet, she can sell $50 - 0.8x$ wallets per month. Write a function for Lorraine's monthly profit.

68. Maximum area. A farmer plans to use 110 feet of fencing to build a rectangular chicken pen. To maximize the area he can enclose with the available fencing, he will use his barn as one side of the pen. Write a function for the area of the pen if x is the side of fencing opposite to the barn.

CHAPTER 2

Limits, Slopes, and the Derivative

Did You Know?

Sir Isaac Newton (1642 – 1727) is credited with inventing calculus. He was recognized in his own lifetime as the greatest physical and mathematical scientist in the world. He coined the word "calculus" from the Latin word meaning to calculate, and he used it in his masterwork *Principia Mathematica* (1687), which provided descriptions of physical principles obtained from his method of calculation. He was able to describe phenomena in the physical world with new and insightful terminology and was able to calculate related numerical values and the way these values changed over time.

Today, we recognize the value of his concept of a functional relation that changes with time and the value of computing the rates of these changes. These concepts, as well as methods of calculations, applications, and extensions, are the central core of calculus even today.

Newton supported himself financially at Cambridge from 1661 to 1665. He had no formal mathematics class until 1663 when Isaac Barrow arrived at Cambridge to become the first of the Lucasian Professors of Mathematics. After graduation, Newton returned to the family farm to manage his mother's estate and to escape from the bubonic plague which was ravaging cities in Britain at the time. There he created his universal law of gravitation, discovered what is called the Binomial Theorem, explained the color spectrum of the rainbow and the nature of light, and invented calculus. In 1669 he replaced his professor and mentor, Isaac Barrow, as Lucasian Professor of Mathematics at Cambridge. In 1696, at the height of his fame, Newton left the academic world for a government post, Warden of the Mint, leaving most of his mathematical discoveries unpublished until the 20th century. The last 30 years of his life were devoted to his government work and to obscure research into alchemy and early Christianity.

Newton was buried in Westminster Abbey, where previously only royalty and the nobility had been interred. Voltaire, the famous French philosopher, attended the funeral and said, "I have seen a professor of mathematics, only because he was great in his vocation, buried like a king who had done good to his subjects."

Shortly before his death, Newton told some friends, "I do not know what I may appear to the world; but to myself I seem to have been only like a boy playing on the seashore, and diverting myself in now and then finding a smoother pebble or prettier shell than ordinary, while the great ocean of truth lay all undiscovered before me."

The principal concept developed in this chapter is the idea of the slope of a curve at a point, or the derivative at the point. We calculate this quantity using the modern concept of the limit developed in Section 2.1. The limit concept is one of the foundations of calculus and is part of what distinguishes calculus from previous courses in mathematics. While the idea is abstract, the approach presented in the text is on an informal, intuitive level with graphs and many examples.

With the limit concept as background, we then discuss rates of change, which leads directly to our first specific calculus topic, the definition of the derivative of a function. From the definition we develop rules and formulas for finding derivatives. These rules are much easier to apply than the definition. More rules for differentiation will be developed in later chapters.

2.1 Limits

Objectives

After completing this section, you will be able to:

1. *Determine left- and right-hand limits of functions.*
2. *Determine if a limit of a function exists as x approaches a real number a.*

One-Sided Limits

This discussion will be on an intuitive basis, with a table of values and informal definitions.

For example, consider the function $f(x) = \dfrac{x^2 + x - 6}{x - 2}$.

Using a graphing utility with a window [–6, 3] by [–4, 8], we obtain a graph like the one in Figure 2.1.1. Using the TRACE function, the student should make a table of (x, y)-values using the x-values

$$1, 1.5, 1.9, 1.99, 1.999$$

without having to retype the expression (See Section 1.7).

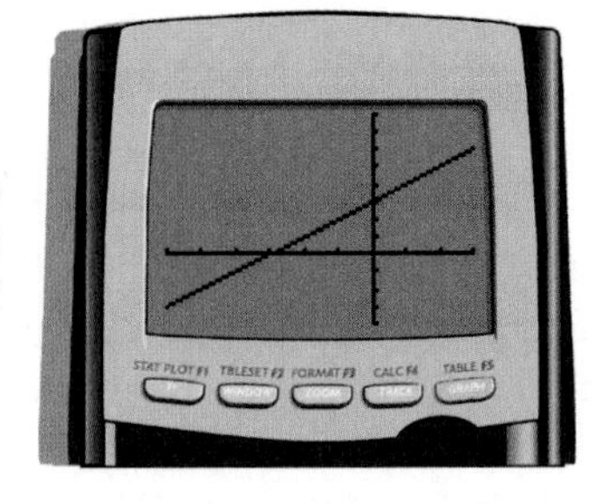

Figure 2.1.1

The results should be like those shown in the table below. The formula does not define a y-value for the x-input of $x = 2$ or, in other words, $f(2)$ is undefined. However, the table shows what happens to $f(x)$ for x-values less than 2 but close to 2.

x	1	1.5	1.9	1.99	1.999	2
$f(x)=\dfrac{x^2+x-6}{x-2}$	4	4.5	4.9	4.99	4.999	5

Reading the table from left to right, we see that as the values of x become closer and closer to 2, the values for $f(x)$ become closer and closer to 5. We say that, **as x approaches 2 from the left, $f(x)$ approaches 5**. The number 5 is called the **left-hand limit**. Symbolically, we write

The function whose limit is to be found.

$$\lim_{x \to 2^-}\left(\frac{x^2+x-6}{x-2}\right)=5.$$

← The left-hand limit.

The arrow signifies "approaches". The minus sign indicates x-values approaching 2 from the left.

The values of the function $f(x)=\dfrac{x^2-1}{x-1}$ in the next table are calculated as x approaches 1 from the right.

x	2	1.5	1.3	1.1	1.01	1.001	1
$f(x)=\dfrac{x^2-1}{x-1}$	3	2.5	2.3	2.1	2.01	2.001	2

We say that, **as x approaches 1 from the right, $f(x)$ approaches 2**. In this case 2 is a **right-hand limit**. We write

$$\lim_{x \to 1^+}\left(\frac{x^2-1}{x-1}\right)=2.$$

Read "The limit of $\dfrac{x^2-1}{x-1}$, as x approaches 1 from the right, is 2."

For the function $f(x)=\dfrac{x^2-1}{x-1}$, the tables have shown that the left-hand limit $\left(\text{as } x \to 1^-\right)$ and the right-hand limit $\left(\text{as } x \to 1^+\right)$ are the same number, 2. The graph of the function is given in Figure 2.1.2 on the following page.

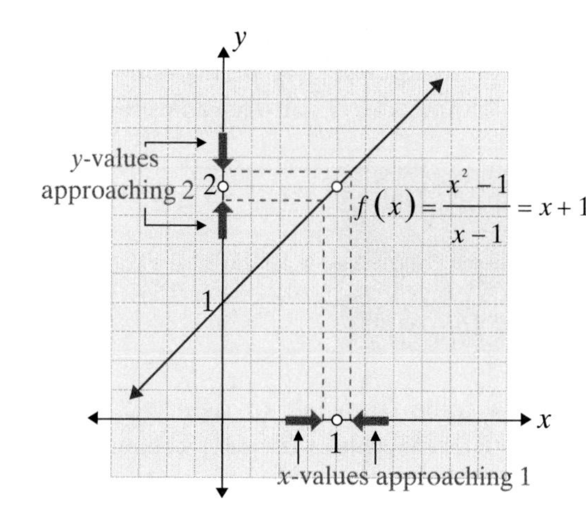

Figure 2.1.2

The open circle at the point (1, 2) indicates that this point is not included in the graph and $x = 1$ is not in the domain of this function.

Left-hand limits and right-hand limits need not be the same. For example, consider another function defined as follows:

$$g(x) = \begin{cases} \frac{1}{2}x & \text{for } 0 \le x \le 4 \\ x - 1 & \text{for } x > 4 \end{cases}.$$

This function is defined in pieces, and 4 is a key value for x since the function is represented by a different expression for values of x on each side of 4. The following table shows what happens to $g(x)$ as x approaches 4 first from the left and then from the right.

Let $g(x) = \frac{1}{2}x$ for $x < 4$ and let $g(x) = x - 1$ for $x > 4$:

x approaches 4 from the left

x	3	3.5	3.8	3.9	3.99	4
$g(x) = \frac{1}{2}x$	1.5	1.75	1.9	1.95	1.995	2

x approaches 4 from the right

4	4.01	4.1	4.2	4.5	5	x
3	3.01	3.1	3.2	3.5	4	$g(x) = x - 1$

Thus

$$\lim_{x\to 4^-} g(x) = \lim_{x\to 4^-}\left(\frac{1}{2}x\right) = 2 \quad \text{and} \quad \lim_{x\to 4^+} g(x) = \lim_{x\to 4^+}(x-1) = 3.$$

If we look at the graph of $y = g(x)$ in Figure 2.1.3, we can see what happens to the y-values for x-values near 4. The fact that the value of the function at $x = 4$ is 2 $\left(\text{since } g(4) = \frac{1}{2}\cdot 4 = 2\right)$ has no bearing on the values of the left-hand and right-hand limits.

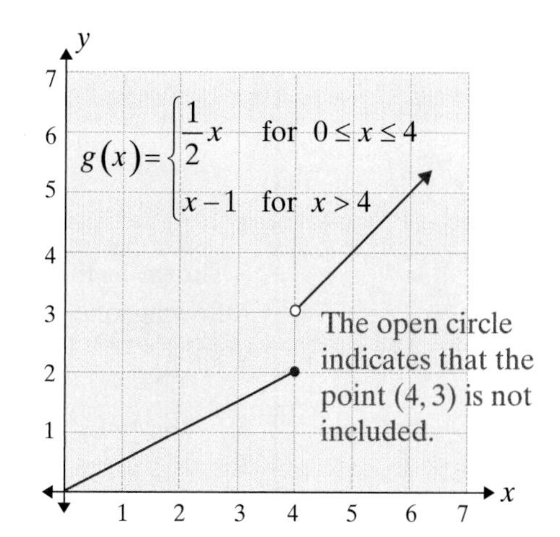

$g(x) = \begin{cases} \frac{1}{2}x & \text{for } 0 \le x \le 4 \\ x-1 & \text{for } x > 4 \end{cases}$
The open circle indicates that the point (4, 3) is not included.

Figure 2.1.3

An intuitive definition of left- and right-hand limits, also called **one-sided limits**, follows.

One-Sided Limits Defined Informally

1. ***Left-hand Limits***

   If the values of a function $y = f(x)$ get closer and closer to some number K as values of x, that are smaller than some number a, get closer and closer to a, then we say that K is the ***limit of $f(x)$ as x approaches a from the left****. We write*

$$\lim_{x\to a^-} f(x) = K.$$

2. ***Right-hand Limits***

   If the values of a function $y = f(x)$ get closer and closer to some number M as values of x, that are larger than some number a, get closer and closer to a, then we say that M is ***the limit of $f(x)$ as x approaches a from the right****. We write*

$$\lim_{x\to a^+} f(x) = M.$$

To help remember one-sided limit notation, think of a^- as representing a **minus** small amounts, that is, numbers near a but slightly less than a. Similarly, think of a^+ as representing a **plus** small amounts, that is, numbers near a but slightly more than a.

In Example 1, the graph of the function is used to find one-sided limits.

Example 1: Finding One-Sided Limits

The graph of the function

$$f(x)=\begin{cases} x^2 & \text{if } 0 \le x \le 2 \\ 2 & \text{if } 2 < x < 4 \\ x & \text{if } 4 \le x \le 6 \end{cases}$$

is shown in the figure at the right. Find the following one-sided limits by inspecting the graph.

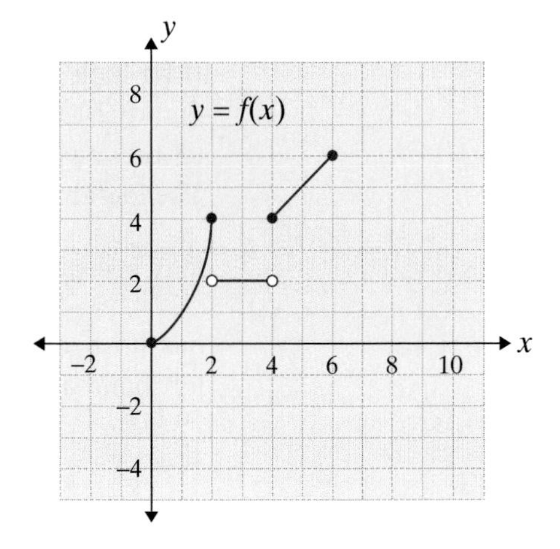

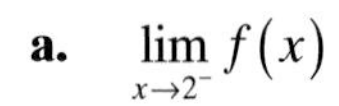

a. $\lim\limits_{x\to 2^-} f(x)$

Solution: $\lim\limits_{x\to 2^-} f(x)=4$

b. $\lim\limits_{x\to 2^+} f(x)$

Solution: $\lim\limits_{x\to 2^+} f(x)=2$

c. $\lim\limits_{x\to 4^-} f(x)$

Solution: $\lim\limits_{x\to 4^-} f(x)=2$

d. $\lim\limits_{x\to 4^+} f(x)$

Solution: $\lim\limits_{x\to 4^+} f(x)=4$

e. $\lim\limits_{x\to 0^+} f(x)$

Solution: $\lim\limits_{x\to 0^+} f(x)=0$

Remarks concerning the graph:

1. The value of the left and right-hand limits as x approaches 2 is independent of the actual y-value at $x = 2$, if there is one.

2. There is no $\lim\limits_{x\to 0^-} f(x)$ and $\lim\limits_{x\to 6^+} f(x)$ since the function is not defined for x-values approaching these numbers from the indicated directions.

Polynomial Functions

In the discussion of one-sided limits thus far, we have dealt with functions whose graphs are missing single points or "jump" from one piece of the curve to another. That is, if you tried to trace any of these curves with a pencil, you would need to lift the pencil in some places. Now we will discuss limits of **polynomial functions**. Since the graphs of polynomial functions behave "nicely" (i.e., the graphs are smooth curves with no jumps or missing points), we will see that finding one-sided limits for these functions is particularly easy.

Polynomial Function

*A **polynomial function** is a function of the form*

$$p(x) = a_n x^n + a_{n-1}x^{n-1} + \dots + a_2 x^2 + a_1 x + a_0$$

where each exponent is a positive integer and $a_n, a_{n-1}, \dots, a_0$ are real numbers.

Two familiar forms of polynomial functions are $p(x) = a_1 x + a_0$ (called **linear functions**) and $p(x) = a_2 x^2 + a_1 x + a_0$ (called **quadratic functions**).

The values for one-sided limits of polynomial functions as $x \to a^-$ or $x \to a^+$ can be found by simply substituting *a* for *x* in the function. The following remark emphasizes this.

One-Sided Limits of Polynomial Functions

If p is a polynomial function and a is a real number, then

$$\lim_{x\to a^-} p(x) = p(a) \quad \text{and} \quad \lim_{x\to a^+} p(x) = p(a).$$

This important fact, actually a theorem, makes polynomial functions particularly easy to deal with, and it will be discussed further in Section 2.5.

Example 2: Finding One-Sided Limits

Find the values of each of the following one-sided limits.

a. $\lim_{x\to 3^-}\left(x^2 + 3x - 7\right)$ **b.** $\lim_{x\to 2^+}\left(x^3 - 5x\right)$ **c.** $\lim_{x\to 0^-}\left(4x^2 - 8x + 3\right)$

Solutions: Since each function is a polynomial function, the one-sided limits can be found by substituting *a* for *x*.

a. $\lim_{x\to 3^-}\left(x^2 + 3x - 7\right) = 3^2 + 3\cdot 3 - 7 = 11$ Substitute $a = 3$ for x.

b. $\lim_{x\to 2^+}\left(x^3 - 5x\right) = 2^3 - 5\cdot 2 = -2$ Substitute $a = 2$ for x.

c. $\lim_{x\to 0^-}\left(4x^2 - 8x + 3\right) = 4\cdot 0^2 - 8\cdot 0 + 3 = 3$ Substitute $a = 0$ for x.

Unbounded Limits

Consider the function $f(x)=\dfrac{1}{x-2}$. This function is undefined at $x = 2$. However, in discussing limits, we are more interested in what happens to $f(x)$ as x gets close to 2, rather than at $x = 2$. The following table shows what happens to $f(x)$ as $x \to 2^-$.

x	1.5	1.8	1.9	1.99	1.999	2
$f(x)=\dfrac{1}{x-2}$	–2	–5	–10	–100	–1000	?

The table shows that, as $x \to 2^-$, the values of $f(x)$ are negative and become smaller and smaller (negative, but large in absolute value). In fact, $f(x)$ decreases without bound. We say that $f(x)$ approaches negative infinity ($-\infty$). (**Note:** The symbol, $-\infty$, is not a number. It is used to indicate unboundedness in the negative y direction.) We write

$$\lim_{x\to 2^-}\left(\frac{1}{x-2}\right)=-\infty.$$

The next table shows what happens to $f(x)$ as $x \to 2^+$.

x	2.5	2.2	2.1	2.01	2.001	2
$f(x)=\dfrac{1}{x-2}$	2	5	10	100	1000	?

The table shows that, as $x \to 2^+$, the values of $f(x)$ are positive and become larger and larger. In fact, $f(x)$ increases without bound. We say that $f(x)$ approaches positive infinity ($+\infty$). (**Note:** The symbol, $+\infty$, is not a number. It is used to indicate unboundedness in the positive y direction.) We write

$$\lim_{x\to 2^+}\left(\frac{1}{x-2}\right)=+\infty.$$

The graph of the function $f(x)=\dfrac{1}{x-2}$ is shown in Figure 2.1.4. The vertical line $x = 2$ is called a **vertical asymptote** since, as the values of x approach 2, the curve approaches this line but never touches it.

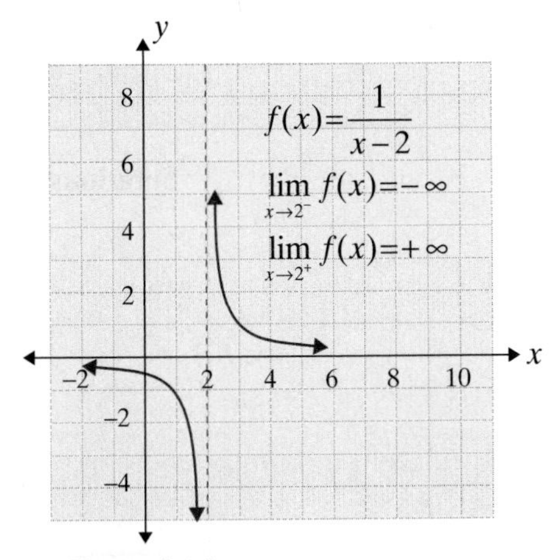

Figure 2.1.4

Making a Table of y-values with the Table Function

To display y-values in table form:

1. Press Y= and enter the function $f(x)$ into Y1.
2. Press 2ND WINDOW to access table settings. Then change "Indpnt:" to "Ask"
3. Press 2ND GRAPH to display the table screen. Finally, enter each x-value followed by ENTER.

If a limit is indicated to be $+\infty$ or $-\infty$, then it **does not exist**. These symbols are used only to indicate the direction of unboundedness.

Infinite One-Sided Limits (Unbounded One-Sided Limits)

1. $+\infty$ ***Limits***

If $f(x)$ increases without bound as x approaches a from the left (or from the right), then we say that $f(x)$ approaches positive infinity, $+\infty$. (See Figure 2.1.5(a) and (b).) We write

$$(a)\ \lim_{x\to a^-} f(x) = +\infty \quad \text{or} \quad (b)\ \lim_{x\to a^+} f(x) = +\infty .$$

2. $-\infty$ ***Limits***

If $f(x)$ decreases without bound as x approaches a from the left (or from the right), then we say that $f(x)$ approaches negative infinity, $-\infty$. (See Figure 2.1.5(c) and (d).) We write

$$(c)\ \lim_{x\to a^-} f(x) = -\infty \quad \text{or} \quad (d)\ \lim_{x\to a^+} f(x) = -\infty .$$

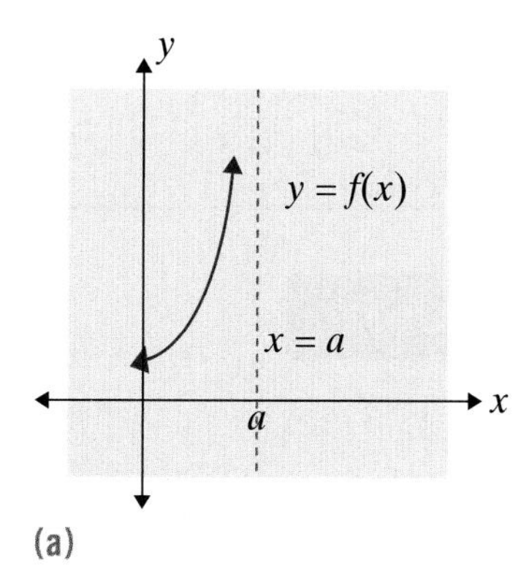

(a)

y
y = f(x)
x = a
x
a

(b)

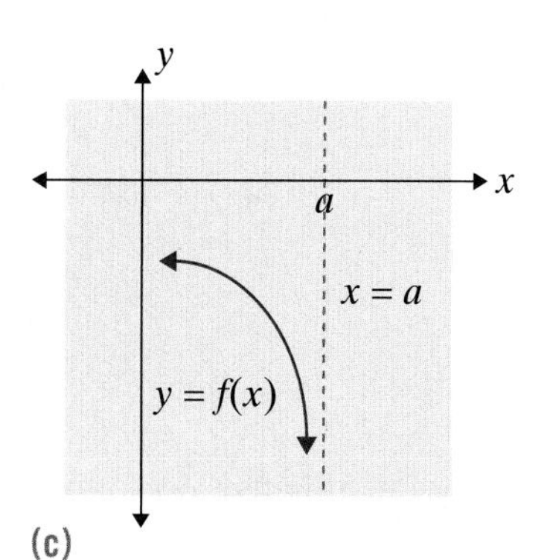

(c)

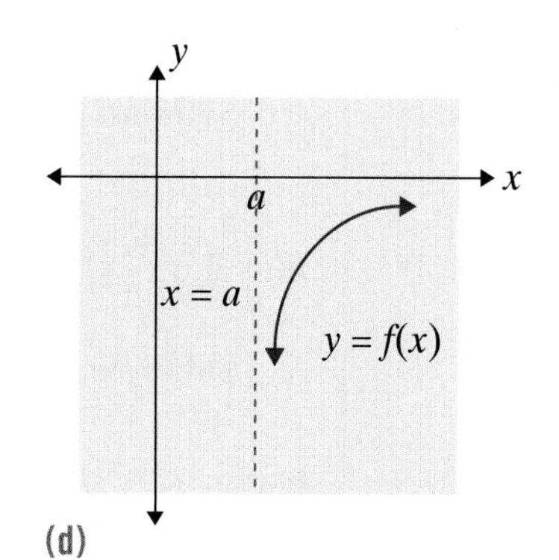

(d)

Figure 2.1.5

Example 3: Finding Infinite One-Sided Limits

a. Find $\lim\limits_{x \to -3^-} \left(\dfrac{x+5}{x+3} \right)$.

Solution: As x approaches -3 from the left, the denominator will always be negative but will approach 0; the absolute value of the denominator will get smaller and smaller. For example, $-3.1 + 3 = -0.1, -3, .01 + 3 = -0.01$, and so on. Meanwhile, the numerator will approach $-3 + 5 = 2$. Thus the fraction $\dfrac{x+5}{x+3}$ will become very large in the negative sense (or unbounded in the negative direction). So

$$\lim_{x \to -3^-} \left(\frac{x+5}{x+3} \right) = -\infty.$$

b. Find $\lim\limits_{x \to -3^+} \left(\dfrac{x+5}{x+3} \right)$.

Solution: Here, as x approaches -3 from the right, $x + 3$ will always be positive. For example, $-2.9 + 3 = +0.1$, $-2.99 + 3 = +0.01$, and so on. Thus the denominator will approach 0 through positive values, and the numerator will approach $-3 + 5 = 2$. Therefore, the fraction $\dfrac{x+5}{x+3}$ will become unbounded in the positive direction, and we have

$$\lim_{x \to -3^+} \left(\frac{x+5}{x+3} \right) = +\infty.$$

Example 4: Finding One-Sided Limits Using a Graph

Study the graph shown for $y = f(x)$ and find the following one-sided limits.

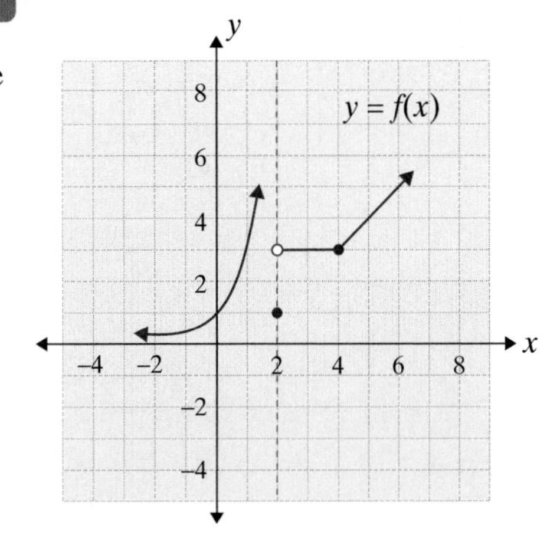

a. 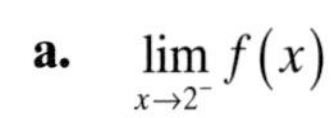 $\lim\limits_{x \to 2^-} f(x)$

Solution: $\lim\limits_{x \to 2^-} f(x) = +\infty$

b. 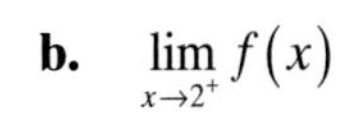 $\lim\limits_{x \to 2^+} f(x)$

Solution: $\lim\limits_{x \to 2^+} f(x) = 3$

Note: $f(2) = 1$ according to the graph, but this fact does not affect either the left or right-hand limits in **a.** and **b.**

c. $\lim_{x \to 4^-} f(x)$

Solution: $\lim_{x \to 4^-} f(x) = 3$

d. $\lim_{x \to 4^+} f(x)$

Solution: $\lim_{x \to 4^+} f(x) = 3$

Limits

We have discussed one-sided limits (i.e., left-hand limits and right-hand limits). With the concepts of one-sided limits as a base, we now develop the concept of a **limit** in general by considering left-hand and right-hand limits simultaneously. That is, for the limit of a function of x to exist as x approaches some number a (from both sides of a), the left-hand limit as $x \to a^-$ and the right-hand limit as $x \to a^+$ must be equal. Thus we say $\lim_{x \to a} f(x) = L$ if and only if $\lim_{x \to a^-} f(x) = L$ and $\lim_{x \to a^+} f(x) = L$.

For example, consider the function

$$f(x) = \frac{x^2 - x - 6}{x - 3}.$$

The following table will help in analyzing the function for values of x close to 3 on both sides of 3.

approaching 3 from the left → 3 ← approaching 3 from the right

x	2	2.5	2.9	2.99	3	3.01	3.1	3.5	4
$f(x) = \frac{x^2 - x - 6}{x - 3}$	4	4.5	4.9	4.99	5	5.01	5.1	5.5	6

The table shows that, as $x \to 3^-$ and as $x \to 3^+$, the value of $f(x)$ gets closer and closer to 5. We write

$$\lim_{x \to 3} \left(\frac{x^2 - x - 6}{x - 3} \right) = 5.$$

The graph of the function is shown in Figure 2.1.6. There is an open circle on the line at $x = 3$ because the function is not defined at this value.

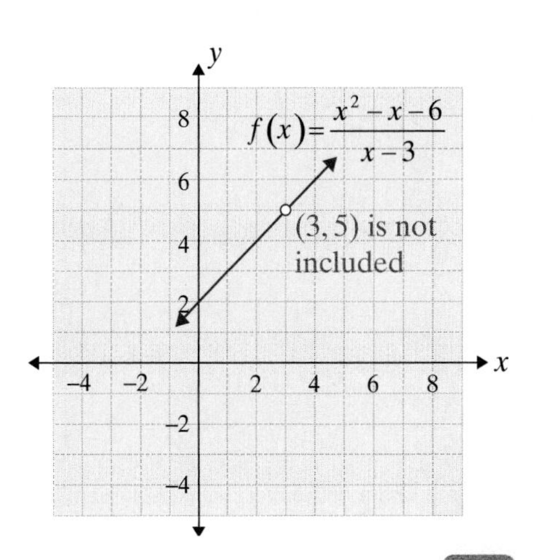

Figure 2.1.6

These ideas lead to the following definitions of a **limit**.

Limit Defined Informally

If the values of f(x) get closer and closer to some number L as values of x that are smaller than some number t and values of x that are larger than t get closer and closer to t (but not equal to t), then ***L is the limit of f(x) as x approaches t.***

Limit Defined Symbolically

For a function f and a real number t, if $\lim\limits_{x \to t^-} f(x) = L$ *and* $\lim\limits_{x \to t^+} f(x) = L$, *where L is a real number, then*

$$\lim_{x \to t} f(x) = L.$$

Existence of a Limit

1. *We say that the limit of a function f as x approaches a real number t* ***exists*** *if and only if* $\lim\limits_{x \to t} f(x) = L$, *where L is a real number.*

2. *If* $\lim\limits_{x \to t^-} f(x) \neq \lim\limits_{x \to t^+} f(x)$, *then* $\lim\limits_{x \to t} f(x)$ ***does not exist.***

3. *If* $\lim\limits_{x \to t} f(x) = +\infty$ *or* $\lim\limits_{x \to t} f(x) = -\infty$, *then* $\lim\limits_{x \to t} f(x)$ ***does not exist.***

The graphs shown in Figure 2.1.7 illustrate a variety of possible situations involving limits. Vertical asymptotes are shown in parts (d), (e), and (f).

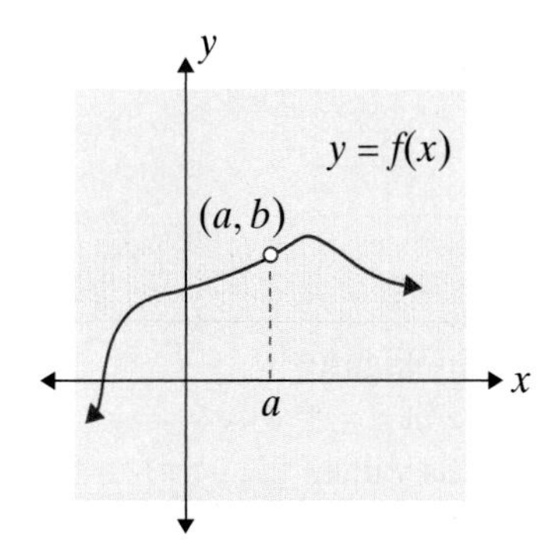

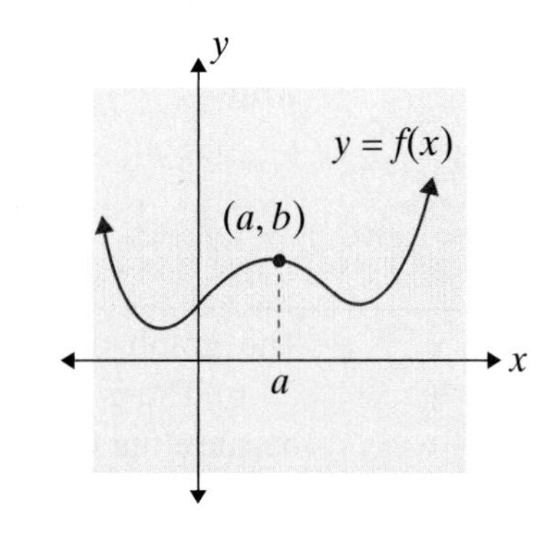

Figure 2.1.7

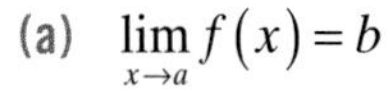

(a) $\lim\limits_{x \to a} f(x) = b$

(b) $\lim\limits_{x \to a} f(x) = b$

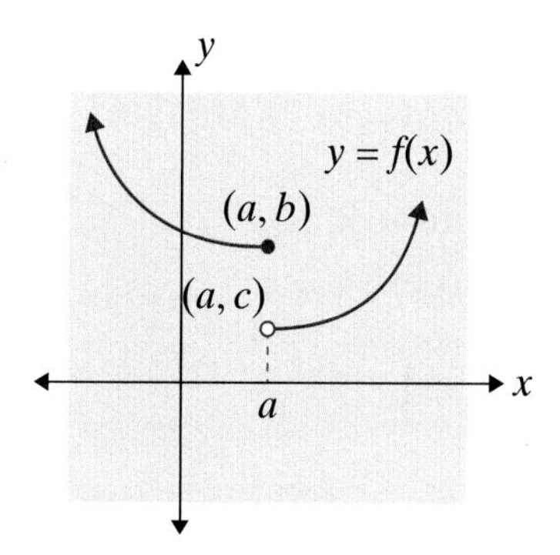

(c) $\lim_{x \to a} f(x)$ does not exist

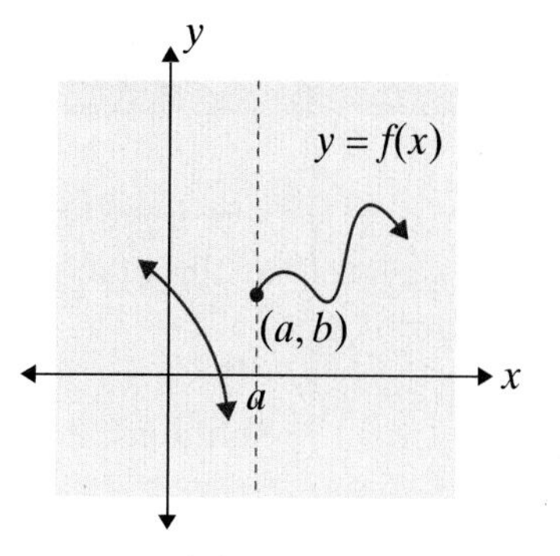

(d) $\lim_{x \to a} f(x)$ does not exist

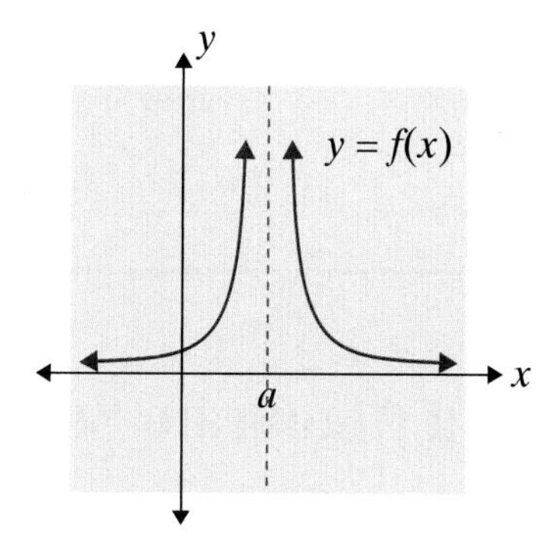

Figure 2.1.7 (e) $\lim_{x \to a} f(x)$ does not exist, or

$\lim_{x \to a} f(x) = +\infty$

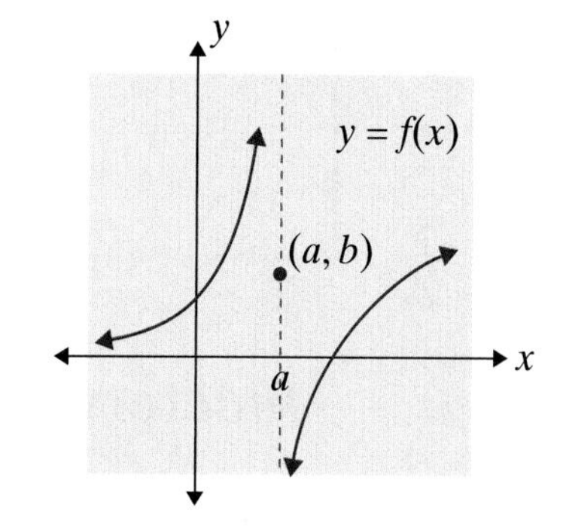

(f) $\lim_{x \to a} f(x)$ does not exist

NOTES Remember, even though the limit does not exist if it is $+\infty$ (or $-\infty$), we still write $+\infty$ (or $-\infty$) for the limit to indicate that the function is unbounded.

Example 5: Finding a Limit Using a Graph

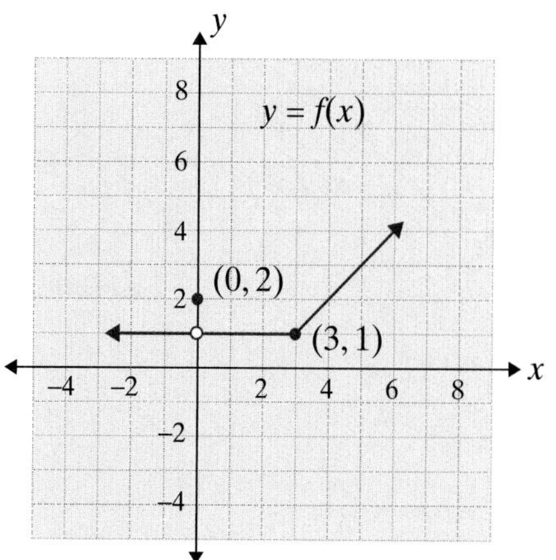

Use the graph of $y = f(x)$ in the figure to find **a.** $\lim_{x \to 3} f(x)$ and **b.** $\lim_{x \to 0} f(x)$.

Solutions: a. We see that $\lim_{x \to 3^-} f(x) = 1$ and $\lim_{x \to 3^+} f(x) = 1$.

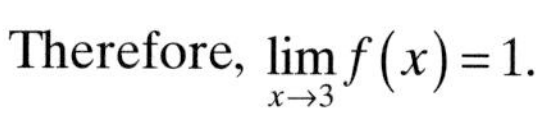
Therefore, $\lim_{x \to 3} f(x) = 1$.

b. We see that $\lim_{x \to 0^-} f(x) = 1$ and $\lim_{x \to 0^+} f(x) = 1$. So, even though $f(0) = 2$, $\lim_{x \to 0} f(x) = 1$.

Example 6: Finding a Limit Using a Graph

Use the graph of $y = g(x)$ in the figure to find **a.** $\lim_{x\to -2} g(x)$ and **b.** $\lim_{x\to 2} g(x)$, if the limits exist.

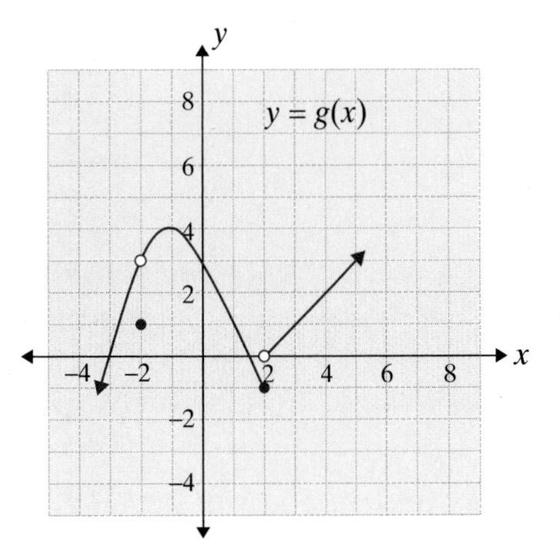

Solutions: a. $\lim_{x\to -2} g(x) = 3$

b. $\lim_{x\to 2^-} g(x) = -1$ and $\lim_{x\to 2^+} g(x) = 0.$

Since $-1 \neq 0$, $\lim_{x\to 2} g(x)$ does not exist.

Indeterminate Forms and Algebraic Solutions

We first point out that, in any limit problem, if the numerator has a limit different from zero and if the limit of the denominator is zero, then the limit of the quotient must fail to exist. However, if **both** the separate limits of a numerator and a denominator are zero, then there may be a limit for the quotient. It is, in fact, exactly this difficult situation which interests us in calculus, and in the next two sections of this text we move from the numerical approach to a more algebraic approach. The familiar factorizations, reviewed in Chapter 1, will be of use in solving limit problems of a special type.

Example 7: Finding Limits Algebraically

a. Determine $\lim_{x\to 2}\left(\frac{10\left(x^2-4\right)}{x-2}\right)$ by using an algebraically equivalent expression.

Solution: We first factor the numerator and simplify the fraction:

$$\frac{10\left(x^2-4\right)}{x-2} = \frac{10(x+2)\cancel{(x-2)}}{\cancel{x-2}} = 10(x+2).$$

The numerical approach, with a table of x-values converging to 2, suggests a limit exists.

Since, for any number x (where $x \neq 2$), the fractions $\frac{10(x^2-4)}{x-2}$ and $10(x+2)$ will be equal, we may replace the given expression with the simplified one. Thus

$$\lim_{x\to 2}\left(\frac{10(x^2-4)}{x-2}\right) = \lim_{x\to 2}\left(\frac{10(x+2)\cancel{(x-2)}}{\cancel{x-2}}\right)$$
$$= \lim_{x\to 2}(10(x+2))$$
$$= 10(2+2)$$
$$= 40.$$

b. Determine $\lim_{x\to 5}\left(\frac{x^3-125}{(x-5)^3}\right)$ algebraically.

Solution: We factor the numerator:

$$\frac{x^3-125}{(x-5)^3} = \frac{x^3-5^3}{(x-5)^3} = \frac{\cancel{(x-5)}(x^2+5x+25)}{(x-5)^{\cancel{3}\,2}}$$
$$= \frac{x^2+5x+25}{(x-5)^2}.$$

The numerator is a difference of two cubes: $A^3 - B^3$ that can be factored into $(A-B)(A^2+AB+B^2)$.

Then the given limit, if it exists, is equivalent to $\lim_{x\to 5}\left(\frac{x^2+5x+25}{(x-5)^2}\right)$.

However, the numerator does not have a limit of zero, but the limit of the denominator is zero. Thus the limit fails to exist.

c. Determine $\lim_{x\to -3}\left(\frac{2x^2+5x-3}{x+3}\right)$ algebraically.

Solution: We factor first:

$$\frac{2x^2+5x-3}{x+3} = \frac{(2x-1)\cancel{(x+3)}}{\cancel{x+3}} = 2x-1.$$

Thus $\lim_{x\to -3}\left(\frac{2x^2+5x-3}{x+3}\right) = \lim_{x\to -3}(2x-1) = 2(-3)-1 = -7.$

Indeterminate Form

A limit expression of the type $\lim\limits_{x \to a}\left(\frac{g(x)}{f(x)}\right)$ *is called an* ***indeterminate form*** *of type* $\frac{0}{0}$ *if* $\lim\limits_{x \to a} g(x) = 0$ *and* $\lim\limits_{x \to a} f(x) = 0$.

A strategy of solving such a problem is given by the two-step method illustrated in the previous example.

1. Replace the quotient with a simplified expression after factoring.
2. Evaluate the new limit problem by substitution if the denominator does not have a limit of 0 as $x \to a$.

2.1 Exercises

In Exercises 1 – 8, determine the limits. In each case, make a suitable table to support your answer. (Take the fourth x-value ± 0.001 *from the indicated a-value.)*

1. $\lim\limits_{x \to 7^-}\left(\frac{x^2 - 49}{x - 7}\right)$

x	y

2. $\lim\limits_{x \to 7^+}\left(\frac{x^2 + 49}{x - 7}\right)$

x	y

3. $\lim\limits_{x \to 3^+}\left(\frac{x^3 - 9x^2 + 27x - 27}{x - 3}\right)$

x	y

4. $\lim\limits_{h \to 0^+}\left(\frac{\sqrt{4 + h}}{h}\right)$

h	y

5. $\lim\limits_{x\to 0^-}\left(\dfrac{4\sin x}{3x}\right)$

(Use a calculator in radian mode.)

x	y

6. $\lim\limits_{a\to 1^+}\left(\dfrac{a^{10}-1}{a-1}\right)$

a	y

7. $\lim\limits_{n\to\sqrt{2}^-}\left(\dfrac{n^2-2}{n-\sqrt{2}}\right)$

n	y

8. $\lim\limits_{\theta\to\pi^-}\left(\dfrac{2\cos\theta-1}{1-\sin\theta}\right)$

(Use a calculator in radian mode.)

θ	y

Given a table for $\lim\limits_{x\to a}(f(x))$ *in Exercises 9 – 11,* **a.** *give the value for a and* **b.** *determine the limit, if there is one.*

9.

x	y
2.500	0.2222
2.100	0.2439
2.010	0.2494
2.001	0.2499

10.

x	y
3.80	15.60
3.900	15.80
3.990	15.98
3.999	15.998

11.

x	y
3.000	3.43
3.100	11.99
3.140	313.90
3.141	843.60

12. **a.** Complete the table.

x	$f(x)=3x-1$
1	
1.4	
1.8	
1.9	
1.99	
1.999	

b. Find $\lim_{x\to 2^-}(3x-1)$.

13. **a.** Complete the table.

x	$f(x)=x^2-2$
0	
−0.4	
−0.8	
−0.9	
−0.99	
−0.999	

b. Find $\lim_{x\to -1^+}(x^2-2)$.

14. **a.** Complete the table.

x	$f(x)=\dfrac{x^2-1}{x+1}$
2	
1.6	
1.2	
1.1	
1.01	
1.001	

b. Find $\lim_{x\to 1^+}\left(\dfrac{x^2-1}{x+1}\right)$.

15. **a.** Complete the table.

x	$f(x)=x^2+3$
2	
2.4	
2.8	
2.9	
2.99	
2.999	

b. Find $\lim_{x\to 3^-}(x^2+3)$.

16. a. Complete the table.

x	$f(x)=\dfrac{1}{x-4}$
3	
3.4	
3.8	
3.9	
3.99	
3.999	

b. Find $\lim_{x\to 4^-}\left(\dfrac{1}{x-4}\right)$.

17. a. Complete the table.

x	$f(x)=\dfrac{x}{x+2}$
−3	
−2.6	
−2.2	
−2.1	
−2.01	
−2.001	

b. Find $\lim_{x\to -2^-}\left(\dfrac{x}{x+2}\right)$.

18. a. Complete the table.

x	$f(x)=\dfrac{x^2-4}{x+2}$
−1	
−1.4	
−1.8	
−1.9	
−1.99	
−1.999	

b. Find $\lim_{x\to -2^+}\left(\dfrac{x^2-4}{x+2}\right)$.

19. a. Complete the table.

x	$f(x)=\dfrac{x-3}{x^2-2x-3}$
4	
3.6	
3.2	
3.1	
3.01	
3.001	

b. Find $\lim_{x\to 3^+}\left(\dfrac{x-3}{x^2-2x-3}\right)$.

In Exercises 20 – 25, use the graph of y = f(x) to find the limits.

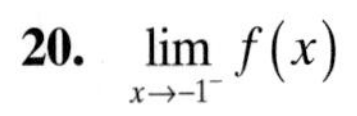

20. $\lim\limits_{x \to -1^-} f(x)$

21. $\lim\limits_{x \to -1^+} f(x)$

22. $\lim\limits_{x \to 2^-} f(x)$

23. $\lim\limits_{x \to 2^+} f(x)$

24. $\lim\limits_{x \to 3^-} f(x)$

25. $\lim\limits_{x \to 3^+} f(x)$

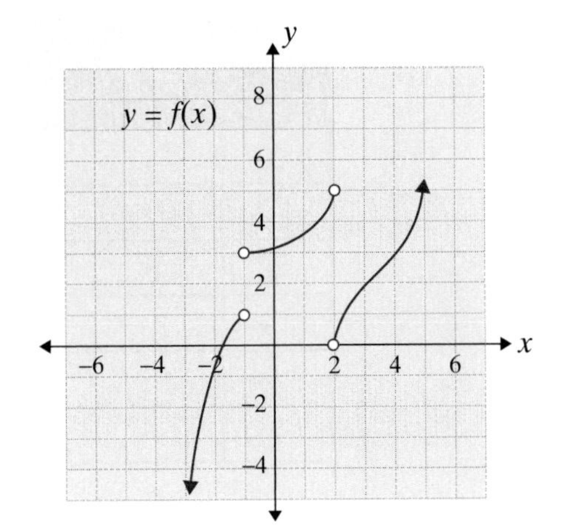

In Exercises 26 – 31, use the graph of y = f(x) to find the limits.

26. $\lim\limits_{x \to -1^-} f(x)$

27. $\lim\limits_{x \to -1^+} f(x)$

28. $\lim\limits_{x \to 0^-} f(x)$

29. $\lim\limits_{x \to 0^+} f(x)$

30. $\lim\limits_{x \to 3^-} f(x)$

31. $\lim\limits_{x \to 3^+} f(x)$

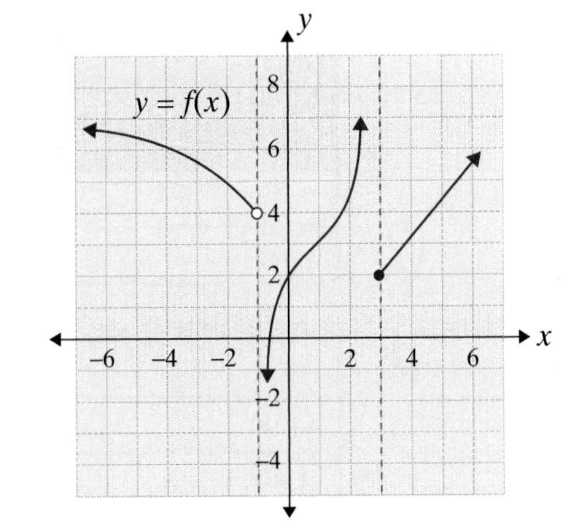

In Exercises 32 – 37, use the graph of y = f(x) to find the limits.

32. $\lim\limits_{x \to 0^-} f(x)$

33. $\lim\limits_{x \to 0^+} f(x)$

34. $\lim\limits_{x \to 4^-} f(x)$

35. $\lim\limits_{x \to 4^+} f(x)$

36. $\lim\limits_{x \to 2^-} f(x)$

37. $\lim\limits_{x \to 2^+} f(x)$

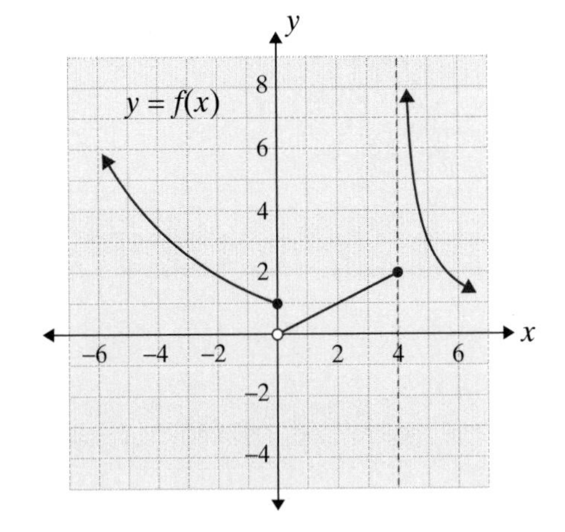

Find the one-sided limits indicated in Exercises 38 – 61.

38. $\lim_{x\to 2^+}(5x-3)$

39. $\lim_{x\to -1^+}(2x+7)$

40. $\lim_{x\to 0^-}(4-3x)$

41. $\lim_{x\to 3^-}(1-6x)$

42. $\lim_{x\to 2^-}\left(x^2-3x+1\right)$

43. $\lim_{x\to -5^+}\left(x^2+4x-2\right)$

44. $\lim_{x\to -4^+}\left(x^2-x+3\right)$

45. $\lim_{x\to -3^-}\left(x^2+2x-3\right)$

46. $\lim_{x\to 10^-}\left(0.01x^2+7x-30\right)$

47. $\lim_{x\to 10^+}\left(0.2x^2-5x+6\right)$

48. $\lim_{x\to 0^+}\left(\frac{x-3}{x}\right)$

49. $\lim_{x\to 0^-}\left(\frac{2x+1}{x}\right)$

50. $\lim_{x\to 1^+}\left(\frac{x-2}{x-1}\right)$

51. $\lim_{x\to 1^-}\left(\frac{x-2}{x-1}\right)$

52. $\lim_{x\to 2^-}\left(\frac{1}{x+2}\right)$

53. $\lim_{x\to 2^+}\left(\frac{1}{x+2}\right)$

54. $\lim_{x\to 3^+}\left(\frac{1}{x+1}\right)$

55. $\lim_{x\to 1^+}\left(\frac{1}{x-5}\right)$

56. $f(x)=\begin{cases}2-3x & \text{if } x<2\\ x-1 & \text{if } x\geq 2\end{cases}$

a. $\lim_{x\to 2^-} f(x)$

b. $\lim_{x\to 2^+} f(x)$

57. $f(x)=\begin{cases}x^2+2 & \text{if } 0\leq x\leq 3\\ 2x+5 & \text{if } x>3\end{cases}$

a. $\lim_{x\to 3^-} f(x)$

b. $\lim_{x\to 3^+} f(x)$

58. $f(x)=\begin{cases}3x+1 & \text{if } 0\leq x\leq 4\\ x^2-3 & \text{if } x>4\end{cases}$

a. $\lim_{x\to 4^-} f(x)$

b. $\lim_{x\to 4^+} f(x)$

59. $f(x)=\begin{cases}x^3 & \text{if } x<2\\ x^2+5 & \text{if } x\geq 2\end{cases}$

a. $\lim_{x\to 2^-} f(x)$

b. $\lim_{x\to 2^+} f(x)$

60. $f(x)=\begin{cases} 3-2x & \text{if } x<1 \\ x & \text{if } 1\le x\le 4 \\ \dfrac{1}{x-4} & \text{if } x>4 \end{cases}$

a. $\lim\limits_{x\to 1^-} f(x)$

b. $\lim\limits_{x\to 1^+} f(x)$

c. $\lim\limits_{x\to 4^-} f(x)$

d. $\lim\limits_{x\to 4^+} f(x)$

61. $f(x)=\begin{cases} x^2-1 & \text{if } 0\le x<2 \\ 3 & \text{if } 2\le x\le 5 \\ \dfrac{1}{x-5} & \text{if } x>5 \end{cases}$

a. $\lim\limits_{x\to 2^-} f(x)$

b. $\lim\limits_{x\to 2^+} f(x)$

c. $\lim\limits_{x\to 5^-} f(x)$

d. $\lim\limits_{x\to 5^+} f(x)$

In Exercises 62– 71, use the graph to find the indicated limits, if they exist.

62.

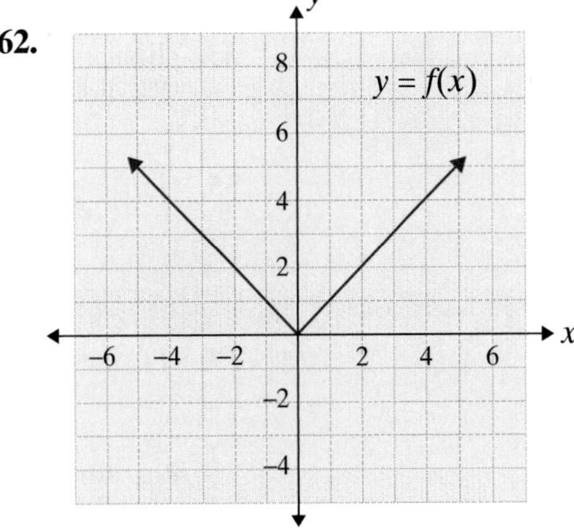

a. $\lim\limits_{x\to 0^-} f(x)$ **b.** $\lim\limits_{x\to 0^+} f(x)$

c. $\lim\limits_{x\to 0} f(x)$ **d.** $\lim\limits_{x\to 2} f(x)$

63.

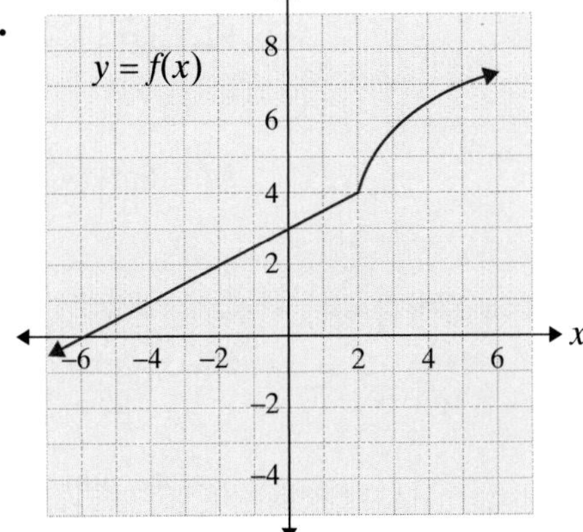

a. $\lim\limits_{x\to 2^-} f(x)$ **b.** $\lim\limits_{x\to 2^+} f(x)$

c. $\lim\limits_{x\to 2} f(x)$ **d.** $\lim\limits_{x\to 0} f(x)$

64.

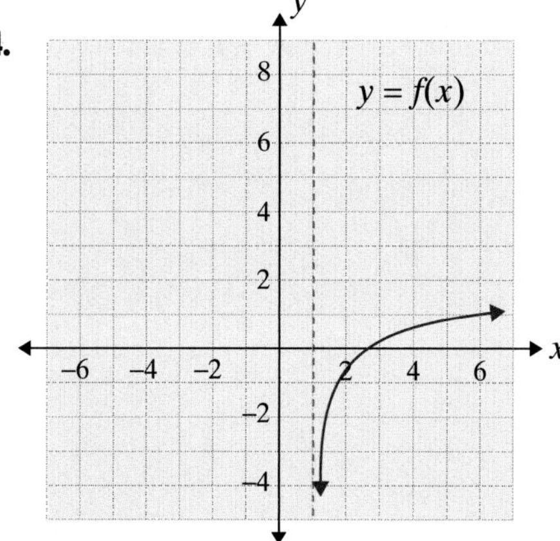

a. $\lim_{x\to 1^+} f(x)$ **b.** $\lim_{x\to 6^-} f(x)$

c. $\lim_{x\to 6^+} f(x)$ **d.** $\lim_{x\to 6} f(x)$

65.

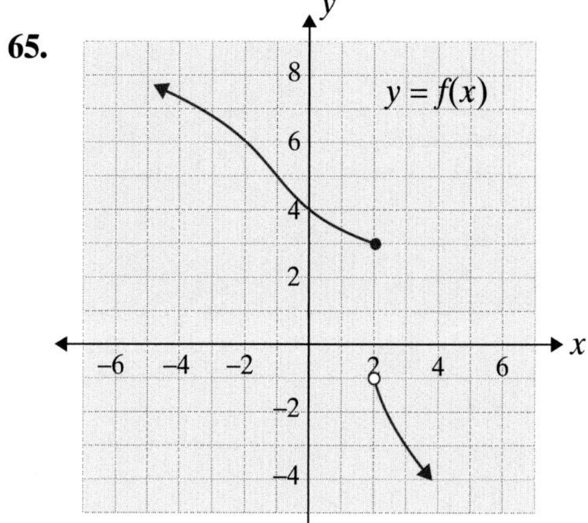

a. $\lim_{x\to 2^-} f(x)$ **b.** $\lim_{x\to 2^+} f(x)$

c. $\lim_{x\to 2} f(x)$ **d.** $\lim_{x\to 0} f(x)$

66.

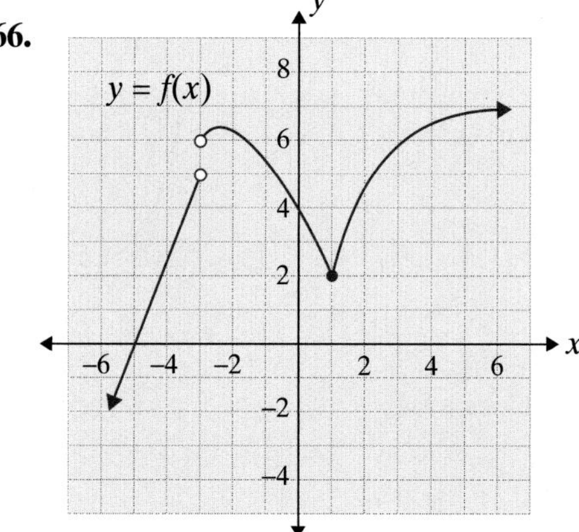

a. $\lim_{x\to -3^-} f(x)$ **b.** $\lim_{x\to -3^+} f(x)$

c. $\lim_{x\to -3} f(x)$ **d.** $\lim_{x\to 1^-} f(x)$

e. $\lim_{x\to 1^+} f(x)$ **f.** $\lim_{x\to 1} f(x)$

67.

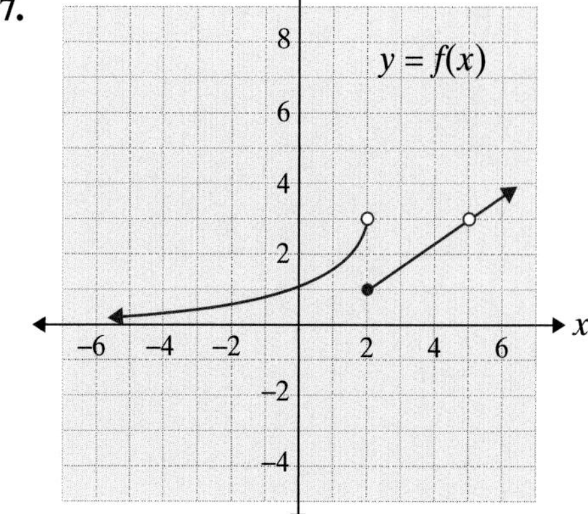

a. $\lim_{x\to 2^-} f(x)$ **b.** $\lim_{x\to 2^+} f(x)$

c. $\lim_{x\to 2} f(x)$ **d.** $\lim_{x\to 5^+} f(x)$

e. $\lim_{x\to 5^-} f(x)$ **f.** $\lim_{x\to 5} f(x)$

68.

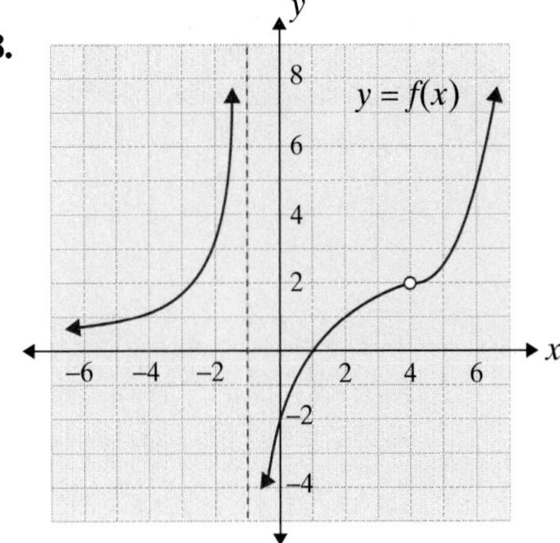

a. $\lim_{x\to -1^-} f(x)$ **b.** $\lim_{x\to -1^+} f(x)$

c. $\lim_{x\to 4} f(x)$ **d.** $\lim_{x\to 1} f(x)$

69.

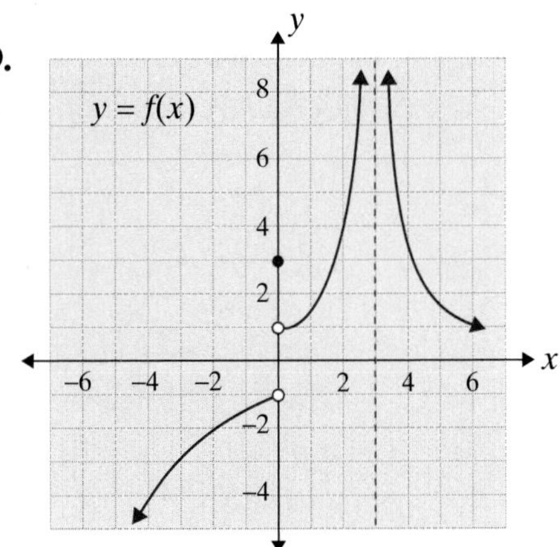

a. $\lim_{x\to 0^-} f(x)$ **b.** $\lim_{x\to 0^+} f(x)$

c. $\lim_{x\to 3^+} f(x)$ **d.** $\lim_{x\to 3^-} f(x)$

70.

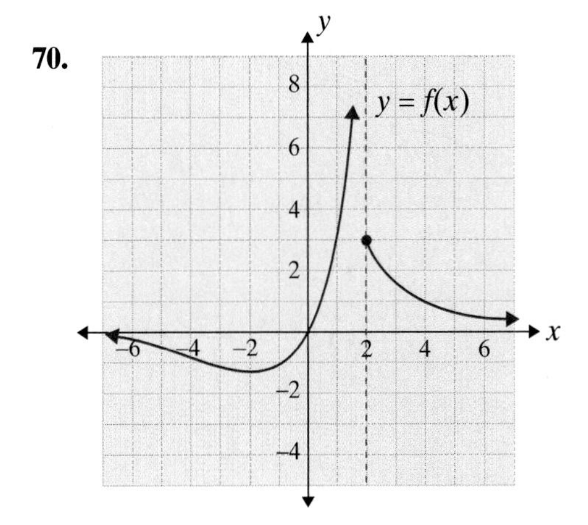

a. $\lim_{x\to 2^-} f(x)$ **b.** $\lim_{x\to 2^+} f(x)$

c. $\lim_{x\to 2} f(x)$ **d.** $\lim_{x\to 0} f(x)$

71.

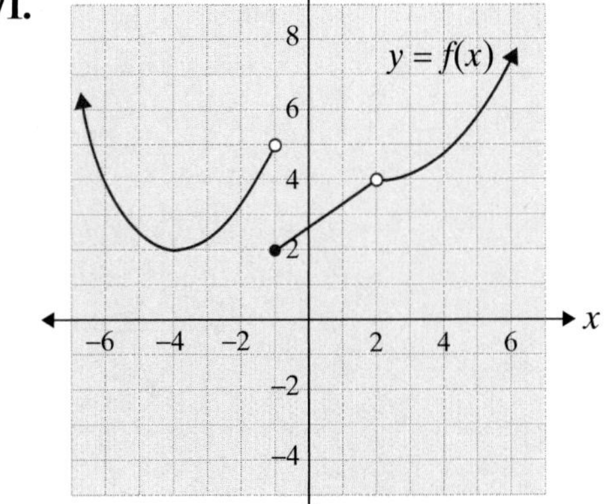

a. $\lim_{x\to -1^-} f(x)$ **b.** $\lim_{x\to -1^+} f(x)$

c. $\lim_{x\to -1} f(x)$ **d.** $\lim_{x\to 2} f(x)$

Solve Exercises 72 – 77 algebraically by factoring first.

72. $\lim_{x\to 3}\left(\frac{3-13x+4x^2}{x-3}\right)$

73. $\lim_{x\to 6}\left(\frac{x^2-36}{x-6}\right)$

74. $\lim_{x\to -7}\left(\frac{x-7}{x^2-49}\right)$

75. $\lim_{h\to 0}\left(\frac{f(3+h)-f(3)}{h}\right),\quad f(x)=x^2-2$

76. $\lim_{h\to 0}\left(\frac{f(2-h)-f(2)}{h}\right),$

$f(x)=1-x+x^2$

77. $\lim_{x\to 4}\left(\frac{x^4-256}{x^2-16}\right)$

78. Salary. Erin is paid a weekly salary of \$12 per hour plus time-and-a-half for overtime (time in excess of 40 hours, but no more than 60 hours). Her salary is given by the function

$$S(t)=\begin{cases}12t & \text{if } 0<t\le 40\\ 480+18(t-40) & \text{if } 40<t\le 60\end{cases}$$

where t is the time in hours, $0<t\le 60$.

a. Find $\lim_{t\to 40^-} S(t)$. **b.** Find $\lim_{t\to 40^+} S(t)$. **c.** Find $\lim_{t\to 40} S(t)$.

79. Parking rates. The rate for parking in the short-term lot (maximum of 24 hours) at the airport is \$1.00 for the first hour plus \$0.75 for each additional hour or part thereof, with a maximum cost of \$7.00. The function for the cost of parking on this lot for t hours (up to 24 hours) is

$$C(t)=\begin{cases}1.00 \text{ for } 0<t\le 1\\ 1.75 \text{ for } 1<t\le 2\\ 2.50 \text{ for } 2<t\le 3\\ 3.25 \text{ for } 3<t\le 4\\ 4.00 \text{ for } 4<t\le 5\\ 4.75 \text{ for } 5<t\le 6\\ 5.50 \text{ for } 6<t\le 7\\ 6.25 \text{ for } 7<t\le 8\\ 7.00 \text{ for } 8<t\le 24\end{cases}$$

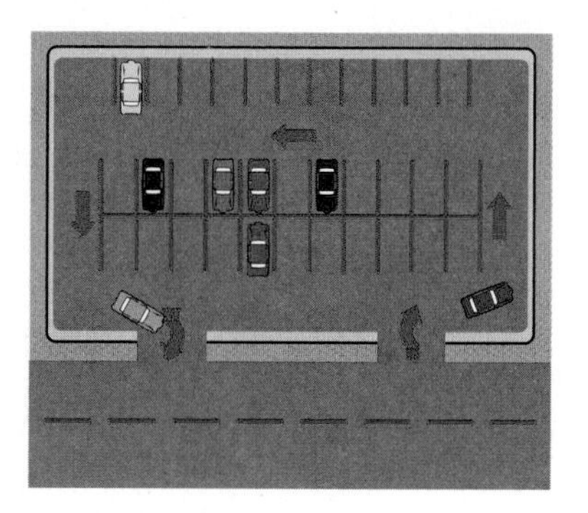

a. Graph the function for $0<t\le 24\text{hr}$.

b. Find $\lim_{t\to 3^-} C(t)$.

c. Find $\lim_{t\to 3^+} C(t)$.

d. Find $\lim_{t\to 8^-} C(t)$.

e. Find $\lim_{t\to 8^+} C(t)$.

80. Suppose $f(x)$ and $g(x)$ are polynomials and $f(t)=0=g(t)=0$ for some t. If $\lim_{x\to t^-}\frac{f(x)}{g(x)}=L$, must $\lim_{x\to t^+}\frac{f(x)}{g(x)}$ also be L?

81. Suppose $f(x)$ and $g(x)$ are equal for all x-values except $x=t$.

a. Is $\lim_{x\to t^-} f(x)=\lim_{x\to t^-} g(x)$ true?

b. What about $\lim_{x\to t^+} f(x)$ and $\lim_{x\to t^+} g(x)$?

c. Is $\lim_{x\to t} f(x)=\lim_{x\to t} g(x)$ necessarily true?

Hawkes Learning Systems: Essential Calculus

Left and Right-hand Limits	p. 96 - 105	Ex. 1 - 61, 79, 80
Limits	p. 105 - 110	Ex. 62 - 78, 81

2.2 Rates of Change and Derivatives

Objectives

After completing this section, you will be able to:

1. *Determine the average rate of change.*
2. *Find the instantaneous rate of change.*
3. *Interpret the graph of a function.*

Suppose that you rode your bicycle 24 miles in 2 hours. Then your average speed (**average rate of change of distance with respect to time**) was $\frac{24 \text{ miles}}{2 \text{ hours}} = 12 \text{mph}$.

This does not mean that you were moving at 12 mph all the time you were riding. At some times you were going slower and at some times faster. For example, if you happened to be riding downhill at a particular instant, then your **instantaneous rate of change** at that moment might have been 18 mph.

These two ideas, average rate of change and instantaneous rate of change, form the basis for our development of the derivative of a function.

Average Rate of Change

The Greek capital letter delta (Δ) is used in calculus to indicate change. Thus Δx is read "delta x" (or "the change in x"), and Δy is read "delta y" (or "the change in y"). For a function $y = f(x)$, as the values of x change from x_1 to x_2, the corresponding values of y change from y_1 to y_2, or from $f(x_1)$ to $f(x_2)$. We write

$$\Delta x = x_2 - x_1$$

and

$$\Delta y = y_2 - y_1 = f(x_2) - f(x_1).$$

Average Rate of Change

If $y = f(x)$ then the ratio

$$\frac{\Delta y}{\Delta x} = \frac{y_2 - y_1}{x_2 - x_1} = \frac{f(x_2) - f(x_1)}{x_2 - x_1}$$

is called the ***average rate of change of y with respect to x*** *as x changes from x_1 to x_2.*

Example 1: Finding the Average Rate of Change

For the function $y = 2x + 1$, find the average rate of change of y with respect to x, $\frac{\Delta y}{\Delta x}$, as x changes from 0 to 3.

Solution: For $x_1 = 0$ and $x_2 = 3$, the corresponding values of y are:

$$y_1 = 2 \cdot x_1 + 1 = 2 \cdot 0 + 1 = 1 \qquad \text{Substitute } x_1 = 0 \text{ into } y \text{ and solve.}$$

and

$$y_2 = 2 \cdot x_2 + 1 = 2 \cdot 3 + 1 = 7. \qquad \text{Substitute } x_2 = 3 \text{ into } y \text{ and solve.}$$

Thus

$$\Delta y = y_2 - y_1 = 7 - 1 = 6,$$

$$\Delta x = x_2 - x_1 = 3 - 0 = 3,$$

and the average rate of change is $\frac{\Delta y}{\Delta x} = \frac{6}{3} = 2.$

Any other distinct choices for x_1 and x_2 will show that $\frac{\Delta y}{\Delta x} = 2$, the slope of the line.

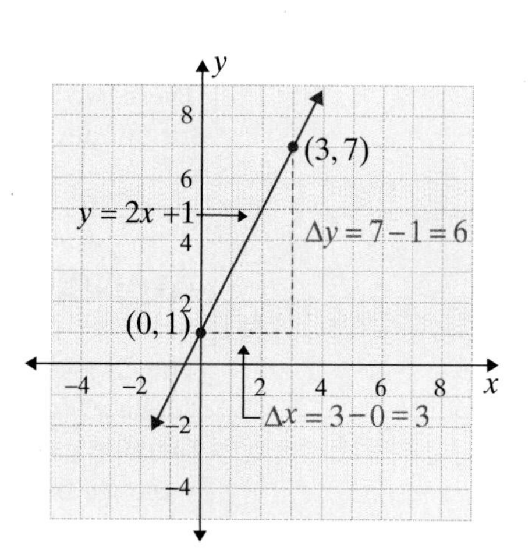

Example 2: Average Rate of Change

Let $f(x)=\frac{2}{x}$. Find and simplify the expression that represents the average rate of change of f between x and $x+h$.

Solution: Here $x_1=x$ and $x_2=x+h$. Therefore, $\Delta x=x_2-x_1=x+h-x=h$.

So
$$\frac{\Delta y}{\Delta x}=\frac{f(x_2)-f(x_1)}{x_2-x_1}=\frac{f(x+h)-f(x)}{h}$$
$$=\frac{\frac{2}{x+h}-\frac{2}{x}}{h}=\frac{\frac{2\cdot x}{(x+h)\cdot x}-\frac{2\cdot(x+h)}{x\cdot(x+h)}}{h}$$
$$=\frac{\frac{2x-2x-2h}{x(x+h)}}{\frac{h}{1}}=\frac{-2\cancel{h}}{x(x+h)}\cdot\frac{1}{\cancel{h}}=-\frac{2}{x(x+h)}.$$

Example 3: A Falling Object

A ball is dropped from an airplane. The distance d that it falls in t seconds is given by the function $d=16t^2$. The table shows the distance the ball falls and the change in distance Δd for 1-second time intervals during the first 5 seconds of its descent.

t (sec)	$d=16t^2$	Δd (ft.)
0	0	—
1	16	16 (d_1-d_0)
2	64	48 (d_2-d_1)
3	144	80 (d_3-d_2)
4	256	112 (d_4-d_3)
5	400	144 (d_5-d_4)

a. How far does the ball fall during the first 3 seconds?

b. What is the average velocity of the ball during the first 3 seconds? (Velocity is the rate of change of distance with respect to time.)

c. How fast is the ball traveling when $t=3$?

Solutions: a. $\Delta d=d_3-d_0=144-0=144$ ft. Using data from the table.

b. $\frac{\Delta d}{\Delta t}=\frac{144\text{ ft.}}{3\text{ sec}}=48$ ft./sec Using $\Delta t=t_3-t_0=3-0=3$ sec.

c. We do not know how fast the ball is traveling when $t=3$ seconds. This is a question of instantaneous velocity, which we will discuss later in this section.

Instantaneous Rate of Change

In this section, through the use of tables, we will discuss the concept of instantaneous rate of change of a function at one specific value of the variable. In the next section we will take a more formal algebraic approach.

Consider the problem in Example 3 of the ball dropped from an airplane. The distance the ball falls is given by the function $d = 16t^2$ where t is time in seconds and d is distance in feet. (See Figure 2.2.1.)

To determine how fast the ball is moving at the instant when $t = 3$ seconds, we analyze the average rate of change $\frac{\Delta d}{\Delta t}$ for smaller and smaller values of Δt, where $t = 3$. Table 2.2.1, on the next page, shows that t_1 is fixed at 3 and t_2 approaches 3. The various values of the function can be found with a calculator.

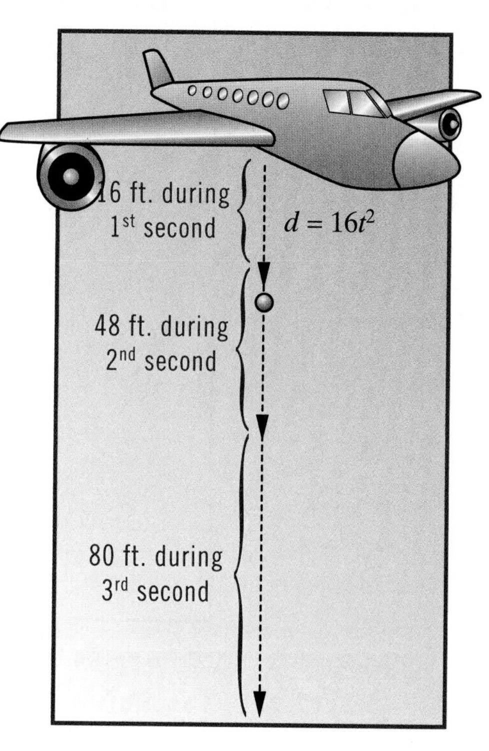

Figure 2.2.1

Notes about Table 2.2.1

1. The values of $t_1 = 3$ and $d_1 = 144$ are fixed.
2. Since t_2 is approaching 3, the difference Δt is approaching 0.
3. The difference Δd is also approaching 0.
4. The average rate of change $\frac{\Delta d}{\Delta t}$ is approaching the **instantaneous rate of change** of d with respect to t when $t = 3$. We can write

$$\lim_{\Delta t \to 0} \frac{\Delta d}{\Delta t} = 96 \text{ ft./sec.}$$

The instantaneous rate of change of distance with respect to time, as illustrated in Table 2.2.1, is called **instantaneous velocity**.

$d = 16t^2$						
t_1	t_2	$\Delta t = t_2 - t_1$	d_1	d_2	$\Delta d = d_2 - d_1$	$\frac{\Delta d}{\Delta t}$
3	2.0	−1.0	144	64.0	−80.0	$\frac{-80.0}{-1.0} = 80$ ft./sec
3	2.1	−0.9	144	70.56	−73.44	$\frac{-73.44}{-0.9} = 81.6$ ft./sec
3	2.5	−0.5	144	100.0	−44.0	$\frac{-44.0}{-0.5} = 88.0$ ft./sec
3	2.8	−0.2	144	125.44	−18.56	$\frac{-18.56}{-0.2} = 92.8$ ft./sec
3	2.9	−0.1	144	134.56	−9.44	$\frac{-9.44}{-0.1} = 94.4$ ft./sec
3	2.99	−0.01	144	143.0416	−0.9584	$\frac{-0.9584}{-0.01} = 95.84$ ft./sec
3	2.999	−0.001	144	143.904016	−0.095984	$\frac{-0.095984}{-0.001} = 95.984$ ft./sec
	↓	↓			↓	↓
	3	0			0	96 ft./sec

Table 2.2.1

Slopes and Rates of Change

Rate of change is a numerical interpretation of $\frac{\Delta y}{\Delta x}$ for some function $y = f(x)$. This is a natural geometric interpretation using the familiar concept of **slope of a line**: for any two points (x_1, y_1) and (x_2, y_2) on the graph of $f(x)$, the number (rate of change) $\frac{\Delta y}{\Delta x} = \frac{y_2 - y_1}{x_2 - x_1}$ is also the slope of the line through the two points.

If we use a graphing utility to graph $y = x^2$ in a window $[-5, 5]$ by $[-10, 20]$, we see this familiar parabola.

Let us consider the points $(x_1, y_1) = (3, 9)$ and $(x_2, y_2) = (4, 16)$.

The slope of the segment containing these points is

$$\frac{\Delta y}{\Delta x} = \frac{16-9}{4-3} = 7.$$

The line $y = mx + b$ through these two points with a y-intercept of b is given by

Using point $(x_1, y_1) = (3, 9)$ and slope = 7. $\quad 9 = 7(3) + b.$

Then, solving for b, we find that

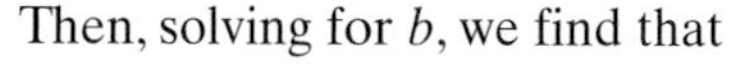

$$b = -12. \qquad \text{Adding } -21 \text{ to both sides.}$$

Therefore, the equation of the line through both points is

$$y = 7x - 12.$$

This line is also called the **secant line** through the two points. Display this line on the same calculator screen as $y = x^2$. To do so, type the equation of this secant line into the position y_2 on the graphing calculator, as shown in Figure 2.2.2(a). Figure 2.2.2(b) is what you should see graphed.

Figure 2.2.2 (a)

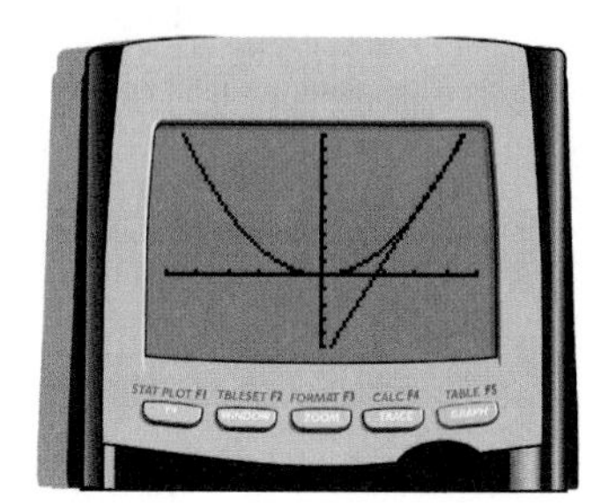

Figure 2.2.2 (b)

Now we want to draw a new view with a smaller Δx. We choose a line through points $(x_1, y_1) = (3, 9)$ and $(x_2, y_2) = (3.1, 9.61)$. The slope of this line is

$$\frac{\Delta y}{\Delta x} = \frac{9.61-9}{3.1-3} = \frac{.61}{.1} = 6.1$$

and the y-intercept is

$$9 = 6.1(3) + b$$

$$9 = 18.3 + b$$

$$-9.3 = b.$$

Therefore, the line through the new points is $y = 6.1x - 9.3$.

Using the window [2.8, 3.3] by [7.5, 9.9], both secant lines are shown graphed in Figure 2.2.3 at the right. Even under high magnification, the two secant lines are difficult to discern.

Figure 2.2.3

We repeat this process a third time with $(x_1, y_1) = (3, 9)$, using an even smaller Δx.

Let $(x_2, y_2) = (3.01, 9.0601)$. Here

$$\frac{\Delta y}{\Delta x} = \frac{9.0601 - 9}{3.01 - 3} = \frac{0.0601}{0.01} = 6.01.$$

As $\Delta x \to 0$, the slope calculated seems to be nearing 6.

That is,

$$\lim_{\Delta x \to 0} \frac{\Delta y}{\Delta x} = 6.$$

One can continue this process, and as $\Delta x \to 0$, the secant line approaches a limiting position. This position is occupied by the line through (3, 9) having slope 6, which has the equation $y = 6x - 9$.

This line touches $y = x^2$ exactly at the point (3, 9) and at no other point. It is called the **tangent line** to $y = x^2$ at the point (3, 9).

As we zoom closer and closer to the point (3, 9), the graph of the function and the graph of the tangent line cannot be separated on the graphing screen (see Figure 2.2.4).

The number 6 is the slope of the tangent line, and it is also the instantaneous rate of change of y at the point (3, 9). At the point (3, 9) the function y is increasing at the rate of 6 units (per 1 unit of change in x). Therefore, from 3 to 4 on the x-axis, we expect y to increase by about 6.

Figure 2.2.4

Slope of a Point on a Curve

The slope of a curve f(x) at the point (a, b) is exactly the slope of the tangent line to the curve at that point. This is also called the ***instantaneous rate of change****.*

Numerically, this slope is

$$\lim_{\Delta x \to 0}\left(\frac{\Delta y}{\Delta x}\right) = \lim_{h \to 0}\frac{f(a+h)-f(a)}{h},$$

where $h = \Delta x$. Technically, the definition given above makes sense only if such a limit exists.

We will use the notation $f'(a)$, read "f-prime of a", to denote the slope at $(a, f(a))$.

Difference Quotient

If $(x, f(x))$ is any fixed point on a curve and $(x+h, f(x+h))$ is another point on the curve, then the difference quotient is the slope of the secant line through these two points:

$$m_{sec} = \frac{f(x+h)-f(x)}{(x+h)-x} = \frac{f(x+h)-f(x)}{h}.$$

The **derivative** $f'(x)$, if it exists, can be interpreted as the slope of the tangent line at the point $(x, f'(x))$. Figure 2.2.5 illustrates this geometric relationship between the derivative and the secant line.

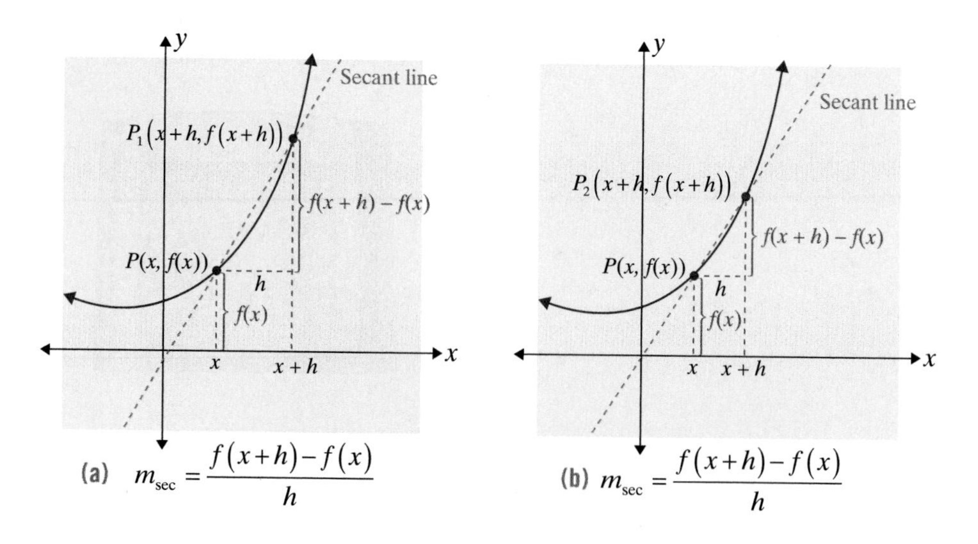

Figure 2.2.5

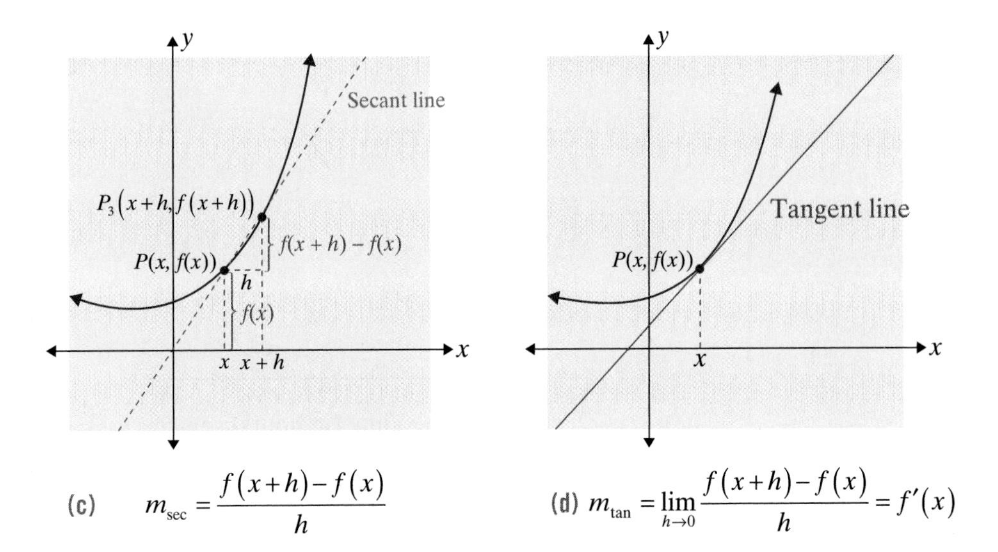

Figure 2.2.5 (c) $m_{\text{sec}} = \dfrac{f(x+h)-f(x)}{h}$ (d) $m_{\tan} = \lim\limits_{h\to 0}\dfrac{f(x+h)-f(x)}{h} = f'(x)$

The derivative $f'(x)$ is the slope of the tangent line at $(x, f(x))$.

For the rest of this book, given a function $y = f(x)$, we will be interested in two "output" numbers corresponding to the input $x = a$, namely the corresponding y-value, $f(a)$, and the slope, $f'(a)$.

NOTES

For an input $x = a$ for a given function $f(x)$:

$f(a)$ = the height of the point, and

$f'(a)$ = the slope of the curve at the point.

Example 4: Temperature

Let $f(t)$ denote the temperature in South Carolina on a typical summer day at time t, where t is the number of hours since midnight. For instance, $t = 1$ corresponds to 1 A.M. and $t = 13$ corresponds to 1 P.M.

a. Is $f'(8)$ positive, negative, or zero?

b. Is $f'(21)$ positive, negative, or zero?

c. Sketch a graph which is approximate for this $f(t)$.

Solutions: a. Since the graph of $f(t)$ will increase from sun up to about mid-afternoon, the slope at $(8, f(8))$ is positive. That is, $f'(8)$ is positive.

b. At 9 P.M. $(t = 21)$, the temperature has begun to decline and so the slope will be negative.

continued on next page ...

c. The graph needs to increase from about 6 A.M. to 4 P.M. and then decline, as described in parts **a.** and **b.** Drawing such a graph will also support the answers found in parts **a.** and **b.**

Example 5: Manufacturing

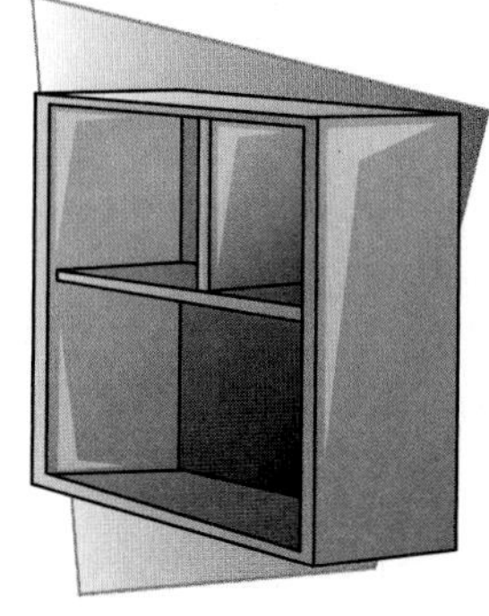

Let $C(x)$ denote the total cost of manufacturing x units of a certain style of metal office bookcase.

a. Given $C(300) = 31{,}000$ and $C'(300) = 90$, sketch a tangent line to $C(x)$ at the point $(300, 31{,}000)$.

b. Estimate the cost of manufacturing 303 bookcases.

Solutions: a. Using the definition of the slope of a curve at a point, $C'(300) = 90$ is the slope of the tangent line at the point $(300, 31{,}000)$. Knowing this, we can draw a line with slope 90 at the point $(300, 31{,}000)$.

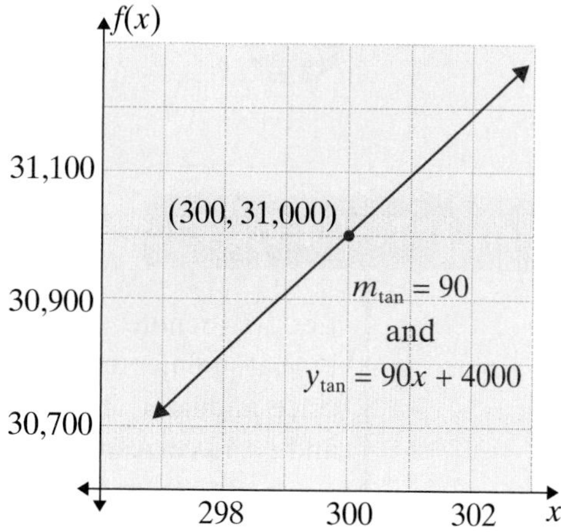

b. First, lets further understand the information given in the example. $C(300) = 31{,}000$ means it costs \$31,000 to manufacture 300 bookcases. Using this, we can determine the average cost of producing the first 300 bookcases:

$$\frac{C(300)}{300} = \frac{31{,}000}{300} \approx \$103.33.$$

$C'(300) = 90$ is the rate of change of y at the point (300, 31,000). This means that y is changing at the rate of \$90 per bookcase at this point. Another way of stating that is \$90 is the cost of an additional bookcase if 301 were produced instead of 300. The average cost is clearly decreasing – a useful fact to be able to calculate for a manufacturing company.

Using these meanings, the cost of producing 3 more bookcases, compared to the cost of producing 300, is about $90 \cdot 3 = 270$ dollars more. So it will cost \$31,000 + \$270 = \$31,270 to manufacture 303 bookcases.

NOTES

We know that $\frac{\Delta y}{\Delta x}$ can be interpreted as the slope of a secant line through two points on a curve. Now, if one point is fixed, then the **tangent line at that point is the limiting position of secant lines**, as illustrated in Figure 2.2.6.

As the points $P_1, P_2, P_3, \ldots, P_n$ approach point P_1, the secant lines $PP_1, PP_2, PP_3, \ldots, PP_n$ approach the tangent at P.

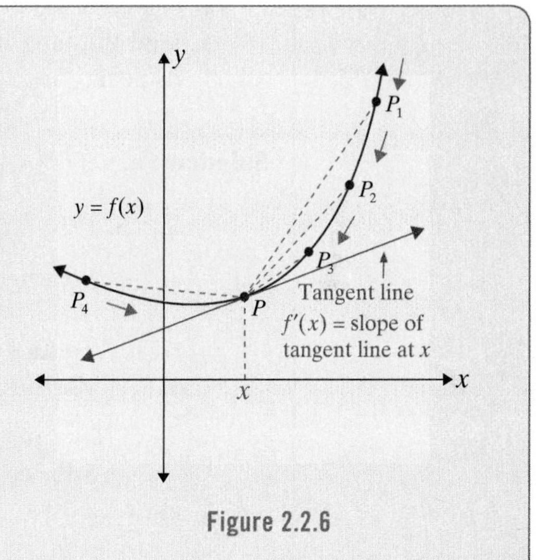

Figure 2.2.6

Example 6: Average Rate of Change

Consider the function $f(x) = \frac{1}{3}x^3$. Find the average rate of change of f as x changes from $x_1 = 0$ to $x_2 = 3$.

Solution: $f(x_1) = f(0) = \frac{1}{3} \cdot 0^3 = 0$

$f(x_2) = f(3) = \frac{1}{3} \cdot 3^3 = 9$

$\frac{\Delta y}{\Delta x} = \frac{f(x_2) - f(x_1)}{x_2 - x_1}$

$= \frac{9-0}{3-0} = \frac{9}{3} = 3$

The average rate of change is the slope of a line secant to both the start and stop points.

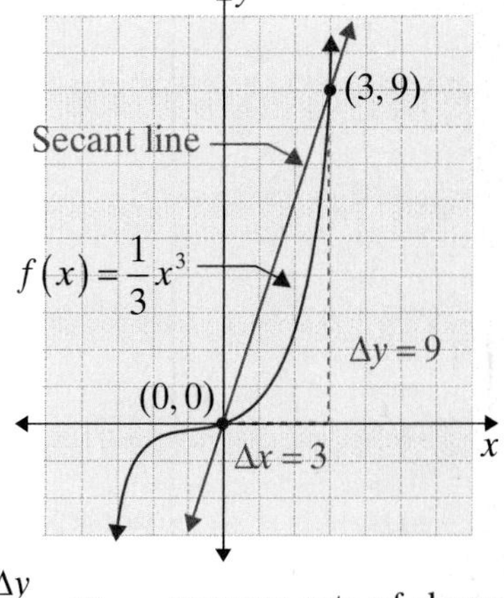

$\frac{\Delta y}{\Delta x} = m_{\text{sec}} =$ average rate of change

In Example 7 we show how the derivative can be interpreted as both the slope of a tangent line to a curve and the instantaneous velocity of an object.

Example 7: Instantaneous Velocity

An arrow is shot into the air and its height in feet after t seconds is given by the function $f(t) = -16t^2 + 80t.$ The graph of the curve $y = f(t)$ is the parabola shown on the right.

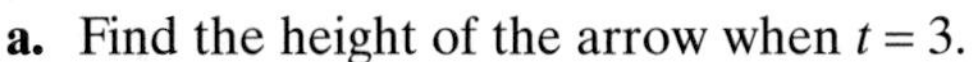

a. Find the height of the arrow when $t = 3$.

b. Find the velocity of the arrow when $t = 3$.

c. Find the slope of the tangent line to the curve at $t = 3$.

d. Find the time it takes the arrow to reach its peak.

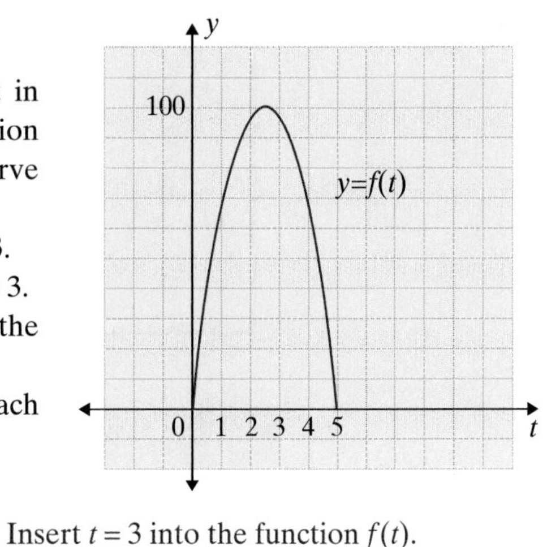

Solutions: a. $f(3) = -16(3)^2 + 80(3)$ Insert $t = 3$ into the function $f(t)$.

$$= -144 + 240 = 96.$$

So, when $t = 3$, the height is 96 feet.

b. To find the instantaneous velocity when $t = 3$, find the derivative $f'(t)$, which is also the slope of the tangent line at $(t, f(t))$, and then evaluate at $t = 3$.

Step 1: Form the difference quotient.

$$\frac{f(t+h) - f(t)}{h} = \frac{\left[-16(t+h)^2 + 80(t+h)\right] - \left[-16t^2 + 80t\right]}{h}$$

Step 2: Simplify the difference quotient.

$$\frac{\left[-16(t+h)^2 + 80(t+h)\right] - \left[-16t^2 + 80t\right]}{h}$$

$$= \frac{-16\left(t^2 + 2th + h^2\right) + 80t + 80h + 16t^2 - 80t}{h}$$

$$= \frac{-16t^2 - 32th - 16h^2 + 80t + 80h + 16t^2 - 80t}{h}$$

$$= \frac{-32ht - 16h^2 + 80h}{h}$$

$$= \frac{\cancel{h}(-32t - 16h + 80)}{\cancel{h}} = -32t - 16h + 80$$

Step 3: Find the limit:

$$\lim_{h \to 0}(-32t - 16h + 80) = -32t + 80 = m_{\tan} = f'(t).$$

Now, using this derivative, insert $t = 3$ and solve:

$$f'(t) = -32t + 80,$$

$$f'(3) = -32 \cdot 3 + 80 = -16.$$

Therefore, when $t = 3$ the arrow is heading towards the ground (this is what the negative sign means) at a rate of 16 ft./sec.

c. From part **b.** we know that $f'(3) = -16$. So, the slope of the tangent line to the curve $f(t) = -16t^2 + 80t$ at $t = 3$ is -16.

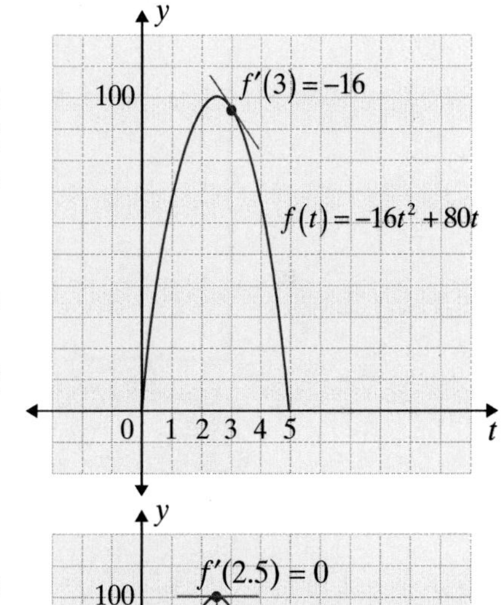

Therefore, we see that the derivative can be interpreted in two ways:

(1) as an instantaneous rate of change and
(2) as the slope of a tangent line.

d. At the instant the arrow hits its peak, its velocity will be 0. That is, the arrow will actually be stopped in midair (it will no longer be traveling upwards, and will not yet be going downwards). **Note:** This also corresponds to the fact that the tangent line at the peak will be horizontal and thus have slope 0 at the highest point of the parabola.

Using this information, we set $f'(t) = 0$ and solve for t.

$$-32t + 80 = 0$$

$$-32t = -80$$

$$t = \frac{80}{32} = \frac{\cancel{16} \cdot 5}{\cancel{16} \cdot 2}$$

$$= \frac{5}{2} = 2.5$$

The arrow hits its peak in 2.5 sec.

Example 8: Interpreting a Graph

Estimate the slopes at the marked points as 1, –1, 0, undefined, positive and greater than 1, positive and less than 1, negative and greater than –1, or negative and less than –1. Keep in mind that a tangent line has slope 1 when it makes a 45° angle with the x-axis, has a slope 0 when it is horizontal, and has an undefined slope when it is vertical.

Solution:

Point	$f'(x)$
A	positive, greater than 1
B	1
C	0
D	negative, greater than –1
E	0
F	positive, less than 1
G	negative, less than –1

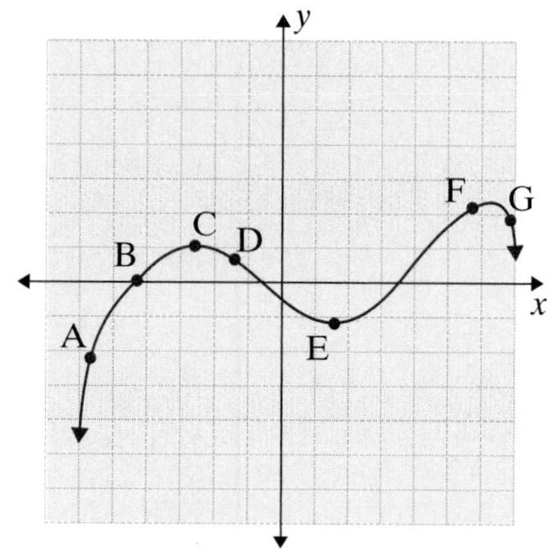

NOTES

We emphasize that $f'(a)$ denotes the slope of a curve $y = f(x)$ at the point $(a, f(a))$. Also, $f'(a)$ is the instantaneous rate of change of $f(x)$ at $(a, f(a))$. The number $f'(a)$ is called the **derivative** of $f(x)$ at $x = a$.

Slopes and Calculators

We investigate the algebra of $f'(x)$ given $f(x)$ in Section 2.3. In this section we continue numerically.

One can use the TI-84 Plus or another graphing calculator to obtain slopes. The TI-84 Plus has several simple mechanisms for obtaining slopes or the equation of a line tangent to $y = f(x)$ at a specified point.

To illustrate the methods, let us create (and complete) the chart shown at the right.

x	$f(x)$	$f'(x)$
–1	1	
0	0	
1	1	
2	4	
3	9	
4	16	

Enter $y = x^2$ in the Y_1 position and graph in a window $[-5, 5]$ by $[-10, 30]$. When the graph appears on screen, press TRACE and then type the first x-value and press ENTER. Note that the values $x = 1$ and $y = 1$ are listed at the bottom of the screen. Type the next x-value and press ENTER. Continue to type the x-values in the first column, one after the other, without having to retype the TRACE command. Without pressing TRACE again, the corresponding y-values can be obtained and written down.

Next, with the graph displayed on the screen, type 2ND TRACE and select item six (type 6, and ENTER). Now type the desired x-value (start with -1). `dy/dx = -2` appears at the bottom of the screen (see Figure 2.2.7). The symbol dy/dx is another notation for *derivative*.

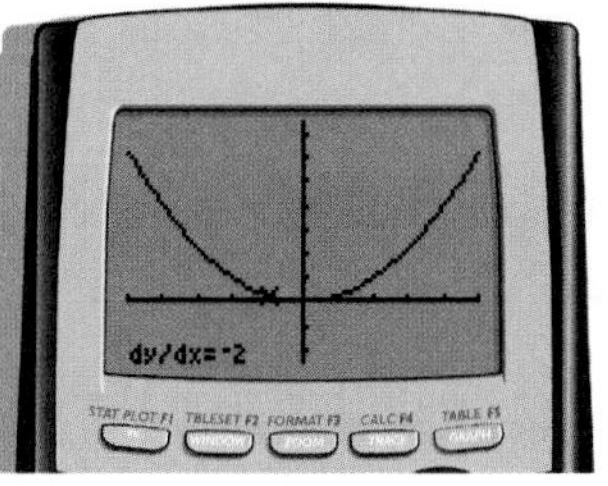

Figure 2.2.7

Retype 2ND TRACE, select item 6 again, and input the next x-value. Continue until all of the x-values in the table have been used and you have written down the appropriate slopes.

The calculator has the ability to draw a tangent line and calculate the line's equation, just as easily performing the previous calculation. With the graph of $y = x^2$ on the screen, press 2ND PRGM (this will activate the `DRAW` function) and select `5: Tangent(` (see Figure 2.2.8).

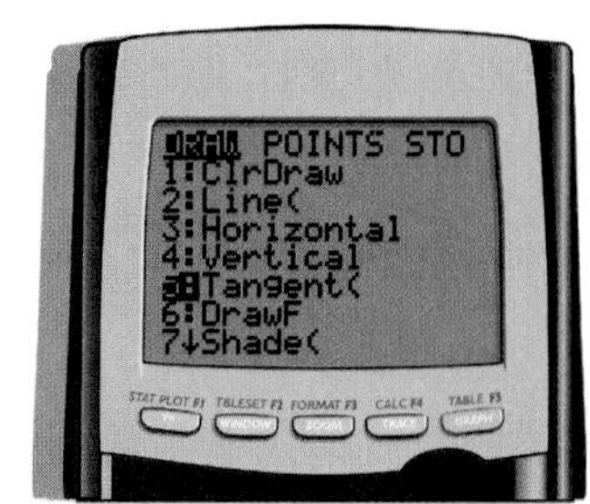

Figure 2.2.8

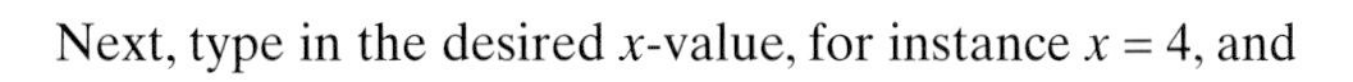

Next, type in the desired x-value, for instance $x = 4$, and press ENTER. The calculator draws a tangent line to the graph at $x = 4$ and displays pertinent information at the bottom of the screen as shown in Figure 2.2.9. The information printed at the bottom is:

`x = 4`
`y = 8x + -16.`

The equation of the tangent line is $y = 8x - 16$, with a slope of 8 and a y-intercept of -16. To solve for the remaining slopes, you need only to repeat the previous commands 2ND, PRGM, 5, and type in the next x-value.

Figure 2.2.9

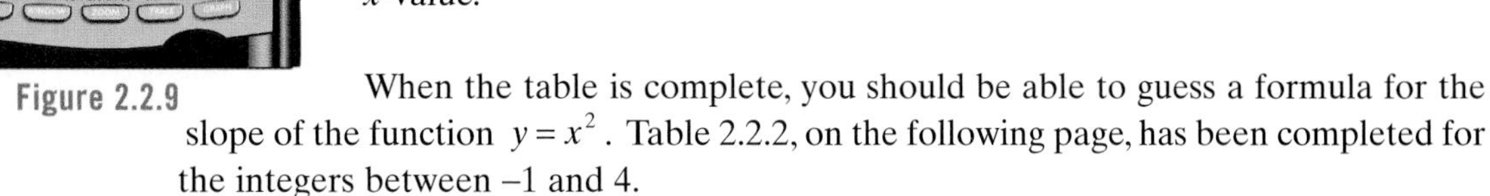

When the table is complete, you should be able to guess a formula for the slope of the function $y = x^2$. Table 2.2.2, on the following page, has been completed for the integers between -1 and 4.

x	$f(x)$	$f'(x)$
–1	1	–2
0	0	0
1	1	2
2	4	4
3	9	6
4	16	8

Table 2.2.2

Example 9: Describing and Graphing Rates of Change

a. Describe $f'(x)$ at the points marked on the graph.

Solution:

Point	$f'(x)$
A	positive
B	0
C	negative
D	negative
E	0

b. Sketch a possible graph of $f'(x)$.

Solution: Using the description found in part **a.**, we can draw a possible graph of $f'(x)$. At A, the slope is positive, and at C, the slope is negative. We know that at point B (and E), $f'(x) = 0$ since the tangent line is horizontal. Therefore, the graph of $y = f'(x)$ is above the x-axis at $x = a$ (positive), passes through the x-axis at $x = b$ (0), and lies below the x-axis at $x = c$ (negative). From point C to point D the slopes continue to decrease, since $f'(x)$ is negative, but from points D to E the slopes increase because we know that at point E the slope is equal to 0.

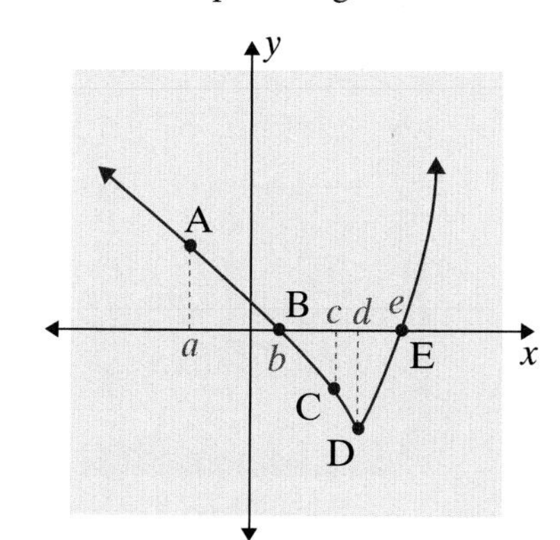

2.2 Exercises

In Exercises 1 – 10, determine what the slope f' represents in terms of the subject in the problem.

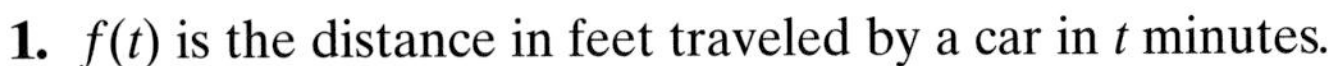

1. $f(t)$ is the distance in feet traveled by a car in t minutes.
2. $f(s)$ is the total money spent in a department store by s customers.
3. $f(n)$ is the number of birds nesting in woods with n trees per acre.
4. $f(x)$ is the total cost of manufacturing x toasters.
5. $f(u)$ is the total revenue from the sale of u car radios.
6. $f(t)$ is the speed of a race car after t seconds.
7. $f(v)$ is the total amount of information in bytes fed into a server at Castle Manufacturing Company in v seconds.
8. $f(x)$ is the vertical distance in meters traveled by a test rocket in x seconds.
9. $f(s)$ is the grade point average of freshmen at Sullivan Technical College where s is the average SAT score of the freshman class.
10. $f(t)$ is the cost of calculus books at college book stores in the United States where t is the time in years since 1980.

Use the graph to the right to solve Exercises 11 – 15. t is the number of hours since midnight and f(t) is the temperature at time t.

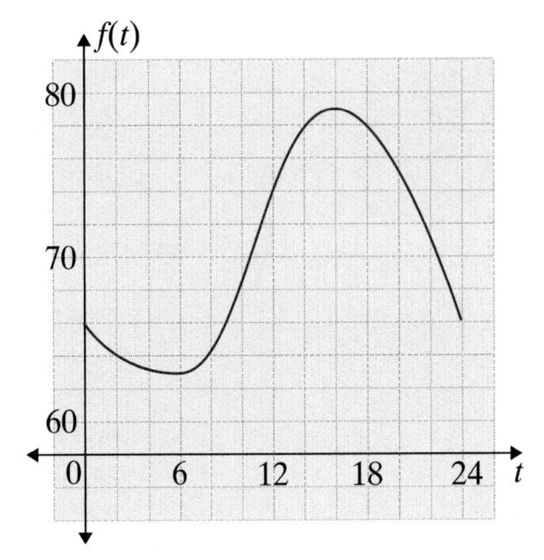

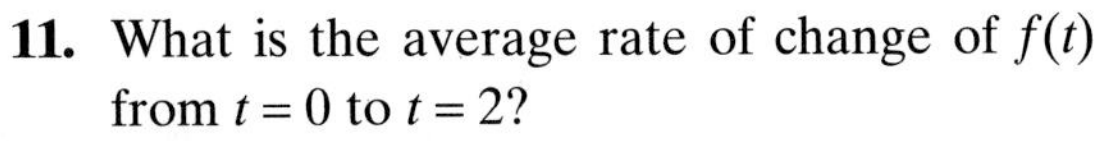

11. What is the average rate of change of $f(t)$ from $t = 0$ to $t = 2$?
12. What is the average rate of change of $f(t)$ from $t = 6$ to $t = 12$?
13. Roughly estimate the instantaneous rate of change of $f(t)$ at 3 P.M. (**Hint:** Extend an imaginary tangent line so as to come close to or to intersect points with integer coordinates.)

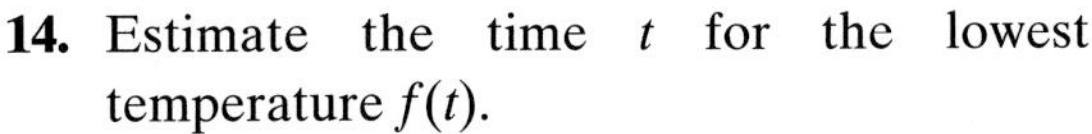

14. Estimate the time t for the lowest temperature $f(t)$.
15. Estimate the time t for the fastest increase in temperature $f(t)$.

In Exercises 16 – 23, find the average rate of change of the given functions between the given values of x_1 and x_2.

16. $f(x) = 5x + 3;\ x_1 = 1,\ x_2 = 3$

17. $f(x) = 3x + 8;\ x_1 = -2,\ x_2 = 1$

18. $f(x) = 2x^2 - x - 3;\ x_1 = 2,\ x_2 = 2.5$

19. $f(x) = 3x^2 - 2x - 1;\ x_1 = 1,\ x_2 = 1.5$

20. $f(x) = \dfrac{-2}{2x-1};\ x_1 = 3,\ x_2 = 3.5$

21. $f(x) = \dfrac{2}{3x+2};\ x_1 = 0.5,\ x_2 = 1$

22. $f(x) = \sqrt{x};\ x_1 = 1,\ x_2 = 2.5$

23. $f(x) = \sqrt{x-3};\ x_1 = 4,\ x_2 = 4.44$

Use a graphing utility in Exercises 24 – 27 to find the slope of f(x) at the given point. Sketch the graph of f(x) and tangent line at the given point on your paper.

24. $f(x)=\dfrac{4+2x}{\sqrt{x}}$; $(16,9)$

25. $f(x)=x^3$; $(3,27)$

26. $f(x)=2-3x+x^2$; $(1,0)$

27. $f(x)=10^x$; $(2,100)$

28. For the function $f(x)=4-x-2x^2$, find a window including the point $(0,4)$ so that the graph of the function and the tangent at $(0,4)$ are indistinguishable.

29. For the function $f(x)=x^3$, make a table with headings a, $f(a)$, and $f'(a)$. Then substitute numbers using $a=-1,0,1,2,3,4$. Give a formula for $f'(x)$.

30. Using the methods described in Chapter 1, locate (with a graphing utility) the x- and y-coordinates of the lowest point on the graph $f(x)=x^2-6x+11$. What is the slope at the lowest point?

31. Interpret the meaning of $f(3)=14$ and $f'(3)=7$ for the function $f(x)=2+x+x^2$.

32. For the function $y=x^2$, add a column to the table on page 134 to include the y-intercepts of the tangent line. What curiosity do you observe in the table?

33. Find the LN button on your graphing calculator and sketch a graph of $y=\ln x$ on your calculator using the window $[-2,8]$ by $[-.5,2]$. What is the slope at the point $(1,0)$? Sketch the graph and tangent on your paper.

34. On $f(x)=\sqrt{x}$, locate the x- and y-coordinates of the point at which the slope is exactly 1. (**Hint:** Find $(a,\sqrt{a})$ so that $f'(a)=1$.)

35. Suppose $f(x)$ is the number of gallons of gas used by a car after it has traveled x miles.

a. Suppose the car gets 20 mi./gal. What is $f(100)$?
b. Is $f'(100)$ positive or negative?
c. Would $f'(100)$ be greater for a subcompact car or for an SUV?

36. Suppose $f(x)$ denotes the production units for input x in labor units (man-hours). Suppose $f(500)=2000$ and $f'(500)=3$.

a. Interpret $f(500)=2000$ and $f'(500)=3$.
b. Estimate the increased production if x is increased from 500 to 501.

37. Suppose $f(x)$ is the total number of students on a college campus that have the flu and x is the number of days after the first case is reported. Interpret $f(8) = 9$ and $f'(8) = 3$.

38. Suppose $f(x)$ is the cost of a Ford Taurus and x is the age of the car.

a. Is $f'(x)$ positive or negative?

b. Interpret the meaning of $f'(3) = -2500$.

39. Average prices for new construction have steadily risen in Charleston, S.C. since 2000, according to local reports. Suppose $f(x) = 3000x + 72{,}000$ is the cost of a new house of 2200 square feet and x is the number of years since 2000.

a. Interpret $f(0) = 72{,}000$.

b. Interpret $f'(x) = 3000$.

c. Interpret $f(3) = 81{,}000$.

40. A bacteria culture in a lab grows according to the formula $y = 1600(2^t)$ where t is time in hours and y is the quantity of bacteria.

a. Interpret the meaning of $f'(1)$.

b. Determine $f'(1)$ using a graphing utility.

41. Sketch $f(x) = (x+10)(x-5)(x-10)$ on a graphing utility. Give the window used.

a. Locate the points $(a, f(a))$ for which $f'(a) = 0$.

b. What is the value of $f'(2)$?

42. Sketch $y = \frac{1}{x}$ on a graphing utility. Find the x- and y-coordinates of any point with slope -25.

43. Sketch one graph so that each of the following is true.

(1) $f'(x)$ is positive for $-2 \le x \le 6$.

(2) $f'(6) = 0$

(3) $f'(x) < 0$ for $x > 6$

(4) $f(6) = 10$ and $f(0) = 1$

44. Suppose $f(x)$ denotes the weight of a cancerous tumor x weeks after discovery. Interpret $f(3) = 4$ grams and $f'(3) = 0.4$ grams/week.

45. Water boils at $212°$F and at $100°$C. Water freezes at $32°$F and $0°$C. Let F denote temperature in degrees Fahrenheit and let x be temperature in degrees Celsius.

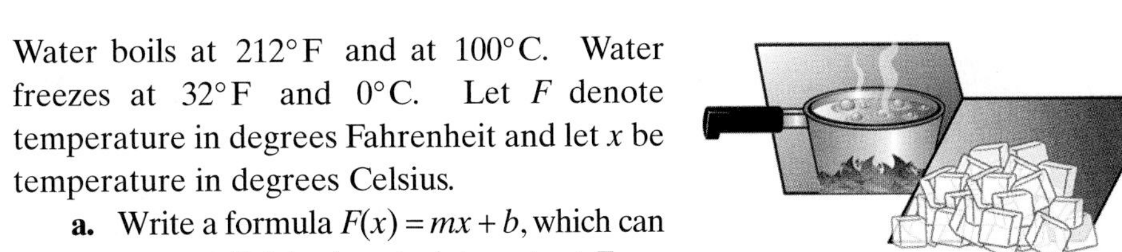

a. Write a formula $F(x) = mx + b$, which can convert Celsius input x into output F.

b. Determine m and input $F'(x)$ in terms of m.

46. Suppose $f(x)=2x^2$. Create a table of values for the slope of $f'(x)$ (see Table 2.2.2). Guess a formula for $f'(x)$.

47. Birth Rate. The fertility decline in many countries can be modeled by an appropriate equation. In Bangladesh, from 1970 to 2000, patterns of fertility changed according to the equation $y=-0.11x+6.45$, where x is the time in years beginning in 1970 and y is the average number of children per woman. (***Source*:** Lori Ashford, "World Population Highlights 2004," *BRIDGE Population Reference Bureau*, (August 2004).)

a. What number is $f(20)$ and what does it represent?

b. What number is $f'(20)$ and what does it represent?

48. Birth Rate. Patterns of fertility changed in India from 1970 to 2000 according to the equation $y=-0.068x+5.22$, where x is the number of years after 1970 and y is the average number of children per woman. (***Source*:** Lori Ashford, "World Population Highlights 2004," *BRIDGE Population Reference Bureau*, (August 2004).)

a. What number is $f(30)$ and what does it represent?

b. What number is $f'(30)$ and what does it represent?

49. Birth Rate. In China, from 1964 to the present, the death rate has remained nearly constant at about 8 deaths per 1000 persons. However, the yearly birth rate, in births per 1000 people, has declined in most years according to the formula $f(x)=-0.641x+35.8$, where x is the number of years since 1964. (***Source*:** Nancy E. Riley, "China's Population: New Trends and Challenges," *Population Bulletin*, Vol. 20, No. 2, (June 2004).)

a. What number is $f(30)$ and what does it represent?

b. What number is $f'(30)$ and what does it represent?

(**Note:** See Exercise 57 in Section 2.3 for a similar problem using a more accurate model than the linear model given here. Both models are based on the same data.)

Hawkes Learning Systems: Essential Calculus

Average Rate of Change	p. 121 - 123	Ex. 11, 12, 16 - 23
Instantaneous Rate of Change and Interpreting Graphs	p. 124 - 136	Ex. 1 - 10, 24 - 50

2.3 Slope and Rate of Change Considered Algebraically

Objectives

After completing this section, you will be able to:

1. *Use the Three-Step Method to find a derivative.*
2. *Determine if a function is or is not differentiable.*
3. *Find derivatives of functions by using the Power Rule.*
4. *Apply the Constant Times a Function Rule and the Sum and Difference Rule.*

We have considered numerical calculation of slope and the interpretation of the number as slopes of tangent lines or as ratio of change. In this section we want to begin to concentrate on the derivative as a function, important in its own right, and on the algebraic connections between $f(x)$ and $f'(x)$.

We begin with the formal definition:

Derivative

For any given function $y = f(x)$, the derivative of f at x is defined to be

$$f'(x) = \lim_{h \to 0} \left(\frac{f(x+h) - f(x)}{h} \right),$$

provided this limit exists. (Note: $\Delta x = (x+h) - x = h$.)

If $f'(x)$ exists, then f is said to be **differentiable at x**. Finding the formula for $f'(x)$ given $f(x)$ is called **differentiating**, and the process itself is called **differentiation**. We can also use $f'(x)$, $\frac{dy}{dx}$, y', or just f' to denote the derivative of $f(x)$.

It is customary for students of calculus to be able to determine the formula for $f'(x)$ given the formula for f. This is done in one of two ways. The first is by application of some rule, or theorem. For instance, there is already one rule that was suggested numerically in the previous section.

$$\text{If } f(x) = x^2, \text{ then } f'(x) = 2x.$$

In the absence of such rules, we have two choices − directly applying the definition of the derivative or using an algorithm we will call the Three-Step Method.

The Three-Step Method for Finding a Derivative

1. *Form the ratio* $\dfrac{f(x+h)-f(x)}{h}$, *called the* ***difference quotient***.
2. *Simplify the difference quotient algebraically.*
3. *Calculate* $f'(x)=\lim_{h\to 0}\left(\dfrac{f(x+h)-f(x)}{h}\right)$, *if it exists.*

Example 1: Finding a Derivative

Use the Three-Step Method to find $f'(x)$ for $f(x)=x^3-5x$.

Solution: **Step 1:** Form the difference quotient.

$$\frac{f(x+h)-f(x)}{h}=\frac{\left[(x+h)^3-5(x+h)\right]-\left(x^3-5x\right)}{h}$$

Step 2: Simplify the difference quotient.

$$\frac{\left[(x+h)^3-5(x+h)\right]-\left(x^3-5x\right)}{h}$$

$$=\frac{x^3+3x^2h+3xh^2+h^3-5x-5h-x^3+5x}{h}$$

$$=\frac{3x^2h+3xh^2+h^3-5h}{h}=\frac{\cancel{h}\left(3x^2+3xh+h^2-5\right)}{\cancel{h}}$$

$$=3x^2+3xh+h^2-5$$

Step 3: Find the limit.

$$\lim_{h\to 0}\left(3x^2+3xh+h^2-5\right)=3x^2+3x\cdot 0+0^2-5$$

$$=3x^2-5$$

Therefore, $f'(x)=3x^2-5$.

Existence of Derivatives

Since derivatives are limits and limits do not always exist, we can infer that derivatives do not always exist. That is, if $y=f(x)$ and $x=a$, then

$$f'(a)=\lim_{h\to 0}\frac{f(a+h)-f(a)}{h},$$

and if this limit exists, we say that $f'(a)$ **exists**. If the limit fails to exist, then $f(x)$ **is not differentiable at** $x = a$. We cannot possibly list all the ways in which a function can fail to be differentiable. However, the following three conditions are commonly discussed.

Conditions of a Non-Differentiable Function

A function $f(x)$ is not differentiable at $x = a$ if any of the following is true:

1. *$f(x)$ is discontinuous at $x = a$.*
2. *The graph of $f(x)$ has a sharp corner at $x = a$.*
3. *The tangent line at $x = a$ is a vertical line.*

Example 2 shows three different functions, each of which is not differentiable due to one of the conditions defined above.

Example 2: Non-Differentiable Functions

a. $f(x) = \frac{1}{x}$ is not defined at $x = 0$. Thus there cannot be a tangent line at $x = 0$ and similarly the limit does not exist.

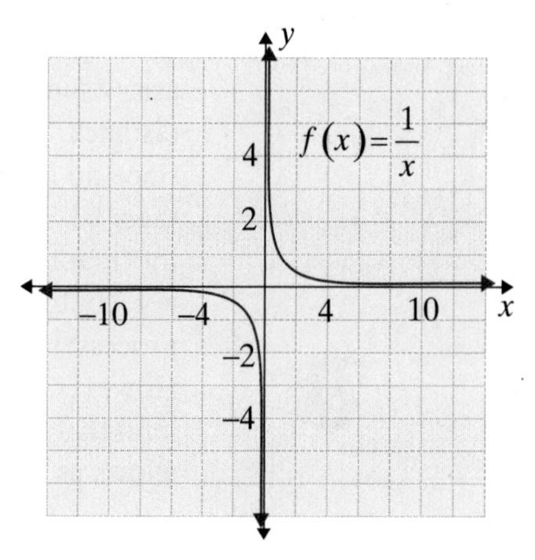

b. $f(x) = (x-3)x^{\frac{2}{3}}$ has a sharp point at $x = 0$ and so is not differentiable at $x = 0$.

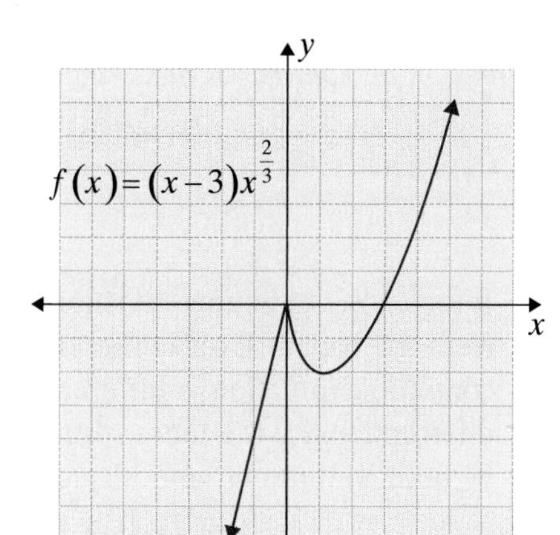

Note: the limit of a set of secant lines from the right of $x = 0$ will have negative slope, but from the left, the limit will have positive slope.

continued on next page ...

c. $f(x) = \sqrt[3]{x}$ has a vertical tangent at $x = 0$. Since vertical lines have no slope, $f'(0)$ is not defined.

For this function, $f'(x)$ exists at every point except for $x = 0$. In general, the functions usually studied in introductory and intermediate algebra or college algebra are "well-behaved" from the calculus point of view. They usually are differentiable except at obvious points.

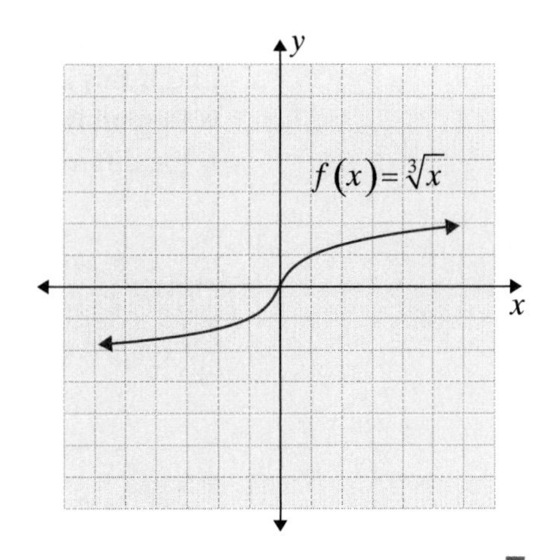

The Power Rule

The idea that a physical process, such as movement from one place to another, is continuous was a difficult idea to accept since it involved an infinite process. Infinity is a tricky concept mathematically. Zeno's Paradox is an ancient paradox that states an arrow shot into the air cannot land! It must go halfway, and then half of the remaining distance, and then half that, and so on. Zeno of Elea (490 BC – 425 BC) claimed that since no infinite process can ever end, no physical motion could be continuous. It can then be argued (by Zeno, not us) that the mathematical functions we use are not continuous either, as they describe processes with infinitely many steps. However, it is the concept of numerical limit that sweeps away contradictions and allows us to resolve this ancient dilemma.

For this discussion we note that a function $f(x)$ is said to be continuous at $x = a$ provided that the left- and right-hand limits agree:

Continuity at $x = a$

The function f is ***continuous at x = a*** *if and only if*

$$\lim_{x\to a^-}\left(f(x)\right) = f(a) = \lim_{x\to a^+}\left(f(x)\right).$$

NOTES

Informally, continuous functions can be drawn without lifting one's pencil from the paper. Every differentiable function is continuous, but not every continuous function is differentiable (see Example 2b). Put another way, differentiability is a more restrictive condition – it requires that a function be both continuous, smooth, and without vertical tangents.

In Section 2.5, we will go into more depth about continuous functions and explore the limit concept further.

The limit concept is viewed today as the perfect tool for giving a formal structure to the ideas of calculus. From the student point of view, the converse idea is even more important: if a function $f(x)$ is continuous at $x = a$, then the limit problem at $x = a$ is solved by substitution! (No calculation or table of values is necessary.)

Continuous Function on an Interval

A function f is **continuous on an interval** (x_1, x_2) *provided the function is continuous at* $x = a$ *for every number a with* $x_1 < a < x_2$.

Recall the fact that $(a+b)^3 = a^3 + 3a^2b + 3ab^2 + b^3$. Now we can get a formula for y' given $y = x^3$. For this function we consider the point $(x, f(x)) = (x, x^3)$ and a nearby point $\left(x+h, (x+h)^3\right)$. Then, with $h = \Delta x$:

Step 1: Form the difference quotient.

$$\frac{\Delta y}{\Delta x} = \frac{(x+h)^3 - x^3}{(x+h) - x}$$

Step 2: Simplify the difference quotient.

$$\begin{aligned} \frac{\Delta y}{\Delta x} = \frac{(x+h)^3 - x^3}{x+h-x} &= \frac{x^3 + 3x^2h + 3xh^2 + h^3 - x^3}{h} \\ &= \frac{3x^2h + 3xh^2 + h^3}{h} \\ &= \frac{\not{h}\left(3x^2 + 3xh + h^2\right)}{\not{h}} \\ &= 3x^2 + 3xh + h^2 \end{aligned}$$

Step 3: Find the limit.

$$\lim_{h \to 0}\left(3x^2 + 3xh + h^2\right) = 3x^2 + 0 + 0 = 3x^2$$

Example 3: Using the Definition of Derivative

Use the definition of derivative to find $f'(x)$ for $f(x)=x^2$.

Solution: Step 1: Form the difference quotient, $\frac{\Delta y}{\Delta x}$.

Here we begin with the two points $(x, f(x))$ and $(x+h, f(x+h))$.

So $\Delta y = f(x+h)-f(x)$ and $\Delta x = (x+h)-x = h$.

Inserting these expressions into the difference quotient, we get:

$$\frac{\Delta y}{\Delta x} = \frac{f(x+h)-f(x)}{h} = \frac{(x+h)^2-x^2}{h}.$$

Step 2: Simplify the difference quotient.

$$\begin{aligned}\frac{\Delta y}{\Delta x} = \frac{(x+h)^2-x^2}{h} &= \frac{x^2+2xh+h^2-x^2}{h} \\ &= \frac{2xh+h^2}{h} \\ &= \frac{\cancel{h}(2x+h)}{\cancel{h}} \\ &= 2x+h\end{aligned}$$

Step 3: Compute or determine the limit as $h \to 0$.

$$\lim_{h\to 0}\left(\frac{\Delta y}{\Delta x}\right) = \lim_{h\to 0}(2x+h) = 2x+0 = 2x$$

It is easy to see that, at any point on the line $y = x$, the slope is 1. We now put several results in a table and we see a pattern:

y	y'
x	1
x^2	$2x$
x^3	$3x^2$

The pattern suggests that if $f(x)=x^4$, then $f'(x)=4x^3$. This is exactly what happens for $y=x^4$. Similarly, if $y=x^5$, then $y'=5x^4$. We are now able to state one of the most useful and striking formulas in mathematics.

The Power Rule

For the function $f(x)=x^r$, where r is any real number, then

$$f'(x)=rx^{r-1}.$$

Especially noteworthy is the fact that the exponent r does not have to be a whole number, or even positive! Also, since 0^0 is undefined, the expression 0^0 is the one case not allowed in the Power Rule.

The second method of obtaining derivatives is by direct application of the definition of derivative itself. Usually, this method is cumbersome and is therefore only used to establish a logical basis for formulae like the Power Rule. In fact, we wish to illustrate our Three-Step Method to justify this very formula.

Our goal for the rest of this section is to provide a way to determine the formula for $f'(x)$ for any polynomial or polynomial-like function $f(x)$. Most of our results here will be proved using the Three-Step Method.

Basic Rule of the Derivative of a Linear Function

For $y = mx + b$, $y' = m$.

This means that we have defined slope for points on a curve consistently since, if the "curve" is a linear function, slope still means slope! The proof is an easy application of the definition. If $(x_1, mx_1 + b)$ and $(x_2, mx_2 + b)$ are any two points on the line, then:

Step 1: Form the difference quotient.

$$\frac{\Delta y}{\Delta x}=\frac{(mx_2+b)-(mx_1+b)}{x_2-x_1}$$

Step 2: Simplify the difference quotient.

$$\frac{\Delta y}{\Delta x}=\frac{(mx_2+b)-(mx_1+b)}{x_2-x_1}=\frac{mx_2+b-mx_1-b}{x_2-x_1}$$

$$=\frac{mx_2-mx_1}{x_2-x_1}=\frac{m(x_2-x_1)}{x_2-x_1}=m$$

Step 3: Determine the limit.

$$\lim_{\Delta x\to 0}\left(\frac{\Delta y}{\Delta x}\right)=m$$

Two Special Cases of the Basic Rule

1. *If $f(x) = c$, then $f'(x) = 0$.* *Has slope 0 (constant functions are horizontal lines).*
2. *If $f(x) = x$, then $f'(x) = 1$.* *Has slope 1.*

Polynomial and Polynomial-Like Functions

Polynomial Functions

*A **polynomial function** is a function of the form*

$$f(x) = a_0 + a_1x + a_2x^2 + \ldots + a_nx^n,$$

where $a_0, a_1, a_2, \ldots, a_n$ are real numbers and the exponents are positive integers.

Also, for any real number a, $\lim_{x \to a} f(x) = f(a)$, and therefore, $f(x)$ is continuous everywhere.

Consider the following functions:

a. $f(x) = 3 + x^{\frac{1}{2}}$,

b. $f(x) = 3 + x^2 - x^3 + x^{\frac{2}{3}}$, and

c. $f(x) = \dfrac{5}{x^2} = 5x^{-2}$.

Each has the form of a polynomial except that the exponents are not always positive integers. In function **c.**, the expression must be revised using a law of exponents. Each function is a sum or difference of terms of the form bx^r where b is some real number and r is some real number. Such functions are called **polynomial-like**. The derivatives of these, and standard polynomial functions, are the easiest type to determine.

Power Rule (General Case)

For the function $f(x) = c \cdot x^r$ where c and r are any real numbers,

$$f'(x) = c \cdot r \cdot x^{r-1}.$$

Example 4: Finding the Derivatives of Functions

Find the derivative of each of the following functions.

a. $f(x)=5$

Solution: $f(x)=5$ is a constant function, so $f'(x)=0$.

This is the first special case of the Basic Rule.

b. $f(x)=5x^3$

Solution: $f'(x)=5\cdot 3\cdot x^{3-1}=15x^2$

Here we use the general Power Rule, where $c=5$ and $r=3$.

c. $f(x)=x^{\frac{1}{2}}$

Solution: $f'(x)=\frac{1}{2}x^{\frac{1}{2}-1}=\frac{1}{2}x^{-\frac{1}{2}}$

We can use the Power Rule with this polynomial-like function because $r=\frac{1}{2}$ is a real number.

Example 5: Rewriting a Fraction to Find the Derivative

Find the derivative of $g(x)$ given that $g(x)=\frac{1}{x}$.

Solution: $g(x)=\frac{1}{x}=x^{-1}$

$g'(x)=-1x^{-1-1}=-x^{-2}$ or $-\frac{1}{x^2}$

We first rewrite the fraction in a form with a negative exponent and then apply the Power Rule.

At the beginning of this section, we listed a few other ways to denote the derivative. A more complete list of commonly used notations is as follows:

$$f'(x),\ y',\ \frac{dy}{dx},\ \frac{d}{dx}f(x),\ f_x,\ D_x\left(f(x)\right),\ \text{and } f'.$$

Example 6: Using Derivative Notation

Find $D_x(h(x))$ where $h(x)=\dfrac{18}{x^{\frac{2}{3}}}$.

Solution: $h(x)=\dfrac{18}{x^{\frac{2}{3}}}=18x^{-\frac{2}{3}}$

Rewrite the expression to be a polynomial-like function.

$$D_x(h(x))=D_x\left(18x^{-\frac{2}{3}}\right)=h'(x)=18\cdot\left(-\frac{2}{3}\cdot x^{-\frac{2}{3}-1}\right)$$

$$=\frac{-18(2)}{3}x^{-\frac{5}{3}}$$

$$=-12x^{-\frac{5}{3}}$$

Use the Power Rule where $r=-\dfrac{2}{3}$ and $c=18$ to find the derivative.

The Sum and Difference Rule

Now we list two more general rules and then show how these rules, along with the Power Rule, can be implemented to find the derivatives of a variety of functions.

Constant Times a Function Rule

If f(x) is a differentiable function, c is a real constant, and $y=c\cdot f(x)$, *then*

$$\frac{dy}{dx}=c\cdot f'(x).$$

In words, the derivative of a constant times a function is the constant times the derivative of the function. (The power rule is a special case of this rule.)

The Sum and Difference Rule

If f(x) and g(x) are differentiable functions and $y=f(x)\pm g(x)$, *then*

$$\frac{dy}{dx}=f'(x)\pm g'(x).$$

In words, the derivative of a sum (or difference) of two functions is the sum (or difference) of their derivatives.

Example 7: Using the Rules

Find the derivative of each of the following functions.

a. $y = 5x^2$

Solution:

$$\frac{dy}{dx} = \frac{d}{dx}\left(5x^2\right)$$
$$= 5 \cdot \frac{d}{dx}\left(x^2\right)$$

Rewrite the equation of the derivative by pulling the constant, 5, out in front. By doing so, we have used the Constant Times a Function Rule.

$$= 5\left(2x^{2-1}\right)$$

Apply the Power Rule to x^2.

$$= 10x$$

b. $y = -4x$

Solution:

$$\frac{dy}{dx} = \frac{d}{dx}(-4x)$$
$$= -4 \cdot \frac{d}{dx}\left(x^1\right)$$

Rewrite the equation of the derivative by pulling the constant, –4, out in front. By doing so, we have used the Constant Times a Function Rule.

$$= -4\left(1x^{1-1}\right)$$

Apply the Power Rule to x^1.

$$= -4x^0 = -4$$

c. $y = 5x^2 - 4x$

Solution: Notice that this function is actually the sum of the functions in Examples 7**a.** and 7**b.** Therefore, to find the derivative of this function we will use the results from 7**a.** and 7**b.**

$$\frac{dy}{dx} = \frac{d}{dx}\left(5x^2 - 4x\right)$$
$$= \frac{d}{dx}\left(5x^2\right) + \frac{d}{dx}(-4x)$$

Rewrite the equation of the derivative as a sum of the two differentiable functions so we can use the rule.

$$= 10x - 4$$

Using the Sum and Difference Rule, we can take the sum of the derivatives of the functions (found in Examples 7**a.** and 7**b.**).

continued on next page ...

d. $y = x^3 + 8\sqrt{x} + \frac{2}{x}$

Solution:

$$y = x^3 + 8x^{\frac{1}{2}} + 2x^{-1}$$

$$\frac{dy}{dx} = \frac{d}{dx}\left(x^3\right) + \frac{d}{dx}\left(8x^{\frac{1}{2}}\right) + \frac{d}{dx}\left(2x^{-1}\right)$$

$$= \frac{d}{dx}\left(x^3\right) + 8 \cdot \frac{d}{dx}\left(x^{\frac{1}{2}}\right) + 2 \cdot \frac{d}{dx}\left(x^{-1}\right)$$

$$= 3x^2 + \overset{4}{\cancel{8}}\left(\frac{1}{\cancel{2}}x^{\frac{1}{2}-1}\right) + 2\left(-1x^{-1-1}\right)$$

$$= 3x^2 + 4x^{-\frac{1}{2}} - 2x^{-2} \quad \text{or} \quad 3x^2 + \frac{4}{\sqrt{x}} - \frac{2}{x^2}$$

First, re-write using exponents. Next, use the Sum and Difference Rule to rewrite the equation of the derivative. Then apply the Constant Times a Function Rule to extract the constants. Lastly, use the Power Rule.

As illustrated in Examples 5 and 7**d**., there can be more than one correct algebraic form for a derivative. In Example 8 we show algebraic manipulations before the derivative is found.

Example 8: Algebraic Manipulations

a. Find $f'(u)$ if $f(u) = u^{\frac{1}{2}}\left(u^2 + 2u\right)$.

Solution:

$$f(u) = u^{\frac{1}{2}}\left(u^2 + 2u\right) = u^{2+\frac{1}{2}} + 2u^{1+\frac{1}{2}} = u^{\frac{5}{2}} + 2u^{\frac{3}{2}}$$

Multiply and simplify.

$$f'(u) = \frac{d}{dx}\left(u^{\frac{5}{2}}\right) + 2 \cdot \frac{d}{dx}\left(u^{\frac{3}{2}}\right)$$

$$= \frac{5}{2}u^{\frac{5}{2}-1} + \cancel{2}\left(\frac{3}{\cancel{2}}u^{\frac{3}{2}-1}\right)$$

$$= \frac{5}{2}u^{\frac{3}{2}} + 3u^{\frac{1}{2}}$$

Use both the Sum and Difference Rule and the Constant Times a Function Rule.
Apply the Power Rule.

b. If $F(v) = \frac{v^2 + 1}{\sqrt{v}}$, find $F'(v)$.

Solution:

$$F(v) = \frac{v^2 + 1}{\sqrt{v}} = \frac{v^2 + 1}{v^{\frac{1}{2}}}$$

$$= \frac{v^2}{v^{\frac{1}{2}}} + \frac{1}{v^{\frac{1}{2}}} = v^{\frac{3}{2}} + v^{-\frac{1}{2}}$$

Simplify by first dividing each term in the numerator by $v^{\frac{1}{2}}$.

$$F'(v) = \frac{d}{dx}\left(v^{\frac{3}{2}}\right) + \frac{d}{dx}\left(v^{-\frac{1}{2}}\right)$$ Use the Sum and Difference Rule.

$$= \frac{3}{2}v^{\frac{1}{2}} - \frac{1}{2}v^{-\frac{3}{2}}$$ Apply the Power Rule.

Or, we can factor and write the answer in fraction form with a single denominator, as shown below:

$$F'(v) = \frac{1}{2}v^{-\frac{3}{2}}\left(3v^2 - 1\right) = \frac{3v^2 - 1}{2v^{\frac{3}{2}}}.$$

Applications of Tangent Lines and Velocity

We know that a derivative of a function can be interpreted as the slope of a line tangent to the graph of the function. In the following example we show how to find the equation of a tangent line at a particular point.

Example 9: Tangent Lines

For the function $f(x) = 8x - x^2$:

a. Find the slope of the tangent lines at $x = 2$, $x = 4$, and $x = 5$.

b. Find the equation of the tangent lines at $x = 2$, $x = 4$, and $x = 5$.

c. Sketch the curve and the three tangent lines.

Solutions: a. First, find the derivative of the function:

$$f'(x) = 8 \cdot 1 \cdot x^{1-1} - 2x^{2-1} = 8 - 2x.$$

Then evaluate the derivative at each point to find the slope of the tangent line at that point.

At $x = 2 \rightarrow f'(2) = 8 - 4 = 4.$
At $x = 4 \rightarrow f'(4) = 8 - 8 = 0.$
At $x = 5 \rightarrow f'(5) = 8 - 10 = -2.$

b. To find the equation of the tangent lines at each of the points, use the point-slope formula for the equation of a line, $y - y_1 = m(x - x_1)$. Insert each value of x into the function to find its corresponding y-value.

At $x = 2 \rightarrow x_1 = 2$, so $y_1 = f(2) = 8 \cdot 2 - 2^2 = 12$. From part **a.**, we know that $f'(2) = 4 = m$.

continued on next page ...

Putting this information into the point-slope formula, we get:

$$y - 12 = 4(x - 2)$$
$$y = 4x + 4.$$

At $x = 4 \rightarrow x_1 = 4$, so $y_1 = f(4) = 8 \cdot 4 - 4^2 = 16$. From part **a.**, we know that $f'(4) = 0 = m$.

Putting this information into the point-slope formula, we get:

$$y - 16 = 0(x - 4)$$
$$y = 16.$$

At $x = 5 \rightarrow x_1 = 5$, so $y_1 = f(5) = 8 \cdot 5 - 5^2 = 15$. From part **a.**, we know that $f'(5) = -2 = m$.

Putting this information into the point-slope formula, we get:

$$y - 15 = -2(x - 5)$$
$$y = -2x + 25.$$

c. The graph is a parabola that opens downward. The vertex of the parabola occurs at (4, 16), where the tangent line has slope 0.

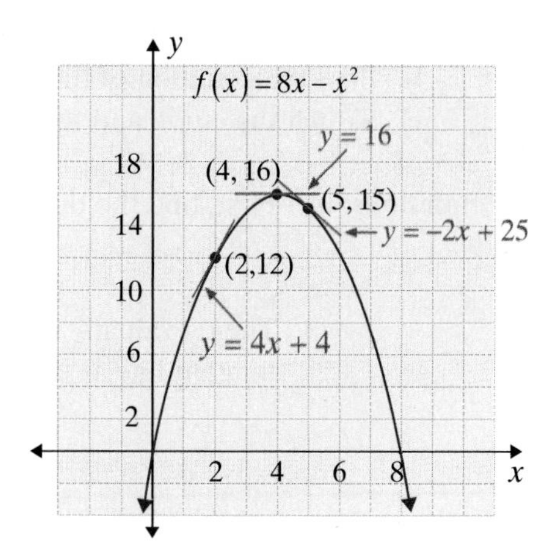

As we stated earlier in Section 2.2, velocity is the instantaneous rate of change of distance with respect to time. Mathematicians generally use the letter s to represent distance. Thus a typical distance function might look like

$$s(t) = t^2 - 2t \qquad \text{where } t \geq 0.$$

Notations for velocity are

$$v(t) = s'(t) = \frac{ds}{dt}.$$

Example 10: Velocity

Suppose that a sailboat is observed, over a period of 5 minutes, to travel a distance from a starting point according to the function $s(t) = t^3 + 60t$, where t is time in minutes and s is the distance traveled in meters.

a. How fast is the boat moving at the starting point?
b. How fast is the boat moving at the end of 3 minutes?

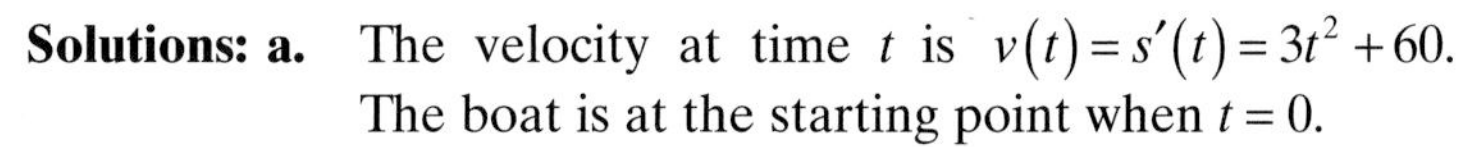

Solutions: a. The velocity at time t is $v(t) = s'(t) = 3t^2 + 60$. The boat is at the starting point when $t = 0$.

$$v(0) = s'(0) = 3 \cdot (0)^2 + 60 = 60 \text{ m/min}.$$

b. When $t = 3$,

$$v(3) = s'(3) = 3 \cdot (3)^2 + 60 = 27 + 60 = 87 \text{ m/min}.$$

Example 11: Identifying Derivatives in Real-Life

Determine which of the following represent (or might represent) the value of a derivative at a point, and if so, give a suitable description of a function f.

a. 30 miles per hour.
b. 6 fish per day.
c. 23 ears of corn.
d. 19 neutrons per millisecond.

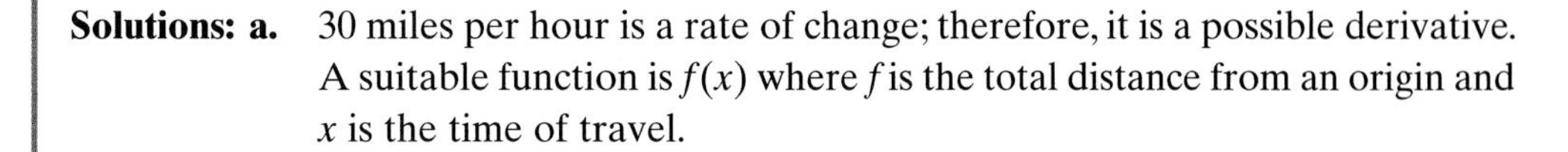

Solutions: a. 30 miles per hour is a rate of change; therefore, it is a possible derivative. A suitable function is $f(x)$ where f is the total distance from an origin and x is the time of travel.

b. 6 fish per day is also a rate of change, and therefore is a possible derivative. A suitable function $f(x)$ is the total number of fish sold by a pet store during June and x is the number of days since June 1.

c. 23 ears of corn is a quantity rather than a rate. Thus it is not a derivative.

d. 19 neutrons per millisecond is a rate of change; therefore, it is a possible derivative. A suitable function $f(x)$ is the total number of neutrons emitted by a nuclear sample, and x is the time since the sample arrived at the college physics laboratory.

2.3 Exercises

Use the various rules of differentiation to find the derivative for each of the functions in Exercises 1 – 30.

1. $f(x) = 4$
2. $f(x) = 3x$
3. $f(x) = 7x - 2$
4. $y = 12$
5. $y = 4x^2$
6. $y = 8x^2$
7. $y = \dfrac{7}{x}$
8. $y = \dfrac{4}{x^5}$
9. $y = \dfrac{1}{2x^3}$
10. $g(x) = \dfrac{4}{3x^2}$
11. $g(x) = 3\sqrt{x}$
12. $h(x) = 2\sqrt[3]{x}$
13. $h(t) = t^{2.3}$
14. $h(t) = t^{-1.4}$
15. $f(x) = 3x^{0.8}$
16. $f(u) = 2u^{0.1}$
17. $f(u) = \dfrac{1}{\sqrt{u}}$
18. $f(x) = \dfrac{2}{\sqrt[4]{x}}$
19. $f(x) = -5x^{\frac{3}{4}}$
20. $f(x) = 6x^{-\frac{2}{3}}$
21. $y = x^3 - 7x$
22. $y = 4x^2 - 9x + 2$
23. $y = 0.3x^2 - 4x + 6$
24. $y = 120 + 8x - 0.2x^2$
25. $y = x^3 - 6x^2 + 5x + 2$
26. $y = \dfrac{1}{3}x^3 - \dfrac{1}{2}x^2 - 3x + 4$
27. $y = 2x^{\frac{3}{2}} + 4x^{\frac{1}{2}} - 5$
28. $y = 2x^{-\frac{2}{3}} + 3x^{\frac{1}{3}} + 7$
29. $f(t) = 2t^{-\frac{1}{2}} + t^{\frac{1}{2}} + t$
30. $f(x) = 3x^{-\frac{1}{3}} - 2x^{-\frac{1}{2}} + 1$

In Exercises 31 – 40, use algebraic techniques to rewrite each function as a sum or difference; then find the derivative.

31. $y = (x+1)(2x-3)$
32. $y = \sqrt{x}\left(2x^2 + x - 3\right)$
33. $f(v) = v^{\frac{3}{2}}\left(v^2 + 2v - 1\right)$
34. $f(v) = v^{\frac{1}{3}}\left(6 - 4v + v^2\right)$

35. $f(x)=\dfrac{x^4+5x^3}{x^2}$

36. $f(x)=\dfrac{6x^2+1}{x^3}$

37. $g(t)=\dfrac{t-2}{\sqrt{t}}$

38. $g(t)=\dfrac{t^2+3}{\sqrt[3]{t}}$

39. $g(x)=\dfrac{4x+5x^{\frac{1}{2}}-1}{\sqrt{x}}$

40. $g(x)=\dfrac{3\sqrt{x}+4x-2}{x^2}$

In Exercises 41 – 46, confirm your results with a graphing utility.

41. For the function $f(x)=x^3+2x-4$, find:
- **a.** The slope of the tangent line at $x=-1$.
- **b.** The equation of the tangent line at $x=-1$.

42. For the function $f(x)=2x^3-x^2-3x$, find:
- **a.** The slope of the tangent line at $x=2$.
- **b.** The equation of the tangent line at $x=2$.

43. For the function $g(x)=x^2+6x+5$, find:
- **a.** The slope of the tangent line at $x=-1, x=-3$, and $x=-4$.
- **b.** The equations of the tangent lines at $x=-1, x=-3$, and $x=-4$.
- **c.** Sketch the graphs of the curve and the three tangent lines.

44. For the function $g(x)=x^2-8x+12$, find:
- **a.** The slope of the tangent line at $x=2, x=4$, and $x=5$.
- **b.** The equations of the tangent lines at $x=2, x=4$, and $x=5$.
- **c.** Sketch the graphs of the curve and the three tangent lines.

45. For the function $f(x)=8-x^2$, find:
- **a.** The slopes of the tangent lines at $x=-2, x=0$, and $x=1$.
- **b.** The equations of the tangent lines at $x=-2, x=0$, and $x=1$.
- **c.** Sketch the graphs of the curve and the three tangent lines.

46. For the function $f(x)=10-3x-x^2$, find:
- **a.** The slopes of the tangent lines at $x=-3, x=-2$, and $x=0$.
- **b.** The equations of the tangent lines at $x=-3, x=-2$, and $x=0$.
- **c.** Sketch the graphs of the curve and the three tangent lines.

47. Velocity of a rocket. A model rocket is fired vertically upward. The height after t seconds is $s(t)=192t-16t^2$ feet.
- **a.** Find the velocity at $t=0$ seconds.
- **b.** Find the velocity at $t=4$ seconds.
- **c.** When will the velocity be zero?

48. Velocity of a particle. A particle moving in a straight line is at a distance of $s(t) = 2.5t^2 + 18t$ feet from its starting point after t seconds, where $0 \le t \le 12$.

a. Find the velocity at $t = 6$.

b. Find the velocity at $t = 9$.

49. Population. A city's population t years from now can be estimated from the formula $P(t) = 9000 + 500t - 72\sqrt{t}$.

a. Find the rate at which the city is growing after 4 years.

b. Find the rate at which the city is growing after 9 years.

50. Cost. The total cost of producing x units of a product is given by $C(x) = 4000 + 25x - 0.2x^2$ dollars, where $0 \le x \le 50$.

a. Find the rate of change in the cost when $x = 10$.

b. Find the rate of change in the cost when $x = 30$.

51. Gravity. A student dropped a pillow from the top floor of his dorm and it fell according to the formula $S(t) = -16t^2 + 8t^{0.5}$, where t is the time in seconds and $S(t)$ is the distance in feet from the top of the building.

a. If the pillow hit the ground in exactly 2.5 seconds, how high is the building?

b. What was the average speed for the trip?

c. What was the instantaneous velocity at $t = 1$ sec?

52. Fuel Consumption. When a factory operates from 6 A.M. to 6 P.M., its total fuel consumption varies according to the formula $f(t) = 0.9t^2 - 0.3t^{0.5} + 20$, where t is the time in hours after 6 A.M. and $f(t)$ is number of barrels of fuel oil.

a. How much fuel oil is consumed by noon?

b. What is the rate of consumption of fuel at 10 A.M.?

c. What is the average rate of consumption from 6 A.M. to 2 P.M.?

53. Population. The population of bacteria in a lab experiment for BIOL 403 at Nevada Tech is given by $f(x) = 2.2x^{1.5} - 0.7x + 2$, where x is the time in hours after 2 P.M. and $f(x)$ is population in suitable units.

a. What is the population at 2 P.M.?

b. The lab is over at 4 P.M. What is the new population of bacteria?

c. What is the average rate of change of bacteria from 2 P.M. to 4 P.M.?

d. What is the instantaneous rate of change of bacteria at 3 P.M.?

54. Electrical Charge. The electrical charge on a new cell phone declines according to the formula $C(t) = 15 - 0.1t^2 - 0.5t$, where t is the time in hours following a full charge and $C(t)$ is a measure of the charge.

a. To the nearest hour, how long does one have until the charge is fully depleted?

b. What is the instantaneous rate of change, in charge units per hour, at $t = 4$?

c. What is the average rate of change from $t = 0$ to $t = 4$?

55. **Spreading a Rumor.** The number of college students at Salis Techical College who have not heard a new rumor is approximated by the formula $N(x) = 300(1 - 0.004x^2)$, where x is the number of days following the start of a new rumor.
    - **a.** How many days does it take for 90 percent of the students to hear the rumor?
    - **b.** What is the instantaneous rate of change in students after one day? Interpret the meaning of this number.
    - **c.** Why is the slope negative in this exercise?

56. **Birth Rate.** The fertility decline in many countries can be modeled by a quadratic equation. In China, from the late 1960's to the present, the number of births per woman has declined according to the formula $f(x) = 0.00675x^2 - 0.3215x + 5.585$, where x is the number of years after 1969. (***Source*:** Nancy E. Riley, "China's Population: New Trends and Challenges," *Population Bulletin*, Vol. 20, No. 2, (June 2004).)
    - **a.** What was the number of births per woman in 1969?
    - **b.** What was the number of births per woman in 1999?
    - **c.** What was the rate of change of this fertility rate in 1979?

57. **Birth Rate.** In China, from 1964 to the present, the death rate has remained nearly constant at approximately 8 deaths per 1000 people. The yearly birth rate, in births per 1000 people, has declined in most years according to the formula $f(x) = -0.00191x^3 + 0.134x^2 - 3.16x + 44.5$, where x is the number of years since 1964. (***Source*:** Nancy E. Riley, "China's Population: New Trends and Challenges," *Population Bulletin*, Vol. 20, No. 2, (June 2004).)
    - **a.** What number is $f(30)$ and what does it represent?
    - **b.** What number is $f'(30)$ and what does it represent?
    - **c.** In what year, according to the model, will the number of new births equal the number of deaths?

58. **Population Growth.** The older population in China (age 60 and over) has grown since the 1950's according to the formula $f(x) = 0.003727x^2 - 0.105x + 7.063$, where x is the number of years since 1953. (***Source*:** Nancy E. Riley, "China's Population: New Trends and Challenges," *Population Bulletin*, Vol. 20, No. 2, (June 2004).)
    - **a.** What was the percentage of older persons in China in 1953?
    - **b.** What is the percentage of older persons projected to be in the year 2025?
    - **c.** At what rate was the percentage changing in 2000?

59. Population Growth. The population in billions of people in the lesser developed countries varied according to the formula $P(x)=\frac{4.953}{10^4}x^2-0.007352x+1.7748$, where x is the number of years since 1900. (***Source*****:** Population Reference Bureau, "Transition's in World Population," *Population Bulletin*, Vol. 59, No. 1, 5, (March 2004).)

- **a.** What was the population in 1950?
- **b.** At what rate was the population changing in 1950?
- **c.** What population is projected for 2020 in these countries?

60. Death Rate. The death rate in Mexico has varied since 1920 according to the formula $M(x)=-\frac{2.076}{10^4}x^3+0.033x^2-1.785x+43.07$, where x is the number of years since 1920. (***Source*****:** Population Reference Bureau, "Transition's in World Population," *Population Bulletin*, Vol. 59, No. 1, 5, (March 2004).)

- **a.** What was the death rate in 1950?
- **b.** At what rate was the death rate changing in 1950?
- **c.** What was the death rate in 2000?
- **d.** In what year does the formula predict that no one will die?

61. Birth Rate. The function $f(x)=0.00375x^2-0.2355x+5.595$ gives the average number of births per woman in Thailand where x denotes the number of years since 1970. (***Source*****:** Lori Ashford, "World Population Highlights 2004," *BRIDGE Population Reference Bureau*, (August 2004).)

- **a.** What number is $f(10)$ and what does it represent?
- **b.** What number is $f'(10)$ and what does it represent?

62. Birth Rate. In Argentina, from 1970 to 2000, the average number of births per woman is given by the function $f(x)=-0.001x^2+0.014x+3.14$, where x is the number of years after 1970. (***Source*****:** Lori Ashford, "World Population Highlights 2004," *BRIDGE Population Reference Bureau*, (August 2004).)

- **a.** What number is $f(30)$ and what does it represent?
- **b.** What number is $f'(30)$ and what does it represent?

Hawkes Learning Systems: Essential Calculus

Definition of the Derivative and Power Rule	p. 141 - 150	Ex. 1 - 20
Slope and Rate of Change Considered Algebraically	p. 150 - 155	Ex. 21 - 62

2.4 Applications: Marginal Analysis

Objectives

After completing this section, you will be able to:

Calculate marginal cost, marginal revenue, marginal profit, and marginal average cost.

Recall from our previous discussions about topics related to business that

$C(x)$ represents the total cost of producing x items,
$R(x)$ represents the total revenue when x items are sold, and
$P(x) = R(x) - C(x)$ represents the profit from selling all x items produced.

Each of these functions depends on x and changes as x changes. The term **marginal** is used in Business and Economics to indicate a rate of change and, therefore, a derivative. In this section we are going to investigate the use of calculus in marginal analysis and the three derivatives:

$C'(x)$, called **marginal cost**;

$R'(x)$, called **marginal revenue**; and

$P'(x)$, called **marginal profit**.

We emphasize that $C'(x)$ is an estimate for the increase in $C(x)$ if x is increased by 1.

Marginal Cost

If the total cost of producing x items is $C(x)$, then the total cost of producing $x + 1$ items is $C(x + 1)$.

For example, suppose $C(x) = 225 + 2x^2$. Then

$C(10) = 225 + 2 \cdot 10^2 = 225 + 200 = \$425$ is the cost of producing 10 items, and

$C(11) = 225 + 2 \cdot 11^2 = 225 + 242 = \$467$ is the cost of producing 11 items.

The difference $C(11) - C(10) = 467 - 425 = \$42$ is the cost of producing the eleventh item.

If we differentiate $C(x)=225+2x^2$, we get

$$C'(x)=4x \text{ and } C'(10)=4\cdot 10=\$40,$$

a very close approximation to the cost of producing the eleventh item. This is exactly what marginal cost at x is – an approximation of the cost of producing the $(x+1)^{\text{th}}$ item.

Marginal Cost

***Marginal cost** is the rate of change of total cost per unit change in quantity x. This concept is illustrated geometrically in Figure 2.4.1.*

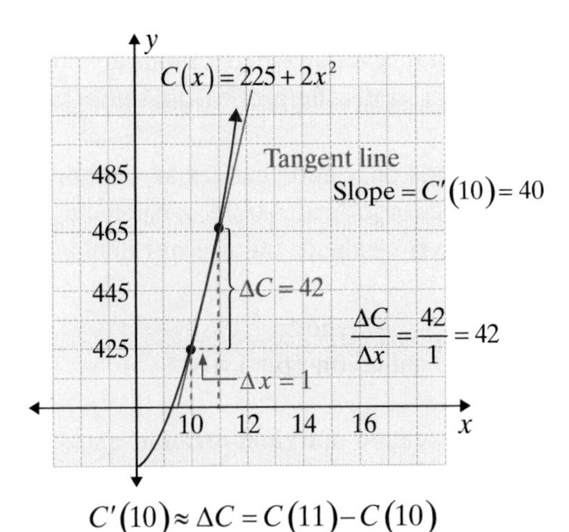

Figure 2.4.1

$C'(10)\approx \Delta C = C(11)-C(10)$

Example 1: Marginal Cost

A specialty item manufacturer determines that the cost of producing x ballpoint pens is $C(x) = 500 + 3x$ in dollars.

a. Find $C(101) - C(100)$, the cost of producing the 101st pen.

b. Find $C'(100)$, the marginal cost at $x = 100$ pens.

Solutions: a. $C(101) = 500 + 3(101) = 500 + 303 = 803$

$C(100) = 500 + 3(100) = 500 + 300 = 800$

$C(101) - C(100) = 803 - 800 = 3$

So, when the production level is increased from 100 to 101 pens, the total cost is increased by \$3. Therefore, the cost of producing the 101st pen is \$3.

b. $C(x) = 500 + 3x$ The given cost function.
$C'(x) = 3$ Marginal cost is the derivative of the cost function.
$C'(100) = \$3$ The marginal cost at $x = 100$.

In Example 1 the cost function is linear and the graph is a straight line with slope 3. Therefore, increasing output by one item always leads to an increase in total cost of \$3. In fact, the marginal cost is \$3 per pen, regardless of how many are produced.

Marginal Revenue

Marginal Revenue

***Marginal revenue** is the rate of change of the total revenue per unit change in sales when the level of sales is x items.*

As with marginal cost, marginal revenue is an approximation of the growth or decline in revenue per unit sold. (Note that revenue can decline even if more items are sold because of possible changes in the price per item.)

Remember that if $p = D(x)$ is the **demand function**, then the revenue function is formed as the product of the demand function (price per item) and the number of x items sold:

$$R(x) = x \cdot p \quad \text{where} \quad p = D(x).$$

Example 2: Marginal Revenue

Suppose that a manufacturer has determined that the price of certain custom-made tables he produces can be determined by the demand function $p = D(x) = 117 - \frac{x}{4}$, where x is the number of tables produced and sold.

a. Find the revenue function.
b. Determine the marginal revenue when $x = 16$ tables.

Solutions: a. $R(x) = x \cdot p = x\left(117 - \frac{x}{4}\right) = 117x - \frac{x^2}{4}$

b.

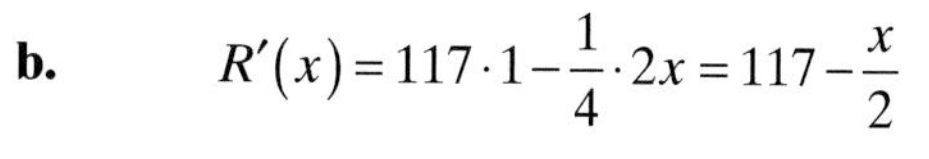

$$R'(x) = 117 \cdot 1 - \frac{1}{4} \cdot 2x = 117 - \frac{x}{2}$$

So $R'(16) = 117 - \frac{16}{2} = 117 - 8 = \$109.$

Thus the revenue is increasing at a rate of \$109 per table as the sixteenth table is sold.

Marginal Profit

Marginal Profit

Marginal profit *is the rate of change of the profit per unit change in sales when x items are produced and sold.*

As long as revenue is greater than cost, there is a profit. When revenue is equal to cost (a break-even point), the profit is 0. As with revenue and cost, when items are produced and sold, profit changes (increases or decreases). A basic question for any person in business is, "What production and sales level will yield maximum profit?" Calculus techniques for answering such a question will be developed in Chapters 3 and 4.

Example 3: Marginal Profit

If the table manufacturer in Example 2 has a cost function of $C(x)=225+2x^2$ along with his revenue function of $R(x)=117x-\frac{x^2}{4}$, find

a. All break-even points.
b. The marginal profit when $x = 10, x = 20, x = 30$.

Solutions: a. Break-even points occur where $C(x) = R(x)$.

$$225+2x^2 = 117x-\frac{x^2}{4}$$
$$900+8x^2 = 468x-x^2$$
$$9x^2-468x+900=0$$
$$9\left(x^2-52x+100\right)=0$$
$$9(x-2)(x-50)=0$$
$$x-2=0 \quad x-50=0$$
$$x=2 \quad x=50$$

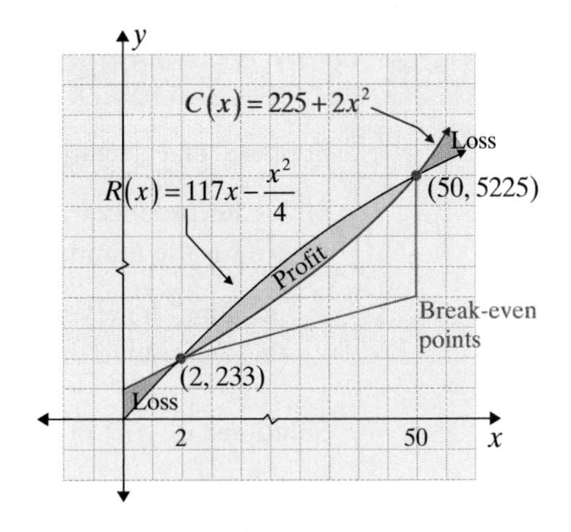

Break-even points occur when $x = 2$ and $x = 50$.

b. $P(x)=R(x)-C(x)$

$$=\left(117x-\frac{x^2}{4}\right)-\left(225+2x^2\right)$$

Marginal profit is rate of change of profit, so first we must find the profit function $P(x)$.

$$= 117x - \frac{x^2}{4} - 225 - 2x^2$$

$$= -225 + 117x - \frac{9}{4}x^2$$

The profit function $P(x)$.

$$P'(x) = 0 + 117 \cdot 1 - \frac{9}{4} \cdot 2x$$

$$= 117 - \frac{9}{2}x$$

Now we can find the derivative of $P(x)$ to calculate the various marginal profits.

$$P'(10) = 117 - \frac{9}{2}(10) = 117 - 45 = \$72$$

$$P'(20) = 117 - \frac{9}{2}(20) = 117 - 90 = \$27$$

$$P'(30) = 117 - \frac{9}{2}(30) = 117 - 135 = -\$18$$

Note that $P'(x)$ is actually getting smaller as x gets larger. Marginal analysis shows that the rate of growth of profit per table is actually decreasing at a rate of \$18 per table when 30 tables are manufactured and sold. This does not necessarily mean there is a loss, but as production increases, the profit per table is growing less because costs are increasing faster than revenue.

Marginal Average Cost

Suppose we let $C(x)$ represent the total cost of producing x items. If we divide the total cost by the number of items, we have the **average cost per item** $\overline{C}(x)$. That is,

$$\overline{C}(x) = \frac{C(x)}{x}.$$

For example, if $C(x) = 0.02x^2 + 100x + 2000$ represents the cost in dollars of producing x radios, then

$$\overline{C}(x) = \frac{C(x)}{x} = \frac{0.02x^2 + 100x + 2000}{x} = 0.02x + 100 + \frac{2000}{x}$$

represents the average cost per radio.

Marginal Average Cost

Marginal average cost *$\overline{C}'(x)$ is the rate of change of the average cost per unit change in production. This is an approximation of the change in average cost when one more item is produced.*

Example 4: Marginal Average Cost

Suppose $C(x)=0.02x^2+100x+2000$ represents the cost in dollars of producing x radios.

a. Find the average cost function.

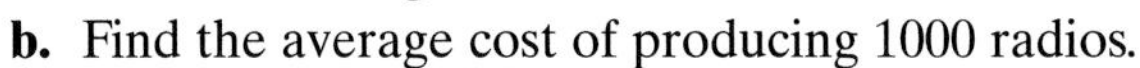

b. Find the average cost of producing 1000 radios.

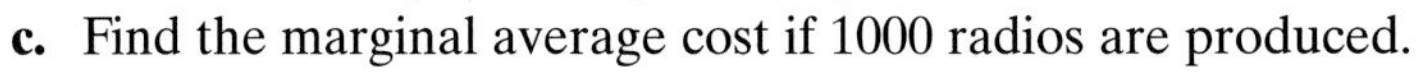

c. Find the marginal average cost if 1000 radios are produced.

Solutions: a. $\overline{C}(x)=\dfrac{C(x)}{x}=\dfrac{0.02x^2+100x+2000}{x}=0.02x+100+\dfrac{2000}{x}$

b. $\overline{C}(1000)=0.02(1000)+100+\dfrac{2000}{1000}=20+100+2=\$122$ Use $x=1000$.

c. $\overline{C}(x)=0.02x+100+2000x^{-1}$ Rewrite the equation so that there are no fractions.

$\overline{C}'(x)=0.02+0+2000(-1)x^{-1-1}$

$=0.02-2000x^{-2}$

$=0.02-\dfrac{2000}{x^2}$

Find the rate of change of the average cost to find the marginal average cost.

$\overline{C}'(1000)=0.02-\dfrac{2000}{1000^2}$

$=0.02-0.002$

$=\$0.018$

Use $x=1000$ in the function for marginal average cost.

The average cost per radio is increasing at a rate of 1.8 cents per radio when 1000 radios are produced.

Marginal Propensity to Consume

Economists study consumption functions for countries in attempts to relate different parts of a nation's economy. Consumption functions give an estimate of a country's total annual consumption of consumer goods. The rate of change of such a function is called the marginal propensity to consume.

Suppose x represents total national income. If $C(x)$ is the amount of consumption and the national savings is $S(x)$, then

$$x = C(x) + S(x)$$

and

$$1 = C'(x) + S'(x).$$

$C'(x)$ is the marginal propensity to consume and $S'(x)$ is the marginal propensity to save.

It follows that

$$C'(x) = 1 - S'(x).$$

Example 5: Marginal Propensity

If the marginal propensity to save is $S'(x) = 0.02x$, what is the marginal propensity to consume?

Solution: Since $C'(x) = 1 - S'(x)$,

$$C'(x) = 1 - 0.02x.$$

We replace $S'(x)$ with $0.02x$.

2.4 Exercises

In Exercises 1 – 10, create an appropriate function of the type indicated.

1. The cost of producing x leather belts is given by $C(x) = 220 + 0.4x$. Determine the average cost function.

2. The cost of manufacturing a certain class of screws for wall hangers is given by $C(x) = 800 + 0.0005x + 0.00002x^2$. Determine the average cost function.

3. A vendor charges $3 for a hot dog. What is the demand function? What is the revenue function?

4. The Knoll Industrial Supply Company charges for 55-gallon drums using the demand function $p(x) = 22.5 - 0.5x$. What is the revenue function?

5. A supplier of souvenir t-shirts charges street vendors for each order based on a set up fee of \$50 and an item charge of \$1.15 per t-shirt. What is the cost function for the vendor?

6. A supplier of souvenir t-shirts has set up costs of \$25 and shirts cost \$0.40 each. What is the cost function?

7. The answer to Exercise 5 is also a revenue function for the supplier of t-shirts. Using the answers to Exercises 5 and 6, determine a profit function for the supplier of t-shirts.

8. For the supplier in Exercise 6, what is the average cost function?

9. A department store's cost estimate for a line of rocker-recliner chairs is given by $C(x) = 100 + 150x$. The store sells them for \$450 each.

- **a.** What is the average cost function?
- **b.** What is the demand function?
- **c.** What is the revenue function?
- **d.** What is the profit function?

10. A shoe store estimates a certain line of dress shoes has costs given by $C(x) = 1500 + 30x + 0.05x^2$. The store charges \$75 per pair.

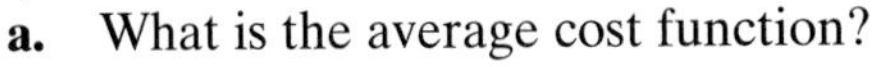

- **a.** What is the average cost function?
- **b.** What is the demand function?
- **c.** What is the revenue function?
- **d.** What is the profit function?

11. The weekly cost of producing x electric drills is given by the function $C(x) = 2400 + 28x + 0.25x^2$.

- **a.** Find $C(10)$, $C(20)$, and $C(30)$.
- **b.** Find the marginal cost function.
- **c.** Find $C'(10)$, $C'(20)$, and $C'(30)$.
- **d.** Find the average cost function and the marginal average cost function.
- **e.** Find the marginal average cost when $x = 10$, $x = 20$, and $x = 30$.

12. The total cost function for producing x units of a product is given by $C(x) = \frac{1}{3}x^3 - \frac{1}{2}x^2 + 7x + 18$.

- **a.** Find $C(3)$, $C(4)$, and $C(6)$.
- **b.** Find the marginal cost function.
- **c.** Find $C'(3)$, $C'(4)$, and $C'(6)$.
- **d.** Find the average cost function and the marginal average cost function.
- **e.** Find the marginal average cost when $x = 3$, $x = 4$, and $x = 6$.

13. The total cost of producing x units of a commodity is given by $C(x) = 60 + 10x - 0.5x^2$.

- **a.** Find $C(4)$, $C(6)$, and $C(9)$.
- **b.** Find the marginal cost function.
- **c.** Find $C'(4)$, $C'(6)$, and $C'(9)$.
- **d.** Find the average cost function.
- **e.** Find the marginal average cost when $x = 4, x = 6$, and $x = 9$.

14. The total cost of producing x clock radios is given by the function $C(x) = 300 + 24x - 0.4x^2 + 0.1x^3$.

- **a.** Find $C(2)$, $C(3)$, and $C(5)$.
- **b.** Find the marginal cost function.
- **c.** Find $C'(2)$, $C'(3)$, and $C'(5)$.
- **d.** Find the average cost function and the marginal average cost function.
- **e.** Find the marginal average cost if $x = 4$.

15. A manufacturer has determined that the revenue from the sale of x cordless telephones is given by $R(x) = 94x - 0.03x^2$ dollars. The cost of producing x telephones is $C(x) = 10{,}800 + 34x$ dollars.

- **a.** Find the profit function $P(x)$.
- **b.** Find $P(200)$, $P(400)$, and $P(600)$.
- **c.** Find the marginal profit function $P'(x)$.
- **d.** Find $P'(200)$, $P'(400)$, and $P'(600)$.
- **e.** Find any break-even points.

16. The revenue from the sale of x fire extinguishers is estimated to be $R(x) = 54x - 0.4x^2$ dollars. The total cost of producing x fire extinguishers is $C(x) = 400 + 30x - 0.2x^2$ dollars.

- **a.** Find the profit function $P(x)$.
- **b.** Find $P(20)$, $P(40)$, and $P(60)$.
- **c.** Find the marginal profit function $P'(x)$.
- **d.** Find $P'(20)$, $P'(40)$, and $P'(60)$.
- **e.** Find any break-even points.

17. A company that produces and sells compact refrigerators has found that the revenue from the sale of x refrigerators is $R(x) = 100x - 0.1x^2$ dollars. The cost function is given by $C(x) = 2070 + 25x + 0.1x^2$ dollars.

- **a.** Find the profit function $P(x)$.
- **b.** Find $P(60)$, $P(80)$, and $P(100)$.
- **c.** Find the marginal profit function $P'(x)$.
- **d.** Find $P'(60)$, $P'(80)$, and $P'(100)$.
- **e.** Find any break-even points.

18. A manufacturer has determined that the cost and the revenue of producing and selling x telescopes are $C(x)=x^2+20x+1050$ dollars and $R(x)=140x-0.5x^2$ dollars, respectively.

a. Find the profit function $P(x)$.
b. Find $P(30)$, $P(35)$, and $P(40)$.
c. Find the marginal profit function $P'(x)$.
d. Find $P'(30)$, $P'(35)$, and $P'(40)$.
e. Find any break-even points.

19. The owner of a leather craft shop has determined that he can sell x attaché cases if the price is $p=D(x)=46+0.25x$ dollars. The total cost for these cases is $C(x)=0.15x^2+6x+190$ dollars.

a. Find the revenue function $R(x)$.
b. Find the profit function $P(x)$.
c. Find $P(25)$, $P(30)$, and $P(40)$.
d. Find the marginal profit function $P'(x)$.
e. Find $P'(25)$, $P'(30)$, and $P'(40)$.

20. A firm can sell x items of a product when the price is $p=D(x)=3.00-0.001x$ dollars. The total production costs are $C(x)=0.002x^2+0.72x+260$ dollars.

a. Find the revenue function $R(x)$.
b. Find the profit function $P(x)$.
c. Find $P(300)$, $P(375)$, and $P(400)$.
d. Find the marginal profit function $P'(x)$.
e. Find $P'(300)$, $P'(375)$, and $P'(400)$.

21. A local publishing company prints a special magazine each month. It has been determined that x magazines can be sold monthly when the price is $p=D(x)=5.50-0.0004x$. The total cost of producing the magazine is $C(x)=0.0002x^2+x+4650$ dollars.

a. Find the revenue function $R(x)$.
b. Find the profit function $P(x)$.
c. Find $P(3000)$, $P(3500)$, and $P(4000)$.
d. Find the marginal profit function $P'(x)$.
e. Find $P'(3000)$, $P'(3500)$, and $P'(4000)$.

22. A sales representative for a company that produces skateboards can sell x units of their deluxe model if the price is $p=D(x)=79.9-0.03x$ dollars. The total cost for these skateboards is given by $C(x)=0.08x^2+5.1x+5800$ dollars.

a. Find the revenue function $R(x)$.
b. Find the profit function $P(x)$.
c. Find $P(320)$, $P(340)$, and $P(350)$.
d. Find the marginal profit function $P'(x)$.
e. Find $P'(320)$, $P'(340)$, and $P'(350)$.

23. Suppose that total national consumption is given by a function $C(x) = 200 - 0.6x - 0.05x^{0.6}$, where x is the total national income.

a. Determine the marginal propensity to consume.

b. Determine the marginal propensity to save.

24. If the marginal propensity to save of a certain country is given by $S'(x) = 0.4x + 0.3$, determine the marginal propensity to consume.

25. The average cost $\overline{C}(x)$ of a product is $\dfrac{C(x)}{x}$, where x is the total cost function.

a. What is the average cost function if its total cost function is $C(x) = 30 + 2x + 0.003x^2$?

b. What is the rate of change of average cost?

c. What value of x results in a minimum average cost?

26. The average cost of a product is given by $A(x) = 20x^{-1} + 3$.

a. Determine the cost function for the product.

b. Determine the marginal cost function.

27. An economics professor claimed that the average cost was a minimum if the average cost equaled the marginal cost. Do you agree? Explain why or why not.

28. "The average revenue $\dfrac{R(x)}{x}$ is not usually studied in the context of business economics." Argue for or against this statement.

29. A certain model car has a valuation in dollars given by the formula $f(x) = 12{,}519.3 - 1391.1x$, for $0 \le x \le 7$, where x is the age of the car in years. $x = 0$ corresponds to this calendar year.

a. What is $f(0)$? Interpret this number.

b. What is the marginal valuation? Interpret this number.

30. Based on averaging results at a certain state supported college, a relationship between grades and SAT scores was found to be $f(s) = 1.36 + 0.00141s$, where s is a student's SAT score and $f(s)$ is the student's graduating GPA (GPA based on 4.0 maximum score).

a. What is the expected GPA for a student with an SAT score of 1000?

b. What is the marginal GPA? Interpret this number.

Exercises 31 – 33 deal with projections of world population based on estimates of fertility around the world. Use the following background information:

> *The Total Fertility Rate (TFR) is the average number of children a woman will have. The United Nations projects world population according to assumptions about TFR values. In general, lower values promote economic well-being and lower world population.*

(***Source*:** Population Reference Bureau, "Transition's in World Population," *Population Bulletin*, Vol. 59, No. 1, 36, (March 2004).)

31. Total Fertility Rate. Using a TFR of 1.5, the U.N. projects total world population, in billions, will be modeled by $F(t)$ where t is the number of years after 2000 and F is the function $F(t) = -\frac{2.1}{10^6}t^4 + \frac{1.6}{10^4}t^3 - 0.00457t^2 + 0.0994t + 6.03.$

- **a.** Give the marginal population function $F'(t)$.
- **b.** Determine the estimate of world population in 2030.
- **c.** What is the marginal population in 2030? Interpret this number.

32. Total Fertility Rate. Using a TFR of 2.0, the U.N. projects total world population, in billions, will be modeled by $G(t)$ where t is the number of years after 2000 and G is the function $G(t) = -\frac{2.1}{10^6}t^4 + \frac{1.66}{10^4}t^3 - 0.0045t^2 + 0.113t + 6.03.$

- **a.** Give the marginal population function $G'(t)$.
- **b.** Determine the estimate of world population in 2030.
- **c.** What is the marginal population in 2030? Interpret this number.

33. Total Fertility Rate. Using a TFR of 2.5, the U.N. projects total world population, in billions, will be modeled by $H(t)$ where t is the number of years after 2000 and H is the function $H(t) = -\frac{2.6}{10^6}t^4 + \frac{1.87}{10^4}t^3 - 0.00396t^2 + 0.116t + 6.05.$

- **a.** Give the marginal population function $H'(t)$.
- **b.** Determine the estimate of world population in 2030.
- **c.** What is the marginal population in 2030? Interpret this number.

Hawkes Learning Systems: Essential Calculus

Applications: Marginal Analysis

2.5 More About Limits and Continuity

Objectives

After completing this section, you will be able to:

1. *Use the properties of limits to determine limits of functions.*
2. *Find limits of rational functions.*
3. *Identify continuous functions.*

Properties of Limits

Because graphs provide a visual tool that is helpful in an informal approach to limits, we have used them extensively in the discussion thus far. However, as graphs are not always accurate or easy to find, we need a more formal approach to evaluating limits in our development of calculus. The following list of properties of limits, stated here without proof, allows us to study limits on a more rigorous level. Applications of the properties are illustrated in the examples that follow.

Properties of Limits as $x \to a$

Suppose that c is any constant, n is a positive integer, a is a real number, $\lim_{x\to a} f(x) = L_1$, *and* $\lim_{x\to a} g(x) = L_2$, *where* L_1 *and* L_2 *are real numbers.*

Property	Comments
1. $\lim_{x\to a} c = c$	*The limit of a constant is the constant itself.*
2. $\lim_{x\to a} x = a$	
3. $\lim_{x\to a} x^n = a^n$	

continued on next page ...

Properties of Limits as $x \to a$ (continued)

Property	Comments
4. $\lim_{x\to a}\left[f(x)\pm g(x)\right]$ $=\lim_{x\to a}f(x)\pm\lim_{x\to a}g(x)=L_1\pm L_2$	*The limit of a sum (or difference) of two functions is the sum (or difference) of the limits.*
5. $\lim_{x\to a}\left[f(x)\cdot g(x)\right]$ $=\left[\lim_{x\to a}f(x)\right]\cdot\left[\lim_{x\to a}g(x)\right]=L_1\cdot L_2$	*The limit of a product is the product of the limits provided these limits exist.*
6. $\lim_{x\to a}\left[\frac{f(x)}{g(x)}\right]=\frac{\lim_{x\to a}f(x)}{\lim_{x\to a}g(x)}=\frac{L_1}{L_2}$ where $L_2\neq 0$	*The limit of a quotient is the quotient of the limits as long as the limit of the denominator is not 0.*
7. $\lim_{x\to a}\left[c\cdot f(x)\right]=c\cdot\lim_{x\to a}f(x)=c\cdot L_1$	*The limit of a constant times a function can be found by first finding the limit of the function and then multiplying this limit by the constant.*
8. $\lim_{x\to a}\sqrt[n]{f(x)}=\sqrt[n]{\lim_{x\to a}f(x)}$ $=\sqrt[n]{L_1}=\left(L_1\right)^{1/n}$ *where* $\left(L_1\right)^{1/n}$ *is defined.*	*The limit of the nth root of a function can be found by calculating the limit of the function first and then taking the nth root of the limit.* *(**Note:** f(x) must be defined on intervals to the left of a and to the right of a.)*
9. *If p is a polynomial function, then* $\lim_{x\to a}p(x)=p(a)$.	*The limit of a polynomial can be found by substituting a for x. This result is particularly useful and follows directly from Properties 1 through 5.*

Example 1: Properties of Limits

Use the properties of limits to find each of the following limits.

a. $\lim_{x \to 4} 10$

Solution: $\lim_{x \to 4} 10 = 10$ — Property 1.

b. $\lim_{x \to 3}\left(x^2 + 5x + 7\right)$

Solution: $\lim_{x \to 3}\left(x^2 + 5x + 7\right) = 3^2 + 5(3) + 7$ — Property 9. Substitute $x = 3$.

$$= 9 + 15 + 7$$

$$= 31$$

c. $\lim_{x \to 2} \sqrt[3]{\dfrac{x-3}{x+6}}$

Solution: $\lim_{x \to 2} \sqrt[3]{\dfrac{x-3}{x+6}} = \sqrt[3]{\lim_{x \to 2}\left(\dfrac{x-3}{x+6}\right)}$ — Property 8.

$$= \sqrt[3]{\frac{\lim_{x \to 2}(x-3)}{\lim_{x \to 2}(x+6)}}$$ — Property 6.

$$= \sqrt[3]{\frac{(2-3)}{(2+6)}}$$ — Property 9. Substitute $x = 2$.

$$= \sqrt[3]{\frac{-1}{8}} = -\frac{1}{2}$$

In the special case when the numerator and denominator of a fractional expression both have value 0 at $x = a$, we say that the expression $\frac{0}{0}$ is an **indeterminate form** (see Section 2.1). In this text we will consider such cases only when $x - a$ is a factor of both the numerator and denominator and the expression for $f(x)$ can be simplified before the limit is found. In more advanced texts, this is called a removable discontinuity. This procedure is allowed because we are concerned with what happens to the function as x approaches a and not at $x = a$. For example, if

$$f(x) = \frac{x^2 - 9}{x - 3}, \quad \text{then} \quad f(3) = \frac{3^2 - 9}{3 - 3} = \frac{0}{0},$$

which is undefined and an indeterminate form. However, by factoring and reducing, we find

$$\lim_{x \to 3} f(x) = \lim_{x \to 3}\left(\frac{x^2 - 9}{x - 3}\right) = \lim_{x \to 3}\frac{(x+3)(x-3)}{x-3} = \lim_{x \to 3}(x+3) = 6.$$

Example 2: Indeterminate Form

If $f(x)=\dfrac{x^3-1}{x-1}$, find $\lim\limits_{x\to 1} f(x)$.

Solution: $f(1)=\dfrac{1^3-1}{1-1}=\dfrac{0}{0}$, so we simplify $f(x)$ first and then find the limit.

$$\lim_{x\to 1} f(x)=\lim_{x\to 1}\left(\frac{x^3-1}{x-1}\right)=\lim_{x\to 1}\left[\frac{\cancel{(x-1)}\left(x^2+x+1\right)}{\cancel{x-1}}\right]$$

$$=\lim_{x\to 1}\left(x^2+x+1\right)=1^2+1+1=3$$

In Example 2 we were able to calculate the limit because we found a common factor. This is no coincidence. The Remainder Theorem in college algebra tells us that, for any polynomial $p(x)$, if $p(a)=0$, then $x-a$ divides evenly into $p(x)$ or, equivalently, that $x-a$ is a factor of $p(x)$. If, for example, $p(x)=x^3-1$, we observe $p(1)=0$. Thus $x-1$ is a factor of x^3-1.

From the other point of view, suppose $q(x)$ satisfies $q(a)=0$ and $p(a)\neq 0$, for two polynomials $p(x)$ and $q(x)$. Then $\lim\limits_{x\to a}\dfrac{p(x)}{q(x)}$ will not exist since, as $x\to a$, the denominator approaches 0 but the numerator does not. There is no factor $x-a$ which will cancel.

Example 3: Undefined Limit

For $f(x)=\dfrac{1}{x-5}$, find $\lim\limits_{x\to 5} f(x)$.

Solution: For $f(x)=\dfrac{1}{x-5}$, $f(5)=\dfrac{1}{0}$, which is not defined, and further, is not of the form $\dfrac{0}{0}$. The expression $\dfrac{1}{x-5}$ cannot be simplified as the expression in Example 2 was. When the limit of the denominator is 0 and the limit of the numerator is some number other than 0, then the limit of the function does not exist. To show this and to determine the nature of the graph of the function near $x=5$, we analyze the left- and right-hand limits:

$$\lim_{x\to 5^-}\left(\frac{1}{x-5}\right)=-\infty \quad\text{and}\quad \lim_{x\to 5^+}\left(\frac{1}{x-5}\right)=+\infty.$$

Thus we see that $\lim_{x\to 5} f(x)$ does not exist and that the graph of the function is unbounded in the negative direction as $x \to 5^-$ and unbounded in the positive direction as $x \to 5^+$.

The following example illustrates the technique for finding a limit of a function defined piecewise.

Example 4: Piecewise Function

The function $f(x)$ is defined piecewise as follows:

$$f(x)=\begin{cases} 2x & \text{if } x \le 3 \\ x+3 & \text{if } x>3 \end{cases}.$$

Find $\lim_{x\to 3} f(x)$ if it exists.

Solution: $\lim_{x\to 3^-} f(x) = \lim_{x\to 3^-} 2x = 2(3) = 6$ Use $f(x)=2x$ since $x \le 3$ when $x \to 3^-$.

$\lim_{x\to 3^+} f(x) = \lim_{x\to 3^+} (x+3) = 3+3 = 6$ Use $f(x)=x+3$ since $x>3$ when $x \to 3^+$.

Since $\lim_{x\to 3^-} f(x) = \lim_{x\to 3^+} f(x) = 6$, we have $\lim_{x\to 3} f(x) = 6$.

Limits of Rational Functions as $x \to +\infty$ or $x \to -\infty$

In this section we will discuss rational functions and their behavior as x increases (or decreases) without bound (that is, as $x \to +\infty$ and $x \to -\infty$).

Rational Function

*A **rational function** is a function of the form*

$$f(x)=\frac{p(x)}{q(x)},$$

where $p(x)$ and $q(x)$ are polynomials.

We begin by analyzing $f(x)=\frac{1}{x}$ as x increases without bound. If we let x = 10, 100, 1000, 10,000, 50,000, 1,000,000, and so on, then $\frac{1}{x} = \frac{1}{10}, \frac{1}{100}, \frac{1}{1000}, \frac{1}{10{,}000}, \frac{1}{50{,}000}, \frac{1}{1{,}000{,}000}$, and so on.

Since $\frac{1}{10} > \frac{1}{100} > \frac{1}{1000} > \frac{1}{10{,}000} > \frac{1}{50{,}000} > \frac{1}{1{,}000{,}000} > \ldots > 0$, the value of $\frac{1}{x}$ becomes smaller and smaller and approaches 0 as x becomes larger and larger. With this intuitive notation as background, we make the following statement:

$$\lim_{x \to +\infty} \frac{1}{x} = 0.$$

A similar analysis will show that

$$\lim_{x \to -\infty} \frac{1}{x} = 0.$$

The graph of the function $y = f(x) = \frac{1}{x}$ is shown in Figure 2.5.1. The line $y = 0$ (the x-axis) is a **horizontal asymptote**.

This table illustrates the behavior of $\frac{1}{x}$ as x increases.

x	$\frac{1}{x}$
10	0.1
100	0.01
1000	0.001
$+\infty$	0

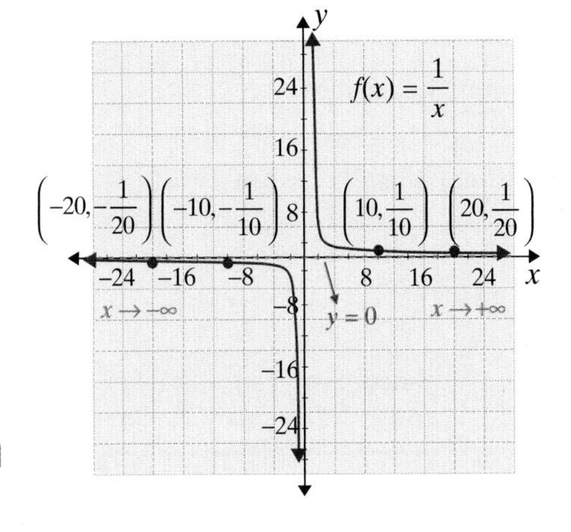

Figure 2.5.1

Earlier in this section we stated the properties of limits as $x \to a$ where a is a real number. These same properties are valid if $x \to +\infty$ or $x \to -\infty$, and we apply them in the following examples.

Example 5: Properties of Limits

Find $\lim_{x \to +\infty} \left(\frac{3x+5}{2x-1} \right)$, if it exists.

Solution: Use a graphing utility and graph the function $f(x) = \frac{(3x+5)}{(2x-1)}$. On the TI-84 Plus, use a window of [–10, 10] by [–3, 3]. As x gets larger we see the y-values level off just above the x-axis (see Figure 2.5.2).

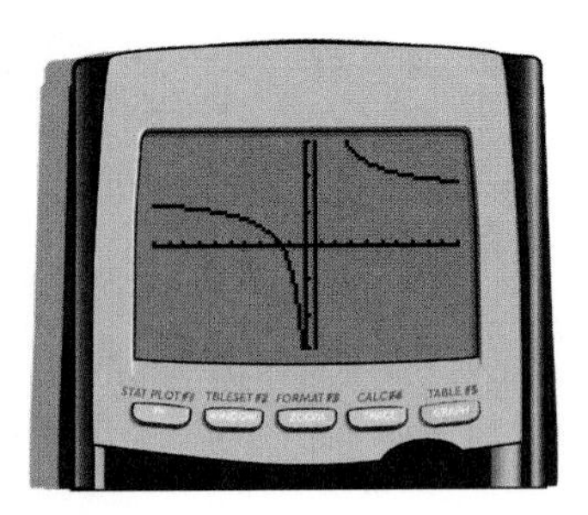

Figure 2.5.2

There is a natural way to investigate this algebraically.

Individually, both the numerator and denominator get very large. That is, they both approach $+\infty$.

$$\lim_{x\to+\infty}(3x+5)=+\infty \quad \text{and} \quad \lim_{x\to+\infty}(2x-1)=+\infty.$$

Thus this approach leads to an expression of the form $\frac{+\infty}{+\infty}$, which is called an indeterminate form. We still do not know what the limit is or even if the limit exists.

Now, if we divide each term in the numerator and denominator by x, we will have an expression with fractions whose limits we can find.

$$\lim_{x\to+\infty}\left(\frac{3x+5}{2x-1}\right)=\lim_{x\to+\infty}\left(\frac{\frac{3x}{x}+\frac{5}{x}}{\frac{2x}{x}-\frac{1}{x}}\right)$$ Divide each term by x.

$$=\lim_{x\to+\infty}\left(\frac{3+\frac{5}{x}}{2-\frac{1}{x}}\right)$$ Simplify.

$$=\frac{\lim_{x\to+\infty}\left(3+\frac{5}{x}\right)}{\lim_{x\to+\infty}\left(2-\frac{1}{x}\right)}$$ Use limit Property 6.

$$=\frac{3+0}{2-0}=\frac{3}{2}$$ Use limit Properties 4 and 7. Also, use $\lim_{x\to+\infty}\left(\frac{1}{x}\right)=0$.

$\left[\textbf{Note:} \lim_{x\to+\infty}\left(\frac{5}{x}\right)=5\cdot\lim_{x\to+\infty}\left(\frac{1}{x}\right).\right]$

Since the constants $+5$ and -1 will have little effect on the fraction for very large values of x (i.e., in the millions), then $\frac{3x+5}{2x-1}$ is approximately the same as $\frac{3x}{2x}=\frac{3}{2}$. Thus the limit of $\frac{3}{2}$ seems quite reasonable from this point of view. The graph shows that the line $y=\frac{3}{2}$ is indeed a horizontal asymptote.

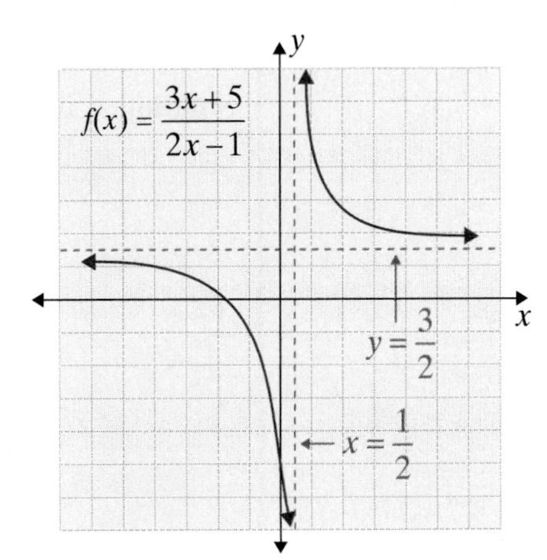

Example 6: Properties of Limits

Find $\lim_{x\to+\infty}\left(\frac{2x+3}{x^3-8}\right)$, if it exists.

Solution: In Example 5 we divided each term in the numerator and denominator by x. In this example, we will divide each term by x^3, since the technique is to divide by the highest power of x present in the function.

$$\lim_{x\to+\infty}\left(\frac{2x+3}{x^3-8}\right)=\lim_{x\to+\infty}\left(\frac{\frac{2x}{x^3}+\frac{3}{x^3}}{\frac{x^3}{x^3}-\frac{8}{x^3}}\right)$$

Divide each term by x^3.

$$=\lim_{x\to+\infty}\left(\frac{\frac{2}{x^2}+\frac{3}{x^3}}{1-\frac{8}{x^3}}\right)$$

Simplify.

$$=\frac{0+0}{1-0}=\frac{0}{1}=0$$

Since the numerator is 0 and the denominator is not 0, the fraction has the value 0.

Example 7: Properties of Limits

Find $\lim_{x\to-\infty}\left(\frac{x^3+x^2-x+1}{x^2-4}\right)$, if it exists.

Solution:

$$\lim_{x\to-\infty}\left(\frac{x^3+x^2-x+1}{x^2-4}\right)=\lim_{x\to-\infty}\left(\frac{\frac{x^3}{x^3}+\frac{x^2}{x^3}-\frac{x}{x^3}+\frac{1}{x^3}}{\frac{x^2}{x^3}-\frac{4}{x^3}}\right)$$

Divide each by the highest power of x present, which in this function is x^3.

$$=\lim_{x\to-\infty}\left(\frac{1+\frac{1}{x}-\frac{1}{x^2}+\frac{1}{x^3}}{\frac{1}{x}-\frac{4}{x^3}}\right)$$

Simplify.

$$=\frac{1}{0}$$

Since $\frac{1}{0}$ is undefined, the limit is either $+\infty$ or $-\infty$. (See Example 3.)

Investigating the expression shows that the highest power is x^3. This term will dominate for very large values of x and will be negative for negative values of x. Thus

$$\lim_{x\to-\infty}\left(\frac{x^3+x^2-x+1}{x^2-4}\right)=-\infty.$$

Summary of Limits for Rational Functions as $x \to +\infty$ or $x \to -\infty$

Consider the function

$$f(x)=\frac{a_n x^n + a_{n-1}x^{n-1}+\ldots+a_0}{b_m x^m + b_{m-1}x^{m-1}+\ldots+b_0},$$

where $a_n \neq 0$ *and* $b_m \neq 0$.

Case 1: *For* $m=n$, $\lim_{x\to+\infty} f(x)=\frac{a_n}{b_m}$.

Case 2: *For* $m>n$, $\lim_{x\to+\infty} f(x)=0$.

Case 3: *For* $m<n$, $\lim_{x\to+\infty} f(x)=+\infty$. *(Or* $-\infty$ *depending on the signs of* a_n *and* b_m*.)*

Continuity

The concept of a continuous function can be explained on an intuitive basis similar to the discussion of polynomial functions in Section 2.3. This is reasonable since all polynomial functions are continuous functions. If you can trace the graph of a function without lifting your pencil from the paper (i.e. there are no "holes" or "jumps" or vertical asymptotes), then the function is **continuous**. (See Figure 2.5.3.)

Figure 2.5.3

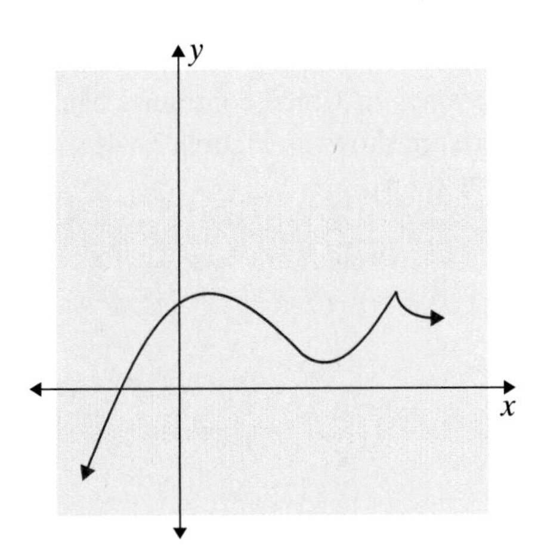

(a) Continuous function

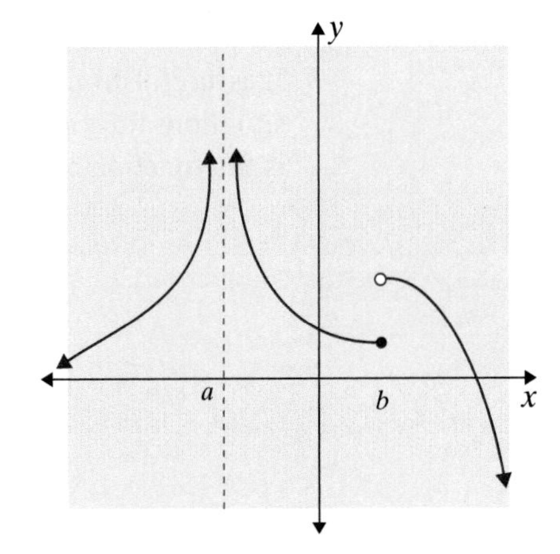

(b) Function not continuous at $x = a$ and at $x = b$

This approach helps to give a mental picture of a continuous function, but it does not take the place of a formal definition that can be applied algebraically.

The concept of continuity is closely related to limits. In our discussion of limits as $x \to a$, we were concerned with the function values for x close to a but not for $x = a$. In some cases, $f(a)$ did **exist** (was **defined**), and in other cases, $f(a)$ did **not exist** (was **not defined**). In no case did the value of $f(a)$ have any effect on the limit. We will find, however, that the value of $f(a)$ is an integral part of the definition of continuity at $x = a$. That is, we discuss continuity in terms of continuity at a point.

Continuity

A function $y = f(x)$ is ***continuous*** *at $x = a$ if all of the following conditions are true:*

1. *$f(a)$ exists.* *($f(a)$ is a real number.)*
2. *$\lim_{x\to a} f(x)$ exists.* *(The limit is a real number.)*
3. *$\lim_{x\to a} f(x) = f(a)$.* *(The limit is equal to the value of the function at $x = a$.)*

If a function fails to satisfy any one of the three conditions of the definition, then it is not continuous at $x = a$ and is said to be ***discontinuous*** *at $x = a$.*

The formal definition not only tells us the technical meaning of "continuous at $x = a$," but it also says something important about computing limits. If a function is known to be continuous at $x = a$, then the limit problem $\lim_{x\to a} f(x)$ is solved merely by substitution of $x = a$ into the formula! For continuous functions, limit problems are trivial. Unfortunately, the important limit problems that occur most often in calculus involve $\frac{\Delta y}{\Delta x}$, which is not continuous at $\Delta x = 0$.

It is useful to consider the ways in which continuity can fail. A variety of possible situations for $y = f(x)$ at $x = a$ are shown in Figures 2.5.4(a) – (e). Only in Figure 2.5.4(a) is the function continuous at $x = a$.

Figure 2.5.4

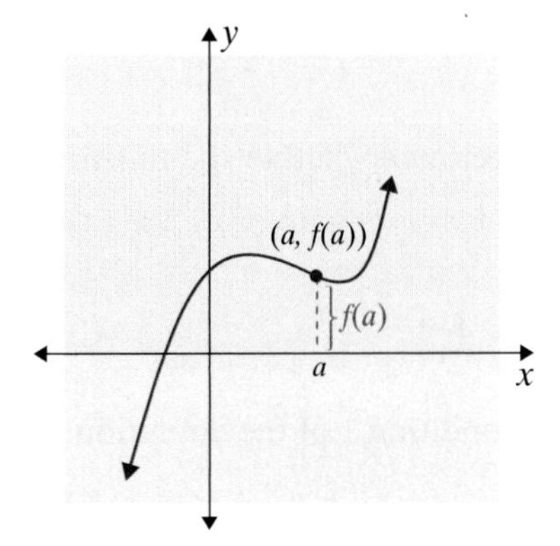

(a) $\lim_{x \to a} f(x) = f(a)$

f is continuous at $x = a$.

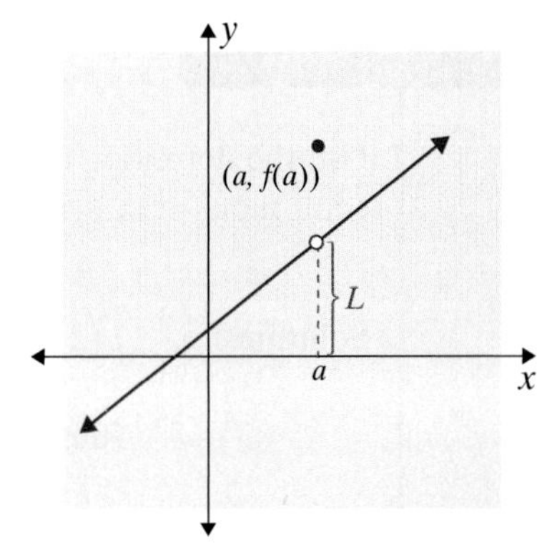

(b) $\lim_{x \to a} f(x) = L \neq f(a)$

There is a "hole" in the curve.
This is called a **removable discontinuity**.

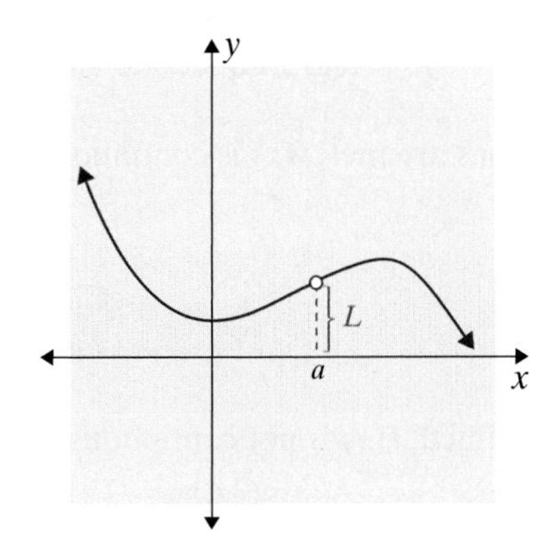

(c) $\lim_{x \to a} f(x) = L$

$f(a)$ does not exist.
This is a **removable discontinuity**.

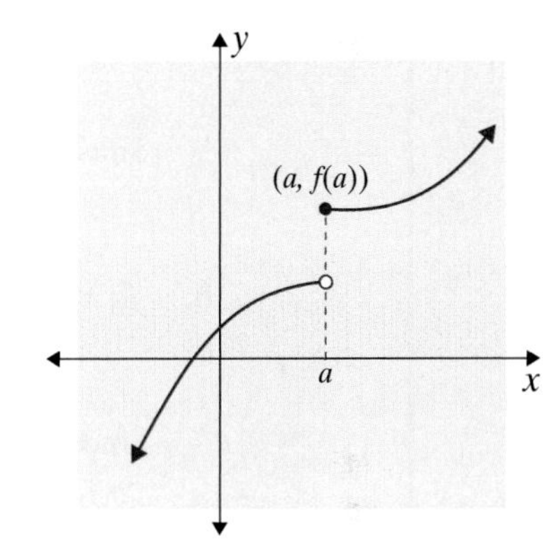

(d) $\lim_{x \to a^-} f(x) \neq \lim_{x \to a^+} f(x)$

$\lim_{x \to a} f(x)$ does not exist.
This is called a **jump discontinuity**.

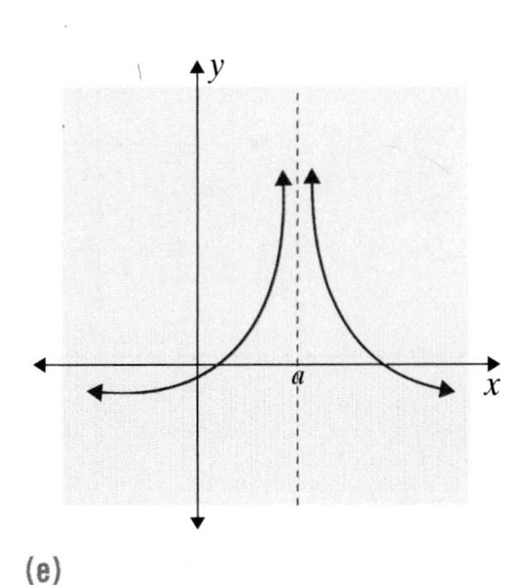

(e)

$\lim_{x \to a} f(x) = +\infty$

f is discontinuous at $x = a$ since the limit does not exist. Also, $f(a)$ does not exist. This is called an **infinite discontinuity** or a **non-removable discontinuity**.

Example 8: Continuity

Use the definition of continuity to determine whether the function $f(x)=\frac{x^2-x-6}{x-3}$ is continuous at **a.** $x = 2$, and **b.** $x = 3$.

Solutions: a. $f(2)=\frac{2^2-2-6}{2-3}=\frac{-4}{-1}=4$ Check Condition 1 by substituting $x = 2$.

Thus $f(2)$ exists, and Condition 1 of the definition is met.

$\lim_{x\to 2} f(x)=\lim_{x\to 2}\frac{(x-3)(x+2)}{x-3}=\lim_{x\to 2}(x+2)=4$ Check Condition 2 by finding a limit at $x = 2$.

The limit exists, so Condition 2 of the definition is met.

$\lim_{x\to 2} f(x)=4=f(2)$ Check Condition 3 by seeing if Conditions 1 and 2 equal one another.

Since all three conditions are met, $f(x)$ is continuous at $x = 2$.

b. $f(3)=\frac{3^2-3-6}{3-3}=\frac{0}{0}$ Check Condition 1 by substituting $x = 3$.

Since $f(3)=\frac{0}{0}$ is undefined, $f(x)$ is not continuous at $x = 3$.

The graph to the right shows a missing point discontinuity at $x = 3$.

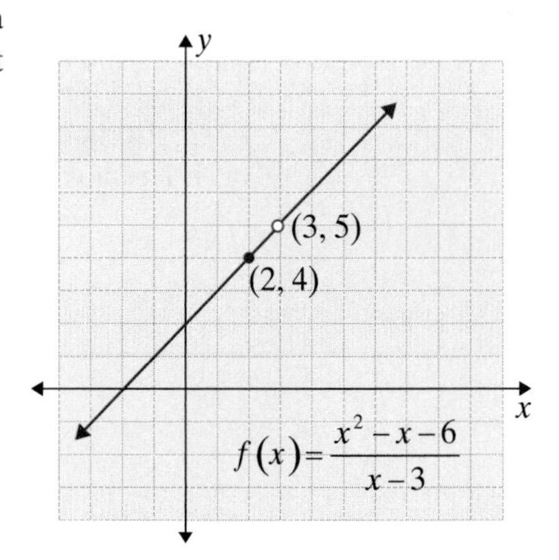

Example 9: Graphing a Discontinuous Funtion

a. Use the definition of continuity to determine whether the function $g(x) = \frac{|x|}{x}$ is continuous at $x = 0$.

Solution: Since $g(0) = \frac{|0|}{0} = \frac{0}{0}$ is undefined, g is not continuous at $x = 0$.

b. Draw the graph of $g(x)$.

Solution: The graph can be found by using the definition of $|x|$ (see Section 1.1) and rewriting the function as follows:

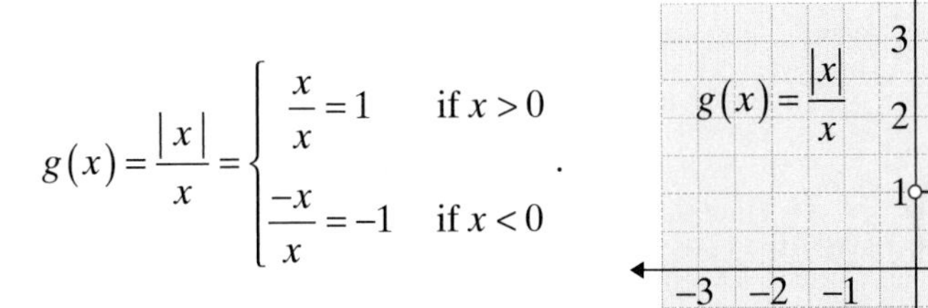

$$g(x) = \frac{|x|}{x} = \begin{cases} \frac{x}{x} = 1 & \text{if } x > 0 \\ \frac{-x}{x} = -1 & \text{if } x < 0 \end{cases}.$$

The graph shows that $g(x)$ has a jump discontinuity at $x = 0$.

A function is **continuous** if it has no point of discontinuity. If a function is discontinuous at any one point, then it is a **discontinuous** function.

NOTES

Analysis shows, polynomial functions have no points of discontinuity. That is, for a polynomial function p, $\lim_{x \to a} p(x) = p(a)$ for every real number a.

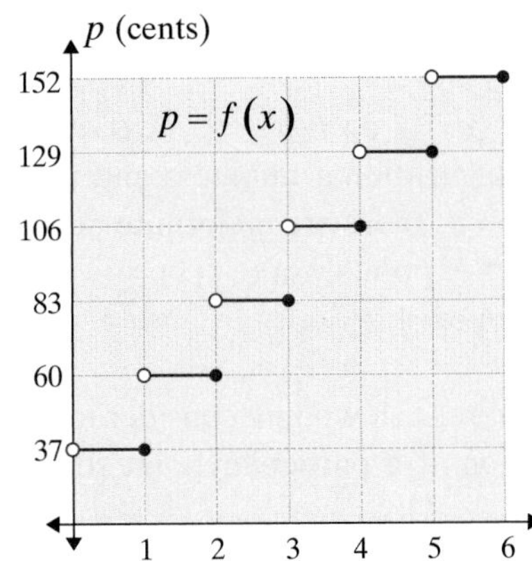

Figure 2.5.5

U.S. Post Office Fees for First-Class Delivery (2005)

An interesting function that has several points of discontinuity is the *post office function*, which relates the weight of an item to the price of delivery by the Postal Service. It is an example of a **step function**, and its graph is shown in Figure 2.5.5.

Another step function is the **greatest-integer function**. It is a particularly interesting example of a function with an infinite number of discontinuities. Each of these discontinuities is a jump discontinuity and occurs at an integer value of x.

Greatest-Integer Function

The function $h(x)=[x]$ *is called the greatest-integer function and is defined as follows:*

$$h(x)=[x]=\text{the greatest integer} \leq x.$$

Informally, $h(x)=[x]$ *is the integer closest to x but still less than or equal to x.*

Thus, for any x, the value of $h(x)$ is an integer. As examples, we show the following:

$$h(3.1)=[3.1]=3,$$
$$h(4.6)=[4.6]=4,$$
$$h(0.8)=[0.8]=0, \text{ and}$$
$$h(-1.7)=[-1.7]=-2.$$

The graph of $h(x)=[x]$ is shown in Figure 2.5.6.

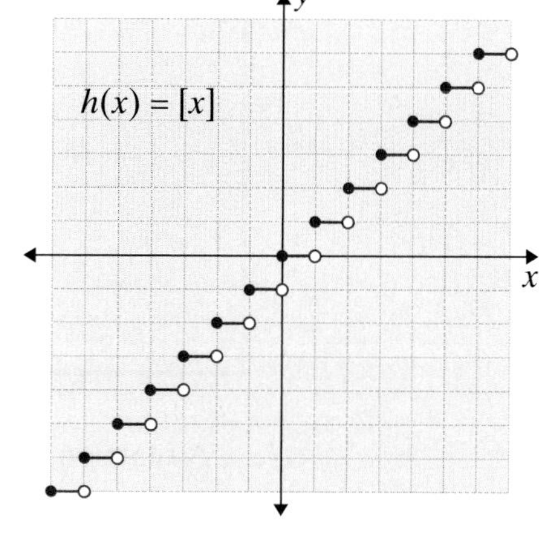

Figure 2.5.6

Another famous function is the **Dirichlet function**. This function is defined as

$$f(x)=\begin{cases} 1 & \text{if } x \text{ is irrational} \\ 0 & \text{if } x \text{ is rational} \end{cases}$$

and is discontinuous at every point. The reasoning is as follows:

If a is a real number, then $f(a)=1$ or $f(a)=0$. Every interval about $x=a$ contains an infinite number of irrational numbers and an infinite number of rational numbers. Thus, as $x \to a$, there are an infinite number of points where $f(x)=1$ and an infinite number of points where $f(x)=0$. Neither 1 nor 0 can be $\lim_{x\to a} f(x)$, so $\lim_{x\to a} f(x)$ does not exist.

For a look at continuity from a different viewpoint, consider the problem of determining the form or the value of a function at a particular point so that the function will be continuous at that point.

Example 10: Making a Function Continuous

Find a value for k so that the function

$$f(x) = \begin{cases} x^2 & \text{if } x < 2 \\ -3x + k & \text{if } x \geq 2 \end{cases}$$

will be continuous at $x = 2$.

Solution: f is defined in pieces, and each piece is a polynomial function. We must determine a value for k so that both pieces will have the same value when $x = 2$.

Set $x^2 = -3x + k$ and substitute $x = 2$:

$$2^2 = -3(2) + k$$
$$4 = -6 + k$$
$$10 = k.$$

Thus the function f is continuous at $x = 2$.

Graphically, we see that the two pieces "fit" together at the point (2, 4).

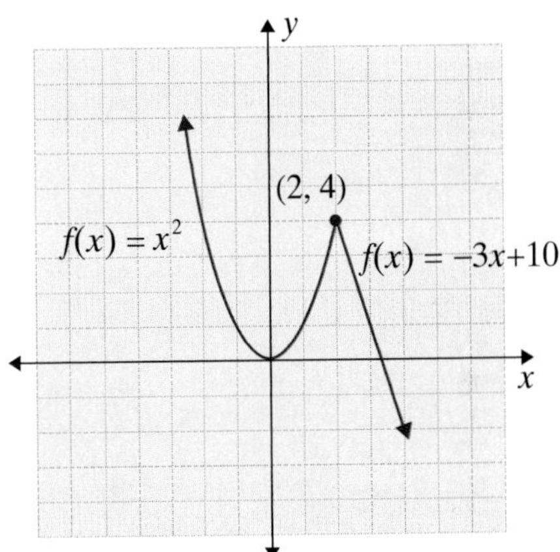

2.5 Exercises

In Exercises 1 – 28, find the indicated limit, if it exists.

1. $\lim_{x \to -2} 6$
2. $\lim_{x \to 4} 2x$
3. $\lim_{x \to 3^-} (x^2 + 1)$
4. $\lim_{x \to -3^+} (5 - 2x^2)$
5. $\lim_{x \to 1^+} \left(\frac{x+2}{x-1}\right)$
6. $\lim_{x \to 1^-} \left(\frac{x+2}{x-1}\right)$
7. $\lim_{x \to \frac{1}{3}} \left(\frac{3x+1}{x+2}\right)$
8. $\lim_{x \to 0^+} \left(\frac{x+4}{x-4}\right)$
9. $\lim_{x \to 0^-} \left(\frac{x}{x^2+2x}\right)$
10. $\lim_{x \to 0^-} \left(\frac{2x^2+x}{x}\right)$
11. $\lim_{x \to +\infty} \left(\frac{x}{x^2+3}\right)$
12. $\lim_{x \to +\infty} \left(\frac{2x^2+7}{3x^2-2}\right)$

13. $\lim_{x\to-\infty}\left(\frac{x^3+64}{x^2-2x+1}\right)$ **14.** $\lim_{x\to-\infty}\left(\frac{4x-x^3}{x^2+2x-7}\right)$ **15.** $\lim_{x\to2}\left(\frac{x^2-x-2}{x^2-4}\right)$

16. $\lim_{x\to-3}\left(\frac{x^2-9}{x^2+2x-3}\right)$ **17.** $\lim_{x\to0^+}\left(4-\frac{3}{x}\right)$ **18.** $\lim_{x\to1^-}\left(2x+\frac{5}{x-1}\right)$

19. $\lim_{x\to+\infty}\left(8+\frac{1}{x}\right)$ **20.** $\lim_{x\to-\infty}\left(11-\frac{2}{x^2}\right)$ **21.** $\lim_{x\to4}\sqrt{x+5}$

22. $\lim_{x\to2}\sqrt{3x+10}$ **23.** $\lim_{x\to1}\left(\sqrt{x}-3\right)$ **24.** $\lim_{x\to4}\left(\sqrt{x}+6\right)$

25. a. $\lim_{x\to4}\left(\frac{\sqrt{x}-2}{x-4}\right)$ [**Hint :** $x-4=\left(\sqrt{x}+2\right)\left(\sqrt{x}-2\right)$]

b. $\lim_{x\to9}\left(\frac{x-9}{\sqrt{x}-3}\right)$ [**Hint :** $x-9=\left(\sqrt{x}+3\right)\left(\sqrt{x}-3\right)$]

26. a. $\lim_{h\to0}\frac{\sqrt{2+h}-\sqrt{2}}{h}$ [**Hint :** multiply by $\frac{\sqrt{2+h}+\sqrt{2}}{\sqrt{2+h}+\sqrt{2}}$, simplify the numerator, and calculate the limit.]

b. $\lim_{h\to0}\frac{\sqrt{5x+h}-\sqrt{5x}}{h}$

27. $\lim_{x\to+\infty}\left(\frac{x^2+3x-4}{x^2+5x-9}\right)$ **28.** $\lim_{x\to-\infty}\left(\frac{x^2+x+1}{2x^3+3x^2+x-2}\right)$

29. For the function

$$f(x)=\begin{cases}-3 & \text{if } x\le1\\ x-4 & \text{if } x>1\end{cases},$$

find the following:

a. $\lim_{x\to1^-}f(x)$ **b.** $\lim_{x\to1^+}f(x)$

c. $\lim_{x\to1}f(x)$ **d.** $\lim_{x\to2}f(x)$

30. For the function

$$f(x)=\begin{cases}2x+1 & \text{if } x<0\\ x^2+1 & \text{if } x\ge0\end{cases},$$

find the following:

a. $\lim_{x\to0^-}f(x)$ **b.** $\lim_{x\to0^+}f(x)$

c. $\lim_{x\to0}f(x)$ **d.** $\lim_{x\to-2}f(x)$

31. For the function

$$f(x)=\begin{cases}5-x & \text{if } x\le2\\ x^2-1 & \text{if } x>2\end{cases},$$

find the following:

a. $\lim_{x\to2^-}f(x)$ **b.** $\lim_{x\to2^+}f(x)$

c. $\lim_{x\to2}f(x)$ **d.** $\lim_{x\to0}f(x)$

32. For the function

$$f(x)=\begin{cases}x^3+4 & \text{if } x\le-2\\ \sqrt{x^2+5} & \text{if } x>-2\end{cases},$$

find the following:

a. $\lim_{x\to-2^-}f(x)$ **b.** $\lim_{x\to-2^+}f(x)$

c. $\lim_{x\to-2}f(x)$ **d.** $\lim_{x\to-1}f(x)$

In Exercises 33 – 40, use the graph of $y = f(x)$ to answer the questions regarding the function.

33.

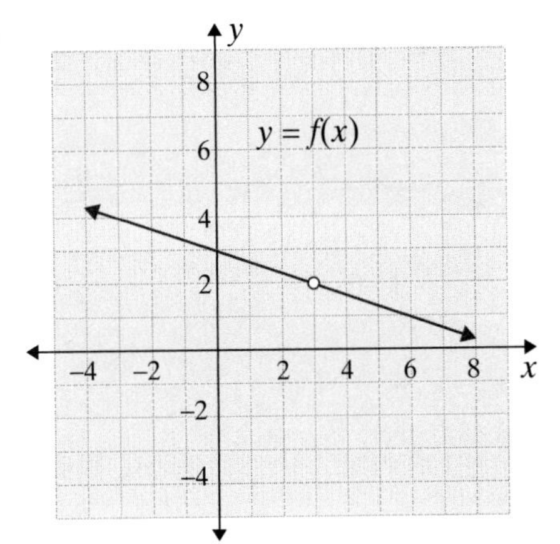

a. Find $\lim_{x \to 3^-} f(x)$.

b. Find $\lim_{x \to 3^+} f(x)$.

c. Find $f(3)$.

d. Is $f(x)$ continuous at $x = 3$? Explain.

34.

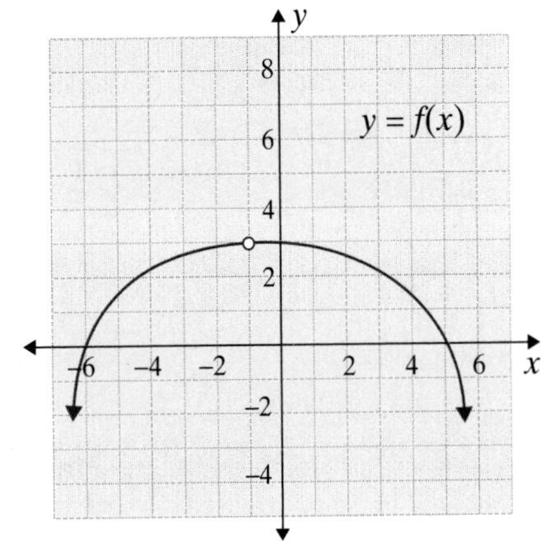

a. Find $\lim_{x \to -1^-} f(x)$.

b. Find $\lim_{x \to -1^+} f(x)$.

c. Find $f(-1)$.

d. Is $f(x)$ continuous at $x = -1$? Explain.

35.

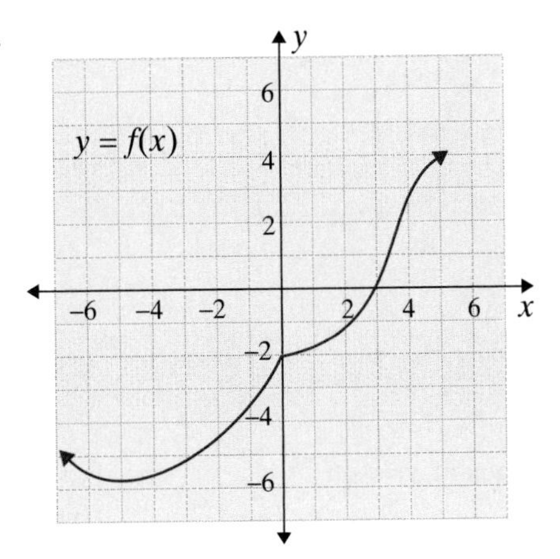

a. Find $\lim_{x \to 0^-} f(x)$.

b. Find $\lim_{x \to 0^+} f(x)$.

c. Find $f(0)$.

d. Is $f(x)$ continuous at $x = 0$? Explain.

36.

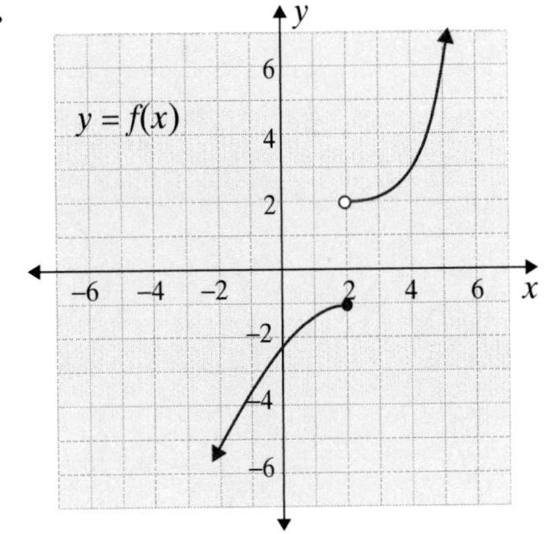

a. Find $\lim_{x \to 2^-} f(x)$.

b. Find $\lim_{x \to 2^+} f(x)$.

c. Find $f(2)$.

d. Is $f(x)$ continuous at $x = 2$? Explain.

37.

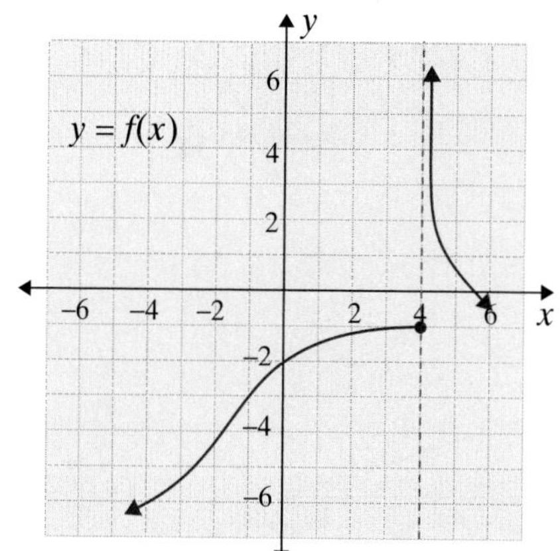

a. Find $\lim_{x \to 4^-} f(x)$.

b. Find $\lim_{x \to 4^+} f(x)$.

c. Find $f(4)$.

d. Is $f(x)$ continuous at $x = 4$? Explain.

38.

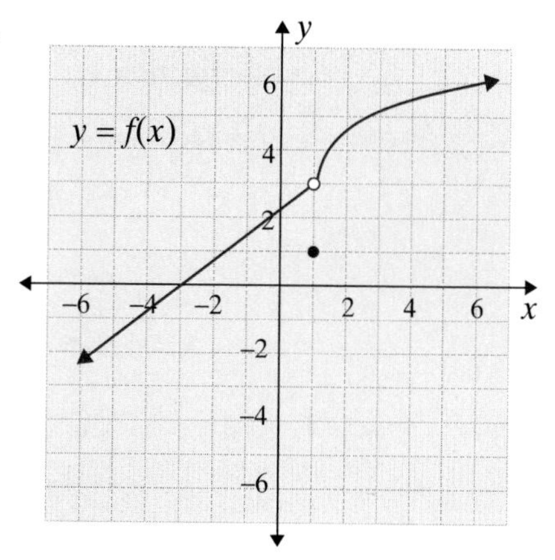

a. Find $\lim_{x \to 1^-} f(x)$.

b. Find $\lim_{x \to 1^+} f(x)$.

c. Find $f(1)$.

d. Is $f(x)$ continuous at $x = 1$? Explain.

39.

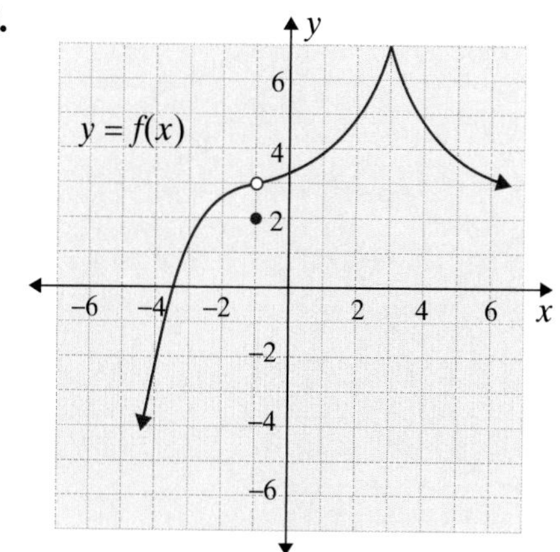

a. Is $f(x)$ continuous at $x = -1$? Explain.

b. Is $f(x)$ continuous at $x = 3$? Explain.

40.

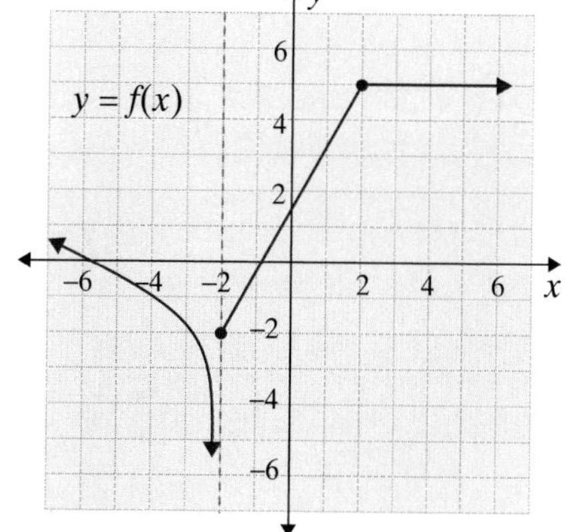

a. Is $f(x)$ continuous at $x = 2$? Explain.

b. Is $f(x)$ continuous at $x = -2$? Explain.

In Exercises 41 – 44, use the graph of y = f(x) to find the points of discontinuity, if any exist. Determine what type of discontinuity each is.

41.

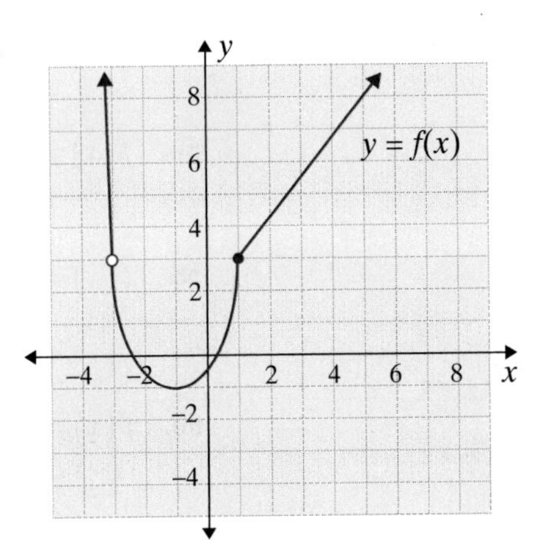

42.

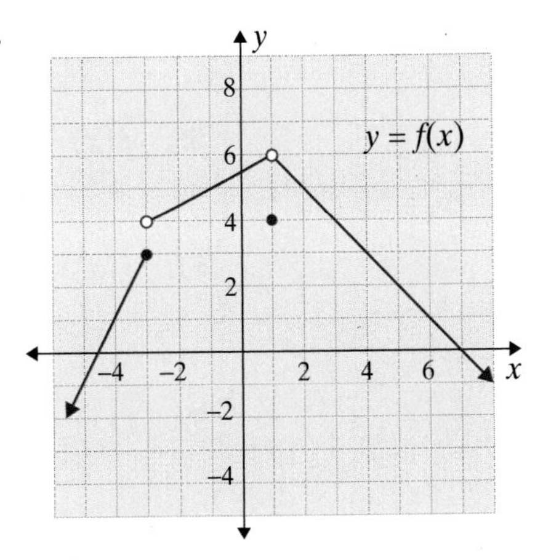

43.

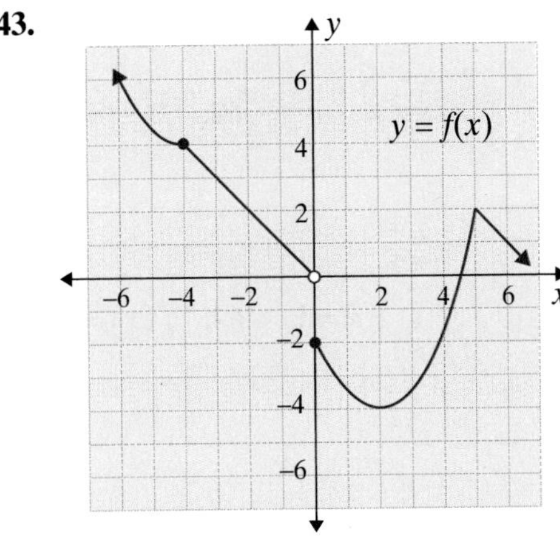

44.

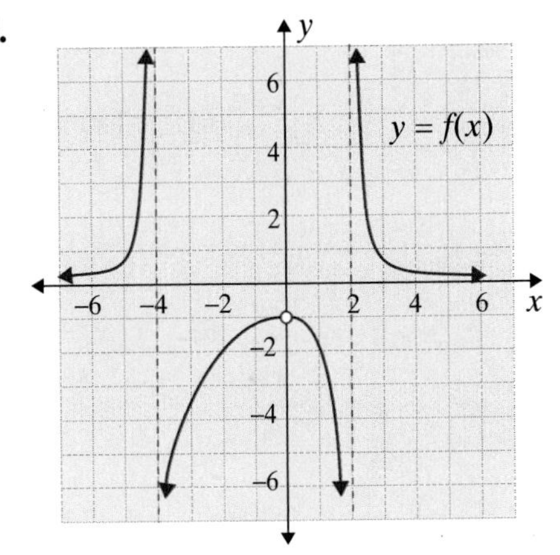

In Exercises 45 – 52, use the definition of continuity to determine whether or not the function is continuous at the given value of x.

45. $f(x) = 3 - 2x;\ x = 1$

46. $f(x) = 5x - x^2;\ x = 0$

47. $f(x) = \dfrac{x^2 - x - 2}{x - 2};\ x = 2$

48. $f(x) = \dfrac{x+1}{x^2 - 1};\ x = 0$

49. $f(x) = \begin{cases} 2x+1 & \text{if } x \le 1 \\ 3 & \text{if } x > 1 \end{cases};\ x = 1$

50. $f(x) = \begin{cases} x^2 & \text{if } x \le 2 \\ 2x & \text{if } x > 2 \end{cases};\ x = 2$

51. $f(x) = \begin{cases} 1 - 3x & \text{if } x < 0 \\ 4x & \text{if } x \ge 0 \end{cases};\ x = 0$

52. $f(x) = \begin{cases} \dfrac{x}{x-3} & \text{if } x \ne 3 \\ 2 & \text{if } x = 3 \end{cases};\ x = 3$

In Exercises 53 – 60, find the points of discontinuity for each function, if any exist. Determine what type of discontinuity each is.

53. $f(x)=2x^2+3x-1$

54. $f(x)=3x^2-x+7$

55. $f(x)=\dfrac{5}{x+3}$

56. $f(x)=\dfrac{x+8}{x}$

57. $f(x)=\dfrac{x}{x^2-9}$

58. $f(x)=\dfrac{2}{x^2-4x}$

59. $f(x)=\begin{cases}2+3x & \text{if } x\le 1\\ x^2+4 & \text{if } x>1\end{cases}$

60. $f(x)=\begin{cases}x^2+1 & \text{if } x\le 2\\ 2x-1 & \text{if } x>2\end{cases}$

In Exercises 61 – 64, find a value for k so that the given function will be continuous at the indicated value for x.

61. $f(x)=\begin{cases}3x & \text{if } x\le 2\\ x^2+k & \text{if } x>2\end{cases};\ x=2$

62. $f(x)=\begin{cases}7 & \text{if } x<-3\\ k-2x & \text{if } x\ge -3\end{cases};\ x=-3$

63. $f(x)=\begin{cases}3x-k & \text{if } x\le 1\\ \dfrac{x^2-3x+2}{x-1} & \text{if } x>1\end{cases};\ x=1$

64. $f(x)=\begin{cases}\dfrac{x^2+3x}{x} & \text{if } x<0\\ 2x^2-k & \text{if } x\ge 0\end{cases};\ x=0$

65. Utility costs. The Municipal Gas Company uses the following function for computing their customers' monthly gas bills:

$$C(x)=\begin{cases}0.37x+3.00 & \text{if } 0<x\le 24\\ 0.78x-6.84 & \text{if } x>24\end{cases},$$

where x is the number of therms (thermal units) used and $C(x)$ is the cost in dollars.

a. Find $\lim_{x\to 24^-} C(x)$. **b.** Find $\lim_{x\to 24^+} C(x)$. **c.** Find $\lim_{x\to 24} C(x)$.

66. Income tax. A federal income tax schedule can be given by the function

$$T(x) = \begin{cases} 0.15x & \text{if } 0 < x \le 23{,}900 \\ 0.28x - 3107 & \text{if } 23{,}900 < x \le 61{,}650 \\ 0.33x - 6189.50 & \text{if } 61{,}650 < x \le 123{,}790 \end{cases},$$

where x is the taxable income in dollars, $0 < x \le 123{,}790$, and $T(x)$ is in dollars.

a. Find $\lim_{x \to 23{,}900^-} T(x)$. **b.** Find $\lim_{x \to 23{,}900^+} T(x)$.

c. Find $\lim_{x \to 23{,}900} T(x)$. **d.** Find $\lim_{x \to 61{,}650} T(x)$.

67. Average cost. A manufacturer of golf clubs estimates that if x sets of golf clubs are produced, then the average cost of producing each set is $A(x) = 73 + \frac{5780}{x}$ dollars. What will be the average cost of producing each set in the long run $\left(\lim_{x \to +\infty} A(x)\right)$?

68. Dictation rate. It has been determined that after t weeks of class, a certain student in an intermediate shorthand class can take dictation at a rate of $W(t) = 60 + \frac{70t^2}{t^2 + 15}$ words per minute. What will be this student's rate of taking dictation in the long run $\left(\lim_{t \to +\infty} W(t)\right)$?

69. Pricing. A leather craft store has the following pricing policy for a belt buckle:

$$C(x) = \begin{cases} 0.79x & \text{if } 0 < x < 12 \\ 0.71x & \text{if } 12 \le x < 50 \\ 0.67x & \text{if } x \ge 50 \end{cases},$$

where x is the number of buckles and $C(x)$ is in dollars.

a. Graph the function $C(x)$.

b. Is $C(x)$ a continuous function? Explain your answer.

70. Salary. A salesperson's weekly salary is determined by the function

$$S(x) = \begin{cases} 550 & \text{if } 0 < x < 10{,}000 \\ 0.06x & \text{if } x \ge 10{,}000 \end{cases},$$

where x is the weekly sales.

a. Graph the function $S(x)$.

b. Is $S(x)$ a continuous function? Explain your answer.

71. Cost of telephone call. The cost of an overseas call is given by the function

$$C(x) = \begin{cases} 9.00 & \text{if } 0 < x \leq 3 \\ 0.95x + 6.15 & \text{if } x > 3 \end{cases},$$

where $C(x)$ is in dollars and x is in minutes.

a. Graph the function $C(x)$.

b. Is $C(x)$ a continuous function? Explain your answer.

72. Revenue. The revenue from the sale of a particular model of radio is given by the function

$$R(p) = \begin{cases} 84p & \text{if } 0 < p \leq 20 \\ 144p - 3p^2 & \text{if } p > 20 \end{cases},$$

where p is the price in dollars.

a. Graph the function $R(p)$.

b. Is $R(p)$ a continuous function? Explain your answer.

Hawkes Learning Systems: Essential Calculus

More about Limits	p. 173 - 181	Ex. 1 - 32, 65 - 68
More about Continuity	p. 181 - 187	Ex. 33 - 64, 69 - 72

Chapter 2 Index of Key Ideas and Terms

Section 2.1 Limits

One-Sided Limits pages 96 - 100

Left-hand Limit

If the values of a function $y = f(x)$ get closer and closer to some number K as values of x, that are smaller than some number a, get closer and closer to a, then we say that K is the limit of $f(x)$ as x approaches a from the left.

$$\lim_{x \to a^-} f(x) = K$$

Right-hand Limit

If the values of a function $y = f(x)$ get closer and closer to some number M as values of x, that are larger than some number a, get closer and closer to a, then we say that M is the limit of $f(x)$ as x approaches a from the right.

$$\lim_{x \to a^+} f(x) = M$$

Polynomial Functions pages 100 - 101

A polynomial function is a function of the form

$$p(x) = a_n x^n + a_{n-1} x^{n-1} + \dots + a_2 x^2 + a_1 x + a_0 ,$$

where n is a whole number and a_n, a_{n-1}, ..., a_0 are real numbers.

Linear Function: $p(x) = mx + b$

Quadratic Function: $p(x) = a_2 x^2 + a_1 x + a_0$

One-Sided Limits of Polynomials: $\lim_{x \to a^-} p(x) = p(a)$ and $\lim_{x \to a^+} p(x) = p(a)$.

Unbounded Limits pages 102 - 105

+∞ Limits

If $f(x)$ increases without bound as x approaches a from the left (or from the right), then we say that $f(x)$ approaches positive infinity, $+\infty$.

$$\lim_{x \to a^-} f(x) = +\infty \quad \text{or} \quad \lim_{x \to a^+} f(x) = +\infty$$

−∞ Limits

If $f(x)$ decreases without bound as x approaches a from the left (or from the right), then we say that $f(x)$ approaches negative infinity, $-\infty$.

$$\lim_{x \to a^-} f(x) = -\infty \quad \text{or} \quad \lim_{x \to a^+} f(x) = -\infty$$

continued on next page ...

Section 2.1 Limits (continued)

Limit pages 105 - 108

For a function f and a real number a, if $\lim_{x\to a^-} f(x) = L$ and $\lim_{x\to a^+} f(x) = L$, where L is a real number, then

$$\lim_{x\to a} f(x) = L.$$

Existence of a Limit:

1. We say that the limit of a function f as x approaches a real number a **exists** if and only if $\lim_{x\to a} f(x) = L$, where L is a real number.
2. If $\lim_{x\to a^-} f(x) \neq \lim_{x\to a^+} f(x)$, then $\lim_{x\to a} f(x)$ **does not exist**.
3. If $\lim_{x\to a} f(x) = +\infty$ or $\lim_{x\to a} f(x) = -\infty$, then $\lim_{x\to a} f(x)$ **does not exist**.

Indeterminate Form and Algebraic Solutions pages 108 - 110

A limit expression of the type $\lim_{x\to a}\left(\frac{g(x)}{f(x)}\right)$ is called an indeterminate form of type $\frac{0}{0}$ if $\lim_{x\to a} g(x) = 0$ and $\lim_{x\to a} f(x) = 0$.

Steps for Solving:

1. Replace the quotient with a simplified expression after factoring.
2. Evaluate the new limit problem by substitution if the denominator does not have a limit of 0 as $x \to \infty$.

Section 2.2 Rates of Change and Derivatives

Average Rate of Change pages 121 - 123

If $y = f(x)$ then the ratio

$$\frac{\Delta y}{\Delta x} = \frac{y_2 - y_1}{x_2 - x_1} = \frac{f(x_2) - f(x_1)}{x_2 - x_1}$$

is called the average rate of change of y with respect to x as x changes from x_1 to x_2.

continued on next page ...

Section 2.2 Rates of Change and Derivatives (continued)

Instantaneous Rate of Change page 124 - 134

Slope page 128

The slope of a curve $f(x)$ at the point (a, b) is exactly the slope of the tangent line to the curve at that point. This is also called the **instantaneous rate of change**.

Derivative page 134

$f'(a)$ denotes the slope of a curve $y = f(x)$ at the point $(a, f(a))$. Also, $f'(a)$ is the instantaneous rate of change of $f(x)$ at $(a, f(a))$. The number $f'(a)$ is called the **derivative** of $f(x)$ at $x = a$.

Difference Quotient page 128

If $(x, f(x))$ is any fixed point on a curve and $(x + h, f(x + h))$ is another point on the curve, then the difference quotient is the slope of the secant line through these two points:

$$m_{\text{sec}} = \frac{f(x+h)-f(x)}{(x+h)-x} = \frac{f(x+h)-f(x)}{h}.$$

Instantaneous Velocity: The instantaneous rate of change of distance with respect to time. page 124

Using a Graphing Calculator to Find Slopes pages 134 - 136

Section 2.3 Slope and Rate of Change Considered Algebraically

Derivative pages 141 - 142

For any given function $y = f(x)$, the derivative of f at x is defined to be

$$f'(x) = \lim_{h \to 0}\left(\frac{f(x+h)-f(x)}{h}\right),$$

provided this limit exists.

Three-Step Method to Find a Derivative:

1. Form the ratio $\frac{f(x+h)-f(x)}{h}$, called the difference quotient.
2. Simplify the difference quotient algebraically.
3. Calculate $f'(x) = \lim_{h \to 0}\left(\frac{f(x+h)-f(x)}{h}\right)$, if it exists.

continued on next page ...

Section 2.3 Slope and Rate of Change Considered Algebraically (continued)

Non-Differentiable Function pages 143 - 144

A function $f(x)$ is not differentiable at $x = a$ if any of the following is true:

1. $f(x)$ is discontinuous at $x = a$.
2. The graph of $f(x)$ has a sharp corner at $x = a$.
3. The tangent line at $x = a$ is a vertical line.

Continuous Function pages 144 - 145

Continuous at a Point $x = a$

The function $f(x)$ is continuous at $x = a$ if and only if

$$\lim_{x \to a^-}\left(f(x)\right) = f(a) = \lim_{x \to a^+}\left(f(x)\right).$$

Continuous on an Interval

A function $f(x)$ is continuous on an interval (x_1, x_2) provided the function is continuous at $x = a$ for every number a with $x_1 < a < x_2$.

Informally, continuous functions can be drawn without lifting one's pencil from the paper. Every differentiable function is continuous, but not every continuous function is differentiable. Put another way, differentiability is a more restrictive condition – it requires that a function be both continuous, smooth, and without vertical tangents.

Power Rule (General Case) pages 144 - 148

For the function $f(x) = c \cdot x^r$ where c and r are any real numbers, $f'(x) = c \cdot r \cdot x^{r-1}$.

Basic Rule of a Linear Function: For $y = mx + b$, $y' = m$.

Two Special Cases:

1. If $f(x) = c$, then $f'(x) = 0$. Has slope 0 (constant functions are horizontal lines).
2. If $f(x) = x$, then $f'(x) = 1$. Has slope 1.

Constant Times a Function Rule page 150

If $f(x)$ is a differentiable function, c is a real constant, and $y = c \cdot f(x)$, then $\frac{dy}{dx} = c \cdot f'(x)$. In words, the derivative of a constant times a function is that constant times the derivative of the function.

continued on next page ...

Section 2.3 Slope and Rate of Change Considered Algebraically (continued)

Sum and Difference Rule pages 150 - 153

If $f(x)$ and $g(x)$ are differentiable functions and $y = f(x) \pm g(x)$, then $\frac{dy}{dx} = f'(x) \pm g'(x)$. In words, the derivative of a sum (or difference) of two functions is the sum (or difference) of their derivatives.

Applications pages 153 - 155

Tangent Lines
Velocity

Section 2.4 Applications: Marginal Analysis

Marginal Cost pages 161 - 163

Marginal cost, $C'(x)$, is the rate of change of total cost per unit change when the production level is x units.

Marginal Revenue page 163

Marginal revenue, $R'(x)$, is the rate of change of the total revenue per unit change in sales when the level of sales is x items.

Marginal Profit page 164 - 165

Marginal profit, $P'(x)$, is the rate of change of the profit per unit change in sales when x items are produced and sold.

Marginal Average Cost pages 165 - 166

Marginal average cost, $\overline{C}'(x)$, is the rate of change of the average cost per unit change in production $\left(\overline{C}(x) = \frac{C(x)}{x}\right)$. This is an approximation of the change in average cost when one more item is produced.

Marginal Propensity to Consume page 167

The rate of change of consumption functions is called the marginal propensity to consume.

Section 2.5 More About Limits and Continuity

Properties of Limits as $x \to a$ pages 173 - 177

Suppose that c is any constant, n is a positive integer, a is a real number, $\lim_{x \to a} f(x) = L_1$, and $\lim_{x \to a} g(x) = L_2$, where L_1 and L_2 are real numbers.

1. $\lim_{x \to a} c = c$
2. $\lim_{x \to a} x = a$
3. $\lim_{x \to a} x^n = a^n$
4. $\lim_{x \to a} [f(x) \pm g(x)] = \lim_{x \to a} f(x) \pm \lim_{x \to a} g(x) = L_1 \pm L_2$
5. $\lim_{x \to a} [f(x) \cdot g(x)] = \left[\lim_{x \to a} f(x)\right] \cdot \left[\lim_{x \to a} g(x)\right] = L_1 \cdot L_2$
6. $\lim_{x \to a} \left[\dfrac{f(x)}{g(x)}\right] = \dfrac{\lim_{x \to a} f(x)}{\lim_{x \to a} g(x)} = \dfrac{L_1}{L_2}$ where $L_2 \neq 0$
7. $\lim_{x \to a} [c \cdot f(x)] = c \cdot \lim_{x \to a} f(x) = c \cdot L_1$
8. $\lim_{x \to a} \sqrt[n]{f(x)} = \sqrt[n]{\lim_{x \to a} f(x)} = \sqrt[n]{L_1} = (L_1)^{1/n}$ where $(L_1)^{1/n}$ is defined.
9. If p is a polynomial function, then $\lim_{x \to a} p(x) = p(a)$.

Indeterminate Form, Undefined Limit, Piecewise Function

Summary of Limits for Rational Functions as $x \to \pm\infty$ page 181

Consider the function $\dfrac{a_n x^n + a_{n-1} x^{n-1} + ... + a_0}{b_m x^m + b_{m-1} x^{m-1} + ... + b_0}$, where $a_n \neq 0$ and $b_m \neq 0$.

Case 1: For $m = n$, $\lim_{x \to +\infty} f(x) = \dfrac{a_n}{b_m}$.

Case 2: For $m > n$, $\lim_{x \to +\infty} f(x) = 0$.

Case 3: For $m < n$, $\lim_{x \to +\infty} f(x) = +\infty$.

(Or $-\infty$ depending on the signs of a_n and b_m.)

continued on next page ...

Section 2.5 More About Limits and Continuity (continued)

Continuity pages 181 - 187

A function $y = f(x)$ is **continuous at $x = a$** if each of the following conditions is true:

1. $f(a)$ exists. ($f(a)$ is a real number.)
2. $\lim_{x \to a} f(x)$ exists. (The limit is a real number.)
3. $\lim_{x \to a} f(x) = f(a)$. (The limit is equal to the value of the function at $x = a$.)

If a function fails to satisfy any one of the three conditions of the definition, then it is not continuous at $x = a$ and is said to be **discontinuous at $x = a$.**

Step Function page 185
Greatest-Integer Function page 186
Dirichlet Function page 186

Chapter 2 Review

For a review of the topics and problems from Chapter 2, look at the following lessons from *Hawkes Learning Systems: Essential Calculus.*

Left and Right-Hand Limits
Limits
Average Rate of Change
Instantaneous Rate of Change and Interpreting Graphs
Definition of the Derivative and Power Rule
Slope and Rate of Change Considered Algebraically
Applications: Marginal Analysis
More About Limits
More About Continuity

Chapter 2 Test

In Exercises 1 – 4, determine the limits. In each case, make a table to support your answer.

1. $\lim_{x\to 3^-}\left(\dfrac{x^2-9}{x+3}\right)$

x	y

2. $\lim_{x\to 3^+}\left(\dfrac{x^2-9}{x}\right)$

x	y

3. $\lim_{x\to 3^+}\left(\dfrac{x^3-27}{x-3}\right)$

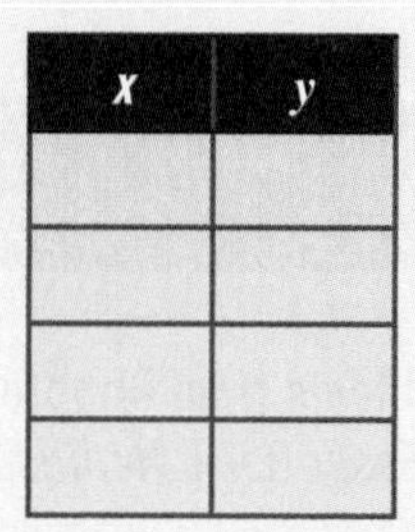

x	y

4. $\lim_{x\to 3^-}\left(\dfrac{x^3-45}{x-3}\right)$

x	y

In Exercises 5 – 8, find a formula for the derivative of the given function.

5. $f(x)=2-30x+16x^2$

6. $f(x)=\dfrac{2-30x+16x^2}{x}$

7. $f(x)=12x^{\frac{2}{3}}-30x^{\frac{1}{2}}+\dfrac{1}{x^5}$

8. $f(x)=\left(2-30x+16x^2\right)\left(x^2\right)$

Use the following information for Exercises 9 – 12.

Given $R(x)=\dfrac{100x}{b+x}$ *for* $0 < x < 80$. *The function* $R(x)$ *denotes the response of a frog to acetylcholine, a chemical compound that diminishes the force with which a muscle contracts. The input* x *represents the concentration of acetylcholine and* $R(x)$ *is a percentage of the maximum response to the drug.*
Suppose $b = 20$.

9. What is the response $R(x)$ if $x = 60$?

10. Sketch a graph of the function by using your graphing calculator. Show the window you used for your sketch.

11. What concentration of x produces a 60% response?

12. Interpret the meaning of $R'(20)$.

For Exercises 13 – 16, $y = f(x) = \frac{1}{3}x^3 - 2x^2$.

13. Sketch a graph using the window $[-4, 9]$ by $[-15, 15]$. Show the dimensions of your window on the drawing.

14. What is y when $x = 5$?

15. What is the slope of this tangent line when $x = 5$? Sketch a tangent line at the point $(5, f(5))$.

16. What is the smallest y-value between $x = 0$ and $x = 6$?

Use the following scenario for Exercises 17 – 19.

The Castle Co. makes and sells designer mugs to stores and colleges. Its total cost function C is given by $C(x) = 3000 + 0.20x + 0.0005x^2$ *for x mugs.*

Sketch a graph with $X_{min} = -1000$, $X_{max} = 5000$, $Y_{min} = -1000$, *and* $Y_{max} = 12{,}000$. *Draw a tangent line at* $x = 1000$.

17. What number is $C(0)$? Interpret its meaning.

18. What number is $C'(1000)$? Interpret its meaning.

19. What is the equation of the tangent line at $x = 1000$? (Use any method you know.)

For Exercises 20 – 23, use $y = f(x) = 2x - \sqrt{x}$.

20. Find and simplify an expression that represents the average rate of change of $y = f(x)$ between $x_1 = 4$ and $x_2 = 4 + h$.

21. Using your answer to Exercise 20, determine the slope if $h = 0.001$.

22. Determine a formula for $f'(x)$.

23. Using your answer to Exercise 22, calculate $f'(4)$.

For Exercises 24 – 26, use $f(x) = x^3 - 3x + 2$.

24. Find a formula for $\frac{dy}{dx}$.

25. Use your answer to Exercise 24 to determine the equation of the tangent line to f at $x = 2$. (Check your answer with a graphing utility.)

26. Determine the largest y-value that corresponds to a negative x-value.

For Exercises 27 – 28, use $S(t) = 2 + 4t - 3t^2$. *This equation gives the distance S in meters of a moving particle with respect to a given origin at time t seconds.*

27. How far from the origin is the particle initially (at $t = 0$ seconds)?

28. The particle passes through the origin at $t = 2$ seconds. What is its speed at that time?

For Exercises 29 – 30, use the following:

A local distributor of cell phones has determined that the monthly cost of maintaining an office is given by $C(x) = 8000 + 4x + 0.05x^2$. *Here x is the number of new cell phones issued each month. The cell phones offered all sell for an average of $135 each.*

29. Determine the revenue function $R(x)$.

30. What is the profit function $P(x)$?

Find the limits, if they exist, in Exercises 31 – 33.

31. $\lim_{x\to\infty}\left(\dfrac{4x-10000}{2x-2000}\right)$

32. $\lim_{x\to 0^-}\left(\dfrac{x^3+2x}{2x^2+x}\right)$

33. $\lim_{x\to 1^+}(F(x))$, where $F(x)=\begin{cases} 3, & x\le 1 \\ 4-x, & x>1 \end{cases}$

34. Population growth. The population growth of African Americans between the years 1900 and 2000 is approximated by the formula

$$P(x)=-\frac{8.813}{10^4}x^3+2.784x^2+0.3917x+8893.47,$$

according to the U.S. Census. Here x is the number of years since 1900 and $f(x)$ is the population in thousands. (***Source*:** U.S. Census Bureau, "Statistical Abstract of the United States: 2003," Mini Historical Statistics Figure No. HS-2, 3 – 5, (2003).)

a. What was the African American U.S. population in 1970?

b. What number is $P'(70)$ and what does it represent?

Algebraic Differentiation Rules

CHAPTER 3

Did You Know?

While Newton (1642 – 1727) was developing calculus, another individual, a man by the name of Gottfried Wilhelm von Leibniz (1646 – 1716), was busily discovering and recording many of the same principles. Both men's findings were publicized around the same time. This, unfortunately, caused Newton's followers to accuse Leibniz of theft and Leibniz's followers to accuse Newton of the same. Both groups were completely outraged at each other, including Newton and Leibniz, who viewed each other with bitterness until their deaths. Today, most scholars agree that they arrived at their conclusions independently rather than through thievery.

Both groups, however, would have been quite surprised to discover that a Greek scientist had used a system very similar to that of both Newton and Leibniz almost two thousand years before. Recent discoveries have led to the theory that Archimedes (287 – 212 B.C.) may have used a variety of calculus in the third century B.C. During the middle ages, a monk in need of manuscript paper tore pages out of Archimedes' text, "cleaned" the pages, and copied new text onto the paper. Now, scientists using modern equipment are trying to recover Archimedes' original notes.

The study of calculus helps us understand rates of change. The physical world around us is constantly changing. Many situations arise for which algebra or trigonometry are not sufficiently sophisticated to describe. Calculus, in contrast, exists to describe changes in the world. For example, with calculus, it is possible to examine the function $P(t) = -4.9t^2 - 45t + 130$ (an equation for a falling object) and find the object's velocity at any instant. Furthermore, calculus can use one rate of change, such as change in radius of an inflating balloon, to describe another related rate, like the volume of that balloon. Calculus is even instrumental to manufacturing processes; corporations can use differential calculus to maximize profits and minimize costs.

Before Isaac Newton and Wilhelm Leibniz were born, a general method for finding slopes of functions was unknown. Both Newton and Leibniz discovered such general methods. Combining their ideas with those developed in the 1700's, the calculus student of today is able to find formulas for the derivative function given almost any of the functions encountered in college mathematics.

Determination of algebraic formulas for many of these derivatives is the highlight of this chapter. We expand the list of rules of differentiation to include the Product Rule, the Quotient Rule, the Chain Rule, and the General Power Rule. Using these four rules as tools, we will be able to solve more complicated applications.

3.1 Product and Quotient Rules

Objectives

After completing this section, you will be able to:

1. *Use the Product Rule and its various forms to differentiate a function.*
2. *Find the derivative of a rational function using the Quotient Rule.*

Product Rule

Suppose that $f(x) = x^2 + 5$ and $g(x) = 2x - 5$. Now consider the function formed by the product of f and g:

$$y = f(x) \cdot g(x) = (x^2 + 5)(2x - 5).$$

Is the derivative of y equal to the product of the derivatives of f and g? Symbolically, is $\frac{dy}{dx} = f'(x) \cdot g'(x)$ a true statement? To answer this question, we first note $f'(x) \cdot g'(x) = 2x \cdot 2 = 4x$. Next, we multiply f and g to express y more simply.

$$\begin{aligned} y &= (x^2 + 5)(2x - 5) \\ &= 2x^3 - 5x^2 + 10x - 25 \end{aligned}$$

Then we differentiate term by term to obtain

$$\frac{dy}{dx} = 6x^2 - 10x + 10,$$

which is a result considerably different from $4x$. Thus we see that the derivative of a product is **not** the product of the derivatives. Nevertheless, simply multiplying the factors of a function and then differentiating the result term-by-term is not always possible. Therefore, we need a method to deal with derivatives of products. Hence we introduce the **Product Rule**.

Product Rule

If f(x) and g(x) are differentiable functions and $y = f(x)\cdot g(x)$, *then*

$$\frac{dy}{dx} = f(x)\cdot g'(x) + g(x)\cdot f'(x).$$

In words, the derivative of the product of two functions is equal to the first function times the derivative of the second plus the second function times the derivative of the first.

Applying the Product Rule to $y = (x^2+5)(2x-5)$ with $f(x) = x^2+5$ and $g(x) = 2x-5$, we get

$\frac{dy}{dx}$	=	$f(x)$	$\cdot$	$g'(x)$	+	$g(x)$	$\cdot$	$f'(x)$
		↑ First Function ↓		↑ Derivative of Second ↓		↑ Second Function ↓		↑ Derivative of First ↓
$\frac{dy}{dx}$	=	(x^2+5)	$\cdot$	(2)	+	$(2x-5)$	$\cdot$	$2x$

$$\frac{dy}{dx} = 2(x^2+5) + 2x(2x-5) = 2x^2+10+4x^2-10x$$

$$= 6x^2 - 10x + 10.$$

This is the same result we obtained earlier by multiplying $f(x)\cdot g(x)$ and then differentiating. The Product Rule can be written in several forms by incorporating the various notations for derivatives.

Other Forms of the Product Rule

If f(x) and g(x) are differentiable functions and $y = f(x)\cdot g(x)$, *then*

1. $\frac{dy}{dx} = f(x)\cdot\frac{d}{dx}[g(x)] + g(x)\cdot\frac{d}{dx}[f(x)]$,
2. $\frac{d}{dx}[f(x)\cdot g(x)] = f(x)\cdot\frac{d}{dx}[g(x)] + g(x)\cdot\frac{d}{dx}[f(x)]$,
3. $y' = f(x)\cdot g'(x) + g(x)\cdot f'(x)$,
4. $D_x[f(x)\cdot g(x)] = f(x)\cdot D_x[g(x)] + g(x)\cdot D_x[f(x)]$, *and*
5. $y' = f\cdot g' + g\cdot f'$.

Example 1: Using the Product Rule

Use the Product Rule to find $\frac{dy}{dx}$ given $y=(x^2+3x-1)(x^3-8x)$.

Solution: $\frac{dy}{dx}=(x^2+3x-1)\cdot\frac{d}{dx}[x^3-8x]+(x^3-8x)\cdot\frac{d}{dx}[x^2+3x-1]$ Apply the Product Rule.

$$=(x^2+3x-1)(3x^2-8)+(x^3-8x)(2x+3)$$

$$=3x^4+9x^3-3x^2-8x^2-24x+8+2x^4-16x^2+3x^3-24x$$

$$=5x^4+12x^3-27x^2-48x+8$$

In Example 1, $\frac{dy}{dx}=(x^2+3x-1)(3x^2-8)+(x^3-8x)(2x+3)$ is a perfectly satisfactory answer. Removing parentheses in an algebraic expression does not necessarily improve appearance or understanding. A rule of thumb we like is to remove parentheses if the end result will be a polynomial of degree 2 or less.

Example 2: Using the Product Rule

Use the Product Rule to find the derivative of the function $f(x)=(\sqrt{x}+x^{-1})(x^2+1)$.

Solution: In this case we have used $f(x)$ to represent the entire function rather than one of the functions in the product. Before differentiating, convert radicals to exponential form.

$$f'(x)=\left(x^{\frac{1}{2}}+x^{-1}\right)\cdot\frac{d}{dx}[x^2+1]+(x^2+1)\cdot\frac{d}{dx}\left[x^{\frac{1}{2}}+x^{-1}\right]$$ Apply the Product Rule.

$$=\left(x^{\frac{1}{2}}+x^{-1}\right)(2x)+(x^2+1)\left(\frac{1}{2}x^{-\frac{1}{2}}-x^{-2}\right)$$

Note: You may choose to use radicals and fractions in your answer:

$$f'(x)=\left(\sqrt{x}+\frac{1}{x}\right)(2x)+(x^2+1)\left(\frac{1}{2\sqrt{x}}-\frac{1}{x^2}\right).$$

Remember, however, that the fractional and negative exponents make differentiating easier.

Quotient Rule

Suppose that $f(x)=x^2$ and $g(x)=5x+1$. Then the function formed by the quotient of these two functions is

$$y=\frac{f(x)}{g(x)}=\frac{x^2}{5x+1}.$$

We can use the following **Quotient Rule** to find $\frac{dy}{dx}$.

Quotient Rule

If f(x) and g(x) are differentiable functions and $y=\frac{f(x)}{g(x)}$, *then*

$$\frac{dy}{dx}=\frac{g(x)\cdot f'(x)-f(x)\cdot g'(x)}{(g(x))^2}.$$

In words, the derivative of a quotient is the denominator times the derivative of the numerator minus the numerator times the derivative of the denominator, all divided by the square of the denominator.

Example 3: Using the Quotient Rule

Use the Quotient Rule to find the derivative of the function $y=\frac{x^2}{5x+1}$.

Solution: Here $f(x)=x^2$ and $g(x)=5x+1$. So $f'(x) = 2x$ and $g'(x) = 5$. Substituting these into the Quotient Rule, we get:

$$\frac{dy}{dx}=\frac{(5x+1)\cdot(2x)-(x^2)\cdot(5)}{(5x+1)^2}$$

$$=\frac{10x^2+2x-5x^2}{(5x+1)^2}=\frac{5x^2+2x}{(5x+1)^2}.$$

In Example 3, since the numerator will be a degree 2 polynomial, removing parentheses and simplifying is appropriate. The student will never find it necessary to square a binomial expression in the denominator.

NOTES

WARNING!
Because of the minus sign in the Quotient Rule, the order of terms in the numerator is critical.

In Example 4 we show two techniques for finding the derivative of the same function. Many times the choice of technique is based on what our experience and practice tell us is the easiest to use in a particular situation.

Example 4: Using the Quotient Rule

For $f(x)=\dfrac{x^2-x+1}{x^2}$, find $f'(x)$ by **a.** using the Quotient Rule and **b.** simplifying algebraically and using negative exponents.

Solutions: a. Here $f(x)$ represents the entire quotient. In this case, we can think of the statement of the Quotient Rule in terms of numerator and denominator rather than in terms of $f(x)$ and $g(x)$.

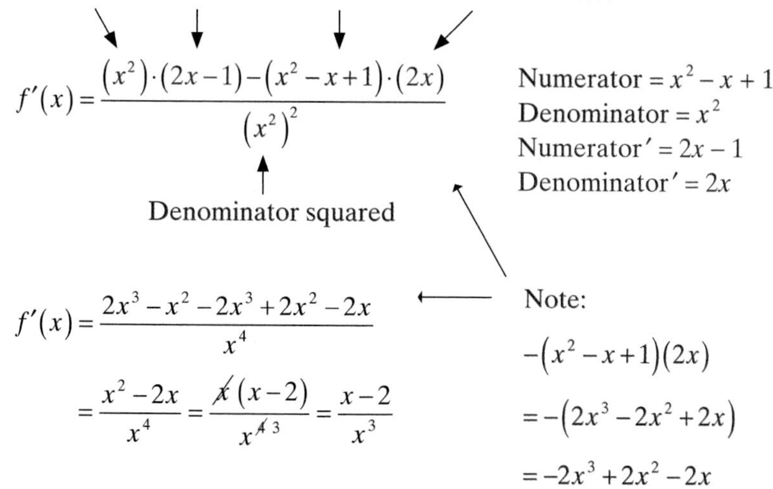

b. Since the denominator is a single term, we can simplify algebraically as follows:

$$f(x)=\frac{x^2-x+1}{x^2}=\frac{x^2}{x^2}-\frac{x}{x^2}+\frac{1}{x^2}$$
$$=1-\frac{1}{x}+\frac{1}{x^2}=1-x^{-1}+x^{-2}.$$

Therefore,

$$f'(x)=0-(-1)x^{-1-1}+(-2)x^{-2-1}$$
$$=x^{-2}-2x^{-3}.$$

Use the Power Rule on each of the terms.

Note that the answers in parts **a.** and **b.** are different forms of the same answer:

$$x^{-2}-2x^{-3}=\frac{1}{x^2}-\frac{2}{x^3}=\frac{x-2}{x^3}.$$

In Example 4, without experience, the student might not realize the enormous simplification which will result. The first expression for f' in solution **a.** is correct; but, by inspection, one can notice two things. First, the cubic terms will subtract out; and second, each term has x as a factor so there will be a cancellation.

Example 5: Growth of Bacteria

Several years of study have shown that t hours after a bacterium is introduced to a particular culture, the number of bacteria is given by $N(t)=\dfrac{t^2+t}{\sqrt{t}+1}$. Find the rate of growth of the bacteria after 4 hours.

Solution: $N(t)=\dfrac{t^2+t}{\sqrt{t}+1}=\dfrac{t^2+t}{t^{\frac{1}{2}}+1}$ Rewrite to remove the square root sign.

$$N'(t)=\frac{\left(t^{\frac{1}{2}}+1\right)\frac{d}{dt}\left[t^2+t\right]-\left(t^2+t\right)\frac{d}{dt}\left[t^{\frac{1}{2}}+1\right]}{\left(t^{\frac{1}{2}}+1\right)^2}$$

Use the Quotient Rule.

$$=\frac{\left(t^{\frac{1}{2}}+1\right)(2t+1)-\left(t^2+t\right)\left(\frac{1}{2}t^{-\frac{1}{2}}\right)}{\left(t^{\frac{1}{2}}+1\right)^2}$$

$$N'(4)=\frac{\left(4^{\frac{1}{2}}+1\right)(2\cdot 4+1)-\left(4^2+4\right)\left(\frac{1}{2}\cdot 4^{-\frac{1}{2}}\right)}{\left(4^{\frac{1}{2}}+1\right)^2}$$

Substitute $t = 4$ into the formula for the derivative.

$$=\frac{(2+1)(9)-(20)\left(\frac{1}{2}\cdot\frac{1}{2}\right)}{(2+1)^2}=\frac{(3)(9)-(20)\left(\frac{1}{4}\right)}{(3)^2}$$

$$=\frac{27-5}{9}=\frac{22}{9}$$

The rate of growth when $t = 4$ is $\frac{22}{9}$ bacteria per hour.

Note that we did not bother to simplify the derivative $N'(t)$ because we were interested in its value at only one time, $t = 4$, rather than a simplified formula. Also, no cancellation is apparent and the rule of thumb suggests this form is as good as any other.

Example 6: Elementary Operations on Functions

You are given that f and g are differentiable functions and that $f(5) = 10$, $g(5) = -3$, $f'(5) = 1$, $g'(5) = 8$. Determine $h'(5)$ given that:

a. $h(x) = \dfrac{f(x)}{g(x)}$ **b.** $h(x) = \left(f(x) + 3x - 1\right)g(x)$

Solutions: a. $h'(x) = \dfrac{g(x)f'(x) - f(x)g'(x)}{\left(g(x)\right)^2}$, so

$$h'(5) = \frac{(-3)(1) - 10(8)}{(-3)^2} = \frac{-3-80}{9} = \frac{-83}{9}.$$

b. $h'(x) = \left(f'(x) + 3\right)g(x) + g'(x)\left[f(x) + 3x - 1\right]$, so

$$h'(5) = (1+3)(-3) + 8[10 + 15 - 1]$$
$$= 4(-3) + 8(24) = -12 + 192 = 180$$

We end the section with the following remarks about the possible algebraic forms of answers. The appropriateness of these remarks will be apparent in the homework as the student gains mastery of the techniques of differentiation and as the related algebraic expressions become more complicated.

NOTES

CAUTION!

The algebraic form of the derivative of a function depends on which rule of differentiation is applied, and there may be several correct forms.

For example, if $f(x) = \dfrac{x^3 + 3x + 1}{(2x+5)}$, then we will find (in this section) that the following two forms for $f'(x)$ are correct.

1. $f'(x) = \dfrac{(2x+5)\left(3x^2+3\right) - \left(x^3 + 3x + 1\right)(2)}{(2x+5)^2}$

2. $f'(x) = \left(x^3 + 3x + 1\right)(-1)(2x+5)^{-2}(2) + (2x+5)^{-1}\left(3x^2 + 3\right)$

Both of these forms can be simplified to the form

3. $f'(x) = \dfrac{4x^3 + 15x^2 + 13}{(2x+5)^2}.$

If the rules of differentiation have been followed, then an answer may be correct even though it does not "look like" the answer given in the Answers section at the end of the text. Be sure to check with your instructor to see how much algebraic simplification is expected and whether or not one form is preferred over another.

3.1 Exercises

In Exercises 1 – 5, find f′(x) two ways: (1) multiply the factors first, then find the derivative, and (2) use the Product Rule.

1. $f(x)=x^2\left(1+3x-2x^2\right)$
2. $f(x)=(x+3)(x-1)$
3. $f(x)=x^{\frac{1}{2}}\left(1+3x^2\right)$
4. $f(x)=x^{\frac{1}{2}}\left(1+x^{\frac{1}{2}}-x^{\frac{3}{2}}\right)$
5. $f(x)=(2x+3)(2x-3)$

In Exercises 6 – 10, find g′(x) two ways: (1) divide the factors first, then find the derivative; and (2) use the Quotient Rule and simplify the answer.

6. $g(x)=\dfrac{1+5x+x^2}{x}$
7. $g(x)=\dfrac{2+\sqrt{x}}{\sqrt{x}}$
8. $g(x)=\dfrac{x^2+1}{x^5}$
9. $g(x)=\dfrac{30x^2-10x^6}{5x}$
10. $g(x)=\dfrac{3x^{\frac{1}{2}}-5x^{\frac{3}{2}}+7x^{\frac{5}{2}}-9x^{\frac{7}{2}}}{x^{\frac{1}{2}}}$

In Exercises 11 – 34, use the Product Rule or Quotient Rule to find the derivative of each of the functions. Simplify your answers.

11. $f(x)=x^3\left(x^2+5\right)$
12. $f(x)=x^5\left(2x-x^3\right)$
13. $f(t)=t^{\frac{1}{2}}(4t+3)$
14. $f(t)=t^{\frac{2}{3}}\left(4t^2+1\right)$
15. $y=x^2\left(\sqrt{x}+\dfrac{1}{\sqrt{x}}\right)$
16. $y=x^{-2}\left(3x+x^{\frac{1}{3}}\right)$
17. $g(u)=\left(2u^2+3\right)(5-3u)$
18. $g(u)=\left(3u^2-8\right)\left(u^2+u\right)$
19. $g(t)=\left(5+\dfrac{1}{t}\right)\left(t^2+\dfrac{1}{5}\right)$
20. $f(t)=\left(1-\dfrac{3}{t^2}\right)\left(2t^2+t-1\right)$
21. $f(x)=\dfrac{3x}{x+6}$
22. $f(x)=\dfrac{7x^2}{2x-1}$
23. $f(x)=\dfrac{x+8}{x-7}$
24. $f(x)=\dfrac{x^2+2x-3}{x+2}$
25. $y=\dfrac{x^3-5}{x^2+1}$
26. $y=\dfrac{2x^2+3x}{x^3+6}$
27. $g(x)=\dfrac{\sqrt{x}}{x+9}$
28. $g(x)=\dfrac{6\sqrt{x}}{3x-4}$
29. $f(u)=\dfrac{u^2}{\sqrt{u}+1}$
30. $f(u)=\dfrac{7}{1-\sqrt[3]{u}}$
31. $f(t)=\dfrac{4-\sqrt{t}}{t^2+3}$
32. $f(t)=\dfrac{3-t}{4-5\sqrt{t}}$

33. $f(x)=\dfrac{x^2-5x}{1+2\sqrt[3]{x}}$

34. $f(x)=\dfrac{x\left(1+3\sqrt{x}\right)}{\sqrt{x}+6}$

In Exercises 35 – 44, you are given that f(x) and g(x) are differentiable functions and that f(2) = 3, f′(2) = –1, g(2) = –11, and g′(2) = 6. In each exercise, find the value of h′(2).

35. $h(x)=x\cdot f(x)$

36. $h(x)=\dfrac{f(x)}{2x+1}$

37. $h(x)=\dfrac{f(x)+3x}{f(x)-3x}$

38. $h(x)=\dfrac{g(x)}{f(x)}$

39. $h(x)=\dfrac{g(x)}{3x+10}$

40. $h(x)=(3x+5)\cdot f(x)$

41. $h(x)=\dfrac{16x+1}{f(x)-11x+1}$

42. $h(x)=f(x)\cdot g(x)$

43. $h(x)=\dfrac{f(x)}{g(x)}$

44. $h(x)=g(x)\cdot(1+3x)$

In Exercises 45 – 50, find the equation of the line tangent to the graph f(x) at the (x, y) coordinate indicated.

45. $f(x)=\left(x+5x^{\frac{1}{2}}\right)\left(6x^2-12x+2\right)$; (4, 700)

46. $f(x)=\dfrac{\left(11x^2-3x+2\right)}{x^2+1}$; (1, 5)

47. $f(x)=\dfrac{2-3x}{5+2x}$; (0, 0.4)

48. $f(x)=\left(x^5-5\right)\left(x^3-x-1\right)$; (0, 5)

49. $f(x)=\dfrac{20}{17x+3}$; (1, 1)

50. $f(x)=\dfrac{\sqrt{x}+2}{x^2-1}$; $\left(9,\dfrac{1}{16}\right)$

51. Given $f(x)=(1-x)\left(16-x^2\right)$, find the (x, y)-coordinates on the graph where the tangent line is horizontal.

52. Given $g(x)=(x-10)\left(x^2+2x+1\right)$, find any (x, y)-coordinates on g(x) for which the tangent line is horizontal.

53. Find any point or points on the graph of $y=(x-5)(x+10)$ so that the slope equals 25. Sketch a graph of y and the tangent line or lines.

54. Find any point or points on the graph of $G(x)=(2x+1)(x-3)$ so that the slope is –20. Sketch a graph of G and the tangent line or lines.

55. Sketch a graph of $F(x)=\dfrac{30x}{2x^2+5}$ on the x-interval [–5, 10]. Determine the (x, y)-coordinates of any point with a horizontal tangent line, and sketch this (or these) horizontal tangent(s). Round to the nearest hundredth.

56. Bacterial growth. It is estimated the population of a bacterial culture after t hours is approximately $N(t)=\dfrac{t^2-2t}{3\sqrt{t}+2}$, where $N(t)$ is in thousands and $2\le t\le 10$. Find the rate of growth after 4 hours.

57. Marginal revenue. The demand function for a particular item is given by $D(x)=\dfrac{115}{3x+1}$. Find the marginal revenue when $x = 3$.

58. Marginal profit. The profit from the sale of x items is given by $P(x)=(2-0.5x)(0.5x-5)$, where $P(x)$ is in hundreds of dollars and $2\le x\le 10$. Find the marginal profit when $x = 5$.

59. Marginal cost. The cost of producing x items of a product is given by $C(x)=(0.1x+100)(0.1x+20)-600$. Find the marginal cost when $x = 60$.

60. Velocity of a particle. A particle is moving slowly along a line. Its position after t seconds is $S(t)=\dfrac{t}{t^2+4}$ feet. Find the velocity when the particle has been moving for 3 seconds.

61. Population growth. It is estimated that t years from now the population of a city will be $P(t)=(0.6t-7)(0.5t+6)+85$ in thousands. How fast will the population be growing in 10 years?

Hawkes Learning Systems: Essential Calculus

Product and Quotient Rules

3.2 The Chain Rule and the General Power Rule

Objectives

After completing this section, you will be able to:

1. *Use the Chain Rule to find the derivative of a function.*
2. *Use the General Power Rule to find the derivative of a function.*
3. *Apply several rules in combination to find the derivative of a complex function.*

The Chain Rule

Suppose that at a certain production level, production costs are \$18 per unit and 40 units are produced each hour. How fast is the total cost changing per hour? The answer is found by multiplying:

$$\$18 \text{ per unit} \cdot 40 \text{ units per hour} = \$720 \text{ per hour.}$$

The total cost is changing at a rate of \$720 per hour: each hour 40 more units are produced and since each unit costs \$18 to produce, the total cost per hour is \$720. Notice that, in order to get this rate of change of cost, we multiplied the two other rates.

In this example the total cost is a function of the number of units produced, $y = C(x)$, and the number of units produced is a function of time, $x = g(t)$. We find that

$$\frac{dy}{dx} \cdot \frac{dx}{dt} = \frac{dy}{dt}.$$

$$\uparrow \qquad\qquad \uparrow \qquad\qquad \uparrow$$

$$\begin{pmatrix}\text{rate of change}\\ \text{of cost with respect}\\ \text{to unit production}\end{pmatrix} \cdot \begin{pmatrix}\text{rate of change}\\ \text{of unit production}\\ \text{with respect to time}\end{pmatrix} = \begin{pmatrix}\text{rate of change}\\ \text{of cost with}\\ \text{respect to time}\end{pmatrix}$$

$$\uparrow \qquad\qquad \uparrow \qquad\qquad \uparrow$$

$$\begin{pmatrix}\$18 \text{ increase}\\ \text{per additional}\\ \text{unit produced}\end{pmatrix} \cdot \begin{pmatrix}\text{40 additional}\\ \text{units produced}\\ \text{per hour}\end{pmatrix} = \begin{pmatrix}\$720 \text{ increase}\\ \text{per hour}\end{pmatrix}$$

These ideas lead to the **Chain Rule**.

The Chain Rule

If $y = f(u)$ and $u = g(x)$, then

$$\frac{dy}{dx} = \frac{dy}{du} \cdot \frac{du}{dx},$$

provided that $\frac{dy}{du}$ and $\frac{du}{dx}$ both exist.

Example 1: Using the Chain Rule

Suppose that $y = u + \sqrt{u}$ and $u = x^3 + 17$. Use the Chain Rule to find $\frac{dy}{dx}$. Then evaluate $\frac{dy}{dx}$ at $x = 2$. This evaluation is denoted more briefly by $\left.\frac{dy}{dx}\right|_{x=2}$.

Solution: $y = u + \sqrt{u}$ $\qquad u = x^3 + 17$

$= u + u^{\frac{1}{2}}$ — Rewrite y using exponents.

$$\frac{dy}{du} = 1 + \frac{1}{2}u^{-\frac{1}{2}} = 1 + \frac{1}{2\sqrt{u}} \qquad \frac{du}{dx} = 3x^2$$

Find the derivatives of each function by using the Power Rule from Section 2.3.

So, by the Chain Rule,

$$\frac{dy}{dx} = \frac{dy}{du} \cdot \frac{du}{dx} = \left(1 + \frac{1}{2\sqrt{u}}\right) \cdot 3x^2.$$

To evaluate this derivative at $x = 2$, we must first substitute it into the equation for u, and then substitute u and x into the equation of the derivative:

$$u = (2)^3 + 17 = 25$$

Substitute $x = 2$ into u.

$$\left.\frac{dy}{dx}\right|_{x=2} = \left(1 + \frac{1}{2\sqrt{25}}\right) \cdot 3(2)^2 = \left(1 + \frac{1}{10}\right) \cdot 12 = \frac{11}{\cancel{10}_5} \cdot \frac{\cancel{12}^6}{1} = \frac{66}{5}$$

Substitute $u = 25$ and $x = 2$ into the formula for the derivative.

Example 2: Ripples on a Pond

A pebble is tossed into a still pond, and concentric circles are formed on the pond's surface. The area of each circle depends on the radius of the circle, and each radius depends on the elapsed time. If the radius is changing at the rate of $\frac{1}{2}$ ft./sec, how fast is the area of the circle growing when the radius is 3 feet?

Solution: Recall that the area of a circle is given by the formula $A = \pi r^2$, where $\pi \approx 3.14$. Differentiating A with respect to r gives

$$\frac{dA}{dr} = 2\pi r.$$

Since the radius is changing at the rate of $\frac{1}{2}$ ft./sec, we have

$$\frac{dr}{dt} = \frac{1}{2} \text{ ft./sec}.$$

Thus, by the Chain Rule,

$$\frac{dA}{dt} = \frac{dA}{dr} \cdot \frac{dr}{dt} = 2\pi r \cdot \frac{1}{2} = \pi r.$$

Therefore, when $r = 3$,

$$\left.\frac{dA}{dt}\right|_{r=3} = \pi \cdot 3 = 3\pi \text{ ft.}^2\text{/sec}.$$

The area of the circle is growing at the rate of 3π ft.^{2}/sec $\left(\approx 9.42 \text{ ft.}^2\text{/sec}\right)$.

In Example 2, as with so many mathematical models of real life situations, there are certain realistic limitations on the validity of the model. In this example, we know (even though it is not explicitly stated) that r is restricted by the size of the pond. Furthermore, if the pebble is small, the circle may dissipate even before the radius becomes 3 feet. In business, such practical considerations can affect a model for profit, revenue, or cost on a yearly, monthly, or weekly basis.

Now we will look at the Chain Rule from the point of view of the composition of functions (see Section 1.6). For example, if

$$y = f(u) = \sqrt{u} \quad \text{and} \quad u = g(x) = x^2 + 9,$$

then

$$y = f(g(x)) = \sqrt{g(x)} = \sqrt{x^2 + 9}.$$

By the Chain Rule,
$$\frac{dy}{dx}=\frac{dy}{du}\cdot\frac{du}{dx}=f'(u)\cdot g'(x)=\frac{1}{2}u^{-\frac{1}{2}}\cdot 2x.$$

Substituting $u = g(x)$, we have
$$\frac{dy}{dx}=f'(g(x))\cdot g'(x)=\frac{1}{2}(x^2+9)^{-\frac{1}{2}}\cdot 2x.$$

This leads to a second form of the Chain Rule.

The Chain Rule (Second Form)

If $y=f(g(x))$, *then*
$$\frac{dy}{dx}=f'(g(x))\cdot g'(x),$$
provided that $f'(g(x))$ *and* $g'(x)$ *both exist.*

Example 3: Second Form of the Chain Rule

Use the second form of the Chain Rule to find $\frac{dy}{dx}$ if $y=(x^2+3x-7)^{\frac{5}{2}}$.

Solution: Let $g(x)=x^2+3x-7$. Then we can write
$$y=f(g(x))=[g(x)]^{\frac{5}{2}}.$$

Using the second form of the Chain Rule, we get

$$\frac{dy}{dx}=f'(g(x)).g'(x)$$
$$=\frac{5}{2}[g(x)]^{\frac{3}{2}}(2x+3)$$
Substitute $f'(g(x))=\frac{5}{2}[g(x)]^{\frac{3}{2}}$ and $g'(x)=2x+3$ into the formula for the second form of the Chain Rule.

$$=\frac{5}{2}(x^2+3x-7)^{\frac{3}{2}}(2x+3).$$
Substitute $g(x)=x^2+3x-7$.

Example 4: Second Form of the Chain Rule

Determine $h'(5)$ if $h(x)=(25-2x+x^2)^3$.

Solution: Set $h(x)=f(g(x))$ where $g(x)=25-2x+x^2$. Then
$$h'(x)=f'(g(x))\cdot g'(x)$$
$$=3(g(x))^2\cdot g'(x)=3(25-2x+x^2)^2(-2+2x)$$

continued on next page ...

Now $x = 5$, so

$$h'(5) = 3\left(25 - 2(5) + 5^2\right)^2 \left(-2 + 2(5)\right)$$
$$= 3(40)^2 (8) = 38400$$

Note:
$h'(5) = f'\left(g(5)\right) \cdot g'(5).$

The General Power Rule

In Example 3 and Example 4, the Chain Rule is applied to a function in the form $f\left(g(x)\right) = \left[g(x)\right]^n$. That is, the function f is a power function, and $g(x)$ is raised to a power. At this time, we state the **General Power Rule**, which is the special case of the Chain Rule where the function f is a power function.

The General Power Rule

If $y = \left[g(x)\right]^n$, *then*

$$\frac{dy}{dx} = n\left[g(x)\right]^{n-1} \cdot g'(x),$$

provided that $g'(x)$ *exists.*

Another form of the General Power Rule is

$$\frac{dy}{dx} = n\left[g(x)\right]^{n-1} \cdot \frac{d}{dx}\left(g(x)\right).$$

NOTES

When using the General Power Rule, remember the order of steps:

1. First apply the Power Rule to $\left[g(x)\right]^n$.
2. Then multiply by the derivative of $g(x)$.

Example 5: The General Power Rule

Using the General Power Rule, find the derivative of each of the following functions.

a. $y = \left(x^2 + 8x\right)^{10}$

Solution: Here $g(x) = x^2 + 8x$ and $g'(x) = 2x + 8$. Thus

$$\frac{dy}{dx} = n\left[g(x)\right]^{n-1} \cdot g'(x)$$
$$= 10\left(x^2 + 8x\right)^9 \cdot (2x + 8).$$

Substitute the following values into the General Power Rule:
$n = 10,$
$g(x) = x^2 + 8x,$
$n - 1 = 9,$
$g'(x) = 2x + 8.$

b. $y = 6\sqrt[3]{2x+5}$

Solution: Rewrite the radical with a fractional exponent; then apply the General Power Rule.

$$y = 6\sqrt[3]{2x+5} = 6(2x+5)^{\frac{1}{3}}$$

Here $g(x) = 2x+5$ and $g'(x) = 2$. So,

$$\frac{dy}{dx} = n\left[g(x)\right]^{n-1} \cdot g'(x)$$

$$= 6 \cdot \frac{1}{3}(2x+5)^{-\frac{2}{3}} \cdot 2$$

$$= 4(2x+5)^{-\frac{2}{3}} \quad \text{or} \quad \frac{4}{(2x+5)^{\frac{2}{3}}}.$$

Substitute the following values into the General Power Rule:

$c = 6$, $n = \frac{1}{3}$,

$g(x) = 2x + 5$, $n - 1 = -\frac{2}{3}$,

$g'(x) = 2$.

Example 6: The General Power Rule

Use the General Power Rule to find $f'(u)$ if $f(u) = \frac{1}{\sqrt{7-u}}$.

Solution: $f(u) = \frac{1}{\sqrt{7-u}} = (7-u)^{-\frac{1}{2}}$ Rewrite $f(u)$ with exponents.

Here $g(u) = 7 - u$ and $g'(u) = -1$. Thus

$$f'(u) = n\left[g(u)\right]^{n-1} \cdot g'(u)$$

$$= -\frac{1}{2}(7-u)^{-\frac{3}{2}} \cdot (-1)$$

$$= \frac{1}{2}(7-u)^{-\frac{3}{2}} \quad \text{or} \quad \frac{1}{2(7-u)^{\frac{3}{2}}}.$$

Substitute the following values into the General Power Rule:

$n = -\frac{1}{2}$,

$g(u) = 7 - u$,

$n - 1 = -\frac{3}{2}$,

$g'(u) = -1$.

Using the Rules in Combination with Each Other

The following examples use a variety of notations and illustrate how the differentiation rules (Sum and Difference Rule, Product Rule, Quotient Rule, General Power Rule, and so on) can be used in combination in one problem. Try to follow each example through, step by step. By careful analysis of these examples, you will be able to work similar problems in the exercises.

Example 7: Using Multiple Rules

Find $\frac{dy}{dx}$ if $y = (x^2+1)\sqrt{2x-3}$.

Solution: Treat y as the product of two functions, $f(x) = (x^2+1)$ and $g(x) = (2x-3)^{\frac{1}{2}}$, and use the Product Rule. In using the Product Rule, differentiate $(2x-3)^{\frac{1}{2}}$ by using the General Power Rule.

continued on next page ...

$f(x)$ ↓ $g'(x)$ ↓ $g(x)$ ↓ $f'(x)$ ↓

$$\frac{dy}{dx}=\left(x^2+1\right)\cdot\frac{d}{dx}\left[(2x-3)^{\frac{1}{2}}\right]+(2x-3)^{\frac{1}{2}}\cdot\frac{d}{dx}\left[x^2+1\right]$$ Product Rule

$$\frac{dy}{dx}=\left(x^2+1\right)\cdot\frac{1}{2}(2x-3)^{-\frac{1}{2}}(2)+(2x-3)^{\frac{1}{2}}(2x)$$

$$=\left(x^2+1\right)(2x-3)^{-\frac{1}{2}}+(2x-3)^{\frac{1}{2}}(2x)$$

$$=(2x-3)^{-\frac{1}{2}}\left[\left(x^2+1\right)+(2x-3)(2x)\right]$$

$$=(2x-3)^{-\frac{1}{2}}\left(x^2+1+4x^2-6x\right)$$

$$=(2x-3)^{-\frac{1}{2}}\left(5x^2-6x+1\right)$$

To find $g'(x)$, substitute the following values into the General Power Rule:

$g(x)=h(x)^{\frac{1}{2}}$

$n=\frac{1}{2},\quad h(x)=2x-3,$

$n-1=-\frac{1}{2},\quad h'(x)=2.$

Also substitute $f'(x)=2x$.

Factor out $(2x-3)^{-\frac{1}{2}}$.

Simplify.

Using radical notation we have

$$\frac{dy}{dx}=\frac{5x^2-6x+1}{\sqrt{2x-3}}.$$

In Example 7 the useful factoring of the radical term illustrates the simplification of any derivative expression when y is a product of a polynomial and a radical. If the expression in the radical is linear, and the other polynomial is of low degree, then the simplification should be made.

Example 8: Using Multiple Rules

Find $D_x\left[\sqrt[3]{\frac{2x-7}{3x+4}}\right]$.

Solution: Apply the General Power Rule and the Quotient Rule.

$$D_x\left[\sqrt[3]{\frac{2x-7}{3x+4}}\right]=D_x\left[\left(\frac{2x-7}{3x+4}\right)^{\frac{1}{3}}\right]$$ Rewrite using exponents.

$$=\frac{1}{3}\left(\frac{2x-7}{3x+4}\right)^{-\frac{2}{3}}\cdot D_x\left[\frac{2x-7}{3x+4}\right]$$ Apply the General Power Rule.

$$=\frac{1}{3}\left(\frac{2x-7}{3x+4}\right)^{-\frac{2}{3}}\left(\frac{(3x+4)\cdot 2-(2x-7)\cdot 3}{(3x+4)^2}\right)$$ Use the Quotient Rule to find $D_x\left[\frac{2x-7}{3x+4}\right]$.

$$= \frac{1}{3}\left(\frac{2x-7}{3x+4}\right)^{-\frac{2}{3}}\left(\frac{6x+8-6x+21}{(3x+4)^2}\right)$$

$$= \frac{1}{3}\left(\frac{2x-7}{3x+4}\right)^{-\frac{2}{3}}\left(\frac{29}{(3x+4)^2}\right)$$ Simplify.

Example 9: Air Pollution

Suppose that the average measure of air pollution in Sootville is given by the formula

$$P(t) = \frac{t^2}{\sqrt{t^2+36}},$$

where t is time in years measured from 1980. How fast was this measure changing in 1988?

Solution: Find the derivative of $P(t)$ by using the Quotient Rule and the General Power Rule.

$$P(t) = \frac{t^2}{\left(t^2+36\right)^{\frac{1}{2}}}$$ Rewrite using exponents.

$$P'(t) = \frac{\left(t^2+36\right)^{\frac{1}{2}} \cdot 2t - t^2 \cdot \frac{1}{2}\left(t^2+36\right)^{-\frac{1}{2}} \cdot 2t}{\left(t^2+36\right)}$$ Here the Quotient Rule and the General Power Rule are used in combination.

$$= \frac{\left(t^2+36\right)^{-\frac{1}{2}}\left[\left(t^2+36\right) \cdot 2t - t^3\right]}{\left(t^2+36\right)}$$ Factor out $\left(t^2+36\right)^{-\frac{1}{2}}$.

$$= \frac{t^3+72t}{\left(t^2+36\right)^{\frac{3}{2}}}$$ Simplify.

To find how fast this measure was changing in 1988 (8 years since 1980), evaluate the derivative for $t = 8$.

$$P'(8) = \frac{8^3+72 \cdot 8}{\left(8^2+36\right)^{\frac{3}{2}}} = \frac{512+576}{(100)^{\frac{3}{2}}} = \frac{1088}{1000} = 1.088$$

In 1988, the air pollution measure was increasing at a rate of 1.088 units per year.

3.2 Exercises

Find the derivative for each function given in Exercises 1 – 40 and simplify your answer.

1. $f(x)=(2x-5)^4$ **2.** $f(x)=(7x+2)^3$ **3.** $f(x)=(1-4x)^3$

4. $f(x)=(3-5x)^5$ **5.** $g(x)=(x^2+4)^{-2}$ **6.** $g(x)=(x^2-8)^{-1}$

7. $h(t)=(2t^2+3t)^{-3}$ **8.** $h(t)=(4t^2-t)^{-2}$ **9.** $y=(2x^2+5x-7)^2$

10. $y=(4x^2+9x-3)^3$ **11.** $y=(x^3+1)^{\frac{1}{2}}$ **12.** $y=(2x^3-5)^{\frac{1}{3}}$

13. $y=\sqrt[3]{4x^2+1}$ **14.** $y=\sqrt{7+4x^2}$ **15.** $y=\sqrt[4]{1-2x^3}$

16. $y=\sqrt[3]{5x^3-4}$ **17.** $f(t)=5t(t^3+3)^4$ **18.** $f(x)=-7x(x^4-2)^3$

19. $f(x)=2x^3(x^2-8)^3$ **20.** $f(t)=t(4-3t^2)^2$ **21.** $g(x)=\dfrac{1}{\sqrt{x^2-6}}$

22. $g(x)=\dfrac{5}{\sqrt{x^3+4}}$ **23.** $g(t)=\dfrac{t}{\sqrt{t^2+8}}$ **24.** $g(x)=\dfrac{x^2}{\sqrt[3]{x^2+6}}$

25. $h(x)=\dfrac{\sqrt[3]{2x+3}}{x^2}$ **26.** $h(x)=\dfrac{\sqrt{5x-2}}{x^3}$ **27.** $y=(2x+1)\sqrt{3x-4}$

28. $y=(4x+3)\sqrt{x^2+3}$ **29.** $y=(3x-2)^2(5x+1)^{-2}$ **30.** $y=(2x+7)^3(3x+1)^{-4}$

31. $f(t)=\dfrac{5t+1}{(t-1)^{\frac{2}{3}}}$ **32.** $f(t)=\dfrac{t^2+t+1}{\sqrt{t^4-1}}$ **33.** $g(t)=\left(\dfrac{2t+5}{t+1}\right)^3$

34. $g(t)=\left(\dfrac{5t+4}{t^2-3}\right)^4$ **35.** $y=\left(\dfrac{x+3}{4-2x}\right)^{\frac{1}{2}}$ **36.** $y=\left(\dfrac{x^2}{4x+1}\right)^{\frac{1}{3}}$

37. $y=\sqrt{\dfrac{x+2}{3x-1}}$ **38.** $y=\sqrt{\dfrac{x^2+6}{x^3}}$ **39.** $y=\dfrac{x^2+x}{\sqrt{7-2x}}$

40. $y=\dfrac{(x^2+2)^2}{\sqrt{5x-3}}$

In Exercises 41 – 50, find $\dfrac{dy}{du}$, $\dfrac{du}{dx}$, and $\dfrac{dy}{dx}$. *Then evaluate* $\dfrac{dy}{dx}$ *for the given value of x.*

41. $y=u^2+2,\ u=3x^2+1;\ x=-1$ **42.** $y=\sqrt{u+4},\ u=x^2+x-1;\ x=2$

43. $y=\dfrac{1}{u^2},\ u=2x^3-3x+3;\ x=1$ **44.** $y=\sqrt[3]{u},\ u=2x^3-4x;\ x=2$

45. $y = u^{\frac{3}{2}},\ u = x^3 - 2x^2;\ x = 3$

46. $y = \frac{1}{u^3},\ u = 3x + 1;\ x = 1$

47. $y = \sqrt[3]{u},\ u = 7x^2 + 1;\ x = -3$

48. $y = u^2 + 3u + 4,\ u = x^3 - 5x - 2;\ x = -2$

49. $y = 2u^2 - 5u + 3,\ u = 5x + 6;\ x = 2$

50. $y = 2u^3 - 3u + 1,\ u = x^3 + 8;\ x = 1$

In Exercises 51 – 56, use the given information to find $h'(2)$:

$f(2) = 3$, $f'(2) = -1$, $g(2) = 4$, $g'(2) = 10$,
$g(3) = 8$, and $g'(3) = 7$.

51. $h(x) = \left(f(x) + 1\right)^3$

52. $h(x) = \left(\frac{f(x)}{g(x)}\right)^2$

53. $h(x) = g\left(f(x)\right)$

54. $h(x) = \sqrt{f(x) + g(x)}$

55. $h(x) = \left(g(x)\right)^3$

56. $h(x) = \left(2 + 3 \cdot f(x)\right)\left(g(x)\right)^3$

Determine the equation of the tangent line for $f(x)$ at the x-value indicated in Exercises 57 – 60.

57. $f(x) = (3x - 1)^3;\ x = 1$

58. $f(x) = \left(\frac{x^2 + 1}{x + 1}\right)^3;\ x = 2$

59. $f(x) = \left(3x^2 + 2x + 8\right)^{\frac{1}{2}};\ x = 4$

60. $f(x) = \sqrt{10x + 1};\ x = 8$

61. Marginal revenue. A dealer of microwave ovens estimates that he can sell x ovens per month when the demand function (price) is $p = D(x) = 20\sqrt{280 - 4x}$ dollars.

a. Find the revenue function $R(x)$.
b. Find $R(21)$.
c. Find the marginal revenue function $R'(x)$.
d. Find $R'(21)$.

62. Marginal revenue. The demand function (price) for a particular product is given by $D(x) = 8\sqrt{25 - 5x + 0.25x^2}$ dollars, where x is the number of units (in hundreds) sold.

a. Find the revenue function $R(x)$.
b. Find $R(2)$.
c. Find the marginal revenue function $R'(x)$.
d. Find $R'(2)$.

63. Population growth. It is estimated that t years from now the population of Castle City will be $P(t) = 10(40 + 2t)^2 - 1600t$.

a. What will the population be in 8 years?
b. Find the rate of change in population in 8 years.

64. Air Pollution. It is estimated that t years from now the level of air pollution in Bohrberg will be $P(t)=\dfrac{0.6\sqrt{8t^2+11t+60}}{(t+1)^2}$ parts per million. Find the rate of change in the pollution level in 7 years.

65. Pollution. After a sewage spill, the level of pollution in San Remo Bay is estimated by $P(t)=\dfrac{200t^2}{\sqrt{t^2+11}}$, where t is the time in days since the spill occurred. How fast is the level changing after 5 days? Round to the nearest whole number.

66. Bacterial growth. It is estimated that in t hours the population of bacteria in a culture will be $P(t)=\dfrac{8000}{\sqrt{8-0.5t}}$.

a. What will be the population in 8 hours?
b. Find the rate of change in the population in 8 hours.

67. Rate of change of cost. A manufacturer of vacuum cleaners estimates that the total cost of producing x vacuum cleaners is given by $C(x)=-0.5x^2+56x+800$ dollars. Records show that after t hours on a typical day, the number of units produced is given by $x=5\sqrt{t^2+5t}$. Find the rate of change of total cost with respect to time at the end of

a. 4 hours.
b. 5 hours.

68. Rate of change of profit. A manufacturer has determined that the weekly profit from the sale of x items is given by $P(x)=-x^2+280x-4000$ dollars. It is estimated that after t days in any week, $x=0.5t^2+5t$ items will have been produced. Find the rate of change of profit with respect to time at the end of

a. 4 days.
b. 5 days.

69. Pollution. Studies show that the average level of certain pollutants in the air is given by $L=1+0.2x+0.001x^2$ parts per million when the population is x thousand people. It is estimated that t years from now the population will be $x=\dfrac{200}{\sqrt{7-0.5t}}$ in thousands. Find the rate of change of the level of pollutants after

a. 6 years.
b. 12 years.

70. Security costs. The annual cost for campus security is given by $C(x)=3x^2-32x+16$ in thousands of dollars. It is estimated that the enrollment in t years will be $x=16+0.5t+0.02t^2$ in thousands. Find the rate of change in security costs after

a. 3 years.
b. 4 years.

71. Baseball attendance. The average home attendance per week at a Class AA baseball park varied according to the formula $N(t)=(3+0.2t)^{\frac{1}{2}}$, where t is the number of weeks into the season $(0 \le t \le 12)$ and N is in thousands of persons.

- **a.** What was the attendance during the first week into the season?
- **b.** Determine the number $N'(5)$.
- **c.** Interpret the meaning of $N'(5)$.

72. Weekly attendance. The semester after its student team won an intercollegiate Duplicate Bridge championship, the average weekly attendance at the University Union Building varied according to the formula $B(t)=100-50\left(1-\frac{t}{16}\right)^{\frac{3}{2}}$, where t is the number of weeks after the championship $(0 \le t \le 16)$ and B is the number of persons.

- **a.** What is $B(0)$ and what does it represent?
- **b.** Determine $B'(t)$.
- **c.** What is $B'(7)$? Interpret the meaning of this number.

73. Class registration. The annual registration in university calculus classes varies according to the formula $P(t)=1+\left(1-\frac{t}{30}\right)^{2.5}$, where t is the number of years since 1995 and P is in millions of students.

- **a.** Determine $P(0)$ and explain its meaning.
- **b.** Determine $P'(t)$.
- **c.** Calculate $P'(10)$ and explain its meaning.

74. Travel. The number of passengers traveling from California to Central America and back on cruise ships is given by $C(t)=(10t+50)^{1.5}$, where t is the number of years since 1990 and C is passenger count in thousands.

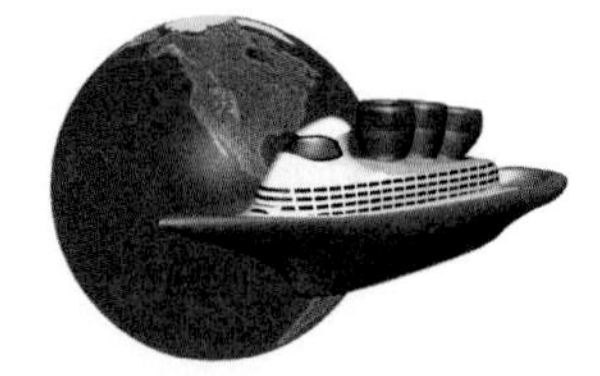

- **a.** When did the number of passengers hit 1,000,000 people ($y = 1000$)?
- **b.** Compute $C'(t)$.
- **c.** Calculate $C'(12)$ and intrepret its meaning.

Hawkes Learning Systems: Essential Calculus

The Chain Rule and the General Power Rule

3.3 Implicit Differentiation and Related Rates

Objectives

After completing this section, you will be able to:

1. *Use the method of implicit differentiation to solve for a derivative.*
2. *Apply the method of implicit differentiation to solve for real-life rates.*

Implicit Differentiation

When a function is expressed in the form $y = f(x)$, we say that y is an **explicit function** of x or that the function is in **explicit form**. For example, the functions

$$y = 2x + 3 \text{ and } y = \frac{1}{x+1}$$

are in explicit form.

If we want to find $\frac{dy}{dx}$ in the equation $xy + y = 1$, where the variable y is **implied** to be a function of x, we can first represent this function in explicit form by solving for y.

$xy + y = 1$	Implicit form.
$y(x+1) = 1$	Factor out y.
$y = \frac{1}{x+1}$	Explicit form (solved for y).

Now that we have the function in explicit form, we can differentiate as follows:

$y = (x+1)^{-1}$	Rewrite y using exponents.
$\frac{dy}{dx} = -(x+1)^{-2} = -\frac{1}{(x+1)^2}.$	Differentiate.

However, in the equation

$$x^2 + y^2 = 4,$$

we cannot solve for y explicitly in terms of x. In fact, the equation represents a circle with radius 2 and center at the origin, and does not represent a function. (See Figure 3.3.1.)

To find $\frac{dy}{dx}$ in the equation $x^2+y^2=4$, we use the method of **implicit differentiation** where y is implied to be a function of x. For example, the upper semicircle $\left(y=\sqrt{4-x^2}\right)$ might be the implied function, or the lower semicircle $\left(y=-\sqrt{4-x^2}\right)$ might be the implied function. However, we do not need to know which function can be implied.

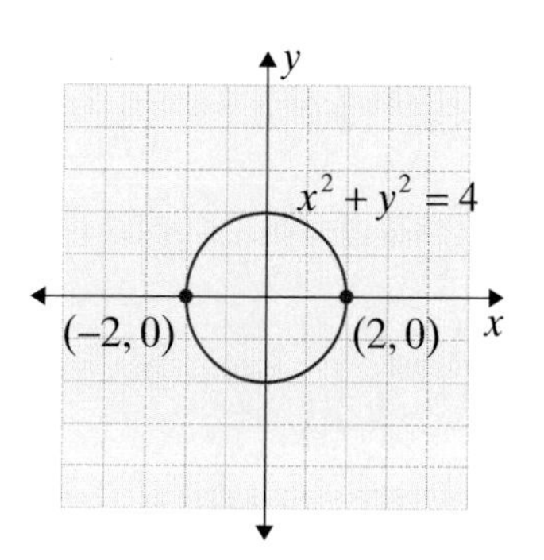

Figure 3.3.1

NOTES There are some cases, such as $x^2+y^2=-1$, where there are no real solutions and no function can be implied. However, the process of implicit differentiation may lead to apparently meaningful results. Such cases are not discussed in this text.

In the procedure of **implicit differentiation**, we use the General Power Rule (or, in more general cases, the Chain Rule). Remember that we are differentiating with respect to x.

For example, suppose that we want to find

$$\frac{d}{dx}\left(y^2\right),$$

where y is implied to be a differentiable function of x. By denoting $y=g(x)$ and $\frac{dy}{dx}=g'(x)$, we can write

$$\frac{d}{dx}\left(y^2\right)=\frac{d}{dx}\left[g(x)^2\right]$$

$$=2\cdot g(x)\cdot g'(x) \qquad \text{By the General Power Rule.}$$

$$=2y\frac{dy}{dx}. \qquad \text{Substitute } y=g(x) \text{ and } \frac{dy}{dx}=g'(x).$$

Or, $\frac{d}{dx}\left(y^2\right)=\left(y^2\right)'=2y\cdot y'$. (Note that $\frac{dy}{dx}=y'$.)

Similarly, by using implicit differentiation, we have

$$\frac{d}{dx}\left(y^3\right)=3y^2\frac{dy}{dx} \quad \text{and} \quad \frac{d}{dx}\left(y^4\right)=4y^3\frac{dy}{dx}.$$

Or, $\left(y^3\right)'=3y^2\cdot y'$ and $\left(y^4\right)'=4y^3\cdot y'$.

Now, to find $\frac{dy}{dx}$ in the equation $x^2+y^2=4$, we differentiate both sides of the equation with respect to x and then solve for $\frac{dy}{dx}$.

$$x^2+y^2=4$$

$$\frac{d}{dx}\left(x^2+y^2\right)=\frac{d}{dx}(4)$$ Use the Sum and Difference Rule.

$$\frac{d}{dx}\left(x^2\right)+\frac{d}{dx}\left(y^2\right)=\frac{d}{dx}(4)$$

$$2x+2y\frac{dy}{dx}=0$$

Differentiate. Use the Power Rule on the first term, implicit differentiation on the second term, and the rule that the derivative of a constant is 0 on the last term.

$$2y\frac{dy}{dx}=-2x$$ Subtract $2x$ from both sides.

$$\frac{dy}{dx}=\frac{-\cancel{2}x}{\cancel{2}y}=-\frac{x}{y}$$ Solve for $\frac{dy}{dx}$ and simplify.

Note that the derivative has both x and y in it. To evaluate such a derivative, we need to know both x and y at a point (x, y) on the curve.

NOTES In general, we do not need to know the explicit representation for y in order to obtain $\frac{dy}{dx}$.

Example 1: Derivative at a Point

Given the equation $x^2+y^2=25$, find $\frac{dy}{dx}$ and then evaluate the derivative at $(4, 3)$ and $(0, 5)$.

Solution:

$$x^2+y^2=25$$

$$\frac{d}{dx}\left(x^2+y^2\right)=\frac{d}{dx}(25)$$ Use the Sum and Difference Rule.

$$\frac{d}{dx}\left(x^2\right)+\frac{d}{dx}\left(y^2\right)=\frac{d}{dx}(25)$$

$$2x+2y\frac{dy}{dx}=0$$

Differentiate. Use the Power Rule on the first term, implicit differentiation on the second term, and the rule that the derivative of a constant is 0 on the last term.

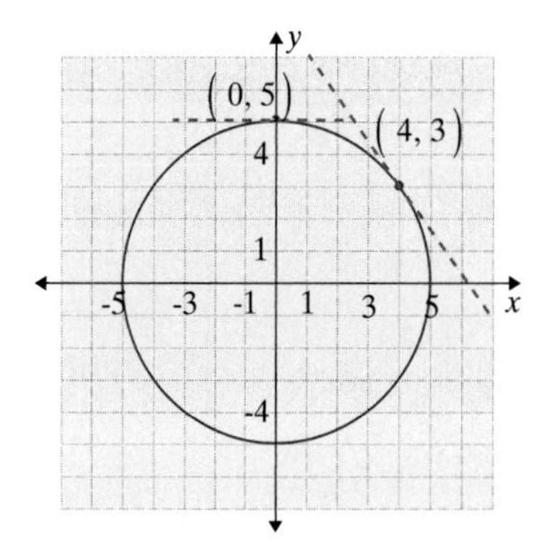

Next, solve for $\frac{dy}{dx}$ and simplify.

$$2y\frac{dy}{dx} = -2x$$

$$\frac{dy}{dx} = -\frac{x}{y}$$

Finally, substitute the given x and y values of each point to evaluate the derivative at that point.

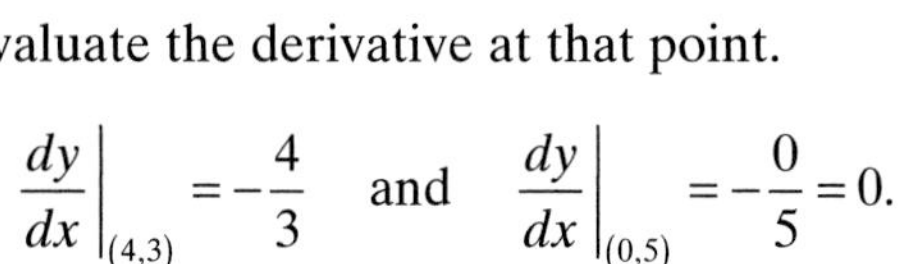

$$\left.\frac{dy}{dx}\right|_{(4,3)} = -\frac{4}{3} \quad \text{and} \quad \left.\frac{dy}{dx}\right|_{(0,5)} = -\frac{0}{5} = 0.$$

Note that $-\frac{4}{3}$ is the slope of the tangent line to the circle at $(4, 3)$.

Example 2: Implicit Differentiation

If $xy + x^2y^3 - 1 = 0$, find $\frac{dy}{dx}$.

Solution: Note that both xy and x^2y^3 are products and that the Product Rule must be used.

$$xy + x^2y^3 - 1 = 0$$

$$\frac{d}{dx}\left(xy + x^2y^3 - 1\right) = \frac{d}{dx}(0)$$

$$\frac{d}{dx}(xy) + \frac{d}{dx}\left(x^2y^3\right) + \frac{d}{dx}(-1) = \frac{d}{dx}(0)$$

Use the Sum and Difference Rule.

$$\left(x\cdot\frac{dy}{dx} + y\cdot 1\right) + \left(x^2\cdot 3y^2\cdot\frac{dy}{dx} + y^3\cdot 2x\right) + 0 = 0$$

$$x\frac{dy}{dx} + y + 3x^2y^2\frac{dy}{dx} + 2xy^3 = 0$$

$$x\frac{dy}{dx} + 3x^2y^2\frac{dy}{dx} = -y - 2xy^3$$

Differentiate. Use the Product Rule in combination with the General Power Rule on the first and second terms and the rule that the derivative of a constant is 0 on the remaining terms.

$$\left(x + 3x^2y^2\right)\frac{dy}{dx} = -y - 2xy^3$$

Solve for $\frac{dy}{dx}$.

$$\frac{dy}{dx} = \frac{-y - 2xy^3}{x + 3x^2y^2}$$

Example 3: Implicit Differentiation

Find $\frac{dy}{dx}$ if $\sqrt{x}-\sqrt{y}=16$.

Solution:

$$\sqrt{x}-\sqrt{y}=16$$

$$x^{\frac{1}{2}}-y^{\frac{1}{2}}=16$$ Rewrite using exponents.

$$\frac{d}{dx}\left(x^{\frac{1}{2}}-y^{\frac{1}{2}}\right)=\frac{d}{dx}(16)$$ Use the Sum and Difference Rule.

$$\frac{d}{dx}\left(x^{\frac{1}{2}}\right)-\frac{d}{dx}\left(y^{\frac{1}{2}}\right)=\frac{d}{dx}(16)$$ Differentiate using the General Power Rule and the rule that the derivative of a constant is 0.

$$\frac{1}{2}x^{-\frac{1}{2}}-\frac{1}{2}y^{-\frac{1}{2}}\cdot\frac{dy}{dx}=0$$

$$-\frac{1}{2}y^{-\frac{1}{2}}\frac{dy}{dx}=-\frac{1}{2}x^{-\frac{1}{2}}$$

$$\frac{dy}{dx}=\frac{x^{-\frac{1}{2}}}{y^{-\frac{1}{2}}}=\frac{y^{\frac{1}{2}}}{x^{\frac{1}{2}}}$$ Solve for $\frac{dy}{dx}$.

$$=\frac{\sqrt{y}}{\sqrt{x}}=\sqrt{\frac{y}{x}}$$ Rewrite using radicals.

We point out that the following notation suggests that x and y are treated differently:

$$\left(x^n\right)'=nx^{n-1} \text{ but } \left(y^n\right)'=ny^{n-1}\cdot y'.$$

When the notation $\frac{d}{dx}$ is used for differentiation with respect to x, the rule is seen to be "equal" since $\frac{d}{dx}\left(x^n\right)=nx^{n-1}\cdot\frac{dx}{dx}$ and $\frac{d}{dx}\left(y^n\right)=ny^{n-1}\cdot\frac{dy}{dx}$.

We actually use, implicitly, the result that $\frac{d}{dx}(x)=\frac{dx}{dx}=1$.

Related Rates

Now we will consider problems in which two or more variables are dependent on time, and we will discuss the use of implicit differentiation to determine their rates of change with respect to time. For example, if x and y are both dependent on time t and

$$y = 5x + 1,$$

then differentiating both sides of the equation with respect to t gives

$$\frac{d}{dt}(y) = \frac{d}{dt}(5x+1)$$
$$\frac{dy}{dt} = 5\frac{dx}{dt}.$$

Thus we see that the rates $\frac{dy}{dt}$ and $\frac{dx}{dt}$ are **related**, and y is changing 5 times as fast as x.

Example 4: Balloon Inflation

Suppose that a spherical balloon is inflated by a continuous flow of air from an air compressor at a rate of 2 in.^{3}/sec. How fast is the radius of the balloon growing when the radius is 3 in.?

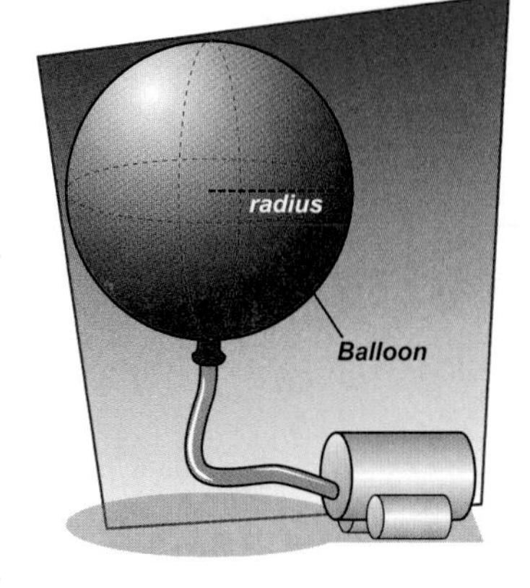

Solution: We know that $\frac{dV}{dt} = 2 \text{ in.}^3/\text{sec}$, and we want to find $\frac{dr}{dt}$.

The formula for the volume of a sphere is $V = \frac{4}{3}\pi r^3$, so

$$V = \frac{4}{3}\pi r^3$$ Write an equation.

$$\frac{d}{dt}(V) = \frac{d}{dt}\left(\frac{4}{3}\pi r^3\right)$$ Differentiate both sides of the equation with respect to t.

$$\frac{dV}{dt} = \frac{4}{\not{3}}\pi \cdot \not{3}r^2 \cdot \frac{dr}{dt}$$ Use implicit differentiation.

$$\frac{dV}{dt} = 4\pi r^2 \frac{dr}{dt}$$

continued on next page ...

Now substitute $\frac{dV}{dt} = 2$ and $r = 3$.

$$2 = 4\pi(3)^2 \frac{dr}{dt}$$ Solve for $\frac{dr}{dt}$.

$$\frac{\overset{1}{\cancel{2}}}{{}_{18}\cancel{36}\pi} = \frac{dr}{dt}$$

Therefore,

$$\frac{dr}{dt} = \frac{1}{18\pi} \text{ in./sec} \approx 0.018 \text{ in./sec.}$$

Example 5: Airplane Holding Pattern

An airplane is flying in a circular "holding" pattern with a radius of 2 miles. The path of the plane can be described by the equation $x^2 + y^2 = 4$. When the plane is at the point $x = \sqrt{3}$ miles and $y = 1$ mile, as shown in the figure, it is traveling north at 300 mph. How fast is it traveling west?

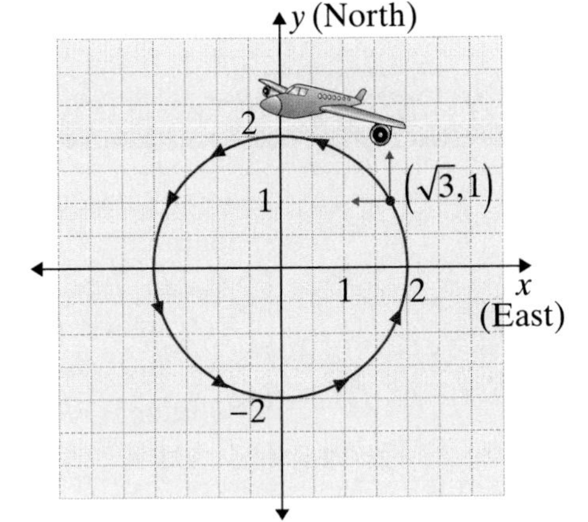

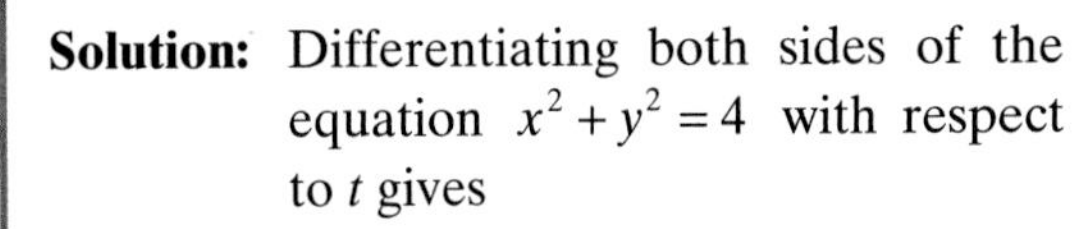

Solution: Differentiating both sides of the equation $x^2 + y^2 = 4$ with respect to t gives

$$\frac{d}{dt}\left(x^2 + y^2\right) = \frac{d}{dt}(4)$$

$$\frac{d}{dt}\left(x^2\right) + \frac{d}{dt}\left(y^2\right) = \frac{d}{dt}(4)$$ Use the Sum and Difference Rule.

$$2x\frac{dx}{dt} + 2y\frac{dy}{dt} = 0.$$ Apply implicit differentiation.

Substituting $x = \sqrt{3}$, $y = 1$, and $\frac{dy}{dt} = 300$, we have

$$2\sqrt{3}\frac{dx}{dt} + 2 \cdot 1 \cdot 300 = 0$$

$$\frac{dx}{dt} = -\frac{\overset{300}{\cancel{600}}}{{}_{1}\cancel{2}\sqrt{3}} = -\frac{300}{\sqrt{3}} \cdot \frac{\sqrt{3}}{\sqrt{3}} = -100\sqrt{3} \text{ mph.}$$ Solve for $\frac{dx}{dt}$.

The plane is traveling west (in the negative x direction) at a rate of $100\sqrt{3}$ mph ≈ 173.2 mph.

Example 6: Moving Cars

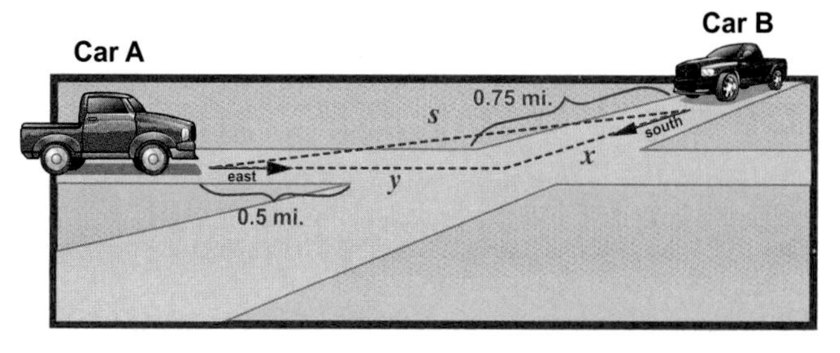

Two cars, A and B, are traveling toward the same intersection, and neither driver plans to stop. Car A is 0.5 miles from the intersection moving east at 40 mph, and Car B is 0.75 miles from the intersection moving south at 60 mph. How fast is the distance between the two cars changing?

Solution: From the figure, $s^2 = x^2 + y^2$.

Differentiating with respect to t, we obtain

$$\frac{d}{dt}\left(s^2\right) = \frac{d}{dt}\left(x^2 + y^2\right)$$

$$2s\frac{ds}{dt} = 2x\frac{dx}{dt} + 2y\frac{dy}{dt}$$

$$\frac{ds}{dt} = \frac{x\frac{dx}{dt} + y\frac{dy}{dt}}{s}.$$

Note: as x and y are decreasing, their velocities are negative.

Now, substituting $x = 0.75$, $\frac{dx}{dt} = -60$, $y = 0.5$, $\frac{dy}{dt} = -40$, and

$$s = \sqrt{x^2 + y^2} = \sqrt{(0.75)^2 + (0.5)^2} = \sqrt{0.8125} \approx 0.9,$$

we find

$$\frac{ds}{dt} = \frac{x\frac{dx}{dt} + y\frac{dy}{dt}}{s} \approx \frac{0.75(-60) + (0.5)(-40)}{0.9} = -72.2 \text{ mph}.$$

The distance between the cars is decreasing at a rate of approximately 72.2 mph.

The costs of a production process can be related to two categories: labor and capital. The cost of labor is essentially the payroll, and the cost of capital is the cost of the physical items such as buildings and machines used in the production process. One relationship frequently used by economists, involving units produced, labor, and capital, is the **Cobb-Douglas Production Formula**.

Cobb-Douglas Production Formula

$$P = Cx^a y^{1-a}$$

where

P = number of units produced,
x = units of labor, and
y = units of capital.
C and a are constants, $0 < a < 1$.

Example 7: Production

Suppose that a firm's level of production is given by $P = 20x^{\frac{1}{4}}y^{\frac{3}{4}}$, where x represents the units of labor and y represents the units of capital. Currently, the company is using 16 units of labor and 81 units of capital. If labor is increasing by 4 units per month, what must the change in units of capital per month be to maintain the current level of production?

Solution: The current level of production is

$$P = 20(16)^{\frac{1}{4}}(81)^{\frac{3}{4}} \qquad \text{Use } x = 16 \text{ and } y = 81.$$

$$= 20(2)(27) = 1080 \text{ units.}$$

Since the current level of production is to be maintained, P is to remain constant at 1080. Thus we have

$$1080 = 20x^{\frac{1}{4}}y^{\frac{3}{4}}, \quad \text{which gives} \quad 54 = x^{\frac{1}{4}}y^{\frac{3}{4}}.$$

We want to find $\frac{dy}{dt}$. Use implicit differentiation and differentiate both sides of the simplified equation with respect to t.

$$\frac{d}{dt}(54) = x^{\frac{1}{4}} \cdot \frac{d}{dt}\left(y^{\frac{3}{4}}\right) + y^{\frac{3}{4}} \cdot \frac{d}{dt}\left(x^{\frac{1}{4}}\right) \qquad \text{By the Product Rule.}$$

continued on next page ...

Thus

$$0 = x^{\frac{1}{4}} \cdot \frac{d}{dt}\left(y^{\frac{3}{4}}\right) + y^{\frac{3}{4}} \cdot \frac{d}{dt}\left(x^{\frac{1}{4}}\right).$$

The Chain Rule gives $\frac{d}{dt}\left(x^{\frac{1}{4}}\right) = \frac{1}{4}x^{-\frac{3}{4}} \cdot \frac{dx}{dt}$, and $\frac{d}{dt}\left(y^{\frac{3}{4}}\right) = \frac{3}{4}y^{-\frac{1}{4}} \cdot \frac{dy}{dt}$; so

$$0 = x^{\frac{1}{4}} \cdot \left(\frac{3}{4}y^{-\frac{1}{4}}\frac{dy}{dt}\right) + y^{\frac{3}{4}} \cdot \left(\frac{1}{4}x^{-\frac{3}{4}}\frac{dx}{dt}\right).$$

We are given that $x = 16$, $y = 81$, and $\frac{dx}{dt} = 4$. Substituting in these values, we get:

$$0 = (16)^{\frac{1}{4}} \cdot \frac{3}{4}(81)^{-\frac{1}{4}}\frac{dy}{dt} + (81)^{\frac{3}{4}} \cdot \frac{1}{4}(16)^{-\frac{3}{4}}(4)$$

$$0 = \cancel{2} \cdot \frac{\cancel{3}}{{}_2\cancel{4}} \cdot \frac{1}{\cancel{3}} \cdot \frac{dy}{dt} + 27 \cdot \frac{1}{\cancel{4}} \cdot \frac{1}{8} \cdot \cancel{4}$$

$$0 = \frac{1}{2} \cdot \frac{dy}{dt} + \frac{27}{8}$$

$$-\left(\frac{1}{2} \cdot \frac{dy}{dt}\right) = \frac{27}{8}$$

$$\frac{dy}{dt} = \left(-\cancel{2}\right)\frac{27}{\cancel{8}_4}$$

$$\frac{dy}{dt} = -\frac{27}{4} = -6.75$$

Because the derivative was negative, the capital must **decrease** at a rate of $\frac{27}{4} = 6.75$ units per month to keep the current level of production.

3.3 Exercises

Use implicit differentiation to find $\frac{dy}{dx}$ *for each of the equations in Exercises 1 – 20.*

1. $2x^2+y^2=4$

2. $x^3+y^3=5$

3. $2x^3+y^3=8$

4. $x^2-y^2=16$

5. $\sqrt{x}+\sqrt{y}=1$

6. $x-\sqrt{y}=2$

7. $x^2y=2$

8. $xy^2=-1$

9. $x^2+xy+y^2=-1$

10. $x^3+2xy-y^2=3$

11. $4x^2+3xy+y^2=2x$

12. $x^3+y^3=3xy$

13. $\frac{1}{x}+\frac{x}{y}=2x$

14. $x^2+\frac{2x}{y}=\frac{1}{x^2}$

15. $x^2+\sqrt{xy}=2y^2$

16. $2y+\sqrt{xy}=5x^2$

17. $x^2y^2+xy^3=x^4$

18. $x^3y+xy^3=3x^3$

19. $x^2+(y-2)^2=16$

20. $x^2+4(y+3)^2=9$

In Exercises 21 – 30, use implicit differentiation to find $\frac{dy}{dx}$ *for the given equations; then find the slope of the tangent line at the given point.*

21. $4x^2-8y^3=24;\ (2,-1)$

22. $3x^3+5y^2+x=1;\ (-1,1)$

23. $x^2y-y^2+4x+8=0;\ (1,4)$

24. $5x^2+xy^2+2x=8;\ (-2,2)$

25. $x^3+2xy-y^2=0;\ (3,-3)$

26. $4x^2-3xy+y^2=7;\ (2,3)$

27. $\frac{1}{x}+\frac{x}{y^2}=2x;\ (1,-1)$

28. $3x-\frac{2x}{y^2}+x^2=12;\ (3,1)$

29. $2y+\sqrt{xy}=5x^2;\ (2,8)$

30. $x^2-\sqrt{xy}=2y^2+3x;\ (4,1)$

In Exercises 31 – 40, x and y are functions of a third variable, t. Use implicit differentiation to find an expression for $\frac{dy}{dt}$.

31. $x^2-4y^2=16$

32. $3x^2+y^4=4$

33. $x^3+5y^2=2x$

34. $6x^2+5x=2y^2$

35. $\sqrt{xy}=4$

36. $\sqrt{x}+\sqrt{y}=3$

37. $x^2+xy+y^2=3$

38. $x+2xy-y^2=6$

39. $2xy+x^2=y^3$

40. $x^2y^2-y^3+4x=0$

41. Retail sales. The manager of an audio electronics store has determined that the number of stereo receivers and the number of speaker systems sold weekly are related by the equation $0.9y^2 = 10x + xy$, where x is the number of receivers and y is the number of speaker systems. Find $\frac{dy}{dx}$ if $x = 12$ and $y = 20$, and interpret your answer.

42. Retail sales. The number of pairs of trousers x and the number of shirts y sold at a department store are related by the equation $36x = 11y + 0.01x^2y$. Find $\frac{dy}{dx}$ when $x = 10$ and $y = 30$ and interpret your answer.

43. Cobb-Douglas production. The level of production of a company is given by $P = 30x^{\frac{1}{3}}y^{\frac{2}{3}}$ units monthly, where x is the units of labor and y is the units of capital. The company is currently utilizing 64 units of labor and 27 units of capital. If labor is increased by 2 units per week, what will be the change in units of capital per week to maintain the current level of production?

44. Cobb-Douglas production. The level of production of a company is given by $P = 18x^{0.3}y^{0.7}$ units monthly, where x is the units of labor and y is the units of capital. The company is currently utilizing 35 units of labor and 24 units of capital. If capital is increased by 3 units per week, what will be the change in units of labor per week to maintain the current level of production?

45. Rate of increase in cost. The cost of producing x units of a product is given by the function $C(x) = 0.02x^3 - x^2 + 8x + 200$. The factory is currently producing 60 units per week but plans to increase production at a rate of 3 units per week. What will be the rate of increase in the total cost?

46. Rate of decrease in cost. The cost of producing x units of commodity is given by the function $C(x) = x^2 - 2x^{\frac{3}{2}} + 7x + 180$. Currently, the production level is 36 units per day. The company plans to decrease production at a rate of 2 units per day. What will be the rate of decrease in the total cost?

47. Sliding ladder. A 17-ft. ladder is leaning against a wall. The bottom of the ladder is pulled away from the wall at a rate of 3 ft./sec. How fast is the top of the ladder moving down the wall when the top is 8 feet above the ground?

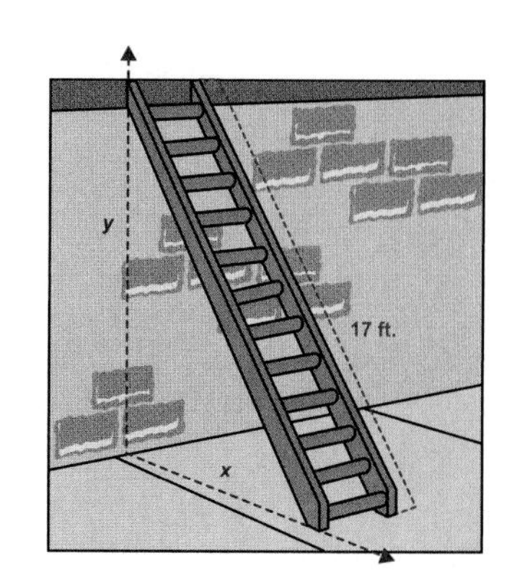

48. Velocity. Marijean is standing on a boat dock pulling in her boat by means of a rope attached to a boat at water level. Her hands are 6 feet above the water and she is pulling in the rope at a rate of 1.5 ft./sec. How fast is the boat approaching the dock if there are 10 feet of rope still out?

49. Driving. A car traveling south at 30 ft./sec. crosses an intersection. When the car is 90 feet past the intersection, a bicyclist crosses the intersection traveling east at a rate of 20 ft./sec. How fast is the distance between the car and the bicycle increasing 5 seconds after the bicycle crosses the intersection?

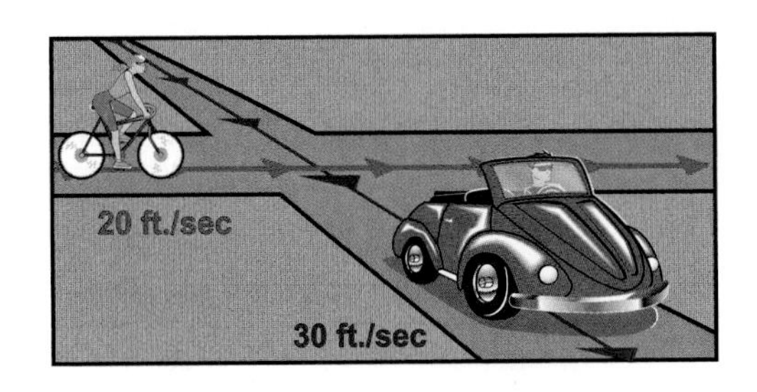

50. Distance to an airplane. An airplane traveling at a height of 3000 feet crosses directly over an observer. The speed of the plane is 400 ft./sec. How fast is the distance between the observer and the plane changing after 10 seconds?

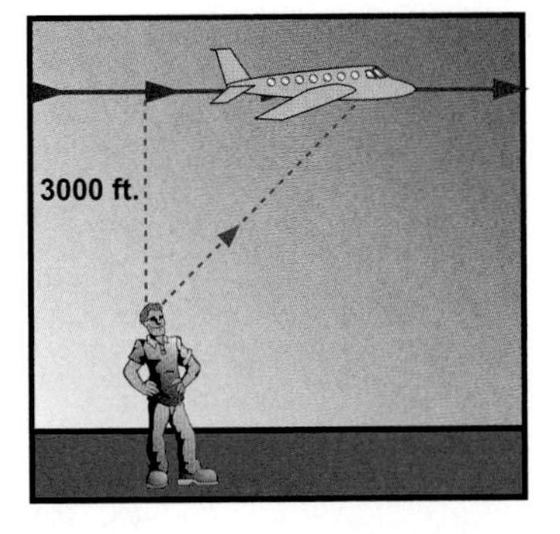

51. Baseball. A base runner heads towards first base with a speed of 20 ft./sec. A baseball diamond is a square, 90 feet on each side.

a. How long will it take for the runner to reach first base?

b. What is the runner's rate of change of distance from the umpire standing on third base (we will call this $\frac{dy}{dt}$)? (**Hint:** Use the diagram to get a relation between x and y.)

c. What is the runner's speed when he arrives at first base?

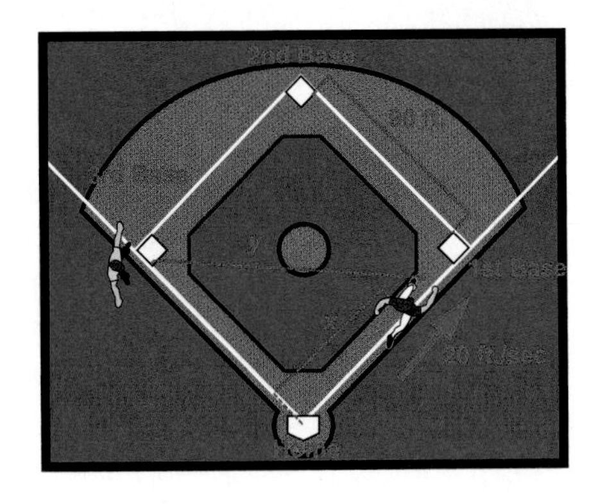

52. Sailing. A Coast Guard radar monitoring station on shore observed a sailboat on its radar grid. It was determined that the boat's East-West distance along the shoreline changed by the formula $x = 12 + 0.1t$ and its seaward distance (North-South) changed by the formula $y = 20 - 0.3t$. Here, x and y are in miles and t is in minutes.

a. What was the sailboat's position at $t = 0$ minutes? Sketch this situation.

b. Determine $\frac{dx}{dt}$ and $\frac{dy}{dt}$.

c. What is $\frac{dy}{dx}$ at $t = 10$? Interpret this number.

d. Where and when will the sailboat hit the shore (assuming it keeps its present heading)?

53. Cartons of oranges. Suppose the price in Miami of cartons of oranges satisfies a demand equation of $xp + 20p = 1040$, where x is the number of cartons supplied and p is the demand (unit price) in dollars.

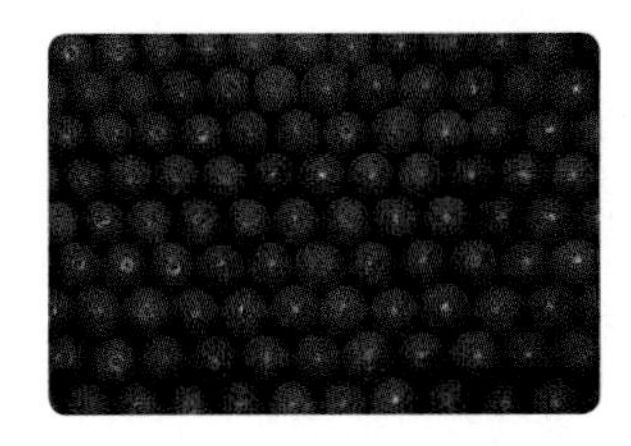

a. If 500 cartons are available today, what is the unit price?

b. If the supply is increasing at the rate of 100 cartons per day, at what rate is the price changing?

54. Flowing water. The radius of a large pan of water is 7 cm and water from a tap flows in at the rate of 10 cubic centimeters per minute.

a. Determine $\frac{dh}{dt}$, where h is the height of water in the pan.

b. How long will it take to fill the pan to a height of 7 cm?

55. Demand. The Arrow Marketing Group sells teddy bears with college logos to college and university gift shops. Their demand equation is $px - 2800 = x$, where x is the quantity of bears and p is the price in dollars that each sells for.

a. How many bears can be sold (or "demanded") at \$4.50 apiece?

b. Determine a formula $\frac{dx}{dp}$ and evaluate for p and x as in part **a.**

56. Electricity costs. Electricity costs per semester at Mount State University are calculated by the formula $C = 0.05x + 0.03y - 0.08xy$, where x is a measure of student size and activity, y is dependent on the usage of various buildings and their efficiencies, and C is in millions of dollars.

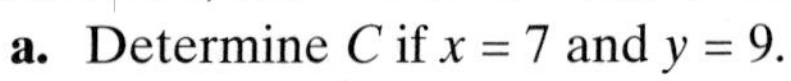

a. Determine C if $x = 7$ and $y = 9$.

b. Determine a general formula for $\frac{dy}{dx}$ and evaluate for $x = 7$ and $y = 9$.

57. Race track. A circular race track has a radius of 840 feet. At a certain point in time, t_0, an observer in a maintenance pit 504 feet from the center of the track clocks a car (when it is directly North of him) traveling counter-clockwise along the track at a speed of 50 miles per hour from his right to left $\left(\frac{dx}{dt}\right)$. (Use 1 mile = 5280 feet.)

a. Determine the location of the car on the track. Assume the center of the track is at the origin.

b. Determine a general equation for $\frac{dy}{dx}$.

c. What is $\left.\frac{dy}{dx}\right|_{t=t_0}$? Interpret its meaning.

d. What is $\frac{dy}{dt}$ at $t = t_0$?

e. What is the speedometer reading $\left(\frac{ds}{dt}\right)$ on the car at $t = t_0$? (**Note:** Assume this speed satisfies $\frac{ds}{dt}$ as in Example 6.)

f. How long will it take the car to go around the track?

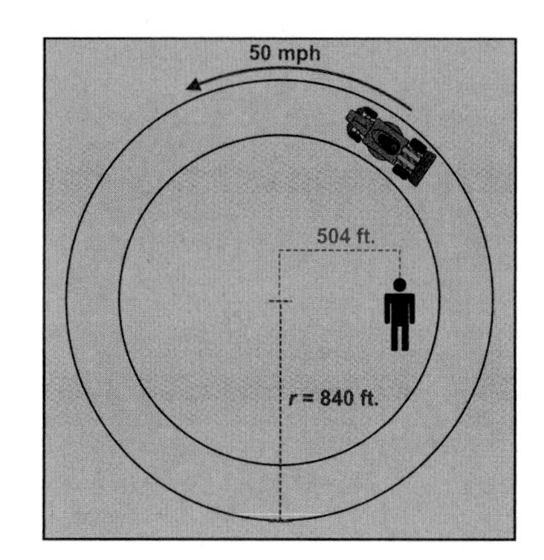

58. Chlorination costs. Chlorination costs for the swimming pool at a spa are given by $C = 0.27x + 2y - 0.001xy^2$, where x is the number of weekly swimmers, y is the number of special functions, and C is in dollars.

a. Determine C if $x = 500$ and $y = 4$.

b. Determine a formula for $\frac{dy}{dx}$ and evaluate at $x = 500$ and $y = 4$.

59. Probability measurement. The standard deviation, S, of a binomial random variable is given by $S = \sqrt{np(1-p)}$, where n is the number of trials or repetitions of an experiment and p is the probability of success on one outcome. S is a measure of how the data tends to vary from the center (or mean).

a. For $n = 768$ and $p = 0.25$, determine S.

b. Suppose we consider S as a fixed quantity. Determine $\frac{dp}{dn}$ for n and p as in part **a.**

Hawkes Learning Systems: Essential Calculus

Implicit Differentiation and Related Rates

3.4 Local Extrema and the First Derivative Test

Objectives

After completing this section, you will be able to:

1. *Define the increasing and decreasing intervals of a function.*
2. *Determine the critical values of a function.*
3. *Use the First Derivative Test to find local extrema of a function and use this information to sketch a graph of the function.*

Increasing and Decreasing Functions

Suppose $y = f(x)$ is a function defined on the interval (a, b). We are interested in what happens to the y-values (functional values) as the values of x increase (i.e., move from left to right on the x-axis). Four basic cases are illustrated in Figure 3.4.1.

Figure 3.4.1

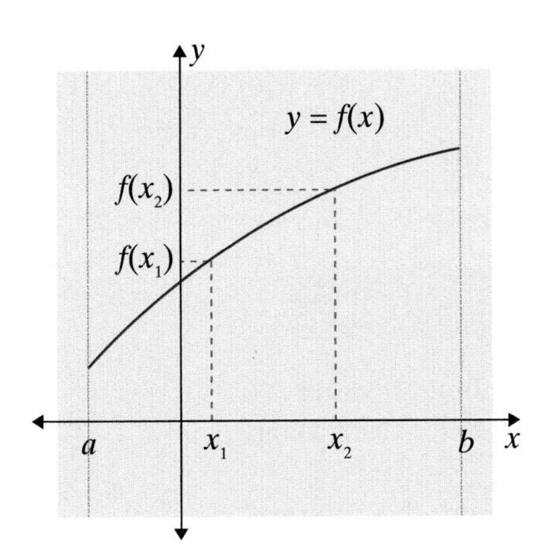

(a) For $x_1 < x_2$, $f(x_1) < f(x_2)$. The function $y = f(x)$ is **increasing**.

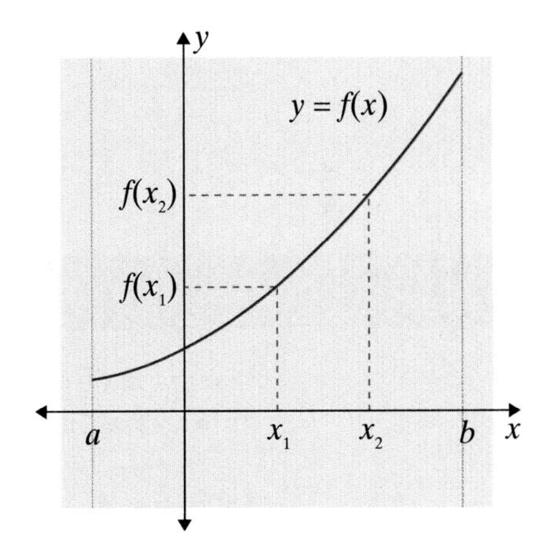

(b) For $x_1 < x_2$, $f(x_1) < f(x_2)$. The function $y = f(x)$ is **increasing**.

continued on next page ...

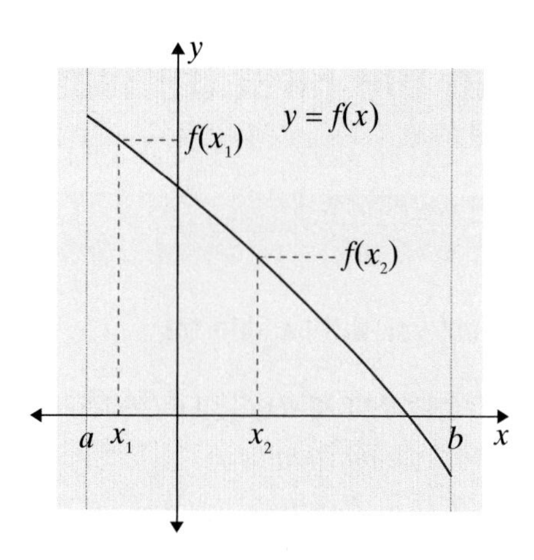

(c) For $x_1 < x_2$, $f(x_1) > f(x_2)$.
The function $y = f(x)$ is **decreasing**.

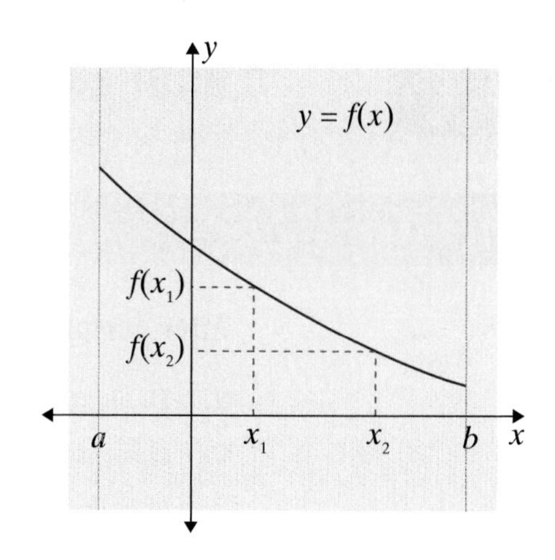

(d) For $x_1 < x_2$, $f(x_1) > f(x_2)$.
The function $y = f(x)$ is **decreasing**.

Increasing and Decreasing Intervals of a Function

Suppose that a function f is defined on the interval (a, b).

1. *If $x_1 < x_2$ implies $f(x_1) < f(x_2)$ for every x_1 and x_2 in (a, b), then $f(x)$ is increasing on (a, b).*
2. *If $x_1 < x_2$ implies $f(x_1) > f(x_2)$ for every x_1 and x_2 in (a, b), then $f(x)$ is decreasing on (a, b).*

Example 1: Increasing and Decreasing Intervals

The graph of $y = f(x)$ is given. Find the intervals on which **a.** f is increasing, and **b.** f is decreasing.

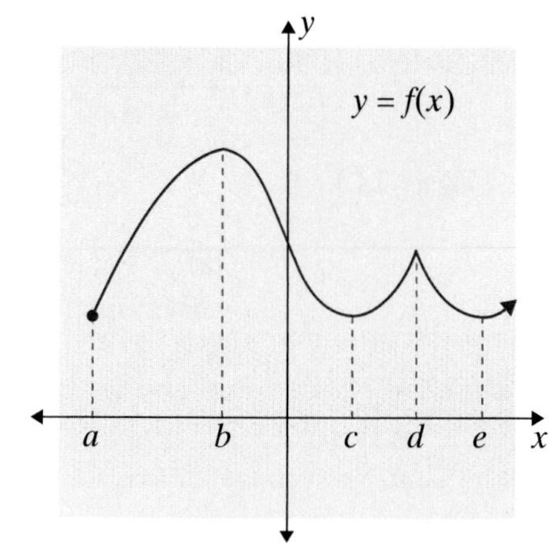

Solutions: a. f is increasing on the intervals (a, b), (c, d), and $(e, +\infty)$.

b. f is decreasing on the intervals (b, c) and (d, e).

The derivative can be used to determine whether a function is increasing or decreasing on an interval. In Figure 3.4.2 we illustrate how the slope of a tangent line (the value of the derivative) at a point can indicate whether the graph of the function is rising or falling at that point. Essentially, we are relating three ideas: the derivative of a function, the slopes of lines tangent to the graph of the function, and the graph of the function.

Figure 3.4.2

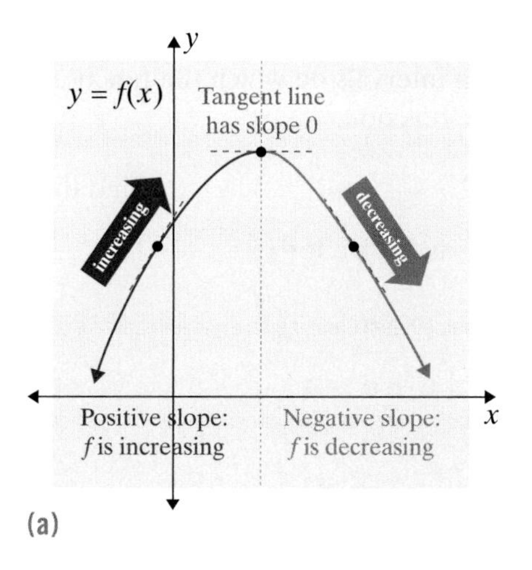

(a)

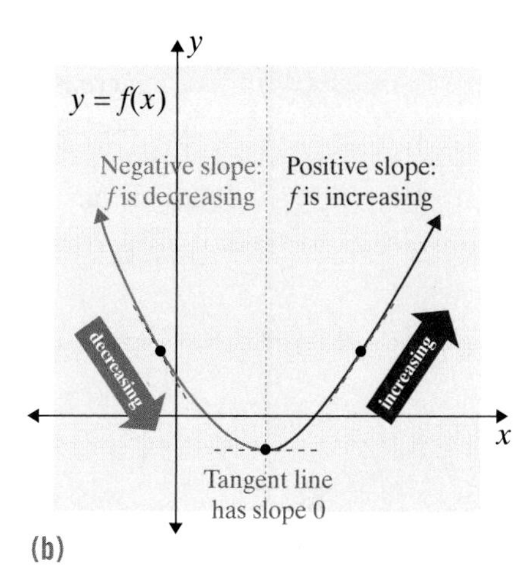

(b)

NOTES Positive values for the derivative indicate positive slopes for the tangent lines which in turn imply that the function is increasing. Negative values for the derivative indicate negative slopes for the tangent lines which imply that the function is decreasing.

The following theorem states these ideas more formally.

Theorem

Suppose that $f(x)$ is differentiable on the interval (a, b).

If $f'(x)$ is positive ($f'(x) > 0$) for all x in (a, b), then f is increasing on (a, b).
If $f'(x)$ is negative ($f'(x) < 0$) for all x in (a, b), then f is decreasing on (a, b).

Note: *The values of x for which a function is increasing (or decreasing) are always presented as open intervals. These intervals are part (or all) of the domain of the function.*

The following examples illustrate how the derivative can be used to determine the intervals on which the function is increasing or decreasing and how this information can be applied to sketch the graph.

Example 2: Graphing a Function

For the function $f(x) = x^3 - 12x + 1$,

a. Find all values of x that correspond to horizontal tangent lines. These values occur where the derivative is 0.

b. Use the values from part **a.** to find the open intervals on which the function is increasing and the open intervals on which the function is decreasing.

c. Sketch the graph of the function.

Solutions: a. $f'(x) = 3x^2 - 12$ Find the derivative of $f(x)$.

We set $f'(x) = 0$ and solve for x:

$$3x^2 - 12 = 0$$

$$3(x^2 - 4) = 0$$

$$3(x+2)(x-2) = 0$$

$$x = -2, 2$$

Horizontal tangent lines occur at $x = -2$ and $x = 2$.

b. We can use a number line to help determine the intervals on which $f'(x) > 0$ (and so $f(x)$ is increasing) and those on which $f'(x) < 0$ (and so $f(x)$ is decreasing).

Mark the values for x where $f'(x) = 0$ on a number line. In this case, they are $x = 2$ and $x = -2$ (this was determined in part **a.**). Then select any one test point in each interval and find the sign of $f'(x)$ at these test points by substituting the value into $f'(x)$.

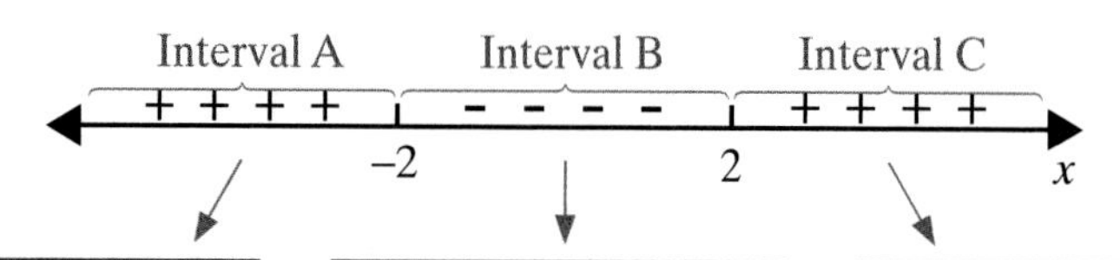

Interval A	Interval B	Interval C
We will use the test point $x = -5$.	We will use the test point $x = 0$.	We will use the test point $x = 3$.
$f'(-5) = 3(-5)^2 - 12$	$f'(0) = 3(0)^2 - 12$	$f'(3) = 3(3)^2 - 12$
$= 75 - 12$	$= 0 - 12$	$= 27 - 12$
$= 63$	$= -12$	$= 15$
$63 > 0$	$-12 < 0$	$15 > 0$

Since $f'(-5)$ is positive, mark interval A with + signs. Since $f'(0)$ is negative, mark interval B with – signs. Since $f'(3)$ is positive, mark interval C with + signs. Note that positive values for the derivative indicate that the function is increasing, and negative values for the derivative indicate that the function is decreasing.

c. The goal for the student is to use the information from parts **a.** and **b.**, to sketch the curve without using a calculator. A secondary goal is to sketch the graph using as few (x, y)-coordinates as possible. We do need the important y-values corresponding to the maximum and to the minimum points (which have the horizontal tangents). We have $f(-2) = 17$ and $f(2) = -15$. We recommend in general that the student note the value of the y-intercept, $f(0)$. Here $f(0) = 1$. With only these three coordinates, a smooth curve can be drawn (on one's paper).

The analysis shows that f increases left-to-right until $x = -2$. At this value f reaches a local maximum of 17. Then f declines until, for $x = 2$, the value -15 is reached. Then f increases without bound.

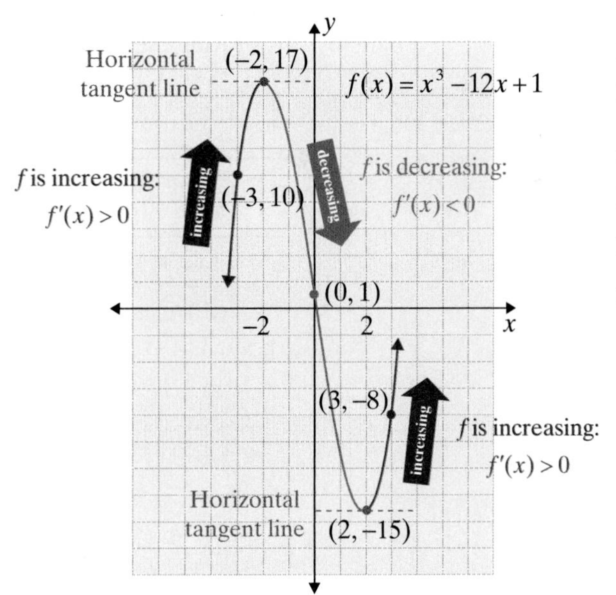

Notice how the pattern of the arrows from the final line graph in part **b.** outlines the general shape of the curve sketched in part **c.**

Example 3: Graphing a Function

For the function $y = x^3 + 1$,

a. Find all values of x where the derivative is 0.

b. Determine where the function is increasing and where it is decreasing.

c. Sketch its graph.

Solutions: a. $\frac{dy}{dx} = 3x^2$ Find the derivative of y.

Set $\frac{dy}{dx} = 0$ and solve for x:

$$3x^2 = 0$$
$$x = 0.$$

A horizontal tangent line occurs at $x = 0$.

b. Mark the values for x where $\frac{dy}{dx} = 0$ on a number line. In this case, there is only one such point: $x = 0$ (this was determined in part **a.**). Select a test point in each of the two intervals and find the sign of $\frac{dy}{dx}$ at these test points by substituting the value into $\frac{dy}{dx}$.

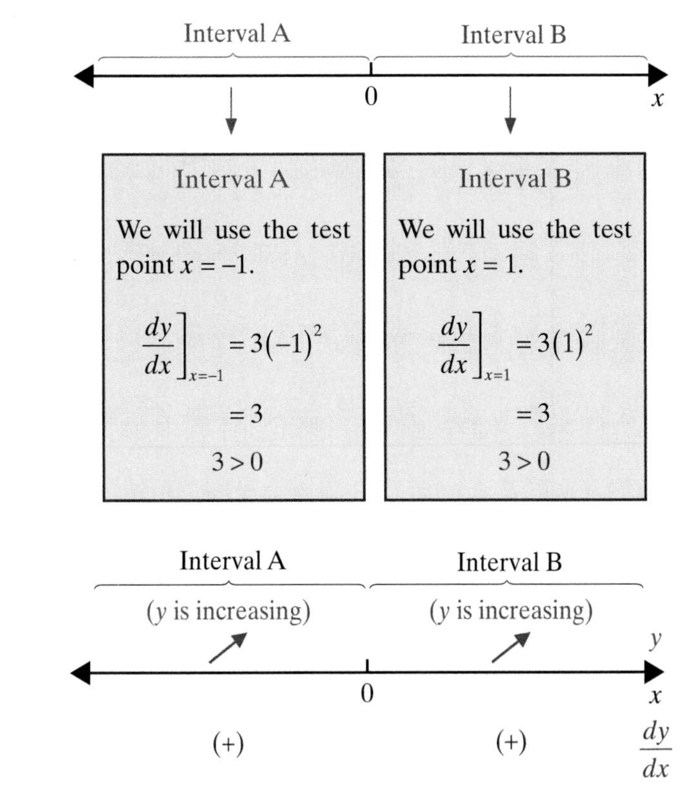

$\frac{dy}{dx} = 3x^2 \geq 0$ for all x.

c. Part **a.** determined a horizontal tangent at $x = 0$, and part **b.** showed that y is increasing on the interval $(-\infty, +\infty)$. (**Note:** since y is increasing on either side of 0, it is also increasing on any interval that contains 0.)

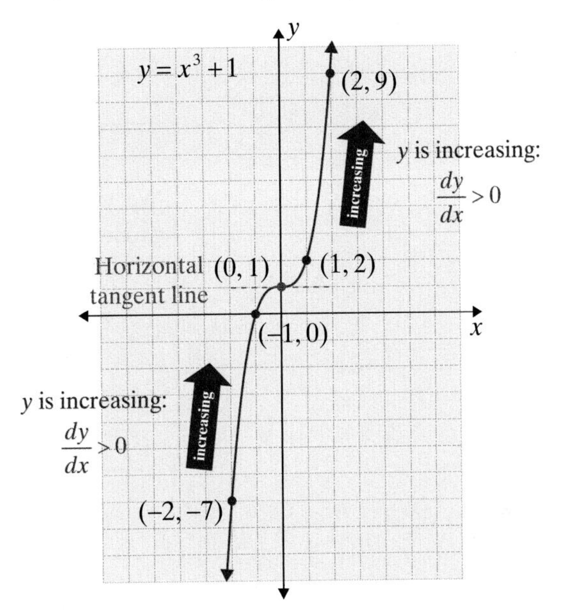

Example 4: Graphing a Function

For the function $f(x) = \frac{1}{x}$,

a. Find all values of x where the derivative is 0.
b. Determine where the function is increasing and where it is decreasing.
c. Sketch its graph.

Solutions: a. $f(x) = \frac{1}{x} = x^{-1}$ — Rewrite $f(x)$ using exponents.

$f'(x) = -x^{-2} = -\frac{1}{x^2} \neq 0$ — Find the derivative of $f(x)$.

There are no horizontal tangent lines. There are no x-values which will make the slope equal to 0 in the formula for f'. However, we note there is no slope defined for $x = 0$. It turns out that we need to put on our slope analysis line not only x-values which make $f'(x) = 0$ but also any x-values for which f' is not defined. In this problem, $x = 0$ is the only point in either category, and we put $x = 0$ as the only point on the analysis line.

continued on next page ...

b. Note that we still investigate the derivative on either side of 0 to help determine the nature of the graph.

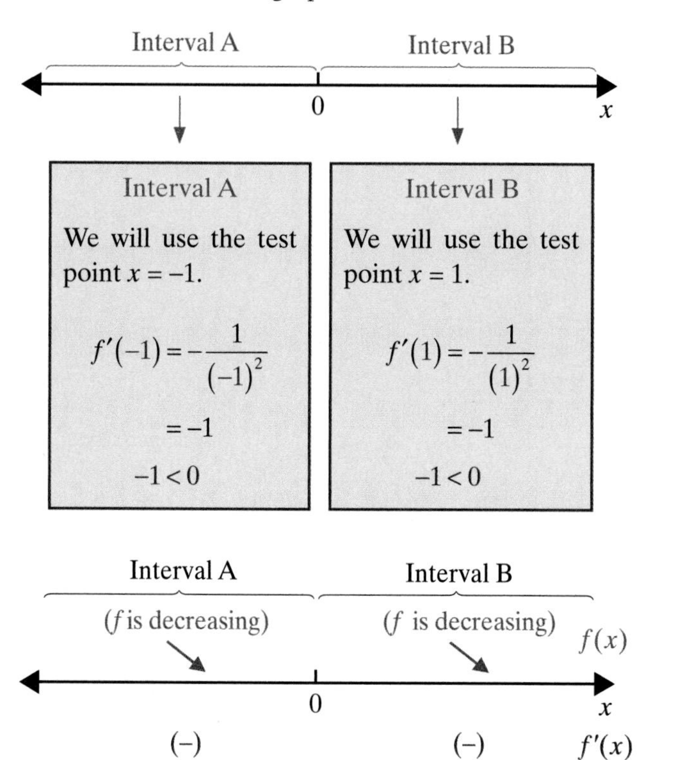

$$f'(x) = -\frac{1}{x^2} < 0 \text{ for all } x \neq 0.$$

c.

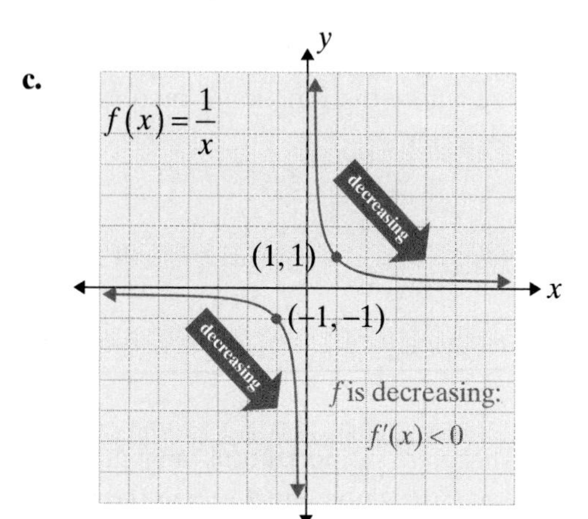

The line $x = 0$ is a **vertical asymptote**. The curve approaches this line without touching it. There is also a horizontal asymptote at $y = 0$.

In this example, $f(0)$ is not defined and so certainly $f'(0)$ is not defined. It can happen that $f(a)$ is defined, but $f'(a)$ is not defined.

Example 5: Graphing $f'(x)$

The graph for $y = f(x)$ is given below. Mark the portions which show where $f'(x)$ is negative, positive, or zero. Draw a possible $f'(x)$ by estimating the absolute values of the slopes on $f(x)$.

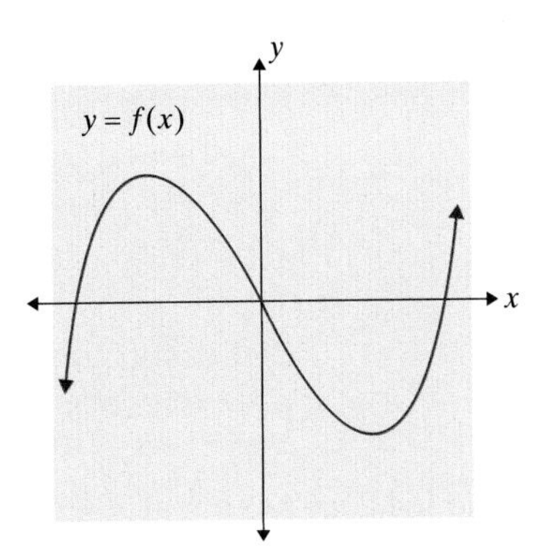

Solution: Let $x = a$ and $x = b$ (say $a < b$) denote the two x-values corresponding to the two local extremes. Then $f'(a) = f'(b) = 0$. In between a and b, the y-values decrease on the interval (a, b); therefore y' is negative. At $x = b$, the y-values start to increase (and y' becomes positive). Thus the graph of y' includes points $(a, 0)$ and $(b, 0)$ and lies below the x-axis (as y' is negative) from $x = a$ to $x = b$. For $x > b$ and $x < a$, the graph of y' lies above the x-axis. We draw a smooth curve from $(a, 0)$ to $(b, 0)$ and extend it in both directions above the x-axis. The low point for y' seems to occur near $x = 0$ because the graph of y is steepest near $(0, 0)$.

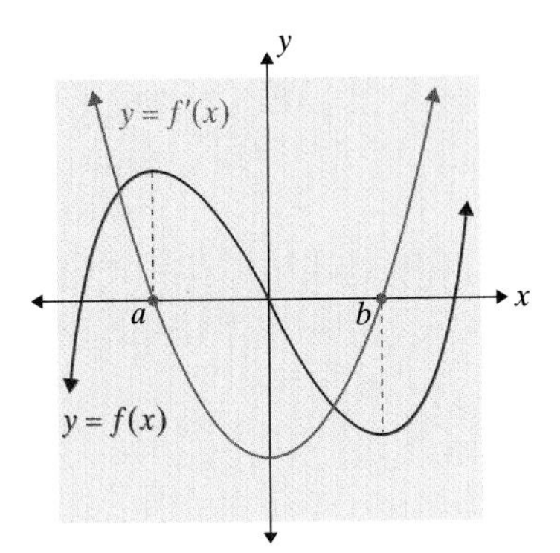

Critical Values

We have discussed the concept of a function increasing or decreasing on an interval. Now, we are interested in locating the points where a function changes from increasing to decreasing, or vice versa. These points are called **local extrema** (singular, **local extremum**). Our examples and discussion include continuous functions and discontinuous functions which have a jump discontinuity (say for a vertical asymptote). Figure 3.4.3 illustrates the basic terms **local maximum** and **local minimum**.

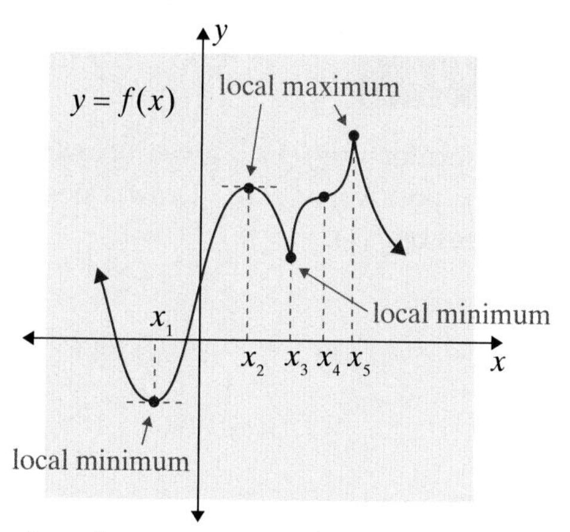

Figure 3.4.3

Local extrema occur for x_1, x_2, x_3, and x_5.
There is no local extremum for x_4.

There is no local extremum at $x = x_4$ since y is increasing on each side of $x = x_4$. However, a tangent line at $x = x_4$ would be horizontal and would cross the graph at $(x_4, f(x_4))$. Such unusual tangents are discussed in Chapter 4.

There is a local minimum at $(x_3, f(x_3))$ even though there is no tangent line at that point. Note that immediately to the left of x_3, slopes are large negative numbers, but just to the right of x_3, slopes are large positive numbers.

Local Maximum and Local Minimum

Let f be a function defined at x = c.

1. *f(c) is called a local maximum (or relative maximum) if there exists some interval (a, b) that contains c such that*

$$f(x) \le f(c)$$

for all x in (a, b). (We say a local maximum occurs at x = c.)

2. *f(c) is called a local minimum (or relative minimum) if there exists some interval (a, b) that contains c such that*

$$f(x) \ge f(c)$$

for all x in (a, b). (We say a local minimum occurs at x = c.)

Note: *We also say that a local extremum occurs at the point (c, f(c)) or that the point (c, f(c)) is a local extremum, with the understanding that the y-value, f(c), is actually the local extremum.*

Finding local extrema without having a graph to refer to can be done with the help of the derivative. We must be careful, however, because there can be several possibilities to consider.

Theorem of Local Extrema

If a function f is continuous on the interval (a, b) and c is in (a, b) and f(c) is either a local maximum or a local minimum, then

1. $f'(c) = 0$, *or*
2. $f'(c)$ *does not exist.*

Critical Values

Critical values of x are those values c in the domain of f where $f'(c) = 0$ (indicating horizontal tangent lines) or $f'(c)$ does not exist (indicating vertical tangent lines or sharp points).

For $f(x) = \frac{1}{x}$ (Example 4), $x = 0$ is not defined for f; however, it is treated as a critical point – that is, it needs to be one of the points marked on an analysis line.

As the theorem implies, **local extrema occur only at critical values**; **however, a critical value does not guarantee a local extremum.** Figure 3.4.4 illustrates some of the possibilities.

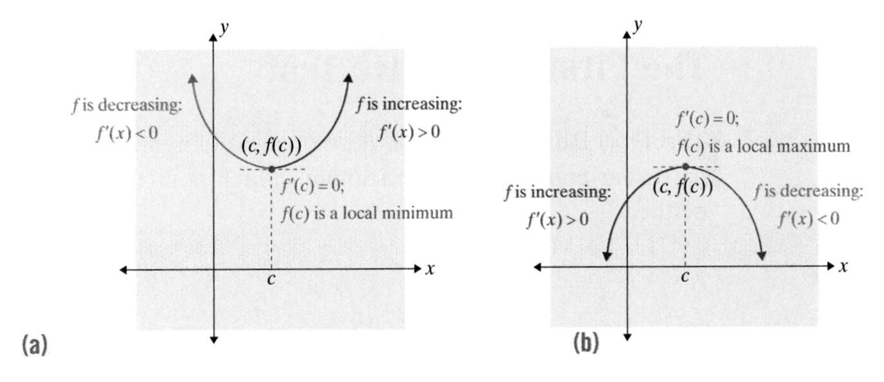

In both (a) and (b), $f'(c) = 0$ and the derivative changes sign from one side of $x = c$ to the other. Under these conditions, $f(c)$ is a local extremum.

Figure 3.4.4

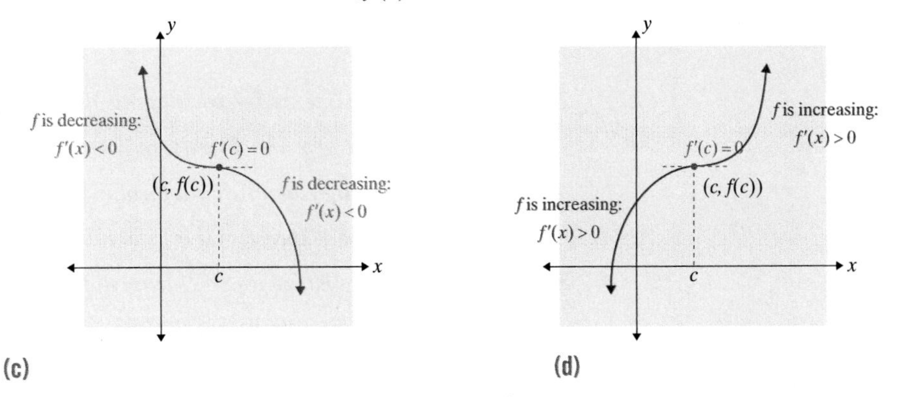

In both (c) and (d), $f'(c) = 0$. However, the derivative does not change sign, so the function is increasing (or decreasing) on both sides of $x = c$. Thus $f(c)$ is not a local extremum.

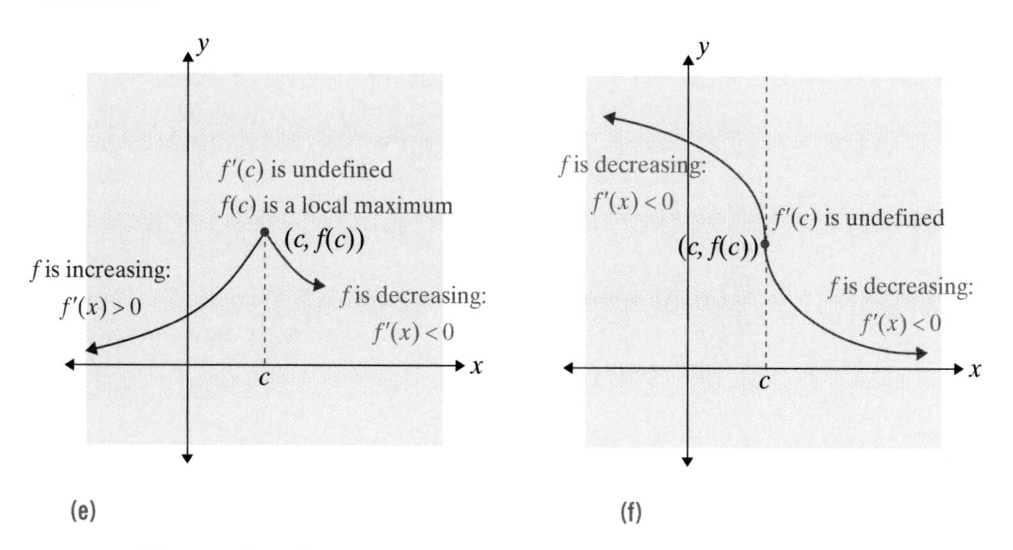

(e) (f)

In both (e) and (f), $f'(c)$ is undefined. In (e), $f(c)$ is a local extremum because the derivative changes sign from one side of $x = c$ to the other. In (f), the derivative does not change sign and the function is decreasing on both sides of $x = c$, indicating that $f(c)$ is not a local extremum.

The First Derivative Test

The First Derivative Test is developed as a numerical test to determine whether a particular point is a local maximum (a high point), a local minimum (a low point), or neither.

The First Derivative Test for Local Extrema

To determine the local extrema for a nonconstant function f continuous on the interval (a, b):

1. *Find the critical values, x-values c in (a, b) such that $f'(c) = 0$ or $f'(c)$ is undefined.*
2. *Draw an analysis line, and mark all critical x-values on it. Check the sign of $f'(x)$ on each side of $x = c$. If the sign changes,*
   - **i.** *from + to –, then f(c) is a local maximum.*
   - **ii.** *from – to +, then f(c) is a local minimum.*
3. *If there is no sign change for $f'(x)$ from the left side of $x = c$ to the right side, then f(c) is not a local extremum.*

The following examples show how to use the First Derivative Test to find local extrema and sketch the graph of a function.

Example 6: Using the First Derivative Test

Let $f(x) = -\frac{1}{3}x^3 + x^2 + 3x + 1.$

a. Find the critical values of f.
b. Use the First Derivative Test to find any local extrema.
c. Sketch the graph of f.

Solutions: a. $f'(x) = -\frac{1}{\not{3}} \cdot \not{3}x^2 + 2x + 3$ Find the derivative of $f(x)$.

$= -x^2 + 2x + 3$

Set $f'(x) = 0$ to find the critical values.

$-x^2 + 2x + 3 = 0$

$x^2 - 2x - 3 = 0$ Multiply both sides by –1.

$(x+1)(x-3) = 0$ Factor.

$x = -1, 3$

Thus the critical values are $x = -1$ and $x = 3$.

b. Step 1 of the First Derivative Test was completed in part **a.** Now we need to check the sign of $f'(x)$ on either side of the critical values (Step 2). Use the technique developed in Examples 1 – 4.

Interval A	Interval B	Interval C
We will use the test point $x = -2$.	We will use the test point $x = 0$.	We will use the test point $x = 5$.
$f'(-2) = -(-2)^2 + 2(-2) + 3$	$f'(0) = -(0)^2 + 2(0) + 3$	$f'(5) = -(5)^2 + 2(5) + 3$
$= -4 - 4 + 3$	$= 0 + 0 + 3$	$= -25 + 10 + 3$
$= -5$	$= 3$	$= -12$
$-5 < 0$	$3 > 0$	$-12 < 0$

continued on next page ...

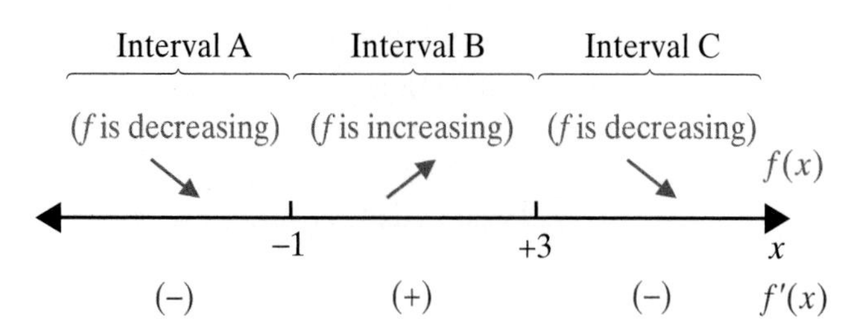

For both of the critical values, there is a sign change from one side of the value to the other. Therefore, local extrema occur at both $x = -1$ and $x = 3$. (Step 3).

We compute the y-values:

At $x = -1$:

$$f(-1) = -\frac{1}{3}(-1)^3 + (-1)^2 + 3(-1) + 1$$
$$= -\frac{2}{3}$$

$\left(-1, -\frac{2}{3}\right)$ is a local minimum.

At $x = 3$:

$$f(3) = -\frac{1}{\not{3}}(3)^{\not{3}\,2} + (3)^2 + 3(3) + 1$$
$$= 10$$

$(3, 10)$ is a local maximum.

c. Use the following summary of information (from parts **a.** and **b.**) to sketch the curve.

1. f is decreasing on $(-\infty, -1)$.
2. f is increasing on $(-1, 3)$.
3. f is decreasing on $(3, +\infty)$.
4. A local minimum occurs at the point $\left(-1, -\frac{2}{3}\right)$.
5. A local maximum occurs at the point $(3, 10)$.

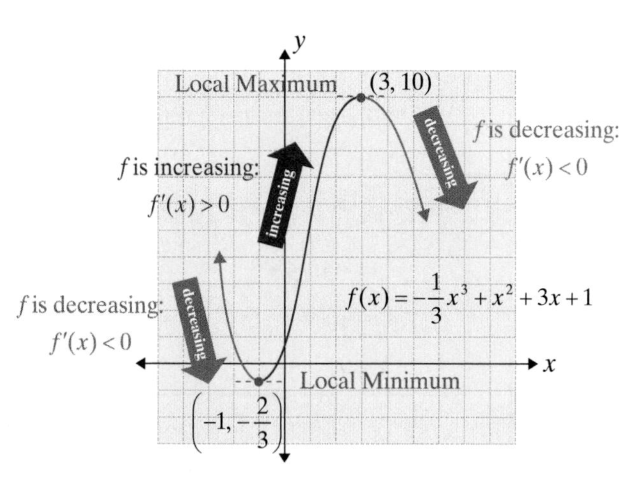

Example 7: Using the First Derivative Test

Let $f(x)=(x-2)^3+1$.

a. Find the critical values of f.
b. Use the First Derivative Test to find any local extrema.
c. Sketch the graph of f.

Solutions: a. $f'(x)=3(x-2)^2$ Find the derivative of $f(x)$.

Set $f'(x) = 0$ to find the critical values.

$$3(x-2)^2=0$$
$$x=2$$

The only critical value is $x = 2$.

b. Step 1 of the First Derivative Test was completed in part **a.** Now we need to check the sign of $f'(x)$ on either side of the critical value (Step 2).

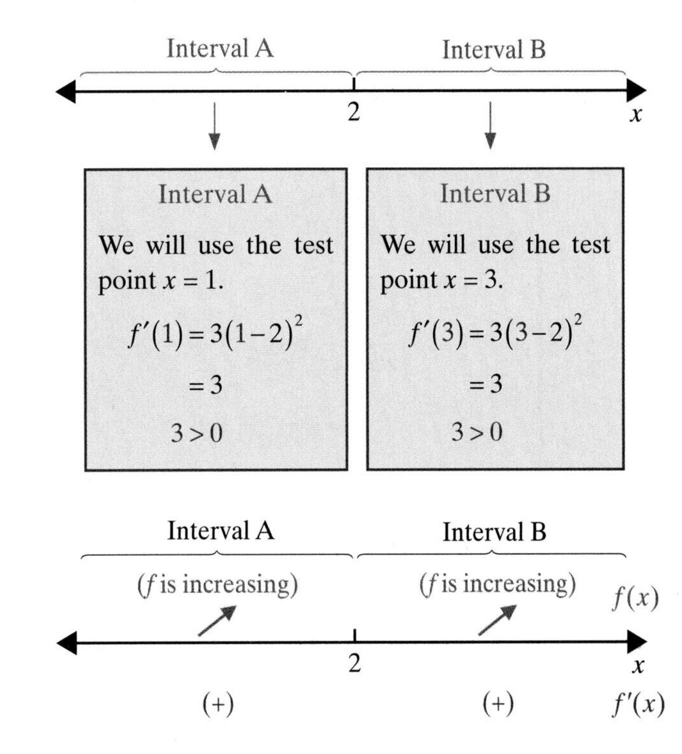

Now we have, $f'(x)=3(x-2)^2 \geq 0$ for all x. Since there is no sign change from one side of the critical value to the other, $x = 2$ is not a local extremum (Step 3).

continued on next page ...

c. Use the following summary of information (from parts **a.** and **b.**) to sketch the curve.

1. f is increasing on $(-\infty, +\infty)$.

2. $f'(2) = 0$, but the point $(2, 1)$ is not a local extremum.

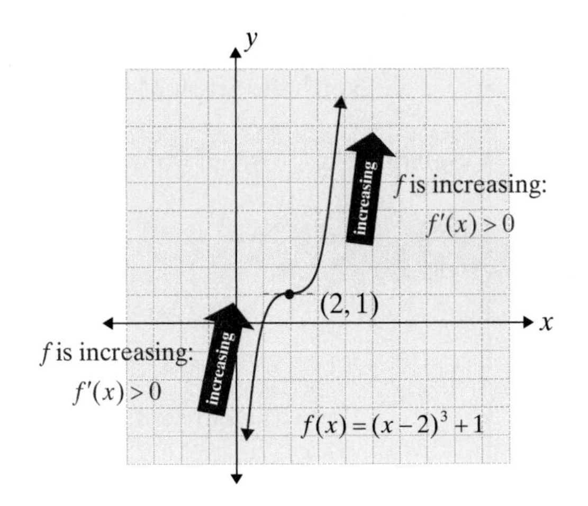

Example 8: Critical Values

Find the critical values for the function $y = \dfrac{1}{x-2}$.

Solution: $y = \dfrac{1}{x-2} = (x-2)^{-1}$ — Rewrite y using exponents.

$\dfrac{dy}{dx} = -(x-2)^{-2}$ — Find the derivative of y.

$= -\dfrac{1}{(x-2)^2} \neq 0$ — y' cannot equal 0. Recall, an algebraic fraction can be zero if and only if its numerator is zero.

Note that $\dfrac{dy}{dx}$ is undefined at $x = 2$ (and defined for all other values of x). The original function, $y = \dfrac{1}{x-2}$, is also not defined at $x = 2$.

Therefore, $x = 2$ is **not** in the domain of the function and is **not** a critical value. There are no critical values. However, $x = 2$ is the only x-value "marked" on the analysis line because the derivative can change signs from positive to negative or vice versa only at $x = 2$.

We will analyze and graph functions similar to that in Example 8 in more detail in Chapter 4.

Example 9: Critical Values

Find the critical values for the function $y = x^{\frac{2}{3}}$.

Solution:

$$y' = \frac{2}{3}x^{-\frac{1}{3}} = \frac{2}{3x^{\frac{1}{3}}} = \frac{2}{3\sqrt[3]{x}} \neq 0$$

Find the derivative of y. Note: y' cannot equal 0 since the numerator cannot equal 0.

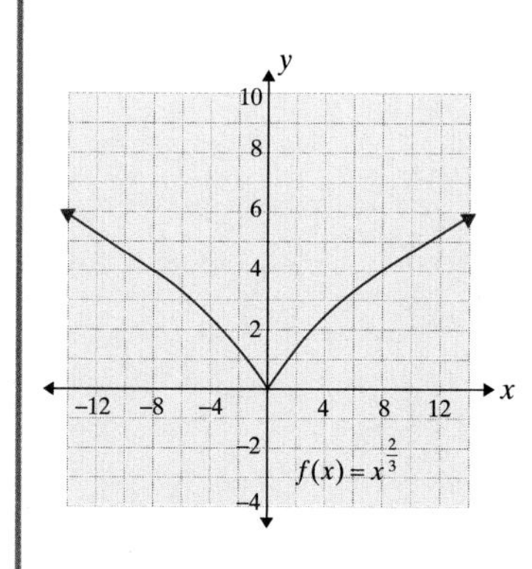

In this case, y' is not defined at $x = 0$. However, since $x = 0$ **is** in the domain of the original function, $x = 0$ **is** a critical value. Also, because y' is never equal to 0, there are no other critical values; only $x = 0$. To the left of $x = 0$, f' is negative (y decreases). To the right of $x = 0$, f' is positive (y increases). Near $x = 0$, the absolute values of slope are huge. We conclude there is a sharp point at $x = 0$, where $x = 0$ is a critical value.

3.4 Exercises

In Exercises 1 – 10, find the open intervals on which **a.** *f is increasing, and* **b.** *f is decreasing.*

1.

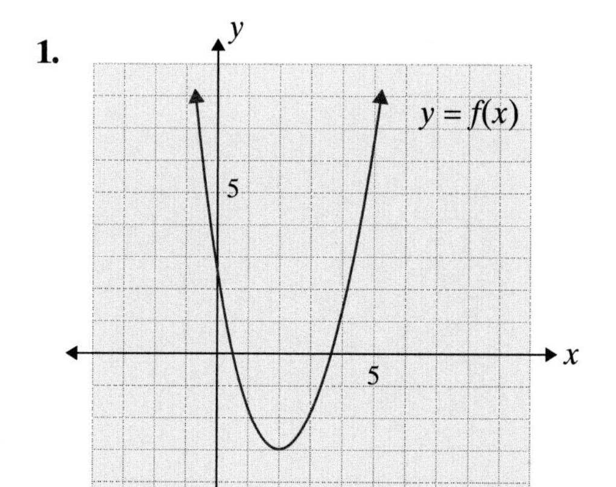

2.

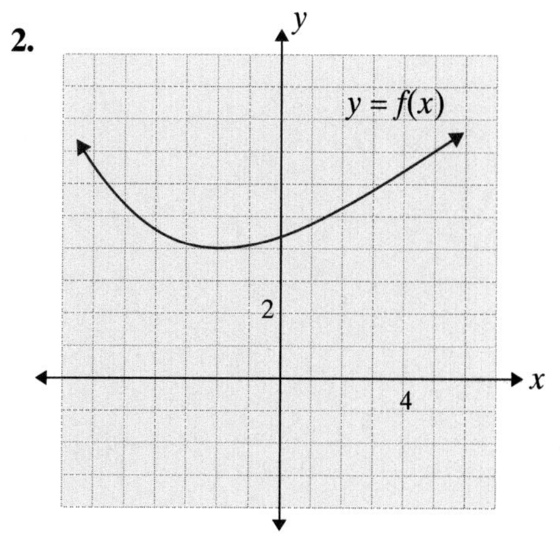

3.

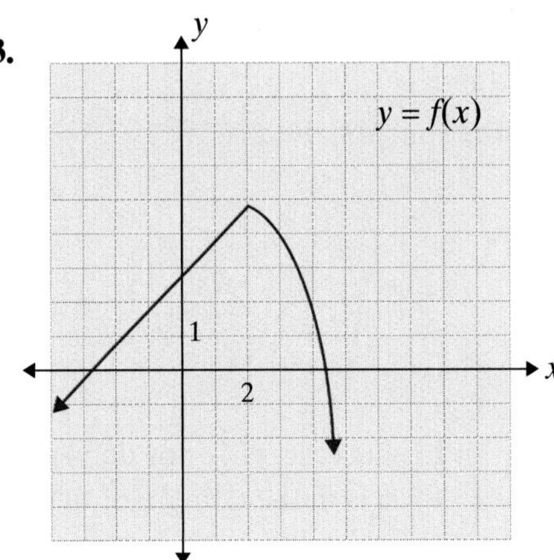

4.

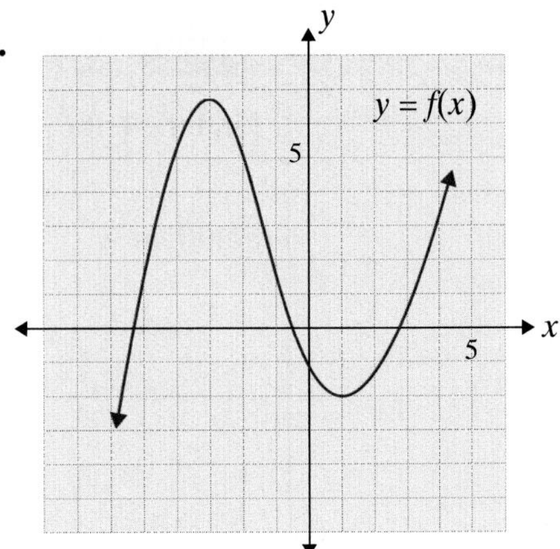

5.

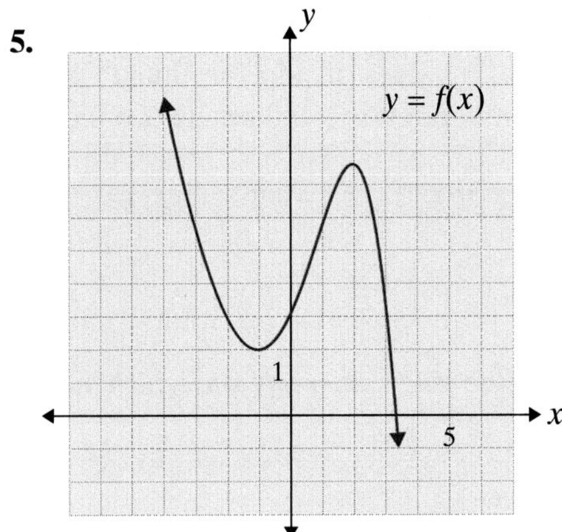

6.

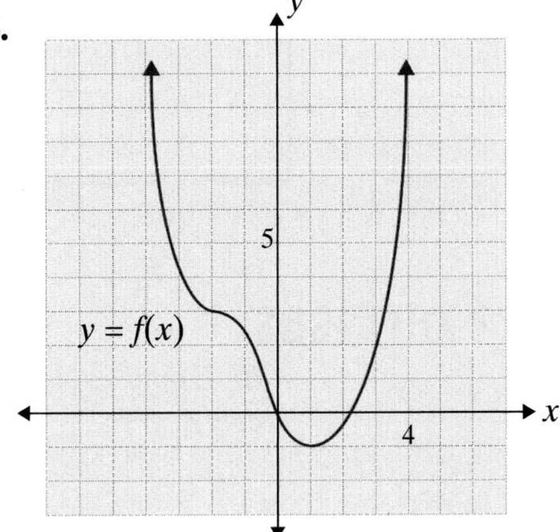

7.

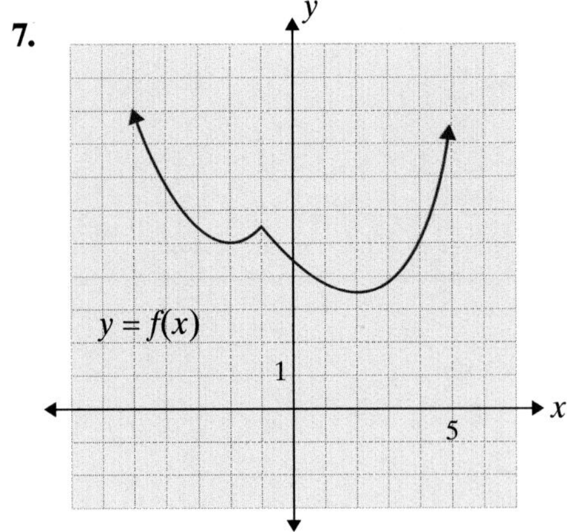

8.

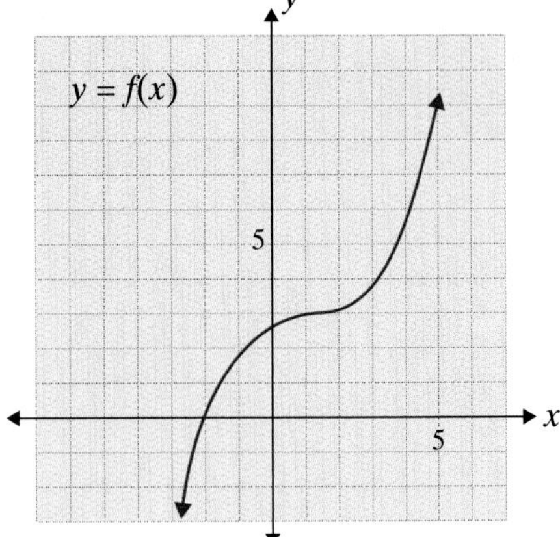

9.

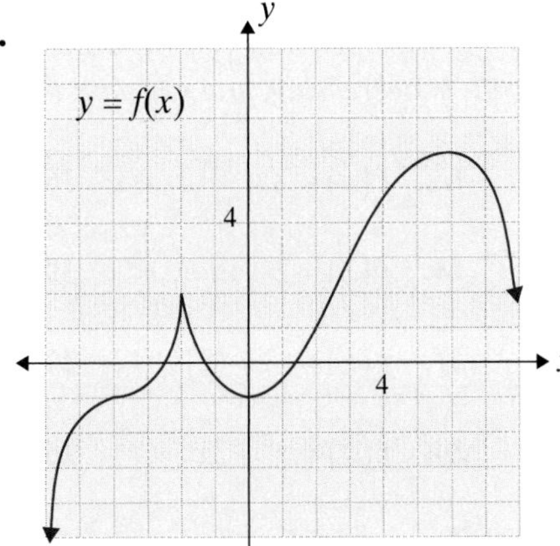

10.

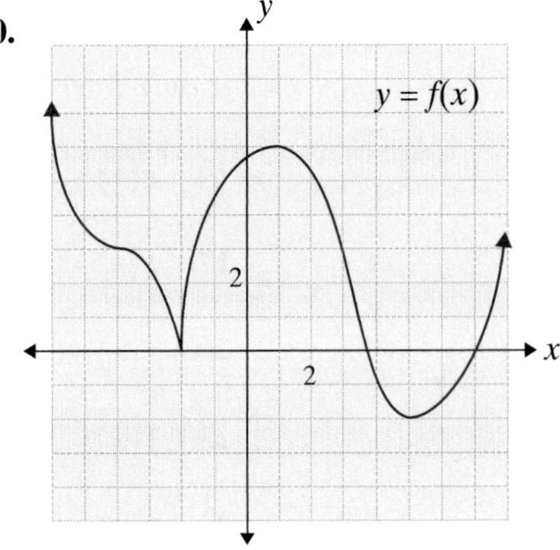

For each of the graphs of $f(x)$ in Exercises 11 – 13, **a.** *determine the open intervals on which the function is increasing and the open intervals on which it is decreasing, and* **b.** *sketch a possible graph of $f'(x)$.*

11.

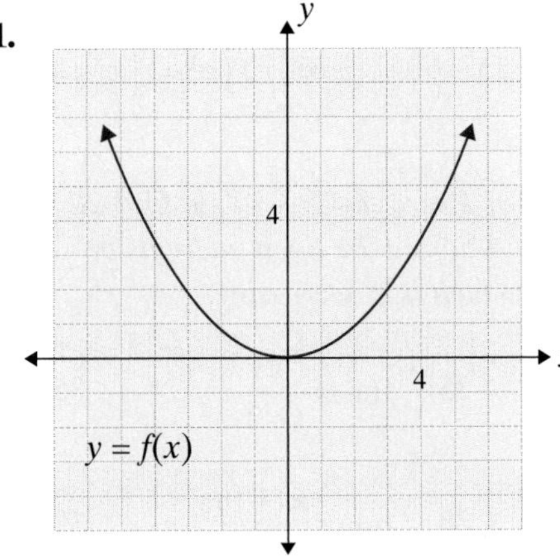

12.

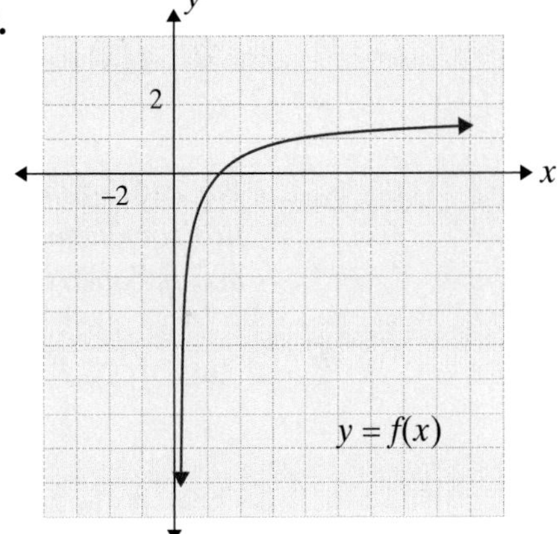

13.

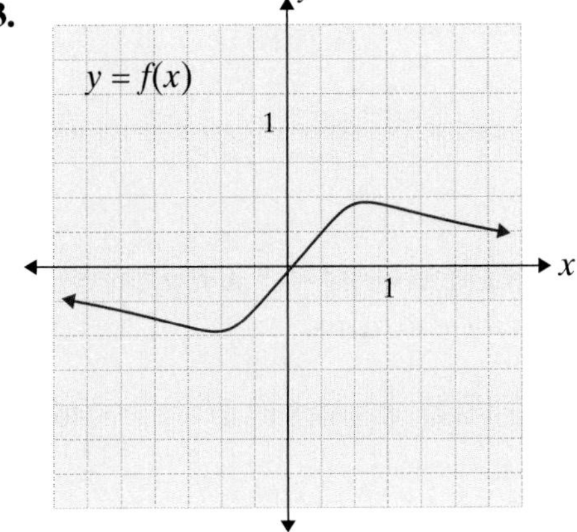

For each of the functions in Exercises 14 – 33, **a.** *find all values of x that correspond to horizontal tangent lines,* **b.** *find the open intervals on which the function is increasing and the open intervals on which it is decreasing, and* **c.** *graph the function.*

14. $f(x)=x^2-8x+3$ **15.** $f(x)=2x^2+12x-1$ **16.** $f(x)=5-3x-x^2$

17. $f(x)=7x-2x^2$ **18.** $f(x)=2-4x-2x^2$ **19.** $f(x)=3x^2-4x+2$

20. $f(x)=(2x+3)^2$ **21.** $f(x)=(3x-2)^2$ **22.** $f(x)=2x^3-5$

23. $f(x)=3x^3+4$ **24.** $f(x)=x^3-3x^2+7$ **25.** $f(x)=x^3-6x^2-4$

26. $f(x)=x^3-3x^2-9x+12$ **27.** $f(x)=x^3-x^2-x$

28. $f(x)=x^3+\frac{1}{2}x^2-2x+3$ **29.** $f(x)=x^3-x^2-5x+2$

30. $f(x)=\frac{1}{3}x^3-x^2-8x+10$ **31.** $f(x)=\frac{1}{3}x^3-2x^2+3x-6$

32. $f(x)=\frac{1}{3}x^3-\frac{3}{2}x^2+2x-6$ **33.** $f(x)=\frac{1}{3}x^3+\frac{5}{2}x^2+4x+11$

For each of the functions in Exercises 34 – 43, **a.** *find all values of x that correspond to horizontal tangent lines, and* **b.** *find the open intervals on which the function is increasing and the open intervals on which it is decreasing.*

34. $f(x)=\frac{x-1}{x}$ **35.** $f(x)=\frac{x}{x-2}$ **36.** $f(x)=\frac{x^2-4}{x}$

37. $f(x)=\frac{x^2-9}{x}$ **38.** $f(x)=\frac{x^3+16}{x}$ **39.** $f(x)=\frac{2x^3-27}{x^2}$

40. $f(x)=\frac{x+2}{x-1}$ **41.** $f(x)=\frac{x-5}{x+3}$ **42.** $f(x)=2x-\frac{125}{x^2}$

43. $f(x)=x^2+\frac{128}{x}$

For each of the functions in Exercises 44 – 63, **a.** *find the critical values, and* **b.** *use the First Derivative Test to find any local extrema.*

44. $f(x)=4x-x^2$ **45.** $f(x)=9x-x^2$ **46.** $f(x)=x^2+6x-2$

47. $f(x)=x^2-10x+12$ **48.** $f(x)=\frac{1}{2}x^2-4x+3$ **49.** $f(x)=-\frac{1}{2}x^2+3x-2$

50. $f(x) = x^3 + x^2 - x + 3$

51. $f(x) = x^3 + 2x^2 + x - 2$

52. $f(x) = -x^3 - \frac{3}{2}x^2 + 18x + 6$

53. $f(x) = 2x^3 + x^2 - 4x$

54. $f(x) = x^3 - 3x + 6$

55. $f(x) = x^3 + 3x^2 - 4$

56. $f(x) = x + \frac{9}{x}$

57. $f(x) = x - \frac{4}{x}$

58. $f(x) = \frac{x^2 - 16}{x}$

59. $f(x) = \frac{4x^2 - 9}{x}$

60. $f(x) = 9x + x^{-1}$

61. $f(x) = 25x + x^{-1}$

62. $f(x) = 16x + x^{-2}$

63. $f(x) = 54x - x^{-2}$

64. Revenue. A store manager has determined that the revenue from the sale of x units of a product is given by $R(x) = 32x - 0.4x^2$ dollars, where $0 \le x \le 80$. On what interval of sales is the revenue increasing, and on what interval of sales is it decreasing?

65. Revenue. A producer of computer software has determined that the revenue from the production and sale of x units is given by $R(x) = 48x - 0.003x^2$ dollars, where $0 \le x \le 10{,}000$. For what interval of production is the revenue increasing, and for what interval is it decreasing?

66. Profit. The revenue from the sale of x coffeemakers is given by $R(x) = 40x - 0.4x^2$ dollars. The total cost is given by $C(x) = 370 + 16x - 0.2x^2$ dollars, where $0 \le x \le 100$. Determine the interval(s) where the profit is increasing and where it is decreasing.

67. Profit. The revenue from the sale of x 50-gallon aquariums is given by $R(x) = 54x - 0.3x^2$ dollars. The total cost function is given by $C(x) = 0.1x^2 + 4x + 200$ dollars, where $0 \le x \le 100$. Determine the interval of sales for which the profit is increasing and the interval for which it is decreasing.

68. Population. The population of the inner-city district of a city is given in thousands by $P(t) = 24 - 0.3t + 0.01t^2$, where t is the number of months after the implementation of an urban renewal project. How long will it be before the population starts to increase?

69. Wildlife management. In an attempt to naturally control the elk population in a national park, the U.S. Fish and Game Department has reintroduced the wolf into the area. It is estimated that the population of the elk herd will be $P(t) = 600 + 12t - 4t^{\frac{3}{2}}$, where t is the number of years after the reintroduction of the wolf. How long will it be before the elk population begins to decrease?

70. Average cost. The cost of producing x clock radios is given in dollars by $C(x) = 320 + 30x + 0.2x^2$, where $x \geq 0$. Determine the interval of production for which the average cost function is increasing.

71. Average cost. The cost of producing x units of a product is given in dollars by $C(x) = 250 + 45x - 0.2x^2$, where $x \geq 0$. Show that the average cost function is always decreasing. (This case corresponds to situations in which increased production distributes the cost so that the cost per unit decreases.)

72. Rate of dictation. It has been determined that after t weeks of class, the average students in an intermediate shorthand class can take dictation at a rate of $W(t) = 60 + \frac{70t^2}{t^2 + 15}$ words per minute. Show that the rate of dictation is an increasing function which approches an upper bound.

73. Court reporting. A typical student in an intermediate court reporting class can reach a level of recording $W(t) = 40 + \frac{35t^2}{t^2 + 20}$ words per minute after t hours of instruction and practice. Show that the number of words recorded per minute increases up to a certain level.

74. Marathon running speed. The speed at which a marathon runner travels varies over time. The function $F(t) = 5 + \frac{5t^2}{t^2 + 200}$ describes the velocity of a particular runner at time t ($F(t)$ is in miles/hour). Determine the intervals over which the function is increasing or decreasing. Is this particular runner an experienced runner? (Experienced runners run faster near the end of the race.)

75. Heating. A frozen pizza is placed in the oven at $t = 0$. The function $F(t) = 30 + \frac{320t^2}{t^2 + 100}$ approximates the temperature of the pizza at time t. Show that the temperature approaches an upper bound. (The pizza will approach oven temperature over time.)

76. Cooling. A cup of hot coffee is placed in a cold room. The temperature in degrees Fahrenheit of the coffee is approximated by the function $F(t) = 190 - \frac{100t^2}{t^2 + 40}$ where t is the number of minutes the coffee has been in the room. Show that the temperature function is a decreasing function, and find the temperature of the room. (The coffee approaches room temperature over time.)

77. Velocity and acceleration. The velocity of a car varies according to the function $F(t) = \frac{t^3}{1875} - \frac{119t^2}{1500} + \frac{53t}{15} + 5$, where t is time ($0 < t < 100$). Determine any relative maximum and minimum velocities, and the time at which they occur. (Be sure to specify whether each is a maximum or mininum.)

78. **Braking.** The velocity of a skydiver after parachute deployment is given by the formula $F(t) = 180 - \frac{165t^2}{t^2+10}$, where t is the time after deployment. Show that the function is decreasing, and determine the terminal velocity for the diver-parachute system. (The velocity approaches terminal velocity as t goes to infinity.)

79. **Pressure.** The pressure in a pressure cooker is given by the function $F(T) = 1 + \frac{T^2}{T^2+180}$ where T is the temperature inside the kettle. Show that the function is increasing, and determine the upper bound for pressure (atm.).

80. **Space probe closing speed.** In order to minimize travel time, a probe on its way to a distant star accelerates for part of the voyage, and then decelerates to enter orbit safely. The closing speed of the probe with the star is given by the function $F(t) = \frac{t^2}{25} - 4t + 0.5$, where t is the number of years after launch and $F(t)$ is measured in percentage of light speed. At what time does the probe begin decelerating? What is the probe's maximum closing speed? (**Hint:** Closing speed is negative.)

81. **Computation speed.** An experimental super computer is undergoing testing to determine whether it will meet necessary specifications. The number of computations it can perform per second is found to be modeled by the function $F(t) = \frac{89t^3}{441000} - \frac{757t^2}{22050} + \frac{2227t}{1470}$, where $0 < t < 100$ is the number of minutes after startup and $F(t)$ is in millions of calculations/sec. Find all relative maximums and minimums for the function over the interval. Round to the nearest tenth. If the number of calculations is zero at any time after startup, the computer crashes. Does the computer crash during testing?

82. **Fuel economy.** The fuel consumption of an automobile is not constant. Fuel economy depends largely on the speed of the vehicle. The function $F(v) = -\frac{8v^3}{19125} + \frac{28v^2}{85} - \frac{288v}{85} + 80$ ($F(v)$ is in miles per gallon) describes the fuel consumption of a new hybrid vehicle, where $0 < v < 85$ is the velocity of the vehicle. On what intervals is the consumption increasing? Which velocities yield maximum efficiencies? (Remember, high efficiency means low consumption. Give answers to tenths place.)

Hawkes Learning Systems: Essential Calculus

Local Extrema p. 245 - 253 Ex. 1 - 43, 64 - 71
Critical Points and the First Derivative Test p. 253 - 261 Ex. 44 - 63, 72 - 82

3.5 Absolute Maximum and Minimum

Objectives

After completing this section, you will be able to:

1. *Find the absolute maximum and absolute minimum of a function.*
2. *Determine maxima and minima in real-life applications.*

An engineer wants to design a carburetor that will allow a car to attain maximum acceleration. Another engineer wants to design a carburetor that will use minimum amount of gasoline yet maintain a respectable acceleration capability. Manufacturers want to minimize costs and maximize profits. These situations illustrate how the concepts of maximum and minimum affect activities in business and manufacturing. Whenever the ideas under consideration can be modeled (or represented) with functions, calculus is a useful tool for determining maximum and/or minimum quantities.

In Section 3.4 we used the derivative to determine where a function is increasing or decreasing and to find local extrema. As was illustrated, a function may have several local extrema. However, a function can have only one largest value or **absolute maximum** and only one smallest value or **absolute minimum**.

Absolute Extrema

If c is in the domain of f and, for all x in the domain of f, we have

1. $f(x) \leq f(c)$, *then* $f(c)$ *is called the* ***absolute maximum*** *of* f.
2. $f(x) \geq f(c)$, *then* $f(c)$ *is called the* ***absolute minimum*** *of* f.

Figure 3.5.1 shows some possible situations involving absolute extrema.

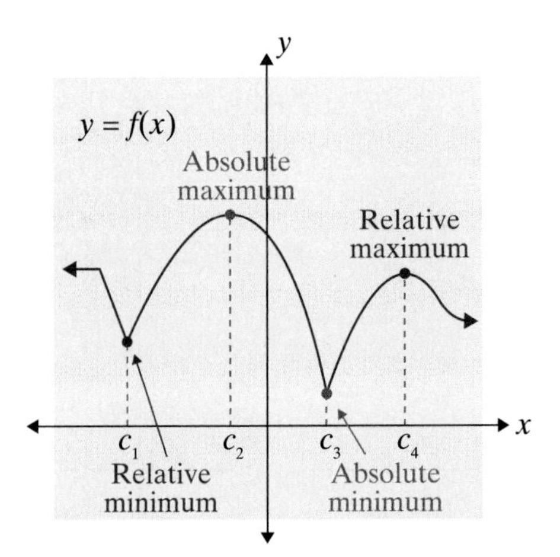

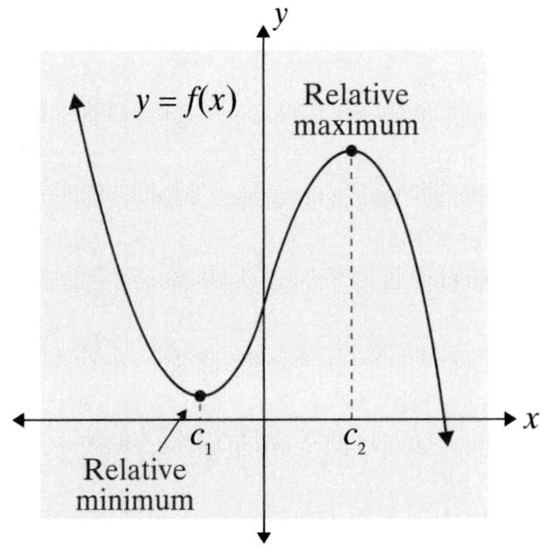

Figure 3.5.1

(a) $f(c_2)$ is the absolute maximum of f. $f(c_3)$ is the absolute minimum of f.

(b) $f(x)$ has no absolute maximum and no absolute minimum.

Generally, we are concerned with absolute extrema on a closed interval $[a, b]$ (i.e., an interval in which the endpoints a and b are included). The following theorem guarantees that a continuous function will indeed have an absolute maximum and an absolute minimum on a closed interval. The proof of this theorem is given in more advanced courses.

Theorem of Absolute Extrema of a Continuous Function

If a function f is continuous on a closed interval $[a, b]$, then f will have an absolute maximum value and an absolute minimum value on $[a, b]$.

NOTES

Although a particular discontinuous function may have an absolute maximum and an absolute minimum on a closed interval, such extrema cannot be guaranteed for all discontinuous functions or for open intervals. For example, $f(x) = \frac{1}{x}$ on the interval $\left(\frac{1}{2}, 3\right)$ has no extrema.

Now that we are assured of the existence of absolute extrema for a function continuous on a closed interval $[a, b]$, we want to know how to find these values. It can be proved that absolute extrema occur at the endpoints (a or b) or at the critical values c, where $f'(c) = 0$ or $f'(c)$ does not exist. Figure 3.5.2 illustrates some possible cases.

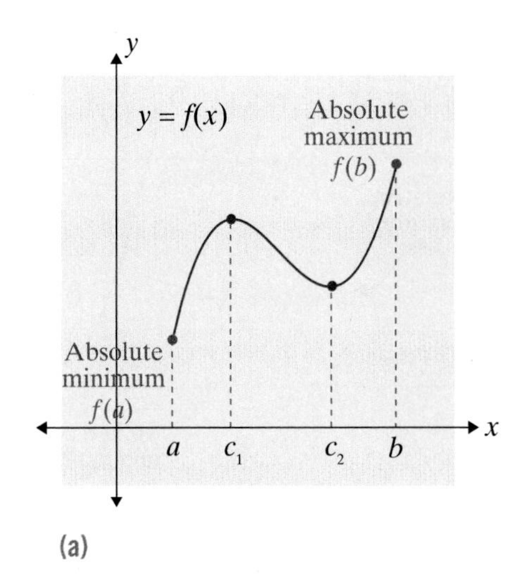

(a)

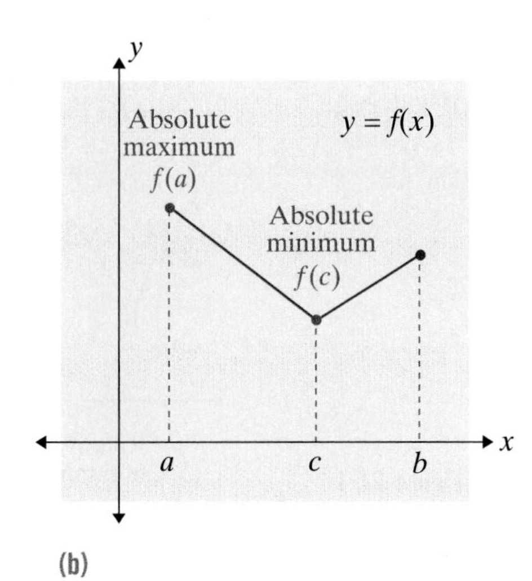

(b)

Figure 3.5.2

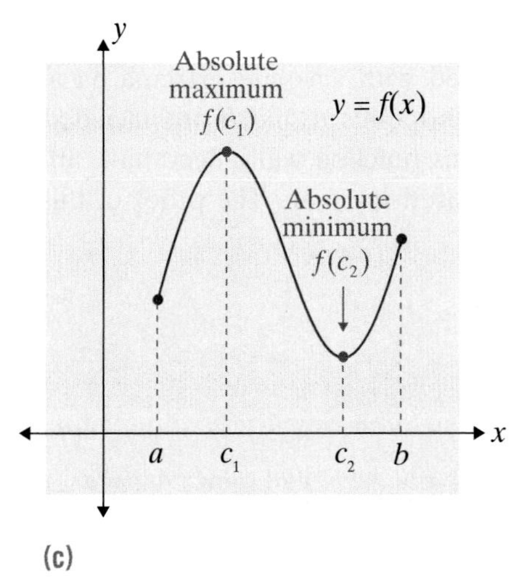

(c)

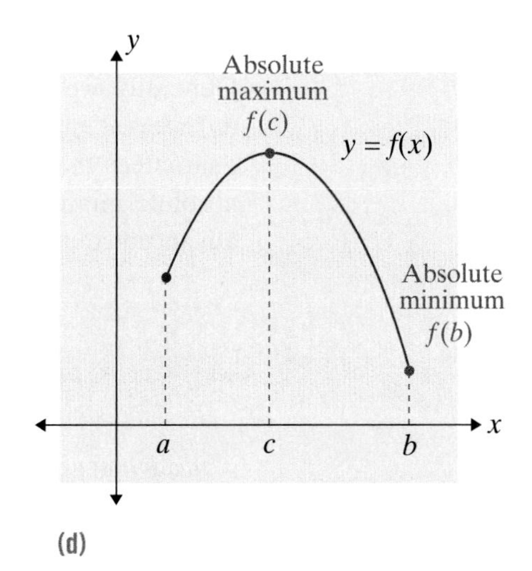

(d)

Finding the Absolute Extrema of a Continuous Function f on the Closed Interval $[a, b]$

1. *Find all the critical values for f in $[a, b]$. That is, find all c in $[a, b]$ where*
   - **a.** $f'(c) = 0$ *or*
   - **b.** $f'(c)$ *is undefined.*
2. *Evaluate $f(a)$, $f(b)$, and $f(c)$ for all critical values c.*
3. *The largest value found in step 2 is the absolute maximum. The smallest value found in step 2 is the absolute minimum.*

Example 1: Finding Absolute Extrema

Find the absolute extrema for $f(x) = 2x^3 - 15x^2 + 24x + 6$ on the interval [0, 5].

Solution: **Step 1:** Find all the critical values for $f(x)$ in [0, 5].

$$f'(x) = 6x^2 - 30x + 24$$
$$0 = 6x^2 - 30x + 24$$
$$0 = 6(x^2 - 5x + 4)$$
$$0 = 6(x-1)(x-4)$$
$$x = 1, 4$$

Find the critical values of f by setting $f'(x) = 0$ and solving for x.

Make sure both critical numbers $x = 1, 4$ are in the interval [0, 5].

Step 2: Evaluate $f(a), f(b)$, and $f(c)$ for all c in the given interval.

Both the critical values, $x = 1$ and $x = 4$, are in the interval [0, 5]. Therefore, evaluate $f(x)$ at $x = 0, x = 1, x = 4$, and $x = 5$.

$x = 0$	$x = 1$
$f(0) = 2(0)^3 - 15(0)^2 + 24 \cdot 0 + 6$	$f(1) = 2(1)^3 - 15(1)^2 + 24 \cdot 1 + 6$
$= 6$	$= 17$ Absolute max

$x = 4$	$x = 5$
$f(4) = 2(4)^3 - 15(4)^2 + 24 \cdot 4 + 6$	$f(5) = 2(5)^3 - 15(5)^2 + 24 \cdot 5 + 6$
$= -10$ Absolute min	$= 1$

Step 3: The largest value found in Step 2 is the absolute maximum and the smallest value is the absolute minimum.

So the absolute maximum is $f(1) = 17$ and occurs at $x = 1$. The absolute minimum is $f(4) = -10$ and occurs at $x = 4$.

continued on next page ...

Note: The full graph of f on [0, 5] is shown to help your understanding of the problem. **Note:** the full graph is easy to draw using only the analysis and the four (x, y)-coordinates (without a calculator).

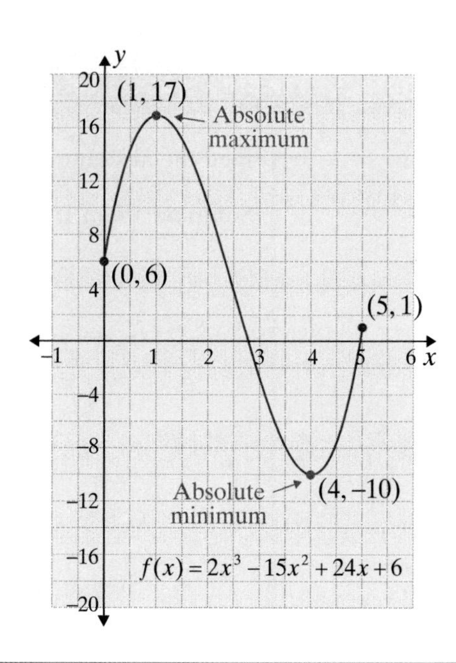

Example 2: Finding Absolute Extrema

Find the absolute extrema for $f(x) = -2x + 10$ on the interval $[1, 3]$.

Solution: Step 1: Find all the critical values for $f(x)$ in $[1, 3]$.

$$f'(x) = -2$$

$$0 \neq -2$$

Find the critical values of f by setting $f'(x) = 0$ and solving for x.

There are no critical values. Furthermore, since $f'(x) = -2 < 0$, f is decreasing for all x.

Step 2: Evaluate $f(a), f(b)$, and $f(c)$ for all c in the given interval.

Again, there are no critical values, c. Therefore, only the endpoints of the interval, $x = 1$ and $x = 3$, need to be evaluated.

$x = 1$	$x = 3$
$f(1) = -2(1) + 10$ $= 8$ Absolute max	$f(3) = -2(3) + 10$ $= 4$ Absolute min

Step 3: The largest value found in Step 2 is the absolute maximum and the smallest value is the absolute minimum.

So the absolute maximum is $f(1) = 8$ and the absolute minimum is $f(3) = 4$.

The graph of f on $[1, 3]$ is a straight line.

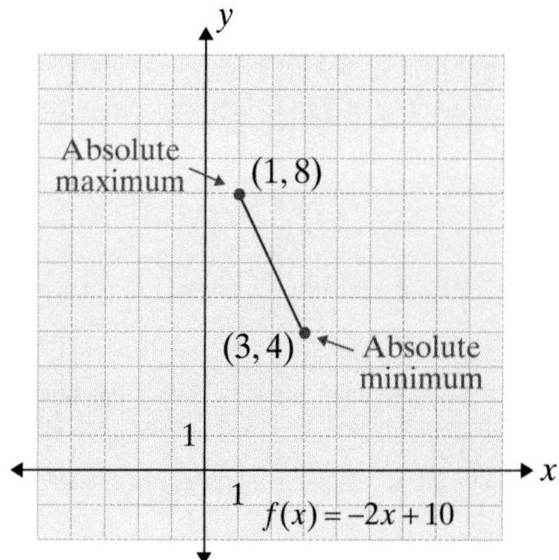

Example 3: Finding Absolute Extrema

Find the absolute extrema for $f(x) = x^{\frac{2}{3}} + 1$ on the interval $[-8, 8]$.

Solution: Step 1: Find all the critical values for $f(x)$ in $[-8, 8]$.

$$f'(x) = \frac{2}{3}x^{-\frac{1}{3}}$$

$$= \frac{2}{3x^{\frac{1}{3}}}$$

Find the critical values of f by setting $f'(x) = 0$ and solving for x.

Since $f'(0)$ is undefined, $x = 0$ is a critical value. In this case, $x = 0$ is the only critical value.

Step 2: Evaluate $f(a), f(b)$, and $f(c)$ for all c in the given interval.

Now evaluate $f(x)$ at $x = -8$, $x = 0$, and $x = 8$.

continued on next page ...

$x = -8$	$x = 0$
$f(-8) = (-8)^{\frac{2}{3}} + 1$ $= 5$ Absolute max	$f(0) = (0)^{\frac{2}{3}} + 1$ $= 1$ Absolute min

$x = 8$
$f(8) = (8)^{\frac{2}{3}} + 1$ $= 5$ Absolute max

Step 3: The largest value found in Step 2 is the absolute maximum and the smallest value is the absolute minimum.

In this case, there are two x-values that result in the absolute maximum. These are $x = -8$ and $x = 8$ (note that these are the end points) and the value of the absolute maximum is 5. The absolute minimum is $f(0) = 1$.

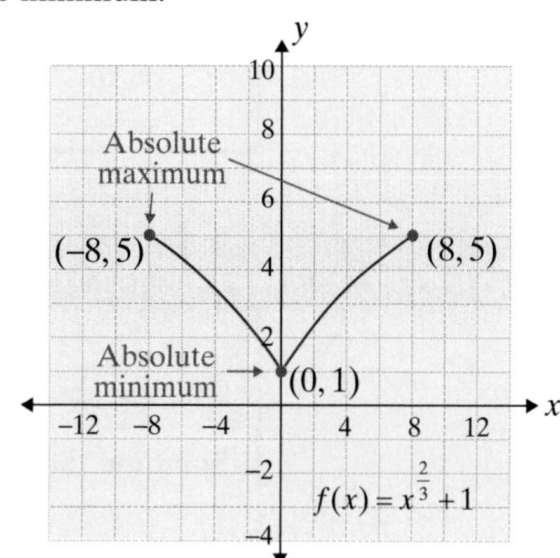

Example 4: Finding Absolute Extrema

Find the absolute extrema for $f(x) = \frac{1}{3}x^3 - 9x$ on the interval [0, 4].

Solution: Step 1: Find all the critical values for $f(x)$ in [0, 4].

$$f'(x) = x^2 - 9$$

$$0 = x^2 - 9$$

$$0 = (x+3)(x-3)$$

$$x = -3, 3$$

Find the critical values of f by setting $f'(x) = 0$ and solving for x.

Although $x = -3$ and $x = 3$ are critical values for the function, $x = -3$ is not in the interval [0, 4], and must be disregarded.

Step 2: Evaluate $f(a), f(b)$, and $f(c)$ for all c in the given interval.

Since the critical value $x = -3$ is not in the interval, evaluate $f(x)$ only at $x = 0, x = 3$, and $x = 4$.

x = 0

$$f(0) = \frac{1}{3}(0)^3 - 9(0)$$

$= 0$ Absolute max

x = 3

$$f(3) = \frac{1}{3}(3)^3 - 9(3)$$

$= -18$ Absolute min

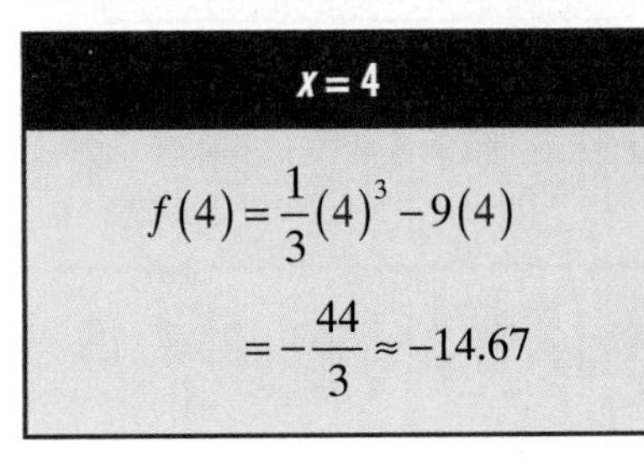

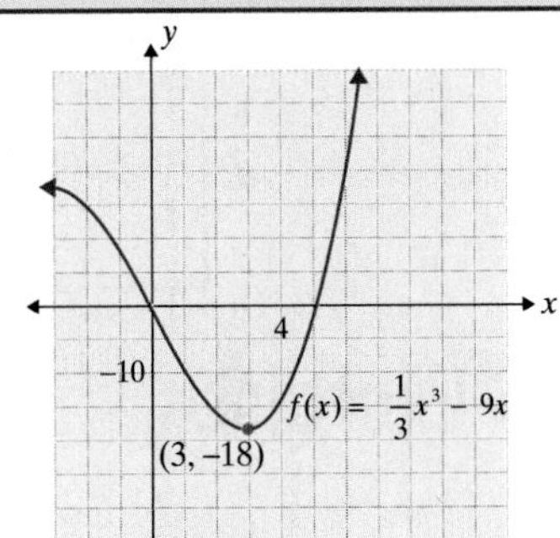

Step 3: The largest value found in Step 2 is the absolute maximum and the smallest value is the absolute minimum. So the absolute maximum is $f(0) = 0$ and the absolute minimum is $f(3) = -18$.

Example 5: Maximizing Profits

A company finds that its profit in dollars for producing x units of a product in one week is given by $P(x) = -2x^2 + 1600x$. If the company is set up so that no more than 500 units can be manufactured in any one week, how many units should the company produce to maximize profit?

Solution: The production restrictions indicate that x is in the closed interval [0, 500]. Find the critical values by setting $P'(x) = 0$ and solving for x (Step 1).

$$P'(x) = -4x + 1600$$
$$0 = -4x + 1600$$
$$4x = 1600$$
$$x = 400$$

Now evaluate $P(x)$ for $x = 0, x = 400$, and $x = 500$ (Step 2).

continued on next page ...

$x = 0$

$$f(0) = -2(0)^2 + 1600 \cdot 0$$
$$= 0$$

$x = 400$

$$f(400) = -2(400)^2 + 1600 \cdot 400$$
$$= 320{,}000 \quad \text{Absolute max}$$

$x = 500$

$$f(500) = -2(500)^2 + 1600 \cdot 500$$
$$= 300{,}000$$

The profit is maximized at $320,000 when 400 units are produced (Step 3).

3.5 Exercises

In Exercises 1 – 8, find the absolute extrema for each graph of f(x) on the given interval.

1. [3, 5]

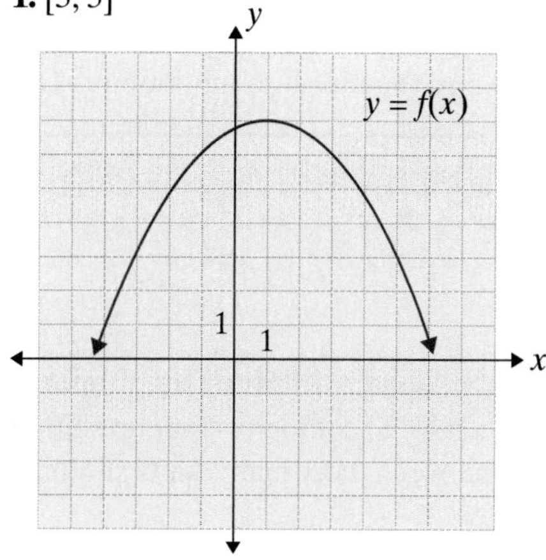

2. [–1, 4]

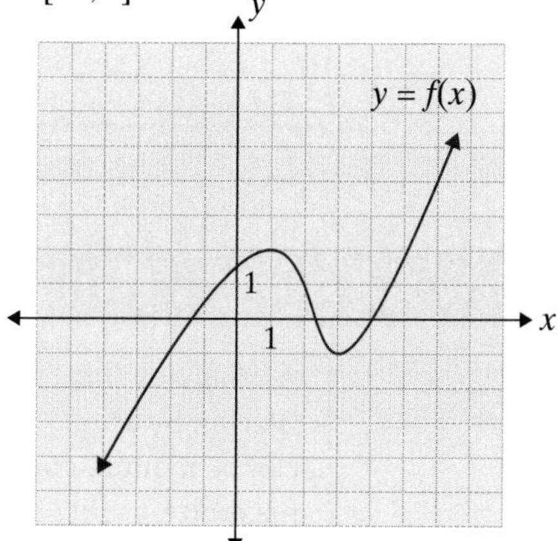

3. [–5, 3]

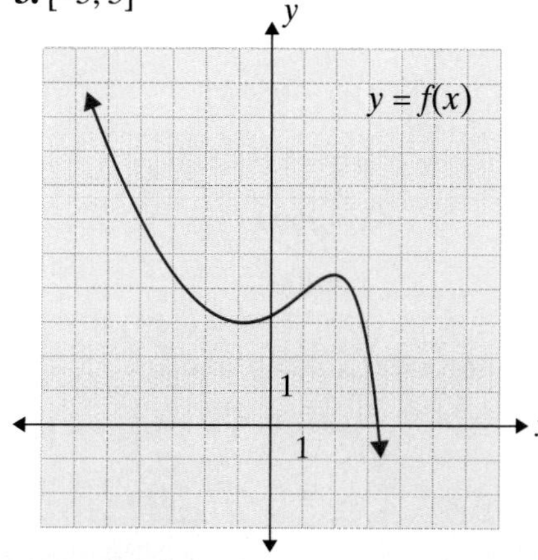

4. [–4, 4]

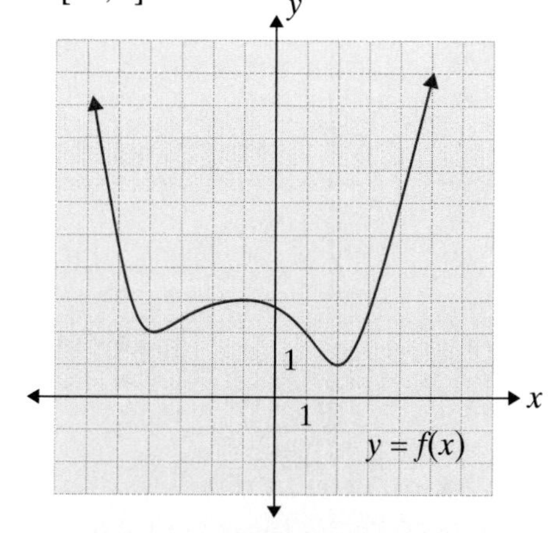

5. [1, 4]

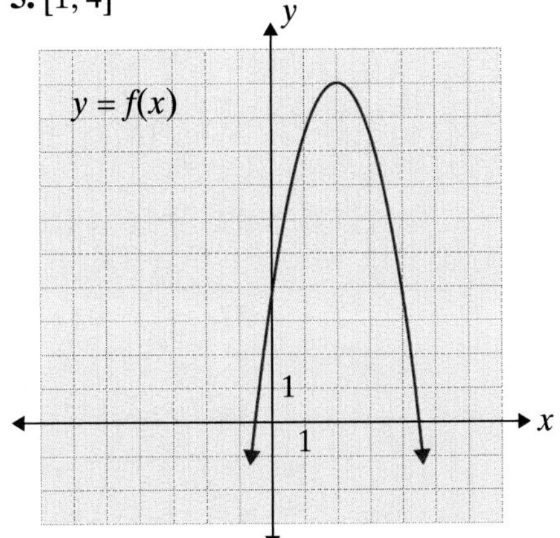

6. [–2, 2]

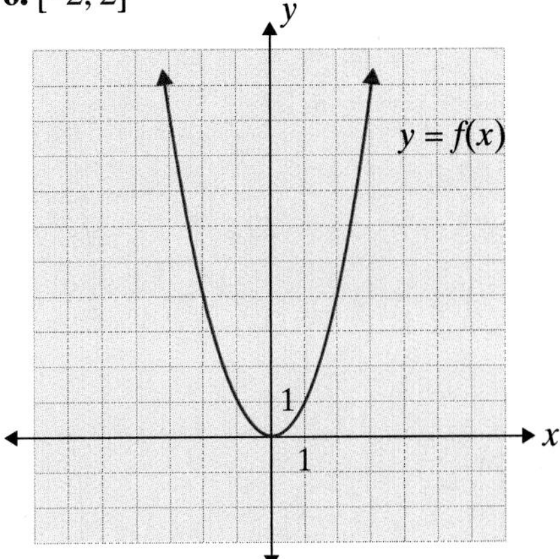

7. [–3, 5]

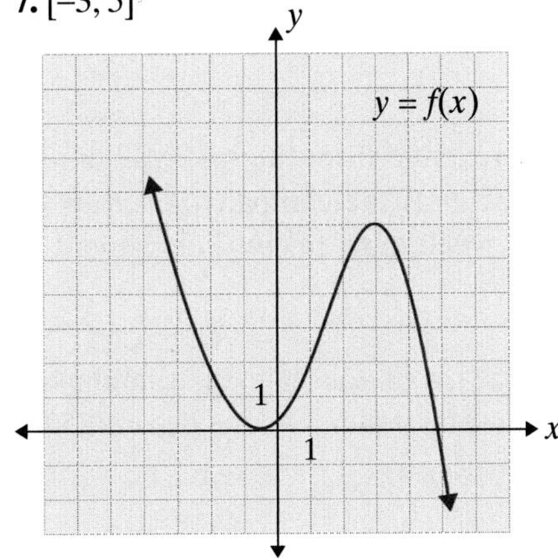

8. [–4, 5]

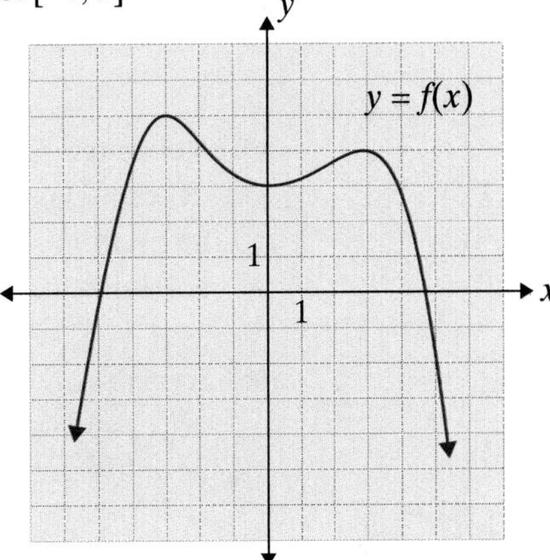

In Exercises 9 – 38, find the absolute extrema for each function on the given interval.

9. $f(x) = x^2 - 8x;\ [0, 5]$

10. $f(x) = 3x^2 - 12x;\ [0, 4]$

11. $f(x) = 6 + 10x - x^2;\ [3, 6]$

12. $f(x) = 11 - 4x - x^2;\ [-3, 0]$

13. $f(x) = 14 - 3x;\ [0, 4]$

14. $f(x) = 7 + \frac{1}{2}x;\ [-2, 4]$

15. $f(x) = 8 - x^3;\ [-1, 3]$

16. $f(x) = x^3 + 4;\ [-2, 2]$

17. $f(x) = x^3 - 12x;\ [-3, 4]$

18. $f(x) = x^3 - 3x;\ [-2, 3]$

19. $f(x) = 2x^3 - x^2;\ [0, 2]$

20. $f(x) = x^3 + 2x^2;\ [0, 3]$

21. $f(x) = 9x - 3x^2 - x^3;\ [-4, 2]$

22. $f(x) = x^3 - 3x^2 - 24x;\ [-3, 3]$

23. $f(x) = 2x^3 - 3x^2 - 12x - 10;\ [0, 4]$

24. $f(x) = x^3 + 3x^2 - 24x;\ [0, 3]$

25. $f(x)=x^{\frac{2}{3}}-4;\ [-1, 8]$

26. $f(x)=3x^{\frac{2}{3}}+2;\ [-1, 4]$

27. $f(x)=3x^{\frac{1}{3}}-4x;\ [0, 1]$

28. $f(x)=3x^{\frac{2}{3}}+x;\ [-9, 1]$

29. $f(x)=\sqrt{x^2+4};\ [-1, 2]$

30. $f(x)=\sqrt{9-x^2};\ [-1, 2]$

31. $f(x)=\sqrt[3]{x^2-1};\ [-2, 2]$

32. $f(x)=\left(x^2-1\right)^{\frac{2}{3}};\ [-2, 2]$

33. $f(x)=x+\frac{4}{x};\ \left[\frac{1}{2}, 3\right]$

34. $f(x)=2x+\frac{18}{x};\ [1, 4]$

35. $f(x)=4x+\frac{9}{x};\ [1, 3]$

36. $f(x)=9x+\frac{16}{x};\ [1, 2]$

37. $f(x)=x^2+\frac{16}{x};\ \left[\frac{1}{2}, 4\right]$

38. $f(x)=x^2+\frac{2}{x};\ [1, 4]$

39. Revenue. The weekly revenue from the sale of x units of a product is given by $R(x)=24x-0.5x^2$ dollars. If the company is set up so that it can produce no more than 40 units per week, how many units should the company produce to maximize revenue?

40. Revenue. The revenue from the sale of x units of a product is given by $R(x)=12x-0.04x^2$ thousand dollars, where $0 \le x \le 250$. How many units should be sold to maximize the revenue?

41. Profit. A manufacturer of telescopes has determined that the revenue from the production and sale of x telescopes is $R(x)=140x-0.5x^2$ dollars. The cost function is given by $C(x)=x^2+20x+1050$ dollars. Find the level of production and sales that will maximize the profit if $0 \le x \le 70$.

42. Profit. A marketing analyst for a company that produces skateboards has determined that if the company sells x units of the deluxe model, the revenue function is $R(x)=79.9x-0.03x^2$ dollars and the cost function is $C(x)=0.08x^2+5.1x+5800$ dollars. Find the sales level that will yield maximum profit if $0 \le x \le 380$.

43. Average cost. The cost of producing x compact refrigerators is given by $C(x)=2880+35x+0.2x^2$ dollars. Find the value of x that minimizes the average cost function if $0 \le x \le 150$.

44. Average cost. The cost of producing x electronic games is given by $C(x)=1080+42x+0.3x^2$ dollars. Find the value of x that minimizes the average cost function if $0 \le x \le 90$.

45. Air quality. The Air Quality Management District monitors the level of pollution in the air. On a good day, the level is approximately $P(t) = 35 + \dfrac{126t}{0.5t^2 + 18}$ PSI (Pollution Standard Index), where t is the number of hours after 7:00 A.M. and $0 \le t \le 11$. At what time will the pollution level be a maximum?

46. Air quality. On a moderately smoggy day, the level of nitrogen dioxide in the air is approximately $N(t) = 0.126 + \dfrac{0.36t}{2t^2 + 40.5}$ ppm (parts per million), where t is the number of hours after 8:00 A.M. and $0 \le t \le 10$. At what time will the level of nitrogen dioxide reach its maximum?

47. Bacteria. It is estimated that t hours after a particular bacterium is introduced into a culture, the population of bacteria in the culture will be $P(t) = \dfrac{4800}{\sqrt{12 - 0.5t}}$, where $0 \le t \le 6$. What will be the absolute maximum population? At what time t will this occur?

48. Population. It is estimated that t years from now the population of a small community will be $P(t) = \dfrac{5000}{\sqrt{25 + 0.4t}}$ people, where $0 \le t \le 10$. What will be the maximum population?

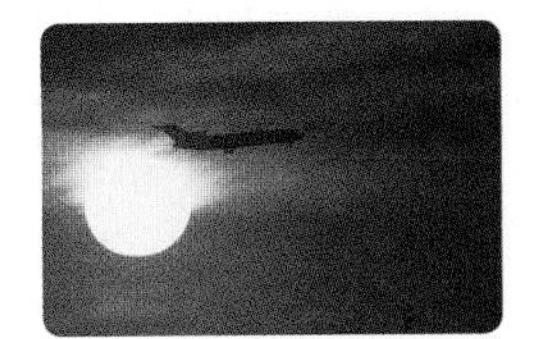

49. Altitude. The altitude of an airplane following a certain flight path is given by the function $F(t) = \dfrac{13t}{20} - \dfrac{3t^2}{200}$, where t is in minutes, and $F(t)$ is in thousands of feet. Find the absolute maximum of the function over the interval $[0, 43]$.

50. Velocity of a car. The velocity of a car in miles per hour varies according to the function $F(x) = \dfrac{x^2}{100} - \dfrac{19x}{25} + \dfrac{5349}{100}$, where x denotes time in seconds. Find the absolute maximum and minimum velocity over the interval $[0, 100]$. Find the car's velocity at the end points.

51. Tire distortion. When a car accelerates, its tires are distorted by the force exerted on them by the motor. For a certain model car, the distortion after the driver floors the accelerator is modeled by the function $F(t) = 0.00026t^3 - 0.0533t^2 + 2.74014t$, where $0 < t < 100$ is the number of milliseconds after acceleration begins, and $F(t)$ is a percentage of the tire's maximum flexibility (at $F(t) = 100$, the tire tears into pieces). Find the time when the maximum distortion occurs, and find the percentage that the tires are distorted. Do the tires survive the acceleration?

52. Pollution. The amount of pollution (measured in parts per million) in a small river is given by the function $F(t)=\frac{450t^3}{169}-\frac{18,225t^2}{169}+\frac{23,625t}{23}+\frac{260,675}{169}$, where t is the number of years after 1980. Sometime after 1980, an environmental protection law was enacted to reduce the amount of pollution in the river, and in the same year the pollution levels began to decrease. Find the maximum pollution for $0 \le t \le 20$, and determine the year the law was enacted. (The largest integer less than the t value specifies the year of enactment.)

53. Wind resistance. An aeronautics company is testing a new airframe. The company wishes to determine an ideal cruising speed, and one step in the process is to find the speed at which the airframe experiences the least wind resistance (other than when it is not moving). After performing a number of tests, the engineers determine that the wind resistance can be modeled by a function $G(v)=\frac{v^3}{3,645,000}-\frac{8v^2}{6075}+\frac{350v}{243}+\frac{614,500}{729}$, where $300 \le v \le 4000$ is the wind velocity in feet per second and $G(v)$ is in Newtons. Determine the ideal cruising speed for the airframe based on this model.

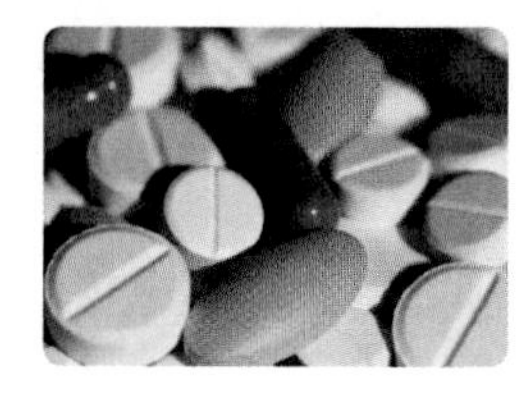

54. Bacteria. A bacteriologist doing research on antibiotics has discovered that a certain type of disease causing bacteria can be effectively treated using a cocktail of two different antibiotics administered at specific intervals. The population changes in response to the antibiotics according to the function $F(t)=-0.0224t^3+4.5676t^2-252.4610t+5000$, where t is in the interval [0, 200]. The maximums of the function correspond to the times when the antibiotics were administered. At what times, t, are the antibiotics administered and what is the population of bacteria at those times?

55. Gravitational pull. A rocket traveling to the moon is affected by the gravity of the earth and of the moon. The total gravitational force exerted on the rocket is approximated by the function $F(h)=\frac{43,750h^2}{3}-\frac{70,000h}{3}+10,000$, where $0 < h < 1$ is the height of the object, given in percentage of the distance between the earth and the moon, and $F(h)$ is measured in Newtons. Find the height at which the minimum occurs and the gravitational force at that altitude. Additionally, there is an onboard experiment which can only be performed if the gravitational force falls below 1000 Newtons. Can this experiment be performed?

56. Thrown object. A cell phone is thrown into the air. The position of the phone is given by the function $F(t)=-\frac{49t^2}{10}+\frac{297t}{10}$, where $0 < t < 10$ is the number of seconds after the phone is thrown. and $F(t)$ is measured in feet. Find the maximum height the phone attains. Also find the time when the phone hits the ground. The phone will shatter if it hits the ground with a speed greater than 32 fps (feet per second). Does the phone shatter?

57. Acceleration. Due to many variable factors in car engines, acceleration is never constant. The acceleration of a particular car in a particular test is approximated by the function $F(t) = -\frac{260t^3}{137} + \frac{2418t^2}{137} - \frac{7371t}{137} + 75$, where $0 \le t \le 5$ is the number of seconds after acceleration begins. What is the maximum acceleration of the car during this test? When does it occur? What is the minimum acceleration, and when does it occur?

58. Power Consumption. A certain piece of equipment in a chemistry lab draws power according to the function

$$G(m) = \left(-6.087 \times 10^{-6}\right)m^3 + \left(8.641 \times 10^{-3}\right)m^2 - 1.751m + 1000,$$

where $0 < m < 1000$ is the mass of the sample to be analyzed in grams and G is power consumption in Watts. Find the sample mass which causes the equipment to draw the most power. How much power does the equipment draw for a sample of this mass?

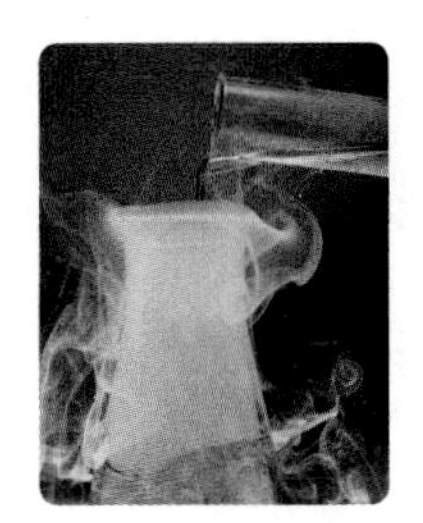

59. Chemistry. A certain chemical procedure requires the addition of reactants and catalysts at precise times to maintain reaction rates. The rate of reaction is given by the function $H(t) = \left(9.423 \times 10^{-7}\right)t^3 - \left(1.230 \times 10^{-4}\right)t^2 + \left(4.384 \times 10^{-3}\right)t + 0.001$, where $0 < t < 100$ is the number of seconds after beginning the reaction, and $H(t)$ is measured in number of moles formed per second. Find the time at which a relative maximum reaction rate is reached and the number of moles per second being formed at that time. At what time was the second reactant/catalyst mixture added? (The reactant/catalyst mixture is added at a relative minimum.) Round to two significant digits.

60. Phone traffic. Phone traffic varies greatly over the course of a day. A long distance provider estimates that the number of long distance calls active per minute on a certain holiday is approximated by the function $P(t) = -.001t^3 + t^2 + 50t + 150000$, where $0 < t < 1000$ is the number of minutes after 6 a.m. and $P(t)$ is the number of active long distance calls. For what t does maximum phone traffic occur? How many calls are active at this time?

61. Stored energy. While a rubber ball is moving, it has kinetic energy. When the ball impacts a hard surface, the kinetic energy is converted to potential energy, and then back to kinetic energy. This happens quite rapidly. The amount of potential energy after impact is approximated by the function $E(t) = -\frac{3t^2}{80} + \frac{3t}{2}$, where t is the number of nanoseconds after impact and t is in $[0, 40]$. When does the ball have maximum potential energy? How much potential energy does it have at its maximum?

62. Photosynthesis. Plants absorb different amounts of light, depending on the wavelength of the light. A study of Rhododentradonzil bushes show that they absorb light according to the formula

$$L(w) = \frac{-2.22w^3}{10^{10}} + \frac{4.44w^2}{10^6} - \frac{2w}{10^3} + 0.444$$

where $100 < w < 650$ is the wavelength of light in nanometers, and $L(w)$ is the proportion of the light shone on the bush. What wavelength does Rhododentradonzil absorb the best? What proportion of light of this wavelength does Rhododentradonzil absorb?

63. Package Delivery. A study says that the package flow in the Southeast USA during the month of August follows the function $D(t) = \frac{7t^3}{9300} - \frac{7t^2}{248} + \frac{7t}{31} + 1$ where $1 \le t \le 31$ is the day of the month, and $D(t)$ is given in millions of packages. On which day are the most packages delivered? How many packages are delivered on this day?

Hawkes Learning Systems: Essential Calculus

Absolute Maximum and Minimum

Chapter 3 Index of Key Ideas and Terms

Section 3.1 Product and Quotient Rules

Product Rule pages 208 - 210

If $f(x)$ and $g(x)$ are differentiable functions and $y = f(x) \cdot g(x)$, then

$$\frac{dy}{dx} = f(x) \cdot g'(x) + g(x) \cdot f'(x).$$

In words, the derivative of the product of two functions is equal to the first function times the derivative of the second plus the second function times the derivative of the first.

Other Forms of the Product Rule page 209

Quotient Rule pages 211 - 213

If $f(x)$ and $g(x)$ are differentiable functions and $y = \frac{f(x)}{g(x)}$, then

$$\frac{dy}{dx} = \frac{g(x) \cdot f'(x) - f(x) \cdot g'(x)}{(g(x))^2}.$$

In words, the derivative of a quotient is the denominator times the derivative of the numerator minus the numerator times the derivative of the denominator, all divided by the square of the denominator.

Section 3.2 The Chain Rule and the General Power Rule

The Chain Rule pages 218 - 222

First Form

If $y = f(u)$ and $u = g(x)$, then

$$\frac{dy}{dx} = \frac{dy}{du} \cdot \frac{du}{dx},$$

provided that $\frac{dy}{du}$ and $\frac{du}{dx}$ both exist.

Second Form

If $y = f(g(x))$, then

$$\frac{dy}{dx} = f'(g(x)) \cdot g'(x),$$

provided that $f'(g(x))$ and $g'(x)$ both exist.

continued on next page ...

Section 3.2 The Chain Rule and the General Power Rule (continued)

The General Power Rule pages 222 - 223

If $y = [g(x)]^n$, then

$$\frac{dy}{dx} = n[g(x)]^{n-1} \cdot g'(x)$$

provided that $g'(x)$ exists.

Another form is $\frac{dy}{dx} = n[g(x)]^{n-1} \cdot \frac{d}{dx}(g(x))$.

Using the Rules in Combination pages 223 - 225

Section 3.3 Implicit Differentiation and Related Rates

Implicit Differentiation pages 230 - 234

Differentiate both sides of the equation with respect to x and then solve for $\frac{dy}{dx}$.

The general power rule can be written in the following form:

$(y^n)' = ny^{n-1} \cdot y'$.

Finding the derivative at a given point using implicit differentiaton.

Related Rates pages 235 - 239

Using implicit differentiation to determine the rates of change with respect to time in problems in which two or more variables are dependent on time.

Suppose $y = f(t)$ and $x = g(t)$. Then we can differentiate and get $\frac{dy}{dt} = f'(t)$ and $\frac{dx}{dt} = g'(t)$. Now, provided $g'(t)$ is not zero, we can say $\frac{dy}{dx} = \frac{f'(t)}{g'(t)}$.

That is,

$$\frac{dy}{dx} = \frac{\frac{dy}{dt}}{\frac{dx}{dt}},$$

and this ratio expresses the rate of change of y with respect to x.

Section 3.4 Local Extrema and the First Derivative Test

Increasing and Decreasing Functions

Increasing and Decreasing Intervals of a Function page 245 - 247

Suppose that a function f is defined on the interval (a, b).

1. If $x_1 < x_2$ implies $f(x_1) < f(x_2)$ for every x_1 and x_2 in (a, b), then $f(x)$ is increasing on (a, b).
2. If $x_1 < x_2$ implies $f(x_1) > f(x_2)$ for every x_1 and x_2 in (a, b), then $f(x)$ is decreasing on (a, b).

Theorem page 247

Suppose that $f(x)$ is differentiable on the interval (a, b).

1. If $f'(x)$ is positive ($f'(x) > 0$) for all x in (a, b), then f is increasing on (a, b).
2. If $f'(x)$ is negative ($f'(x) < 0$) for all x in (a, b), then f is decreasing on (a, b).

Graphing a Function page 248 - 253

Critical Values pages 253 - 256

Let f be a function defined at $x = c$.

1. $f(c)$ is called a local maximum (or relative maximum) if there exists some interval (a, b) that contains c such that $f(x) \le f(c)$ for all x in (a, b). (We say a local maximum occurs at $x = c$.)
2. $f(c)$ is called a local minimum (or relative minimum) if there exists some interval (a, b) that contains c such that $f(x) \ge f(c)$ for all x in (a, b). (We say a local minimum occurs at $x = c$.)

Note: We also say that a local extremum occurs at the point $(c, f(c))$ or that the point $(c, f(c))$ is a local extremum, with the understanding that the y-value, $f(c)$, is actually the local extremum.

Theorem of Local Extrema page 255

If a function f is continuous on the interval (a, b) and c is in (a, b) and $f(c)$ is either a local maximum or a local minimum, then

1. $f'(c) = 0$, or
2. $f'(c)$ does not exist.

continued on next page ...

Section 3.4 Local Extrema and the First Derivative Test (continued)

The First Derivative Test for Local Extrema pages 256 - 261

To determine the local extrema for a nonconstant function f continuous on the interval (a, b):

1. Find the critical values c for f in (a, b) such that $f'(c) = 0$ or $f'(c)$ is undefined.
2. Check the sign of $f'(x)$ on either side of $x = c$. If the sign changes,
   i. from + to –, then $f(c)$ is a local maximum.
   ii. from – *to* +, then $f(c)$ is a local minimum.
3. If there is no sign change for $f'(x)$ from the left side of $x = c$ to the right side, then $f(c)$ is not a local extremum.

Section 3.5 Absolute Maximum and Minimum

Absolute Extrema pages 268 - 269

If c is in the domain of f and, for all x in the domain of f, we have

1. $f(x) \leq f(c)$, then $f(c)$ is called the **absolute maximum** of f.
2. $f(x) \geq f(c)$, then $f(c)$ is called the **absolute minimum** of f.

Theorem of Absolute Extrema of a Continuous Function page 269

If a function f is continuous on a closed interval $[a, b]$, then f will have an absolute maximum value and an absolute minimum value on $[a, b]$.

Finding the Absolute Extrema of a Continuous Function f on the Closed Interval $[a, b]$ pages 270 - 276

1. Find all the critical values for f in $[a, b]$. That is, find all c in $[a, b]$ where
   a. $f'(c) = 0$ or
   b. $f'(c)$ is undefined.
2. Evaluate $f(a), f(b)$, and $f(c)$ for all critical values c.
3. The largest value found in Step 2 is the absolute maximum. The smallest value found in Step 2 is the absolute minimum.

Chapter 3 Review

For a review of the topics and problems from Chapter 3, look at the following lessons from *Hawkes Learning Systems: Essential Calculus.*

Product and Quotient Rules
The Chain Rule and the General Power Rule
Implicit Differentiation and Related Rates
Local Extrema
Critical Points and the First Derivative Test
Absolute Maximum and Minimum

Chapter 3 Test

In Exercises 1 – 18, use the rules for differentiation to find the derivative for each of the functions.

1. $f(x)=(5x+6)(4x^2+5)$

2. $f(x)=(4-x)(x^3+7)$

3. $y=\sqrt{x}\left(5x^2-2x+3\right)$

4. $y=x^{\frac{3}{2}}\left(6x^2+1\right)$

5. $y=\dfrac{x^3+3x+1}{x+5}$

6. $y=\dfrac{x^2-5x+6}{2x-3}$

7. $f(x)=\dfrac{\sqrt{x}}{x-4}$

8. $f(x)=\dfrac{x+10}{x^{\frac{3}{2}}}$

9. $f(t)=(4-7t)^{-3}$

10. $f(x)=\left(5x^2-6\right)^{-\frac{1}{2}}$

11. $y=x^2\sqrt{2x-9}$

12. $f(x)=\dfrac{\sqrt[3]{5x+2}}{x^4}$

13. $y=(2x+7)^2(4x-3)^{-2}$

14. $y=(6x+1)\sqrt{x^2-3}$

15. $y=\dfrac{3-2x}{\sqrt{x-7}}$

16. $y=\dfrac{8x+9}{\sqrt[3]{x^2+6}}$

17. $f(x)=\left(\dfrac{4x-1}{x^2+2}\right)^4$

18. $f(x)=\sqrt{\dfrac{2x-7}{5x+2}}$

In Exercises 19 – 22, use the Chain Rule to find $\frac{dy}{dx}$. Then evaluate $\frac{dy}{dx}$ for the given value of x.

19. $y=u^3+2u,\ u=2x^2+3;\ x=1$

20. $y=4u-\sqrt{u},\ u=x^2-2x+1;\ x=3$

21. $y=\sqrt{2u+3},\ u=2x^2+x-3;\ x=-2$

22. $y=5u-2u^2,\ u=3\sqrt{2x^2-1};\ x=-1$

In Exercises 23 – 26, use implicit differentiation to find $\frac{dy}{dx}$. Then find the slope of the tangent line at the given point.

23. $x^2+2y^2+3xy=4;\ (2, 0)$

24. $2x^3-5x+y^3=5;\ (1, 2)$

25. $x\sqrt{y}+x^2-y=3x+2;\ (3, 1)$

26. $\dfrac{y}{x}+\dfrac{x}{y}=\dfrac{13}{6};\ (2, 3)$

27. **Demand.** The price of a product is related to the number of units available (supply) by the equation $px + 3p - 16x = 234$, where p is the price in dollars and x is the number of units available. Find the rate at which the price is changing if there are 90 units available and the supply is increasing at the rate of 15 units per week.

28. **Tax revenue.** The revenue from taxes for a city is estimated to be $R(x) = 0.016(15 + 1.3x + 0.02x^2)$ million dollars, where x is the population in thousands. The population t years from now is estimated to be $x = 60 + 2t + 0.1t^2$ thousand people.
    a. Find the revenue 5 years from now.
    b. Find the marginal revenue with respect to time 5 years from now.

In Exercises 29 and 30, find the open intervals on which **a.** *f is increasing and* **b.** *f is decreasing.*

29.

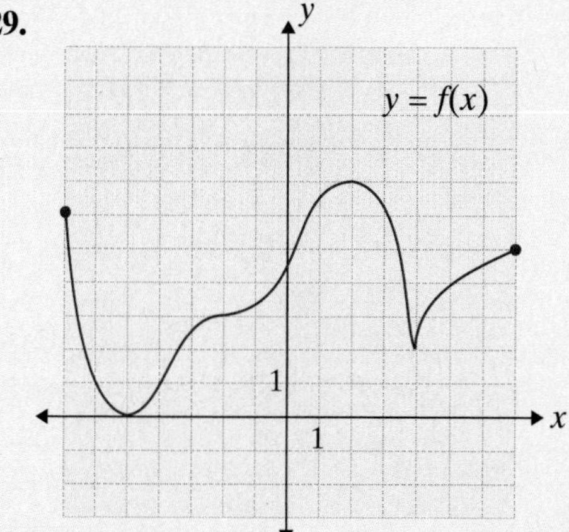

30.

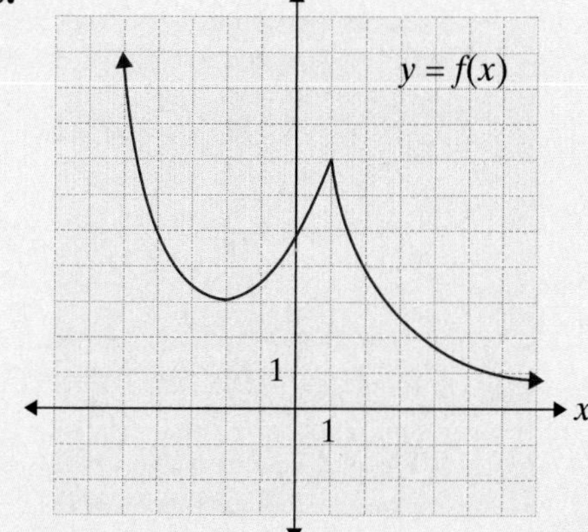

In Exercises 31 – 34, for each function, find **a.** *the values for x that correspond to horizontal tangent lines and* **b.** *the open intervals on which the function is increasing and the open intervals on which it is decreasing.*

31. $f(x) = 3x^2 - 9x + 2$

32. $f(x) = x^3 - 12x + 3$

33. $f(x) = x^4 - 2x^2 + 5$

34. $f(x) = \dfrac{x+3}{x^2}$

In Exercises 35 – 38, for each function, **a.** *find the critical values, and* **b.** *use the First Derivative Test to find any local extrema.*

35. $f(x) = 4 + 6x - x^2$

36. $f(x) = x^3 + x^2 - 5x + 2$

37. $f(x) = x^4 + x^3 - 5x^2 + 3$

38. $f(x) = \dfrac{4x^2 + 25}{x}$

39. For the graph of $f(x)$,
    a. Determine the open intervals on which the function is increasing and the open intervals on which it is decreasing.
    b. Sketch a possible graph of $f'(x)$.

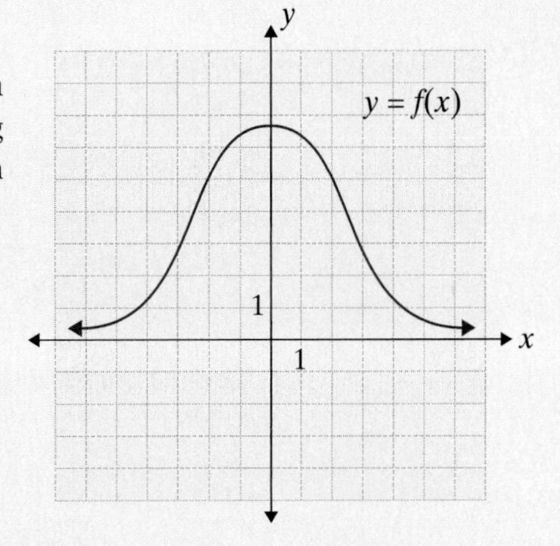

In Exercises 40 – 43, for each function, find the absolute extrema on the given interval.

40. $f(x) = 2x^2 - 3x + 7;\ [0, 2]$

41. $f(x) = 4 + x + x^2 - x^3;\ [0, 4]$

42. $f(x) = \sqrt{x^2 + 3};\ [0, 2]$

43. $f(x) = 3x + \frac{12}{x};\ [-3, -0.1]$

44. **Manufacturing.** A furniture manufacturer has determined that x rocking chairs can be sold if the price is $p = 180 - 0.03x$ dollars each. For what levels of production is the revenue increasing?

45. **Profit.** The weekly revenue from the sale of a particular style of shirt is $R(x) = 43x - 0.02x^2$ dollars. The total cost for the shirts is $C(x) = 0.04x^2 + 13x + 1200$ dollars. Find the sales level that will yield maximum profit.

46. **Population.** It is estimated that t years from now the population of a town will be $P(t) = 4500 + \frac{400t}{t^2 + 4}$, where $0 \le t \le 5$. Find the maximum population.

Cumulative Review: Chapters 2 – 3

In Exercises 1 – 12, determine the derivative of the given function.

1. $f(x) = 2 + 3x - 6x^4$

2. $f(x) = 3 - 6x^{\frac{1}{2}} + 12x^{\frac{1}{3}}$

3. $f(x) = 4x^3 + \sqrt{x}$

4. $f(x) = 3x + \frac{3}{\sqrt{x}}$

5. $f(x) = \sqrt{3 + 2x}$

6. $f(x) = \sqrt{5 + 17x}$

7. $F(u)=(16u+1)^5$

8. $F(u)=(u^2+u-1)^3$

9. $g(t)=(3t+1)(5t^2+2t-10)$

10. $g(t)=(2t-1)(18+5t-11t^2)$

11. $F(u)=\dfrac{2u-1}{4u^2-4u+1}$

12. $F(u)=\dfrac{3u^2-3u+1}{u^2+u-3}$

Use the graph on the right to answer Exercises 13 – 17.

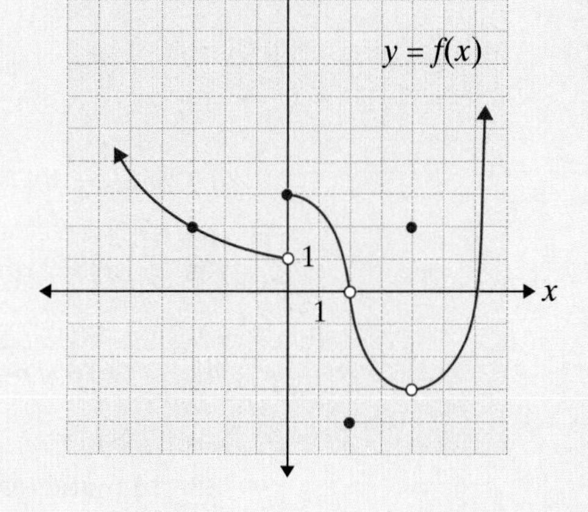

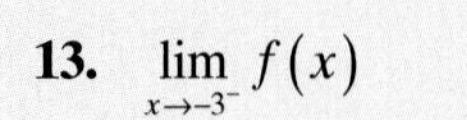

13. $\lim\limits_{x\to -3^-} f(x)$

14. $\lim\limits_{x\to 0^-} f(x)$

15. $\lim\limits_{x\to 0^+} f(x)$

16. $\lim\limits_{x\to 2^+} f(x)$

17. $\lim\limits_{x\to 4^+} f(x)$

18. For each of the following quantities, determine whether or not it is (or could be) the value of a derivative at a point. If the quantity is not a derivative, explain why not. If it is a derivative, give, or briefly describe, a possible corresponding function f and identify what x represents.

a. 40 man-hours.

b. 19 miles per gallon.

c. 16 years of age.

d. 12 hens-a-laying.

e. \$9 per hour.

f. 9 tons per man-hour.

19. For the function $G(x)$ pictured at the right, list the points in order of increasing slope.

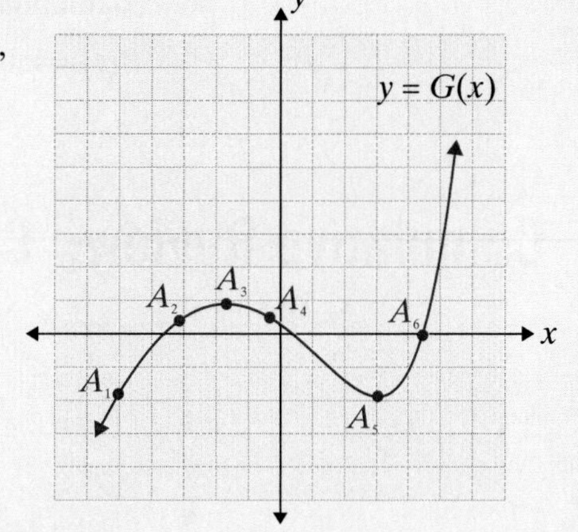

20. For the function $f(x) = 2x + \sqrt{5x-1}$ determine $f'(2)$ using the following methods.

a. Complete the table below.

$h = \Delta x$	0.1	0.01	0.001	0.0001
$\frac{\Delta y}{\Delta x}$				

b. Determine the derivative algebraically and evaluate at $x = 2$.

21. Suppose the TFR (see Section 2.4, Exercise 31 – 33) for Canada is 3 and its population (in millions) is modeled by $P(t) = 250 + 0.8t - 0.002t^2$, where t is the number of years since 2000.

a. Interpret the meaning of $P(0) = 250$.

b. How fast was the population changing in 2005?

c. What population is predicted for 2010?

d. What is the average rate of change of population from 2000 to 2010?

22. Using a graphing calculator and the function $f(x) = x^3 - 20x + 4$:

a. Draw a graph of the function on your paper.

b. Locate the two points on the curve which have a slope equal to 1. (Give their (x, y)-coordinates to the nearest hundredth.)

c. Draw the two tangent lines to $f(x)$ at the two points found in part **b.**

23. Suppose the total cost for a manufacturer of grommets is given by $C(x) = 500 + 0.01x + 0.003x^2$, where x is the number of grommets produced. The total revenue is given by $R(x) = x$.

a. What is the unit price p?

b. Give the marginal cost function.

c. Give the profit function.

d. What is the average cost $A(x)$ at $x = 1000$? (Hint: The average cost is given by $A(x) = \frac{C(x)}{x}$.)

e. Find $(x, A(x))$ for the minimum average cost. Determine the value of $(x, C'(x))$ for the same x-value. Interpret the results.

24. The new floor supervisor at a small manufacturing plant reports that his data charts tell him that the current production results in 40 units of production per man-hour of labor. The Chief Financial Officer (CFO) reports that the average labor costs are \$16 per man-hour.

- **a.** What are the current costs of production?
- **b.** Express the given quantities and the answer to part **a.** as derivatives (use x represents units of production, c represents cost, and y is labor in man-hours).
- **c.** Write a formula which uses the answers to part **b.** but which calculates the number in part **a.**

25. Suppose a company's daily revenue R is given as a function of unit price p, and that p is a function of the quantity x of items sold. When $x = 250$, $\frac{dR}{dp}$ is \$200 for a \$1 increase in unit price, and $\frac{dp}{dx}$ is \$9 per additional item sold per day. Find the marginal revenue when $x = 250$.

26. Determine the equation of the tangent line to $y = 14 + 2x - x^2$ at the point $(0, 14)$. Draw a sketch of the function and the tangent line.

27. Use a graphing calculator to sketch $y = 3^x$ in the window [–1, 3] by [–3, 3]. Determine the slope at the point (0, 1) and get the equation of the tangent line. Draw the graph of the function and the tangent line.

28. Put $f(x) = 2x^2 - 3x - 3$ into a graphing calculator with the window [–4, 4] by [–10, 20]. With the graph on the display, command the calculator to draw a tangent line at $x = -2$. What is the equation of this line? Draw a graph of the function and the tangent line.

29. If $f(t) = 3t + 17 - \frac{225}{t^2}$, determine a formula for $f'(t)$ and use it to find the slope at (1.5, –78.5). Confirm your answer with a graphing calculator.

30. Suppose $G(0) = 53$ and $G'(0) = -1$ for $G'(t)$, where G denotes the population of gorillas in the wilderness of Myst Mountains, Zimolayeo, and t is the time in years since 2000.

- **a.** Interpret $G(0) = 53$.
- **b.** Interpret $G'(0) = -1$.

31. On a psychology test, calculus I students were asked to find derivatives of ten functions. The student responses are modeled by $R(t) = 100(1 - 2^{-t})$, where t is the number of times the test has been given and R is the percentage of students scoring 90% or better on the test. Here, $0 \le t \le 6$ and $0 \le R \le 100$.

- **a.** Use a graphing calculator to sketch the graph.
- **b.** Interpret $R(2) = 75$.
- **c.** Interpret $R'(2) \approx 17.3$.
- **d.** Why is $R(3) \ne 75 + 17.3$?

In Exercises 32 – 33, for the graph of $f(x)$, **a.** *Determine the open intervals on which the function is increasing and the open intervals on which it is decreasing, and* **b.** *sketch a possible graph of $f'(x)$.*

32.

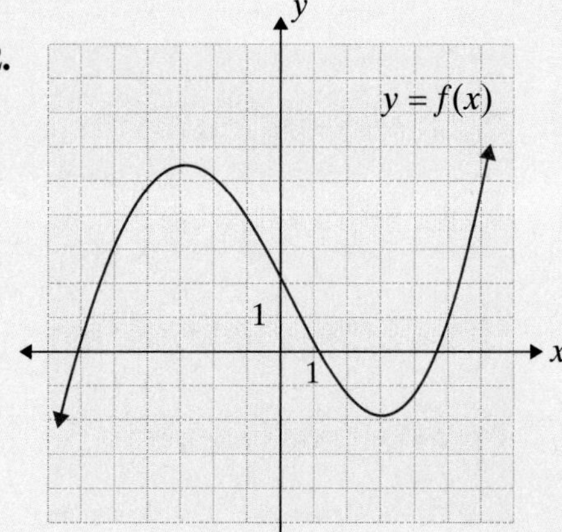

33.

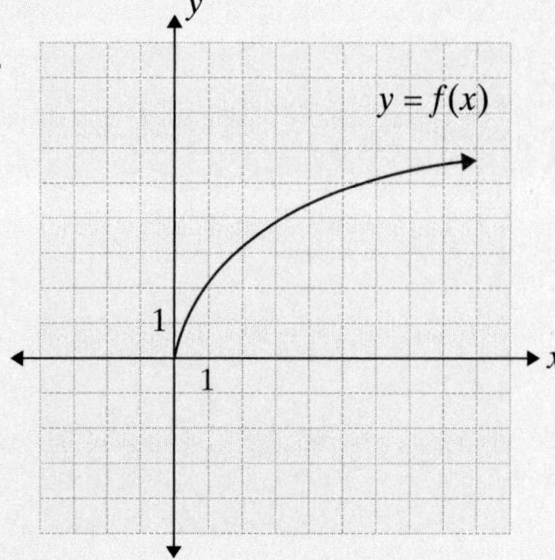

Applications of the Derivative

CHAPTER 4

Did You Know?

In the late 1500's two remarkable astronomers came to the forefront of the scientific community, Tycho Brahe (1546 – 1601) and Galileo Galilei (1564 – 1642). Brahe was the first astronomer to take detailed records of the movements of planetary objects using his invention which was the predecessor of the sextant (his successor, Kepler, used these meticulous records to formulate his laws of planetary motion). Galileo was the first to study far-reaching parts of the galaxy with his improvements upon the telescope (he claimed to have seen mountains on the moon, and to have proved that the Milky Way was made up of tiny stars). These men not only made vital contributions to the scientific community, revolutionizing astronomical measurement and observation instruments, but were also two of history's most colorful scientists, passing many unique and intriguing stories on to posterity.

The popular story associated with Brahe has to do with his nose. Brahe lost his nose in what is said to have been a duel with a fellow student over who was the better mathematician! From then on he wore a prosthetic nose made of a gold and silver alloy.

Galileo, upon learning of an instrument called the telescope, set out to build his own. After successfully building a replica of the reported telescope, he went on to make great improvements to the instrument in future revisions, including increasing the known magnification (which was 4 times) to around 9 times! He then sold the exclusive rights of manufacture of this powerful device (difficult to guarantee even today) to the Venetian senate, effectively pirating another man's invention and swindling the government in one business deal. The senate later froze his salary, finally realizing what Galileo had done.

The lenses in a telescope are called convex lenses. Convex lenses are thicker at the center than at the edges, while concave lenses are thicker at the edges than at the center. Both convex and concave lenses have concavity in the mathematical sense of the word. If one were to draw out the "graph" of these lenses, one would see that they both form a parabola.

These two famous astronomers and their colleague Johannes Kepler (1571-1630) were mathematicians first and used ideas usually thought of as part of calculus in their calculations. Formal trigonometry (Chapter 9) and the new logarithms (Chapter 5) increased the value of mathematics to the newly emerging sciences of the Renaissance period. When the calculus of Newton appeared (1668-1700), it was exactly what was needed to accelerate their developments.

In this chapter, we point out that f' is a function which has its own derivative called the second derivative of f. This second derivative is used to describe and understand $f(x)$ itself. We also develop the "Second Derivative Test" and show how both the first derivative and second derivative can be used in curve sketching. Higher order derivatives are discussed but are not applied here. The ideas in this chapter on sketching are mainly applied to polynomial functions; however, the student should be aware that the ideas presented apply to sketching the graphs of most types of functions.

While we have discussed applications throughout the first three chapters, the emphasis has been on developing rules and techniques for differentiation. Now we focus on the real-world applications of these techniques and show how calculus can be used to maximize profit, minimize inventory cost, reduce time or distance of travel, or, in general, to optimize a function.

The last section on differentials shows how the notation dy and dx can be used to indicate small changes in the variables y and x, respectively. Although the applications of differentials are somewhat limited, understanding the idea of differentials and being able to use corresponding notation will prove valuable when integration is described in Chapters 6 and 7.

4.1 Higher-Order Derivatives: Concavity and the Second Derivative Test

Objectives

After completing this section, you will be able to:

1. *Find the second derivative of a function.*
2. *Determine the concavity of a function.*
3. *Locate the points of inflection of a function by hand and by using a graphing calculator.*

Higher-Order Derivatives

Given a function f, we have defined and discussed the derivative f', called the **first derivative**. We have seen that there may be x-values for which f is defined but for which the derivative does not exist. However, in general, the functions defined and applied in this book have slope defined at most points for which the function is defined. This means that f' is itself a function which has its own derivative defined at most points in its domain, and it too can be differentiated. The **second derivative**, f'', if it exists, is the derivative of the first derivative. The **third derivative**, f''' (or $f^{(3)}$), if it exists, is the derivative of the second derivative, and so on.

For example, if

$$f(x) = \frac{1}{x} = x^{-1} \qquad \text{(where } x \neq 0\text{)}$$

$$f'(x) = -1 \cdot x^{-2},$$

then

$$f''(x) = -1(-2)x^{-3} = 2x^{-3},$$

$$f'''(x) = 2(-3)x^{-4} = -6x^{-4},$$

$$f^{(4)}(x) = -6(-4)x^{-5} = 24x^{-5},$$

and so on.

Although there are important applications that involve an infinite number of derivatives, we will study only first and second derivatives and related applications in this chapter.

Recall, the first derivative of a function $y = f(x)$ can be represented by any of the following symbols:

$$y', \ f'(x), \ \frac{dy}{dx}, \ f_x, \text{ and } D_x[y].$$

Similar notation is used for the second derivative.

Notation for the Second Derivative

The second derivative of the function $y = f(x)$ can be denoted by any of the following symbols:

$$y'', \ f''(x), \ f^{(2)}, \ \frac{d^2y}{dx^2}, \ f_{xx}, \text{ and } D_x^2[y].$$

NOTES

The notation $\frac{d^2y}{dx^2}$ deserves special attention. The expression $\frac{d}{dx}$ is called an **operator** and is meaningless by itself. When applied to a function, it means to find the derivative of that function with respect to x. Thus

$$\frac{d}{dx}\left[\frac{dy}{dx}\right]$$

indicates the derivative of $\frac{dy}{dx}$. This leads to the following simplified notation:

$$\frac{d}{dx}\left[\frac{dy}{dx}\right] = \frac{d^2y}{dx^2}.$$

Example 1: First and Second Derivatives

Find both the first and second derivatives of the function $f(x) = x^3 - 12x + 1$.

Solution: $f'(x) = 3x^2 - 12$ Differentiate $f(x)$.

$f''(x) = 6x$ Differentiate $f'(x)$.

Example 2: First and Second Derivatives

Find $\frac{dy}{dx}$ and $\frac{d^2y}{dx^2}$ if $y = \sqrt{x^2+1}$.

Solution:

$$y = \sqrt{x^2+1} = \left(x^2+1\right)^{\frac{1}{2}}$$ Rewrite y using exponents.

$$\frac{dy}{dx} = \frac{1}{2}\left(x^2+1\right)^{-\frac{1}{2}}(2x)$$ Use the General Power Rule to find the first derivative of y.

$$= x\left(x^2+1\right)^{-\frac{1}{2}}$$ Simplify.

In order to find $\frac{d^2y}{dx^2}$, we must take the derivative of $\frac{dy}{dx}$. Since $\frac{dy}{dx}$ is a product of x and $\left(x^2+1\right)^{-\frac{1}{2}}$, we use the Product Rule to find $\frac{d^2y}{dx^2}$. When using the Product Rule, differentiate $\left(x^2+1\right)^{-\frac{1}{2}}$ by using the General Power Rule. If $f(x) = x$, $g(x) = \left(x^2+1\right)^{-\frac{1}{2}}$ and, $\frac{dy}{dx} = f(x)\cdot g(x)$, we have

$f(x)$ ↓ $g'(x)$ ↓ $g(x)$ ↓ $f'(x)$ ↓

$$\frac{d^2y}{dx^2} = x\cdot\frac{d}{dx}\left[\left(x^2+1\right)^{-\frac{1}{2}}\right] + \left(x^2+1\right)^{-\frac{1}{2}}\cdot\frac{d}{dx}[x]$$ Use the Product Rule.

$$= x\cdot\left(-\frac{1}{\not 2}\right)\left(x^2+1\right)^{-\frac{3}{2}}(\not 2x) + \left(x^2+1\right)^{-\frac{1}{2}}\cdot 1$$ Use the General Power Rule to find $g'(x)$.

$$= -x^2\left(x^2+1\right)^{-\frac{3}{2}} + \left(x^2+1\right)^{-\frac{1}{2}}$$

$$= \left(x^2+1\right)^{-\frac{3}{2}}\left[-x^2 + \left(x^2+1\right)\right]$$ Factor out $\left(x^2+1\right)^{-\frac{3}{2}}$.

$$= \left(x^2+1\right)^{-\frac{3}{2}}(1) = \frac{1}{\left(x^2+1\right)^{\frac{3}{2}}}$$ Simplify.

Example 3: First and Second Derivatives

For the function $f(u)=\frac{u^2}{u^2-9}$, find $f'(u)$ and $f''(u)$.

Solution: In order to find $f'(u)$, we must use the Quotient Rule.

Denominator Numerator′ Numerator Denominator′

$$f'(u)=\frac{(u^2-9)(2u)-(u^2)(2u)}{(u^2-9)^2}$$

Denominator squared

Numerator $= u^2$
Denominator $= u^2-9$
Numerator′ $= 2u$
Denominator′ $= 2u$

$$f'(u)=\frac{2u^3-18u-2u^3}{(u^2-9)^2}$$

$$=-\frac{18u}{(u^2-9)^2}$$ Simplify.

To find $f''(u)$, we must again use the Quotient Rule. When using the Quotient Rule to find the derivative of this denominator, use the General Power Rule.

Numerator $= -18u$ Numerator′ $= -18$
Denominator $= (u^2-9)^2$ Denominator′ $= 2(u^2-9)(2u)$

Denominator Numerator′ Numerator Denominator′

$$f''(u)=\frac{(u^2-9)^2(-18)-(-18u)\left[2(u^2-9)(2u)\right]}{(u^2-9)^4}$$

Denominator squared

$$f''(u)=\frac{\left[-18\cancel{(u^2-9)}\right]\left[(u^2-9)-4u^2\right]}{(u^2-9)^{\cancel{4}\,3}}$$ Factor out $-18(u^2-9)$ and cancel the common factor.

$$=\frac{-18(-3u^2-9)}{(u^2-9)^3}=\frac{54(u^2+3)}{(u^2-9)^3}$$ Simplify.

Concavity

Earlier, we discussed how the first derivative can be used to indicate the intervals on which a function is increasing or decreasing. When $f'(a)$ is positive, we know that f is increasing on an interval containing the point $(a, f(a))$. The meaning for $f''(a)$ is exactly analogous: if $f''(a)$ is positive, then f' is increasing on an interval containing $x = a$. However, it becomes important then, in using information about f'', to be able to visualize an example $f(x)$ about which one can say that "the slope of f is increasing at $x = a$."

There are visually (and numerically) two ways in which the increasing slope of a function can be recognized. One of these occurs in Figure 4.1.1(a) for the graph of $f(x) = x^2$. For example, at $x = 1$, the slope is increasing. To the left of $x = 1$, at $x = .75$ say, the slope is 1.5. But $f'(1) = 2$, and $f'(1.5) = 3$. Thus the slope is increasing from small positive numbers to bigger positive numbers. But if slope increased, then its derivative is positive – that is, $f''(1)$ is positive. Visually, we observe the curve is "steeper" from left to right.

The graph of $y = x^2$ at $x = -1$ is also significant. Notice that the slopes increase here too. At the point $(-1, 1)$, the slope is -2. But just to the left, at $x = -1.5$, the slope is -3, and, just to the right at $x = -0.5$, we have $f'(-0.5) = -1$. The slope increased from the negative value -3 to the less negative value -1. Thus we expect $f''(-1)$ to be positive as well. Although "steeper" does not seem to describe the graph here, say between $(-1, 1)$ and $(0, 0)$, the second derivative $f''(x)$ is positive and the slope is increasing in this region.

The graph of $f(x) = x^2$ at all of its points can also be characterized by the fact that the tangent lines lie below the curve.

On the other hand, the graph of $f(x) = \sqrt{x}$ pictured in Figure 4.1.1(b) has the property that tangent lines lie above the curve. Notice, as well, that the slopes for $f(x) = \sqrt{x}$ are decreasing.

Figure 4.1.1

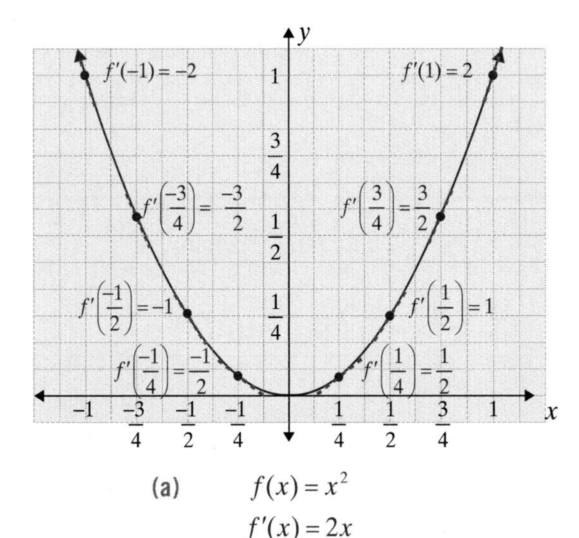

(a) $f(x) = x^2$

$f'(x) = 2x$

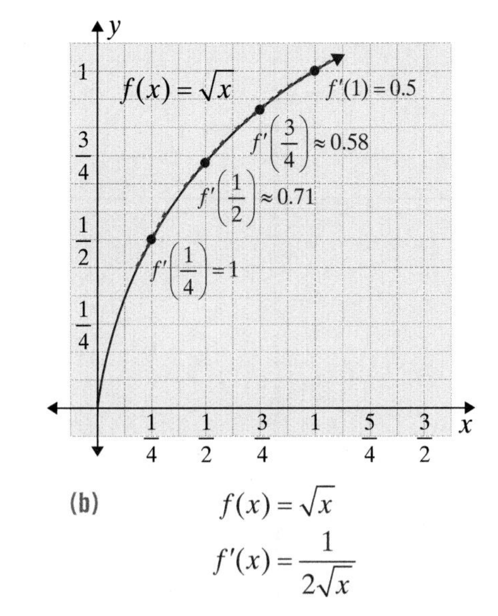

(b) $f(x) = \sqrt{x}$

$f'(x) = \dfrac{1}{2\sqrt{x}}$

This property of curves which we are discussing is called **concavity**. As illustrated in Figure 4.1.1(a), a curve that is **concave upward** lies above the lines tangent to the curve. Conversely, as in Figure 4.1.1(b), a curve that is **concave downward** lies below the lines tangent to the curve.

The following definition tells us how the first derivative can be used to determine concavity.

Concavity

Suppose that f is differentiable on the interval (a, b).

1. *If f' is increasing on (a, b), then the graph of f is* **concave upward** *on (a, b).*
2. *If f' is decreasing on (a, b), then the graph of f is* **concave downward** *on (a, b).*

NOTES If the graph is concave upward, we also say that f is concave upward. If the graph is concave downward, we say that f is concave downward. Alternatively, we can think of a function being "cupped up"where it is concave up and we may say a function is "cupped down" where it is concave down.

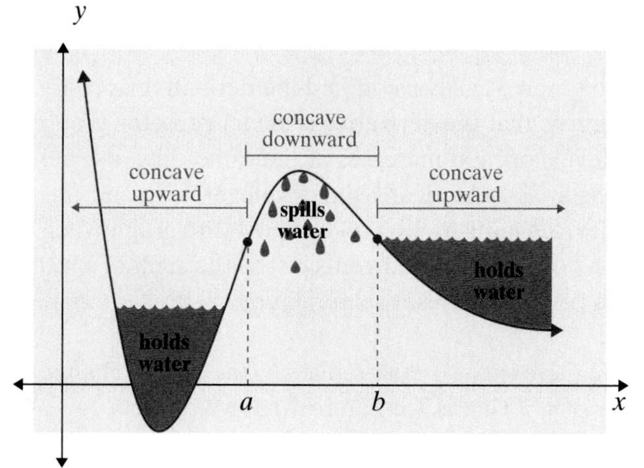

Figure 4.1.2

Intuitively, we say that the curve will "hold water" on an interval where the curve is concave upward and will "spill water" on an interval where the curve is concave downward. (See Figure 4.1.2.)

To determine the x-axis intervals for which a given graph is concave up or concave down, as in Figure 4.1.2, one uses the second derivative because it is precisely the tool for determining whether f' is increasing or decreasing.

One of the more interesting tasks the calculus student is asked to do is to think about how the slope of $y = f(x)$ is changing when, in past years, one has considered only how the magnitude or size of y-values might be changing.

In Figure 4.1.2, one observes that, at the minimum point, say (x_1, y_1), to the left of $x = a$, the graph is certainly concave up. Thus, at x_1,we have $f'(x_1)=0$ and $f''(x_1)>0$. At the local maximum, say (x_2, y_2), the graph is certainly concave down, and $f'(x_2)=0$ and $f''(x_2)<0$. Somewhere between x_1 and x_2, the concavity changed from concave up to concave down, and somewhere in between x_1 and x_2, the values of $f''(x)$ changed from positive to negative. These changes occur at the same point, namely $(a, f(a))$.

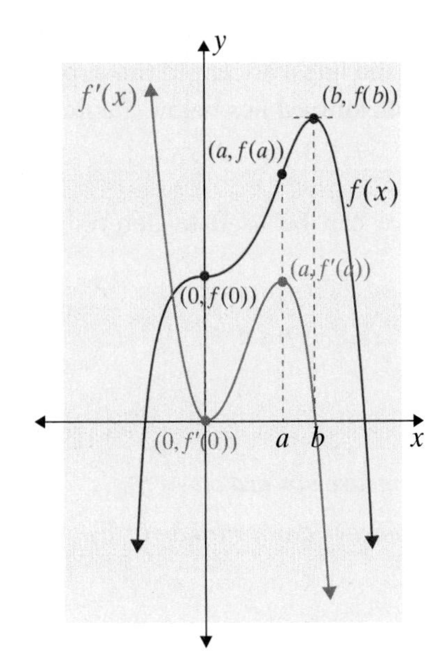

Figure 4.1.3

In similar fashion, to the right of the local maximum at (x_2, y_2), the graph begins to level off (slope decreases). Thus, to the right of (x_2, y_2), there is another point where the concavity changes, namely $(b, f(b))$. The student will be asked both to estimate (visually) the location of such points where the concavity changes and also to determine the exact x-values where such changes occur.

In Figure 4.1.3, we have marked the point $(a, f(a))$. Slopes decrease for all x-values less than $x = 0$. For the negative x-values not shown, the slopes are large positive numbers which decrease to small positive numbers, and then, at about $x = 0$, the slope is zero. The slopes then increase, becoming positive and larger until $x = a$, and then the slopes start to decrease. They decrease to zero at $x = b$. Then, exactly at $(b, f(b))$, the tangent to $y = f(x)$ is horizontal, and $f'(b) = 0$. The maximum point for f at $(b, f(b))$ lies directly above the point $(b, f'(b)) = (b, 0)$. Following $x = b$, the slopes continue to decrease and do so for all other positive x-values.

Let us examine more closely the graph of $y = f(x)$ in Figure 4.1.3. We have also graphed $f'(x)$ on the same axes. These graphs are very revealing. For negative values of the input x, the output values of y are increasing (left to right). We expect f' to be positive. This is shown by the fact that the graph of f' lies above the x-axis. Algebraically, we can write

$$f'(x) > 0 \text{ when } x < 0, \text{ and thus } y \text{ is increasing.}$$

As x approaches zero from the left, the graph of y levels off and the derivative appears to be zero at the point $(0, f(0))$. We observe that the derivative is in fact zero; the graph of y' hits the x-axis at $x = 0$. Since y levels off and increases at the same time, its rate of increase declines. Therefore, we reason y' to be decreasing for negative x, and this is also visible in the graph of y'. Further, tangents to $y = f(x)$ lie above the graph when $x < 0$. So $f(x)$ is concave down to the left of $x = 0$. But tangents just to the right of $x = 0$ will clearly lie below the graph. So y is concave up just to the right of $x = 0$.

Since y' decreases (until $x = 0$) and then increases, we can predict y'' (the rate of change of y') to be zero at the minimum value of y'. That is, for Figure 4.1.3, we expect

$$f''(x) < 0 \text{ when } x \text{ is negative,}$$

$$f''(0) = 0, \text{ and}$$

$$f''(x) > 0 \text{ for } x \text{ bigger than zero but less than } a, \text{ stated algebraically } 0 < x < a.$$

This is precisely what it means for $(0, f(0))$ to be a point on the graph of y at which the concavity changes.

Does y have yet another point at which concavity changes? In fact, the rate of change of y (that is, y') increases to a maximum from $x = 0$ to $x = a$ and then decreases. Thus, for Figure 4.1.3, we expect y'' is positive to the left of $x = a$ and is negative to the right of $x = a$. This means the point $(a, f(a))$ is another point of change in the concavity of f.

We emphasize that such points occur when y' has a local extrema. Here, for the graph of f', $(0, f'(0)) = (0, 0)$ is a local minimum and $(a, f'(a))$ is a local maximum.

The following test for concavity gives us an analytical method for determining the intervals on which the derivative is increasing or decreasing. This method involves the second derivative, f''. Just as f' tells us where f is increasing or decreasing, f'' tells us where f' is increasing or decreasing.

Interpreting $f''(x)$ to Determine Concavity

Suppose that f is a function and f' and f'' both exist on the interval (a, b).

1. *If $f''(x) > 0$ for all x in (a, b), then f' is increasing and f is* ***concave upward*** *on (a, b).*
2. *If $f''(x) < 0$ for all x in (a, b), then f' is decreasing and f is* ***concave downward*** *on (a, b).*

Example 4: Determining Concavity

In the figure shown, the graph of $y = f(x)$ is given.

a. List the intervals on which f is concave upward.

b. List the intervals on which f is concave downward.

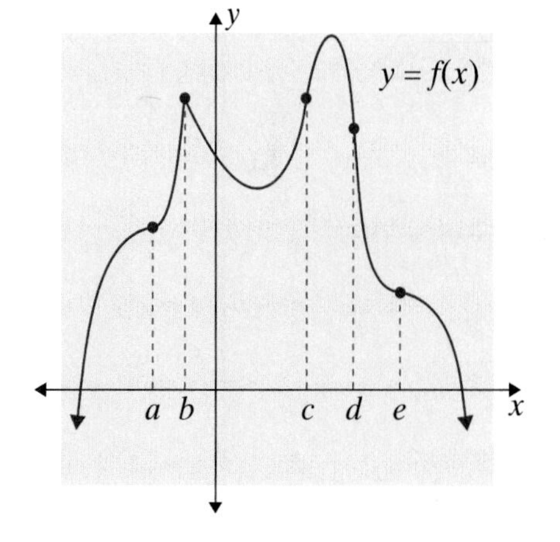

Solutions: a. f is concave upward on the intervals (a, b), (b, c), and (d, e).

b. f is concave downward on the intervals $(-\infty, a)$, (c, d), and $(e, +\infty)$.

NOTES

If one has a graph of y (and possibly y') on a graphing utility, then one can use them to understand the ideas of this section. If a graphing utility is not accessible, one can use calculus to obtain quite accurate graphs of functions, even those not usually discussed in algebra classes.

Example 5: Sketching a Curve Using Concavity

For the function $f(x) = x^4 - 4x^3$,

a. Determine the intervals on which the function is concave upward and the intervals on which it is concave downward.

b. Use the information from part **a.** to draw a rough sketch of the graph of $f(x)$.

Solutions: a. To determine concavity, we must interpret $f''(x)$. First, we find the second derivative and then determine the intervals where it is positive and the intervals where it is negative.

$$f'(x) = 4x^3 - 12x^2$$
$$f''(x) = 12x^2 - 24x$$

To find the second derivative, find the first derivative of $f(x)$ and then find its derivative.

Now that we have determined $f''(x)$, set $f''(x) = 0$ and solve for x to find the endpoints of the intervals to be tested for concavity.

$$12x^2 - 24x = 0$$
$$12x(x-2) = 0$$
$$x = 0, 2$$

So there are three intervals to be considered:

$$(-\infty, 0),\ (0, 2),\ \text{and}\ (2, +\infty).$$

The values of y'' in the interval $(-\infty, 0)$ are all positive or all negative. (If some were positive and some were negative, then there would have to be some other x-value, say b, such that $f''(b) = 0$. But there is no such x-value in that interval.) Similarly, in each interval, all the values of y'' are negative or all are positive. Therefore, for each interval, we choose one test value in the selected interval which seems convenient for calculation. We use this x-value, say $x = a$, to determine whether $f''(a)$ is positive or negative. If $f''(a)$ is positive, then $f''(x)$ is positive for all x in that interval, and the graph of y is concave up for that interval. Likewise, if $f''(a)$ is negative, then $f''(x)$ is negative for all x in that interval, and the graph of y is concave down for that interval.

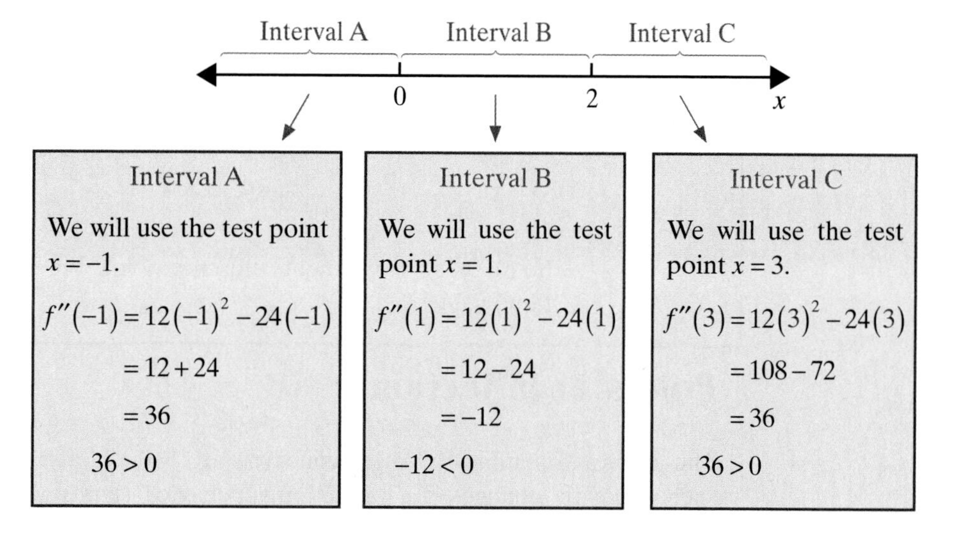

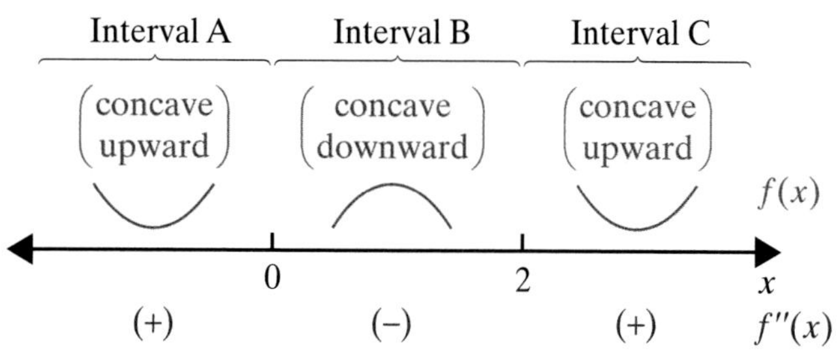

f is concave upward on intervals $(-\infty, 0)$ and $(2, +\infty)$.
f is concave downward on the interval $(0, 2)$.

b. In making a rough sketch of a function using concavity, one wishes to calculate as few (x, y) points as possible. However, the y-intercept, $(0, f(0))$, is usually easy to calculate at a glance (without a calculator). Here $f(0) = 0$, so the origin is a point on the graph. Also, at $x = 0$, the graph changes concavity from concave up to concave down. Near the origin, the graph of $f(x)$ must look like the sketch provided. We have calculated $f(2) = -16$, where the concavity changes from down to up. From the form of the polynomial $f(x) = x^4 - 4x^3$, we can say that eventually all the y-values are positive (since the term of largest degree is positive). This means that there is a global minimum to the right of $(2, -16)$; after this point, y increases without further concavity changes. The x-intercept to the right of $(2, -16)$ is easily found since there is no constant term for y.

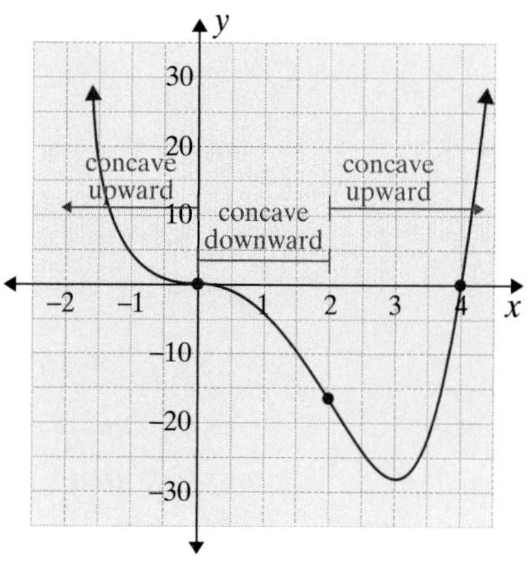

continued on next page ...

We set $f(x) = 0$ and solve by factoring.

$$x^4 - 4x^3 = 0$$

$$x^3(x-4) = 0$$

$$x = 0, \quad x = 4$$

So (4, 0) is a convenient reference point. We have calculated only three (x, y) values.

Points of Inflection

The graph in Example 5 changes concavity at $x = 0$ and at $x = 2$. In most applications, such dramatic changes in a function's behavior signify an important event. We concentrate now on locating such points algebraically. The graphs of three continuous functions are shown in Figure 4.1.4. In each case, the graph changes concavity at a point marked with the letter I. In Figure 4.1.4(a), the tangent lines lie above the graph to the left of I but below the graph to the right of I. At that point I, the tangent line is said to cross through the graph. Such points are called **points of inflection**. Similar situations are presented in parts (b) and (c).

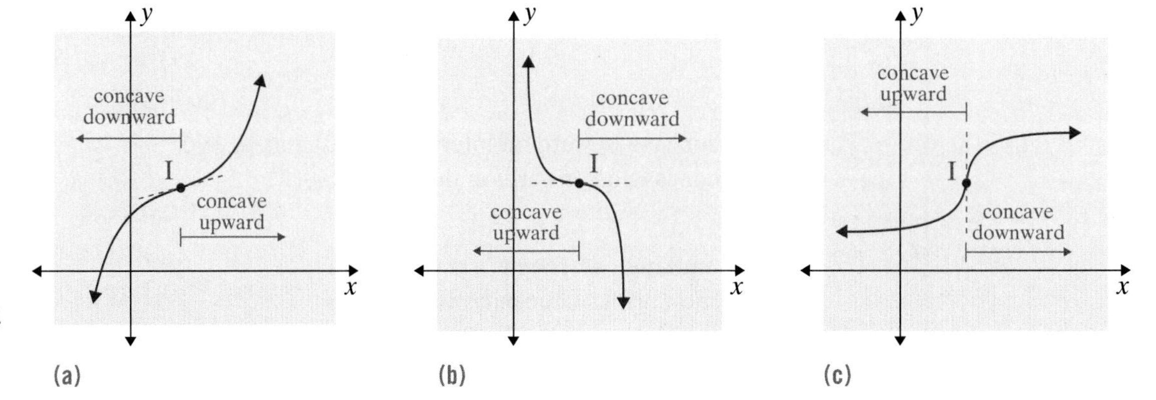

Figure 4.1.4

Point of Inflection

If the graph of a continuous function has a tangent line at a point (possibly a vertical line) and the graph changes concavity at that point, then the point is called a ***point of inflection****.*

Hypercritical values of x are those values $x = c$ in the domain of f where $f''(c) = 0$ or $f''(c)$ does not exist. As we will see, points of inflection occur only at hypercritical values. It turns out, however, that hypercritical values do not guarantee points of inflection. Remember that at a point of inflection, the function must change concavity from up to down, or vice versa.

To Determine Points of Inflection

Suppose that f is a continuous function.

1. *Find $f''(x)$.*
2. *Find the hypercritical values of x. That is, find the values $x = c$ where*
   a. *$f''(c) = 0$, or*
   b. *$f''(c)$ is undefined.*
3. *Using the hypercritical values as endpoints of intervals, determine the intervals where*
   a. *$f''(x) > 0$ and f is concave upward, and*
   b. *$f''(x) < 0$ and f is concave downward.*
4. *Points of inflection occur at those hypercritical values where f changes concavity. (See Figure 4.1.5.)*

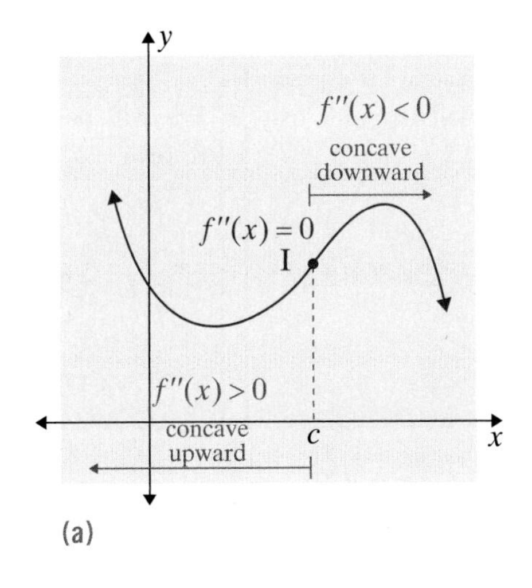

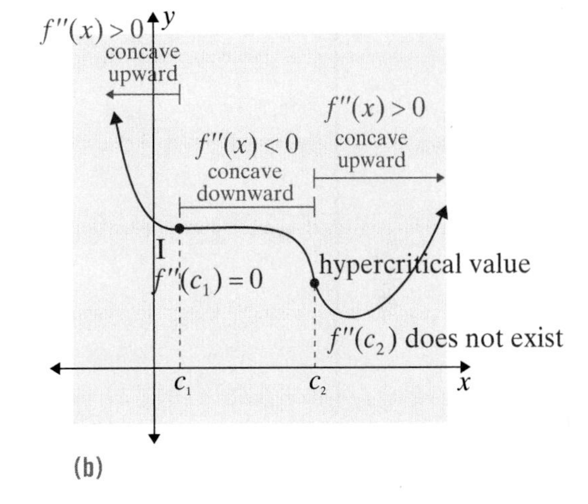

Figure 4.1.5

(a) (b)

Example 6: Finding Points of Inflection

For the function $f(x) = x^3 - \frac{3}{2}x^2 + 5$,

a. Determine the intervals on which *f* is concave upward and the intervals on which it is concave downward.

b. Locate any points of inflection.

Solutions: a. We find the hypercritical values and then determine the concavity on the related intervals.

continued on next page ...

$$f'(x) = 3x^2 - 3x$$

$$f''(x) = 6x - 3$$

To find the second derivative, find the first derivative of $f(x)$ and then find its derivative.

Now set $f''(x) = 0$.

$$6x - 3 = 0$$

$$x = \frac{1}{2}$$

Note that there are no values where $f''(x)$ is undefined.

There are two intervals to be considered:

$$\left(-\infty, \frac{1}{2}\right) \text{ and } \left(\frac{1}{2}, +\infty\right).$$

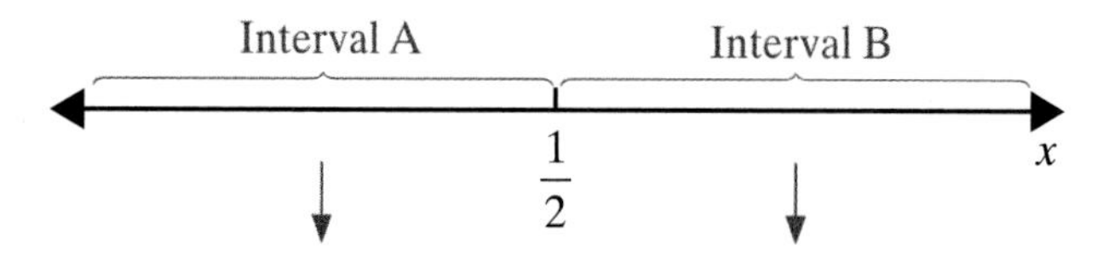

Interval A	Interval B
We will use the test point $x = -1$.	We will use the test point $x = 1$.
$f''(-1) = 6(-1) - 3$	$f''(1) = 6(1) - 3$
$= -6 - 3$	$= 6 - 3$
$= -9$	$= 3$
$-9 < 0$	$3 > 0$

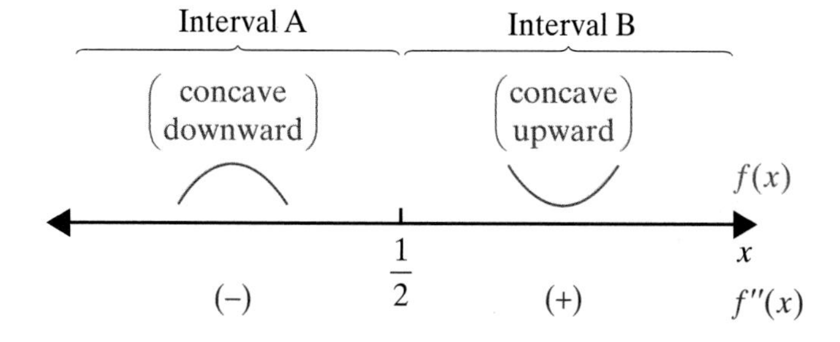

f is concave upward on $\left(\frac{1}{2}, +\infty\right)$.

f is concave downward on $\left(-\infty, \frac{1}{2}\right)$.

b. Since we found in part **a.** that f changes concavity at $x = \frac{1}{2}$, there must be a point of inflection there.

$$f(x) = x^3 - \frac{3}{2}x^2 + 5$$

$$f\left(\frac{1}{2}\right) = \left(\frac{1}{2}\right)^3 - \frac{3}{2}\left(\frac{1}{2}\right)^2 + 5$$

$$= \frac{1}{8} - \frac{3}{8} + 5$$

$$= \frac{19}{4}.$$

Substitute $x = \frac{1}{2}$ into $f(x)$ to find the point of the inflection.

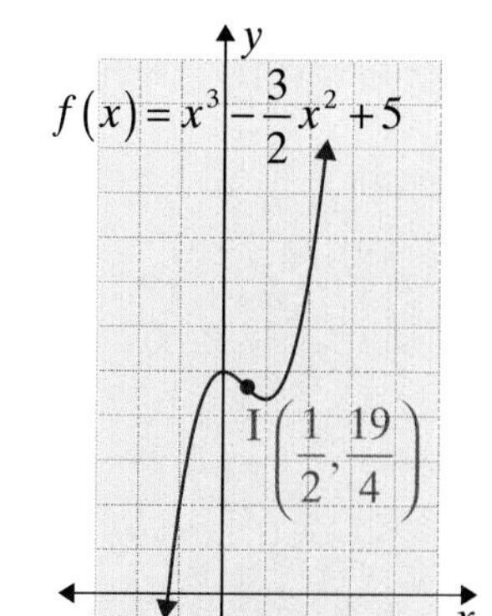

$\left(\frac{1}{2}, \frac{19}{4}\right)$ is a point of inflection of $f(x)$.

Using a Calculator to Find Inflection Points

Powerful hand-held calculators can sketch y, y', and y'', all on the same axes. They can locate coordinates of maximum points, minimum points, and points of inflection.

A TI-84 Plus can be used to graph y' and from this graph of y', one can find inflection points easily. To illustrate this procedure, we use the function $y = \frac{8}{3}x^3 - x^4 + 5$. (This function was actually used to graph Figure 4.1.3. Refer to this figure and it's accompanying text as necessary.)

Step 1: Type $y = \frac{8}{3}x^3 - x^4 + 5$ into the `Y1` position (in the Y= menu).

Step 2: Put the cursor in the `Y2` position. Next, press the MATH button. Arrow down the menu to item `8: nDeriv(` and press ENTER. This inserts "`nDeriv(`" into the `Y2` slot. Next press VARS, and arrow right one position to `Y-VARS`. The menu cursor will be on item `1: Function`. Press enter. When you see the menu of functions, press enter again. This selects `Y1` and puts it into the `nDeriv(` function. Next, input "`,X,X)`".

The full entry in the `Y2` position should look like `nDeriv(Y1,X,X)`. (See Figure 4.1.6.)

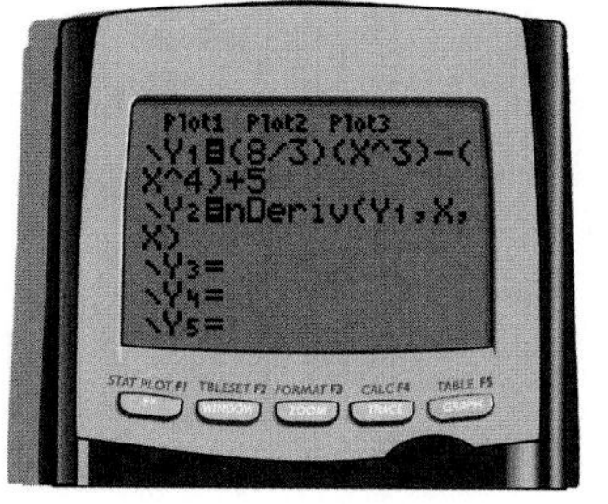

Figure 4.1.6

Step 3: Using a window of [–2, 3] by [–6, 12], press GRAPH. Both graphs should be displayed on the screen as shown in Figure 4.1.9 at left.

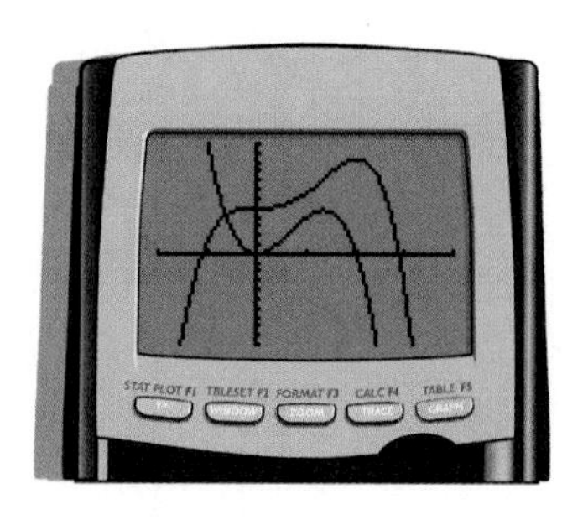

Figure 4.1.7

Note: Typically, you will not want the graph of y' displayed in every window. You can "turn-off" the graph of y' by moving the calculator's cursor, by using the arrow keys, to the equal sign just after Y_2. With the cursor on the equal sign, press ENTER.. This deselects the Y_2 graph, but leaves the typed equation or commands in place, so that the graph of y' can be reselected when desired.

Step 4: Type 2ND TRACE and select item 4: Maximum. If the cursor is blinking on the graph of $f(x)$, press the down arrow once. (This switches graphs so that the graph of $f'(x)$ is now selected.)

Step 5: You are now being prompted to select a left bound for the maximum point on the graph of $f'(x)$. Using the arrow keys, move the blinking cursor to the left of the maximum point, at or to the left of $x = 1$. Press ENTER.

Step 6: Now, using the arrow keys, select a right bound by moving the blinking cursor to the right of the maximum point, say to $x = 1.7$. Press ENTER.

Step 7: Finally, you are prompted "Guess". You should press the left arrow key once and then press ENTER. The (x, y)-values of the maximum point will be displayed at the bottom of the screen. In this case $x = 1.3333304$ and $y = 4.7407381$ will be given. The y-value here means $f'(1.33) = 4.74$.

As might be expected, the actual value of x is $\frac{4}{3}$, but the answer given by the calculator is quite accurate. The computation shows $f''\left(\frac{4}{3}\right)=0$. So $\left(\frac{4}{3}, f'\left(\frac{4}{3}\right)\right)$ is the high point on the graph of f' and thus $\left(\frac{4}{3}, f\left(\frac{4}{3}\right)\right)$ is the inflection point on $f(x)$.

We can confirm the calculator results:

$$f(x) = \frac{8x^3}{3} - x^4 + 5,$$

$$f'(x) = 8x^2 - 4x^3, \text{ and}$$

$$f''(x) = 16x - 12x^2.$$

We set $f''(x) = 0$ and obtain

$$16x - 12x^2 = 0, \text{ so}$$
$$4x(4 - 3x) = 0.$$

Thus $x = 0$ and $x = \frac{4}{3}$ will make $f''(x)$ equal to zero. Both values give inflection points for $f(x)$.

We can use $f'(x) = 0$ to locate the maximum and minimum points for $f(x)$.

$$f'(x) = 0$$
$$8x^2 - 4x^3 = 0$$
$$4x^2(2 - x) = 0$$
$$x = 0, 2$$

In the figure, we observe $x = 0$ corresponds to an inflection point (neither a maximum nor a minimum) and $x = 2$ corresponds to the maximum point $\left(2, \frac{31}{3}\right)$.

In this problem, the algebraic solutions of $f'(x) = 0$ and $f''(x) = 0$ are of a familiar type. When these equations are too difficult for an algebraic solution, a graphing utility is invaluable. When the goal (as in this book) is to teach applications of calculus, curve sketching problems are very important.

Second Derivative Test for Local Extrema

We have seen how the first derivative of a function can be used to locate local extrema and how the second derivative can be used to analyze concavity and locate points of inflection. Now we will show how the second derivative, if it exists, provides a relatively simple test for local extrema.

Second Derivative Test for Local Extrema

Suppose that f is a function, f′ and f″ exist on the interval (a, b), c is in (a, b), and $f'(c) = 0$.

1. *If $f''(c) > 0$, then f(c) is a local minimum.* *(See Figure 4.1.8(a).)*
2. *If $f''(c) < 0$, then f(c) is a local maximum.* *(See Figure 4.1.8(b).)*
3. *If $f''(c) = 0$, then the test fails to give any information about local extrema.*

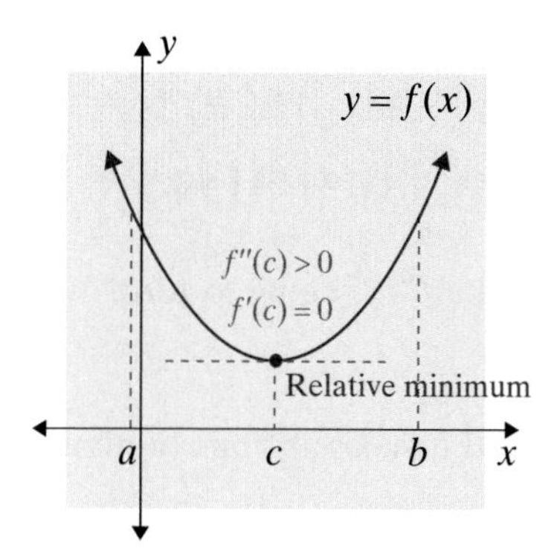

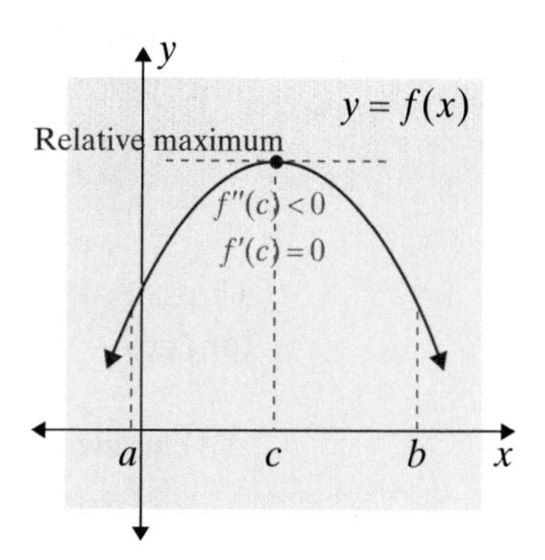

Figure 4.1.8

(a) Figure 4.1.8(a) suggests that $f(x)$ must be concave upward ($f''(c) > 0$) if it has a relative minimum at $x = c$.

(b) Similarly, Figure 4.1.8(b) shows that at a relative maximum, $x = c$, $f(x)$ must be concave downward ($f''(c) < 0$).

NOTES

The principle being applied in the Second Derivative Test is really very simple. If $f''(c)$ is defined at a local maximum $(c, f(c))$, a curve is concave down. If $f''(c)$ is defined at a local minimum $(c, f(c))$, a curve is concave up.

Of course, the primary value of the Second Derivative Test occurs only when the algebra formula is available for analysis. In the pursuit of fully understanding calculus, it is very important that the student is capable of graphing a few functions successfully by using calculus, without the aid of a calculator. The next example will demonstrate this.

Example 7: Using the Second Derivative Test

For the function $f(x) = x^4 - 18x^2$,

a. Locate the local extrema.

b. Locate the points of inflection.

Solutions: a. We will use the Second Derivative Test to find all local extrema of $f(x)$. First, find all values $x = c$ so that $f'(c) = 0$.

$$f'(x) = 4x^3 - 36x$$ Find the derivative of $f(x)$.

$$0 = 4x^3 - 36x$$ Set $f'(x) = 0$.

$$0 = 4x(x^2 - 9)$$

$$0 = 4x(x+3)(x-3)$$

$$x = -3, 0, 3$$

Substituting these x-values into $f(x)$, the corresponding y-values are $f(-3) = -81$, $f(0) = 0$, and $f(3) = -81$.

Now we need to determine how $f''(x)$ responds at these x-values.

$$f''(x) = 12x^2 - 36$$ Find $f''(x)$.

$$f''(-3) = 12(-3)^2 - 36 = 72$$ Substitute each of the values found for c: -3, 0, and 3.

$$f''(0) = 12(0)^2 - 36 = -36$$

$$f''(3) = 12(3)^2 - 36 = 72$$

Use the method for determining concavity in addition to the Second Derivative Test to interpret these results:

$x = -3$	$x = 0$	$x = 3$
$f''(-3) > 0$. Therefore, $f(-3)$ is concave up and $(-3, -81)$ is a minimum point.	$f''(0) < 0$. Therefore, $f(0)$ is concave down and $(0, 0)$ is a maximum point.	$f''(3) > 0$. Therefore, $f(3)$ is concave up and $(3, -81)$ is a minimum point.

b. The points of inflection come from setting $f''(x) = 0$ and solving for x:

$$f''(x) = 12x^2 - 36$$

$$0 = 12x^2 - 36$$ Set $f''(x) = 0$.

$$0 = 12(x^2 - 3)$$

$$0 = 12(x - \sqrt{3})(x + \sqrt{3})$$

$$x = \pm\sqrt{3}$$

continued on next page ...

The corresponding y-values are

$$
\begin{aligned}
f\left(-\sqrt{3}\right) &= \left(-\sqrt{3}\right)^4 - 18\left(-\sqrt{3}\right)^2 \\
&= 9 - 18(3) \\
&= -45 \\
f\left(\sqrt{3}\right) &= \left(\sqrt{3}\right)^4 - 18\left(\sqrt{3}\right)^2 \\
&= 9 - 18(3) \\
&= -45
\end{aligned}
$$

So the points of inflection are $\left(-\sqrt{3}, -45\right)$ and $\left(\sqrt{3}, -45\right)$.

Verifying Your Graph With a Calculator

1. Enter the given function into Y1.
2. Check that your WINDOW is appropriate.
3. Press GRAPH.

We point out that it is easy (without a graphing calculator) to sketch $y = x^4 - 18x^2$. We need only the coordinates of the three extreme points, and we may sketch a smooth curve only knowing which is a maximum and/or a minimum. In this example, the potential inflection points occur between a maximum and a minimum point and so we "know" the concavity changes at these points.

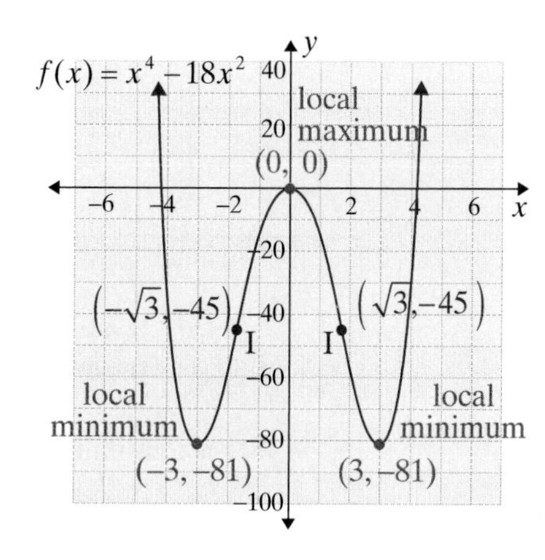

In the next example, we confirm intervals on which the graph of a function is concave up and on which it is concave down.

Example 8: Locating Local Extrema

Locate the local extrema for the function $f(x) = \frac{8}{3}x^3 - x^4$ and sketch a graph of the function.

Solution:

$$
\begin{aligned}
f'(x) &= 8x^2 - 4x^3 \\
0 &= 8x^2 - 4x^3 \\
0 &= 4x^2(2 - x) \\
x &= 0, 2
\end{aligned}
$$

Find $f'(x)$ and all values $x = c$ where $f'(c) = 0$.

Determine the corresponding y-values of $x = 0$ and $x = 2$.

$$f(x) = \frac{8}{3}x^3 - x^4$$

$$f(0) = \frac{8}{3}(0)^3 - (0)^4$$

$$= 0$$

$$f(2) = \frac{8}{3}(2)^3 - (2)^4$$

$$= \frac{16}{3}$$

The critical points are $(0, 0)$ and $\left(2, \frac{16}{3}\right)$.

Now find $f''(0)$ and $f''(2)$ and apply the Second Derivative Test to interpret these points.

$$f''(x) = 16x - 12x^2$$

$$f''(0) = 16(0) - 12(0)^2$$

$$= 0$$

$$f''(2) = 16(2) - 12(2)^2$$

$$= -16$$

$$-16 < 0$$

Substitute $x = 0$ and $x = 2$ into $f(x)$.

Since $f''(0) = 0$, the Second Derivative Test fails at this point.

$x = 0$	$x = 2$
$f''(0) = 0$. The Second Derivative Test cannot provide information about the point $(0, 0)$ as a local extremum.	$f''(2) < 0$. Therefore, f is concave down at $x = 2$ and $\left(2, \frac{16}{3}\right)$ is a maximum point.

A local maximum occurs at the point $\left(2, \frac{16}{3}\right)$.

Because the Second Derivative Test provides no information about the point $(0, 0)$ as a local extremum, we must use the First Derivative Test. Let us check the sign of $f'(x)$ on either side of 0 (Step 2 and Step 3 of the test). Using $x = -1$ and $x = 1$,

$$f'(-1) = 12 \text{ (positive) and } f'(1) = 4 \text{ (also positive).}$$

continued on next page ...

This shows that f increases to the left of $x = 0$ as well as to the right. Since there is no sign change from one side of $f'(0)$ to the other, $x = 0$ does not give a local minimum or a local maximum.

In fact, we have shown that slope is positive to the left of the origin, decreases to zero at the origin, and then increases to the right of the origin. This shows that the concavity changed at (0, 0) which must be a point of inflection.

Since the graph is concave up to the right of (0, 0), and concave down at the maximum $\left(2, \frac{16}{3}\right)$, there must be an inflection point in between these points. Set $y'' = 0$ and solve for x.

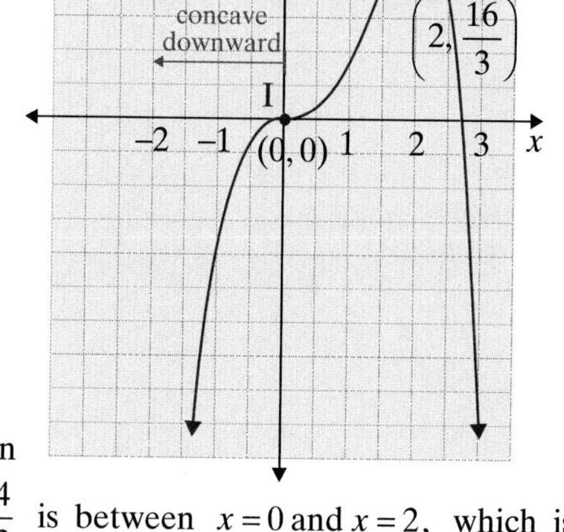

$$\begin{aligned} f''(x) &= 16x - 12x^2 = 0 \\ &= 4x(4 - 3x) = 0 \\ x &= 0, \text{ and } x = \frac{4}{3}. \end{aligned}$$

So $\left(\frac{4}{3}, f\left(\frac{4}{3}\right)\right)$ is the other inflection point. Note as a check that $x = \frac{4}{3}$ is between $x = 0$ and $x = 2$, which is already known.

NOTES The Second Derivative Test requires only one number to be checked. The First Derivative Test requires two numbers to be computed. For polynomial functions, the formula for f'' is one degree lower than that for f', so numbers are easier to calculate. For these reasons, the Second Derivative Test is usually preferred. However, sometimes, as in Example 8, it provides no information, and one must then use the First Derivative Test.

Application: Point of Diminishing Returns

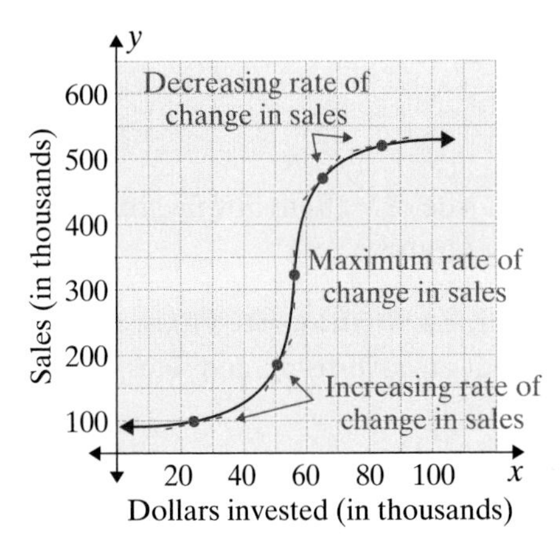

Figure 4.1.9

A company expects that as it spends more dollars on an advertising campaign for a product, the sales of the product will increase. A typical sales curve (sales as a function of advertising dollars) is shown in Figure 4.1.9.

At the beginning of a successful advertising campaign, not only do the sales increase, but the rate at which the sales grow also increases (the sales curve is concave upward). As the campaign continues, usually there is some point at which the rate of growth is a maximum. As more money is spent beyond this point, sales continue to increase, but at a slower rate. In economics, this point of rate change is called the **point of diminishing returns**.

Point of Diminishing Returns

*The **point of diminishing returns** occurs at a point of inflection where the sales curve changes from concave upward to concave downward.*

Example 9: Point of Diminishing Returns

Find the point of diminishing returns for the sales function

$$S(x) = -0.02x^3 + 3x^2 + 100,$$

where x represents thousands of dollars spent on advertising, $0 \le x \le 80$, and S is sales in thousands of dollars for automobile tires.

Solution: To find the point at which concavity changes from concave upward to concave downward, we must first find the hypercritical values of x between 0 and 80. Then we determine whether these points are points of inflection.

$$S(x) = -0.02x^3 + 3x^2 + 100$$

$$S'(x) = -0.06x^2 + 6x \qquad \text{Find } S'(x).$$

$$S''(x) = -0.12x + 6 \qquad \text{Find } S''(x).$$

$$-0.12x + 6 = 0$$

$$x = 50$$

Set $S''(x)$ equal to 0 to find the hypercritical values.

Note: $S''(10) = -0.12(10)+6 = 4.8$ (positive). Also, $S''(60) = -1.2$. Testing shows that

$$S''(x) > 0 \quad \text{for} \quad 0 < x < 50$$

and

$$S''(x) < 0 \quad \text{for} \quad 50 < x < 80.$$

Concavity changes from upward on the left side of $S(50)$ to downward on the other, signifying it is a point of diminishing returns.

Thus the point of diminishing returns is at $(50, S(50)) = (50, 5100)$. At the point of diminishing returns, \$50,000 is spent on advertising, and sales in tires are \$5,100,000.

4.1 Exercises

1. At each point marked on the graph in the figure, determine if f' is positive, negative, or zero. Determine if f'' is positive, negative or zero.

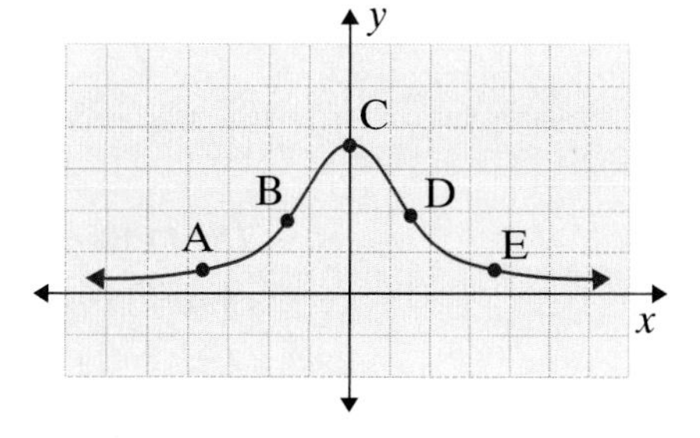

Draw a graph which satisfies the given conditions in Exercises 2 – 5.

2. Given $f(5) = 9,\ f'(5) = 2,\ f''(5) = -2.$

3. Given $f(-5) = -9,\ f'(-5) = 2,\ f''(-5) = 2.$

4. Given $f(5) = -9,\ f'(5) = 0,\ f''(5) = 3.$

5. Given $f(0) = 12,\ f'(0) = 0,\ f''(0) = -3.$

Find both the first and second derivatives for each of the functions in Exercises 6 – 17. Locate any relative maximum or minimum points and any points of inflection. Determine the intervals on which the function is concave upwards or concave downwards.

6. $f(x) = 7x^2 - 28x + 8$

7. $f(x) = 5x^2 - 9x + 2$

8. $f(x) = 2x^3 + 5x - 1$

9. $f(x) = 3x^3 + 6x - 8$

10. $f(x) = x^3 + 2\sqrt{x} + 5$

11. $f(x) = x^4 - 3\sqrt{x} + 2$

12. $f(x) = \left(x^2 + 7\right)^2$

13. $f(x) = \left(2x^2 - 5\right)^2$

14. $f(x) = \sqrt{x^2 + 3}$

15. $f(x) = \sqrt[3]{x^2 + 9}$

16. $f(x) = \dfrac{3x}{x^2 + 1}$

17. $f(x) = \dfrac{2x + 1}{x^2 - 4}$

For Exercises 18 – 25, find $f''(x)$. Then evaluate $f''(0)$, $f''(1)$, and $f''(4)$, if they exist.

18. $f(x) = x^3 + x^2 + 3$

19. $f(x) = x^3 - x^2 + 7$

20. $f(x) = x^2 - 5\sqrt{x} + 1$

21. $f(x) = x^2 + 2\sqrt{x} - 3$

22. $f(x) = \sqrt{x-4}$

23. $f(x) = \sqrt{2x+1}$

24. $f(x) = \dfrac{x}{x+5}$

25. $f(x) = \dfrac{x-2}{x+4}$

In Exercises 26 – 29, find all inflection points. Apply the Second Derivative Test at possible maximum/minimum points. Make a sketch of the graph and confirm your results with a graphing calculator.

26. $f(x) = (x+5)\sqrt[3]{x}$

27. $f(x) = (x^2+1)\sqrt[3]{x}$

28. $f(x) = 2x\sqrt[3]{x+1}$

29. $f(x) = (x+10)\sqrt[3]{x^2+10}$

30. Sketch y and y' on the same coordinate axis using the graph of y (figure at the right). Show how to locate inflection points using y'.

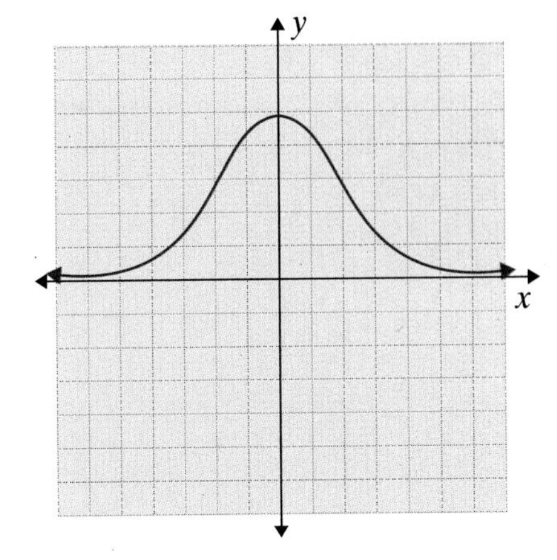

31. Given $f(x)$ as shown, graph a possible $f'(x)$ on the same axis. Show the inflection points.

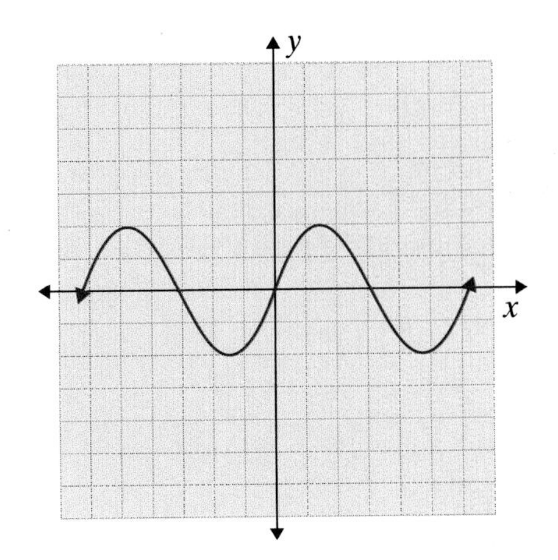

32. On the same axis, sketch $f'(x)$ and locate all inflection points.

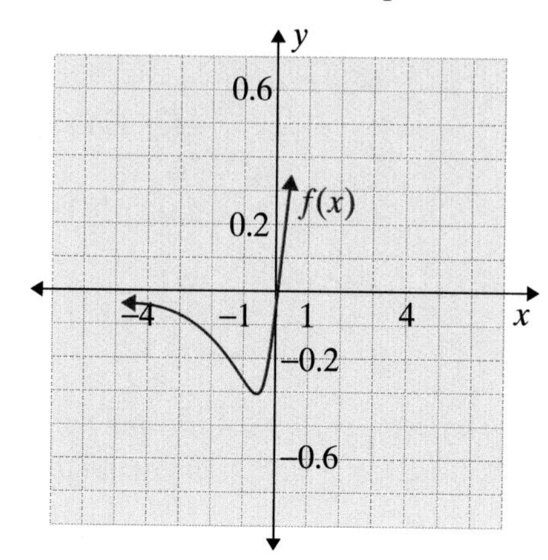

33. On the same axis, sketch $f'(x)$ and locate all inflection points.

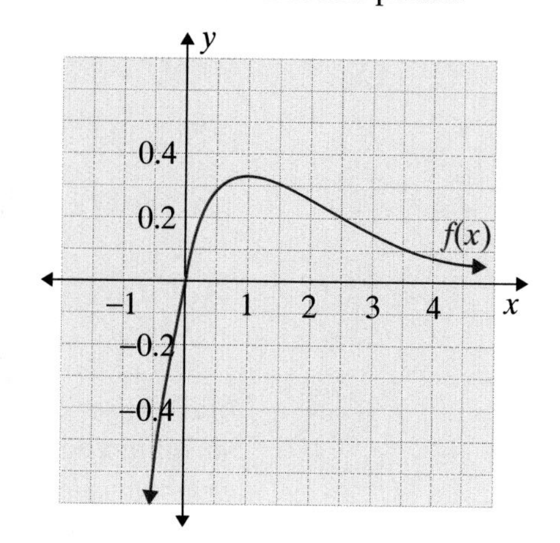

For each of the graphs in Exercises 34 – 37, list the interval(s) **a.** *on which f is concave upward and* **b.** *on which f is concave downward; then* **c.** *locate all points of inflection.*

34.

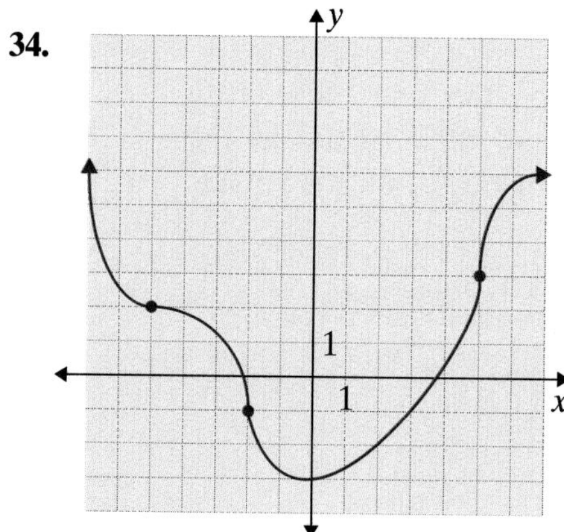

35.

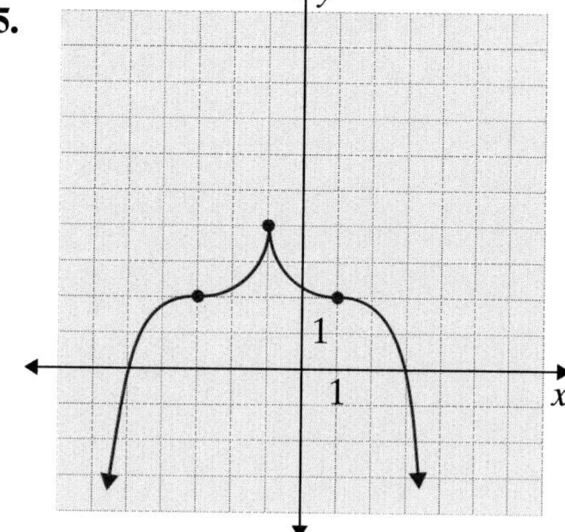

36.

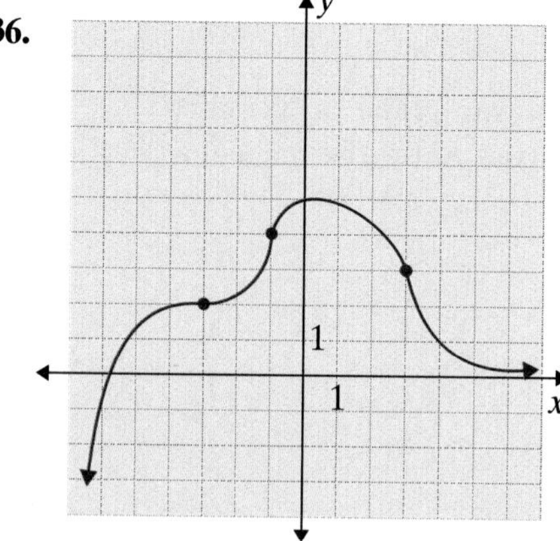

37.

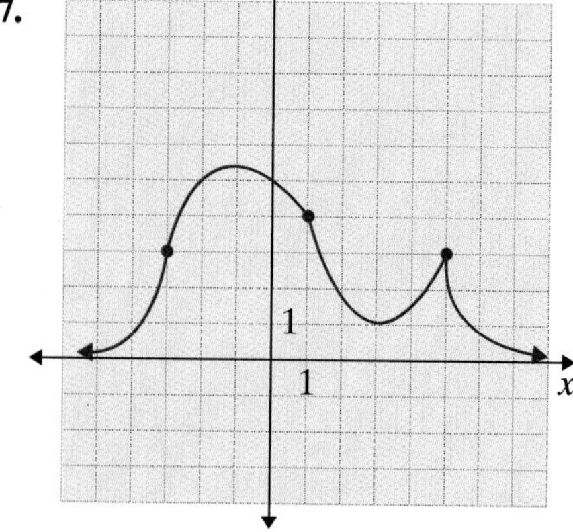

In Exercises 38 – 49, determine the intervals on which each function is **a.** *concave upward and* **b.** *concave downward; then* **c.** *locate all points of inflection. Use the information gathered to sketch the function. Confirm the details with a graphing calculator.*

38. $f(x)=2x^2+5x-9$

39. $f(x)=5x^2+8x-1$

40. $f(x)=x^3-3x^2+7$

41. $f(x)=x^3+6x^2-10$

42. $f(x)=x^3+11x-4$

43. $f(x)=5x^3+7x+2$

44. $f(x)=\frac{1}{3}x^3-2x^2+x-3$

45. $f(x)=\frac{1}{3}x^3+3x^2+2x-5$

46. $f(x)=\sqrt[3]{2x+3}$

47. $f(x)=\sqrt[3]{5x-3}$

48. $f(x)=\frac{x}{x^2-4}$

49. $f(x)=\frac{4x}{x^2-5}$

In Exercises 50 – 63, use the Second Derivative Test to find all local extrema, if the test applies. Otherwise, use the First Derivative Test.

50. $f(x)=x^2-3x+5$

51. $f(x)=8+7x-2x^2$

52. $f(x)=x^3-3x^2+8$

53. $f(x)=x^3+6x^2-10$

54. $f(x)=x^3-12x+3$

55. $f(x)=x^3-3x+4$

56. $f(x)=\frac{2}{3}x^3-x^2-4x-2$

57. $f(x)=\frac{1}{3}x^3+x^2-3x-1$

58. $f(x)=x^4-8x^2+7$

59. $f(x)=x^4-2x^2+3$

60. $f(x)=x^4+2x^3-4$

61. $f(x)=x^4-6x^3+8$

62. $f(x)=2x+\frac{8}{x}$

63. $f(x)=\frac{x^2+9}{x}$

In Exercises 64 – 67, give an example of a polynomial function which satisfies the conditions.

64. $F(5)=15$. $F'(x)$ is nonzero, but $F''(x)=0$ for all x.

65. $G(0)=0$, $G'(0)=0$, and $G''(0)=0$. $G(x)$ is concave upward everywhere and has no point of inflection.

66. $H(4)=0$. $H'(x)$ is positive for $x>4$ and negative for $x<4$. $H(x)$ has no inflection points.

67. $J(4) = 0$. $J'(4)$ is zero but $J'(x)$ is positive if $x \neq 4$. $J''(4) = 0$.

68. Given $f(x) = px^3 + bx + 10$, answer the following questions.

a. Suppose p and b are positive numbers. What can be said about maximum/minimum points and points of inflection?

b. Suppose p and b have opposite parity (one is positive and the other negative). What can be said about maximum/minimum points and points of inflection?

c. If the constant term 10 is changed to some other value, do your responses to parts **a.** and **b.** change?

d. Put your answers to parts **a.**, **b.**, and **c.** together in a "Lab Report" which discusses the coefficients in the given polynomial $y = px^3 + bx + c$.

69. Point of diminishing returns. Find the point of diminishing returns for the sales function $S(x) = 112 + 1.8x^2 - 0.1x^3$, where x represents thousands of dollars spent on advertising, $0 \le x \le 10$, and S is sales in thousand of dollars.

70. Point of diminishing returns. The sales function for a product is given by $S(x) = 204 + 6.3x^2 - 0.25x^3$, where x represents thousands of dollars spent on advertising, $0 \le x \le 12$, and S is sales in thousand of dollars. Find the point of diminishing returns.

71. Marginal cost. The cost function for a particular product is given by $C(x) = 0.1x^3 - 2.4x^2 + 24x + 190$ dollars, where $0 \le x \le 12$. Find the minimum marginal cost.

72. Marginal cost. Find the minimum marginal cost of a product if the cost function is given by $C(x) = 0.0001x^3 - 0.036x^2 + 16.8x + 1900$ dollars, where $0 \le x \le 150$.

73. Law enforcement. Due to the rapid increase in major crimes, the mayor of a large city plans to organize a major crime task force. It is estimated that for every 1000 persons in the city, the numbers of major crimes will be $N(t) = 56 + 3t^2 - 0.8t^{\frac{5}{2}}$, where t is the number of months after the task force is organized and $0 \le t \le 12$.

a. Find the maximum $N(t)$.

b. Find the maximum rate of increase in $N(t)$.

74. Meteorology. Meteorology records for a certain city suggest that for the month of June, the daily temperature between midnight and 6:00 P. M. can be approximated by $T(t) = -0.04t^3 + 1.14t^2 - 7.2t + 66$ degrees, where t is the number of hours after midnight and $0 \le t \le 18$.

a. Find the maximum and minimum temperatures.

b. Find the maximum rate of increase in the temperature.

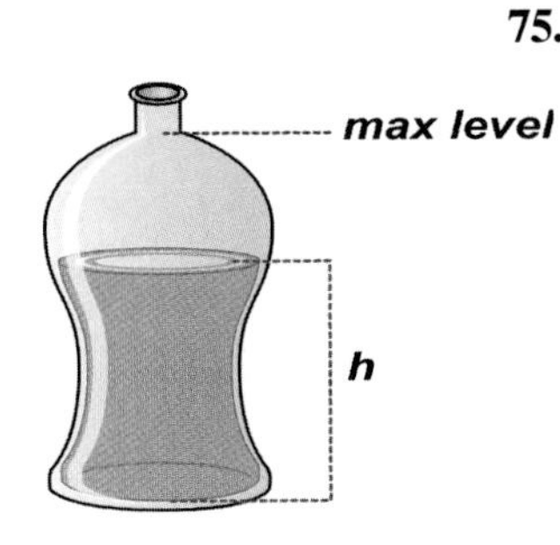

75. **Filtrate.** In a chemistry lab a filtrate drips slowly but continuously at a constant rate into a glass container shaped like the one at the left. The container eventually fills to the base of the neck. Let t denote the passage of time and h be the height of the liquid.
  a. Describe at what points on the bottle that $\frac{dh}{dt}$ will be a maximum and a minimum.
  b. Sketch a graph of $\frac{dh}{dt}$. Are there any inflection points (on a graph of $y = h(t)$)?
  c. Add a sketch of $y = h(t)$ on the same axis for part **b.**

Hawkes Learning Systems: Essential Calculus

Higher-Order Derivatives and Concavity p. 296 - 311, 317, 318
  Ex. 1 - 25, 30 - 49, 64 - 67, 69, 70
Higher-Order Derivatives: the Second Derivative Test p. 311 - 316
  Ex. 6 - 17, 26 - 29, 50 - 63, 68, 71 - 75

4.2 Curve Sketching: Polynomial Functions

Objectives

After completing this section, you will be able to:

1. *Sketch a polynomial function given a list of conditions.*
2. *Create your own list of conditions given a polynomial function, and sketch the graph of the function using this list.*

Curve Sketching

We begin by studying the basic characteristics that graphs must have if they are to satisfy a particular list of conditions. The more information we have (or the more conditions to be satisfied), the more accurate the corresponding graph can be. In each case, we assume that the function is continuous.

List of Given Conditions
1. $f'(x) < 0$ on $(-\infty, 2)$
2. $f'(x) > 0$ on $(2, +\infty)$
3. $f'(2) = 0$

What we know from given conditions is that:

1. f is decreasing on $(-\infty, 2)$,
2. f is increasing on $(2, +\infty)$, and
3. f has a horizontal tangent at $x = 2$.

That is, at $x = 2$, f has a local (and absolute) minimum. We have no idea what the minimum is and have no idea about the concavity of the graph.

Figure 4.2.1 shows three possible graphs that satisfy all the conditions given.

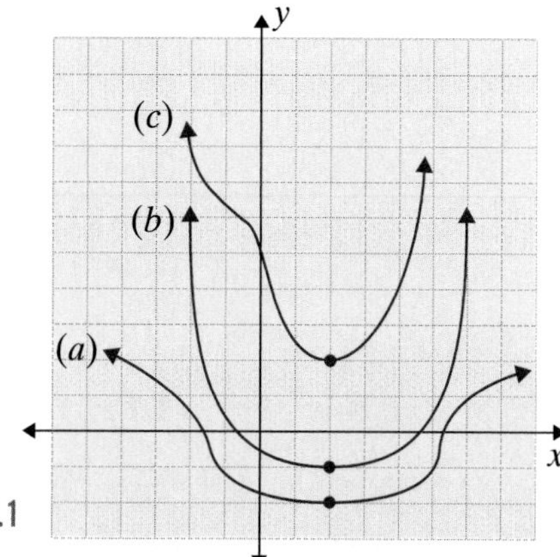

Figure 4.2.1

Possible Graphs of $f(x)$

Let us additionally suppose

4. $f''(x) > 0$ for all x.

This new condition eliminates graphs (a) and (c) since it declares that $f(x)$ is concave up for all x.

Let us add one more condition:

5. $f(2) = -1$.

We know that curves (a) and (c) in Figure 4.2.1 will not satisfy Condition 4 because they are not concave up for all x. Graph (b) is a likely candidate since the point $(2, -1)$ is on the graph of (b); however, there are still an infinite number of possible graphs that satisfy all the given conditions (see Figure 4.2.2). Thus we can sketch only a **general curve** that satisfies all the conditions. We cannot expect to graph a specific curve unless we know more conditions that the function must satisfy, or, even better, the equation that represents the function.

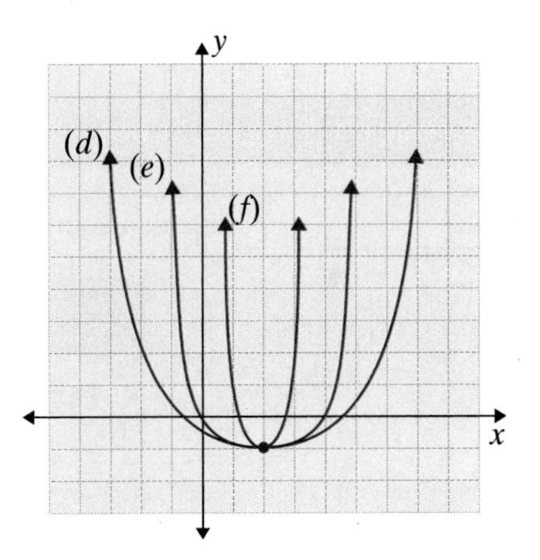

Possible General Graphs of $f(x)$

Figure 4.2.2

Example 1: Curve Sketching

Sketch a continuous function that satisfies all the given conditions.

List of Given Conditions
1. $f'(x) > 0$ for all x.
2. $f''(3) = 0$ and $f(3) = 4$.
3. $f''(x) < 0$ on $(-\infty, 3)$.
4. $f''(x) > 0$ on $(3, +\infty)$.

Solution: Condition 1 means that f is increasing for all x. Conditions 2, 3, and 4 indicate that $(3, 4)$ is a point of inflection and f is concave downward on $(-\infty, 3)$ and concave upward on $(3, +\infty)$.

continued on next page ...

One possible curve that satisfies all the given conditions is shown at right.

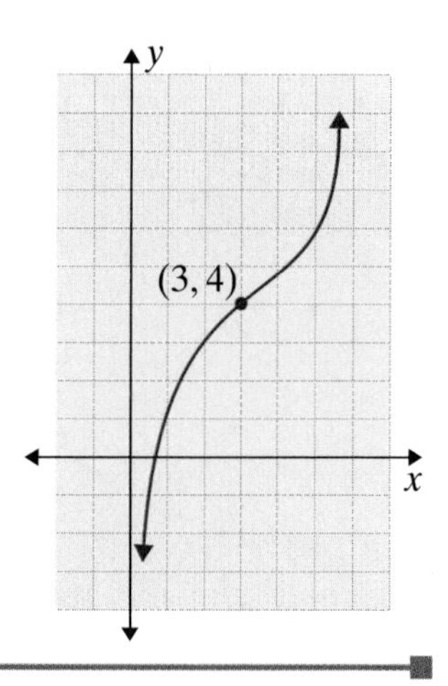

Example 2: Curve Sketching

Sketch a continuous function that satisfies all the given conditions.

List of Given Conditions
1. $g(-1) = 2,\ g(0) = 1,\ g(1) = -2.$
2. $g'(0) = 0,\ g'(1) = 0.$
3. $g''(0) = 0,\ g(.5) = 0$
4. $g''(x) < 0$ if $0 < x < .5$
5. $g''(x) > 0$ if $x < 0$ or $x > .5.$

Solution: In Condition 1, three points are given: (–1, 2), (0, 1), and (1, –2). There are horizontal tangent lines at $x = 0$ and $x = 1$ (Condition 2). Due to changes in concavity inferred from Conditions 3 to 5, points (0, 1) and (0.5, 0) are points of inflection. We also know that a local minimum occurs at the point (1, –2) because the function is concave up by condition 5 but has zero slope by condition 2.

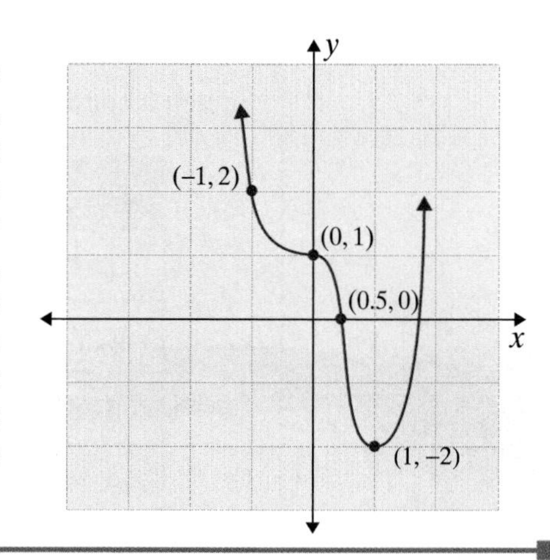

We have previously seen how f' and f'' can be used to describe characteristics of f (whether it is increasing, decreasing, concave up, or concave down) and to locate critical values and hypercritical values of f. The following systematic plan or strategy shows how to collect all this information for a given function in order to obtain an accurate graph of that function.

Strategy for Curve Sketching Given a Function f

To sketch the graph of a function f:

1. *Find $f'(x)$ and $f''(x)$.*
2. *Find the critical values of f. That is, find the values $x = c$ where*
   a. *$f'(c) = 0$, or*
   b. *$f'(c)$ is undefined.*
3. *Using the critical values as endpoints, find intervals*
   a. *where $f'(x) > 0$ (f is increasing), and*
   b. *where $f'(x) < 0$ (f is decreasing).*
4. *Locate the local extrema using the First Derivative Test or the Second Derivative Test.*
5. *Find the hypercritical values of f. That is, find the values $x = c$ where*
   a. *$f''(c) = 0$, or*
   b. *$f''(c)$ is undefined.*
6. *Using the hypercritical values as endpoints, find intervals*
   a. *where $f''(x) > 0$ (f is concave upward), and*
   b. *where $f''(x) < 0$ (f is concave downward).*
7. *Locate all points of inflection. These occur at hypercritical values where the curve changes concavity.*
8. *Using the combined information from Steps 1 – 7 and any other specific points that might be helpful, sketch the graph. (A helpful point would usually be the y-intercept $(0, f(0))$ or an x-intercept: $(a, f(a))$ if $f(a) = 0$.)*

Polynomial Functions

In the following examples we use the plan for curve sketching to sketch the graphs of polynomial functions. Remember that polynomial functions are functions of the form

$$f(x) = a_n x^n + a_{n-1} x^{n-1} + \ldots + a_1 x + a_0,$$

where the coefficients $a_n, a_{n-1}, \ldots, a_0$ are real numbers and the exponents $n, n-1, \ldots, 0$ are positive integers.

NOTES

We make the following important statement without proof:

The graphs of polynomial functions are smooth, continuous curves, and $f'(x)$ and $f''(x)$ are defined for all real x.

Example 3: Using the Strategy for Curve Sketching

Sketch the graph of the polynomial function $f(x) = x^2 - 2x - 3$.

Solution: In this example several steps are listed together because the first and second derivatives are easily found.

Step 1: Find $f'(x)$ and $f''(x)$.

$$f'(x) = 2x - 2$$

$$f''(x) = 2$$

Steps 2 and 3: Find the critical values and the intervals where $f'(x) > 0$ and where $f'(x) < 0$.

$$f'(x) = 2x - 2 = 2(x - 1)$$

$$0 = 2(x - 1)$$

$$x = 1$$

Find the critical values of f by setting $f'(x) = 0$ and solving for x.

The critical value is $x = 1$.

Now we can determine the following:

If $x < 1$, then $f'(x) = 2(x - 1) < 0$ and f is decreasing.
If $x > 1$, then $f'(x) = 2(x - 1) > 0$ and f is increasing.

Step 4: Find all local extrema.

$$f''(1) = 2$$

$$2 > 0 \qquad \text{local minimum}$$

Evaluate $f''(x)$ for the critical value to find the local extrema and use the Second Derivative Test.

Find the corresponding y-value of the critical value.

$$f(1) = (1)^2 - 2(1) - 3$$

$$= -4$$

The point $(1, -4)$ is the local minimum.

Steps 5, 6, and 7: Find the hypercritical values, the intervals where $f''(x) > 0$ and where $f''(x) < 0$, and all points of inflection.

For all x,

$$f''(x) = 2 > 0.$$

Thus f is concave up for all x, and there are no points of inflection.

Step 8: Sketch the graph.

Interval or Value	Derivative	Nature of Graph
$(-\infty, 1)$	$f'(x) < 0$	Decreasing
$(1, +\infty)$	$f'(x) > 0$	Increasing
$x = 1$	$f'(1) = 0$	Local minimum
$(-\infty, +\infty)$	$f''(x) = 2 > 0$	Concave up

Setting $y = x^2 - 2x - 3 = 0$ gives $(x + 1)(x - 3) = 0$. So the x-intercepts are at $x = -1$ or $x = 3$. We can also see from the graph that -4 is the absolute minimum value of f.

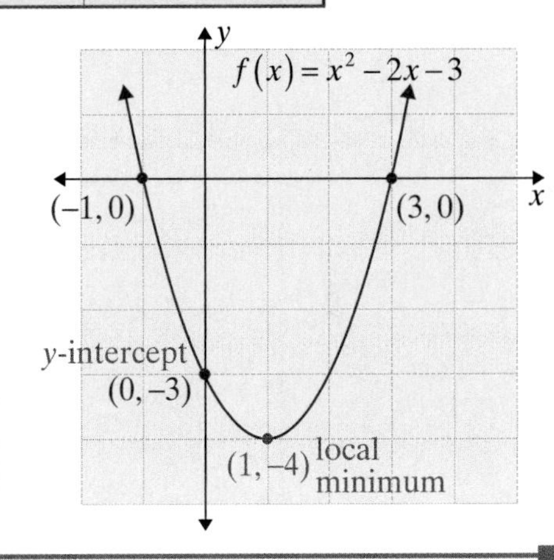

The graph is a parabola, which agrees with the fact that the graph of every second-degree polynomial function is a parabola.

Example 4: Using the Strategy for Curve Sketching

Use the curve sketching strategy to sketch the graph of the function

$$f(x) = \frac{1}{3}x^3 - 2x^2 + 3x + 1.$$

Solution: **Step 1:** Find $f'(x)$ and $f''(x)$.

$$f'(x) = x^2 - 4x + 3$$

$$f''(x) = 2x - 4$$

Step 2: Find the critical values.

$$f'(x) = x^2 - 4x + 3$$

$$x^2 - 4x + 3 = 0$$

$$(x - 3)(x - 1) = 0$$

$$x = 1, 3$$

Find the critical values of f by setting $f'(x) = 0$ and solving for x.

The critical values are $x = 1$ and $x = 3$.

continued on next page ...

Step 3: Use the critical values $x = 1$ and $x = 3$ as endpoints of intervals where $f'(x) > 0$ and where $f'(x) < 0$.

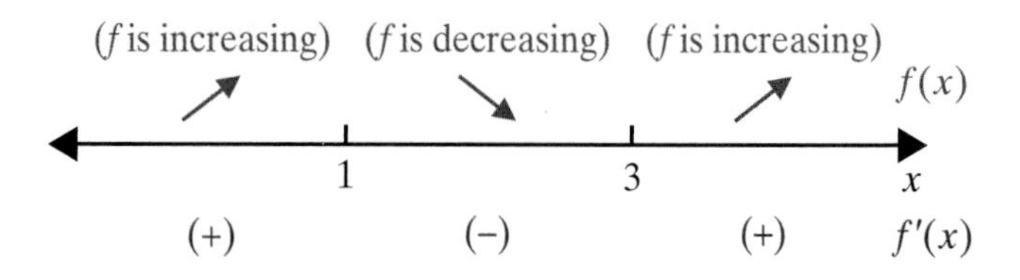

Step 4: Find all local extrema.

Evaluate $f''(x)$ for the critical values to find the local extrema and use the Second Derivative Test.

$$f''(1) = 2(1) - 4$$
$$= -2$$
$$-2 < 0 \qquad \text{local maximum}$$

$$f''(3) = 2(3) - 4$$
$$= 2$$
$$2 > 0 \qquad \text{local minimum}$$

Find the corresponding y-value for each of the critical values.

$$f(1) = \frac{1}{3}(1)^3 - 2(1)^2 + 3(1) + 1$$
$$= \frac{7}{3}$$

$$f(3) = \frac{1}{3}(3)^3 - 2(3)^2 + 3(3) + 1$$
$$= 1$$

Therefore, $\left(1, \frac{7}{3}\right)$ is the local maximum and $(3, 1)$ is the local minimum.

Step 5: Find all hypercritical values.

Find the hypercritical values of f by setting $f''(x) = 0$ and solving for x.

$$f''(x) = 2x - 4$$
$$2x - 4 = 0$$
$$2(x - 2) = 0$$
$$x = 2$$

There is one hypercritical value, $x = 2$.

Step 6: Using the hypercritical value $x = 2$ as an endpoint of intervals, find the intervals where $f''(x) > 0$ and where $f''(x) < 0$.

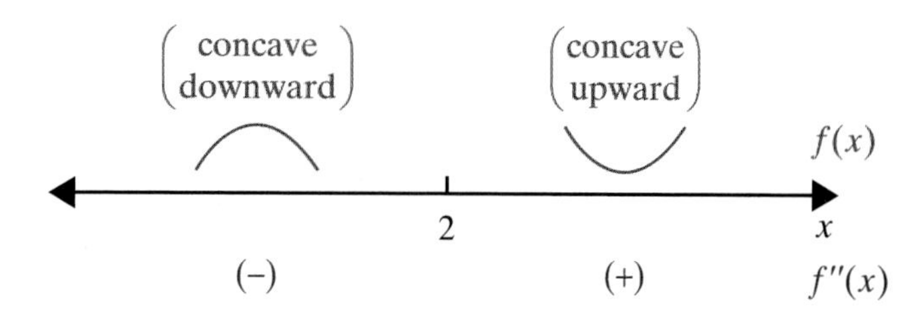

f is concave down on the interval $(-\infty, 2)$ and concave up on the interval $(2, +\infty)$.

Step 7: Find all points of inflection.

$$f(2) = \frac{1}{3}(2)^3 - 2(2)^2 + 3(2) + 1$$
$$= \frac{5}{3}$$

Find the corresponding y-value of the hypercritical values.

Since f changes concavity at $x = 2$, the point $\left(2, \frac{5}{3}\right)$ is a point of inflection.

Step 8: Sketch the graph.

The following table summarizes the information found in Steps 1 – 7.

Interval or Value	Derivative	Nature of Graph
$(-\infty, 1)$ or $(3, +\infty)$	$f'(x) > 0$	Increasing
$(1, 3)$	$f'(x) < 0$	Decreasing
$x = 1$	$f'(1) = 0$	Local maximum
$x = 3$	$f'(3) = 0$	Local minimum
$(-\infty, 2)$	$f''(x) < 0$	Concave down
$(2, +\infty)$	$f''(x) > 0$	Concave up
$x = 2$	$f''(2) = 0$	Point of inflection

continued on next page ...

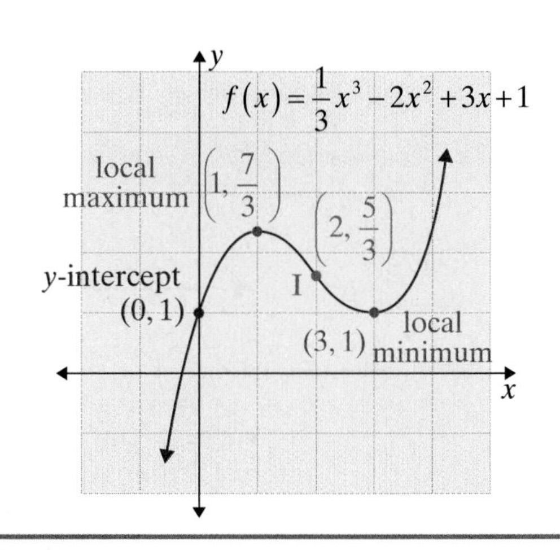

Example 5: Graphing the Derivative

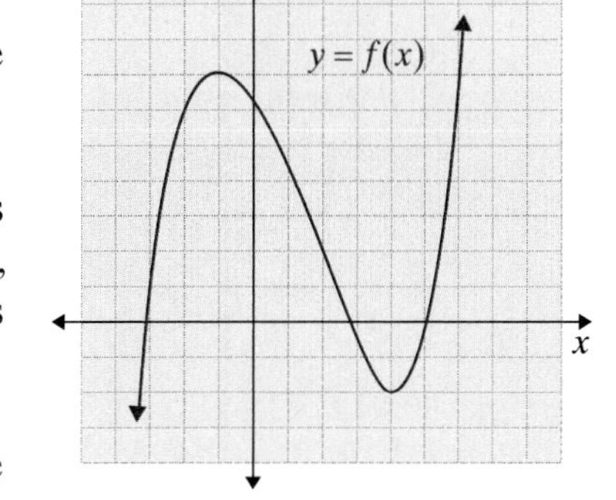

For the function graphed at right:

a. Identify the local extrema and locate the point(s) of inflection.

b. Determine the intervals on which $f(x)$ is increasing and on which $f(x)$ is decreasing, and identify the intervals on which $f(x)$ is concave upwards and concave downwards.

c. Sketch on the same coordinate axis a possible graph of $f'(x)$.

Solutions: a. A local maximum is located at $(-1, 7)$, and a local minimum is located at $(4, -2)$. There is a point of inflection at about $(1.5, 3.5)$.

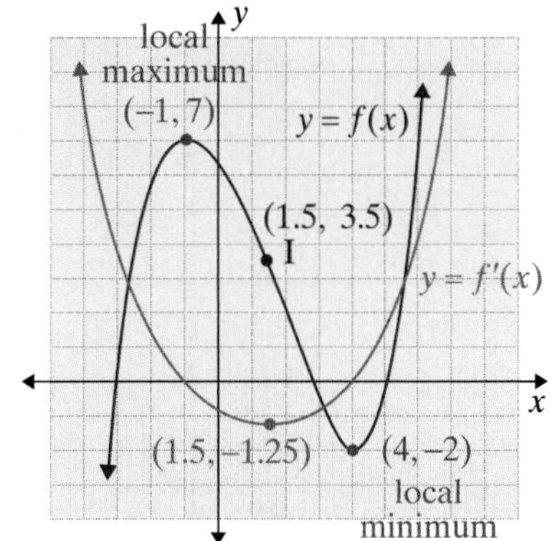

b. The function is increasing on the intervals $(-\infty, -1)$ and $(4, +\infty)$. It is decreasing on the interval $(-1, 4)$. The function is concave downwards on the interval $(-\infty, 1.5)$ and concave upwards on $(1.5, +\infty)$.

c. The slope of $f(x)$ is positive but decreasing from $-\infty < x < -1$, and is 0 at $x = -1$. The slope then becomes negative and continues decreasing until $x = 1.5$, at which point slope is a minimum. Then the slope starts to increase from some negative value (which we estimate to be -1.25) to 0 (at $x = 4$). It then becomes positive and continues increasing.

4.2 Exercises

In Exercises 1 – 16, sketch the graph of a continuous function that satisfies all the given conditions.

1. Given conditions:
- **a.** $f(-1) = 2$.
- **b.** $f'(-1) = 0$.
- **c.** $f'(x) < 0$ if $x < -1$.
- **d.** $f'(x) > 0$ if $x > -1$.
- **e.** $f''(x) > 0$ for all x.

2. Given conditions:
- **a.** $f(3) = 4$.
- **b.** $f'(3) = 0$.
- **c.** $f'(x) < 0$ if $x > 3$.
- **d.** $f'(x) > 0$ if $x < 3$.
- **e.** $f''(x) < 0$ for all x.

3. Given conditions:
- **a.** $f(-2) = 4,\ f(-1) = 1,\ f(1) = -1$.
- **b.** $f'(-2) = 0,\ f'(1) = 0$.
- **c.** $f'(x) < 0$ if $-2 < x < 1$.
- **d.** $f'(x) > 0$ if $x < -2$ or $x > 1$.
- **e.** $f''(-1) = 0$.
- **f.** $f''(x) < 0$ if $x < -1$.
- **g.** $f''(x) > 0$ if $x > -1$.

4. Given conditions:
- **a.** $f(0) = -2,\ f(2) = 0,\ f(3) = 3$.
- **b.** $f'(0) = 0,\ f'(3) = 0$.
- **c.** $f'(x) < 0$ if $x < 0$ or $x > 3$.
- **d.** $f'(x) > 0$ if $0 < x < 3$.
- **e.** $f''(2) = 0$.
- **f.** $f''(x) < 0$ if $x > 2$.
- **g.** $f''(x) > 0$ if $x < 2$.

5. Given conditions:
- **a.** $f(-3) = 5,\ f(-1) = 2,\ f(0) = -1$.
- **b.** $f'(-3) = 0,\ f'(0) = 0$.
- **c.** $f'(x) < 0$ if $x < 0$ and $x \neq -3$.
- **d.** $f'(x) > 0$ if $x > 0$.
- **e.** $f''(-3) = 0,\ f''(-1) = 0$.
- **f.** $f''(x) < 0$ if $-3 < x < -1$.
- **g.** $f''(x) > 0$ if $x < -3$ or $x > -1$.

6. Given conditions:
- **a.** $f(1) = 2,\ f(2) = 3,\ f(4) = 4,$ $f(6) = 2$.
- **b.** $f'(1) = 0,\ f'(4) = 0$.
- **c.** $f'(x) < 0$ if $x > 4,\ x < 1$.
- **d.** $f'(x) > 0$ if $1 < x < 4$.

7. Given conditions:
- **a.** $f(x) = ax^3 + bx + c$.
- **b.** $f(0) = 0$.
- **c.** $f(1) = 15$.
- **d.** $f'(-1) = 0$ and $x = -1$ is a local max.

8. Given conditions:
- **a.** $f(10) = 5$.
- **b.** $f'(5) = 0$.
- **c.** $f''(5) = 10$.
- **d.** $f''(x) < 0$ if $x > 10$.
- **e.** $f''(x) > 0$ if $x < 10$.

9. Given conditions:
- **a.** $f''(x) > 0$ if $x < 5$.
- **b.** $f''(5) = 0$.
- **c.** $f''(x) < 0$ if $x > 5$.
- **d.** $f'(x) > 0$ for all x.

10. Given conditions:
- **a.** $f(4) = 8$.
- **b.** $f'(4) = 0$.
- **c.** $f''(4) = 8$.

11. Given conditions:
 a. $f(-5)=4$.
 b. $f'(-5)=0$.
 c. $f''(-5)=-2$.

12. Given conditions:
 a. $f(x)=ax^2+bx+c$.
 b. $f'(-3)=0$.
 c. $f''(-3)=2$.

13. Given conditions:
 a. $f(x)=ax^3+bx^2+cx+d$.
 b. $f(0)=25$.
 c. $f'(4)=0,\ f'(-4)=0$.
 d. $f''(4)=48,\ f''(-4)=-48$.

14. Given conditions:
 a. $f(x)=ax^2+bx+c$.
 b. $f(0)=79$.
 c. $f'(5)=0$.
 d. $f''(x)=6$.

15. Given conditions:
 a. $f(x)=ax^3+bx^2+cx+d$.
 b. $f(0)=2$.
 c. $f'(0)=5$.
 d. $f''(0)=4$.
 e. $f''(1)=12$.

16. Given conditions:
 a. $f(x)=ax^3+bx$.
 b. $f'(0)=-12$.
 c. $f'(2)=0$.

For each of the functions in Exercises 17 – 36, determine $f'(x)$ and $f''(x)$. Then complete a summary table like those in Examples 3 and 4. Use this table to sketch the graph of the function. (If available, use a graphing utility or calculator to obtain a suitable window and confirm the accuracy of your calculations.)

17. $f(x)=x^2-4x+7$

18. $f(x)=x^2+6x-8$

19. $f(x)=6+5x-x^2$

20. $f(x)=2+3x-2x^2$

21. $f(x)=x^3+3x^2-6$

22. $f(x)=\frac{1}{3}x^3-4x+3$

23. $f(x)=\frac{1}{3}x^3+x^2-3x+5$

24. $f(x)=2x^3-3x^2-12x+5$

25. $f(x)=x^4-2x^2+4$

26. $f(x)=x^4-8x^2-3$

27. $f(x)=\frac{1}{4}x^4-x^3+5$

28. $f(x)=x^4+4x^3+12$

29. $f(x)=x^4-4x+7$

30. $f(x)=3x^4-4x^3+3$

31. $f(x)=(x+5)(x-3)^2$

32. $f(x)=(x+1)^2(x-10)^2$

33. $f(x)=(2x+1)(x-8)^3$

34. $f(x)=(x-5)(x-10)(x+3)$

35. $f(x)=2x(5x+8)^3$

36. $f(x)=16x(21+x)^3$

37. Suppose that $f(x)=mx^2+6x+4$. Determine a value of m so that $f(x)$ has a minimum at $x=-1$.

38. Given $y=4x^2+nx+8$, determine a value for n so that y has a minimum at $x=2$.

39. Determine a value for m such that at $x=1$ the tangent to the function $f(x)=mx^2+6x+1$ has an equation of $y=12x-2$.

40. Determine a value for m so that $y=4x^3+mx^2$ has an inflection point at $x=-10$.

41. Is it possible for a polynomial $y=ax^2+bx+c$ to have an inflection point?

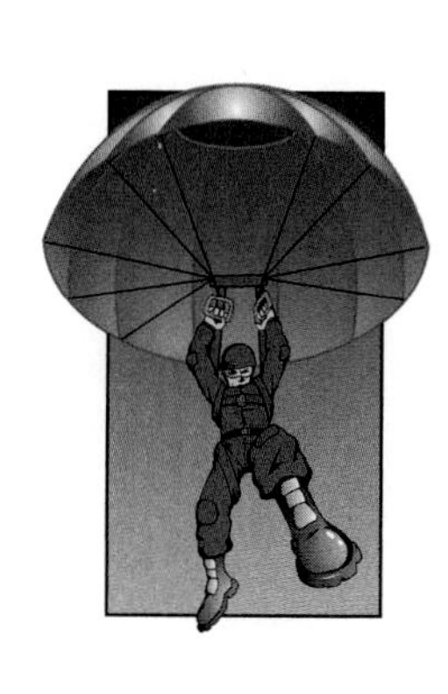

42. In an action movie, the hero is seen fighting the villain inside a plane which has a large hole in its side. The hero (actually a movie stunt man) is then thrown from the plane. He falls quickly and soon reaches a constant velocity. The hero opens his parachute but it deploys slowly, as if he is having trouble, but finally, in triumph, all is well and he drifts steadily and slowly to the ground.

- **a.** Draw a graph of the hero's vertical distance to the ground, represented by y, versus time t (in seconds).
- **b.** Describe any interesting points on the graph with points in the movie narrative.

43. The effectiveness of a certain medical injection is modeled by $E(t)=.01t(100-t)$, where t is time in minutes and E is a measure of concentration in the bloodstream. Effectiveness readings above 9.0 are satisfactory and readings above 30 are dangerous.

- **a.** If an injection is given at midnight, when are the readings satisfactory? When do they become too low?
- **b.** How high do the effectiveness readings get?
- **c.** The supervising nurse and the resident pharmacologist must assign a schedule for injections. For the next week, assuming injected dosages are additive, give a reasonable schedule for injections so that the patient's E-reading stays at or above 9 but never exceeds 30.

44. The productivity rating of an individual worker at the Cruz Corporation assembly line is based on the number of tasks accomplished, mistakes made, and responsiveness to difficulties encountered. The average of all scores allows the company to use a simple model based on time on the floor given by $\text{PR}=-0.4x^3+2x^2+10x+5$, where x is in hours at work. A PR score of 20 is acceptable and a score of 40 is highly unusual.

- **a.** When are workers' scores the highest?
- **b.** Design an 8-hour day where workers do the most demanding jobs for about 6 hours and have 2 hours for less stressful work. Explain your reasoning.

Hawkes Learning Systems: Essential Calculus

Curve Sketching: Polynomial Functions

4.3 Curve Sketching: Rational Functions

Objectives

After completing this section, you will be able to:

1. *Find vertical, horizontal, and oblique asymptotes of a rational function.*
2. *Sketch the graph of a given rational function.*

In this section we will apply the graphing strategies discussed in Section 4.2, along with a few new ideas, to **rational functions**. Examples of rational functions are

$$f(x)=\frac{x-1}{x},\ g(x)=\frac{x+2}{x-2}\ \text{and}\ h(x)=\frac{2x}{x^2+1}.$$

Rational Function

*A **rational function** is a function of the form*

$$f(x)=\frac{P(x)}{Q(x)},$$

where $P(x)$ and $Q(x)$ are polynomials and $Q(x)\neq 0$.

If, in the definition of a rational function, $Q(x)$ is a constant, then the rational function is just a polynomial function. Since we have already discussed polynomial functions in Section 4.2, the analysis here will focus on rational functions that are not polynomial functions. Also, we restrict the discussion to rational functions that are completely reduced with no common factors in the numerator and denominator.

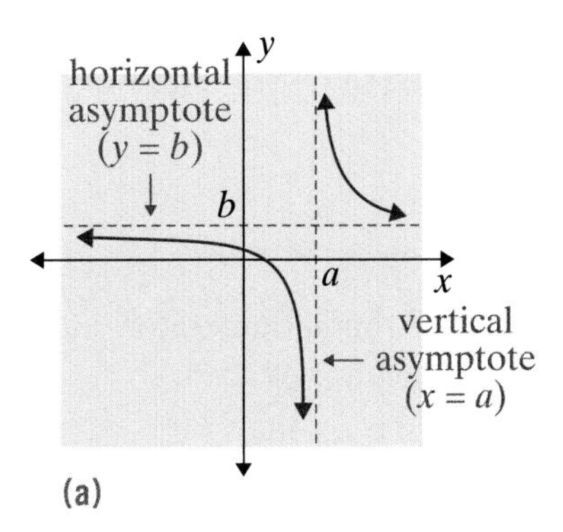

(a)

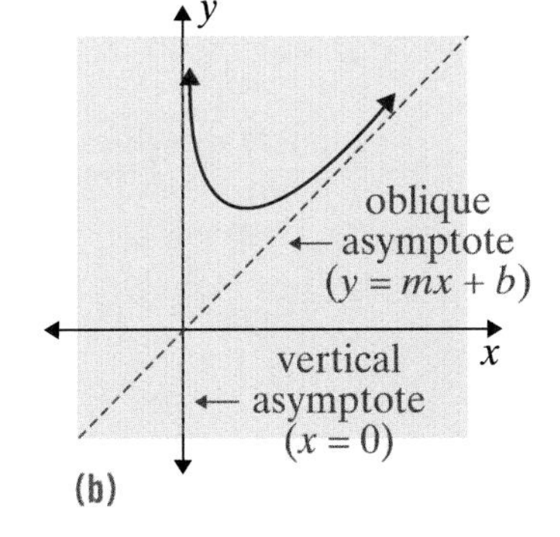

(b)

Figure 4.3.1

A distinguishing characteristic of the graphs of the rational functions we will investigate is that they have one or more asymptotes. Intuitively, an **asymptote** is a line that a curve "approaches" or "gets close to" in a limiting sense. Figure 4.3.1 illustrates three kinds of asymptotes: **horizontal**, **vertical**, and **oblique**.

Asymptotes can be found by taking limits, as outlined in the following list.

Asymptotes for Rational Functions $f(x)=\frac{P(x)}{Q(x)}$

1. ***Vertical asymptotes*** *are of the form x = a and occur where the denominator is* 0 *and the numerator is not* 0, *(that is,* $Q(a)=0$ *and* $P(a)\neq 0$*).*

2. ***Horizontal asymptotes*** *are of the form y = b and occur where*

$$\lim_{x\to+\infty} f(x)=b \qquad or \qquad \lim_{x\to-\infty} f(x)=b.$$

3. ***Oblique asymptotes*** *are of the linear form y = mx + b and occur when the numerator is one degree larger than the denominator and f(x) can be written as*

$$f(x)=mx+b+\frac{R(x)}{Q(x)},$$

where

$$\lim_{x\to+\infty}\frac{R(x)}{Q(x)}=0 \qquad or \qquad \lim_{x\to-\infty}\frac{R(x)}{Q(x)}=0.$$

Example 1: Locating Vertical and Horizontal Asymptotes

For the function $f(x)=\frac{2x}{x-5}$, find **a.** the vertical asymptotes and **b.** the horizontal asymptotes.

Solutions: a. A vertical asymptote occurs when the denominator is 0 (i.e., where $x-5=0$). Thus the line $x=5$ is a vertical asymptote.

b. We find the horizontal asymptotes by finding $\lim_{x\to+\infty} f(x)$ and $\lim_{x\to-\infty} f(x)$.

$$\lim_{x\to+\infty} f(x)=\lim_{x\to+\infty}\frac{2x}{x-5}$$

$$=\lim_{x\to+\infty}\frac{\frac{2x}{x}}{\frac{x}{x}-\frac{5}{x}}$$ Divide both numerator and denominator by x.

$$=\lim_{x\to+\infty}\frac{2}{1-\frac{5}{x}}$$ From Section 2.5, we know that $\lim_{x\to+\infty}\frac{a}{x}=0$.

$$=\frac{2}{1-0}=2$$

The line $y = 2$ is a horizontal asymptote. Since $\lim_{x\to-\infty}\left(\frac{5}{x}\right)=0$, we see that $\lim_{x\to-\infty} f(x)=2$ also.

Example 2: Locating Vertical and Horizontal Asymptotes

For the function $f(x)=\frac{1}{x^2+3}$, find **a.** the vertical asymptotes and **b.** the horizontal asymptotes.

Solutions: a. A vertical asymptote occurs when the denominator is 0. Since the denominator, $x^2 + 3$, is never 0, there are no vertical asymptotes.

b. We find the horizontal asymptotes by finding $\lim_{x\to+\infty} f(x)$ and $\lim_{x\to-\infty} f(x)$.

$$\lim_{x\to+\infty}\frac{1}{x^2+3}=0 \quad \text{and} \quad \lim_{x\to-\infty}\frac{1}{x^2+3}=0.$$

Therefore, the line $y = 0$ is a horizontal asymptote.

Example 3: Locating Oblique Asymptotes

Find the oblique asymptote for the function $f(x)=\frac{2x^2+7}{3x}$.

Solution: Since the numerator is one degree larger than the denominator, we can write $f(x)$ in the following form:

$$f(x)=\frac{2x^2+7}{3x}=\frac{2x^2}{3x}+\frac{7}{3x}=\frac{2}{3}x+\frac{7}{3x}.$$

Now we find that $\lim_{x\to+\infty}\frac{7}{3x}=0.$

So the remaining term gives the oblique asymptote:

$$y=\frac{2}{3}x.$$

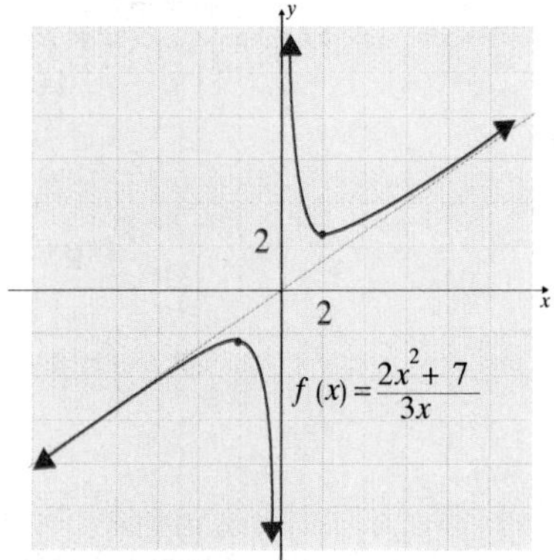

In this example $x = 0$ is not in the domain of the function f. Thus $x = 0$ is technically neither a critical value nor a hypercritical value of f. Nevertheless, in the sketch, we notice that f is concave down to the immediate left of $x = 0$. To the immediate right of $x = 0$, f is concave up. Although x-values which determine vertical asymptotes are not critical points, they are "critical" in the sense of having relevance to the analysis. (See Example 5 in this section.)

In the following example, we apply the structured curve sketching strategy from Section 4.2, along with our knowledge of asymptotes, to sketch the graphs of some rational functions. One useful fact (that will not be proven) is that **a rational function can have at most one horizontal asymptote**. In the next chapter, we will encounter functions with two horizontal asymptotes.

Example 4: Sketching the Graph of a Rational Function

Sketch the graph of the rational function $f(x)=\frac{x-1}{x-2}$. Include any related asymptotes.

Solution: Step 1: Find $f'(x)$ and $f''(x)$.

$$f'(x)=\frac{(x-2)\frac{d}{dx}(x-1)-(x-1)\frac{d}{dx}(x-2)}{(x-2)^2}=-\frac{1}{(x-2)^2}=-1(x-2)^{-2}$$

$$f''(x)=-1(-2)(x-2)^{-3}\frac{d}{dx}(x-2)=\frac{2}{(x-2)^3}$$

Steps 2, 3, and 4: Find the critical values, the intervals where $f'(x)>0$ and where $f'(x)<0$, and the local extrema.

The first derivative, $f'(x)=-\frac{1}{(x-2)^2}$, is never 0. Note that f' is undefined at $x=2$, but $x=2$ is not a critical value since 2 is not in the domain of f. Thus there are no critical values of f, and f has no local maxima or local minima.

Also, $f'(x)=-\frac{1}{(x-2)^2}<0$ for all $x\neq 2$ and f is decreasing for all x in its domain.

Steps 5, 6, and 7: Find all hypercritical values, intervals where $f''(x)>0$ and where $f''(x)<0$, and points of inflection.

Since $f''(x)=\frac{2}{(x-2)^3}\neq 0$ and is undefined only for $x=2$, there are no hypercritical values and no points of inflection.

However, if $x<2$, then $f''(x)=\frac{2}{(x-2)^3}<0$, and f is concave down, and if $x>2$, then $f''(x)=\frac{2}{(x-2)^3}>0$, and f is concave up.

Asymptotes: The line $x=2$ is a vertical asymptote because $x-2$ is not a factor of the numerator and the value of 2 for x will make the denominator 0.

We now find the horizontal asymptotes by finding $\lim_{x\to+\infty} f(x)$ and $\lim_{x\to-\infty} f(x)$.

$$\lim_{x\to+\infty} f(x) = \lim_{x\to+\infty} \frac{x-1}{x-2}$$

$$= \lim_{x\to+\infty} \frac{\frac{x}{x}-\frac{1}{x}}{\frac{x}{x}-\frac{2}{x}}$$

Divide both the numerator and the denominator by the highest power of x present in the function.

$$= \lim_{x\to+\infty} \frac{1-\frac{1}{x}}{1-\frac{2}{x}} = \frac{1-0}{1-0} = 1$$

We know that $\lim_{x\to+\infty} \frac{a}{x} = 0$.

Proceeding as above, we can also find that $\lim_{x\to-\infty} f(x) = 1$.

Therefore, the line $y = 1$ is a horizontal asymptote.

We also know that this function does not have an oblique asymptote because the degree of the numerator is not one degree higher than the denominator.

Step 8: Sketch the graph.

Interval or Value	Derivative	Nature of Graph
$(-\infty, 2)$ or $(2, +\infty)$	$f'(x) < 0$	Decreasing
$(-\infty, 2)$	$f''(x) < 0$	Concave down
$(2, +\infty)$	$f''(x) > 0$	Concave up
$x = 2$	NA	Vertical asymp.
$y = 1$	NA	Horizontal asymp.

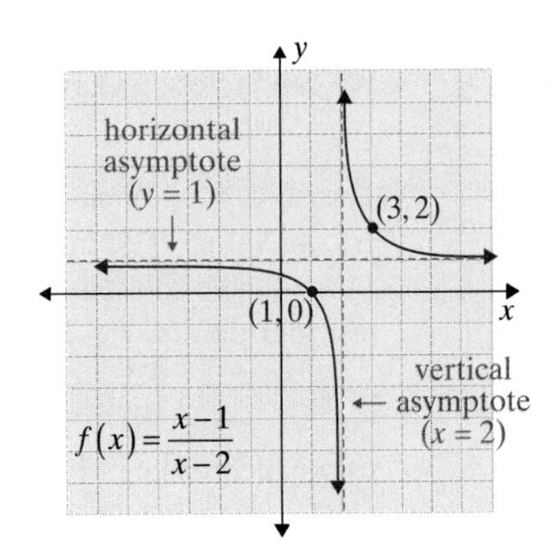

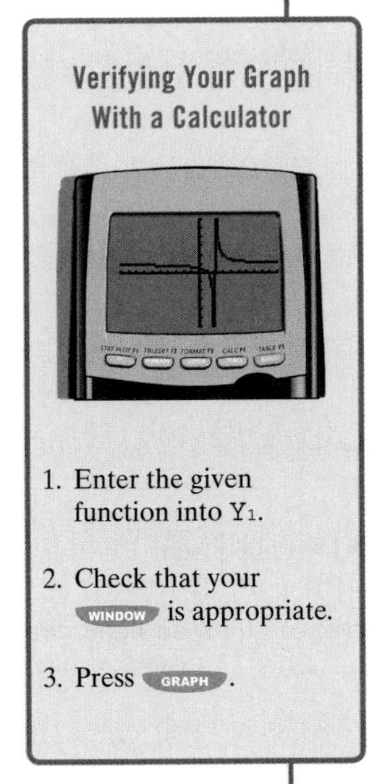

Verifying Your Graph With a Calculator
1. Enter the given function into Y1.
2. Check that your WINDOW is appropriate.
3. Press GRAPH.

In Example 5 we show all the analysis tools available to the student.

Example 5: Sketching the Graph of a Rational Function

Sketch the graph of the rational function $f(x)=\dfrac{100+x^2}{2x}$.

Solution: Here we will first identify the function's asymptotes. First, it is easy in this case to see by inspection of the formula that $x = 0$ makes the denominator 0 but not the numerator. Thus $x = 0$ is a vertical asymptote.

$$\begin{aligned} f(x) &= \frac{100+x^2}{2x} \\ &= \frac{x^2}{2x}+\frac{100}{2x} \\ &= \frac{x}{2}+\frac{50}{x} \end{aligned}$$

Since $\lim\limits_{x\to+\infty}\dfrac{50}{x}=0$, we observe $y=\dfrac{x}{2}$ is an oblique asymptote.

Unlike the function in Example 4, this function's numerator is one degree larger than the denominator, so we know immediately that this function will have an oblique asymptote. Using the above work (equivalent to division of polynomials), the line $y=\dfrac{x}{2}$ is seen to be an oblique asymptote.

Now we will analyze the derivative for $x > 0$.

$$f'(x)=\frac{2x(2x)-(100+x^2)(2)}{4x^2}=\frac{x^2-100}{2x^2}$$

$$f''(x)=\frac{2x^2(2x)-(x^2-100)4x}{4x^4}=\frac{100}{x^3}$$

We can factor $f'(x)$ even further: $f'(x)=\dfrac{x^2-100}{2x^2}=\dfrac{(x+10)(x-10)}{2x^2}$.

This shows $x = 10$ and $x = -10$ are critical values, and we must treat $x = 0$ as a critical value. The intervals to consider are $(-\infty,-10),(-10,0),(0,10)$, and $(10,\infty)$. We select a convenient point in each interval and find the slope. We test four points: $f'(-11) = .087, f'(-1) = -49.5, f'(1) = -49.5$, and $f'(11) = .087$.

For $-\infty<x<-10$, $f'(x)>0$ so y is increasing. For $-10 < x < 0$, $f'(x) < 0$ so y is decreasing. For $0 < x < 10$, $f'(x) < 0$ so y is decreasing. For $x > 10$, $f'(x) > 0$ so y is increasing. Note: $f(10) = 10$, and $f(-10) = -10$. The analysis shows $(-10, -10)$ is a local maximum and $(10, 10)$ is a local minimum.

$$f''(x)=\frac{100}{x^3}>0 \text{ for all } x>0, \text{ so } f \text{ is concave up.}$$

Since $f'(10) = 0$ and $f''(10) > 0$, by the Second Derivative Test, we confirm that the point $(10, f(10)) = (10, 10)$ is a local minimum. Similarly $f''(-10) < 0$ and $(-10, -10)$ is a local maximum.

Also, because $f''(x) \neq 0$ and there are no points in the domain of f where $f''(x)$ is undefined, there are no hypercritical values and no points of inflection.

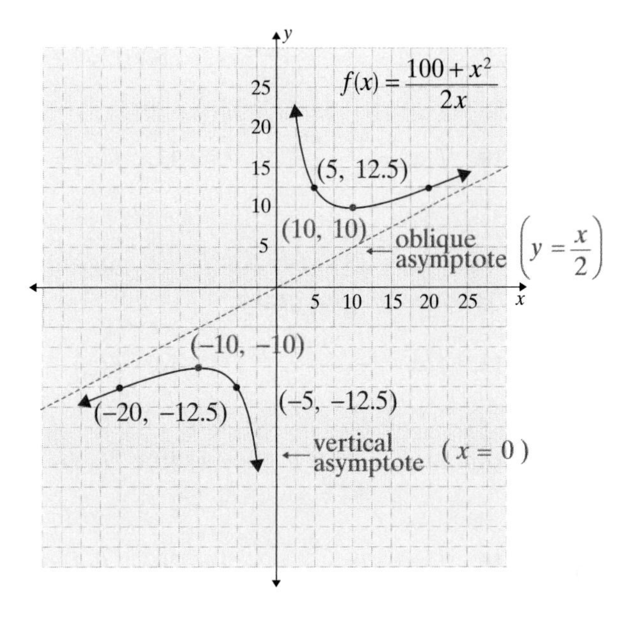

The oblique asymptotes that we have studied thus far have been linear. It is possible for a rational function to have a nonlinear asymptote. See the graph of $f(x)=x^2+\frac{1}{x}$ in Figure 4.3.2. This function has an asymptote of $y=x^2$. However, for analysis purposes, we will only consider linear asymptotes in this book.

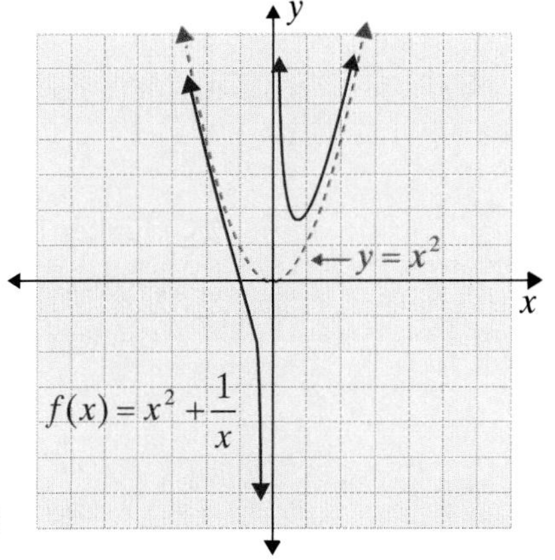

Figure 4.3.2

4.3 Exercises

For each of the rational functions in Exercises 1 – 12, find **a.** *any vertical asymptotes,* **b.** *any horizontal asymptotes, and* **c.** *any oblique asymptotes.*

1. $f(x)=\frac{1}{x-4}$ **2.** $f(x)=-\frac{3}{x+6}$ **3.** $f(x)=\frac{2x}{x+8}$

4. $f(x)=\frac{5x}{2x+1}$ **5.** $f(x)=\frac{x+2}{x^2+1}$ **6.** $f(x)=\frac{x-7}{x^2+3}$

7. $f(x)=\frac{5x^2}{3x^2-2x-1}$ **8.** $f(x)=\frac{2x^2}{x^2+3x}$ **9.** $f(x)=\frac{x^2-4}{x}$

10. $f(x)=\frac{3x^2+2}{x}$ **11.** $f(x)=\frac{x^2+1}{x+1}$ **12.** $f(x)=\frac{x^2-5}{x-2}$

In Exercises 13 – 24, sketch the graph of each rational function. Show any related asymptotes on each graph.

13. $f(x)=-\frac{2}{x+5}$ **14.** $f(x)=\frac{4}{x-3}$ **15.** $f(x)=\frac{2x}{x+1}$

16. $f(x)=\frac{3x}{x-2}$ **17.** $f(x)=\frac{x-2}{x-1}$ **18.** $f(x)=\frac{x+4}{2x+1}$

19. $f(x)=2x+\frac{2}{x}$ **20.** $f(x)=3x+\frac{12}{x}$ **21.** $f(x)=\frac{3x^2+6}{x}$

22. $f(x)=\frac{2x^2+1}{3x}$ **23.** $f(x)=\frac{x^3-3x^2+1}{(x-10)(x-3)}$

24. $f(x)=\frac{x^3+x^2-13x+100}{(x-3)(x-10)}$

25. Junker Renovation completely overhauls junked or abandoned cars. Data shows their 1970's models hold their value quite well. The value $F(x)$ of one of these cars is given by $F(x)=70-\frac{15x}{x+1}$ where x is the number of years since repurchase and F is in hundreds of dollars.

a. What is the initial resale price of a car?
b. Find all asymptotes.
c. Sketch the function.
d. What is the long term value of one of these cars?

26. The average cost $A(x)$ is the total cost $C(x)$ divided by the quantity x. Thus $A(x) = \frac{C(x)}{x}$. If the total cost function for a product is $C(x) = 3x + 12$, graph the average cost $A(x)$. If there are any asymptotes, locate them and interpret their meaning.

27. A product's total costs are given by $C(x) = 0.03x^2 + 24x + 10$.

a. Graph the average cost function, locating any asymptotes.
b. What is the meaning of the asymptotes for average cost?

28. The Polar Pollution Control Company removes debris from old motors. Suppose the cost $F(x)$ of removing x percent of the pollutants is given by $F(x) = \frac{100000}{100 - x}$, where x is a percentile, $0 \le x < 100$, and F is in dollars.

a. Determine $\lim_{x \to 0^+} F(x)$ and $\lim_{x \to 100^-} F(x)$ and interpret their meanings.
b. What percentile can be removed at a cost of \$3000?
c. Show that $F(x)$ is always increasing. Does this make sense in the context of the problem?

29. The cost of camels in Tunisia is modeled by the function $F(x) = 40 - \frac{20x}{x+3}$, where x is the number of years since 2000 and F is a national average cost in dinars.

a. What was the cost of a camel in 2002?
b. What are the asymptotes for F and which is significant in the problem?
c. When will the average cost be 26 dinars?

30. The sugar level concentration in the bloodstream of a certain diabetes patient is modeled by $S(t) = 1 + \frac{0.2t}{t^2 + 2}$, where S is in suitable units and t is the time in hours following a meal of allowed carbohydrate content.

a. Which asymptotes play a role here?
b. For $0 \le t \le 6$, what is the highest S-value and when does it occur? (If this level exceeds 4, the patient will become ill.)
c. Are there any inflection points? What is the meaning in the context of the problem of an inflection point?

31. Data suggests a professional football team will win $F(x)$ games (out of 16) if the salaries of the superstar players increase. For one team, the function F is given by $F(x) = 8 + \frac{6x}{0.125x^2 + 2}$, where x is the average salary (in millions) of the superstars (players earning at least one million dollars).

a. Are there any asymptotes of consequence in the problem?
b. What average salary gives the biggest return on games won, according to this model? (Here return is total games won.)

32. If administrative assistants at Bookworm Publications make phone follow-ups after textbook reviews, more colleges and universities will adopt a new statistics book. The publisher noticed that new book sales varied in the second year according to $S(x) = A_0 - \frac{(x+200)}{x^2}$, where S is total sales and x is the number of phone calls made to colleges which have adopted the book. A_0 denotes the sales from the previous year.

a. Assuming $A_0 = 2500$, what sales can Bookworm Publications expect if they make 100 follow-up phone calls?

b. What is the horizontal asymptote and what is its significance?

Hawkes Learning Systems: Essential Calculus

Curve Sketching: Rational Functions

4.4 Business Applications

Objectives

After completing this section, you will be able to:

1. *Minimize inventory costs of a business.*
2. *Determine the price a company should charge for their product in order to maximize revenue.*
3. *Find a company's maximum profit assuming the demand function is linear.*

Minimizing Inventory Costs

Retailers' costs of maintaining an inventory of goods can be considered to consist of two categories:

1. Storage costs (warehouse fees, insurance, etc), and
2. Ordering costs (paperwork and shipping).

A large inventory can tie up valuable space in a store or warehouse. On the other hand, frequent ordering can run up a large shipping bill. We want to find a way to minimize the costs related to storage and shipping. These costs are called **inventory costs.**

For the mathematical model (function) that represents inventory costs to be manageable, the following "ideal" conditions are to be assumed:

1. The total sales for the year are known.
2. The same numbers of items are sold each day until all items in the inventory are sold. That is, the inventory is depleted linearly.
3. Orders are for the same number of items at regular time periods throughout the year.

Figure 4.4.1 illustrates these conditions graphically.

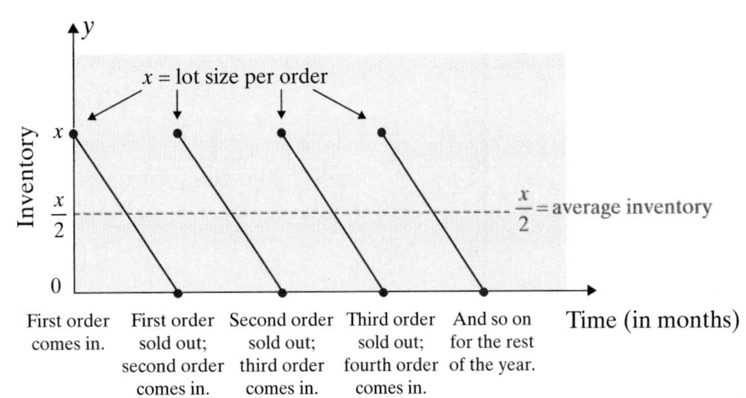

Figure 4.4.1

The mathematical model has the following components:

$$x = \text{lot size};$$

$$\frac{x}{2} = \text{average number of items inventory};$$

$$\frac{\text{total ordered}}{x} = \text{number of orders per year; and}$$

$$\underset{\left(\begin{array}{c}\text{inventory}\\ \text{costs}\end{array}\right)}{C(x)} = \underset{\left(\begin{array}{c}\text{storage}\\ \text{costs}\end{array}\right)}{(\text{storage cost per item})\cdot\left(\frac{x}{2}\right)} + \underset{\left(\begin{array}{c}\text{ordering}\\ \text{costs}\end{array}\right)}{(\text{cost per order})\cdot\left(\frac{\text{total ordered}}{x}\right)}.$$

Example 1: Minimizing Inventory Costs

A furniture dealer sells 500 desks per year. The desks take up floor space and warehouse space, and the dealer estimates his storage costs at \$6 per desk. The distributor charges the dealer a \$60 fee for each order. How many times per year and in what lot size should the dealer order to minimize inventory costs?

Solution: Using the given information, we can determine a function for the inventory costs, $C(x)$. Let x = lot size. Then $\frac{500}{x}$ is the number of orders per year, and $\frac{x}{2}$ is average inventory.

$$C(x) = (\text{storage cost per item})\cdot\frac{x}{2} + (\text{cost per order})\cdot\left(\frac{500}{x}\right)$$

$$C(x) = 6\left(\frac{x}{2}\right) + 60\left(\frac{500}{x}\right)$$

The number of desks ordered is between 1 and 500. At the extremes, one order for 500 desks would cost

$$C(500) = 6\left(\frac{500}{2}\right) + 60\left(\frac{500}{500}\right) = \$1560,$$

and 500 orders for one desk at a time would cost

$$C(1) = 6\left(\frac{1}{2}\right) + 60\left(\frac{500}{1}\right) = \$30,003.$$

Now we need to differentiate $C(x)$ so we can determine the local minima.

Evaluating a Function With a Calculator

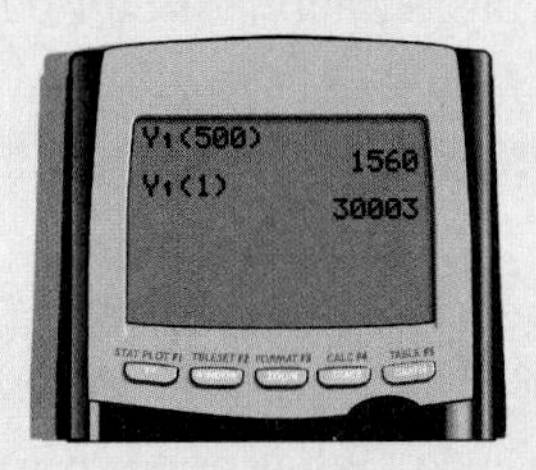

To evalute the function $C(x)$ in Example 1:

1. Enter the function $C(x)$ that was found into Y1.
2. Now, on the main screen, press VARS, scroll right to `Y-VARS`, select `1: Function`, and then `1: Y1` to display Y1 on the main screen.
3. With your cursor after Y1, type an opening parenthesis and then the x-value we are evaluating, 500. Type a closing parenthesis.
4. Hit Enter to calculate the corresponding y-value.
5. This process can be repeated for the next x-value, 1.

Differentiating $C(x)$ gives the following results.

$$C(x) = 6\left(\frac{x}{2}\right) + 60\left(\frac{500}{x}\right)$$

$$= 3x + 30{,}000x^{-1} \quad \text{Rewrite } C(x) \text{ using exponents.}$$

$$C'(x) = 3 - 30{,}000x^{-2}$$

$$= 3 - \frac{30{,}000}{x^2}$$

Setting $C'(x) = 0$ gives

$$3 - \frac{30{,}000}{x^2} = 0$$

$$3x^2 = 30{,}000$$

$$x^2 = 10{,}000$$

$$x = \pm 100.$$

$x = -100$ is not in the interval $[1, 500]$, so $x = 100$. Substituting this value into the equation for the number of orders per year, we get:

$$\frac{500}{x} = \frac{500}{100} = 5 \text{ orders per year.}$$

To minimize inventory costs, the dealer should order 100 desks at a time, 5 times per year. The minimum inventory costs are

$$C(100) = 6\left(\frac{100}{2}\right) + 60\left(\frac{500}{100}\right)$$

$$= 6(50) + 60(5)$$

$$= 300 + 300 = \$600.$$

Substitute $x = 100$ into $C(x)$ to find the minimum inventory costs.

Since we are dealing with the closed interval $[1, 500]$ and $C(1) = \$1560$ and $C(500) = \$30{,}003$, $C(100) = \$600$ is also the absolute minimum.

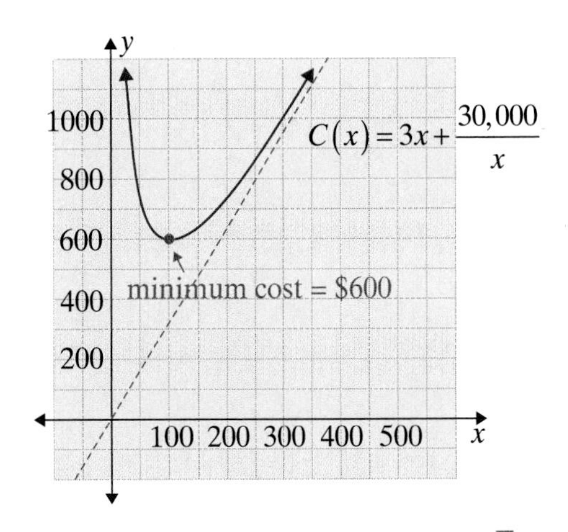

Maximizing Revenue

Consider the following situation. The owner of a business can afford to lower (or raise) prices somewhat. This will have a direct effect on revenue. Generally, raising prices lowers sales and lowering prices increases sales. Sometimes past experience can be a guide in creating appropriate functions, and then calculus can be applied.

Example 2: Maximizing Revenue

Susan knows that she can sell 440 tacos a day at 50 cents per taco. What price should she charge to maximize her revenue if, for every increase of 10 cents in price, she sells 40 fewer tacos? What will this revenue be?

Solution: Let x = number of 10-cent increases in price. (The choice of x here does not represent the unknown price but the number of price increases. This choice for x makes the equation relatively easy to set up.) Then

$$0.50+0.10x = \text{new price, and}$$
$$440-40x = \text{new sales.}$$

Therefore,

$$\begin{aligned}R(x) &= (\text{price})\cdot(\text{sales})\\ &= (0.50+0.10x)(440-40x)\\ &= 220+44x-20x-4x^2\\ &= 220+24x-4x^2\end{aligned}$$

Differentiating $R(x)$, we get

$$R'(x) = 24 - 8x.$$

Setting $R'(x) = 0$ gives

$$\begin{aligned}24-8x &= 0\\ -8x &= -24\\ x &= 3.\end{aligned}$$

We can verify that $x = 3$ is a maximum by the Second Derivative Test: $R''(x) = -8 < 0$ for all x. Therefore, $x = 3$ does indeed give a maximum revenue.

Thus there should be three 10-cent increases in price. The new price should be

$$\$0.50 + \$0.10(3) = \$0.80 \text{ per taco.}$$

The sales will then be

$$440 - 40(3) = 320 \text{ tacos},$$

and the maximum revenue will be

$$R(3) = (0.80)(320) = \$256.$$

Note that no mention of profit is made in Example 2. Do you think that the profit will be likely to increase or decrease with an increase in price? What factors should be taken into account?

Linear Demand Functions

Suppose that two points relating sales and price are given. If the demand function is assumed to be linear, then the demand function can be found by using these two points and the slope and line equations reviewed in Chapter 1. (See Figure 4.4.2.)

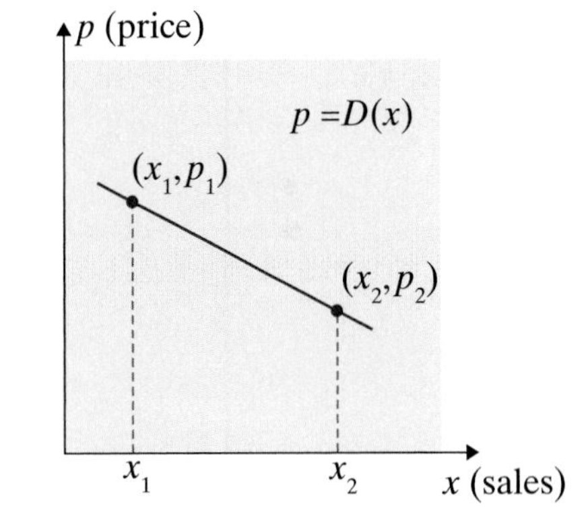

Figure 4.4.2

Once the demand function has been determined, then the revenue function $[R(x) = x \cdot D(x)]$ can be set up and differentiated to find what sales and price will yield maximum revenue.

Example 3: Finding Maximum Revenue With a Linear Demand Function

A company sells 3000 calculators per month when the price is \$7 per calculator. When the price is lowered to \$4 per calculator, 6000 are sold. The maximum number of calculators the company can manufacture is 7000 per month. Assuming that the demand function is linear, determine how many calculators the company should produce and sell per month in order to maximize revenue.

Solution: The two points (3000, 7) and (6000, 4) can be used to find the linear demand function.

First, we will use these points in the formula for slope, $m = \frac{y_2 - y_1}{x_2 - x_1}$, and then we will use the point-slope form for the equation of a line, $y - y_1 = m(x - x_1)$. For this example we use the points in the form (x, p) instead of (x, y).

continued on next page ...

$$m=\frac{4-7}{6000-3000}=\frac{-3}{3000}=-\frac{1}{1000}$$

$$p-7=-\frac{1}{1000}(x-3000)$$

$$p-7=-\frac{1}{1000}x+3$$

$$p=-\frac{1}{1000}x+10=D(x)$$

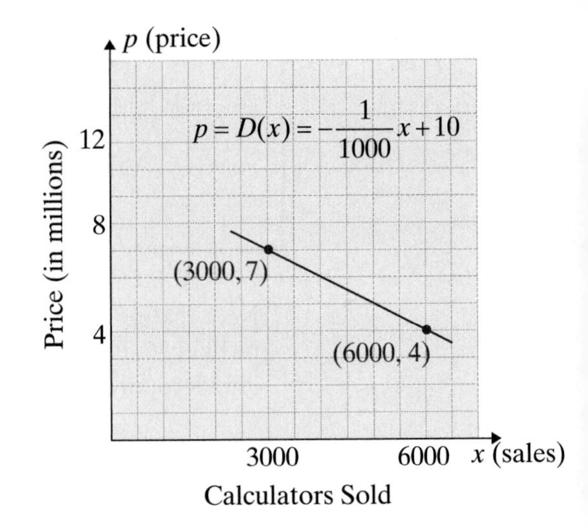

$D(x)$ is the demand function where $x =$ number of calculators sold, and $p =$ price per calculator.

The revenue function is then,

$$R(x)=x\cdot D(x)$$
$$=x\cdot\left(-\frac{1}{1000}x+10\right)$$
$$=-\frac{1}{1000}x^2+10x$$

Differentiating $R(x)$ gives

$$R'(x)=-\frac{2}{1000}x+10.$$

Setting $R'(x)=0$ and solving for x gives

$$-\frac{2}{1000}x+10=0$$
$$-2x=-10{,}000$$
$$x=5000$$

The company should produce 5000 calculators and sell them at a price of

$$p=-\frac{1}{1000}(5000)+10=\$5\text{ each}$$

for the maximum revenue of $R(5000) = 5000 \cdot 5 = \$25{,}000$ per month.

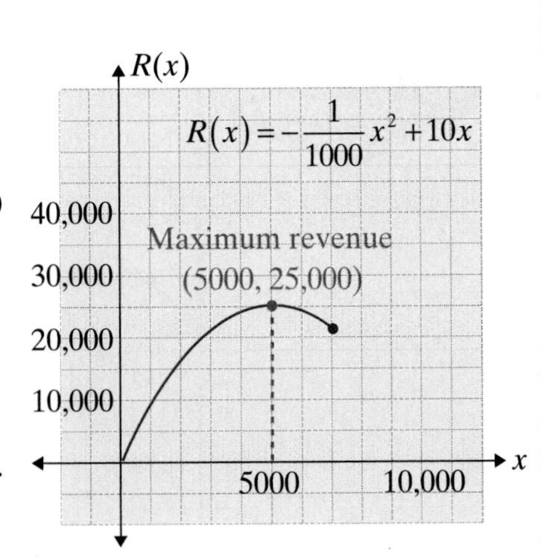

Example 4: Finding Maximum Profit With a Linear Demand Function

Suppose that the company in Example 3 has a cost function $C(x) = 5000 + 3x$. How many calculators should the company produce and sell to maximize profit?

Solution: Find the profit function:

$$
\begin{aligned}
P(x) &= R(x) - C(x) \\
&= -\frac{1}{1000}x^2 + 10x - (5000 + 3x) \\
&= -\frac{1}{1000}x^2 + 10x - 5000 - 3x \\
&= -\frac{1}{1000}x^2 + 7x - 5000
\end{aligned}
$$

Differentiate $P(x)$:

$$P'(x) = -\frac{2}{1000}x + 7.$$

Set $P'(x)$ equal to 0, and solve for x:

$$
\begin{aligned}
-\frac{2}{1000}x + 7 &= 0 \\
-2x &= -7000 \\
x &= 3500
\end{aligned}
$$

Thus, a maximum profit occurs if the company produces and sells 3500 calculators, which is only half its production capabilities. The price for each calculator would be

$$p = -\frac{1}{1000}(3500) + 10 = \$6.50.$$

The graph at right illustrates the relationship between profit, revenue, and cost.

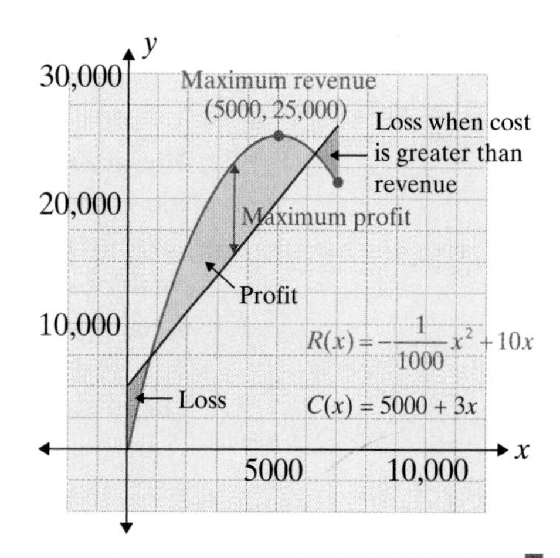

Note that maximum revenue and maximum profit do not necessarily occur at the same level of production and sales.

4.4 Exercises

1. **Minimizing inventory costs.** An appliance store owner estimates that he will sell 125 vacuum cleaners of a particular model. It costs \$12 to store one vacuum cleaner for one year. There is a fixed cost of \$30 for each order. Find the lot size and the number of orders per year that will minimize inventory costs.

2. **Minimizing inventory costs.** A hardware store sells 96 chainsaws per year. It costs \$5 to store one chainsaw for one year. There is a fixed reordering cost of \$15. Find the lot size and the number of orders per year that will minimize inventory costs.

3. **Minimizing inventory costs.** An art gallery owner expects to sell 90 copies of a limited-edition print during the next year. It costs \$1.50 to store one copy for one year. For each order she places, there is a fixed cost of \$7.50, plus \$0.50 for each copy. Find the lot size and the number of times the gallery owner should order per year to minimize her inventory costs.

4. **Minimizing inventory costs.** The owner of Lamps-4-U expects to sell 180 brass lamps during the year. For each order he places, there is a fixed cost of \$18, plus \$2 for each lamp ordered. It costs \$5 to store one lamp for one year. In what lot size and how many times per year should he reorder to minimize the inventory costs?

5. **Minimizing inventory costs.** A tee-shirt company sells 4000 sweatshirts per year. To reorder, there is a fixed cost of \$6 plus \$0.80 for each sweatshirt. It costs \$1.20 to store one sweatshirt for one year. In what lot size and how many times per year should an order be placed to minimize inventory costs?

6. **Minimizing inventory costs.** A computer software store sells 7500 computer diskettes per year. It costs \$0.15 to store one diskette for one year. To reorder diskettes, there is a fixed cost of \$22.50, plus \$0.10 for each diskette. In what lot size and how many times per year should an order be placed to minimize inventory costs?

7. **Minimizing inventory costs.** A snowmobile dealer in Minnesota expects to sell 960 snowmobiles during the next year. It costs \$9 to store one snowmobile for one year. To reorder, there is a fixed cost of \$67.50, plus \$7.50 for each snowmobile. In what lot size and how many times per year should an order be placed to minimize inventory costs?

8. **Minimizing inventory costs.** A car dealer expects to sell 1320 cars during the next year. It costs \$660 to store one car for one year. To reorder, there is a fixed cost of \$225, plus \$304 for each car. Find the lot size and the number of orders that should be placed so inventory costs will be minimized.

9. **Maximizing revenue.** A discount store sells 84 radios per month at \$20 each. The owners estimate that for each \$1 increase in price, they will sell 3 fewer radios per month. How much should they charge for their radios to maximize their revenue?

10. **Maximizing revenue.** A farmer estimates that if he plants 30 grapefruit trees per acre, the average yield per tree will be 480 pounds. For each additional tree planted per acre, the yield per tree will be reduced by 12 pounds. How many trees should be planted per acre to maximize the yield?

11. **Maximizing revenue.** Sam operates a small hotdog stand. He estimates that he can sell 600 hotdogs per day if he charges 75 cents each. Sam determines that for each 10-cent reduction in price, he will sell an additional 80 hotdogs per day. How much should he charge for his hotdogs to maximize his revenue?

12. **Maximizing revenue.** A sporting goods store sells 200 baseball gloves per month at \$36 each. The owner estimates that for each \$2 increase in price, he will sell 5 fewer gloves. Find the price that will maximize revenue.

13. **Maximizing revenue.** A sports arena has 40 roaming soda salespeople, each of whom sells 200 sodas per event. Management estimates that for each additional salesperson, the yield per salesperson decreases by 4. How many additional salespeople should management hire to maximize the number of sodas sold?

14. **Maximizing revenue.** Ms. Wills owns a 16-unit apartment complex. The unit rent is currently \$400 per month, and all units are rented. Each time rent is increased by \$20, one tenant will move out. Find the rental price that will maximize Ms. Wills' revenue.

15. **Linear demand function.** A local amusement park found that if the admission was \$7, about 1000 customers per day were admitted. When the admission was dropped to \$6, the park had about 1200 customers per day. Assuming a linear demand function, determine the admission price that will yield maximum revenue.

16. **Linear demand function.** A department store manager has determined that when the price of a necktie was \$12, she sold 100 neckties per month. However, only 80 neckties were sold per month when the price was raised to \$14. Assuming a linear demand function, determine the price that would maximize the revenue.

17. **Linear demand function.** The cost of producing x units of an item is $C(x) = 10x + 20$. When the selling price is \$20, twenty-one items are sold. However, when the price is \$16, twenty-three items are sold. Assuming the demand function is linear, determine the price per unit and the number of units sold that will maximize the profit.

18. Linear demand function. The manager of an electronics store knows he can sell 60 blank compact disks (CDs) when the price is \$1.20 each. If the price is \$1.50, only 48 CDs are sold. The total cost function for x CDs is $C(x) = 0.70x + 15$ dollars. Assuming a linear demand function, determine the price per CD and the number of CDs sold that will maximize the profit.

19. Linear demand function. The manufacturer of color televisions can sell 800 sets to his dealers at \$392 each. If the price is \$380, he can sell 1000 sets. The total cost of producing x television sets is $C(x) = 3600 + 250x - 0.01x^2$ dollars. Assuming the demand function is linear, find the price per set and the number of sets sold that will maximize profit.

20. Linear demand function. A candy store can sell 180 candy bars at 62 cents each. The store can sell 220 candy bars if the price is 54 cents each. The total cost of producing x candy bars is $C(x) = 3050 - 10x + 0.04x^2$ cents. Find the number of candy bars that should be produced to maximize profit.

21. Profit. Suppose $P(x)$ represents profit on the sales of x telephones. Suppose $P(25{,}000) = 12{,}000$, $P'(25{,}000) = 2$, and $P''(25{,}000) = -3$.

- **a.** Is the company making money or losing it? How much?
- **b.** If sales are increased, will the profits rise or fall? By how much?
- **c.** What is the meaning of $P''(25{,}000) = -3$?

22. Profit. Suppose the monthly marginal profit from John's newspaper route is $P'(x) = 3$, where x is the number of subscribers.

- **a.** The profit function is (choose one): linear, quadratic, or a polynomial of degree 3 or higher.
- **b.** Suppose monthly costs are $C(x) = 9x + 20$. Assuming initial revenue is 0, what is the revenue function?

23. Average cost. $A(x) = \dfrac{C(x)}{x}$ gives average cost.

- **a.** Calculate a formula for $A'(x)$.
- **b.** Set $A'(x) = 0$ and solve for $C'(x)$.
- **c.** If average costs are minimal , describe a relationship between average cost and marginal cost. That is, interpret the result of part **b.**

24. Earnings. Suppose a company's earnings are given by $E(x) = P(x) + I(x)$, where x is the number of years since 2000, $P(x)$ is the annual profit function, and $I(x)$ is the intangible growth (the growth in value of the company's intangible assets such as its good name). If $P(x) = 1.3x + 2$ and $I(x) = 0.25x + 1$ for a certain company, determine the following.

- **a.** The marginal earnings for year x.
- **b.** The actual earnings for 2002.
- **c.** The average earnings formula (earnings per year since 2000).

25. **Average cost.** Suppose that a company's average cost is $A(x) = 0.2x + 3$.
    a. Determine the cost function.
    b. Determine the marginal average cost.
    c. Determine the marginal cost.

26. **Profit.** Suppose the demand for a product is \$12 and the total costs are $C(x) = 0.3x^2 + 2x + 5$.
    a. What is the revenue function?
    b. What is the profit function?
    c. What is the maximum value of the profit?

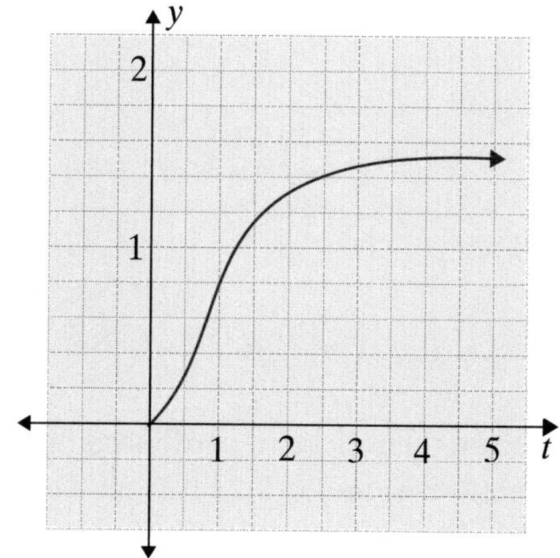

27. **Sales function.** A sales function $S(t)$ for a new product is at left. $S(t)$ is total sales (quantity of items) and t is time in months since the products release. Copy this graph onto your paper and add a graph of a possible $S'(t)$. Locate approximately the point of inflection on your curve for S.

28. **Cost function.** A certain cost function $C(x)$ satisfies $C(10) = 20$, $C(20) = 40$, and $C(30) = 60$. Suppose $C''(10) = -2$, $C''(20) = 0$, and $C''(30) = 2$. Draw a suitable function.

29. **Value.** The value of a new business franchise grows according to the formula $V(x) = 10 + \frac{10x}{1 + 0.5x}$. Here V is in thousands of dollars and x is the number of years after 2000.
    a. What is the expected value in 2010?
    b. Is the value V increasing or decreasing in 2010?
    c. Is the rate of increase in value changing? What has this to do with V' or V''?

30. **Value.** The timber value of a small stand of pine trees is given by $V(x) = 50\left(1 - \frac{1}{x+2}\right)$, where x is the number of years after 1990 and V is in dollars.
    a. What was the value in 1990?
    b. At what rate was the value changing in 2000?
    c. What asymptotes are present and what is their significance in the problem?

Hawkes Learning Systems: Essential Calculus

Business Applications

4.5 Other Applications

Objectives

After completing this section, you will be able to:

Use the concepts of maximum and minimum in applications related to geometry and science including the Pythagorean Theorem, volume, area, distance, and velocity.

Geometry

Example 1: Volume of a Box

A rectangular box with no top is to be made from a rectangular piece of cardboard that is 18 in. long by 12 in. wide. Small squares (all the same size) are to be cut from each corner and the sides folded to form the box. What size should the squares be to give a maximum volume to the box?

Solution: Let x = length of one side of the square to be cut.

$$\text{Volume} = \text{length} \cdot \text{width} \cdot \text{height}$$

$$V = (18-2x)\cdot(12-2x)\cdot x$$

$$= 216x - 60x^2 + 4x^3$$

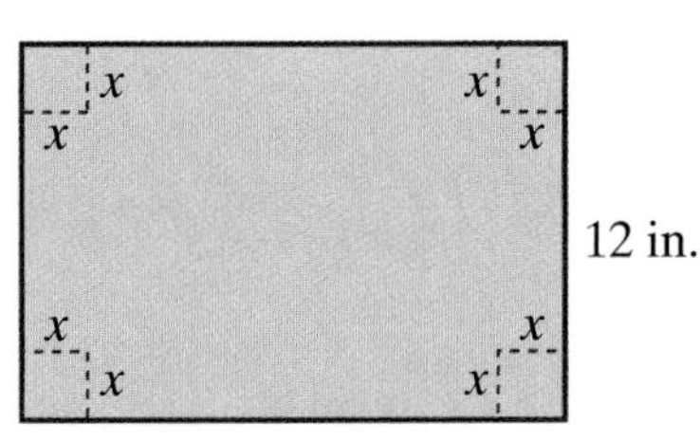

Differentiating V, we obtain

$$\frac{dV}{dx} = 216 - 120x + 12x^2$$

$$= 12\left(x^2 - 10x + 18\right).$$

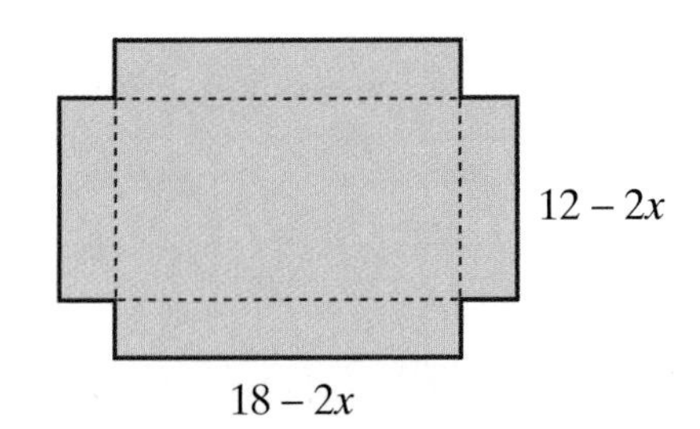

Setting $\frac{dV}{dx} = 0$ gives

$$12\left(x^2 - 10x + 18\right) = 0$$

$$x^2 - 10x + 18 = 0$$

$$x = \frac{10 \pm \sqrt{100-72}}{2}$$

x
18 – 2x
12 – 2x

$$x = \frac{10 \pm \sqrt{28}}{2} = \frac{10 \pm 2\sqrt{7}}{2} = 5 \pm \sqrt{7}.$$

We know that the size of the square cannot be any larger than 6 in. or smaller than 0 in.; that is, x must be in the closed interval $[0, 6]$.

Note, $x = 5+\sqrt{7}$ is too large to apply in the problem since $5+\sqrt{7} > 6$. Therefore, we check $x = 5-\sqrt{7} \approx 2.35$ in with the second derivative test. $V'' = 12(2x-10)$ and $V''(2.35) = 12(5.7-10) < 0$. So V is concave down at $x = 5-\sqrt{7}$ which means this x gives a maximum volume.

Example 2: Surface Area of a Box

A rectangular box is to be made so that the top and bottom are squares. The volume is to be 250 cm³. Material for the top and bottom costs \$2 per square centimeter, but the material for the sides costs only \$1 per square centimeter. What dimensions will give a minimum cost for the materials? What is the minimum cost?

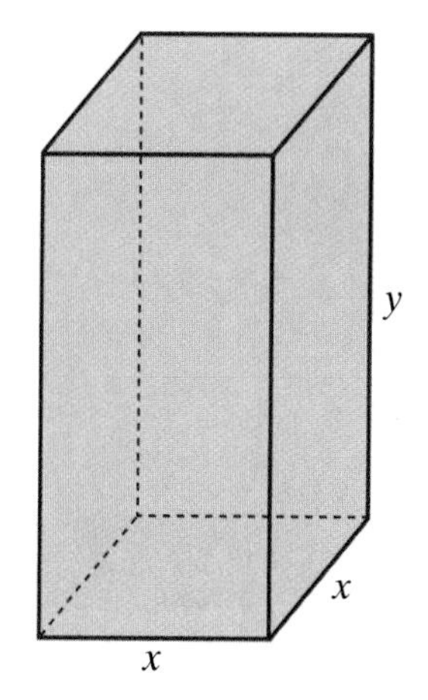

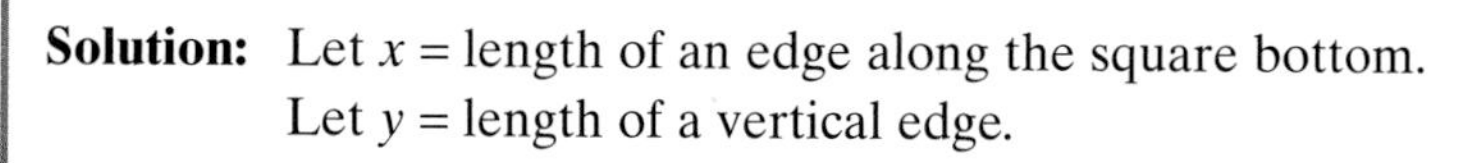

Solution: Let x = length of an edge along the square bottom.
Let y = length of a vertical edge.

$$\text{Volume} = x^2 y = 250 \text{ cm}^3.$$

Note: We are assuming that cost only depends on surface area. The cost, $C(x)$, of the materials is found by multiplying the areas of the faces of the box by \$2 and \$1 as follows:

$$C(x) = \$2x^2 + \$2x^2 + \$1(4xy)$$
$$= 4x^2 + 4xy.$$

Cost of the top, bottom, and 4 sides.

Now, using the volume, solve for y and substitute into the cost function.

$$y = \frac{250}{x^2}$$

Solve volume for y.

$$C(x) = 4x^2 + 4x\left(\frac{250}{x^2}\right)$$
$$= 4x^2 + \frac{1000}{x}$$

Substitute y in the equation for $C(x)$.

continued on next page ...

$$C(x) = 4x^2 + 1000x^{-1}$$ Rewrite $C(x)$ using exponents.

Differentiating $C(x)$, we have

$$C'(x) = 8x - 1000x^{-2}.$$

Setting $C'(x) = 0$ and solving for x gives

$$8x - \frac{1000}{x^2} = 0$$

$$8x = \frac{1000}{x^2}$$

$$8x^3 = 1000$$

$$x^3 = 125$$

$$x = 5 \text{ cm}$$

So

$$y = \frac{250}{5^2} = 10 \text{ cm}.$$

Therefore, the dimensions of the box are 5 cm by 5 cm by 10 cm, and the minimum cost is

$$C(5) = 4(5)^2 + \frac{1000}{5} = 100 + 200 = \$300.$$

Example 3: Fencing

A farmer wants to enclose a rectangular area next to his barn. If he has 200 feet of fencing to use, what dimensions of the rectangle will maximize the area? What is the maximum area? (Note that the fence is only three sides of the rectangle, since the barn serves as the fourth side of the enclosure.)

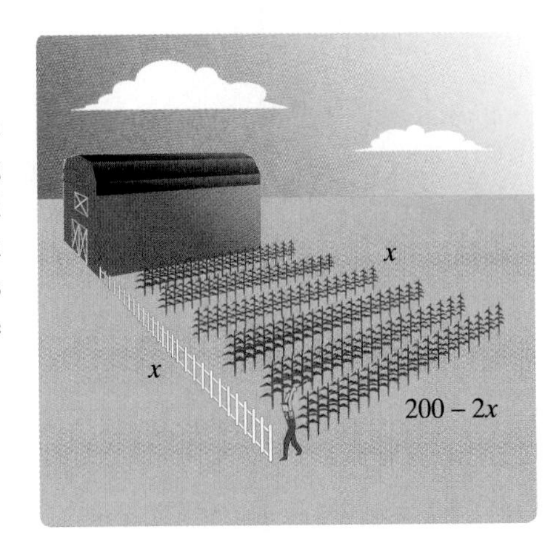

Solution: Let x = length of one of the two equal sides of the rectangle. Then, since there is only 200 feet of fencing available, the length of the third side is $200 - 2x$.

The area of the rectangle is to be maximized; so we need a function representing the area:

$$A(x) = x(200 - 2x)$$
$$= 200x - 2x^2.$$

Then

$$A'(x) = 200 - 4x.$$

Setting $A'(x) = 0$ and solving for x, we have

$$200 - 4x = 0$$
$$4x = 200$$
$$x = 50$$

We apply the Second Derivative Test:
$A'' = -4$. So $A''(50) = -4$ and A is concave down at $x = 50$. Thus the critical value gives a maximum.

The rectangle should be 50 ft. by 100 ft., and the maximum area is

$$A(50) = 50(200 - 2 \cdot 50) = 50(100) = 5000 \text{ ft.}^2$$

Example 4: Oil Pipeline

An oil company wants to lay a pipeline from its offshore drilling rig to a storage tank on shore, as illustrated in the accompanying figure. The rig (point R) is three miles offshore (point A), and the storage tank (point B) is 8 miles down the shoreline. The costs of laying pipe underwater are \$800 per mile and along the shoreline are \$400 per mile. Point P is the point on shore where the underwater pipe connects with the shoreline pipe. Where should point P be located so as to minimize the cost of laying pipe?

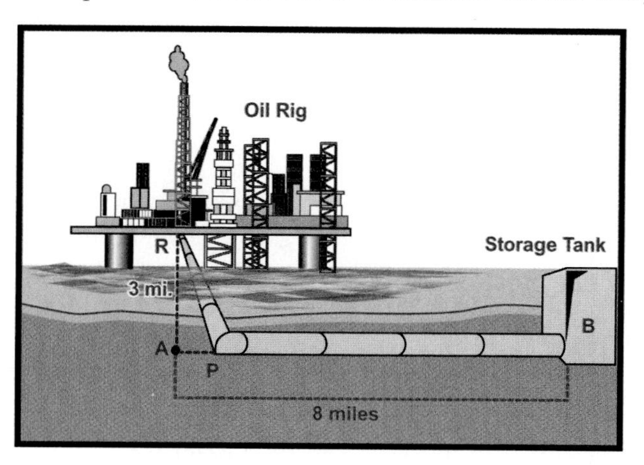

continued on next page ...

Solution: Let x be the distance from A to P.

Since $\overline{AB} = 8$, $\overline{PB} = 8 - x$.

By the Pythagorean Theorem,

$$\overline{RP} = \sqrt{x^2 + 9}.$$

Oil Rig
Storage Tank
R
3
$\sqrt{x^2+9}$
A
x
P
B
8 – x

The total cost of laying pipe is

$$C(x) = 800\sqrt{x^2+9} + 400(8-x)$$
$$= 800\left(x^2+9\right)^{\frac{1}{2}} + 3200 - 400x.$$

Differentiating $C(x)$ gives

$$\frac{dC}{dx} = 800 \cdot \frac{1}{2}\left(x^2+9\right)^{-\frac{1}{2}}(2x) - 400$$
$$= \frac{800x}{\sqrt{x^2+9}} - 400.$$

(Note: $\frac{d^2C}{dx^2} = C'' = 7200\left(x^2+9\right)^{-\frac{3}{2}}$)

Setting $\frac{dC}{dx} = 0$ and solving for x, we have

$$\frac{800x}{\sqrt{x^2+9}} - 400 = 0$$
$$\frac{800x}{\sqrt{x^2+9}} = 400$$
$$2x = \sqrt{x^2+9}$$
$$4x^2 = x^2 + 9$$
$$3x^2 = 9$$
$$x^2 = 3$$
$$x = \pm\sqrt{3}$$

Note that $x = -\sqrt{3}$ is not a reasonable solution because distance cannot be negative. We have assumed x to be in the interval [0, 8]. Now, $C''\left(\sqrt{3}\right) = 173.2 > 0$, so $x = \sqrt{3}$ gives a minimum.

So the minimum cost is incurred if point P is located $\sqrt{3}$ miles (approximately 1.732 miles) along the shoreline from point A (or $8 - \sqrt{3} \approx 6.268$ miles from the storage tank at point B.)

We should check the endpoints $x = 0$ and $x = 8$ to be sure that they do not give an absolute minimum:

$$C(0) = 800\sqrt{0^2 + 9} + 400(8-0) = \$5600$$

and

$$C(8) = 800\sqrt{8^2 + 9} + 400(8-8) \approx \$6835.$$

Thus $C\left(\sqrt{3}\right) = 800\sqrt{\left(\sqrt{3}\right)^2 + 9} + 400\left(8 - \sqrt{3}\right) \approx \$5278$ is indeed the minimum cost.

Distance and Velocity

Velocity is the rate of change of distance with respect to time. Customary notation in science and mathematics is $s(t)$ for distance and $v(t)$ for velocity, since both distance and velocity depend on time t. We also have

$$\frac{ds}{dt} = s'(t) = v(t).$$

Example 5: Helicopter Lift-Off

Suppose that a helicopter is rising from the ground in such a way that its distance s in feet from the ground is given by $s(t) = 5t^2$, where t is time after take off in seconds and $0 \le t \le 6$. How fast is the helicopter rising 3 seconds after takeoff? How high is the helicopter at that time?

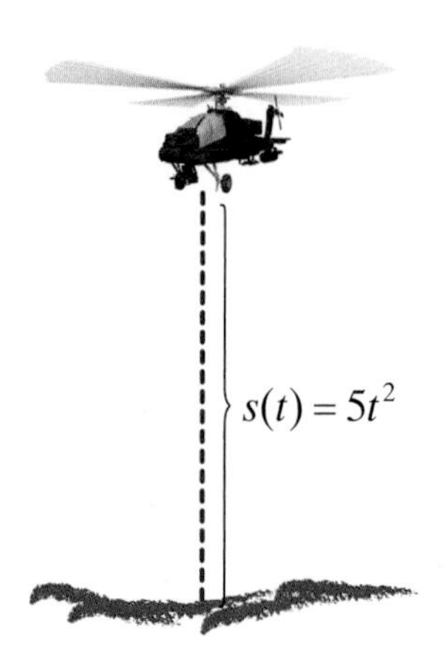

Solution: Differentiating the distance function with respect to time gives the velocity function:

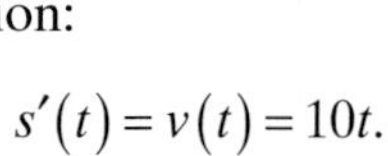

$$s'(t) = v(t) = 10t.$$

$v(3) = 10(3) = 30$ ft./sec — Substitute $t = 3$ into $v(t)$ to find the velocity.

Three seconds after lift-off, the helicopter is rising at 30 ft./sec.

$s(3) = 5(3)^2 = 45$ ft. — To find the height of the helicopter after 3 seconds, substitute $t = 3$ into $s(t)$.

In 3 seconds the helicopter has risen to an altitude of 45 feet.

Example 6: Vertical Projectile

A projectile is fired vertically (straight up) from the top corner of the 176-foot tall building. The height of the projectile at time t (in seconds) is given by the function $s(t) = -16t^2 + 160t + 176$.

a. What is the maximum height of the projectile?

b. When does the projectile hit the ground?

c. How fast is the projectile moving when it hits the ground?

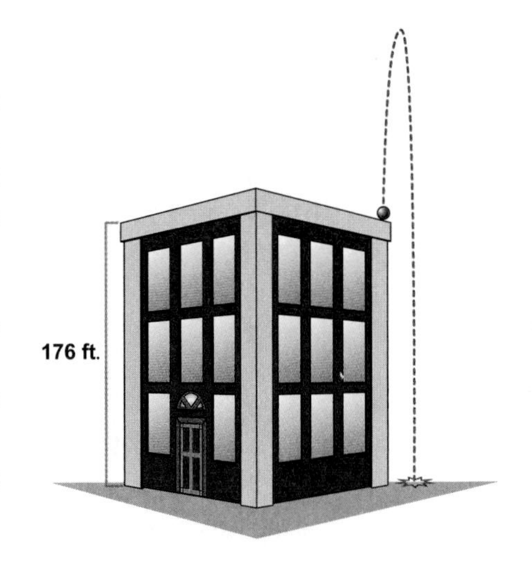

Solutions: a. The maximum height of the projectile occurs at the point where its velocity is 0.

$$s(t) = -16t^2 + 160t + 176$$

$$v(t) = s'(t) = -32t + 160$$

$$-32t + 160 = 0$$

$$-32t = -160$$

$$t = 5$$

Differentiate $s(t)$ to determine $v(t)$. Then set $v(t)$ equal to 0 and solve for t.

Since $s'' = -32$ (a constant), the curve $s(t)$ is concave down everywhere. By the Second Derivative Test, the maximum height occurs when $t = 5$ seconds. Insert this t-value into the $s(t)$ to calculate the maximum height.

$$s(5) = -16(5)^2 + 160 \cdot 5 + 176$$

$$= -400 + 800 + 176$$

$$= 576 \text{ ft.}$$

b. The projectile hits the ground when $s(t) = 0$. (This fact is not related to calculus.)

$$-16t^2 + 160t + 176 = 0$$

Solve $s(t) = 0$ for t.

$$-16(t^2 - 10t - 11) = 0$$

$$-16(t - 11)(t + 1) = 0$$

$$t = 11 \text{ or } \cancel{t = -1}$$

Time cannot be negative, so we discard the $t = -1$.

Thus the projectile hits the ground 11 seconds after it is fired.

c. The projectile's velocity when it hits the ground is $v(11)$. From part **a.** we know

$$v(t) = -32t + 160.$$

So

$$v(11) = -32(11) + 160$$
$$= -352 + 160$$
$$= -192 \text{ ft./sec.}$$

The negative sign indicates the projectile is falling.

The projectile will be falling at a rate of 192 ft./sec when it hits the ground.

4.5 Exercises

1. **Volume of a box.** A rectangular box with no top is to be made from a piece of cardboard that is 24 in. by 24 in. Equal squares are to be cut from each corner and the sides folded to form the box. What size should the squares be to maximize the volume of the box?

2. **Volume of a box.** A rectangular box with no top is to be made from a piece of cardboard that is 20 in. by 20 in. Equal squares are to be cut from each corner and the sides folded to form the box. What size should the squares be to maximize the volume of the box?

3. **Volume of a box.** Equal squares are to be cut from each corner of a rectangular piece of thin sheet metal, and the sides are to be folded to form a box. If the piece of metal is 8 in. by 15 in., find the dimensions of the box having maximum volume.

4. **Volume of a box.** Equal squares are to be cut from each corner of a rectangular piece of thin sheet metal, and the sides are to be folded to form a box. If the piece of metal is 10 in. by 16 in., find the dimensions of the box having maximum volume.

5. **Surface area.** A rectangular box is designed to have a square base and an open top. The volume is to be 864 in.^{3}
   a. What dimensions will give a minimum surface area?
   b. What is the minimum surface area?

6. **Surface area.** A rectangular box is designed to have a square base and an open top. The volume is to be 256 in.^{3}
   a. What dimensions will give a minimum surface area?
   b. What is the minimum surface area?

7. **Container design.** A container manufacturer is asked to design a closed rectangular shipping crate with a square base. The volume is 10 ft.^{3} The material for the top and sides costs \$2 per square foot and the material for the bottom costs \$3 per square foot. Find the dimensions of the box that will minimize the total cost.

8. **Container design.** A container manufacturer is asked to design a closed rectangular shipping crate with a square base. The volume is 36 ft.^{3} The material for the top costs \$1 per square foot, the material for the sides costs \$0.90 per square foot, and the material for the bottom costs \$1.40 per square foot. Find the dimensions of the box that will minimize the total cost of material.

9. **Container design.** An investor plans to manufacture rectangular box containers whose bottom and top measure x by $3x$. The box must contain 18 cubic feet. The top and bottom will cost \$2 per square foot, and the four sides will cost \$3 per square foot. What should the height h be so as to minimize costs?

10. **Area.** A rectangular plot is to be enclosed with an existing block wall as one side. If there are 680 ft. of fencing available for the other three sides, find the dimensions that will maximize the area.

11. **Area.** A rectangular play area is to be enclosed with the side of a house as one of the sides. If there are 74 ft. of fencing available for the other three sides, find the dimensions that will maximize the area.

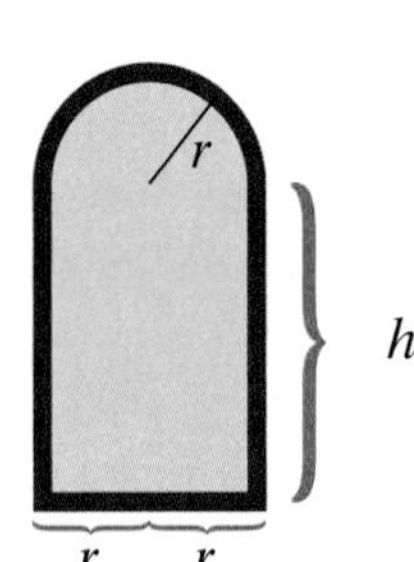

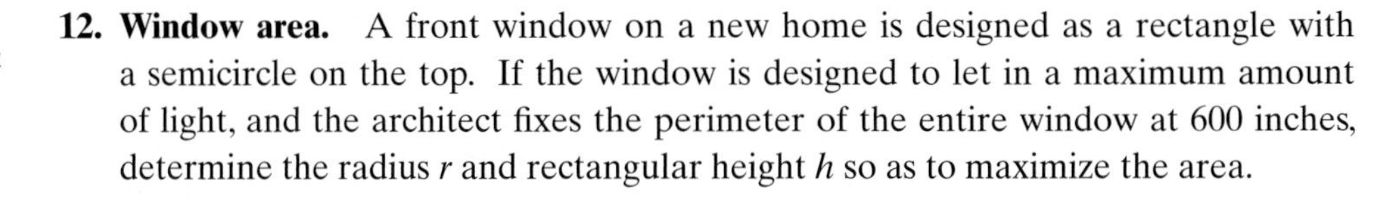

12. **Window area.** A front window on a new home is designed as a rectangle with a semicircle on the top. If the window is designed to let in a maximum amount of light, and the architect fixes the perimeter of the entire window at 600 inches, determine the radius r and rectangular height h so as to maximize the area.

13. **Wall construction.** An old stone wall makes two legs of a right angle, one 40 feet long and the other 20 feet long. A contractor is told to add 220 feet of new stone fence to complete a rectangular fence. How should he complete the fence so as to maximize the enclosed area? Determine the maximum enclosed area he may obtain.

14. **Construction.** A warehouse is being constructed with a total floor area of 2200 ft.^{2} A single partition is built to divide the building into storage area and office space. The exterior walls cost \$160 per foot, and the interior wall costs \$120 per foot. Find the dimensions of the warehouse that will minimize the cost.

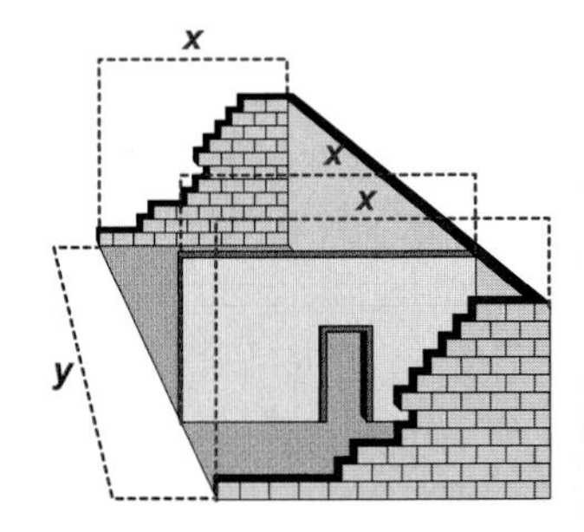

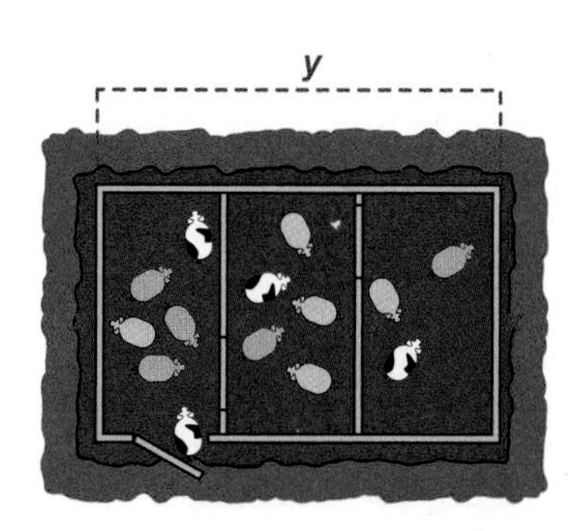

15. Construction. A farmer wants to build a rectangular pen and then divide it with two interior fences. The total area is to be 2484 ft.^{2} The exterior fence costs \$18 per foot, and the interior fence costs \$16.50 per foot. Find the dimensions of the pen that will minimize the cost.

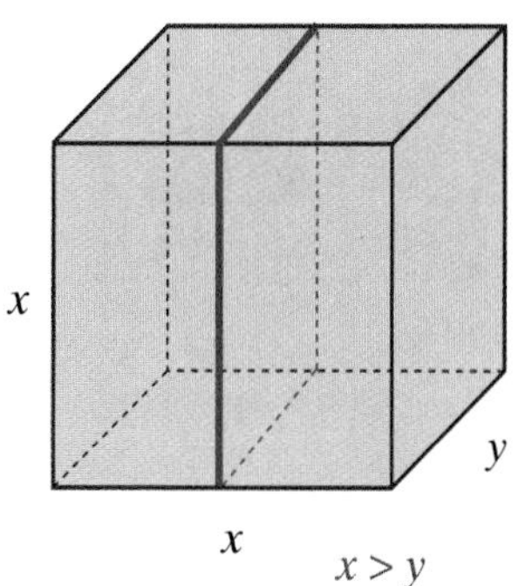

16. Shipping. The Postal Service has a limit of 108 in. on the combined length and girth of a rectangular package to be sent by parcel post. Find the dimensions of the package of maximum volume that has a square cross section. (**Hint:** There are two different answers, depending on the shape of the box. The two shapes are shown here. The **girth** is defined to be the smallest perimeter of a rectangular cross section of the box.)

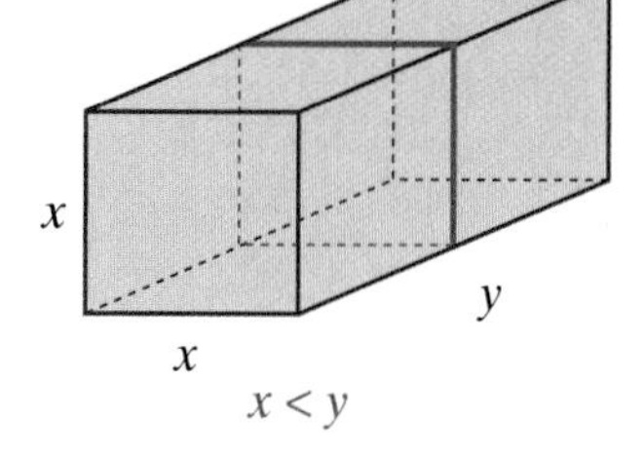

17. Shipping. An independent parcel service has a limit of 130 in. on the combined length and girth of a rectangular package it will ship. Find the dimensions of the package of maximum volume that has a square cross section. (**Hint:** There are two different answers. The **girth** is defined to be the smallest perimeter of a rectangular cross section of the box.)

18. Pipeline construction. An oil company wishes to run a pipeline from a drilling platform located 5 miles offshore to a shipping terminal 16 miles down the coast. The costs are \$130,000 per mile to lay the pipeline underwater and \$120,000 per mile to lay the pipeline over land. Find the location of point P (as illustrated in the diagram) so that the total cost of laying pipe will be minimized.

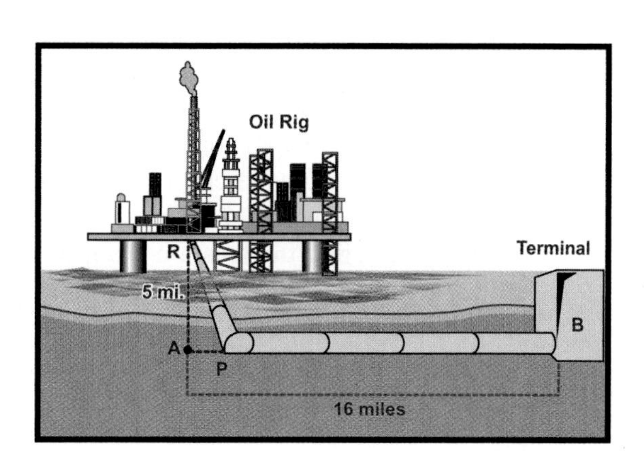

19. Power-line construction. The U.S. Forest Service wishes to run a power line to a fire lookout tower located in a wooded area. The tower is 4 miles from the nearest road and the power source is 7 miles down that road. It costs \$5000 per mile to run the line through the forest and \$3000 per mile to run the line along the road. Find the location of point P (as illustrated in the diagram) so that the total cost of running the power line will be minimized.

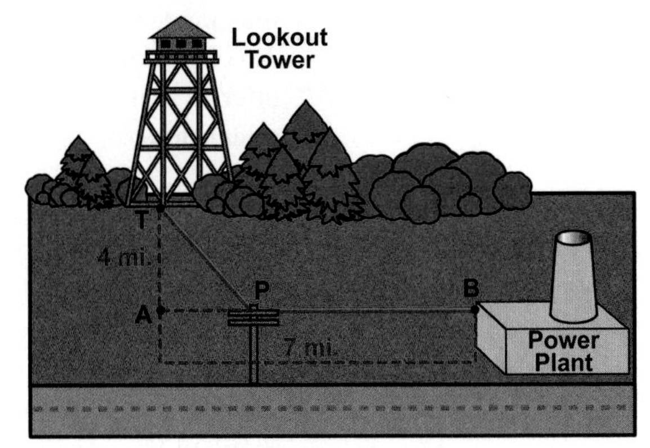

20. Minimum time. The Off-Roaders, an all-terrain vehicle club, were driving their four-wheelers in the desert when one member had a serious accident. At the time, they were 12 miles from the nearest paved road. The nearest hospital was located 11 miles down the paved road. If they can average 20 mph on the desert and 52 mph on the road, locate point P on the road toward which they should drive in order to minimize the time needed to get to the hospital.

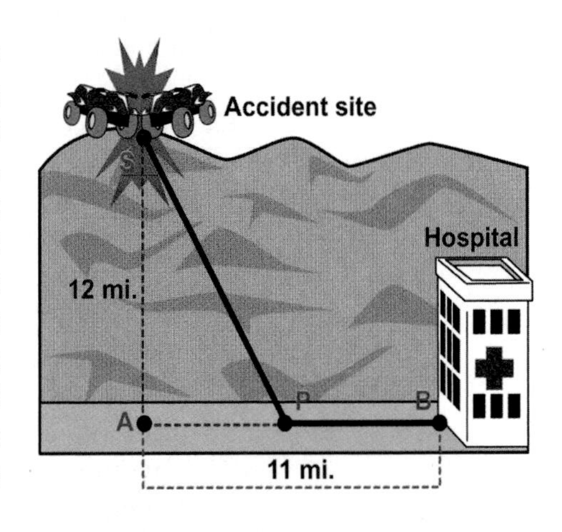

21. Minimum time. A man is on the bank of a river that is 0.6 miles wide. He wants to reach a point on the opposite shore that is 1.5 miles downstream. If he can row a boat across the river at 4 mph and walk at 5 mph, find the location P, on the opposite shore, to which he should row in order to minimize the total time he would need to reach the point downstream.

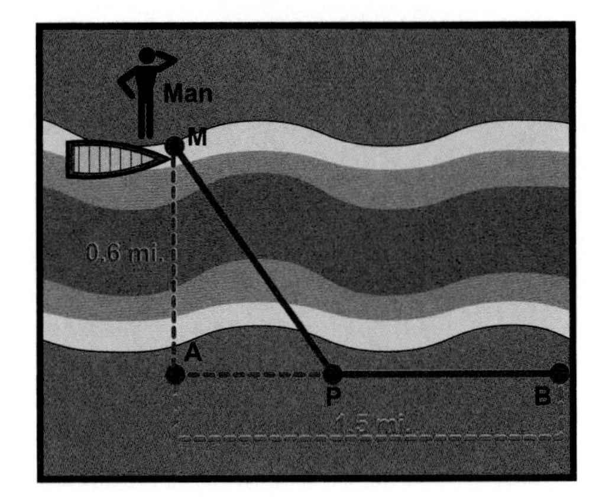

22. Distance and velocity. A particle is moving along a straight line such that the distance traveled at the end of t seconds is given by $s(t) = 7t^2 + 30t$ feet.

   a. Find the velocity if $t = 2$ seconds.

   b. How far has the particle traveled?

23. Distance and velocity. A ball is rolled down an incline. The distance (in feet) of the ball from the starting point after t seconds is given by $s(t) = 19t + 8t^2$.

   a. Find the velocity after 3 seconds.

   b. How far has the ball traveled in 3 seconds?

24. Distance and velocity. A projectile is fired vertically, and its height (in feet) after t seconds is given by $s(t) = 104t - 16t^2$.

   a. Find the maximum height of the projectile.

   b. When does the projectile hit the ground?

   c. How fast is the projectile moving when it hits the ground?

25. Distance and velocity. A stone is projected vertically. The height (in feet) of the stone at time t (in seconds) is given by $s(t) = -16t^2 + 112t + 128$.

   a. What is the maximum height of the stone?

   b. When will the stone hit the ground?

   c. What is the speed of the stone when it hits the ground?

26. Distance and velocity. A child rolls a hoop down a hilly street with the distance traveled given by $S(t) = 4t + t^2$.

   a. How far has the hoop traveled in 3 seconds?

   b. What was the speed at 3 seconds?

   c. At what rate was the speed changing at $t = 3$?

Hawkes Learning Systems: Essential Calculus

Other Applications

4.6 Differentials

Objectives

After completing this section, you will be able to:

1. *Find the differential of a function.*
2. *Use differentials to approximate changes in volume, revenue, and other real-life applications.*

For the function $y = f(x)$, the expressions y and $f(x)$ both represent the distance from the x-axis to the point $(x, f(x))$ on the curve. (See Figure 4.6.1.)

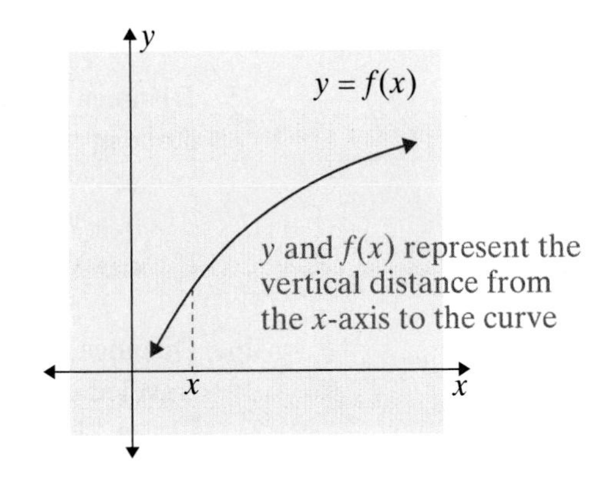
y
y = f(x)
y and f(x) represent the vertical distance from the x-axis to the curve
x
x

Figure 4.6.1

If x is changed by a small amount h, then the corresponding y is also changed. The value of y becomes $f(x + h)$, and the change in y becomes $f(x + h) - f(x)$. (See Figure 4.6.2.)

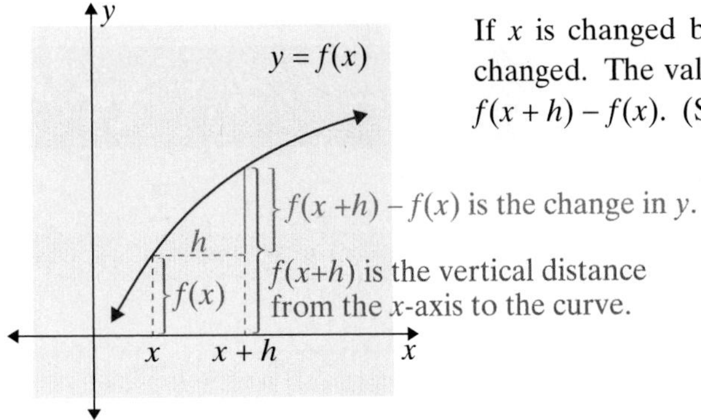
y
y = f(x)
f(x +h) − f(x) is the change in y.
h
f(x)
f(x+h) is the vertical distance from the x-axis to the curve.
x
x + h
x

Figure 4.6.2

As we discussed in Chapter 2, the change in y can be denoted by Δy (delta y):

$$\Delta y = f(x+h) - f(x).$$

Now $\Delta x = h$ and the **difference quotient** is given by

$$\frac{\Delta y}{\Delta x} = \frac{f(x+h) - f(x)}{h} \quad \text{or} \quad \frac{\Delta y}{\Delta x} = \frac{f(x+\Delta x) - f(x)}{\Delta x}.$$

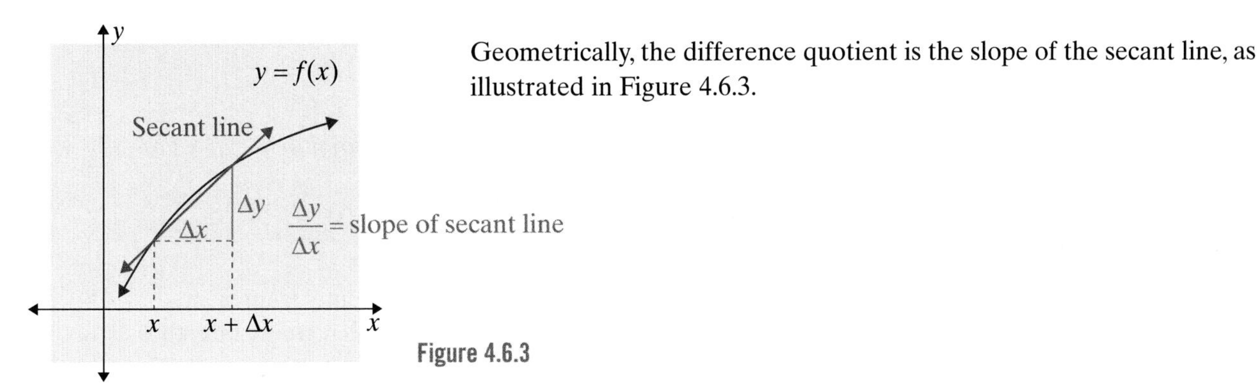

Geometrically, the difference quotient is the slope of the secant line, as illustrated in Figure 4.6.3.

Figure 4.6.3

The slope of the tangent is the first derivative $\frac{dy}{dx}$. Now we take some liberty with the notation $\frac{dy}{dx}$ and think of it as being composed of two parts, dy and dx. (See Figure 4.6.4.)

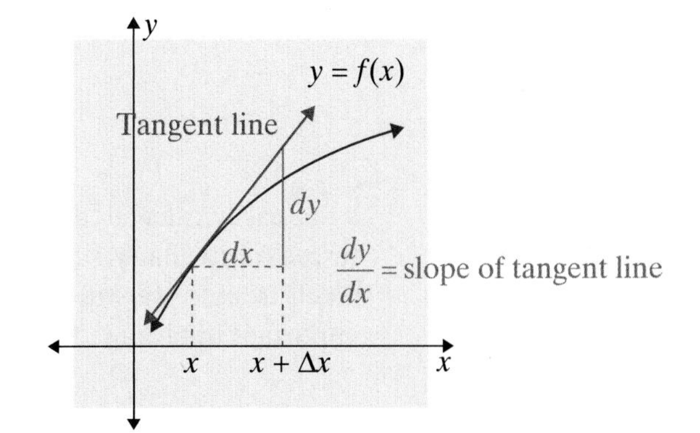

Figure 4.6.4

The symbols dy and dx are called **differentials**. We let $h = \Delta x = dx$. Then, as illustrated in Figure 4.6.5, the difference quotient and the derivative $\frac{dy}{dx}$ (quotient of the differentials) are approximately equal for small values of Δx.

$$\frac{dy}{dx} \approx \frac{\Delta y}{\Delta x}.$$

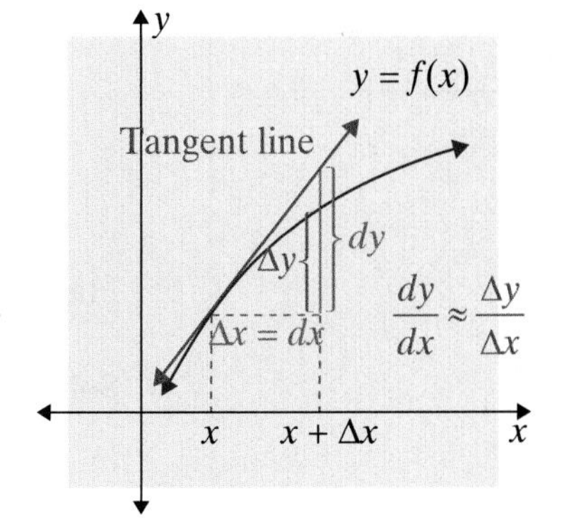

Figure 4.6.5

From Figure 4.6.5, we can see that the differential dy and the change in y, Δy, are almost equal. In fact, the smaller Δx is, the closer dy and Δy will be to each other. We make the following observations:

$$\frac{dy}{dx} \approx \frac{\Delta y}{\Delta x},$$

$$dy \approx \frac{\Delta y}{\Delta x} dx, \text{ and}$$

$$dy = \lim_{\Delta x \to 0}\left(\frac{\Delta y}{\Delta x}\right) \cdot dx = f'(x)\,dx.$$

This last expression becomes the definition for the differential dy.

Differential

If $y = f(x)$ is a function and $y' = f'(x)$ exists, then the **differential** *dy is defined as*

$$\boldsymbol{dy = f'(x) \cdot dx}.$$

As we have seen in the previous development, the change in y (represented by Δy) can be approximated by the differential dy. That is, we can calculate Δy directly as

$$\Delta y = f(x+\Delta x) - f(x)$$

or use

$$dy = f'(x)dx$$

as an easily calculated approximation to Δy. While this use of dy may seem somewhat forced, particularly since most students have calculators and can calculate Δy without much trouble, we will find other uses for differentials later. At this stage, though, it is important to become familiar with both the notation and the terminology.

Example 1: Finding Differentials

a. Find dy for $y = x^2$.

Solution: $f(x) = x^2$ and $f'(x) = 2x$. — Find the derivative of y.

So

$$dy = f'(x) \cdot dx$$
$$= 2x \cdot dx.$$

Substitute $f'(x) = 2x$ into the definition of differential.

b. Find du for $u = (x^2+5)^3$.

Solution: The differential du is found in the same way as dy,

$$f(x) = (x^2+5)^3 \text{ and } f'(x) = 3(x^2+5)^2 \cdot 2x$$
$$= 6x(x^2+5)^2.$$

Find the derivative of u.

So

$$du = f'(x)dx$$
$$= 6x(x^2+5)^2 dx$$

Substitute $f'(x) = 6x(x^2+5)^2$ into the definition of differential.

c. Find dv if $v=\dfrac{x^2}{x^2+9}$.

Solution: $f(x)=\dfrac{x^2}{x^2+9}$ and $f'(x)=\dfrac{(x^2+9)(2x)-x^2(2x)}{(x^2+9)^2}$ Find the derivative of v.

$$=\frac{18x}{(x^2+9)^2}.$$

So

$$dv=\frac{18x}{(x^2+9)^2}dx$$

Substitute $f'(x)=\dfrac{18x}{(x^2+9)^2}$ into the definition of differential.

Example 2: Using Differentials

For $y=x^2$, $x=9$, $\Delta x=0.1$, find **a.** Δy, **b.** dy, and **c.** $\Delta y-dy$.

Solutions: a.

$\Delta y=f(x+\Delta x)-f(x)$ From earlier this section.

$=f(9+0.1)-f(9)$ Substitute $x=9$ and $\Delta x=0.1$.

$=(9.1)^2-(9)^2$ Find $f(9.1)$ and $f(9)$.

$=82.81-81.0$

$=1.81$ Solve the equation.

b. $f(x)=x^2$ and $f'(x)=2x.$ Find the derivative of y.

$dy=f'(x)dx$

$=2xdx$ Substitute $f'(x)=2x$ into the definition of differential.

$=2(9)(0.1)$ Substitute in $x=9$ and $dx=0.1$.

$=1.8$

c. $\Delta y-dy=1.81-1.8=0.01$ Substitute the values found in parts **a.** and **b.** into the equation and solve.

Example 3: Estimating Square Roots

Using differentials, estimate $\sqrt{15}$.

Solution: Here we let $f(x)=\sqrt{x}$, $x=16$, and $\Delta x=-1$.

We have chosen $x = 16$ because 16 is the closest square number to 15. This gives $15 = 16 + \Delta x$, which leads to $\Delta x = -1$.

So

$$f(x)=\sqrt{x}=x^{\frac{1}{2}}$$ Rewrite $f(x)$ using exponents.

$$f'(x)=\frac{1}{2}x^{-\frac{1}{2}}=\frac{1}{2\sqrt{x}}$$ Find the derivative of $f(x)$.

$$dy=f'(x)\,dx$$

$$=\frac{1}{2\sqrt{x}}dx$$ Substitute $f'(x)$ into the definition of differential.

$$dy\Big|_{x=16}=\frac{1}{2\sqrt{16}}(-1)$$ Find dy for $x = 16$: substitute $x = 16$ and $\Delta x = -1$.

$$=-\frac{1}{8}$$

$$=-0.125$$

This means that

$$\sqrt{15}\approx\sqrt{16}+dy$$

$$=4-0.125$$

$$=3.875.$$

Your calculator will give $\sqrt{15}\approx 3.872983346$.

Example 4: Volume

The volume of a cube with edge x is given by $V = x^3$. Using differentials, approximate the change in V if x is

a. Changed from 5 cm to 5.01 cm.

b. Changed from 5 cm to 4.99 cm.

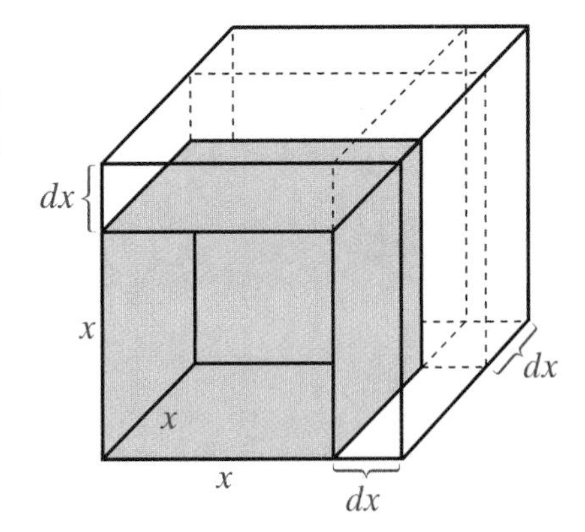

Solutions: a. $V = f(x) = x^3$

$f'(x) = 3x^2$ Find the derivative of $f(x)$.

$dV = f'(x)\,dx = 3x^2 dx$ Find the differential.

If $x = 5$ and $dx = 0.01$, then

$$dV = 3(5)^2(0.01) = 0.75 \text{ cm}^3.$$

Substitute in $x = 5$ (original length of an edge) and $dx = 0.01$ (the change in length).

b. Using $x = 5$ and $dx = -0.01$,

$$dV = 3(5)^2(-0.01) = -0.75 \text{ cm}^3.$$

Here $x = 5$ and $dx = -0.01$.

Note that with a calculator, we can find the volume of each cube:

$V(5) = 5^3 = 125 \text{ cm}^3$,

$V(5.01) = (5.01)^3 = 125.751501 \text{ cm}^3$, and

$V(4.99) = (4.99)^3 = 124.251499 \text{ cm}^3$.

So our values of $dV = \pm\, 0.75$ (calculated in parts **a.** and **b.**) are very close to the actual changes in volume.

Example 5: Change in Revenue

Suppose that a company makes and sells x tennis rackets per week, and the corresponding revenue function (in hundreds of dollars) is $R(x)=20x-\frac{x^2}{30}$, where x is between 0 and 600. Use dR to estimate the approximate change in revenue if production is

a. Increased from 150 to 160 rackets.

b. Increased from 480 to 490 rackets.

Solutions: a.

$$R(x)=20x-\frac{x^2}{30}$$

$$R'(x)=20-\frac{x}{15}$$ Find the derivative of $R(x)$.

$$dR=R'(x)\,dx=\left(20-\frac{x}{15}\right)dx$$ Find the differential.

Let $x=150$ and $\Delta x=dx=160-150=10$.
Then

$$\begin{aligned} dR&=\left(20-\frac{150}{15}\right)10\\ &=(20-10)10\\ &=100. \end{aligned}$$

The change in revenue is approximately 100 hundred dollars (or \$10,000).

b. Let $x=480$ and $\Delta x=dx=490-480=10$.
Then

$$\begin{aligned} dR&=\left(20-\frac{480}{15}\right)10\\ &=(20-32)10\\ &=-120. \end{aligned}$$

The change in revenue is negative 120 hundred dollars, or a loss of \$12,000. In this case, the revenue is actually decreasing at a high level of production and sales because the corresponding price of rackets has been lowered to maintain sales.

4.6 Exercises

Find the differential for each of the functions in Exercises 1 – 14.

1. $y=x^3+5$

2. $y=4x^3+x-7$

3. $u=\left(2t^2+1\right)^2$

4. $u=(5t+9)^2$

5. $A=\pi r^2$

6. $A=x(44-2x)$

7. $V=\frac{4}{3}\pi r^3$

8. $V=x^3$

9. $S=4x^2+\frac{1350}{x}$

10. $S=2\pi r^2+\frac{90\pi}{r}$

11. $C=40+3x+0.4\sqrt{x}$

12. $C=75+10x-0.6\sqrt{2x}$

13. $P=-0.2x^2+75x-2400$

14. $P=-0.3x^2+84x-870$

For each of the functions given in Exercises 15 – 22, use the given values for x and Δx to find **a.** Δy, **b.** dy, *and* **c.** $\Delta y - dy$.

15. $y=x^2-3x+4,\ x=3,\ \Delta x=0.2$

16. $y=x^2+5x-9,\ x=2,\ \Delta x=0.15$

17. $y=\left(2x^2-4\right)^3,\ x=-2,\ \Delta x=0.04$

18. $y=\left(x^2+x-1\right)^3,\ x=-3,\ \Delta x=0.05$

19. $y=20\left(x-\frac{24}{x}\right),\ x=2,\ \Delta x=-0.12$

20. $y=2x^2+\frac{125}{x^2},\ x=5,\ \Delta x=-0.15$

21. $y=\sqrt{3x+4},\ x=7,\ \Delta x=0.5$

22. $y=\sqrt{12-5x},\ x=2,\ \Delta x=0.07$

Use differentials to approximate the indicated roots in Exercise 23 – 30. Express your answer as $a\pm{}^{b}\!/_{c}$*, where a is the nearest integer.*

23. $\sqrt{37}$

24. $\sqrt{65}$

25. $\sqrt[3]{26}$

26. $\sqrt[3]{126}$

27. $\sqrt{50.4}$

28. $\sqrt{79.5}$

29. $\sqrt[3]{62.3}$

30. $\sqrt[3]{218.3}$

31. Cost. A total cost function (in dollars) is given by $C(x)=375+9x+0.01x^2$. Use differentials to estimate the change in cost when the level of production is increased from 60 to 62 units.

32. Cost. A total cost function (in dollars) is given by $C(x)=930+15x+0.2x^2$. Using differentials, estimate the change in cost from $x = 100$ to $x = 101$.

33. **Profit.** The weekly revenue (in dollars) from the sale of x coffeemakers is given by $R(x) = 40x$. The total cost function is given by $C(x) = 370 + 16x + 0.2x^2$. Use differentials to approximate the change in profit if the weekly sales are increased from 25 to 28 coffeemakers.

34. **Profit.** The monthly revenue from the sale of x 50-gallon aquariums is given by $R(x) = 54x - 0.3x^2$ dollars. The total cost function is given by $C(x) = 0.1x^2 + 4x + 200$ dollars. Using differentials, find the approximate change in profit if the monthly sales are increased from 40 to 44 aquariums.

35. **Population.** It is estimated that t years from now the population of a city will be $P(t) = 10(4000 + 2t^2) - 1600t$. Use differentials to estimate the change in population as t changes from 6 to 6.25 years.

36. **Bacterial population.** It is estimated that t hours from now the population of bacteria in a culture will be $P(t) = \frac{8000}{\sqrt{8 - 0.5t}}$. Use differentials to estimate the change in population as t changes from 8 to 8.3 hours.

37. **Volume.** The edge of a cube measures 18 in. with a possible error in measurement of 0.02 in. Use differentials to estimate the possible error in computing the volume.

38. **Fiberglass coating.** A cube is 12 in. on a side. It is to be covered with a fiberglass coating 0.25 in. thick. Use differentials to estimate the volume of the fiberglass coating.

39. **Measurement error.** A manufacturer of cargo containers receives an order for a cube-shaped container. The specifications state that the volume should be 125 ft.^{3} with a maximum error of no more than 1 ft.^{3} Using differentials, find the possible error in the length of the edges.

40. **Melting ice.** A block of ice is in the form of a 10-in. cube. If it melts uniformly until the volume changes to 972 in.^{3}, approximate the change in the length of each edge by using differentials.

41. **Volume of a weather balloon.** A spherical weather balloon is being inflated. Use differentials to find the approximate change in the volume if the radius changes from 20 to 21.5 inches. $\left(\textbf{Hint:}\ V = \frac{4}{3}\pi r^3.\right)$

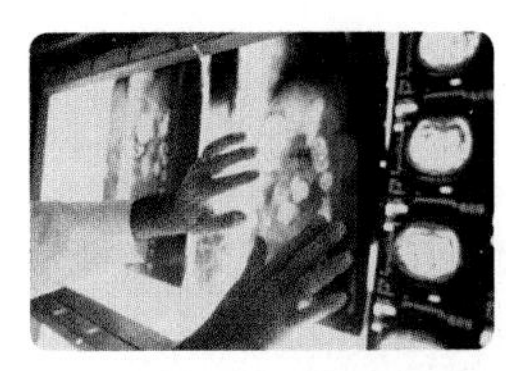

42. Volume of a tumor. A spherical cancer tumor is being treated with an experimental drug. The radius of the tumor has been reduced from 1.6 to 1.4 cm. Use differentials to estimate the change in the volume of the tumor. $\left(\textbf{Hint:}\ V = \frac{4}{3}\pi r^3.\right)$

Hawkes Learning Systems: Essential Calculus

Differentials

Chapter 4 Index of Key Ideas and Terms

Section 4.1 Higher-Order Derivatives: Concavity and the Second Derivative Test

Higher-Order Derivatives pages 296 - 299

For a function f that has slope defined at most points for which the function is defined, f' is itself a function which has its own derivative defined at most points in its domain, and it too can be differentiated. The **second derivative**, f'', if it exists, is the derivative of the first derivative. The **third derivative**, f''', if it exists, is the derivative of the second derivative, and so on.

Notation for the Second Derivative

The second derivative of the function $y = f(x)$ can be denoted by any of the following symbols:

$$y'',\ f''(x),\ \frac{d^2y}{dx^2},\ f_{xx},\ \text{and}\ D_x^2[y].$$

Concavity pages 300 - 306

Suppose that f is differentiable on the interval (a, b).

1. If f' is increasing on (a, b), then the graph of f is **concave upward** on (a, b).
2. If f' is decreasing on (a, b), then the graph of f is **concave downward** on (a, b).

Interpreting $f''(x)$ to Determine Concavity page 303

Suppose that f is a function and f' and f'' both exist on the interval (a, b).

1. If $f''(x) > 0$ for all x in (a, b), then f' is increasing and f is concave upward on (a, b).
2. If $f''(x) < 0$ for all x in (a, b), then f' is decreasing and f is concave downward on (a, b).

Points of Inflection pages 306 - 309

If the graph of a continuous function has a tangent line at a point (possibly a vertical line) and the graph changes concavity at that point, then the point is called a **point of inflection**.

Hypercritical Values

Hypercritical values of x are those values $x = c$ in the domain of f where $f''(c) = 0$ or $f''(c)$ does not exist. Points of inflection occur only at hypercritical values. However, hypercritical values do not guarantee points of inflection.

continued on next page ...

Section 4.1 Higher-Order Derivatives: Concavity and the Second Derivative Test (continued)

To Determine Points of Inflection page 307

Suppose that f is a continuous function.

1. Find $f''(x)$.
2. Find the hypercritical values of x. That is, find the values $x = c$ where
    a. $f''(c) = 0$, or
    b. $f''(c)$ is undefined.
3. Using the hypercritical values as endpoints of intervals, determine the intervals where
    a. $f''(x) > 0$ and f is concave upward, and
    b. $f''(x) < 0$ and f is concave downward.
4. Points of inflection occur at those hypercritical values where f changes concavity.

Using a Calculator to Find Inflection Points pages 309 - 311

Second Derivative Test for Local Extrema pages 311 - 316

Suppose that f is a function, f' and f'' exist on the interval (a, b), c is in (a, b), and $f'(c) = 0$.

1. If $f''(c) > 0$, then $f(c)$ is a local minimum.
2. If $f''(c) < 0$, then $f(c)$ is a local maximum.
3. If $f''(c) = 0$, then the test fails to give any information about local extrema.

Application: Point of Diminishing Returns page 317

The point of diminishing returns occurs at a point of inflection where the sales curve changes from concave upward to concave downward.

Section 4.2 Curve Sketching: Polynomial Functions

Strategy for Curve Sketching Given a Function f page 327

To sketch the graph of a function f:

1. Find $f'(x)$ and $f''(x)$.
2. Find the critical values of f. That is, find the values $x = c$ where
    a. $f'(c) = 0$, or
    b. $f'(c)$ is undefined.
3. Using the critical values as endpoints, find intervals
    a. where $f'(x) > 0$ (f is increasing), and
    b. where $f'(x) < 0$ (f is decreasing).
4. Locate the local extrema using the First Derivative Test or the Second Derivative Test.

continued on next page ...

Section 4.2 Curve Sketching: Polynomial Functions (continued)

Strategy for Curve Sketching Given a Function f (continued)

5. Find the hypercritical values of f. That is, find the values $x = c$ where
   a. $f''(c) = 0$, or
   b. $f''(c)$ is undefined.
6. Using the hypercritical values as endpoints, find intervals
   a. where $f''(x) > 0$ (f is concave upward), and
   b. where $f''(x) < 0$ (f is concave downward).
7. Locate all points of inflection. These occur at hypercritical values where the curve changes concavity.
8. Using the combined information from steps 1 – 7 and any other specific points that might be helpful, sketch the graph.

Section 4.3 Curve Sketching: Rational Functions

Rational Function

pages 337 - 343

A **rational function** is a function of the form

$$f(x)=\frac{P(x)}{Q(x)},$$

where $P(x)$ and $Q(x)$ are polynomials and $Q(x) \neq 0$.

Asymptotes for Rational Functions $f(x)=\frac{P(x)}{Q(x)}$

1. **Vertical asymptotes** are of the form $x = a$ and occur where the denominator is 0 and the numerator is not 0 (that is, $Q(a) = 0$ and $P(a) \neq 0$).

2. **Horizontal asymptotes** are of the form $y = b$ and occur where
   $$\lim_{x\to+\infty} f(x)=b \quad \text{or} \quad \lim_{x\to-\infty} f(x)=b.$$

3. **Oblique asymptotes** are of the linear form $y = mx + b$ and occur when the numerator is one degree larger than the denominator and $f(x)$ can be written as
   $$f(x)=mx+b+\frac{R(x)}{Q(x)},$$
   where
   $$\lim_{x\to+\infty}\frac{R(x)}{Q(x)}=0 \quad \text{or} \quad \lim_{x\to-\infty}\frac{R(x)}{Q(x)}=0.$$

Chapter 4 Review

For a review of the topics and problems from Chapter 4, look at the following lessons from *Hawkes Learning Systems: Essential Calculus.*

Higher-Order Derivatives and Concavity
Higher-Order Derivatives: the Second Derivative Test
Curve Sketching: Polynomial Functions
Curve Sketching: Rational Functions
Business Applications
Other Applications
Differentials

Chapter 4 Test

In Exercises 1 – 4, find the second derivative for each function.

1. $f(x)=x^4+6x^{\frac{3}{2}}-9$

2. $f(x)=\left(2x^2+7\right)^{\frac{2}{3}}$

3. $f(x)=5x+\dfrac{2}{x+1}$

4. $f(x)=\dfrac{7x+2}{x^3}$

In Exercises 5 and 6, find $f''(x)$. Then find $f''(-2), f''(0)$, and $f''(1)$, if they exist.

5. $f(x)=x^3-4x^2-x+8$

6. $f(x)=x^2-\dfrac{14}{x-2}$

In Exercises 7 and 8, **a.** *find the interval(s) on which f is concave upward,* **b.** *find the interval(s) on which f is concave downward, and* **c.** *locate any points of inflection.*

7.

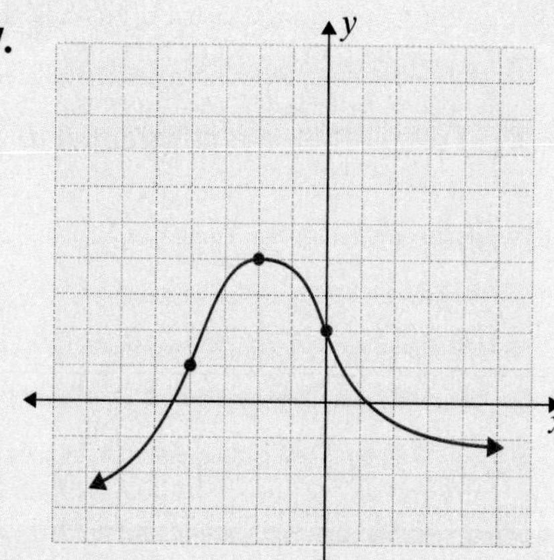

8.

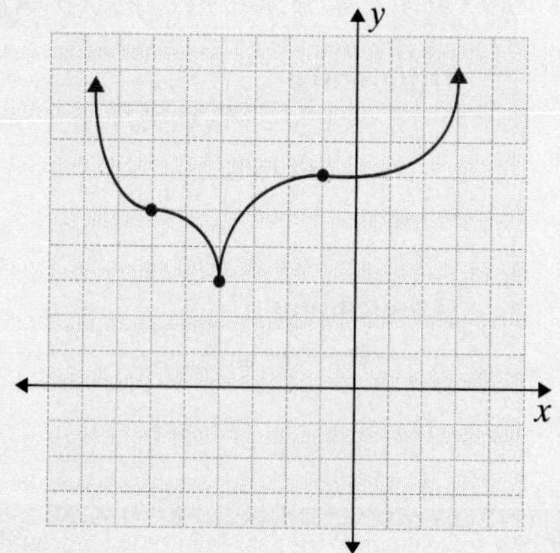

In Exercises 9 – 12, determine the interval(s) on which each function is **a.** *concave upward,* **b.** *concave downward, and* **c.** *locate any points of inflection.*

9. $f(x)=x^3-6x^2+4x+2$

10. $f(x)=\sqrt{x^2+1}$

11. $f(x)=\dfrac{x^3-16}{3x}$

12. $f(x)=\dfrac{1}{3}x^4-6x^2+2$

In Exercises 13 – 16, Use the Second Derivative Test to locate extrema.

13. $f(x)=5-9x+3x^2$

14. $f(x)=\dfrac{1}{3}x^3-3x^2-7x+2$

15. $f(x)=3x+\dfrac{12}{x}$

16. $f(x)=x^4+4x^3-8x^2+2$

17. Sketch the graph of a continuous function that satisfies all the following conditions:

a. $f(2) = -3$.
b. $f'(2) = 0$.
c. $f'(x) < 0$ if $x < 2$.
d. $f'(x) > 0$ if $x > 2$.
e. $f''(x) > 0$ for all x.

18. Marginal cost. Find the minimum marginal cost of a product if the cost function is given by $C(x) = 2x^{\frac{3}{2}} + 54x^{\frac{1}{2}} + 500$ dollars, where x is the number of units produced and $0 \le x \le 64$.

19. Homeless population. The number of homeless people in a midwestern city is estimated by the function $P(t) = 680 - 1.5t^2 + 8t^{\frac{3}{2}}$, where t is the number of months after a special low income housing program has been introduced and $0 \le t \le 36$.

a. Find the maximum number of homeless people in the city.
b. Find the maximum rate of increase in the homeless population.

For each function in Exercises 20 – 23, complete a summary table (as in Section 4.3) and sketch the graph.

20. $f(x) = 3 - 5x - x^2$

21. $f(x) = \frac{2}{3}x^3 + \frac{3}{2}x^2 - 2x - 3$

22. $f(x) = x^4 - 6x^2 + 2$

23. $f(x) = 2x - \frac{1}{x^2}$

In Exercises 24 and 25, use the given values for x and Δx to find **a.** Δy, **b.** dy, *and* **c.** $\Delta y - dy$.

24. $y = (4x^2 - 3)^2$, $x = 1$, $\Delta x = 0.5$

25. $y = \sqrt{5x + 1}$, $x = 3$, $\Delta x = -0.15$

In Exercises 26 and 27, use differentials to approximate the indicated roots.

26. $\sqrt{48}$

27. $\sqrt[3]{65}$

28. Drug concentration. The level of concentration (expressed as a percent) of a certain drug in the bloodstream is given by the function $L(x) = \frac{6x}{5x^2 + 80}$, where x is the number of hours after the drug has been administered.

a. Find the time at which the concentration is a maximum.
b. Find the maximum concentration.

29. Minimizing inventory costs. An auto parts store sells 540 radios per year. It costs \$3 to store one radio for 1 year. The ordering costs are \$10 for each order. In what lot size and how many times per year should the orders be placed to minimize inventory costs?

30. Maximizing Revenue. A travel agent is planning a cruise. She knows that if 30 people go, it will cost $420 per person. However, the cost per person will decrease $10 each time an additional person signs up to go. Find the price that will maximize her revenue.

31. Linear demand. A bicycle shop sells 100 racing bikes of one model when the price is $300. If the price is reduced to $270, the shop sells 120 bikes. The total cost of x bikes is given by $C(x)=0.2x^2+25x+6500$. Assuming the demand function is linear, find

a. The total number of bikes sold to maximize profit.

b. The price per bike that will maximize profit.

32. Cabinet construction. A rectangular cabinet is built with a square top and bottom. The front of the cabinet is left open. If the volume is 4.5 ft.^{3}, find the dimensions that will minimize the material needed.

33. Fence construction. Ann wants to build a rectangular pen for her two dogs. She plans to use her existing yard fence for one of the sides but wants to divide the pen into two sections. What is the area of the largest pen she can build with 120 feet of fencing?

34. Minimum time. A man wishes to travel from an island located 0.8 miles offshore to a town 3 miles down the coast. He plans to take his motorboat to a point P on the shore and then ride his minibike into town. If the boat travels at 15 mph, and the minibike travels at 39 mph, find the location of point P that will minimize his time.

35. Vertical projectile. A projectile is fired vertically, and its height (in feet) t seconds after it is fired is given by $S(t)=176t-16t^2$.

a. Use the Second Derivative test to find the maximum height of the projectile.

b. When does the projectile hit the ground?

c. How fast is the projectile moving when it hits the ground?

36. Cost. A cost function is given by $C(x)=435+22x+0.4x^2$. Use differentials to estimate the change in cost when the level of production is increased from 30 to 32 units.

37. Bacterial population. It is estimated that t hours from now the population of bacteria in a culture will be $P(t)=\dfrac{6400}{\sqrt{6-0.2t}}$. Use differentials to estimate the change in bacterial population as t changes from 10 to 10.4 hours.

38. Volume. The specifications for the radius of a ball are 6 cm with a possible error of 0.02 cm. Use differentials to estimate the error in computing the volume of the ball.

Cumulative Review

In Exercises 1 – 4, use the graph of y = f(x) to determine the limits.

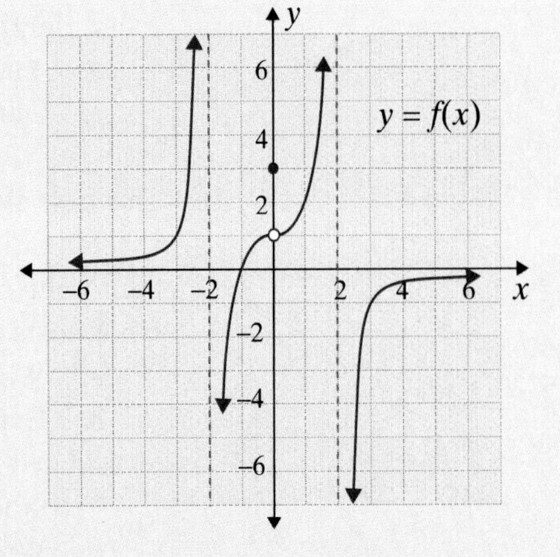

1. 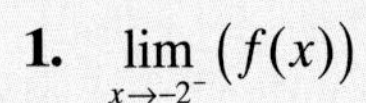 $\lim\limits_{x\to -2^-}(f(x))$

2. $\lim\limits_{x\to -2^+}(f(x))$

3. $\lim\limits_{x\to 0^-}(f(x))$

4. $\lim\limits_{x\to 0^+}(f(x))$

In Exercises 5 – 8, determine the indicated limits (if they exist). In each case, make a table with 5 (x, y) pairs to justify your answer.

5. $\lim\limits_{x\to 4^-}\left(\dfrac{2x^3-8x^2+6x-24}{x-4}\right)$

6. $\lim\limits_{x\to -1^+}\left(\dfrac{2x^2+3\sqrt{x+1}}{\sqrt{x+1}}\right)$

7. $\lim\limits_{x\to 3^-}\left(\dfrac{x^2-9}{x-3}\right)$

8. $\lim\limits_{x\to 4^+}\left(\dfrac{9x-54}{x-6}\right)$

9. Give examples of functions whose left- and right-hand limits (as x approaches 4):

- **a.** Are not equal.
- **b.** Are equal.
- **c.** Whose left-hand limit is 2 and whose right-hand limit is 3.

10. If y represents the number of 5-gallon drums of heating oil consumed by a utility company and x represents the number of weeks since January 1st, what does the derivative represent?

11. Suppose (2.0, 3) and (2.10, 3.3) are two points on the graph of a function $y = f(x)$. Estimate the slope of the tangent line to $f(x)$ at $x = 2$.

In Exercises 12 – 14, determine the formula for the derivative of the given function.

12. $f(x) = 6x^2$

13. $f(x) = 6x^2(x-1)$

14. $f(x) = 14\sqrt{x}$

15. For each of the following quantities, determine whether or not it is (or could be) the value of a derivative at a point. If the quantity is not a derivative, explain why not. If it is a derivative, give, or briefly describe, a possible corresponding function f and identify what x represents.

a. 4 aces per deck. **b.** 2 weekends per semester.

c. 240 calories per serving. **d.** $120 per week.

16. The new floor supervisor at a small manufacturing plant reports that his data charts tell him that the current production results in 40 units of production per man-hour of labor. The Chief Financial Officer (CFO) reports that the average labor costs are $16 per man-hour.

a. What are the current costs of production?

b. Express the given quantities and the answer to part **a.** as derivatives (use x represents units of production, c represents cost in dollars, and y is labor in man-hours).

c. Write a formula which uses the answers to part **b.** but which calculates the number in part **a.**

17. Suppose a company's daily revenue R is given as a function of unit price p, and that p is a function of the quantity x of items sold. When $x = 250$, $\frac{dR}{dp}$ is $200 for a $1 increase in unit price, and $\frac{dp}{dx}$ is $9 per additional item sold per day. Find the marginal revenue when $x = 250$.

In Exercises 18 – 23, determine the derivative of the given function.

18. $f(x) = \frac{23}{\sqrt{17+5x}}$ **19.** $g(t) = \frac{2t-1}{3t+1}$

20. $g(t) = \sqrt{1+2t} + \sqrt{3t+1} + 2 + 5t$ **21.** $g(t) = (100+75t)\sqrt{t}$

22. $H(s) = \sqrt{1+8s+16s^2}$ **23.** $F(u) = \sqrt{\frac{3u-1}{3u+1}}$

24. If $f(t) = 3t + 17 - \frac{225}{t^2}$, determine a formula for $f'(t)$ and use it to find the slope at $(1.5, -78.5)$.

Find the coordinates of any local extrema of the given functions in Exercises 25 – 28.

25. $f(x) = \frac{2050}{250{,}047}x^3 + \frac{820}{27{,}783}x^2 - \frac{17{,}425}{18{,}522}x - \frac{4765}{18{,}522}$

26. $f(x) = 0.5x^3 + 15x^2 - 2.3x - 35$ **27.** $f(x) = 5x^3 - 0.1x + 3.3$

28. $f(x) = 25(x-3)^3 + 5x$

29. A ship traveling parallel to the shore at 45 ft./sec is passing a person standing on the beach. The ship's distance to the shore is 3000 ft. and, at $t = 3$, the ship is directly perpendicular to the person. How fast is the distance between the observer and the ship changing at $t = 13$ seconds?

30. A 25-foot ladder is leaning against a wall. The top of the ladder is being pulled by a rope up the side of the building at a rate of 2.3 ft./sec. How fast is the bottom of the ladder moving towards the wall when the ladder's base is 5 feet from the wall?

31. The size of usable beach on Hurricane Island is given by the function $F(t) = -\frac{60}{7943}t^3 + \frac{2970}{7943}t^2 - \frac{3000}{611}t + 40$, where $0 \le t \le 24$ is the number of years after 1981 and $F(t)$ is given in width (feet) of usable beach. The beach is constantly being eroded by storms, tides, and the destruction of the beach plant life. When was the beach the largest? When was it the smallest? Have efforts to restore the beach been successful? Explain.

In Exercises 32 – 35, find the first, second, third, and fourth derivatives for the given function.

32. $f(x) = 3 + 17x - 11x^2 + 8x^3 - 2x^4$

33. $f(x) = \frac{1}{x}$

34. $f(x) = \frac{2x+5}{x-3}$

35. $f(x) = \sqrt{5+3x}$

In Exercises 36 and 37, sketch a continuous function that satisfies each of the given conditions. For problem 36, draw only the Quadrant I portion of the graph.

36.
a. $f(2) = 6$
b. $f'(2) = 6$
c. $f''(2) = -6$

37.
a. $f'(x) < 0$ if $x > 0$
b. $f''(x) > 0$ if $x < 0$

In Exercises 38 and 39, use the curve sketching plan to graph the functions given.

38. $f(x) = x\sqrt[3]{x-8}$

39. $f(x) = (x-8)(x+12)^2$

In Exercises 40 and 41, identify all asymptotes.

40. $f(x) = x^2 + \frac{100}{x}$

41. $f(x) = \frac{x^2}{100+x^3}$

42. The total cost function for Castle Armored Vest Company is $C(x) = 22{,}000 + 450x + 0.02x^2$. The average cost function is $A(x) = \dfrac{C(x)}{x}$.

a. At what rate is the average cost of an armored vest changing when the company produces 50 vests. Interpret this number.

b. Find the horizontal asymptote for $A(x)$. Interpret the meaning of the asymptote.

43. The census bureau of East Warmongria says the percentage of veteran male workers over 45 is modeled by the function $W(t) = \dfrac{20t+5}{.5t+1}$, where t is the number of years since January 1, 1900.

a. Determine $W(0)$ and interpret this number.

b. To the nearest year and month, when was the percentage of older veteran workers 25%?

c. What percentage does the function predict for the year 2010?

44. A cube is 15 centimeters on a side. It is to be coated with a plate of 1 millimeter thick silver.

a. Use differentials to estimate the volume of the silver coating.

b. Your supervisor at Cubic Industries asks you to determine the cost to the company of the silver used per cube. (The density of silver is 10.5 grams per cm^3 and the present cost is about \$0.23 per gram [www.silverinfo.net].)

Exponential and Logarithmic Functions

CHAPTER 5

Did You Know?

The rapid advances in astronomy in the 16th century were hindered by the difficulties of the computations involved. Some simplifications due to Islamic advances in trigonometry were known, and these were used, in combination with trigonometric tables, to replace certain lengthy multiplications with relatively quick additions. John Napier (1550 – 1617), a Scottish Laird, saw how useful it would be if all multiplication could be reduced to simple addition and began looking for an easier method.

Since Johannes Kepler (1571 – 1630) had recently used his knowledge of mathematics to demonstrate his three laws of planetary motion for Mars, astronomers were eager to do the same for the other planets. The amount of multiplication required, however, was daunting. For navigational computations, multiplication of pairs of eight-digit numbers was common. As an example, let us consider the multiplication $111 \cdot 57$. Technically, this problem consists of adding $(111_1 + 111_2 + 111_3 + 111_4 + \ldots + 111_{57})$. It would be very helpful if the problem could be simplified into a single addition. Since we know that we can write $111 = 10^a$ for some number a, and $57 = 10^b$ for some number b, then we can say that $111 \cdot 57 = 10^a \cdot 10^b = 10^{a+b} = 10^c$ with $c = a + b$. Thus a long multiplication can be simplified by reformulating the problem as an addition of exponents. Mathematicians and astronomers faced the same problem, with much more complex computations. In order to avoid such lengthy calculations, Napier devised the concept of the logarithm.

Napier did not know that logarithmic and exponential functions would become central pillars of modern mathematics. In addition to greatly simplifying calculations, logarithmic and exponential functions describe some very interesting physical events. For example, the exponential function can describe population growth or compound interest, while logarithmic functions model such things as the magnitude of an earthquake or the brightness of stars.

Computers and calculators have reduced the need to simplify calculations, but the exponential and logarithmic functions will continue to be vital in the description of many real-life situations.

From the point of view of this course, and for the calculus student, exponential and logarithmic functions are valuable models for a wide variety of important applications. These functions are so important that even though they are discussed in most intermediate algebra courses, they are reviewed thoroughly in this chapter, before any related calculus is introduced.

The irrational number $e = 2.71828182846...$ plays a singularly important role in most applications involving exponential and/or logarithmic functions. With this understanding, we emphasize functions of the form $y = e^x$ and $y = \ln x$, where e is the base. The student is expected to become skilled with the use of the corresponding keys on a graphing calculator.

In Sections 5.3 and 5.4 we develop methods for differentiating these functions and analyze some applications. In Section 5.5 we discuss further applications as interesting and diverse as compound interest, population growth, radioactive decay and half-life, and the growth of bacteria. We also discuss a topic important in economics: elasticity of demand.

5.1 Exponential Functions

Objectives

After completing this section, you will be able to:

1. *Define exponential functions.*
2. *Discuss characteristics of exponential functions and their graphs.*

Suppose you buy a new car for $20,000 and you know, from past records, that this car will depreciate by 16 percent of its value each year. That is, each year the car will maintain 84 percent of its value of the previous year. What will be its value in 5 years? In n years?

You might analyze the problem as follows:

$$\text{End of Year 1:}\quad \text{Value} = 20{,}000(0.84)$$

$$\text{End of Year 2:}\quad \text{Value} = \big(20{,}000(0.84)\big)(0.84) = 20{,}000(0.84)^2$$

$$\text{End of Year 3:}\quad \text{Value} = \big(20{,}000(0.84)^2\big)(0.84) = 20{,}000(0.84)^3$$

$$\text{End of Year 4:}\quad \text{Value} = \big(20{,}000(0.84)^3\big)(0.84) = 20{,}000(0.84)^4$$

$$\text{End of Year 5:}\quad \text{Value} = \big(20{,}000(0.84)^4\big)(0.84) = 20{,}000(0.84)^5.$$

Using a calculator, we can determine that the car's value at the end of 5 years will be

$$20,000(0.84)^5 = 20,000(0.418212) = \$8364.24.$$

This pattern shows that at the end of n years we expect the value of the car to be

$$\text{Value} = 20,000(0.84)^n.$$

This is an example of an exponential function.

Exponential Functions

We are already familiar with polynomial functions such as $f(x) = x$, $f(x) = x^2$, and $f(x) = x^3$. Note that in each of these examples the base is a variable x and the exponent is constant.

Now, reversing the roles of base and exponent, we will discuss **exponential functions** of the form

$$f(x) = b^x,$$

where the base is a constant and the exponent is a variable.

Consider the problem of calculating values for $f(x) = 2^x$, where x is any real number. We have already defined values for 2^x if x is an integer or other rational number in terms of powers and roots. For example,

$$f(0) = 2^0 = 1, \quad f(1) = 2^1 = 2, \quad f(5) = 2^5 = 32,$$

$$f(-1) = 2^{-1} = \frac{1}{2}, \quad f(-3) = 2^{-3} = \frac{1}{8}, \quad \text{and} \quad f\left(\frac{1}{3}\right) = 2^{\frac{1}{3}} = \sqrt[3]{2}.$$

If x is an irrational number, such as $\sqrt{2}$, $\sqrt{3}$, or π, we approximate x with rational numbers and use a limiting process to define 2^x. For example, approximate values of $\sqrt{2}$ are 1.4, 1.41, 1.414, and so on.

Thus

$$2^{1.4} = 2^{\frac{14}{10}} \approx 2.6390$$

$$2^{1.41} = 2^{\frac{141}{100}} \approx 2.6574$$

$$2^{1.414} = 2^{\frac{1414}{1000}} \approx 2.6647$$

$$\vdots$$

and so on.

These values get closer and closer to $2^{\sqrt{2}} \approx 2.6651$. In this way we can define values for $f(x) = 2^x$ for all real numbers x. This means that the graph of the function $f(x) = 2^x$ contains such points as $\left(\sqrt{2}, 2^{\sqrt{2}}\right)$, $\left(\pi, 2^{\pi}\right)$, and $\left(\sqrt{5}, 2^{\sqrt{5}}\right)$. The graphs of the functions $f(x) = 2^x$ and $f(x) = 3^x$ are shown in Figure 5.1.1. One can observe that the appearance is similar and we expect, for both functions, that the slope is always positive (as y is increasing) and that y'' is always positive also (as the curve is concave up).

Figure 5.1.1

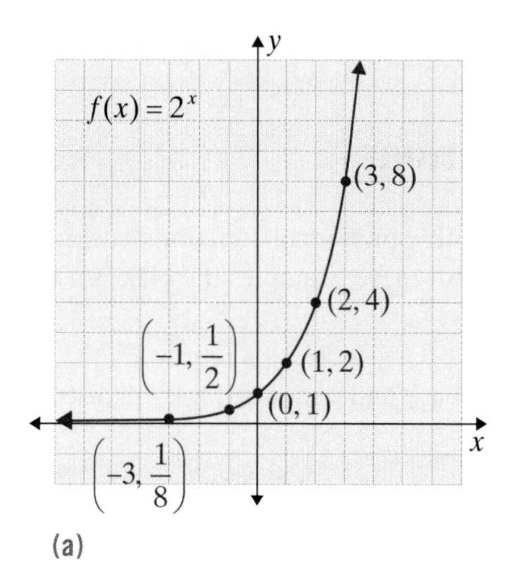

(a)

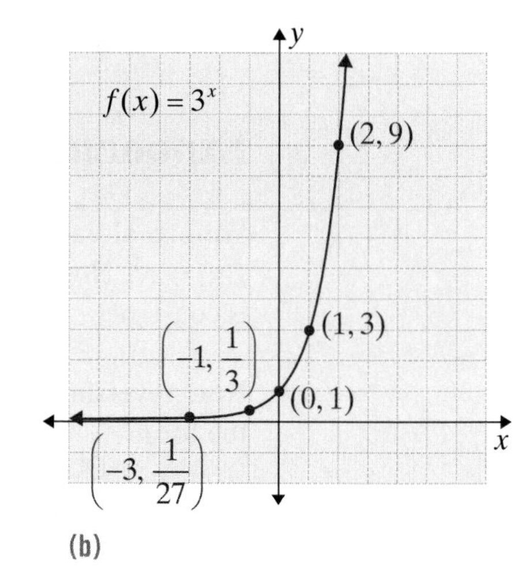

(b)

A similar analysis is valid for any positive base $b > 1$. We omit the case $b = 1$ because $1^x = 1$ for all real x. If $0 < b < 1$, the corresponding graph will have y decreasing and will be concave up (see Figure 5.1.2). If $b < 0$, the function b^x is not defined for many x-values and is neither continuous nor differentiable.

Exponential Function with Base *b*

A function of the form

$$f(x) = c \cdot b^{rx},$$

where b is a positive real number, b ≠ 1, and c and r are real constants, is called an exponential function with base b.

If $0 < \frac{1}{b} < 1$, then the graph of the exponential function $y = \left(\frac{1}{b}\right)^x = \left((b)^{-1}\right)^x = b^{-x}$ is a reflection across the y-axis of the corresponding graph $y = b^x$, where $b > 1$. Figure 5.1.2 illustrates the two cases corresponding to the graphs in Figure 5.1.1.

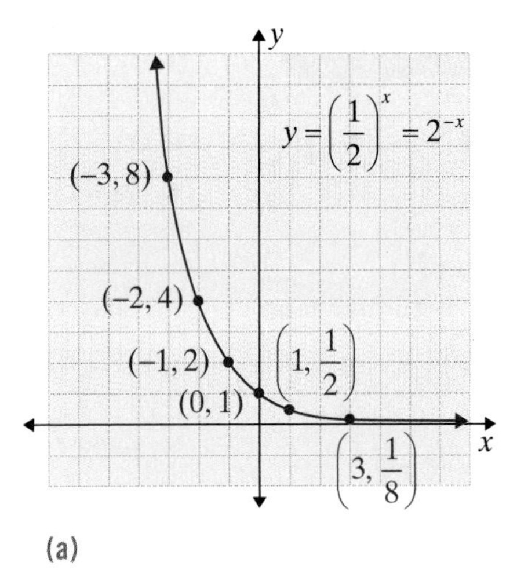

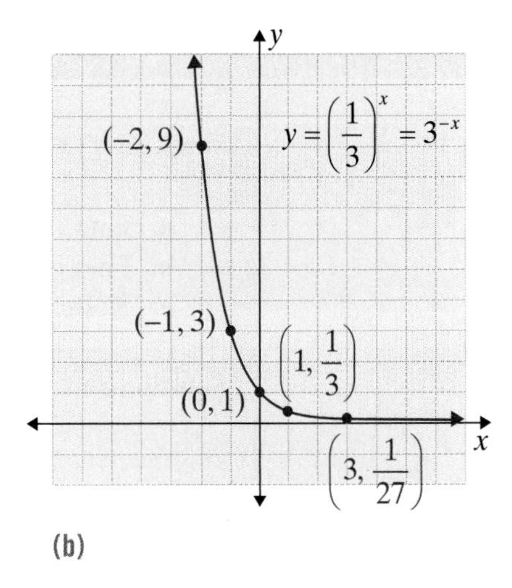

Figure 5.1.2

(a) (b)

The following properties are true for all exponential functions. We state them without proof.

Properties of Exponential Functions

For any exponential function $f(x) = c \cdot b^{rx}$, the following properties are true.

1. *The domain is the set of all real numbers.*
2. *The range is the set of all positive real numbers when $c > 0$.*
3. *The y-intercept is the point $(0, c)$.*
4. *The x-axis is a horizontal asymptote.*
5. *The function is continuous for all real x.*
6. *Suppose the base b is greater than 1. Then, when $c = 1$ and $r = 1$, the slope is positive, and y is increasing and concave up.*
7. *Suppose the base b satisfies $0 < b < 1$. Then, when $c = 1$ and $r = 1$, the slope is negative, and y is decreasing and concave up.*

We can locate the graphs of exponential functions by plotting a few points and applying the properties just listed. As we will see later, the graphs can be analyzed in detail with the aid of calculus.

Example 1: Graphing an Exponential Function

Suppose $f(x)=2^{3x}$ and $g(x)=\frac{1}{3}(2^{3x})$.

a. Find $f(-1)$, $f(0)$, and $f(1)$.
b. Find $g(-1)$, $g(0)$, and $g(1)$.
c. Sketch the graphs of both functions.

Solutions: a. $f(-1)=2^{3(-1)}=2^{-3}=\frac{1}{8}$ — Substitute $x=-1$ into $f(x)$.

$f(0)=2^{3(0)}=2^{0}=1$ — Substitute $x=0$ into $f(x)$.

$f(1)=2^{3(1)}=2^{3}=8$ — Substitute $x=1$ into $f(x)$.

b. $g(-1)=\frac{1}{3}\left(2^{3(-1)}\right)=\frac{1}{3}\left(\frac{1}{8}\right)=\frac{1}{24}$ — Substitute $x=-1$ into $g(x)$.

$g(0)=\frac{1}{3}\left(2^{3(0)}\right)=\frac{1}{3}(1)=\frac{1}{3}$ — Substitute $x=0$ into $g(x)$.

$g(1)=\frac{1}{3}\left(2^{3(1)}\right)=\frac{1}{3}(8)=\frac{8}{3}$ — Substitute $x=1$ into $g(x)$.

c.

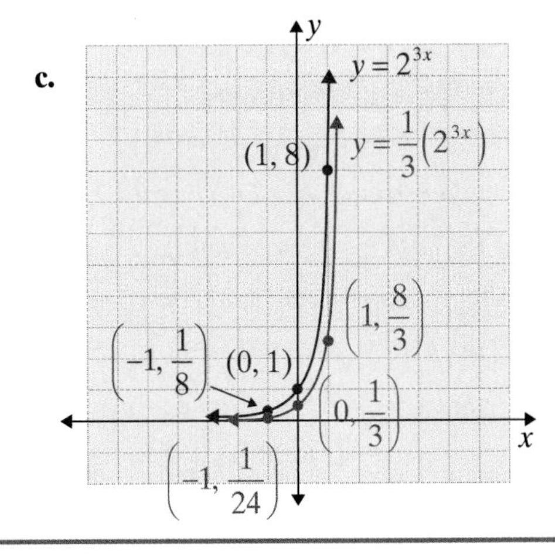

Compound Interest

When the interest earned on an account is added to the account and new interest is paid on the interest earned, we say that interest is **compounded**. Interest is usually compounded at regular intervals of time (called **periods**), most often annually, quarterly, monthly, or daily. No matter what the length of the compounding period is, the general process for calculating compound interest is the same. At the end of each compounding period, the principal and the interest on that principal are added together to form new principal for the next compounding period.

For example, suppose that a principal of $10,000 is invested at 8 percent compounded quarterly. Since the interest quoted (8 percent) is an annual rate, we divide by 4 to find the interest per quarter: $\frac{0.08}{4}=0.02$. The amount A at the end of 1 year can be calculated as follows:

$$\begin{aligned}
\text{1st Quarter:}\quad & 10{,}000+0.02(10{,}000) = 10{,}000(1+0.02)\\
& = 10{,}000(1.02) = \$10{,}200.\\
\text{2nd Quarter:}\quad & 10{,}200+0.02(10{,}200) = 10{,}200(1+0.02)\\
& = 10{,}200(1.02) = \$10{,}404.\\
\text{3rd Quarter:}\quad & 10{,}404+0.02(10{,}404) = 10{,}404(1+0.02)\\
& = 10{,}404(1.02) = \$10{,}612.08.\\
\text{4th Quarter:}\quad & 10{,}612.08+0.02(10{,}612.08) = 10{,}612.08(1+0.02)\\
& = 10{,}612.08(1.02) = \$10{,}824.32.
\end{aligned}$$

Thus A = $10,824.32 if $10,000 is compounded quarterly for 1 year. This same value can be found as follows:

$$A = 10{,}000(1.02)^4 = 10{,}000(1.082432) = \$10{,}824.32.$$

This approach seems quite reasonable once we realize that in the previous list of calculations the amount in each quarter was multiplied by 1.02 to get the amount for the next quarter.

Number of Corresponding Periods in 1 Year	Amount After 1 Year	Amount After *t* Years
1	$A=P(1+r)$	$A=P(1+r)^t$
2	$A=P\left(1+\frac{r}{2}\right)^2$	$A=P\left(1+\frac{r}{2}\right)^{2t}$
3	$A=P\left(1+\frac{r}{3}\right)^3$	$A=P\left(1+\frac{r}{3}\right)^{3t}$
⋮	⋮	⋮
n	$A=P\left(1+\frac{r}{n}\right)^n$	$A=P\left(1+\frac{r}{n}\right)^{nt}$

Table 5.1.1

A general formula for calculating compound interest is shown in Table 5.1.1 with the use of the following notation:

P = principal (amount invested),
r = rate (annual rate of interest in decimal form),
t = time (time in years),
n = number of compounding periods in 1 year, and
A = amount (called the future value of P).

The last equation in the right-hand column of Table 5.1.1 is the general **compound interest formula**.

Compound Interest Formula

For principal P invested at annual rate r (in decimal form) compounded n times per year for t years, the amount A is given by

$$\boldsymbol{A = P\left(1+\frac{r}{n}\right)^{nt}}.$$

Example 2: Compound Interest

Suppose that \$1000 is invested at 12 percent compounded quarterly. What will be the value of the investment in **a.** 1 year? **b.** 3 years?

Solutions: a. Use $P = 1000$, $r = 0.12$, $n = 4$, and $t = 1$.

$$\begin{aligned} A &= 1000\left(1+\frac{0.12}{4}\right)^{4(1)} \\ &= 1000(1.03)^4 \\ &\approx 1000(1.12551) \\ &= \$1125.51 \end{aligned}$$

The amount is \$1125.51 at the end of 1 year.

b. Use $P = 1000$, $r = 0.12$, $n = 4$, and $t = 3$.

$$\begin{aligned} A &= 1000\left(1+\frac{0.12}{4}\right)^{4\cdot 3} \\ &= 1000(1.03)^{12} \\ &\approx 1000(1.42576) \\ &= \$1425.76 \end{aligned}$$

The amount is \$1425.76 at the end of 3 years.

The Number *e*

Now we will define and use a particular irrational number that has special significance because it appears in applications concerning compound interest as well as applications in the life, social, and physical sciences. This number is denoted by the letter e, and

$$e \approx 2.71828182846 \quad \text{(to 12 significant digits).}$$

We define e as a limit and then illustrate how e relates to compound interest.

Definition of e

$$e = \lim_{x \to +\infty} \left(1 + \frac{1}{x}\right)^x$$

To see that this limit does in fact equal 2.71828182846... study the following table of values and use a calculator to generate values for larger x.

x	$\left(1+\frac{1}{x}\right)^x$
1	$\left(1+\frac{1}{1}\right)^1 = (2)^1 = 2$
10	$\left(1+\frac{1}{10}\right)^{10} = (1.1)^{10} \approx 2.59374246$
100	$\left(1+\frac{1}{100}\right)^{100} = (1.01)^{100} \approx 2.704813829$
200	$\left(1+\frac{1}{200}\right)^{200} = (1.005)^{200} \approx 2.711517123$
500	$\left(1+\frac{1}{500}\right)^{500} = (1.002)^{500} \approx 2.715568521$
1000	$\left(1+\frac{1}{1000}\right)^{1000} = (1.001)^{1000} \approx 2.716923932$
↓	↓
$+\infty$	e

The Function $f(x) = e^x$

As we will see in the remainder of the text, one of the most interesting and useful exponential functions is

$$f(x) = e^x \text{ where the base is } e \approx 2.71828182846.$$

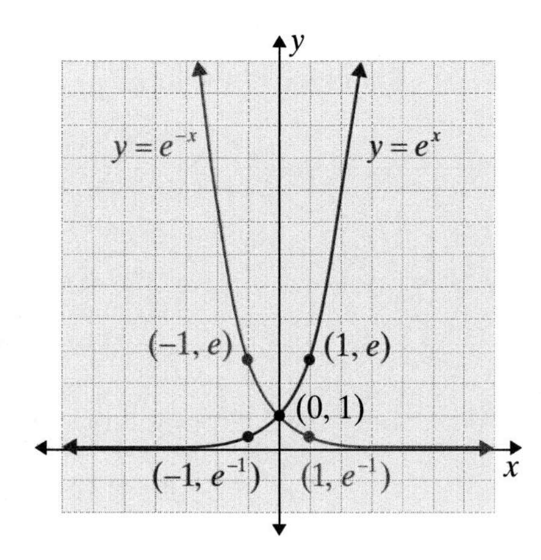

Figure 5.1.3

This function has all the properties of exponential functions listed earlier. Its domain is the set of all real numbers, and its functional values are always positive. The graphs of

$$y = e^x \text{ and } y = e^{-x}$$

are both shown in Figure 5.1.3.

Notice that both graphs stay above the x-axis and never touch the x-axis.

Now we will show that by taking smaller and smaller periods of time for compounding, the number e is related to the concept of **compounding interest continuously**.

Recall the formula for compound interest.

$$A = P\left(1 + \frac{r}{n}\right)^{nt}$$

Let $x = \frac{n}{r}$. Then, as $n \to +\infty$, we also have $x \to +\infty$. Compounding continuously corresponds to $n \to +\infty$, and

$$A = \lim_{n \to +\infty} P\left(1 + \frac{r}{n}\right)^{nt} = \lim_{n \to +\infty} P\left(1 + \frac{1}{\frac{n}{r}}\right)^{\frac{n}{r} \cdot rt}, \quad \left(x = \frac{n}{r}\right)$$

$$= \lim_{x \to +\infty} P\left[\left(1 + \frac{1}{x}\right)^{x}\right]^{rt}$$

$$= Pe^{rt}.$$

Continuous Compound Interest Formula

For principal P invested at annual rate r (in decimal form) compounded continuously for t years, the amount A is given by

$$\mathbf{A = Pe^{rt}}.$$

Example 3: Investment Value

If \$5000 is invested at 8 percent, what will be the value of the investment at the end of 5 years if the interest is

a. compounded quarterly?

b. compounded continuously?

Solutions: a. $A = 5000\left(1 + \frac{0.08}{4}\right)^{4 \cdot 5}$

$= 5000(1.02)^{20}$

$\approx 5000(1.485947396)$

$\approx \$7429.74$

Use the compound interest formula. $P = 5000$, $r = 0.08$, $n = 4$ (for quarterly compounding), and $t = 5$.

b. $A = 5000e^{0.08 \cdot 5}$

$= 5000e^{0.4}$

$\approx 5000(1.491824698)$

$\approx \$7459.12$

Use the continuous compound interest formula. $P = 5000$, $r = 0.08$, and $t = 5$.

A calculator can be used to find the value of $e^{0.4}$. (See sidebar at the left.)

You may find it interesting to note that the difference in interest is only about \$30, even after 5 years.

Finding the Value of *e* to a Power With a Calculator

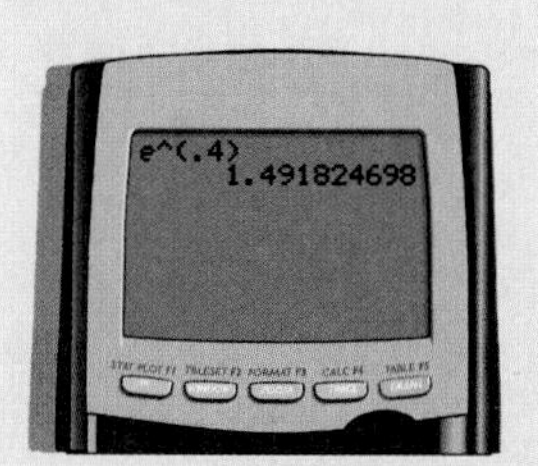

1. Press  .
2. Enter given exponent (in Example 3b, this value would be 0.4).
3. Close the parentheses by pressing).
4. Hit ENTER.

So far, we have used the formula $A = Pe^{rt}$ to calculate the future value A of an account that will result from an investment P in the present. Now suppose we know the future value to be accumulated and would like to determine how much should be deposited today to guarantee that future amount. We solve the formula $A = Pe^{rt}$ for P to obtain

$$P = Ae^{-rt}.$$

We call P the present value of A.

Example 4: Initial Deposit Amount

Suppose that the grandparents of a child would like to make a one-time investment now so that they can help provide for the child's college education in 20 years. How much should they deposit in an account earning 10 percent interest compounded continuously so that the value of the account will accumulate to \$20,000?

Solution: $P = Ae^{-rt}$

$= 20{,}000e^{-0.10 \cdot 20}$

$\approx 20{,}000(0.135335283)$

$\approx \$2706.71$

We use $A = 20{,}000$, $r = 0.10$, and $t = 20$.

continued on next page ...

That is, \$2706.71 (invested at 10 percent compounded continuously for 20 years) is the present value of \$20,000.

5.1 Exercises

Use a calculator to evaluate the expressions in Exercises 1 – 8. Round off your answers to the nearest hundredth.

1. $10^{2.3}$ **2.** $2^{1.5}$ **3.** $e^{-0.5}$ **4.** $e^{2.1}$

5. e^{3} **6.** e^{-1} **7.** $2e^{-1.6}$ **8.** $3e^{2.5}$

9. For $f(x)=9^{\frac{x}{2}}$:
- **a.** Find $f(0)$.
- **b.** Find $f(1)$.
- **c.** Find $f(3)$.

10. For $y=-2e^{x^2+1}$, find the value of y (accurate to 2 decimal places) if
- **a.** $x = 1$.
- **b.** $x = 0$.
- **c.** $x = -1$.

Graph each of the functions in Exercises 11 – 26.

11. $y=4^x$ **12.** $y=5^x$ **13.** $y=\left(\frac{2}{3}\right)^x$ **14.** $y=\left(\frac{1}{4}\right)^x$

15. $y=3^{-2x}$ **16.** $y=2^{-3x}$ **17.** $y=2^{x-1}$ **18.** $y=2^{x+1}$

19. $y=3e^x$ **20.** $y=2e^{-x}$ **21.** $y=e^{-1.5x}$ **22.** $y=e^{2x}$

23. $y=6e^{-0.5x}$ **24.** $y=10e^{0.5x}$ **25.** $y=e^{-x^2}$ **26.** $y=5e^{-x^2}$

27. Compound interest. Suppose that \$5000 is invested at an annual rate of 6 percent for 4 years. Find the total amount available if the interest is compounded:
- **a.** annually.
- **b.** semiannually.
- **c.** quarterly.
- **d.** daily. (using 360 days/year)
- **e.** continuously.

28. Compound interest. Suppose that \$3000 is deposited in a saving account at the rate of 8 percent per year. Find the total amount available at the end of 5 years if the interest is compounded:
- **a.** annually.
- **b.** semiannually.
- **c.** quarterly.
- **d.** daily. (using 360 days/year)
- **e.** continuously.

29. Compound interest. How much money must be deposited in a savings account that pays interest at a rate of 7 percent per year compounded quarterly so that after 5 years the balance will be $12,000?

30. Compound interest. Matt's father is planning to open a savings account to pay for Matt's college education. He has found a bank that will pay 8 percent interest compounded monthly. How much will he need to deposit initially so that in 4 years the balance will be $20,000?

31. Compound interest. Maggie has $8000 to invest for 2 years. She has two different accounts she can use. One account will pay 8 percent interest compounded monthly. The other account will pay 7.5 percent compounded continuously. Which is the better investment, and by how much?

32. Reliability. The reliability of a certain type of flashlight battery is given by $f(t) = e^{-0.05t}$, where f is the fractional part of the batteries that lasts t hours. What fraction of the batteries produced is good after 20 hours of use?

33. Drug concentration. The concentration of a certain drug in the body fluids is given by $C = C_0 e^{-0.4t}$, where C_0 is the initial dose and t is the time in hours elapsed after the dose is administered. If 20 mg of the drug is given, how much of the drug will remain in the body after 8 hours?

34. Bee population. A colony of bees grows according to the formula $P = P_0 e^{0.18t}$, where P_0 is the number present initially and t is the time in days. How many bees will be present after 6 days if there were 1200 present initially?

35. Advertising. A radio station knows that during an intense advertising campaign, the total number of people N who hear a commercial is given by $N = A(1 - e^{-0.03t})$, where A is the number of people in the broadcasting area and t is the number of hours the commercial is run. If there are 500,000 people in the area, how many will hear the commercial during the first 20 hours?

36. Reliability. Studies show that the fractional part P of light bulbs that are still burning after t hours of use is given by $P = 1 - e^{-0.03t}$. What fractional part of light bulbs are burning after 50 hours of use?

37. Depreciation. The value V of a machine at the end of t years is given by $V = C(1 - r)^t$, where C is the original cost of the machine and r is the rate of depreciation. Find the value at the end of 6 years of a machine that costs $8400 if $r = 0.15$.

38. Depreciation. A machine that is 5 years old is valued at $4000. If the rate of depreciation is 12 percent, find the original cost of the machine. (See Exercise 37.)

39. Present value. Mr. and Mrs. Jackson want to accumulate $50,000 for their retirement in 15 years with a one-time deposit now. If the interest is to be figured at 8 percent compounded continuously, what should the Jacksons' initial deposit be?

40. Present value. If a trust fund is designed to have a value of $600,000 in 25 years and interest is compounded continuously at 9 percent, what is the present value of this amount?

Hawkes Learning Systems: Essential Calculus

Exponential Functions

5.2 The Algebra of the Natural Logarithm Function

Objectives

After completing this section, you will be able to:

1. *Define natural logarithms.*
2. *Use the properties of natural logarithms to solve equations.*

The Natural Logarithm Function

Logarithms are exponents. If e raised to a power is equal to a number x, then the exponent on e is called the natural logarithm of x (denoted as $\ln x$).

Natural Logarithm

For natural logarithms,

$$e^y = x \quad \text{if and only if} \quad y = \ln x.$$

Since e^y is always positive, we see that the domain of the natural log function is the set of positive numbers (see Figure 5.2.1)

Since the two equations $y = \ln x$ and $e^y = x$ are equivalent, we can write an exponential equation in an equivalent logarithmic form, and vice versa. For example,

Exponential Equation	Equivalent Logarithmic Equation
$e^0 = 1$	$\ln 1 = 0$
$e^1 = e$	$\ln e = 1$
$e^2 \approx 7.389056099$	$\ln 7.389056099 \approx 2$
$e^{-1} \approx 0.367879441$	$\ln 0.367879441 \approx -1$
$e^{4.1} \approx 60.3402876$	$\ln 60.3402876 \approx 4.1$

For example, ln 2 is the exponent for e which gives 2. That is $\ln 2 \approx 0.6931$, approximately, since $e^{0.6931} \approx 2$, or $e^{\ln 2} = 2$.

Creating a Table With a Calculator

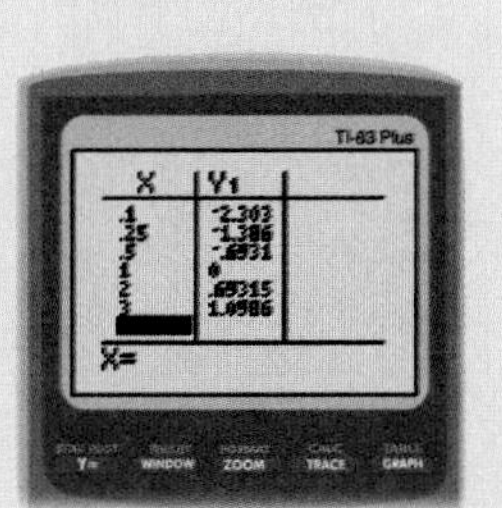

To create a table of x- and y-values for $\ln x$ as shown on the next page.

1. Enter the function $\ln x$ into Y1.
2. Press 2ND WINDOW to access Table Setup.

continued ...

continued ...

3. For this table we will input specific values of x, so `TblStart`, `ΔTbl`, and `Depend` can be ignored.
4. For `Indpnt`, move the blinking cursor over `Ask`. Press ENTER.
5. Press 2ND GRAPH to access the table.
6. With the cursor in the `X` column, enter the first x-value you would like to find y for (use 0.1). Press ENTER.

   `-2.303` should appear in the `Y` column.
7. Continue entering the remaining x-values.

The function $y = \ln x$ is defined for all positive real numbers x, and its graph is a smooth continuous curve. One way to locate the graph is to plot a few points and then sketch the curve through these points, as shown in Figure 5.2.1. The values in the table can be found by using the calculator. The calculus of the function will be explored in the next section. For now, we note that $y = \ln x$ is concave down and has positive slope (and so is increasing).

x	$\ln x = y$
0.1	$\ln(0.1) \approx -2.30$
0.25	$\ln(0.25) \approx -1.39$
0.5	$\ln(0.5) \approx -0.69$
1.0	$\ln(1.0) \approx 0$
2.0	$\ln(2.0) \approx 0.69$
3.0	$\ln(3.0) \approx 1.10$

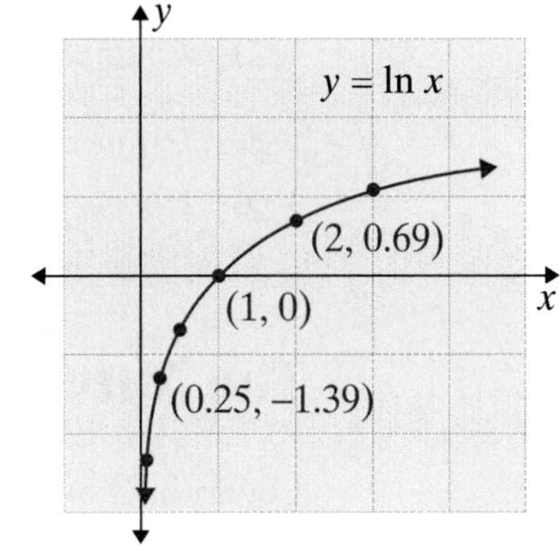

Figure 5.2.1

Another approach to understand the logarithm function is to analyze the relationship between the two functions

$$f(x) = e^x \text{ and } g(x) = \ln x.$$

Forming the composition of these two functions, we get the following important statements:

1. $f(g(x)) = e^{\ln x} = x$ for $x > 0$.
2. $g(f(x)) = \ln e^x = x$ for all real x.

These two relationships between exponential and logarithmic functions lead to equations such as the following.

$e^{\ln x} = x$	$\ln e^x = x$
$e^{\ln 2} = 2$	$\ln e^{-1} = -1$
$e^{\ln 3} = 3$	$\ln e^2 = 2$
$e^{\ln 0.5} = 0.5$	$\ln \sqrt{e} = \ln e^{\frac{1}{2}} = \frac{1}{2}$

Figure 5.2.2

The preceding discussion indicates that the two functions $y = e^x$ and $y = \ln x$ are **inverses** of each other. The graphs of the two functions are reflections (mirror images) of each other across the line $y = x$. For example, if $b = e^a$, then the point (a, b) is on the graph of the function $y = e^x$. Now, from the definition of $\ln x$, we know that $a = \ln b$. Therefore, the point (b, a) is on the graph of the function $y = \ln x$. These relationships are illustrated in Figure 5.2.2.

Properties of Natural Logarithms

Consider the fact that when we multiply two exponential expressions with the same base, such as $e^2 \cdot e^3$, we add the exponents:

$$e^2 \cdot e^3 = e^{2+3} = e^5.$$

More generally, if

$$M = e^x \quad \text{and} \quad N = e^y,$$

then

$$x = \ln M \quad \text{and} \quad y = \ln N.$$

However,

$$MN = e^x \cdot e^y = e^{x+y},$$

and we have

$$\ln MN = \ln\left(e^{x+y}\right) = x + y = \ln M + \ln N.$$

This leads us to a basic property of logarithms:

$$\ln MN = \ln M + \ln N.$$

This property and others that we will use throughout the remainder of the text are summarized in the following box.

Properties of Natural Logarithms

The domain of $f(x) = \ln x$ is the set of positive numbers. For positive real numbers M and N:

1. $\ln (MN) = \ln M + \ln N.$
2. $\ln\left(\dfrac{M}{N}\right) = \ln M - \ln N.$
3. $\ln (M^r) = r \cdot \ln M$
4. $\ln M = \ln N$ *if and only if* $M = N.$
5. $\ln 1 = 0$
6. $\ln e = 1$
7. $e^{\ln x} = x$
8. $\ln e^x = x$

NOTES

Note: There are no rules to simplify $\ln(M+N)$ and $\dfrac{\ln M}{\ln N}$. For instance, $\ln(2+3) = \ln 5 \neq \ln 2 + \ln 3 = \ln 6$. Also, $\dfrac{\ln 10}{\ln 5} \approx 1.43$, but $\ln\left(\dfrac{10}{5}\right) = \ln 2 \approx 0.693.$

Suppose B is a positive number and $B > 1$. The function $f(x) = B^x$ has an inverse function since f is strictly increasing. We denote this inverse by $f^{-1}(x) = \log_B x$. The inverse is the logarithm function with base B. It follows that $B^{\log_B x} = x$ and $\log_B B^x = x$.

Historically, the "natural logarithms" (base e) and "common logarithms" (base 10) have proved sufficient for computations and most theoretical purposes. For this reason tables for base e and base 10 are built into all graphing calculators. The calculus of exponential and logarithmic functions of a general base B is developed in Section 5.5.

Example 1: Properties of Natural Logarithms

Use the properties of logarithms to rewrite each of the following expressions as the logarithm of a single expression.

a. $2 \ln x + \ln y$
b. $\ln (x + 1) + \ln (x - 1)$
c. $\ln (x - 4) - \ln 2$
d. $\ln x + \frac{1}{2}\ln(x+5) - \frac{1}{2}\ln(x-3)$

Solutions: a. $2\ln x + \ln y = \ln x^2 + \ln y$ By Property 3.

$= \ln x^2 y$ By Property 1.

b. $\ln(x+1) + \ln(x-1) = \ln(x+1)(x-1)$ By Property 1.

$= \ln(x^2 - 1)$

c. $\ln(x-4) - \ln 2 = \ln\frac{x-4}{2}$ By Property 2.

d. $\ln x + \frac{1}{2}\ln(x+5) - \frac{1}{2}\ln(x-3)$

$= \ln x + \ln(x+5)^{\frac{1}{2}} - \ln(x-3)^{\frac{1}{2}}$ By Property 3.

$= \ln x(x+5)^{\frac{1}{2}} - \ln(x-3)^{\frac{1}{2}}$ By Property 1.

$= \ln\left(\frac{x(x+5)^{\frac{1}{2}}}{(x-3)^{\frac{1}{2}}}\right)$ By Property 2.

$= \ln\left(\frac{x\sqrt{x+5}}{\sqrt{x-3}}\right)$

Example 2: Properties of Natural Logarithms

Use the properties of logarithms to rewrite each of the following expressions as a sum, difference, or constant multiple of logarithms.

a. $\ln\left(\frac{5x}{y}\right)$ **b.** $\ln\sqrt{2x+3}$ **c.** $\ln(x^2+6x+5)$

Solutions: a. $\ln\left(\frac{5x}{y}\right) = \ln 5x - \ln y$ By Property 2.

$= \ln 5 + \ln x - \ln y$ By Property 1.

b. $\ln\sqrt{2x+3} = \ln(2x+3)^{\frac{1}{2}}$

$= \frac{1}{2}\ln(2x+3)$ By Property 3.

c. $\ln(x^2+6x+5) = \ln(x+1)(x+5)$

$= \ln(x+1) + \ln(x+5)$ By Property 1.

Example 3: Solving Equations

Use the properties of logarithms to solve each of the following equations.

a. $e^{x+1} = 30$
b. $5^{2x-1} = 60$
c. $\ln(x-3) + \ln(x+3) = 0$
d. $\ln 2 + \ln x = \ln 30$

Solutions: a. $e^{x+1} = 30$

$x + 1 = \ln 30$ By the definition of $\ln x$.

$x = -1 + \ln 30$

In decimal form, we can write

$x = -1 + \ln 30$

$\approx -1 + 3.4012 = 2.4012.$

continued on next page ...

b.

$$5^{2x-1} = 60$$

$$\ln 5^{2x-1} = \ln 60 \qquad \text{By Property 4.}$$

$$(2x-1)\ln 5 = \ln 60 \qquad \text{By Property 3.}$$

$$2x-1 = \frac{\ln 60}{\ln 5}$$

$$2x = 1 + \frac{\ln 60}{\ln 5}$$

$$x = \frac{1}{2} + \frac{\ln 60}{2\ln 5}$$

In decimal form, we can write

$$x \approx 0.5 + \frac{4.0943}{3.2189} \approx 1.7720.$$

c.

$$\ln(x-3) + \ln(x+3) = 0$$

$$\ln(x-3)(x+3) = 0 \qquad \text{By Property 1.}$$

$$\ln(x^2-9) = 0$$

$$x^2 - 9 = e^0$$

$$x^2 - 9 = 1 \qquad \text{By Property 5.}$$

$$x^2 - 10 = 0$$

$$(x+\sqrt{10})(x-\sqrt{10}) = 0$$

$$\cancel{x = -\sqrt{10}} \text{ or } x = \sqrt{10}$$

Since $\ln(-\sqrt{10}-3)$ is not defined, $x = -\sqrt{10}$ is not a solution, and the only solution is $x = \sqrt{10}$.

d.

$$\ln(2) + \ln x = \ln 30$$

$$\ln(2x) = \ln 30 \qquad \text{By Property 1.}$$

$$2x = 30 \qquad \text{By Property 4.}$$

$$x = 15$$

Note: In order to avoid extra parentheses, it is customary to write ln x instead of ln(x). Similarly, multiplication and exponentiations may occur in the argument of a function without an extra pair of parentheses. For example:

$$\ln x = \ln(x) \qquad \ln x(x+3) = \ln(x(x+3)) \qquad \ln x^2 = \ln(x^2)$$

$$\ln yz = \ln(yz) \qquad \ln xy^3 = \ln(xy^3) \qquad \ln 5xy = \ln(5xy).$$

Example 5: Rate of Inflation

Suppose the price of a box of popcorn at a movie theatre costs \$0.35 in 1960 and costs \$2.60 in 2005. Assuming the cost of a box of popcorn increased continuously at a constant rate r, determine this rate.

Solution: We are assuming P, the price of a box of popcorn, is an exponential function of t, the number of years since 1960. This means $t = 0$ in 1960 and $t = 45$ in 2005. We write $P = P(t)$ to mean P is a function of t, and in this case we are assuming $P(t) = P_0e^{rt}$, where P_0 and r are two constants which need to be determined.

At $t = 0$, $P(0) = .35$. From the formula for $P(0) = P_0e^{r \cdot 0} = P_0e^0 = P_0(1) = P_0$. Thus we have found that $P_0 = .35$. We conclude that $P(t) = 0.35e^{rt}$. The other data value is used to determine r.

$$P(45) = 0.35e^{45r} = 2.60$$

$$e^{45r} = \frac{2.60}{0.35} \approx 7.428571429$$

$$\ln(e^{45r}) \approx \ln(7.428571429)$$

$$45r \approx 2.00533357$$

$$r \approx .04456 \text{ or } r = 4.46\%$$

Example 6: Continuous Compounding Interest

The formula $A = Pe^{rt}$ represents the amount accumulated when a principal is compounded continuously (as discussed in Section 5.1). How long will it take a principal P to double if interest is compounded continuously at 10 percent?

Solution: If P is doubled, then the amount A is $2P$. Thus

$$Pe^{0.10t} = 2P$$

$$e^{0.10t} = 2$$

$$0.10t = \ln 2$$

$$t = \frac{\ln 2}{0.10} \approx \frac{0.6931}{0.10} = 6.931$$

The principal will double in approximately 6.9 years.

5.2 Exercises

In Exercises 1 – 6, write each exponential equation approximation as a logarithmic equation.

1. $e^{0.5} \approx 1.648721$ **2.** $e^{1.6} \approx 4.953032$ **3.** $e^{-2.1} \approx 0.122456$

4. $\ln 1.822119 \approx 0.6$ **5.** $e^{0.25} \approx 1.284025$ **6.** $e^{-0.42} \approx 0.657047$

In Exercises 7 – 12, write each logarithmic equation as an exponential equation.

7. $\ln 1.822119 \approx 0.6$ **8.** $\ln 12.182494 \approx 2.5$ **9.** $\ln 23.6 \approx 3.161247$

10. $\ln 0.697676 \approx -0.36$ **11.** $\ln 0.069460 \approx -2.6670$ **12.** $\ln 3 \approx 1.098612$

In Exercises 13 – 22, use the inverse relationship between $f(x) = e^x$ and $g(x) = \ln x$ to simplify each expression.

13. $e^{\ln 4.6}$ **14.** $e^{\ln 0.7}$ **15.** $\ln e^3$ **16.** $\ln e^{-0.86}$

17. $\ln \sqrt[3]{e}$ **18.** $\ln \frac{1}{e^2}$ **19.** $e^{\ln 5x}$ **20.** $e^{\ln(x+4)}$

21. $\ln e^{(2t-1)}$ **22.** $\ln e^{(1-0.2t)}$

In Exercises 23 – 34, use the properties of logarithms to rewrite each expression as the logarithm of a single expression. Be sure to use positive exponents and avoid radicals.

23. $\ln x + 3\ln y$

24. $\frac{1}{2}(\ln x + \ln y)$

25. $\ln x - \ln y + \ln z$

26. $\ln x + \ln y - \frac{1}{2}\ln z$

27. $\ln x + \ln(x+3)$

28. $\ln(x+4) + \ln(x-1)$

29. $2\ln x - \ln(x+1)$

30. $\ln(x^2+2) - \frac{1}{2}\ln x$

31. $\ln(x^2-2x+1) - \ln(x-1)$

32. $3\ln x + \frac{1}{2}\ln(x+5)$

33. $\ln(x+3) + \ln(x+1) - \frac{1}{2}\ln x$

34. $\ln(x^2-3x-4) - \ln(x-4)$

In Exercises 35 – 46, use the properties of logarithms to rewrite each expression as a sum, difference, or constant multiple of logarithms. Replace all radicals with exponents.

35. $\ln x^3 y^2$

36. $\ln(xy)^{-2}$

37. $\ln\sqrt[3]{\frac{x}{y}}$

38. $\ln\sqrt[3]{xy^2}$

39. $\ln\left(\frac{15x}{\sqrt{y}}\right)$

40. $\ln\left(\frac{\sqrt{3x}}{y^4}\right)$

41. $\ln\sqrt{4x+1}$

42. $\ln\left(x\sqrt{2x+3}\right)$

43. $\ln\frac{e}{(x+2)^2}$

44. $\ln\left(\frac{x-5}{e^x}\right)^2$

45. $\ln(x^2-16)$

46. $\ln(x^2-3x-10)$

In Exercises 47 – 60, use the properties of logarithms to solve each equation.

47. $2e^{2x+1} = 26$

48. $e^{2x+3} = 3.5$

49. $e^{-0.6t} = 0.8$

50. $6e^{-0.4t} = 48$

51. $3\cdot 5^x = 54$

52. $4^{2x} = 0.016$

53. $10^{3x-1} = 8.6$

54. $10^{1-2x} = 72$

55. $\ln x + \ln 2x = \ln 18$

56. $\ln(x+4) + \ln(x-4) = \ln 9$

57. $\ln(x+1) + \ln(x-1) = 0$

58. $\ln(x+5) + \ln(x-5) = 0$

59. $\ln(x^2+4x-5) - \ln(x+5) = 1$

60. $\ln(x^2+2x-3) - \ln(x-1) = 2$

61. Compound interest. A deposit of \$5000 is made into an account that pays interest at an annual rate of 9.6 percent. Find the time required for the investment to double if the interest is compounded as indicated. Round to the nearest hundredth of a year.

a. annually
b. quarterly
c. monthly

62. Compound interest. A deposit of \$2000 is made into an account that pays 8.4 percent annual interest. Find the time required for the account to be worth \$3600 if the interest is compounded as indicated. Round to the nearest hundredth of a year.

- **a.** annually
- **b.** quarterly
- **c.** monthly

63. Interest compounded continuously. A deposit of \$5000 is made into an account that pays interest at a rate of 8 percent compounded continuously. Find the time required for the account to double. Round to the nearest hundredth of a year.

64. Interest compounded continuously. Anna deposited \$6000 into an account that pays 7.6 percent annual interest compounded continuously. How long will it be before the balance will be \$10,000? Round to the nearest hundredth of a year.

65. Depreciation. The value V of a machine t years after the machine is purchased is given by $V = C(1-r)^t$, where C is the original cost and r is the rate of depreciation. A machine is purchased for \$12,000. If $r = 0.16$, in how many years will the machine be worth \$6000? Round to the nearest hundredth of a year.

66. Depreciation. The owner of the furniture shop purchased a special saw for \$8600. She plans to depreciate the saw at a rate of 12 percent per year. In how many years will the value of the saw be \$2380? Round to the nearest hundredth of a year.

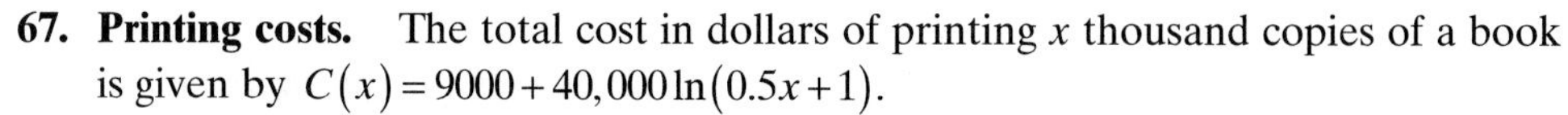

67. Printing costs. The total cost in dollars of printing x thousand copies of a book is given by $C(x) = 9000 + 40{,}000\ln(0.5x+1)$.

- **a.** Find $C(2)$, $C(8)$, and $C(10)$.
- **b.** Find the cost of printing 12,000 copies.
- **c.** If the total cost of printing is \$100,000, how many books are printed?

68. City planning. To plan for the future needs of the city, the Department of Public Works uses the function $P(t) = 36{,}000 + 8400\ln(0.8t+1)$ to estimate the population t years from now.

- **a.** What is the city's current population?
- **b.** Find $P(2)$, $P(6)$, and $P(10)$.
- **c.** In how many years will the population reaches 60,000? Round to the nearest hundredth of a year.

Hawkes Learning Systems: Essential Calculus

The Algebra of the Natural Logarithm Function

5.3 Differentiation of Logarithmic Functions

Objectives

After completing this section, you will be able to:

1. *Differentiate logarithmic functions.*
2. *Use curve sketching techniques to graph logarithmic functions.*

The Derivative of $y = \ln x$

From the graph of $y = \ln x$, we already know that $\frac{d}{dx}\ln x = (\ln x)'$ will be positive since the function increases with increasing x.

Also, we know that $(\ln x)''$ will be negative for all x since the graph is concave down.

It is always useful to consider numerical information which helps in understanding. Let us consider the tangent line to $f(x) = \ln x$ at $x = 9$. The slope of this line is $f'(9)$. In order to calculate this slope approximately, let point $A = (9, \ln 9)$ and point $B = (9.00001, \ln 9.00001)$. The slope of the segment AB is close to that for the tangent line. This slope is

$$\frac{\Delta y}{\Delta x} = \frac{\ln(9.00001) - \ln(9)}{.00001} = 0.11111105.$$

Evidently, the slope of the tangent line is quite close to $\frac{1}{9} = .11111\ldots$. The following table can be quickly constructed in the same way or by using a calculator. The student will be able to guess the formula for $\frac{d}{dx}(\ln x)$.

x	$f(x) = \ln x$	$f'(x)$
2	$\ln 2 \approx .6931$	$.499999 \approx \frac{1}{2}$
3	$\ln 3 \approx 1.098$	$.3333327 \approx \frac{1}{3}$
4	$\ln 4 \approx 1.386$	$.249999 \approx \frac{1}{4}$
9	$\ln 9 \approx 2.197$	$.111111 \approx \frac{1}{9}$

Creating a Table of Multiple Columns With a Calculator

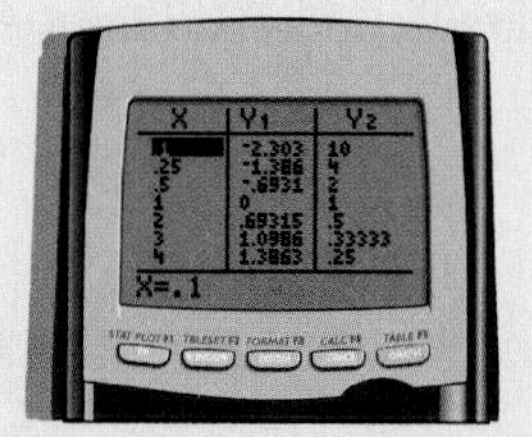

To create a table of x- and y-values for $\ln x$ as shown at right.

1. Enter the function $\ln x$ into Y1.
2. Enter the derivative of $\ln x$ into Y2 by first pressing MATH and selecting item `8:nDeriv(`. At the blinking cursor, type in `Y1,X,X)`.
3. To display a table, follow the steps presented in the calculator sidebar on pages 405 – 406. However, instead of 2 columns appearing, there will be 3.

We are ready to give a simple and clear derivation now for one of the most amazing formulas in calculus. It is not expected that the student would carry out the steps in the proof or would even anticipate any of them, but each is straightforward and all the preliminaries have been given. Before looking at the details, the student might review the definition of the number e, as it will be needed in the following proof.

To find the formula for the derivative of

$$f(x) = \ln x, \text{ for } x > 0,$$

we go directly to the definition of the derivative (see Section 2.3).

$$f'(x) = \lim_{h \to 0} \frac{f(x+h) - f(x)}{h}$$

Step 1: Form the difference quotient.

$$\frac{f(x+h) - f(x)}{h} = \frac{\ln(x+h) - \ln x}{h}$$

Step 2: Simplify the difference quotient. Here we use properties of logarithms.

$$\frac{\ln(x+h) - \ln x}{h} = \left(\frac{\ln\left(\frac{x+h}{x}\right)}{h} \right) \qquad \text{By Property 2.}$$

$$= \frac{1}{h} \cdot \ln\left(1 + \frac{h}{x}\right)$$

$$= \ln\left(1 + \frac{h}{x}\right)^{\frac{1}{h}} \qquad \text{By Property 3.}$$

$$= \frac{1}{x} \cdot x \cdot \ln\left(1 + \frac{h}{x}\right)^{\frac{1}{h}} \qquad \frac{1}{x} \cdot x = 1.$$

$$= \frac{1}{x} \ln\left(1 + \frac{h}{x}\right)^{\frac{x}{h}} \qquad \text{By Property 3.}$$

Step 3: Find the limit. Here we substitute $u = \frac{x}{h}$ and $\frac{1}{u} = \frac{h}{x}$. Then, since x is some fixed positive value, as $h \to 0^+$, we have $u \to +\infty$. (We omit the discussion of what happens as $h \to 0^-$. In effect, the result will be the same as when $h \to 0^+$.) Now,

$$\lim_{h \to 0} \left[\frac{1}{x} \cdot \ln\left(1 + \frac{h}{x}\right)^{\frac{x}{h}} \right] = \lim_{u \to \infty} \left[\frac{1}{x} \cdot \ln\left(1 + \frac{1}{u}\right)^{u} \right] \qquad \text{Substitute } u = \frac{x}{h}.$$

$$= \frac{1}{x} \cdot \lim_{u \to \infty} \left[\ln \left(1 + \frac{1}{u} \right)^u \right]$$ Since x is constant, u is the variable.

$$= \frac{1}{x} \ln \left(\lim_{u \to \infty} \left[1 + \frac{1}{u} \right]^u \right)$$ Since the expression is continuous.

$$= \frac{1}{x} \cdot \ln(e)$$ Using the definition of e on p. 399.

$$= \frac{1}{x} \cdot 1$$ By Property 6.

$$= \frac{1}{x}.$$

Thus the derivative of ln x has been proved.

Derivative of ln x

If $f(x) = \ln x$, *then*

$$f'(x) = \frac{1}{x}.$$

Example 1: Finding the Derivative

Find $\frac{dy}{dx}$ for $y = x^3 + \ln x$.

Solution: $\frac{dy}{dx} = 3x^2 + \frac{1}{x}$ By the Sum and Difference Rule.

Example 2: Finding the Derivative

Find $\frac{dy}{dx}$ for $y = x^2 \ln x$.

Solution: $\frac{dy}{dx} = x^2 \cdot \frac{d}{dx}[\ln x] + \ln x \cdot \frac{d}{dx}\left[x^2\right]$ By the Product Rule.

$$= x^2 \cdot \frac{1}{x} + \ln x \cdot (2x)$$

$$= x + 2x \ln x$$

Example 3: Equation of a Tangent Line

Find the equation of the tangent line to $y = \ln x$ at the point (5, 1.61). (ln 5 is actually equal to 1.609437912... but round off to two decimal places here.)

Solution: For $f(x) = \ln x$, the slope we need is $f'(5) = \frac{1}{5}$. In point-slope form, we have

$$y - 1.61 = \frac{1}{5}(x-5)$$

Use the given (x, y)-coordinate $(5, 1.61)$ in the point-slope formula as well as $f'(5) = \frac{1}{5} = m$.

$$y - 1.61 = 0.2x - 1$$

$$y = 0.2x + 0.61.$$

The Derivative of $y = \ln(g(x))$

Suppose we want to find $\frac{dy}{dx}$ given that

$$y = \ln\left(x^2+1\right).$$

We can let

$$y = \ln u \quad \text{and} \quad u = x^2 + 1.$$

Then

$$\frac{dy}{du} = \frac{1}{u} \quad \text{and} \quad \frac{du}{dx} = 2x.$$

So, by the Chain Rule (Section 3.2), we have

$$\begin{aligned} \frac{dy}{dx} &= \frac{dy}{du} \cdot \frac{du}{dx} \\ &= \frac{1}{u} \cdot 2x \\ &= \frac{1}{x^2+1} \cdot 2x \qquad \text{Substitute } u = x^2+1. \\ &= \frac{2x}{x^2+1}. \end{aligned}$$

Use the same approach for the general case:

$$y = \ln(g(x)),$$

where $g(x)$ is positive and differentiable.

Let

$$y = \ln u \quad \text{and} \quad u = g(x).$$

Then

$$\frac{dy}{du} = \frac{1}{u} \quad \text{and} \quad \frac{du}{dx} = g'(x).$$

Applying the Chain Rule, we get

$$\begin{aligned}\frac{dy}{dx} &= \frac{dy}{du} \cdot \frac{du}{dx} \\ &= \frac{1}{u} \cdot g'(x) \\ &= \frac{1}{g(x)} \cdot g'(x) \qquad \text{Substitute } u = g(x). \\ &= \frac{g'(x)}{g(x)}.\end{aligned}$$

Derivative of ln(g(x))

If $y = \ln(g(x))$, *then*

$$y' = \frac{1}{g(x)} \cdot g'(x) = \frac{g'(x)}{g(x)}.$$

Example 4: Finding the Derivative

a. Find $f'(x)$ for $f(x) = \ln(x^2 + 2x + 10)$.

Solution: Let $g(x) = x^2 + 2x + 10$. Then $g'(x) = 2x + 2$.

So

$$f'(x) = \frac{g'(x)}{g(x)} = \frac{2x+2}{x^2 + 2x + 10}.$$

b. Find $f'(x)$ for $f(x) = \ln\sqrt{x^2 + 3}$.

Solution: We simplify first to obtain

$$f(x) = \ln\sqrt{x^2+3} = \ln(x^2+3)^{\frac{1}{2}} = \frac{1}{2}\ln(x^2+3).$$

Now, differentiating with $g(x) = x^2 + 3$ and $g'(x) = 2x$, we have

$$f'(x) = \frac{1}{2} \cdot \frac{g'(x)}{g(x)} = \frac{1}{\not{2}} \cdot \frac{\not{2}x}{x^2+3} = \frac{x}{x^2+3}.$$

Logarithmic Differentiation

Functions that involve several products and quotients can be very difficult to differentiate directly. However, we know that logarithms of products and quotients can be written as sums and differences. And sums and differences are generally easier to differentiate than products and quotients. Taking the logarithm of both sides of an equation and then differentiating is called **logarithmic differentiation**. The only condition is that both sides must represent positive values.

Example 5: Using Logarithmic Differentiation

Use logarithmic differentiation to find $f'(x)$, given that $f(x)=\dfrac{\left(x^2+1\right)^3\sqrt{x^2-2x}}{(x-5)}$.

Solution: Take the natural logarithm of both sides and simplify.

$$\begin{aligned}\ln f(x)&=\ln\left[\frac{\left(x^2+1\right)^3\sqrt{x^2-2x}}{(x-5)}\right]\\&=\ln\left(x^2+1\right)^3+\ln\left(x^2-2x\right)^{\frac{1}{2}}-\ln(x-5)\\&=3\ln\left(x^2+1\right)+\frac{1}{2}\ln\left(x^2-2x\right)-\ln(x-5)\end{aligned}$$

Now differentiate both sides.

$$\begin{aligned}\frac{f'(x)}{f(x)}&=3\cdot\frac{1}{x^2+1}\cdot 2x+\frac{1}{2}\cdot\frac{1}{x^2-2x}\cdot(2x-2)-\frac{1}{x-5}\\f'(x)&=f(x)\cdot\left(\frac{6x}{x^2+1}+\frac{x-1}{x^2-2x}-\frac{1}{x-5}\right)\\&=\frac{\left(x^2+1\right)^3\sqrt{x^2-2x}}{x-5}\cdot\left(\frac{6x}{x^2+1}+\frac{x-1}{x^2-2x}-\frac{1}{x-5}\right)\end{aligned}$$

We complete this section with two examples that apply logarithmic functions and their derivatives. In Example 6 we find the absolute extrema of a function on a closed interval. In Example 7 we use the technique for graphing functions that was discussed in Chapter 4.

Example 6: Locating Absolute Extrema

Find the absolute extrema for the function $f(x) = x^2 \ln x$ on the interval $[0.1, 2.0]$.

Solution: The function $f(x) = x^2 \ln x$ is a continuous function over the interval $(0, +\infty)$, so it has an absolute maximum and an absolute minimum on the closed interval $[0.1, 2.0]$.

We know from Example 2 that

$$\begin{aligned} f'(x) &= x + 2x \ln x \\ &= x(1 + 2\ln x). \end{aligned}$$

Setting $f'(x) = 0$ gives

$$x(1+2\ln x) = 0$$

$$\cancel{x=0} \text{ or } 1+2\ln x = 0$$

x cannot be 0.

$$2\ln x = -1$$

$$\ln x = -\frac{1}{2} = -0.5$$

$$x = e^{-0.5} \approx 0.61$$

Solve $1 + 2\ln x = 0$ for x to find the critical value.

x	$f(x) = x^2 \ln x$
0.1	$f(0.1) = (0.1)^2 \ln 0.1 \approx -0.02$
0.61	$f(0.61) = (0.61)^2 \ln 0.61 \approx -0.18$ Absolute min
2.0	$f(2.0) = (2.0)^2 \ln 2.0 \approx 2.77$ Absolute max

Evaluate $f(x)$ at the critical values. (Remember, these include the endpoints of the function's interval.)

The absolute minimum is -0.18 and occurs at $x = 0.61$. The absolute maximum is 2.77 and occurs at $x = 2.0$.

Example 7: Graphing Logarithmic Functions

Using the curve sketching techniques discussed in Chapter 4, sketch the graph of the function $f(x) = x \ln x$.

Solution: The domain is the interval $(0, +\infty)$ because $\ln x$ is defined only for positive values of x.

$$f'(x) = x \cdot \frac{1}{x} + \ln x \cdot 1$$

$$= 1 + \ln x$$

Find the critical value(s) by setting $f'(x)$ equal to 0 and solving for x.

$$0 = 1 + \ln x$$

$$\ln x = -1$$

$$x = e^{-1} \approx 0.37$$

Using the critical value $x = e^{-1}$ as well as the endpoint $x = 0$, we have two intervals to test.

By testing values, we find that $f(x)$ is decreasing on the interval $(0, e^{-1})$ and increasing on the interval $(e^{-1}, +\infty)$.

Taking the second derivative of $f(x)$, we find that $f''(x) = \frac{1}{x} > 0$ for all x on the interval $(0, +\infty)$. Therefore, $f(x)$ is concave upward on the interval $(0, +\infty)$.

A local (and absolute) minimum occurs at the critical value $x = e^{-1}$, and

$$f(e^{-1}) = e^{-1} \ln e^{-1}$$

$$= e^{-1}(-1) = -e^{-1}.$$

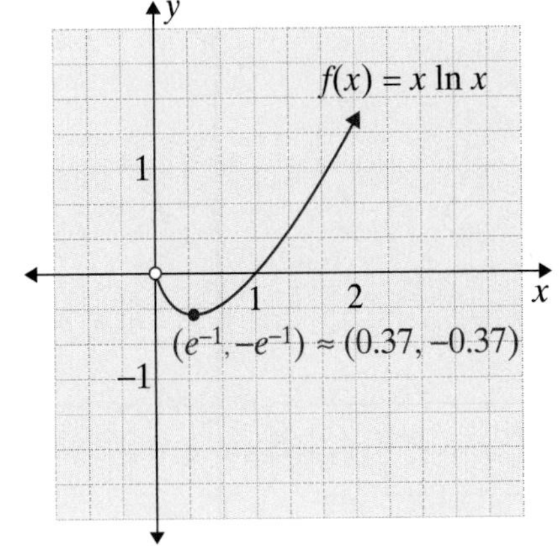

Note: one easy point to get without a calculator is $(1, 0)$ (the student should recall that $\ln 1 = 0$). Therefore, it is possible to notice that $1 \cdot \ln 1 = 0$, and $(1, 0)$ is an x-intercept.

5.3 Exercises

Find the derivative of each of the functions in Exercises 1 – 28.

1. $f(x)=x^2-\ln x$

2. $f(x)=4x^2+\ln x^2$

3. $f(x)=25+x\ln x$

4. $f(x)=(x^2+1)\ln x$

5. $y=\dfrac{\ln x}{x}$

6. $y=\dfrac{x^2}{\ln x}$

7. $y=(\ln x)^3$

8. $y=\sqrt{\ln x}$

9. $f(x)=\ln(5x+3)$

10. $f(x)=\ln(7x-2)$

11. $f(x)=\ln(x^2+2)$

12. $f(x)=\ln(x^2+3x)$

13. $f(x)=\ln\sqrt{2x^2-1}$

14. $f(x)=\ln\sqrt{x^3+4}$

15. $y=\ln(4x+3)^2$

16. $f(x)=\ln(x^2-4)^3$

17. $y=\sqrt{x}\ln(x^2+2)$

18. $y=\dfrac{\ln(5x+2)}{x^3}$

19. $y=\dfrac{\ln(x^2+2x-1)}{x}$

20. $y=x^{-2}\ln(x^2-3x+4)$

21. $f(x)=\ln((3x+1)(x^2+3))$

22. $f(x)=\ln(x^2(4x-1))$

23. $f(x)=\ln((2x-1)^2(x^2-2))$

24. $f(x)=\ln(\sqrt{4x-7}(6x+7))$

25. $y=\ln\left(\dfrac{x+1}{x-2}\right)$

26. $y=\ln\left(\dfrac{3x-1}{x+5}\right)$

27. $f(x)=\ln\left(\dfrac{x^2-5}{2x+9}\right)$

28. $f(x)=\ln\sqrt{\dfrac{4x+3}{x^2-6}}$

For Exercises 29 – 34, determine a formula for $f''(x)$. Use your calculator (if necessary) to solve the equation $f''(x)=0$ and locate any possible inflection points.

29. $f(x)=\dfrac{x}{\ln x}$

30. $f(x)=(2x+x^2)\ln x$

31. $f(x)=3x^2\ln x$

32. $f(x)=\ln(x^3)$

33. $f(x)=(\ln x)^3$

34. $f(x)=\ln x+x^2+3x+2$

Use logarithmic differentiation to find the derivatives of the functions in Exercises 35 – 42.

35. $y=(2x-5)^3\sqrt{x^2-2x}$

36. $y=(4-5x)^4(7x+2)^{-\frac{2}{3}}$

37. $y=(x^2+2)^4(x^2-1)^{-\frac{1}{3}}$

38. $y=(3x-2)^5\sqrt{x^2+x}$

39. $y=\dfrac{(2x+x^2)^3}{(4x-9)^2}$

40. $y=\dfrac{\sqrt[3]{x^2+4}}{(2-5x)^3}$

41. $y=\dfrac{(x+2)^2(3x+4)^2}{\sqrt{x-6}}$

42. $y=\dfrac{(x^2-3)^2\sqrt{x^2+3x}}{x+7}$

In Exercises 43 – 48, find the absolute extrema for each of the functions on the indicated interval.

43. $f(x)=x-\ln x;\ [0.5, 2]$

44. $f(x)=x^2-8\ln x;\ [0.3, 4]$

45. $f(x)=\dfrac{x}{\ln x};\ [1.2, 3]$

46. $f(x)=\dfrac{\ln x}{x^2};\ [1, 2]$

47. $f(x)=x^2-\ln x^3;\ [1, e]$

48. $f(x)=\ln(3+2x-x^2);\ [0, 2.5]$

Using the graph-sketching techniques discussed in Chapter 4, sketch the graph of each function in Exercises 49 – 54.

49. $f(x)=4x\ln x$

50. $f(x)=x^2\ln x$

51. $f(x)=x-3\ln x$

52. $f(x)=4x-\ln x^2$

53. $f(x)=\dfrac{\ln x}{x}$

54. $f(x)=\dfrac{\ln x}{x^2}$

55. Marginal revenue. A retailer has determined that the revenue from the sale of x end tables is given by the function $R(x)=96x+\ln(4x^2+15)$ dollars. Find the marginal revenue.

56. Advertising. An automobile dealer has estimated that he can sell $N(x)=420+72\ln(1+0.5x)$ cars annually, where x (in thousands of dollars) is the amount spent on advertising.

a. Find the number of cars sold if $6000 is spent on advertising.

b. Find the rate of change in number of cars sold if $6000 is spent on advertising.

57. Revenue. The daily demand for a product is given by $p=-8\ln(0.01x)$, where p is the price in dollars when x units are sold and $0<x\le 100$. Find the maximum daily revenue.

58. Revenue. The demand for a product is given by $p = 14 - 6.5\ln x$, where x is the number of units (in thousands) that can be sold at a price p dollars and $2 \leq x \leq 8$. Find the maximum revenue.

59. Insect population. Mediterranean fruit flies are discovered in a citrus orchard. The Department of Agriculture has determined that the population of flies t hours after the orchard has been sprayed with pesticide is approximated by $N(t) = 25 - 5t\ln(0.04t) - t$, where $0 < t \leq 25$.

  a. Find $N(3)$, $N(10)$, $N(25)$.

  b. What will be the maximum number of flies in the orchard?

60. Air Quality. The Air Quality Management Board estimates that t hours after midnight in a major city the level of ozone in the air is about $N(t) = 0.013 - 0.007t\ln(0.026t)$ parts per million, where $0 < t \leq 18$.

  a. Find $N(6)$, $N(10)$, $N(18)$.

  b. Find the maximum level of pollution.

Hawkes Learning Systems: Essential Calculus

Differentiation of Logarithmic Functions

5.4 Differentiation of Exponential Functions

Objectives

After completing this section, you will be able to:

1. *Differentiate exponential functions.*
2. *Use curve sketching techniques to graph exponential functions.*

Slopes and Exponential Functions

For the function $y = 2^x$, let us approximate the slope at $x = 0$ by considering the line through $(0, 2^0)$ and $(.0001, 2^{.0001})$. We determine the slope is

$$\frac{2^{.0001} - 2^0}{.0001 - 0} = \frac{.00006931..}{.0001} = 0.6931\ldots.$$

Here we want to consider the Table 5.4.1. Using the method above or a graphing calculator, we calculate y-values and slopes for $y = 2^x$.

$f(x) = 2^x$		
x	y	y'
0	1	0.6931
1	2	1.386
2	4	2.773
3	8	5.545

Table 5.4.1

Because of the exponential character of the function, as x increases by 1, y doubles. This is not surprising, but what is surprising is that y' also doubles! Put another way, $f'(0) = .6931$ is a number which specifies the slope not only at $x = 0$ but also for other x-values. This is revealed if we add another column to Table 5.4.1 (see Table 5.4.2).

$f(x) = 2^x$			
x	y	y'	y'
0	1	0.6931	1(0.6931)
1	2	1.386	2(0.6931)
2	4	2.773	4(0.6931)
3	8	5.545	8(0.6931)

Table 5.4.2

Table 5.4.2 shows, numerically, that for $y = 2^x$, we have

$$y' = (.6931)2^x = .6931y.$$

We can restate this as:

The slope of $f(x) = 2^x$ is linearly related (directly proportional) to $f(x)$ itself.

$f(x) = 3^x$		
x	$y = 3^x$	$y' = 1.099y$
0	1	1.099
1	3	$3.296 = 3(1.099)$
2	9	$9.888 = 9(1.099)$
3	27	$29.66 = 27(1.099)$

Table 5.4.3

This linear relationship between y-values of exponential functions and the corresponding slopes was understood by the mathematician John Napier (1550 – 1617) and is exactly what he tried to capture with his logarithm tables.

Suppose we let $f(x) = A^x$ and we look at other values of A besides 2. For example, if $A = 3$, we get the numbers shown in Table 5.4.3.

There are two important facts which we want to infer from these tables. First, that for any exponential function $y = A^x$, the slopes are directly proportional to the y-values, that is,

$$\frac{d}{dx}A^x = k \cdot A^x$$

or, $y' = ky$ for some constant k. Secondly, the constant k, called the **constant of proportionality**, is given by

$$k = f'(0).$$

We will eventually prove these facts. Before that, however, we want to consider the set of possible k-values. In exercise #52 we ask that the student calculate k for two other bases. It will appear that as the base A increases, for the function $y = A^x$, the value of k increases. Apparently, however, between $A = 2$ and $A = 3$, there is a value of A which makes $k = 1$. This value $k = 1$ is very nice since k is used to multiply in the derivative formula $y' = ky$.

When $k = 1$, the formula for slope is just $y' = y$. That is, the function has the remarkable property that the slope exactly equals the corresponding y-value, regardless of x. In a real sense this base, for which the slope equals y-values, is a "perfect" base for calculus. A numerical method for determining the perfect base is outlined below and the results are shown in Table 5.4.4.

Derivation for e			
x	$y = B^x$	y'	$\frac{y'}{y}$
0	1	0.99999504	0.99999504
0.5	1.6471677	1.6487086	0.99999504
1	2.718267	2.71825353	0.99999504
1.5	4.4816524	4.48163019	0.99999504
2	7.38897548	7.38893887	0.99999504
2.5	12.1823278	12.1822675	0.99999504
3	20.0852082	20.0851087	0.99999505

When $k = 1$, B is roughly 2.718267.

$$y' \approx \frac{\left(f\left(x+0.000001\right)-f\left(x\right)\right)}{0.000001}$$

Table 5.4.4

We can confirm analytically that, for $y = e^x$, the slopes are equal to the y-values. The key is to use the results for $\ln x$ and the fact that e^x and $\ln x$ are inverses of each other. The details are amazingly simple – use the fact that e^x is always positive as well as your knowledge of logarithmic differentiation.

$$y = e^x$$

$$\ln y = \ln e^x \qquad \text{By Property 4.}$$

$$\ln y = x \qquad \text{By Property 8.}$$

$$\frac{d}{dx}[\ln y] = \frac{d}{dx}x$$

$$\frac{1}{y} \cdot \frac{dy}{dx} = 1$$

Differentiate both sides and treat y as $g(x)$ in the expression $\ln(g(x))$.

$$\frac{dy}{dx} = y = e^x$$

This proves that the derivative of $y = e^x$ is the function itself.

Derivative of the Exponential Function

If $f(x) = e^x$, then

$$f'(x) = e^x.$$

This derivative formula and the corresponding one for $\ln x$ are probably the two most amazing formulas in calculus. We consider $f'(x)$ for $f(x) = B^x$ in Section 5.5.

Example 1: Finding the Derivative

a. Find $\frac{dy}{dx}$ for $y = x^3 + e^x$.

Solution: $\frac{dy}{dx} = 3x^2 + e^x$ By the Sum and Difference Rule.

b. Find $\frac{dy}{dx}$ for $y = x^2 e^x$.

Solution: By the Product Rule.

$$\frac{dy}{dx} = x^2 \cdot \frac{d}{dx}\left[e^x\right] + e^x \cdot \frac{d}{dx}\left[x^2\right]$$

$$= x^2 e^x + e^x \cdot 2x$$

$$= xe^x(x+2)$$

Now consider the function

$$y = e^{-x^2},$$

where the exponent is a function of x other than x itself. To find $\frac{dy}{dx}$, we can proceed as before and take the natural logarithm of both sides.

$$\ln y = \ln e^{-x^2}$$

$$\ln y = -x^2 \ln e$$

$$\ln y = -x^2$$

$$\frac{1}{y} \cdot \frac{dy}{dx} = -2x \qquad \text{Differentiate both sides.}$$

$$\frac{dy}{dx} = -2x \cdot e^{-x^2} \qquad \text{Substitute } e^{-x^2} \text{ for } y.$$

While logarithmic differentiation works well here and is a good technique, another approach is to use the Chain Rule with

$$y = e^u \quad \text{and} \quad u = -x^2.$$

Thus

$$\frac{dy}{du} = e^u \quad \text{and} \quad \frac{du}{dx} = -2x.$$

Therefore,

$$\frac{dy}{dx} = \frac{dy}{du} \cdot \frac{du}{dx} = e^u(-2x) = -2xe^{-x^2}.$$

Applying either technique to the general case $y = e^{g(x)}$, where $g(x)$ is differentiable, we get the following result.

Chain Rule for Exponential Functions

If $f(x) = e^{g(x)}$, then

$$f'(x) = e^{g(x)} \cdot g'(x).$$

Example 2: Using the Chain Rule

Find $\frac{dy}{dx}$ if $y = e^{x^2+3x}$.

Solution: Let $g(x) = x^2 + 3x$. Then $g'(x) = 2x + 3$.

Thus

$$\frac{dy}{dx} = e^{g(x)} \cdot g'(x) = e^{x^2+3x}(2x+3) = (2x+3)e^{x^2+3x}.$$

Example 3: Using the Chain Rule

For $f(x) = xe^{-x}$, find **a.** $f'(x)$, **b.** $f''(x)$, and **c.** the equation of the tangent line to $f(x)$ at the point where $x = -1$.

Solutions: a.

$$f'(x) = x \cdot \frac{d}{dx}\left[e^{-x}\right] + e^{-x} \cdot \frac{d}{dx}[x] \qquad \text{By the Product Rule.}$$

$$= x \cdot e^{-x}(-1) + e^{-x}(1)$$

$$= e^{-x}(-x+1) \qquad \text{Factor out } e^{-x}.$$

$$= e^{-x}(1-x)$$

b.

$$f''(x) = e^{-x} \cdot \frac{d}{dx}[1-x] + (1-x) \cdot \frac{d}{dx}\left[e^{-x}\right] \qquad \text{By the Product Rule applied to } f'(x).$$

$$= e^{-x}(-1) + (1-x)(-1)e^{-x}$$

$$= e^{-x}(-1-1+x) \qquad \text{Factor out } e^{-x}.$$

$$= e^{-x}(x-2)$$

c. Since $f(-1) = (-1)e^{-(-1)} = -e$, we need the equation at the point $(-1, -e)$.

To determine the slope of the tangent line at the point $(-1, -e)$, we need to find $f'(-1)$.

$$f'(-1) = e^{-(-1)}(1-(-1)) = 2e$$

Using the slope-intercept formula, the equation of the tangent line is

$$y-(-e) = 2e(x-(-1))$$

$$y+e = 2ex+2e$$

$$y = 2ex+e$$

$$y \approx 5.437x + 2.718.$$

In Example 4 we will use the results found in Example 3 to analyze the function $f(x) = xe^{-x}$ and sketch its graph.

Example 4: Graphing Exponential Functions

For $f(x) = xe^{-x}$

a. Find any critical values.

b. Find any hypercritical values.

c. Sketch the graph of the function.

Solutions: a. To find critical values, set $f'(x) = 0$.

$$f'(x) = e^{-x}(1-x)$$

We know $f'(x)$ from Example 3.

$$0 = e^{-x}(1-x)$$

$$0 = 1-x$$

$$x = 1$$

e^{-x} is never equal to 0, so we can divide both sides by e^{-x} (that is, cancel e^{-x}).

The only critical value is $x = 1$, since there are no values where $f'(x)$ is undefined.

b. To find hypercritical values, set $f''(x) = 0$.

$$f''(x) = e^{-x}(x-2)$$

We know $f''(x)$ from Example 3.

$$0 = e^{-x}(x-2)$$

$$0 = x-2$$

$$x = 2$$

The only hypercritical value is $x = 2$.

c. Using the critical value and the hypercritical value found in parts **a.** and **b.**, we can find the local extrema and the points of inflection.

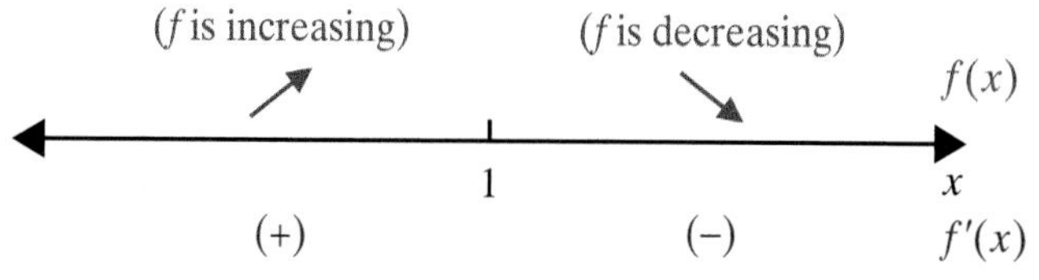

So there is a local max at $(1, e^{-1})$.

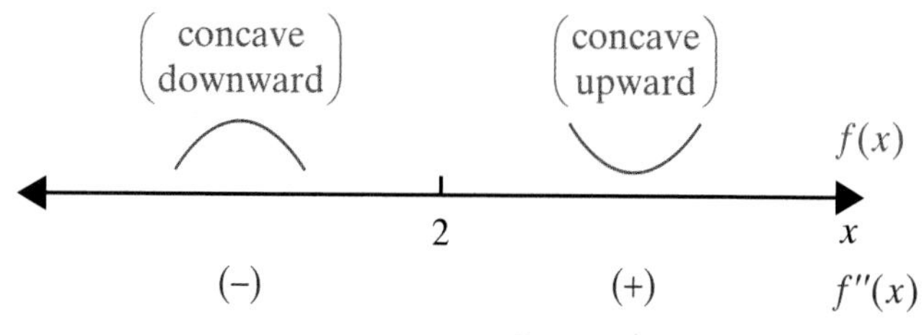

There is a point of inflection at $(2, 2e^{-2})$.

continued on next page ...

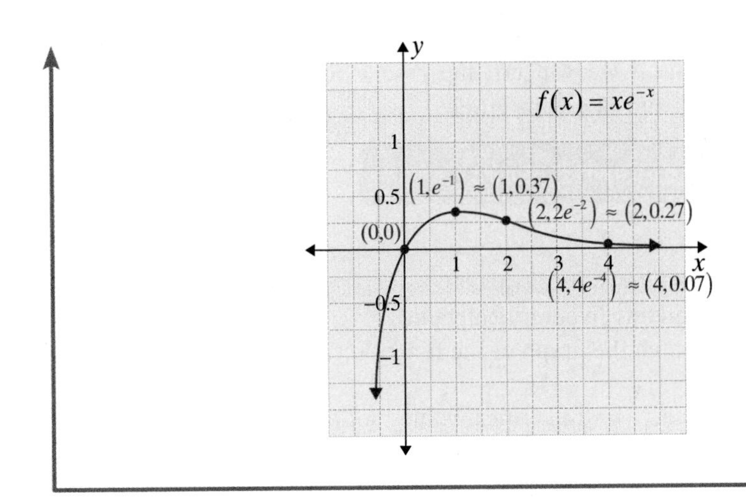

5.4 Exercises

Find the first and second derivative of each of the functions in Exercises 1 – 6.

1. $f(x) = 3e^x$

2. $f(x) = -6e^x$

3. $f(x) = x^2 + 5e^x$

4. $f(x) = 4x^2 - 2e^x$

5. $f(x) = xe^x$

6. $f(x) = -7x^2e^x$

For Exercises 7 – 14, find a formula for $f'(x)$ and determine the slope $f'(a)$ at the point where $x = a$ is given.

7. $f(x) = e^x \ln x;\ x = 1$

8. $f(x) = e^x \ln(x+4);\ x = 1$

9. $f(x) = \dfrac{e^x}{e^x - 1};\ x = 2$

10. $f(x) = \dfrac{e^x}{\ln x};\ x = e$

11. $f(x) = 2e^{4x};\ x = -1$

12. $f(x) = 8e^{-3x};\ x = 4$

13. $f(x) = 3e^{2x+1};\ x = 0$

14. $f(x) = 5e^{\frac{x}{2}};\ x = 0$

For Exercises 15 – 18, find a formula for $f'(x)$ and use it to determine the intervals on which $f(x)$ is increasing or decreasing.

15. $f(x) = e^{1-x^2}$

16. $f(x) = e^{-0.04x^2}$

17. $f(x) = \left(e^{2x} - 4\right)^2$

18. $f(x) = \left(e^{4x} + 2\right)^3$

For Exercises 19 and 20, determine $f'(x)$ and use it to determine the intervals on which $f(x)$ is increasing or decreasing. Determine for each function if there is a horizontal asymptote. Confirm your results with a graphing calculator.

19. $f(x)=\sqrt{e^{-0.2x}+11}$

20. $f(x)=\dfrac{1}{\sqrt{3e^{x}+1}}$

Find $f'(x)$ and use it to argue whether or not there is a diagonal asymptote for each of the functions in Exercises 21 and 22.

21. $f(x)=\ln\left(e^{x}+1\right)$

22. $f(x)=\ln\sqrt{5+e^{2x}}$

Use a graphing calculator to graph $f(x)$ and $f'(x)$. Then locate all extrema and all inflection points, if any.

23. $f(x)=e^{-x^{2}}\ln\left(x^{2}+2\right)$

24. $f(x)=\ln\dfrac{2+3e^{-x}}{x+2}$

For each of the following functions in Exercises 25 – 32, find the absolute extrema on the indicated interval.

25. $f(x)=xe^{2x};\ [-2,1]$

26. $f(x)=xe^{\frac{x}{3}};\ [-4,0]$

27. $f(x)=x^{2}e^{-x};\ [-1,2]$

28. $f(x)=2x^{2}e^{-x};\ [-2,3]$

29. $f(x)=5e^{1-x^{2}};\ [-2,1]$

30. $f(x)=3xe^{-x^{2}};\ [-2,2]$

31. $f(x)=(2x+3)e^{-0.2x};\ [1,4]$

32. $f(x)=(4x-1)e^{-0.5x};\ [0,3]$

For each of the functions in Exercises 33 – 38, **a.** *find any critical values,* **b.** *find any hypercritical values,* **c.** *find all intervals of concavity, and* **d.** *sketch the graph of the function. If available, confirm your results with a graphing utility.*

33. $f(x)=xe^{-0.4x}$

34. $f(x)=2xe^{-0.5x}$

35. $f(x)=4x^{2}e^{-x}$

36. $f(x)=3x^{2}e^{-x}$

37. $f(x)=e^{x}+e^{-x}$

38. $f(x)=\dfrac{e^{x}}{x}$

39. Revenue. The demand for a product is given by $D(x)=140e^{-0.05x}$, where x is the number of units sold each week and $0\le x\le 30$.

a. Find the number of units sold that will yield maximum revenue.

b. Find the price per unit that will yield maximum revenue.

40. **Revenue.** The demand equation for a certain product is given by $D(x) = 210e^{-0.025x}$, where x is the number of units sold each week and $0 \le x \le 60$.

  a. Find the number of units sold that will yield the maximum revenue.

  b. Find the price per unit that will yield maximum revenue.

41. **Advertising.** An automobile manufacturer is planning a television advertisement campaign to introduce a new model for their truck. It is estimated that $N(t) = 600\left(1 - e^{-0.02t}\right)$ people (in thousands) will have seen the advertisement after t days of advertising. How fast is N increasing at the end of 7 days?

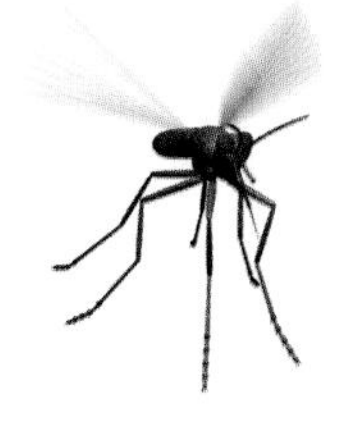

42. **Insect population.** The mosquito population of a pool of water is estimated to be $P(t) = 400 + 1400e^{-0.3t}$, where t is the number of hours after the pool has been treated. Find the rate of change in the population at the end of 5 hours.

43. **Bacterial population.** The population of bacteria in an experimental culture is estimated by $N(t) = \dfrac{10{,}000}{1 + 9e^{-0.14t}}$, where t is the number of hours after the experiment begins. How fast is the population changing at the end of 5 hours?

44. **Disease control.** The elk herd in a national park has been infected by a contagious disease. The number of infected animals is estimated by $N(t) = \dfrac{600}{1 + 49e^{-0.36t}}$, where t is the number of days after the disease was discovered. How fast is the disease spreading after 4 days?

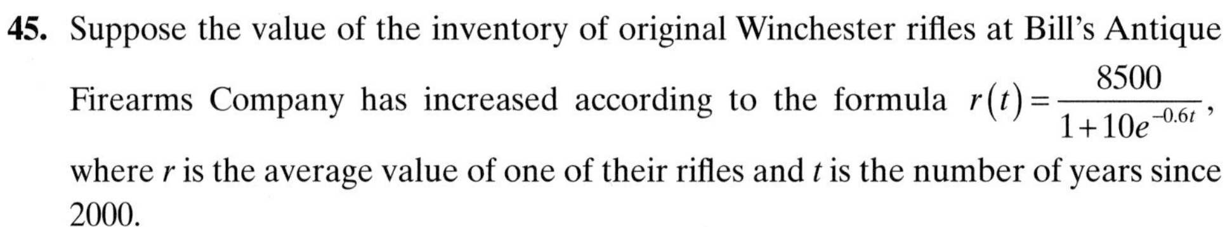

45. Suppose the value of the inventory of original Winchester rifles at Bill's Antique Firearms Company has increased according to the formula $r(t) = \dfrac{8500}{1 + 10e^{-0.6t}}$, where r is the average value of one of their rifles and t is the number of years since 2000.

  a. What was the average value of a rifle in 2000? In 2005?

  b. At what rate is r changing in 2005? In 2006?

  c. If there is an inflection point for $r(t)$, locate it and explain its significance in the application.

  d. When is the rate of increase of r at a maximum?

46. Suppose an advertising campaign for the sale of a new magazine, Dungeons and Creeps, causes sales to vary according to the formula $S(t) = 8\left(1 - 0.3e^{-0.2t+1}\right)$, where S is sales in thousands and t is time in months since the ad campaign started.

  a. What were the monthly sales when the ad campaign started?

  b. What was the rate at which sales were changing after 4 months into the campaign?

  c. What are the long-term sales expectations?

47. The growth cycle of a mass of algae in a pond grows according to $A(t)=1+2te^{-0.5t}$, where A represents the mass-density of algae in the pond in suitable units and t is the time in months ($t=0$ corresponds to April 1st).

- **a.** What day of the year corresponds to a maximum A value?
- **b.** When does the rate of decline in algae reach its maximum?

48. The weekly sales of deluxe cheeseburgers at Cheese House is given by $C(t)=\dfrac{10\ln(2+2t^2)}{1+0.8t}$, where C is the number of cheeseburgers sold each week in thousands and t is the number of weeks following a new radio commercial.

- **a.** Determine the intervals on which C is increasing or decreasing.
- **b.** How long after introduction of the new commercial will it take for sales to reach a maximum?

49. A certain calculus student recalls information according to the formula $p=70e^{-0.6x}+30$, where p is the percentage of information retained after x weeks.

- **a.** After 4 weeks, what percentage of a lesson is retained?
- **b.** Ater 4 weeks, at what rate is the percentage changing?
- **c.** What does the model predict a few months after the Calculus course is over?

50. Inexpensive videos detailing the championship basketball season of Castle High School are sold locally by a civic club to raise money for next year's team. The total sales are given by $S=\dfrac{12{,}500}{1+15e^{-0.5x}}$, where S is the total sales after x weeks.

- **a.** What are the total sales after 3 weeks?
- **b.** What is the rate of change of sales after 3 weeks?
- **c.** After about how many weeks will the total sales begin to level off?
- **d.** When is the sales rate increasing fastest? Illustrate this point graphically.

51. **a.** Find the equation of the tangent line to $f(x)=2e^{-x^2+1}+4$ at the point where $x=2$.

b. Discuss the advantages and disadvantages of using the tangent line to get values of $f(x)$ for $x\geq 2$ rather than the function itself.

52. Determine k in the equations that follow by finding $f'(0)$ (see p. 428).

- **a.** Let $f(x)=10^x$. Determine the value of k in the formula $f'(x)=k\cdot 10^x$.
- **b.** Let $y=f(x)=\pi^x$. Determine k in the formula $f'(x)=ky$.

Hawkes Learning Systems: Essential Calculus

Differentiation of Exponential Functions

5.5 Applications of Exponential Functions

Objectives

After completing this section, you will be able to:

Solve problems involving growth, decay, and elastic demand.

Exponential Growth

Exponential and logarithmic functions can be used to describe (or model) the behavior of real life phenomena. A quantity that increases at a rate of change that is proportional to the amount present is said to **increase exponentially** (or **grow exponentially**). For example, in the study of bacterial cultures in biology and medicine, the number of bacteria increases as time goes on and the rate of growth often increases at any time t in proportion to the number of bacteria present at that time.

Another example of exponential growth was introduced in Section 5.1 when we discussed continuously compounded interest. The related function is the exponential function

$$A = Pe^{rt}.$$

Now suppose that \$1000 is invested in an account that earns 6 percent interest compounded continuously. Then

$$P = \$1000, \quad r = 0.06, \quad \text{and} \quad A = 1000e^{0.06t}.$$

To find the rate of change of the amount A at time t, we differentiate with respect to t.

$$\begin{aligned} \frac{dA}{dt} &= 1000(0.06)e^{0.06t} \\ &= 0.06\left(1000e^{0.06t}\right) \\ &= 0.06A \end{aligned}$$

Thus we see that the rate of change of A, $\frac{dA}{dt}$, is directly proportional to A. The constant 0.06 is called the **constant of proportionality**. For continuous compounding, $r = 0.06 = 6$ percent is the **interest rate** or the **growth rate**. The rate of change of A is found by multiplying A by its growth rate.

More generally, suppose that the mathematical model of a situation is an exponential function of the form

$$y = ce^{kt},$$

where c and k are positive constants.

Then

$$\frac{dy}{dt} = c \cdot ke^{kt}$$
$$= k \cdot ce^{kt}$$
$$= ky$$

and the rate of change in y is directly proportional to y. Therefore, the function

$$y = ce^{kt}$$

is a mathematical model for exponential growth.

Exponential Growth

A function $y = f(t)$ satisfies the equation

$$\frac{dy}{dt} = ky,$$

if and only if

$$y = ce^{kt}.$$

If k and c are positive, we say that ***y grows*** *(or* ***increases****)* ***exponentially****.*

See Figure 5.5.1 for a general graph of exponential growth.

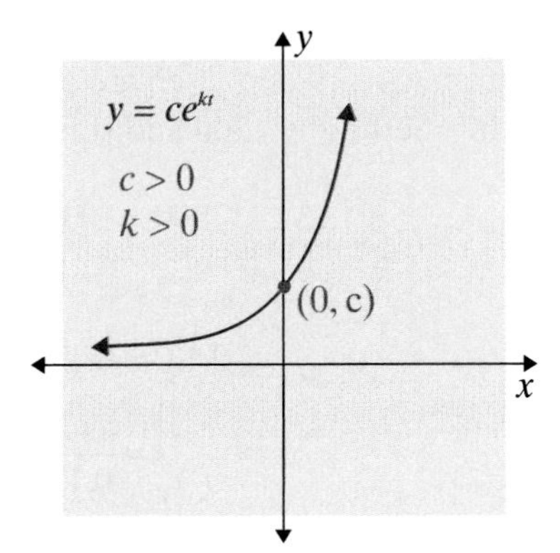

Figure 5.5.1

Example 1: Population Growth

Suppose that a population of rabbits grows exponentially; that is, the rate of population growth depends on the size of the population. Further, suppose that on Monday at 8 A.M. there are 100 rabbits and that at 8 A.M. on Wednesday there are 130 rabbits.

a. Find an exponential function that represents the rabbit population at time t, where t is measured in days from Monday at 8 A.M.

b. Find the time that the population will be 200 rabbits (double the original 100).

Solutions: a. The general form of the function is

$$y = ce^{kt}.$$

When $t = 0$ (Monday at 8 A.M.), $y = 100$.

$$100 = c \cdot e^{k \cdot 0} = c \cdot 1$$
$$100 = c$$

Substitute the given values into the general form of the function and solve for C.

This gives

$$y = 100e^{kt}.$$

Substitute $C = 100$ into the general form of the funtion.

Now, when $t = 2$ (Wednesday at 8 A.M.), $y = 130$.

$$130 = 100e^{k \cdot 2}$$
$$1.30 = e^{2k}$$

Substitute the given values into our formula for y.

$$2k = \ln 1.30$$
$$k = \frac{\ln 1.30}{2} \approx 0.13$$

Use the definition of natural logarithm.

Therefore, the exponential function is $y = 100e^{0.13t}$, where t is measured in days from Monday at 8 A.M.

b. Setting $y = 200$ and solving for t, we have

$$200 = 100e^{0.13t}$$
$$2 = e^{0.13t}$$
$$0.13t = \ln 2$$
$$t = \frac{\ln 2}{0.13} \approx \frac{0.6931}{0.13} \approx 5.3 \text{ days},$$

which is approximately 3 P.M. on Saturday.

Example 2: Continuous Compounding Interest

Suppose that $2000 is invested in an account that earns 8 percent compounded continuously.

a. What will be the balance in 1 year?

b. When will the original investment be doubled?

Solutions: a. We know that the growth is exponential because the interest is compounded continuously. Thus

$$A = Pe^{rt} = 2000e^{0.08t}.$$

Substitute $P = 2000$ and $r = 0.08$ into the formula for continuous compounding interest.

Substituting $t = 1$ gives

$$\begin{aligned} A &= 2000e^{0.08 \cdot 1} \\ &= 2000(1.08329) \\ &= \$2166.57. \end{aligned}$$

$2166.57 is the balance at the end of 1 year.

b. Setting $A = 4000$ and solving for t, we obtain

$$4000 = 2000e^{0.08t}$$

$$2 = e^{0.08t}$$

$$0.08t = \ln 2$$

$$t = \frac{\ln 2}{0.08} \approx \frac{0.6931}{0.08} \approx 8.7 \text{ years.}$$

The original investment of $2000 will double in approximately 8.7 years.

Exponential Decay

If, instead of growing at a rate proportional to the amount present, a quantity declines (or decays) at a rate proportional to the amount present, then the quantity is said to **decline exponentially** (or **decay exponentially**). The function that models this situation is

$$y = ce^{-kt},$$

where c and k are positive constants.

Exponential decline is associated with the study of radioactive substances, food production under adverse conditions, amounts of medicine remaining in the blood stream after an injection, and populations attacked by disease.

Exponential Decline

A function $y = f(t)$ satisfies the equation

$$\frac{dy}{dt} = -ky,$$

if and only if

$$y = ce^{-kt}.$$

If k and c are positive, we say that ***y declines*** *(or* ***decays****)* ***exponentially***.

See Figure 5.5.2 for a general graph of exponential decline.

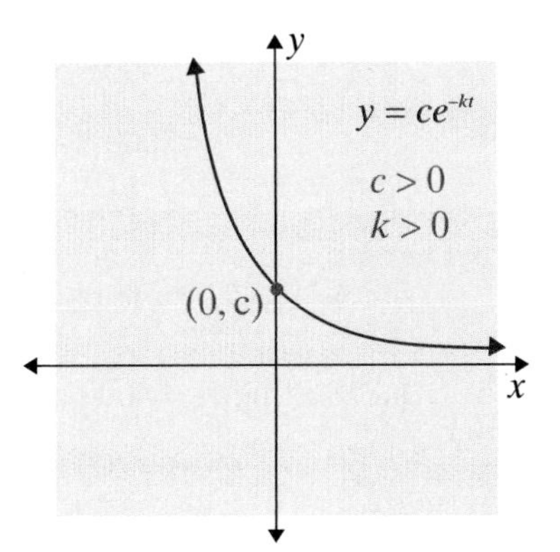

Figure 5.5.2

Example 3: Carbon-14

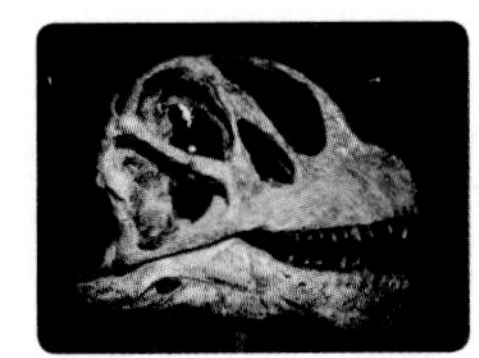

The radioactive substance carbon-14 is known to maintain a constant level in living plants and animals. When a plant or animal dies, the level of carbon-14 decays at a rate proportional to the amount present. That is, carbon-14 decays exponentially. For carbon-14, the constant of proportionality is approximately $k = 0.000124$. Suppose that an animal bone is found at an archeological dig and that it contains 40 percent of its original amount of carbon-14. How old is the bone?

Solution: Using the function

$$y = ce^{-0.000124t}$$

to represent the amount of carbon-14 present, we find that if $t = 0$, then

$$y = ce^{0} = c.$$

That is, c is the amount of carbon-14 present when the animal died. We know that t years after the animal's death there is 40 percent of the original amount of carbon-14 left. The amount of carbon-14 currently present is thus,

$$y = 0.40c.$$

Now, to find the age of the bone t, we substitute $y = 0.40c$ into our original equation, and solve for t.

$$y = ce^{-0.000124t}$$

$$0.40c = ce^{-0.000124t}$$

$$0.40 = e^{-0.000124t}$$

$$-0.000124t = \ln 0.40$$

$$t = \frac{\ln 0.40}{-0.000124} \approx \frac{-0.9163}{-0.000124} \approx 7400 \text{ years}$$

The bone is approximately 7400 years old.

Limited Growth

Even when environmental conditions are good, the growth of populations of wild animals is limited by such factors as a finite food supply and natural enemies or predators. Another example of limited growth occurs in industry. As workers become more efficient, their production level increases, but not more than the capacity of the company as restricted by money, materials, and machinery. This type of limited growth in productivity is called a **learning curve**.

Limited Growth

The mathematical model for limited growth is

$$y = c(1 - e^{-kt}),$$

where c and k are positive constants.

NOTES

1. As t increases, $e^{-kt} = \dfrac{1}{e^{kt}}$ decreases and approaches 0.
2. As t increases, $\left(1 - e^{-kt}\right)$ approaches 1.
3. As t increases, y approaches c, as shown in Figure 5.5.3.

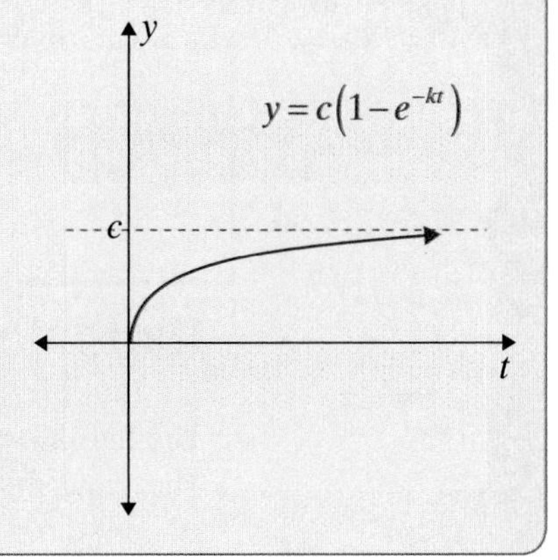

Figure 5.5.3

Example 4: Limited Growth

Suppose that a man learning to make widgets can make $N = 100(1 - e^{-0.3t})$ widgets per week after t weeks on the job.

a. About how many widgets will this man produce during his 4th week on the job?

b. At what rate will his widget-making skills be improving at the end of the 4th week? the 10th week?

Solutions: a. Substitute $t = 4$ into the formula for N.

$$\begin{aligned} N &= 100\left(1 - e^{-0.3\cdot 4}\right) \\ &= 100\left(1 - e^{-1.2}\right) \\ &\approx 100(1 - 0.30) \\ &= 100(0.70) = 70 \end{aligned}$$

He will make approximately 70 widgets during his 4th week on the job.

b. To find his rate of learning (or rate of improvement), we need $\dfrac{dN}{dt}$.

$$N = 100\left(1 - e^{-0.3t}\right) = 100 - 100e^{-0.3t}$$

$$\begin{aligned} \frac{dN}{dt} &= -100(-0.3)e^{-0.3t} && \text{Find } N'. \\ &= 30e^{-0.3t} \end{aligned}$$

At $t = 4$,

$$\left.\frac{dN}{dt}\right|_{t=4} = 30e^{-0.3\cdot 4} = 30e^{-1.2} \approx 30(0.3012) \approx 9.$$

At $t = 10$,

$$\left.\frac{dN}{dt}\right|_{t=10} = 30e^{-0.3\cdot 10} = 30e^{-3} \approx 30(0.0498) \approx 1.5.$$

At 4 weeks, his rate of improvement is about 9 widgets per week. By 10 weeks, he is getting closer to the top of his learning curve as his rate of improvement is slowing to 1.5 widgets per week.

Half-Life

The half-life of a material is the amount of time required for half the material to decay. The concept of half-life is particularly useful when studying radioactive decay. For example, the half-life of carbon-14 is about 5700 years, and the half-life of radium is about 1600 years.

Example 5: Half-Life

Suppose that the decay of a particular radioactive isotope is described by the equation

$$y = 200e^{-0.05t},$$

where y is the amount of the isotope in grams and t is time in years. Find the half-life of this isotope.

Solution: When $t = 0$, we have

$$y = 200e^{0} = 200.$$

Thus, initially, there are 200 grams of the isotope present. We want to find the time it takes for there to be 100 grams left (half of the initial amount of 200 grams).

Therefore, we want to find t in the equation

$$100 = 200e^{-0.05t}.$$

We divide both sides of the equation by 200 and then use the definition of natural logarithm:

$$\frac{100}{200} = \frac{200e^{-0.05t}}{200}$$

$$\frac{1}{2} = e^{-0.05t}$$

$$-0.05t = \ln\frac{1}{2}.$$

Thus

$$t = \frac{\ln\frac{1}{2}}{-0.05} \approx \frac{-0.6931}{-0.05} \approx 13.86.$$

The half-life of the isotope is about 13.86 years.

Logarithmic Differentiation and the General Exponential and Logarithmic Derivatives

In Exercise 52, Section 5.4, students were asked to calculate the slope of $f(x) = 10^x$. The result was $f'(x) = 2.3 \cdot 10^x$.

In general, when we considered any general exponential function $f(x) = B^x$, for some $B > 0$, we discovered that there was a constant, say b, such that $f'(x) = b \cdot B^x$. When $B = e$, then $b = 1$. We can use the technique of logarithmic differentiation to discover easily the nature of the constant b.

Starting with

$$f(x) = B^x$$

we can take the natural logarithm of both sides. This gives

$$\ln(f(x)) = \ln(B^x) = \ln(B) \cdot x.$$

So

$$\frac{d}{dx}\ln(f(x)) = \frac{d}{dx}(\ln(B) \cdot x) = \ln B \cdot \frac{d}{dx}x = \ln B \cdot 1$$

$$\frac{1}{f(x)}\frac{d}{dx}f(x) = \ln B \cdot 1.$$

General Exponential Derivative

1. *If* $f(x) = B^x$, *then* $f'(x) = \ln B \cdot f(x) = (\ln B) \cdot B^x$.
2. *Chain Rule: if* $f(x) = B^{g(x)}$

   then $f'(x) = (\ln B) \cdot B^{g(x)} \cdot g'(x)$.

The mysterious constant is the natural logarithm of the base B. If $x = 0$, then we see that $f'(0) = \ln B$.

Restated, if $f(x) = B^x$ for some $B > 0$, then $f'(x) = \ln B \cdot B^x$. Moreover, $f'(0) = \ln B$.

We can use the same method and the result just obtained to get a general formula for the derivative of any logarithmic function.

Suppose $y = \log(x) = \log_{10}(x)$. Then $10^y = x$.

So

$$\ln(10^y) = \ln x$$

$$y \cdot \ln(10) = \ln x$$

$$y = \frac{1}{\ln 10}\ln x$$

$$y' = \frac{1}{\ln 10} \cdot \frac{1}{x}.$$

General Logarithmic Derivative

1. *If* $y = \log_B(x)$ *then* $y' = \frac{1}{\ln B} \cdot \frac{1}{x}$.
2. *Chain Rule: If* $y = \log_B(g(x))$, *then* $y' = \frac{1}{\ln B} \cdot \frac{1}{g(x)} g'(x)$.

Example 6: Using the Derivatives

Given $f(x)=(x^2-3x+6)\log_2(x)$, determine $f'(x)$.

Solution: Use the Product Rule.

$$\frac{d}{dx}f(x)=(x^2-3x+6)\cdot\frac{d}{dx}\log_2(x)+\log_2(x)\cdot\frac{d}{dx}(x^2-3x+6)$$

$$\frac{d}{dx}f(x)=(x^2-3x+6)\cdot\frac{1}{\ln 2}\cdot\frac{1}{x}+\log_2(x)\cdot(2x-3)$$

Elasticity of Demand

In Section 2.2 we introduced the demand function $p = D(x)$, where p is the price per item at which consumers are willing to buy x items. Recall that revenue is the product of the number of items sold times the price per item,

$$R(x) = x \cdot p = x \cdot D(x).$$

We know that the demand functions are always decreasing, indicating that an increase in supply (items demanded or sold) will result in a decrease in price. Since revenue depends on both price and supply, a natural concern is whether an increase in supply will result in a percent price change so small that revenue increases or a percent price change so great that revenue decreases. Economists have developed a concept called **elasticity of demand** for determining just what effect can be expected.

Intuitively, if demand is "elastic," then an increase in sales will be accompanied by a relatively small decrease in price and will result in an increase in revenue. If demand is "inelastic," then an increase in sales will be accompanied by a relatively large decrease in price and will result in decrease in revenue.

For instance, if computers are currently selling about 10,000 units a year, then an increase of sales of 200 units will result in a **relative rate of change** in demand of $\frac{200}{10{,}000}=0.02=2$ percent per year. The calculation of elasticity of demand involves a comparison of the *relative rate of change of the quantity demanded* with the *relative rate of change in price*.

Consider the following analysis:

1. If Δx is the change in quantity demanded, then $\frac{\Delta x}{x}$ is the percent change in quantity.

2. If Δp is the change in price, then $\frac{\Delta p}{p}$ is the percent change in price.

3. The comparison we want is the ratio

$$\frac{\text{Percent change in } x}{\text{Percent change in } p} = \frac{\frac{\Delta x}{x}}{\frac{\Delta p}{p}} = \frac{p}{x} \cdot \frac{\Delta x}{\Delta p} = \frac{p}{x} \cdot \frac{1}{\frac{\Delta p}{\Delta x}}.$$

4. Now, if we let $\Delta x \to 0$ and assume that $p = D(x)$ is continuous, we have

$$\lim_{\Delta x \to 0}\left[\frac{p}{x} \cdot \frac{1}{\frac{\Delta p}{\Delta x}}\right] = \frac{p}{x} \cdot \frac{1}{\frac{dp}{dx}}.$$

For simplicity, we make the following adjustments in the notation:

$$\frac{p}{x} \cdot \frac{1}{\frac{dp}{dx}} = \frac{1}{x} \cdot \frac{p}{\frac{dp}{dx}} = \frac{1}{x} \cdot \frac{D(x)}{D'(x)}.$$

This last expression will be negative because $D'(x)$ is negative and x and $D(x)$ are positive. So that the measure of elasticity of demand will be a positive number, this expression is multiplied by –1.

Elasticity of Demand

If $p = D(x)$ is the demand function for a product, then the ***elasticity of demand*** *for that product is*

$$\boldsymbol{E = -\frac{1}{x} \cdot \frac{D(x)}{D'(x)}}.$$

Since total revenue, $R(x)$, equals quantity times unit price, we have: $R(x) = x \cdot D(x)$. Differentiating both sides gives (with the product rule) $R'(x) = D(x) + xD'(x)$. Using the definition of Elasticity of Demand, $x\,D'(x) = \frac{-D(x)}{E}$. Now, substituting in the expression for $R'(x)$,

$$R'(x) = D(x) - \frac{D(x)}{E}$$

$$R'(x) = D(x)\left(1 - \frac{1}{E}\right).$$

This relates the rate of change of revenue to elasticity of demand and helps one understand the analysis used in economics.

Properties of Elasticity of Demand

1. *If* $E > 1$, *then* $R'(x) = D(x)\left[1 - \frac{1}{E}\right] > 0$, *the revenue is increasing, and we say that the demand is* ***elastic.***

2. *If* $E < 1$, *then* $R'(x) = D(x)\left[1 - \frac{1}{E}\right] < 0$, *the revenue is decreasing, and we say that the demand is* ***inelastic.***

3. *If* $E = 1$, *then* $R'(x) = 0$, *the revenue is at its maximum, and we say that the demand has* ***unit elasticity.***

Figure 5.5.4 illustrates the various possibilities for E.

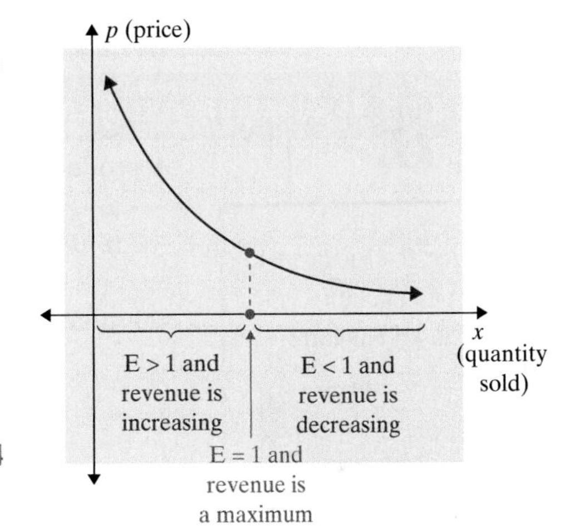

Figure 5.5.4

Example 7: Elasticity of Demand

Suppose that the demand function for a product is $p = D(x) = 300 - 2x$.

a. What is the price per unit if 50 units are sold?
b. Find the function describing the elasticity of demand E.
c. Find $E(50)$ and $E(110)$.
d. What quantity x maximizes revenue, R?

Solutions: a. For $x = 50$,

$$p = D(50) = 300 - 2(50) = 200.$$

If the demand is 50 units, the price per unit is \$200.

b. For $D(x) = 300 - 2x$,

$$D'(x) = -2. \qquad \text{Find } D'(x).$$

Substituting $D(x) = 300 - 2x$ and $D'(x) = -2$ in the formula for $E(x)$ gives

$$E(x) = -\frac{1}{x} \cdot \frac{300 - 2x}{-2} = \frac{150 - x}{x}.$$

continued on next page ...

c. $E(50) = \dfrac{150-50}{50} = \dfrac{100}{50} = 2 > 1$ Demand is elastic.

$E(110) = \dfrac{150-110}{110} = \dfrac{40}{110} \approx 0.36 < 1$ Demand is inelastic.

When the demand level is at 50 units, a change in the quantity sold will result in a relatively small change in price, and revenue will be increasing. When the demand level reaches 110, the revenue will be decreasing.

d. $R = xD(x) = x(300-2x) = 300x - 2x^2$
$R' = 300 - 4x$
If $R' = 0$, $x = 75$
Since $R'' = -4$, (a constant), at $x = 75$, R is concave down at $x = 75$ gives a maximum for R. This is consistent with part **c.**

Example 8: Elasticity of Demand

A product is known to have a demand function $p = D(x) = 100e^{-0.5x}$.

a. Find the value of x for which $E = 1$.
b. Find the value for x that maximizes the revenue.

Note: these should be the same x-value.

Solutions: a. $D(x) = 100e^{-0.5x}$

$D'(x) = 100(-0.5)e^{-0.5x} = -50e^{-0.5x}$ Find $D'(x)$.

Thus

$$E(x) = -\frac{1}{x} \cdot \frac{100e^{-0.5x}}{-50e^{-0.5x}} = \frac{2}{x}.$$

Substitute $D(x)$ and $D'(x)$ into $E(x)$.

Setting $E(x) = 1$ gives

$$\frac{2}{x} = 1$$
$$x = 2.$$

b. The revenue function is $R(x) = x \cdot P = x \cdot 100e^{-0.5x} = 100xe^{-0.5x}$. Find $R'(x)$ and set $R'(x) = 0$ to find the value of x that gives the maximum revenue.

$$R'(x) = 100x \cdot (-0.5)e^{-0.5x} + e^{-0.5x}(100)$$
$$= e^{-0.5x}(-50x + 100)$$
$$0 = e^{-0.5x}(-50x + 100)$$
$$0 = -50x + 100 \quad \text{or} \quad \cancel{0 = e^{-0.5x}}$$
$$50x = 100$$
$$x = 2$$

Thus $x = 2$ gives the maximum revenue.

Verifying the Maximum of a Function With a Calculator

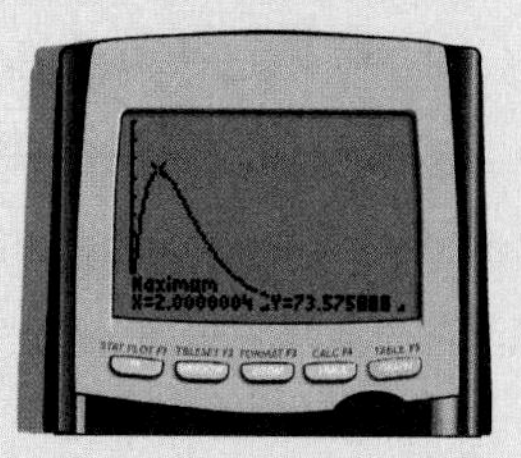

To find the value of x that maximizes the revenue in Example 8.

1. Enter the revenue function into Y1.
2. Graph Y1 in an appropriate window (0 by 20 for x and 0 by 100 for y).
3. With the graph displayed, press 2ND TRACE and select item 4 : maximum.

continued ...

continued ...

4. When asked `Left Bound?` on the graphing screen, use the arrows to move the blinking tracer to the left side of the maximum point. Press ENTER.
5. Do the same for `Right Bound?`. When asked `Guess?`, move the tracer in the middle of the left and right bound points and again press enter.
6. The maximum point will appear in decimal form at the bottom of the screen.

There is also an alternative approach to elasticity of demand, which we briefly outline here. The slope of the demand curve is negative, and as a consequence, the demand function is one - to - one and thus is invertible. Let $F = D^{-1}$ and write $F(p) = x$ whenever $p = D(x)$. If p is the unit price, then x is the quantity demanded at that price. We can differentiate $F(p)$ with respect to x and obtain

$$\frac{d}{dx}F(p) = \frac{d}{dx}(x)$$

$$F'(p)\frac{dp}{dx} = 1.$$

Note: $F = F(p)$ is the quantity demanded when the unit price is p.

Since $\frac{dp}{dx} = D'(x)$, the elasticity of demand becomes, with substitution, ($x = F$, and $p = D$).

$$E = -\frac{1}{x}\frac{D(x)}{D'(x)} = -\frac{1}{F(p)}\cdot\frac{p}{\frac{dp}{dx}}$$

$$E = -\frac{pF'(p)}{F(p)} = \frac{-p}{F(p)\frac{1}{F'(p)}}$$

This gives a useful equivalent formulation of elasticity of demand.

Example 9: Alternative Approach to Elasticity of Demand

A grocery store determines that the demand function for its bakery's bread is $x = 180 - 30p$, where x is the number of loaves of bread it sells daily and p is the unit price.

a. Find the quantity demanded when the price is \$1.70.
b. Find the function describing the elasticity as a function of p.
c. Find the elasticity at $p = \$1.70$.
d. Interpret the resulting elasticity.
e. Determine the revenue function R and find p so that R is a maximum.

Solution: a. The quantity demanded is x and $x = F(p)$. Thus

$$\begin{aligned} x &= F(1.70) \\ &= 180 - 30(1.70) \\ &= 129. \end{aligned}$$

b.
$$\begin{aligned} E &= -\frac{pF'(p)}{F(p)} \\ &= -\frac{p(-30)}{180 - 30p} \\ &= \frac{p}{6 - p} \end{aligned}$$

continued on next page ...

c. $E(1.70) = \dfrac{30(1.70)}{180-30(1.70)}$

$= \dfrac{51}{129} = \dfrac{17}{43}$

d. Since $E < 1$, the demand is inelastic and so an increase in price will bring a decrease in revenue.

e. $R = xp = (180-30p)p$

$= 180p - 30p^2$

Thus $R' = 180 - 60p$. Setting $R' = 0$, we obtain $60p = 180$, so $p = \$3.00$ per loaf.

Since $R'' = -60$, the function R is concave down at $p = 3$ and this price gives a maximum for the revenue.

5.5 Exercises

1. **Population.** The population of a city is growing exponentially at a rate of 3.5 percent per year. The population was 8400 in 2000.
   a. Find an exponential function that represents the population t years after 2000.
   b. What will be the population in the year 2010?
   c. When will the population be 12,800?

2. **Bee population.** A swarm of bees grows exponentially at a rate of 4 percent hourly. Initially, there were 900 bees in the swarm.
   a. Find an exponential function for the number of bees in the swarm after t hours.
   b. How many bees are in the swarm after 6 hours?
   c. How many hours will it take for the swarm to double in size? Round your answer to the nearest tenth.

3. **Cost.** In 1996, the cost of a prime rib dinner was about $9.00. In 1999, the cost was $12.00. If the cost is growing exponentially, predict the cost of a prime rib dinner in 2005?

4. **Ant colony.** A colony of ants is growing exponentially. When first observed, the colony contained about 400 ants. If at the end of 9 days there are about 700 ants, approximately how many ants will be present at the end of 15 days?

5. **Bacterial population.** A bacteria culture grows at a rate proportional to its size. If the population doubles every 6 hours, how long will it take for the population to be three times its initial size?

6. **Demand for oil.** The demand for oil in the United States doubles every 8 years. How long will it take for the demand to triple?

7. **Inflation.** The amount of goods and services that costs \$100 on January 1, 2000 costs \$139.10 on January 1, 2003. Estimate the cost of the same goods and services on January 1, 2010. Assume the cost is growing exponentially.

8. **Interest compounded continuously.** One thousand dollars is deposited in a savings account where the interest is compounded continuously. After 4 years, the balance will be \$1366.15. When will the balance be \$1870.00?

9. **Half-life.** The decay rate for a radioactive isotope is 2.6 percent per year. Find its half-life.

10. **Half-life.** The decay rate of a radioactive isotope is 6.5 percent per year. Find its half-life.

11. **Archaeological dating.** A wooden carving found at an archaeological dig contains about 34 percent of its carbon-14. Approximately how old is the carving?

12. **Archaeological dating.** Bones from the skeleton of an animal have lost 62 percent of their carbon-14. Estimate the age of the bones.

13. **Atmospheric pressure.** As the elevation above sea level is increased, the atmospheric pressure declines exponentially. The pressure at sea level is approximately 15 lbs per sq. in. and the pressure at 3000 feet of elevation is about 13 lbs per sq. in. Find the pressure at 5000 ft.

14. **Drug concentration.** The concentration of a drug in the body fluids is known to decline exponentially. If 20 mg of a drug is administered and 8 mg remains after 3 hours, how much will remain after 5 hours?

15. **Depreciation.** It is determined that the value of a piece of machinery declines exponentially. A machine that was purchased 5 years ago for \$65,000 is worth \$35,000 today. What will be the value of the machine 5 years from now?

16. **Population.** The population of a certain economically depressed union is declining exponentially at a rate of 1.5 percent. If the population in 1990 was 30,000, estimate the population in 2010.

17. **Reliability.** Studies show that the fractional part P of light bulbs that has burned out after t hours of use is given by $P = 1 - e^{-0.03t}$. What fractional part of the bulbs has burned out after 50 hours? How long will it be before half of the bulbs have burned out?

18. Advertising. A radio station estimates that during an intense advertising campaign, the number of people N who will hear a commercial is given by $N = A\left(1 - e^{-0.02x}\right)$, where A is the number of people in the broadcasting area and x is the number of times the commercial is run. If there are 60,000 people in the area,

a. how many people will hear the commercial if it is run 20 times?

b. how many times should the station plan to run the commercial to be certain that at least 30,000 people hear it?

19. Ecology. The Department of Fisheries has begun a reclamation project at a lake where the fish population was nearly destroyed by agricultural chemicals. They estimate that the population of fish in t years will be $P = 6000 - 5200e^{-0.28t}$.

a. What was the initial population?

b. What will be the population after 4 years?

c. How long will it take for the population to be 5000 fish?

20. Advertising. The manager of The Sound Lab has determined that after an intense advertising campaign, the monthly sales of a particular compact disc player can be approximated by $N = 300 + 180e^{-0.04t}$ units, where t is the number of months after the campaign.

a. Find the monthly sales initially.

b. Find the monthly sales when $t = 6$.

c. When will the monthly sales be 400 units?

21. Skills development. Beverly is making a small souvenir to give to each person attending her family reunion. The length of time, in minutes, she takes to make the nth one is given by the function $T(n) = 12 + 30e^{-0.1n}$. How long will it take her to make the 30th souvenir?

22. Dairy farming. The number of dairy farmers in a particular state who are feeding a new supplement to their milking cows is given by the function $W(t) = 340\left(1 - e^{-0.09t}\right)$, where t is the number of months the supplement has been available. How long will it be before 200 farmers are feeding the supplement to their cows?

23. Cost. The total cost function for a local company is given by $C(t) = 12 - ce^{-kt}$ in thousands of dollars, where t is the time in months. The fixed costs are \$5000 and the total cost after 2 months is \$10,200. Find the total cost at the end of 6 months.

24. Skills development. The time that it takes a service attendant to change a tire is given by the function $T(x) = 4.4 + Ce^{-kx}$ minutes, where x is the number of tires the attendant has changed before. It takes Patrick 15 minutes to change the first tire ($x = 0$) and 9.3 minutes to change the seventh tire. How long will it take him to change the eleventh tire?

For each of the demand functions in Exercises 25 – 40, find **a.** *the elasticity function and* **b.** *the value of x that maximizes the revenue.*

25. $p = D(x) = 84 - 3x$

26. $p = D(x) = 144 - 1.5x$

27. $p = D(x) = 520 - 2.6x$

28. $p = D(x) = 480 - 3.2x$

29. $p = D(x) = 200e^{-0.2x}$

30. $p = D(x) = 67e^{-0.1x}$

31. $p = D(x) = 88e^{-0.025x}$

32. $p = D(x) = 130e^{-0.04x}$

33. $p = D(x) = \sqrt{150 - x}$

34. $p = D(x) = \sqrt{180 - 2x}$

35. $p = D(x) = \sqrt{162 - 3x}$

36. $p = D(x) = \sqrt{255 - 2.5x}$

37. $p = D(x) = 18 - \sqrt{x}$

38. $p = D(x) = 21 - 2\sqrt{x}$

39. $p = D(x) = 363 - x^2, \quad x \le 18$

40. $p = D(x) = 600 - 0.5x^2, \quad x \le 34$

41. Maximum revenue. The demand function for an electric pencil sharpener is given by $p = D(x) = 19.2 - 0.4x$ dollars. Find the level of production for which the revenue is maximized.

42. Maximum revenue. The demand function for a popular stereo tuner is given by $p = D(x) = 540 - 0.05x^2$ dollars. Find the level of production for which the revenue is maximized.

43. Elastic demand. The demand function for an exclusive wool blanket is given by $p = D(x) = 33 - 2\sqrt{x}$ dollars, where x is in thousands of blankets. Find the level of production for which the demand is elastic.

44. Elastic demand. Find the levels of production for which the demand is elastic if the demand is given by $p = D(x) = \sqrt{207 - 3x}$ dollars.

45. Elastic demand. An arcade sells video games and determines that $x = 30\left(1 - e^{-\frac{p}{10}}\right)$, where x is the number of video games demanded for a unit price p.

a. Determine the quantity demanded when $p = \$10$ per game.

b. Determine E and interpret the result at $p = \$10$.

c. What revenue is generated at $p = \$10$?

46. Elastic demand. Lucky Blooms sells a new rose variety which has established a demand of $x = f(p) = \dfrac{e^{\frac{p}{3}} + 350}{e^{\frac{p}{2}}}$.

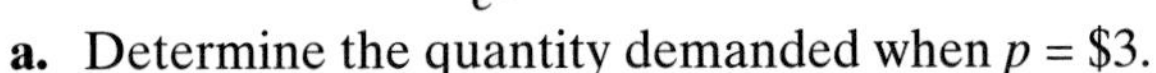

a. Determine the quantity demanded when $p = \$3$.

b. Determine E and interpret the result at $p = \$3$.

47. Elastic demand. The demand for a product is given by $x = F(p) = \dfrac{1800}{10 + \ln(1+p)}$.

a. If $p = 20$, determine E and interpret the results.

b. What is the revenue function R?

c. Use R' to determine if R is increasing at $p = 20$. Is your answer consistent with part **a.**?

48. Elastic demand. Suppose a product has a demand function $x = F(p) = 300e^{-\frac{p}{10}}$.

a. Find the elasticity function.

b. Is the demand elastic or inelastic at $p = \$20$.

c. Determine the unit price which maximizes revenue.

d. Discuss whether or not your answers to **b.** and **c.** are consistent.

49. Elastic demand. Suppose the demand for a product is $p = D(x) = 300e^{-\frac{x^2}{200}}$.

a. Determine the unit price if the quantity $x = 5$.

b. What is formula for $D'(x)$?

c. What is elasticity at $x = 5$?

d. Determine the value of x which maximizes revenue.

Hawkes Learning Systems: Essential Calculus

Applications of Exponential Functions: Growth and Decay
p. 436 - 443 Ex. 1 - 24

Logarithmic Differentiation and Elasticity of Demand
p. 443 - 450 Ex. 25 - 49

Chapter 5 Index of Key Ideas and Terms

Section 5.1 Exponential Functions

Exponential Function with Base b pages 393 - 395

A function of the form

$$f(x) = c \cdot b^{rx},$$

where b is a positive real number, $b \neq 1$, and c and r are real constants, is called an exponential function with base b.

Properties of Exponential Functions page 395

For any exponential function $f(x) = c \cdot b^{rx}$, the following properties are true.

1. The domain is all real numbers.
2. The range is the set of all positive real numbers.
3. The y-intercept is at $(0, c)$.
4. The x-axis is a horizontal asymptote.
5. The function is continuous for all real x.
6. Suppose the base b is greater than 1. Then, when $c = 1$ and $r = 1$, the slope is positive, y is increasing and concave up.
7. Suppose the base b satisfies $0 < b < 1$. Then, when $c = 1$ and $r = 1$, the slope is negative, y is decreasing and concave up.

Compound Interest Formula page 398

For principal P invested at annual rate r (in decimal form) compounded n times per year for t years, the amount A is given by

$$A = P\left(1 + \frac{r}{n}\right)^{nt}.$$

Definition of e pages 398 - 399

$$e = \lim_{x \to +\infty}\left(1 + \frac{1}{x}\right)^{x} = 2.71828182846\ldots$$

Continuous Compound Interest Formula page 400

For principal P, invested at annual rate r (in decimal form), compounded continuously for t years, the amount A is given by

$$A = Pe^{rt}.$$

Section 5.2 The Algebra of the Natural Logarithm Function

Natural Logarithm pages 405 - 406

For natural logarithms,

$$e^y = x \quad \text{if and only if} \quad y = \ln x.$$

Properties of Natural Logarithms page 407 - 412

For positive real numbers M and N:

1. $\ln(MN) = \ln M + \ln N$.
2. $\ln\left(\frac{M}{N}\right) = \ln M - \ln N$.
3. $\ln(M^r) = r \cdot \ln M$
4. $\ln M = \ln N$ if and only if $M = N$.
5. $\ln 1 = 0$
6. $\ln e = 1$
7. $e^{\ln x} = x$
8. $\ln e^x = x$
9. The domain is all positive numbers, and the range is all real numbers.

Section 5.3 Differentiation of Logarithmic Functions

Derivative of ln *x* pages 415 - 418

If $f(x) = \ln x$, then

$$f'(x) = \frac{1}{x}.$$

Derivative of ln(g(*x*)) pages 418 - 419

If $y = \ln(g(x))$, then

$$y' = \frac{1}{g(x)} \cdot g'(x) = \frac{g'(x)}{g(x)}.$$

Logarithmic Differentiation pages 420

Taking the logarithm of both sides of an equations, and then differentiating.

Graphing Logarithmic Functions page 422

Section 5.4 Differentiation of Exponential Functions

Slopes and Exponential Functions pages 426 - 428

Derivative of the Exponential Function pages 428 - 429

If $f(x) = e^x$, then

$$f'(x) = e^x.$$

continued on next page ...

Section 5.4 Differentiation of Exponential Functions (continued)

Chain Rule for Exponential Functions pages 429 - 430

If $f(x)=e^{g(x)}$, then

$$f'(x)=e^{g(x)}\cdot g'(x).$$

Graphing Exponential Functions pages 431 - 432

Section 5.5 Applications of Exponential Functions

Exponential Growth pages 436 - 439

A function $y = f(t)$ satisfies the equation

$$\frac{dy}{dt}=ky,$$

if and only if

$$y=ce^{kt}.$$

If k is positive, we say that y grows (or increases) exponentially.

Exponential Decay pages 439 - 441

A function $y = f(t)$ satisfies the equation

$$\frac{dy}{dt}=-ky,$$

if and only if

$$y=ce^{-kt}.$$

If k is positive, we say that y declines (or decays) exponentially.

Limited Growth pages 439 - 440

The mathematical model for limited growth is

$$y=c(1-e^{-kt}),$$

where c and k are positive constants.

Half-Life page 442 - 443

Logarithmic Differentiation and the General Exponential and Logarthmic Derivatives pages 443 - 444

If $y=B^{x}$, then $\frac{dy}{dx}=\ln B\cdot B^{x}=(\ln B)y$

If $y=\log_{B}(x)$, then $\frac{dy}{dx}=\frac{1}{(\ln B)x}$

Chain Rules

If $y=B^{g(x)}$, then $\frac{dy}{dx}=(\ln B)\cdot B^{g(x)}\cdot g'(x)$

If $y=\log_{B}(g(x))$, then $\frac{dy}{dx}=\frac{g'(x)}{(\ln B)g(x)}$

continued on next page...

Section 5.5 Applications of Exponential Functions (continued)

Elasticity of Demand pages 445 - 450

If $p = D(x)$ is the demand function for a product, then the elasticity of demand for that product is

$$E = -\frac{1}{x} \cdot \frac{D(x)}{D'(x)}.$$

Properties of Elasticity of Demand page 447

1. If $E > 1$, then $R'(x) = D(x)\left[1 - \frac{1}{E}\right] > 0$, the revenue is increasing, and we say that the demand is elastic.
2. If $E < 1$, then $R'(x) = D(x)\left[1 - \frac{1}{E}\right] < 0$, the revenue is decreasing, and we say that the demand is inelastic.
3. If $E = 1$, then $R'(x) = 0$, the revenue is at its maximum, and we say that the demand has unit elasticity.

Elasticity of Demand (Alternate Form) pages 449 - 450

If $x = F(p)$ is the demand function for a product, then the elasticity of demand for that product is

$$E = -\frac{pF'(p)}{F(p)}.$$

Chapter 5 Review

For a review of the topics and problems from Chapter 5, look at the following lessons from *Hawkes Learning Systems: Essential Calculus*.

Exponential Functions
The Algebra of the Natural Logarithmic Function
Differentiation of Logarithmic Functions
Differentiation of Exponential Functions
Applications of Exponential Functions: Growth and Decay
Logarithmic Differentiation and Elasticity of Demand

Chapter 5 Test

Graph each of the functions in Exercises 1 – 4.

1. $y = 3^x$

2. $y = 2^{-0.5x}$

3. $y = 4e^{-0.5x}$

4. $y = 3e^{0.2x}$

5. \$8000 is invested in a savings account at a rate of 7 percent per year. Find the total amount on deposit at the end of 4 years if the interest is compounded

a. annually.
b. semiannually.
c. quarterly.
d. daily.
e. continuously.

6. How much must be deposited in a savings account that pays interest at a rate of 7.8 percent compounded quarterly so that after 4 years, the balance will be \$8600?

Express each equation in Exercises 7 and 8 in logarithmic form.

7. $e^{0.25} = 1.284$

8. $e^m = 14.6$

Express each equation in Exercises 9 and 10 in exponential form.

9. $\ln 1.98 = 0.6831$

10. $\ln P = 1.3741$

In Exercises 11 – 14, use the properties of logarithms to rewrite each expression as a sum, difference, or constant multiple of logarithms. Replace all radicals with exponents.

11. $\ln \frac{7x-2}{5}$

12. $\ln \sqrt[3]{1-12x}$

13. $\ln \frac{x^2}{y^2-1}$

14. $\ln\left(xy^2\right)^3$

Solve each of the equations in Exercises 15 – 20.

15. $e^{0.3x} = 29$

16. $e^{1-2x} = 18.6$

17. $12^{x+2} = 267$

18. $\ln(x+1) = 2.631$

19. $\ln x + \ln(2x-1) = 0$

20. $\ln\left(x^2-3x+2\right) - \ln(x-2) = 2$

Find the derivative for each of the functions in Exercises 21 – 26.

21. $f(x) = x^3 + e^{-2x}$

22. $f(x) = e^x(4x^2 - 6)$

23. $y = (e^{3x} - 7)^5$

24. $y = \dfrac{\ln(x+2)}{e^x}$

25. $y = \ln\left[(2x^2+5)^3(x+3)\right]$

26. $y = \ln\left[\dfrac{6x+1}{(2x-5)^3}\right]$

For each of the functions in Exercises 27 and 28, find the absolute extrema on the indicated interval.

27. $f(x) = 200e^{\sqrt{x}-0.25x}$; $[0, 5]$

28. $f(x) = xe^x - 3e^x$; $[1, 4]$

Using the graph sketching technique discussed in Chapter 4, sketch the graph of each function in Exercises 29 and 30.

29. $f(x) = \sqrt{x}\ln x$

30. $f(x) = e^x - e^{-x}$

31. Interest compounded continuously. How much money should be invested today in an account paying 9 percent compounded continuously so that the account will contain $8000 six years from now?

32. Cell growth. A cell culture grows from 1000 to 1600 cells in 12 hours. If the rate of growth is proportional to the size, when will the culture contain 2000 cells?

33. Depreciation. A machine depreciates continuously at a rate of 15 percent. If the machine costs $9400 new, find its value after 6 years.

For each of the demand functions in Exercises 34 and 35, find **a.** *the elasticity function and* **b.** *the value of x that will maximize revenue.*

34. $D(x) = \sqrt{108 - 1.5x}$

35. $D(x) = 45 - 2\sqrt{x}$

Cumulative Review

In Exercises 1 – 4, use the graph of y = f(x) to determine the limits.

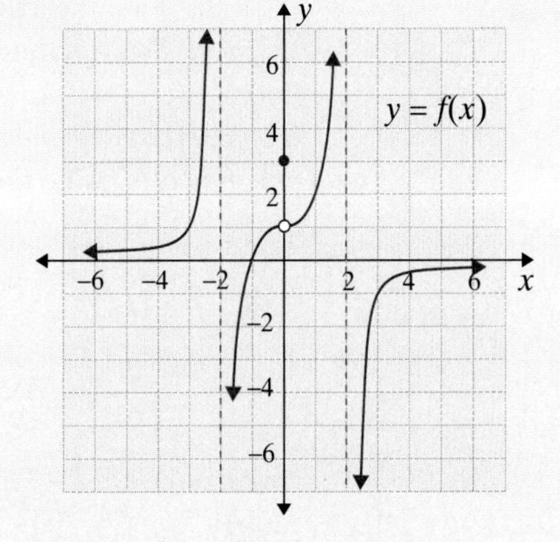

1. $\lim\limits_{x \to 2^-} \left(f(x)\right)$ **2.** $\lim\limits_{x \to 2^+} \left(f(x)\right)$

3. $\lim\limits_{x \to \infty} \left(f(x)\right)$ **4.** $\lim\limits_{x \to -\infty} \left(f(x)\right)$

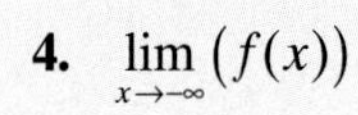

5. Determine the equation of the tangent line to $y = 14 + 2x - x^2$ at the point (0, 14). View the graph of the function and the tangent line with a graphing calculator.

6. In a psychology study, Calculus I students were asked to find derivatives of ten functions. The student responses are modeled by $R(t) = 100\left(1 - 2^{-t}\right)$, where t is the number of times the test has been given and R is the percentage of students scoring 90% or better on the test. Here, $0 \leq t \leq 6$ and $0 \leq R \leq 100$.

- **a.** Use a graphing calculator to sketch the graph.
- **b.** Interpret $R(2) = 75$.
- **c.** Interpret $R'(2) = 17.3$.
- **d.** Why is $R(3) \neq 75 + 17.3$?

7. Determine the derivative of $f(x) = 1 + \dfrac{5}{\sqrt{x}} - \dfrac{6}{\sqrt[3]{x}}$.

8. Suppose the total cost for a manufacturer of grommets is given by $C(x) = 500 + 0.01x + 0.003x^2$, where x is the number of grommets produced. The total revenue is given by $R(x) = x$.

- **a.** What is the unit price p?
- **b.** Determine the profit function, $P(x)$.
- **c.** Give the marginal cost function.
- **d.** Give the marginal profit function.
- **e.** What is the average cost $A(x)$ at $x = 1000$. (Hint: The average cost is given by $A(x) = \dfrac{C(x)}{x}$.)
- **f.** Set $A'(x)$ equal to zero and solve for marginal cost. Determine whether the following is true or false: marginal cost equals average cost when average cost is a minimum.

9. Suppose $C(x) = 0.04x^2 + 90x + 1500$ represents the cost of raising x hogs.

a. Find the average cost function.
b. Find the average cost of producing 200 hogs.
c. What is the marginal average cost when 200 hogs are produced?

For Exercises 10 and 11, find a value for k so that the given function will be continuous at the indicated value for x.

10. $f(x) = \begin{cases} \frac{1}{2}x^2 & \text{if } x < 3 \\ 3e^{kx} & \text{if } x \geq 3 \end{cases}, \quad x = 3$

11. $f(x) = \begin{cases} \frac{1}{x} & \text{if } x \leq 4 \\ \ln(kx) & \text{if } x \geq 4 \end{cases}, \quad x = 4$

For Exercises 12 – 14, calculate f′(x) and f″(x).

12. $f(x) = 14 - 3x + 6x^2$

13. $f(x) = (3 + 16x)(8 - 2x - 4x^2)$

14. $f(x) = \sqrt{100 + 3x}$

For Exercises 15 and 16, use the Chain Rule to find $f'(x)$. Then evaluate $f'(x)$ for the given value of x.

15. $f(x) = (xe^x + x^2)^{\frac{3}{2}}, \quad x = 2$

16. $f(x) = \left(x^{\frac{3}{2}} + 2e^{3x}\right)^{\frac{1}{2}}, \quad x = \frac{1}{2}$

For Exercises 17 and 18, use your knowledge of y′ and y″ to locate all maximum and minimum points and make a rough sketch of f(x). Show all details of your work. Use a graphing calculator only to confirm the analysis you made.

17. $f(x) = 30 + 8x - 2x^2$

18. $f(x) = 2x + \frac{3}{x}$

19. Let $f(x) = \frac{x^2 + 9}{x}$.

a. What are the asymptotes for $f(x)$?
b. Calculate $f'(x)$ and $f''(x)$.
c. Locate the minimum value of f for positive x.

For Exercises 20 and 21, determine the location of all inflection points and determine the intervals on which f(x) is concave up or concave down.

20. $y = 2x^3 - 8x + 11$

21. $y = \frac{1}{1 + x^2}$

22. Suppose the students in a court recorder certification class can take dictation at a rate of $W(t) = 50 + \frac{65t^2}{t^2 + 12}$ words per minute where t is time in weeks of class. For what values of t is W increasing?

For Exercises 23 and 24, complete the following parts:

a. *Determine formulas for* $\frac{dy}{dx}$ *and* $\frac{d^2y}{dx^2}$.

b. *For what a-values is* $f'(a) = 0$?

c. *Use the second derivative test to determine for which a-values f(a) is a maximum or minimum.*

d. *Set* $f''(x) = 0$ *and locate potential inflection points. Determine which are actual inflection points and which are not.*

23. $y = 3x^3 - 144x$

24. $y = 140 + 30x - x^2$

In Exercises 25 and 26, sketch a continuous function that satisfies each of the given conditions. In each case, draw only the portion of the graph for positive x values.

25. **a.** $f(6) = 2$
b. $f'(6) = -2$
c. $f''(6) = 2$

26. **a.** $f(3) = -3$
b. $f'(a) > 0$ for all $a \geq 0$
c. $f''(a) < 0$ if $a < 3$
d. $f''(a) > 0$ if $a > 3$

27. If the cost and revenue functions for crates of tomatoes are $C(x) = 0.002x^2 + 29x + 2400$ and $R(x) = 45x$, determine the following:

a. The marginal cost and marginal revenue functions.
b. The profit functions and marginal profit functions.
c. The value of x which will maximize profits.

28. Suppose a vendor at a flea market sells T-shirts to tourists for \$5.50 each. He sells an average of about 45 T-shirts daily, and they cost \$3 each. His data suggests he can sell about 5 more shirts for each \$0.20 price decrease he makes. What selling price will maximize his profit?

Find the derivative for each of the functions in Exercises 29 – 32.

29. $y = 5x - \ln x$

30. $y = x^2 \ln x$

31. $f(x) = \ln\sqrt{x^2 + 2}$

32. $f(x) = \frac{\ln(2x+1)}{x^2}$

Use logarithmic differentiation to find the derivative for each of the functions in Exercises 33 and 34.

33. $y = \frac{(5x+2)(2x+1)^2}{x+4}$

34. $y = \frac{11x^3}{(2x-9)\sqrt{4x+5}}$

For each of the functions in Exercises 35 and 36, find the absolute extrema on the indicated interval.

35. $f(x) = x - 2\ln(x+1);\ [0, 3]$

36. $f(x) = x(1 - 2\ln x);\ [1, 6]$

Using the graph sketching technique discussed in Chapter 4, sketch the graph of each function in Exercises 37 and 38.

37. $f(x) = 2x - \ln x$

38. $f(x) = 4e^{-x^2}$

39. Half-life. The half-life of a radioactive material is 1480 years. After how many years will only 20 percent of the material remain?

40. Skills development. The time required for a new employee to assemble the nth picture frame is given by $T(n) = 10 + 12e^{-0.09(n-1)}$ minutes.

- **a.** How long will it take the employee to assemble the first frame?
- **b.** How long will it take him to assemble the 20th frame?

41. Sales. It has been determined that t weeks after the introduction of a promising new product, the weekly sales for the product is given by $N(t) = 360 - 280e^{-0.08t}$ units. What is the level of sales after 6 weeks?

For each of the demand functions in Exercises 42 and 43, find **a.** *the elasticity function and* **b.** *the value of x that will maximize revenue.*

42. $D(x) = 63 - 0.9x$

43. $D(x) = 120e^{-0.3x}$

Integration with Applications

CHAPTER 6

Did You Know?

In the early 1600's Johannes Kepler (1571 – 1630) discovered three laws describing the nature of the relationship and relative motion of the Earth, the moon, the other known planets, and the sun. Kepler's first law states that planets travel around the sun in elliptical orbits. His second law states that an imaginary line between a planet and the sun sweeps out equal areas in equal times. His third law helps determine the amount of time it takes a planet to complete one full orbit.

Kepler's Second Law of Motion illustrates an application of integral calculus to physics. The equality of the two areas is possible because the planet orbits at a higher velocity when it is closer to the sun and at a lower velocity when it is farther away. To determine his second law, Kepler calculated the rate of change of the position of the planets using very accurate data supplied by his mentor, Tycho Brahe (1546 – 1601). Next, he determined a position function of the planets by using the sum of the areas involved. The process of using rates (derivatives) and areas to get position functions is a consequence of what is presented in Chapter 6 under the imposing title of "The Fundamental Theorem of Calculus."

Integrals form one subcategory of calculus, but what are integrals and integration? When we take a derivative, we find the rate of change of a function. Integration is exactly the opposite. When we take an integral, we transform the rate of change into its original function. For example, we know that velocity is the derivative of position, so we know that position is the integral of velocity. The same relationship is true of velocity and acceleration.

In engineering, biology, business, and economics, the process of using data about rates of change of y with respect to x in order to infer the form of a function of y with respect to x is now a common approach to problem solving. This chapter will address applications of the integral in these fields. In addition, the study of integration will lead us to powerful applications which are only now emerging.

Our emphasis in Chapter 6 is on understanding the process that is the reverse of differentiation by using many basic applications. In Section 6.1 we discuss only a minimum number of basic formulas and rules for integration. In Section 6.2 these formulas are expanded to cover a broader spectrum of functions with a technique called ***u*-substitution**. Even here, though, we use only the same few basic formulas of integration.

There are two general problems which we will ask the student to complete. The first is to determine a possible function $f(x)$ given $f'(x)$. The second type of problem is to apply this formula to calculate specific values of $f(x)$ which will have geometric or physical meaning in the problem at hand. The many applications will give the student the opportunity to master the technique while developing an appreciation for the wide extent of calculus applications in business, statistics, and science. The culmination and high point (theoretically!) of the book is in Section 6.3 where we explain the connection between definite integrals and what is called the **Fundamental Theorem of Calculus**. In Sections 6.4, 6.5, and 6.6, we apply the theory to a wide range of new applications.

6.1 The Indefinite Integral

Objectives

After completing this section, you will be able to:

1. *Use the definition of the antiderivative.*
2. *Use rules of integration to find integrals.*

The Antiderivative

In Chapter 2 through Chapter 5, we developed several formulas for differentiation and used those formulas to find rates of change, slopes of tangent lines, marginal cost, maxima and minima, and so on. Now we want to consider reversing the process of differentiation. That is, we want to **antidifferentiate**.

If the management of a company knows the current rate of change of cost (i.e., they know the marginal cost function), they can use this knowledge to find the related cost function and proceed to make a reasonably accurate budget projection. With special instruments, scientists can measure the speed of particles along straight lines. That is, they know the derivative of the position function (or velocity), and with this information, they can find the function that will tell the location of the particle at any particular time. In such situations, a derivative is known and the objective is to find the function (or functions) with this derivative.

Antiderivative

If F and f are two functions and $\mathbf{F'(x) = f(x)}$ *for all x in the domain of f, then F is an* ***antiderivative*** *of f.*

Note: *We also say that F(x) is an antiderivative of f(x).*

For example, suppose that

$$P(x) = 4x^2, \quad Q(x) = 4x^2 + 5, \quad \text{and} \quad R(x) = 4x^2 - 100.$$

In each case, we have

$$P'(x) = Q'(x) = R'(x) = 8x.$$

Thus $P(x)$, $Q(x)$, and $R(x)$ are all antiderivatives of $8x$. Any two of these antiderivatives differ from each other by a constant; in fact, all the antiderivatives of $8x$ belong to the family of functions $4x^2 + C$, where C is an arbitrary constant.

Consider the following general analysis. Suppose that $G(x)$ and $F(x)$ are both antiderivatives of $f(x)$. This means that $G'(x) = f(x)$ and $F'(x) = f(x)$. Now let

$$H(x) = G(x) - F(x).$$

Then

$$H'(x) = G'(x) - F'(x) = f(x) - f(x) = 0.$$

If the derivative of $H(x)$ is zero for every x, then we know that $H(x)$ is some constant C because only constant functions have 0 for a derivative. Therefore,

$$G(x) - F(x) = C$$

or

$$G(x) = F(x) + C.$$

Antiderivatives Differ by a Constant

If G and F are both antiderivatives of the same function f, then

$$\mathbf{G(x) = F(x) + C},$$

where C is a constant.

We emphasize that "any two antiderivatives of the same function differ by a constant." Geometrically, this means that the graphs of $y = G(x)$ and $y = F(x)$ are congruent, and the vertical distance between any two points on their graphs is C.

Example 1: Antiderivatives

Show that both $F(x)=x^3+x^2+50$ and $G(x)=x^3+x^2-80$ are antiderivatives of $f(x)=3x^2+2x$.

Solution: Differentiating both F and G, we have

$$F(x)=x^3+x^2+50 \quad \text{and} \quad G(x)=x^3+x^2-80$$
$$F'(x)=3x^2+2x \qquad\qquad G'(x)=3x^2+2x.$$

Thus

$$F'(x)=G'(x)=f(x),$$

and F and G are both antiderivatives of f. Note that $F(x)-G(x)$ is a constant.

Formulas of Integration

Now we want to develop a general approach to find antiderivatives. The **indefinite integral** of a function represents the set of all its antiderivatives. This integral is symbolized by an elongated *s* in the form $\int$.

The Indefinite Integral

If F(x) is any antiderivative of f(x), then the ***indefinite integral*** *of f(x), symbolized by* $\int f(x)dx$, *is defined as*

$$\int f(x)dx=F(x)+C,$$

where C is an arbitrary constant. f(x) is called the ***integrand****, dx is called the* ***differential*** *(see Section 4.6), and C is called the* ***constant of integration****.*

NOTES

The differential dx indicates that the integral is taken with respect to the variable x. At this time it plays a minor role, but it is a necessary part of the notation. Its importance will become apparent in the next two sections.

To **integrate** (or **to find the indefinite integral of**) a function, we develop formulas based on reversing the differentiation formulas that we already know. In this chapter we are interested in only four basic integrals and combinations of these integrals. We develop these integrals on an informal basis with the knowledge that their proofs lie in the fact that their derivatives do give the integrands.

Formulas of Integration

I. $\int \mathbf{k\,dx = kx + C}$

II. $\int \mathbf{x^r dx = \frac{1}{r+1} \cdot x^{r+1} + C},$ *where* $r \neq -1$.

III. $\int \mathbf{\frac{1}{x} dx = \int x^{-1} dx = \ln|x| + C}$ Note: the power rule, formula II, does not apply.

IV. $\int \mathbf{e^x dx = e^x + C}$

I. $\int k dx = kx + C$

Consider the following pattern:

a. If $F(x) = 8x$, then $F'(x) = 8$. Therefore, $\int 8\,dx = 8x + C$.

b. If $F(x) = 3x + 5$, then $F'(x) = 3$. Therefore, $\int 3\,dx = 3x + C$.

c. If $F(x) = 2x - 7$, then $F'(x) = 2$. Therefore, $\int 2\,dx = 2x + C$.

These results imply the following: for any constant *k*,

$$\int k\,dx = kx + C.$$

II. $\int x^r dx = \frac{1}{r+1} \cdot x^{r+1} + C$ (where $r \neq -1$.)

a. If $F(x) = \frac{1}{2}x^2$, then $F'(x) = \frac{1}{\cancel{2}} \cdot \cancel{2}x = x$. Therefore, $\int x\,dx = \frac{1}{2}x^2 + C$.

b. If $F(x) = \frac{1}{3}x^3 + 14$, then $F'(x) = \frac{1}{\cancel{3}} \cdot \cancel{3}x^2 = x^2$.

Therefore, $\int x^2 dx = \frac{1}{3}x^3 + C$.

c. If $F(x) = -\frac{1}{3}x^{-3}$, then $F'(x) = -\frac{1}{\cancel{3}}(-\cancel{3})x^{-4} = x^{-4}$.

Therefore, $\int x^{-4} dx = -\frac{1}{3}x^{-3} + C$.

Thus, for $r \neq -1$,

$$\int x^r dx = \frac{1}{r+1} \cdot x^{r+1} + C.$$

III. $\int \frac{1}{x} dx = \int x^{-1} dx = \ln|x| + C$

We know that if $x > 0$ and $F(x) = \ln x$, then

$$F'(x) = \frac{1}{x}.$$

However, suppose that $x < 0$ and

$$G(x) = \ln(-x).$$

Then, by the Chain Rule,

$$G'(x) = \frac{1}{-x} \cdot \frac{d}{dx}[-x] = -\frac{1}{x}(-1) = \frac{1}{x}.$$

This means that both

$$F(x) = \ln x \quad (\text{for } x > 0)$$

and

$$G(x) = \ln(-x) \quad (\text{for } x < 0)$$

have the same derivative, namely, $f(x) = \frac{1}{x} = x^{-1}$. Since logarithms are defined only for positive numbers, we use $|x|$ in the following integral. This means that the integral $\int \frac{1}{x} dx$ is defined when x is positive or when x is negative.

$$\int \frac{1}{x} dx = \int x^{-1} dx = \ln|x| + C.$$

IV. $\int e^x dx = e^x + C$

Since the function $f(x) = e^x$ is its own derivative, a reasonable conclusion is that it is also its own integral. For example, if

$$F(x) = e^x + 7, \quad \text{then} \quad F'(x) = e^x.$$

Thus we see that

$$\int e^x dx = e^x + C.$$

Individually, the four integral formulas we have just discussed are somewhat limiting in their applications. However, the following two rules, essentially reversing two familiar rules of differentiation, allow us to integrate a wide variety of functions.

Constant Multiple Rule

$$\int kf(x)\,dx = k\int f(x)\,dx$$

In words, the integral of a constant times a function is equal to that constant times the integral of the function.

Sum and Difference Rule

$$\int [f(x) \pm g(x)]\,dx = \int f(x)\,dx \pm \int g(x)\,dx$$

In words, the integral of a sum (or difference) of two functions is the sum (or difference) of their individual integrals.

The following examples illustrate a variety of applications of the formulas and rules that we have discussed for finding indefinite integrals.

Example 2: Finding the Indefinite Integral

Find $\int 3x^5 dx$.

Solution: $\int 3x^5 dx = 3\int x^5 dx$ — By the Constant Multiple Rule.

$= 3 \cdot \frac{1}{5+1} \cdot x^{5+1} + C$ — By Formula II.

$= \frac{\cancel{3}}{\cancel{6}_2} x^6 + C$

$= \frac{1}{2} x^6 + C$

Check: Each integral can be checked by differentiation. Its derivative must be the integrand.

$$\frac{d}{dx}\left[\frac{1}{2}x^6 + C\right] = \frac{1}{2} \cdot 6x^5 + 0 = 3x^5$$

NOTES

Just as the derivative rules apply regardless of the variables used, it is also true that the variable used in integration is a matter of choice. For example, $\int 3w^5 dw = \frac{1}{2}w^6 + C$.

Example 3: Finding the Indefinite Integral of Polynomial Expressions

Find $\int (6t^5 - 4t^3 + 10t^2 + 1)\,dt$.

Solution:
$$\int (6t^5 - 4t^3 + 10t^2 + 1)\,dt = 6\int t^5 dt - 4\int t^3 dt + 10\int t^2 dt + 1\int dt$$
$$= 6\left(\frac{1}{6}t^6\right) - 4\left(\frac{1}{4}t^4\right) + 10\left(\frac{1}{3}t^3\right) + 1(t) + C$$
$$= t^6 - t^4 + \frac{10}{3}t^3 + t + C$$

Check:
$$\frac{d}{dt}\left(t^6 - t^4 + \frac{10}{3}t^3 + t + C\right)$$
$$= 6t^5 - 4t^3 + \frac{10}{3}(3t^2) + 1 + 0$$
$$= 6t^5 - 4t^3 + 10t^2 + 1$$
This is the original integrand.

Example 4: Finding the Indefinite Integral

Find $\int \frac{5}{x}\,dx$.

Solution: $\int \frac{5}{x}\,dx = 5\int \frac{1}{x}\,dx$ By the Constant Multiple Rule.

$= 5\ln|x| + C$ By Formula III.

Example 5: Finding the Indefinite Integral

Find $\int \frac{1}{\sqrt{y}}\,dy$.

Solution: $\int \frac{1}{\sqrt{y}}\,dy = \int \frac{1}{y^{\frac{1}{2}}}\,dy = \int y^{-\frac{1}{2}}\,dy$ Rewrite using exponents.

$= \frac{1}{-\frac{1}{2}+1} \cdot y^{-\frac{1}{2}+1} + C$ By Formula II.

$$= \frac{1}{\frac{1}{2}} y^{\frac{1}{2}} + C$$

$$= 2y^{\frac{1}{2}} + C = 2\sqrt{y} + C$$

Example 6: Using Several Formulas for Integration

Find $\int \left(e^x - x^{\frac{2}{5}} + 3 \right) dx$.

Solution: $\int \left(e^x - x^{\frac{2}{5}} + 3 \right) dx = \int e^x dx - \int x^{\frac{2}{5}} dx + \int 3dx$ By the Sum and Difference Rule.

$$= \left(e^x + C_1 \right) - \left(\frac{1}{\frac{2}{5}+1} x^{\frac{2}{5}+1} + C_2 \right) + \left(3x + C_3 \right)$$ By Formulas IV, II, and I.

$$= e^x - \frac{5}{7} x^{\frac{7}{5}} + 3x + C$$

One use of the constant representative C is sufficient since the sum of three constants is simply another constant.

Example 7: Finding the Indefinite Integral of a Rational Function

Find $\int \frac{t^3 + t + 1}{t^2} dt$.

Solution: The general approach is to replace a quotient with a polynomial-like expression whenever the denominator is a single term:

$$\int \frac{t^3 + t + 1}{t^2} dt = \int \left(t + \frac{1}{t} + \frac{1}{t^2} \right) dt$$

$$= \int t dt + \int \frac{1}{t} dt + \int \frac{1}{t^2} dt$$

$$= \int t dt + \int t^{-1} dt + \int t^{-2} dt$$

$$= \frac{1}{2} t^2 + \ln|t| + \frac{1}{-1} t^{-1} + C$$ By Formulas II and III.

$$= \frac{t^2}{2} + \ln|t| - \frac{1}{t} + C$$

Example 8: Finding the Constant of Integration

If $F'(x) = x - \frac{1}{x}$ and $F(1) = \frac{3}{2}$, find $F(x)$.

Solution: Since $F'(x) = x - \frac{1}{x}$, we know that

$$F(x) = \int \left(x - \frac{1}{x} \right) dx.$$

Thus

$$F(x) = \int \left(x - \frac{1}{x} \right) dx$$

$$= \int x\,dx - \int \frac{1}{x}\,dx$$ By the Sum and Difference Rule.

$$= \frac{x^2}{2} - \ln|x| + C$$ By Formulas II and III.

Now we use the fact that $F(1) = \frac{3}{2}$ to find the value of C.

$$F(1) = \frac{(1)^2}{2} - \ln|1| + C = \frac{3}{2}$$ Substitute the given values into $F(x)$.

$$\frac{1}{2} - 0 + C = \frac{3}{2}$$

$$C = 1$$ Solve for C.

Therefore,

$$F(x) = \frac{x^2}{2} - \ln|x| + 1.$$

Example 9: Evaluating an Antiderivative

Find $\int 25x^4 (6x - 1)\,dx$.

Solution: Multiply the integrand first; then integrate.

$$\int 25x^4 (6x - 1)\,dx = \int \left(150x^5 - 25x^4 \right) dx$$

$$= \frac{150x^6}{6} - \frac{25x^5}{5} + C = 25x^6 - 5x^5 + C$$

Example 10: Cost Function

A company's marginal cost function is given as $f(x) = 3x + 4$, and fixed costs are known to be \$5000. What is the cost function?

Solution: We find the cost function by integrating the marginal cost function, since marginal cost is the derivative of cost. That is, in this case, $f(x) = C'(x)$.

$$C(x) = \int (3x+4)\,dx = \frac{3}{2}x^2 + 4x + C$$

Since fixed costs are \$5000, we know that

$$C(0) = 5000,$$

and this allows us to determine a specific value for C, the constant of integration.

$$C(0) = \frac{3}{2}(0)^2 + 4 \cdot 0 + C = 5000$$

Therefore,

$$C = 5000$$

and

$$C(x) = \frac{3}{2}x^2 + 4x + 5000.$$

Distance, Velocity, and Acceleration

Suppose the distance s of a particle from an origin is given by a function of time t. We indicate this by $s = s(t)$.

In this case the velocity v of the particle is given by $v = \frac{ds}{dt}$, and the acceleration a is $a = s''(t) = \frac{d^2s}{dt^2}$. Suppose only the acceleration function is known. How would we obtain the distance function $s(t)$?

For example, a meteor 50,000 kilometers away is traveling towards the Earth with an acceleration of $-60t + 64$ meters per second per second. Since $a = \frac{dv}{dt}$,

$$dv = a \cdot dt.$$

The term dv is the differential of v and we refer to this step as "multiplying" by dt on both sides.

Now, integrating on both sides,

$$\int dv = \int a \cdot dt.$$

Since $v = \int dv$ and $a = -60t + 64$, by substitution, we have

$$v = \int (-60t + 64)\,dt$$

$$v = -\overset{30}{\cancel{60}}\left(\frac{1}{\cancel{2}}t^2\right) + 64t + C_1$$

$$= -30t^2 + 64t + C_1.$$

Since v is also a function of time t, we see $v(t) = -30t^2 + 64t + C_1$. The constant C_1 can be evaluated only if there is more information. Suppose, with careful observation we measure the speed of the meteor and determine that it is 200 m/s. It is customary and convenient to refer to the time of the initial observation as $t = 0$. Substituting 0 for t, we have $v(0) = -30(0)^2 + 64(0) + C_1$. So $v(0) = C_1$. The constant of integration C_1 is the speed at $t = 0$. Thus $C_1 = -200$ m/s. We use –200 rather than +200 since the direction of the meteor is towards the origin (so s is decreasing due to the speed). Therefore,

$$v(t) = -30t^2 + 64t - 200.$$

Since $v = \frac{ds}{dt}$, we can say $ds = v \cdot dt$. Integrating again, we have

$$\int ds = v \cdot dt = \int (-30t^2 + 64t - 200) \cdot dt.$$

Integrating a second time gives

$$s = -30\left(\frac{1}{3}t^3\right) + 64\left(\frac{1}{2}t^2\right) - 200t + C_2$$

$$s = -10t^3 + 32t^2 - 200t + C_2.$$

Once again there is a constant of integration to evaluate. If at the time of initial measurement the distance of the meteor from the Earth is 50,000 kilometers, then $s(0) = 50{,}000{,}000$ meters. Substituting these values,

$$50{,}000{,}000 = s(0) = -10(0)^3 + 32(0)^2 - 200(0) + C_2 = C_2.$$

Here we see $C_2 = 50{,}000{,}000$ meters, the initial distance of the meteor from the Earth.

It is typical of "acceleration" problems that there are two integrations, two constants of integration to evaluate, and two additional pieces of data necessary for this evaluation. It is common to use the notation v_0 and s_0 to denote initial ($t = 0$) values of velocity and distance.

One case of special interest is a body falling due to the Earth's gravity. In this case the acceleration a is a constant. As before, $a = \dfrac{dv}{dt}$ so $dv = a \cdot dt$. Integrating both sides gives

$$v = at + C.$$

The constant C is v_0, the initial velocity. That is, $v = at + v_0$. Since $\dfrac{ds}{dt} = v$, $ds = v \cdot dt$. One more integration gives

$$s = \int ds = \int (at + v_0)\,dt = a\left(\frac{1}{2}t^2\right) + v_0 t + s_0$$

$$s = \frac{1}{2}at^2 + v_0 t + s_0.$$

This familiar formula from physics also applies to other situations in which the acceleration is constant (for example, a traveling car stopping by a uniform application of the brakes). We emphasize that a, v_0, and s_0 are positive if directed away from the origin and are negative if directed toward the origin.

6.1 Exercises

In Exercises 1 – 12, show that the function F(x) is an antiderivative of the function f(x) by differentiating F.

1. $F(x) = 4x - 1, \ f(x) = 4$

2. $F(x) = 6x, \ f(x) = 6$

3. $F(x) = 3x^2 + 5x + 2, \ f(x) = 6x + 5$

4. $F(x) = \frac{1}{2}x^2 - 4x + e^{2x} - 1, \ f(x) = x - 4 + 2e^{2x}$

5. $F(x) = \ln x - \frac{1}{x} - 4e^{x^2}, \ f(x) = \frac{1}{x} + \frac{1}{x^2} - 8xe^{x^2}$

6. $F(x) = \ln x^3 + \frac{1}{x^2} + 6, \ f(x) = \frac{3}{x} - \frac{2}{x^3}$

7. $F(x) = \left(x^2 + 3\right)^4 - 1, \ f(x) = 8x\left(x^2 + 3\right)^3$

8. $F(x) = 3(5x - 1)^{\frac{2}{3}} + 8, \ f(x) = \frac{10}{\sqrt[3]{5x - 1}}$

9. $F(x)=\frac{5}{3}(e^x-4)^3+e,\ f(x)=5e^x(e^x-4)^2$

10. $F(x)=3e^{x^2-1}-7,\ f(x)=6xe^{x^2-1}$

11. $F(x)=\ln(x^2+5x-3)-\sqrt{5},\ f(x)=\frac{2x+5}{x^2+5x-3}$

12. $F(x)=\ln(e^{3x}-x)+\sqrt{11},\ f(x)=\frac{3e^{3x}-1}{e^{3x}-x}$

Find the indefinite integrals in Exercises 13 – 32.

13. $\int 7dx$

14. $\int \frac{2}{3}dx$

15. $\int 5x^4dx$

16. $\int -2x^{-3}dx$

17. $\int (x^2-3)dx$

18. $\int (x^4+5)dx$

19. $\int \left(\frac{1}{3}-e^t\right)dt$

20. $\int (e^t+t)dt$

21. $\int \left(\frac{1}{y}+y^3\right)dy$

22. $\int \left(\frac{1}{\sqrt{y}}-\frac{1}{y}\right)dy$

23. $\int \left(4x^2+\frac{2}{x}+\frac{1}{x^2}\right)dx$

24. $\int \left(9x-\frac{3}{x}-\frac{1}{\sqrt{x}}\right)dx$

25. $\int (2\sqrt[3]{x}+5\sqrt{x})dx$

26. $\int \left(\sqrt{x}+6e^x-\frac{5}{x}\right)dx$

27. $\int \left(4e^y-2y^5-\frac{1}{5}\right)dy$

28. $\int \left(y^{\frac{3}{2}}+5y^{-\frac{2}{3}}-y^{-1}\right)dy$

29. $\int \left(\frac{2}{\sqrt[3]{t}}+\frac{7}{t^3}\right)dt$

30. $\int \left(2t^{\frac{5}{2}}+\frac{4}{t}-\sqrt{3}\right)dt$

31. $\int \left(\frac{2}{3}e^x+x^{-\frac{3}{2}}-7x^{-1}\right)dx$

32. $\int \left(0.25e^x+4x^{-\frac{1}{4}}\right)dx$

In Exercises 33 – 38, perform the indicated multiplication and then integrate.

33. $\int x^3(2x-1)dx$

34. $\int x^2(3x-5)dx$

35. $\int (3t+2)^2dt$

36. $\int (5x+6)^2dx$

37. $\int \sqrt{y}(y^2+2y-1)dy$

38. $\int \sqrt{y}(4-3y-2y^2)dy$

In Exercises 39 – 44, simplify the indicated quotient and then integrate.

39. $\int \frac{3x^2+5x-4}{x^2}dx$

40. $\int \frac{x^3-6x^2+x}{x^2}dx$

41. $\int \frac{4+\sqrt{x}-3x}{x}dx$

42. $\int \frac{5x^2-2x+3}{\sqrt{x}}dx$

43. $\int \frac{x^{\frac{3}{2}}+6-2xe^x}{x}dx$

44. $\int \frac{4x^2+4\sqrt{x}-7x}{x^2}dx$

45. Find the antiderivative $F(x)$ that satisfies the given condition.

a. $F'(x) = x^2 - e^x,\ F(0) = 1$ **b.** $F'(x) = 6x^2 + x - 10,\ F(0) = 0$

c. $\dfrac{dF}{dx} = \dfrac{10}{\sqrt{x}},\ F(1) = 20$ **d.** $\dfrac{dF}{dx} = 6e^x - 2,\ F(0) = -10$

46. a. Compute the derivative of $y = e^{mx+b}$ where m and b are constants. Use your answer to determine an integration formula for $\int e^{mx+b} \cdot dx$.

b. Compute the derivative of $y = \dfrac{1}{e^{mx+b}}$ where m and b are constants. Use your answer to determine an integration formula for $\int \dfrac{1}{e^{mx+b}} \cdot dx$.

47. Given that $f'(x) = x^2 - 2$, determine the function $f(x)$ with the given constant of integration C. Draw all three on the same coordinate system.

a. $C = -1$. **b.** $C = 1$. **c.** $C = 3$.

48. Given $f'(x) = 6x^2 - 24x$.

a. Determine the x-values at which $f(x)$ has a maximum or minimum.

b. Determine whether there is an inflection point.

c. Given $f(1) = -9$, sketch f and determine if the answers to part **a.** are correct.

49. Cost. The weekly marginal cost of producing x ice cream makers is $28 + 0.05x$ dollars per ice cream maker. Find the cost function if the fixed costs are \$2400.

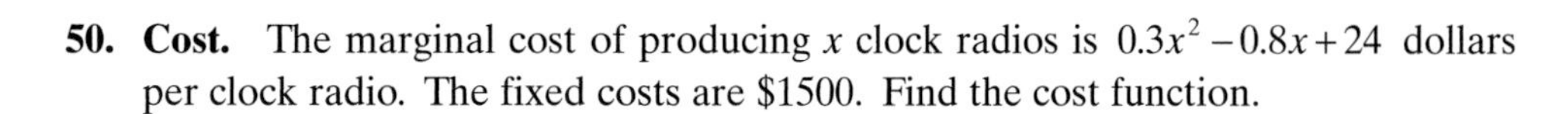

50. Cost. The marginal cost of producing x clock radios is $0.3x^2 - 0.8x + 24$ dollars per clock radio. The fixed costs are \$1500. Find the cost function.

51. Revenue. The marginal revenue from selling x irons is $94 - 0.06x$ dollars per iron. Find the revenue function. (**Hint:** $R(0) = 0$.)

52. Revenue. The marginal revenue from selling x floor lamps is $100 - 0.2x$ dollars per lamp. Find the revenue function. (**Hint:** $R(0) = 0$.)

53. Profit. The marginal profit from the production and sale of x cameras is estimated to be $24 - 0.4x$ dollars per unit.

a. Find the firm's profit function if the profit from the production and sale of 80 units is \$240.

b. What is the profit from the sale of 90 units?

54. Profit. A manufacturer has determined that the marginal profit from the production and sale of x stereo speakers is approximately $120 - 3x$ dollars per speaker.

a. Find the profit function if the profit from the production and sale of 30 speakers is \$1200.

b. What is the profit from the sale of 40 speakers?

55. Population. The population of a community is growing at a rate given by $\frac{dP}{dt} = 120 - 15t^{\frac{1}{2}}$ people per year. Find a function to describe the population t years from now if the present population is 8600 people.

56. Rodent control. Animal control officers have implemented a program to eliminate rats in a community. They estimate that the population of rats is changing at a rate of $\frac{dP}{dt} = 24t^{\frac{1}{2}} - 40t$ rats per month. Find a function for the rat population t months from now if the current population is estimated to be 6300 rats.

57. Air quality. The air quality control office estimates that for a population of x thousand people, the level of pollution in the air is increasing at a rate of $\frac{dL}{dx} = 0.2 + 0.002x$ parts per million per thousand people. Find a function to estimate the level of the pollutants if the level is 5.4 parts per million when the population is 20,000 people.

58. Ecology. Biologists are treating a stream contaminated with bacteria. The level of contamination is changing at a rate of $\frac{dN}{dt} = -\frac{960}{t^2} - 240$ bacteria per cubic centimeter per day, where t is the number of days since the treatment began. Find a function $N(t)$ to estimate the level of contamination if the level after 1 day was about 5720 bacteria per cubic centimeter.

59. Height. An object is projected vertically so that the velocity after t seconds is given by $v(t) = 96 - 32t$ feet per second.

a. Find the height function $s(t)$ if $s(0) = 18$ feet.

b. What will be the height after 3 seconds?

60. Distance. A vehicle travels in a straight line for t minutes with a velocity of $v(t) = 72t - 6t^2$ feet per minute, for $0 \le t \le 10$.

a. Find the distance function $s(t)$ if $s(0) = 0$.

b. How far will the vehicle travel in 5 minutes?

c. How far will the vehicle travel in 10 minutes?

Hawkes Learning Systems: Essential Calculus

The Indefinite Integral

6.2 Integration by Substitution

Objectives

After completing this section, you will be able to:

Use the method of integration by substitution to find integrals.

None of the integration formulas or rules discussed in Section 6.1 can be used directly to find any of the following integrals:

$$\int\left(x^2+4\right)^6 2x\,dx,\quad \int e^{x^2}2x\,dx,\quad \text{or}\quad \int\frac{2t}{t^2+1}\,dt.$$

In this section we develop a method of integration for finding integrals such as these. This method is called **integration by substitution**.

The concept of substitution has been used before in formulas for differentiation. For example, in Section 5.4 we found that if

$$f(x)=e^{g(x)},\quad \text{then}\quad f'(x)=e^{g(x)}g'(x).$$

Therefore, for

$$f(x)=e^{x^2+3x},\quad \text{we have}\quad f'(x)=e^{x^2+3x}(2x+3).$$

In this case we have "substituted" x^2+3x for $g(x)$ and $(2x+3)$ for $g'(x)$.

Similarly, the General Power Rule (discussed in Section 3.2) states that if

$$f(x)=\left[g(x)\right]^n,\quad \text{then}\quad f'(x)=n\left[g(x)\right]^{n-1}g'(x).$$

Thus, for

$$f(x)=\left(x^2+4\right)^7,\quad \text{we have}\quad f'(x)=7\left(x^2+4\right)^6 2x.$$

In this case we have substituted x^2+4 for $g(x)$ and $2x$ for $g'(x)$.

The four formulas for integration from Section 6.1 can be restated with the variable x replaced by $g(x)$ and the differential dx replaced by $g'(x)\,dx$. These formulas can be verified by using the Chain Rule to show that the derivative of each function on the right is equal to the integrand.

Formulas of Integration with Function Notation

I. $\int kg'(x)dx = kg(x) + C$

II. $\int [g(x)]^r g'(x)dx = \frac{1}{r+1}[g(x)]^{r+1} + C,$ *where* $r \neq -1.$

III. $\int \frac{g'(x)dx}{g(x)} = \int \frac{1}{g(x)} g'(x)dx = \int [g(x)]^{-1} g'(x)dx = \ln|g(x)| + C$

IV. $\int e^{g(x)} g'(x)dx = e^{g(x)} + C$

These same formulas for integration can be restated again where substitutions (called ***u*-substitutions**) are made as follows:

$$u = g(x) \text{ and } du = g'(x)dx.$$

Formulas of Integration with *u*-Substitution

I. $\int k\,du = ku + C$

II. $\int u^r du = \frac{1}{r+1} u^{r+1} + C,$ *where* $r \neq -1.$

III. $\int \frac{du}{u} = \int \frac{1}{u} du = \int u^{-1} du = \ln|u| + C$

IV. $\int e^u du = e^u + C$

NOTES

These forms indicate three particularly informative ideas about the formulas as stated in Section 6.1:

1. The letter x is a placeholder. Any other letter, such as u, v, y, or t, can be used.
2. Any substitution made in an integrand requires a related substitution for the differential such as du, dv, dy, or dt.
3. The u-substitution method will be appropriate to use when, in order to check the correctness of the antiderivative found, one must use the chain rule in some form.

Now consider the integral listed first in the introduction to this section:

$$\int (x^2+4)^6\, 2x\, dx.$$

WARNING!
There is no product rule for integration.

Our goal is to make substitutions in the integrand and differential so that the resulting integral will be one that we recognize in the list of integration formulas. Our choice (for it is a choice) is based on this goal. If this substitution does not lead to a recognizable integral form, then we make another choice.

If we let $u = x^2 + 4$, then $du = 2x\,dx$.

From Section 4.6, we know that if $u = g(x)$, then $du = g'(x)\,dx$.

If we make these substitutions, the integral takes the form of one of the formulas we know.

$$\int (x^2+4)^6\, 2x dx = \int u^6 du$$ Substituting u and du.

$$= \frac{1}{7}u^7 + C$$ Using Formula II.

$$= \frac{1}{7}(x^2+4)^7 + C$$ Substitute x^2+4 for u so that the answer is in the terms of the original variable x.

We have just performed **integration by substitution**, which involves the following two substitutions:

1. If $u = g(x)$,
2. then $du = g'(x)\,dx$.

In the discussion above, we let $u = x^2 + 4$ and $du = 2x\,dx$.

The following examples illustrate integration by substitution in a variety of situations. Study these examples carefully.

Example 1: Integration by Substitution with Rational Expressions

Find $\int \frac{2t}{t^2+1}\,dt$.

Solution: Let $u = t^2 + 1$. Then $du = 2t\,dt$.
In this case $2t\,dt$ is in the integral expression just as we need it. So no adjustments are necessary.

$$\int \frac{2t}{t^2+1}\,dt = \int \frac{du}{u}$$ Make the substitutions.

continued on next page ...

$= \int u^{-1} du$	Rewriting emphasizes that the exponent is –1.
$= \ln\lvert u\rvert + C$	By Formula III.
$= \ln\lvert t^2 + 1\rvert + C$	Substitute in $u = t^2 + 1$.
$= \ln\left(t^2 + 1\right) + C$	In this problem, the absolute value symbols are not necessary because $t^2 + 1 > 0$ for all t.

Example 2: Integration by Substitution with Radicals

Find $\int \sqrt{5x+1}\, dx$.

Solution: Choose $u = 5x + 1$. Then $\frac{du}{dx} = 5$ and, multiplying by dx on both sides, we get $du = 5dx$.

Since $du = 5dx$ and the constant factor 5 is not part of the original integrand, multiply the integrand by 1 in the form $\frac{1}{5} \cdot 5$ as follows:

$\int \sqrt{5x+1}\, dx = \int (5x+1)^{\frac{1}{2}} \cdot \frac{1}{5} \cdot 5dx$	Use fractional exponents.
$= \int u^{\frac{1}{2}} \cdot \frac{1}{5} \cdot du$	Substitute.
$= \frac{1}{5} \int u^{\frac{1}{2}}\, du$	By the Constant Multiple Rule.
$= \frac{1}{5} \cdot \frac{1}{\frac{1}{2}+1} \cdot u^{\frac{1}{2}+1} + C$	By Formula II.
$= \frac{1}{5} \cdot \frac{2}{3} \cdot u^{\frac{3}{2}} + C$	Simplify.
$= \frac{2}{15}(5x+1)^{\frac{3}{2}} + C$	Substitute $5x + 1$ for u.

A slightly different approach: "Solve" $du = 5dx$ for dx; we get $dx = \frac{1}{5} du$.

Then substitute this with $u = (5x+1)$, resulting directly in the integral $\int \left(u^{\frac{1}{2}} \cdot \frac{1}{5} du \right)$.

NOTES The technique used in Example 2, multiplying by 1 in the form $\frac{1}{5}\cdot 5$ and then factoring out $\frac{1}{5}$ from the integral, is valid because $\frac{1}{5}$ is a constant. The same technique is not valid with variables. For example, inserting $\frac{1}{x}\cdot x$ is valid, but then factoring out $\frac{1}{x}$ from the integral is not valid.

Example 3: Integration by Substitution with *e*

Find $\int 7xe^{x^2+3}dx$.

Solution: Let $u = x^2 + 3$. Then $du = 2x\,dx$.
We adjust for the 2 by introducing $\frac{1}{2}\cdot 2$ into the integrand.

$$\int 7xe^{x^2+3}dx = 7\int e^{x^2+3}\cdot\frac{1}{2}\cdot 2x\,dx$$ Use the Constant Multiple Rule with 7 and $\frac{1}{2}\cdot 2 = 1$, since we need $2x\,dx$ for du.

$$= \frac{7}{2}\int e^u\,du$$ Substitute and use the Constant Multiple Rule with $\frac{7}{2}$.

$$= \frac{7}{2}e^u + C$$ By Formula IV.

$$= \frac{7}{2}e^{x^2+3} + C$$

Example 4: Integration by Substitution

Find $\int \frac{x^2}{\left(x^3-1\right)^5}dx$.

Solution: Choose $u = x^3 - 1$. Then $du = 3x^2\,dx$.

$$\int \frac{x^2}{\left(x^3-1\right)^5}dx = \int\left(\frac{\frac{1}{3}\cdot 3x^2dx}{\left(x^3-1\right)^5}\right)$$ Multiply the integrand by $\frac{1}{3}\cdot 3 = 1$ to get $3x^2\,dx$.

continued on next page ...

$$= \frac{1}{3}\int \frac{du}{u^5}$$ By the Constant Multiple Rule.

$$= \frac{1}{3}\int u^{-5}du$$

$$= \frac{1}{3}\cdot\frac{1}{(-5+1)}\cdot u^{-5+1} + C$$ By Formula II.

$$= -\frac{1}{12}u^{-4} + C$$

$$= -\frac{1}{12}(x^3-1)^{-4} + C \quad \left[\text{or } -\frac{1}{12(x^3-1)^4} + C\right]$$

NOTES

We have used the letter u throughout this section. This particular letter is commonly used in most textbooks, and the substitution technique is sometimes known as the ***u*-substitution** technique.

A rule of thumb that you may have observed while using this technique is to substitute u for

1. the exponent in an exponential function,
2. an expression in parentheses,
3. a denominator, or
4. an expression under a radical sign.

The following example illustrates **an incorrect choice** for substitution. Remember, if a substitution does not result in an integral that matches one of the basic formulas we know, then try another substitution.

Example 5: Incorrect Integration by Substitution

Find $\int 3x^2\sqrt{x^3+1}\,dx$.

Solution: Let $u = 3x^2$. Then $du = 6x\,dx$.
This substitution will not work because the expression $\sqrt{x^3+1}\,dx$ is not du.

A better substitution is to let u represent the expression under the radical sign. Let $u = x^3+1$. Then $du = 3x^2dx$. Now substitution gives a familiar form.

$$\int 3x^2\sqrt{x^3+1}\,dx = \int \left(x^3+1\right)^{\frac{1}{2}} \cdot 3x^2 dx \qquad \text{Rewrite using exponents.}$$

$$= \int u^{\frac{1}{2}} du \qquad \text{Substitute.}$$

$$= \frac{1}{\frac{1}{2}+1} u^{\frac{1}{2}+1} + C \qquad \text{By Formula II.}$$

$$= \frac{2}{3} u^{\frac{3}{2}} + C \qquad \text{Simplify.}$$

$$= \frac{2}{3}\left(x^3+1\right)^{\frac{3}{2}} + C \qquad \text{Substitute } x^3+1 \text{ for } u.$$

Example 6: Incorrect Integration by Substitution

Find $\int (x^2+5)^{10} dx$.

Solution: Let $u = x^2+5$. Then $du = 2x\,dx$ and $u^{10} = \left(x^2+5\right)^{10}$.

This substitution will not work. The differential is $dx = \frac{du}{2x}$. The factor $2x$ is not present in the original problem.

Actually, in this example, the method cannot be made to work by any u-substitution.

Integrals of the particular form in Example 6 can be worked by a method called trigonometric substitution which is not discussed in this book. We do not suggest expanding the polynomial and integrating term-by-term!

There are many problems which require other methods, and there are some functions which cannot be integrated in "closed form." This means there is no simple formula for the antiderivative. Fortunately, all the problems that follow can be solved by the methods here.

6.2 Exercises

In Exercises 1 – 36, use the technique of substitution to perform each integration.

1. $\int (x+4)^7\,dx$ **2.** $\int (y-6)^{-3}dy$ **3.** $\int 2x\left(x^2-1\right)^{\frac{1}{2}}dx$

4. $\int 3x^2\left(x^3-5\right)^{\frac{1}{3}}dx$ **5.** $\int \frac{1}{t+2}dt$ **6.** $\int \frac{1}{x-11}dx$

7. $\int \frac{3t^2}{t^3+4}dt$ **8.** $\int \frac{1}{y^2-8}2y\,dy$ **9.** $\int e^{y+5}dy$

10. $\int e^{x-9}dx$ **11.** $\int e^{-0.2x}(-0.2)dx$ **12.** $\int e^{0.5t}(0.5)dt$

13. $\int \frac{1}{5x+3}dx$ **14.** $\int (3y-2)^{-2}dy$ **15.** $\int \frac{1}{\sqrt{4x-1}}dx$

16. $\int e^{-4x}dx$ **17.** $\int xe^{2x^2}dx$ **18.** $\int \frac{x}{2x^2+5}dx$

19. $\int \frac{x}{(3x^2-1)^2}dx$ **20.** $\int y^2e^{-2y^3}dy$ **21.** $\int \frac{2t+1}{t^2+t-4}dt$

22. $\int (x^2+3x-1)^4(2x+3)dx$ **23.** $\int 5ye^{-y^2}dy$ **24.** $\int \frac{e^{\sqrt{t}}}{\sqrt{t}}dt$

25. $\int 6x\sqrt{5+2x^2}dx$ **26.** $\int \frac{e^t}{e^t-1}dt$ **27.** $\int \frac{4}{x^2}e^{\frac{1}{x}}dx$

28. $\int 4y\sqrt[3]{3y^2+7}dy$ **29.** $\int e^{2x}(1-3e^{2x})^2dx$ **30.** $\int \frac{4x+10}{x^2+5x+2}dx$

31. $\int \frac{\ln x}{x}dx$ **32.** $\int \frac{\ln 4x}{x}dx$ **33.** $\int \frac{1}{x\ln x}dx$

34. $\int \frac{(\ln x)^2}{x}dx$ **35.** $\int \frac{e^x-e^{-x}}{e^x+e^{-x}}dx$ **36.** $\int \frac{7x}{e^{x^2}}dx$

In Exercises 37 – 40, divide first and then integrate.

37. a. $\int \frac{x-1}{x-2}dx$ $\left(\textbf{Hint: } \frac{x-1}{x-2}=1+\frac{1}{x-2}\right)$

b. $\int \frac{x+3}{x+1}dx$ $\left(\textbf{Hint: } \frac{x+3}{x+1}=1+\frac{2}{x+1}\right)$

38. a. Rework Exercise 37a. and substitute $u = x - 2$.
b. Rework Exercise 37b. and substitute $u = x + 1$.

39. $\int \frac{x+2}{x+4}dx$ **40.** $\int \frac{x+5}{x-3}dx$

In Exercises 41 – 44, find a function f(x) given $f'(x)$ and one (x, y)-value.

41. $f'(x)=10xe^{5x^2}$; (0, 11) **42.** $f'(x)=\frac{3}{x-5}$; (6, 5)

43. $f'(x) = (2x+2)^5$; $(-1, 10)$

44. $f'(x) = \dfrac{12x^2}{(x^3+6)^2}$; $\left(0, -\dfrac{2}{3}\right)$

45. Appreciation. The value V of a painting is increasing at a rate of $4500(25-1.8t)^{-\frac{3}{2}}$ dollars per year. The painting originally sold for \$1000.

- **a.** Write a function for its value t years after the original sale.
- **b.** What will the painting be worth 5 years after the original sale?

46. Skills development. It is estimated that after t weeks of practice, students in a typing class can increase their speed $24e^{-0.2t}$ words per minute for each additional week of practice.

- **a.** If their initial speed was 0 words per minute ($S(0) = 0$), write a function to represent their speed after t weeks of practice.
- **b.** How fast can they type after 10 weeks (to the nearest word)?

47. Price. After the NBA championship basketball game, the price of a souvenir cap changes at the rate of $-\dfrac{1}{2x}$ dollars per cap, where x is the number of caps sold (in thousands). Write a function for the price p if $p(10) = 12.85$.

48. Profit. The marginal profit from the production and sale of x units of a product is estimated to be $\dfrac{100}{1+0.5x} - 4.5$ dollars per unit. If $P(0) = -60$, find the profit function $P(x)$.

49. Daily production. Records show that t hours after starting work on a typical day, an employee can assemble bikes at a rate of $\dfrac{6t+15}{\sqrt{t^2+5t}}$ per hour. Find the daily production function $N(t)$ if $N(0) = 0$.

50. Position of a particle. A particle is moving in a straight line with the velocity $v(t) = \sqrt{2t+7}$ feet per second, where t represents time in seconds. Find the position function $s(t)$ if $s(0) = 0$.

51. Position of a projectile. The velocity of a projectile moving in a straight line t seconds after it is fired is given by $v(t) = 36 + \dfrac{60}{(t+1)^2}$ feet per second. Find the position function $s(t)$ if $s(1) = 10$.

52. The marginal value for a tract of land is $V' = \dfrac{20}{2t+1}$, where t is time in years since 2000 and V is in thousands of dollars.

- **a.** Determine the function V in terms of t if $V(0) = 20$.
- **b.** What will be the value of the land in 2015?

53. A point mass moving on a horizontal axis has a deceleration given by $a(t) = -48(2t+1)^2$, where t is time in seconds and a is in feet per second per second.

- **a.** If $v(0) = 6$ ft./sec, determine a velocity function $v(t)$.
- **b.** If $s(0) = -\dfrac{3}{4}$ feet, determine a distance function $s(t)$.

54. The number of viewers of a new soap opera grew at a rate $V'(t) = 900e^{0.3t-4}$ all summer long, where V is the total number of viewers in thousands and t is the number of weeks since June 1st.

- **a.** Determine $V(t)$ if $V(0) = 100$.
- **b.** At what rate is V changing when $t = 5$.
- **c.** When will the show hit 1,000,000 viewers ($V = 1000$)?

55. Suppose in a medieval country, from 1200 to 1300, the life expectancy of a female serf changed at the rate $f'(t) = \dfrac{0.3}{1+0.01t}$, where t is time in years after 1200 and $f(t)$ is the average age of death.

- **a.** What are the units of $f'(t)$.
- **b.** Determine $f(t)$ if $f(0) = 30$.
- **c.** What was the life expectancy of a female serf in 1300?

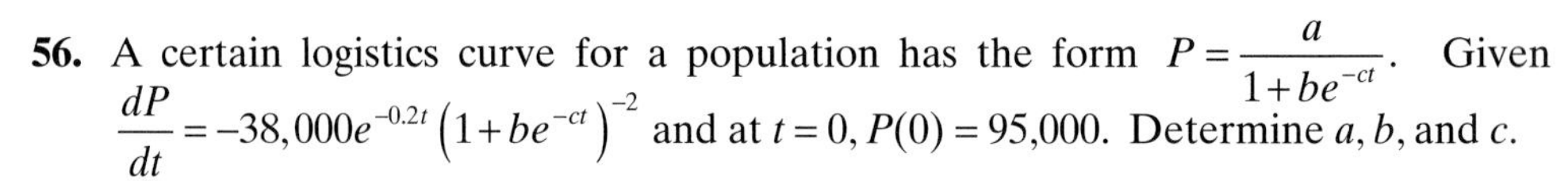

56. A certain logistics curve for a population has the form $P = \dfrac{a}{1+be^{-ct}}$. Given $\dfrac{dP}{dt} = -38{,}000e^{-0.2t}\left(1+be^{-ct}\right)^{-2}$ and at $t = 0$, $P(0) = 95{,}000$. Determine a, b, and c.

57. A new evening school is growing at the rate of $p'(t) = \dfrac{150}{\sqrt{1+0.2t}}$, where $p(t)$ is the total evening school population and t is the time in years after 2000. The initial enrollment was 1500 students.

- **a.** How fast was enrollment changing in 2002?
- **b.** What was the expected enrollment in 2005?

58. A particle moves along an axis with speed given by $v(t) = 3t - 1$, where t is in seconds and v is in ft./sec.

- **a.** Determine the acceleration, $a(t)$.
- **b.** What is the distance function, $s(t)$, if $s(0) = 5$?

59. A meteor falls partly under the influence of the Earth's gravity at a speed given by $v(t) = 200 + 30t + 24t^{\frac{1}{2}}$ for $0 \le t \le 24$, where t is in hours and v is in miles per hour.

- **a.** Determine the acceleration.
- **b.** Determine the distance function if $s(0) = 5000$ miles.

Hawkes Learning Systems: Essential Calculus

Integration by Substitution

6.3 The Fundamental Theorem of Calculus and the Definite Integral

Objectives

After completing this section, you will be able to:

1. *Have a thorough understanding of the Fundamental Theorem of Calculus.*
2. *Evaluate definite integrals.*

We remind the reader that the area of a planar region is defined simply as the number of unit squares (1-by-1 squares) contained inside the region. It follows that the area of a rectangle is the product of its base and height. In this section we will try to calculate, using calculus, areas bounded by continuous curves. Our general method is to consider the problem of "filling" a region in a coordinate plane with rectangles. The idea of using polygons to calculate areas bounded by curves is very old. Archimedes made very accurate estimates of π by filling circles with regular polygons with ever increasing numbers of sides, and he calculated the exact area of a section of a parabola by filling the area with infinitely many triangles (of uniformly decreasing size). Kepler used triangles to determine his famous Second Law of Motion; the ray from the sun to a planet sweeps out equal areas in equal times. Amazingly, he did this 50 years before the invention of calculus.

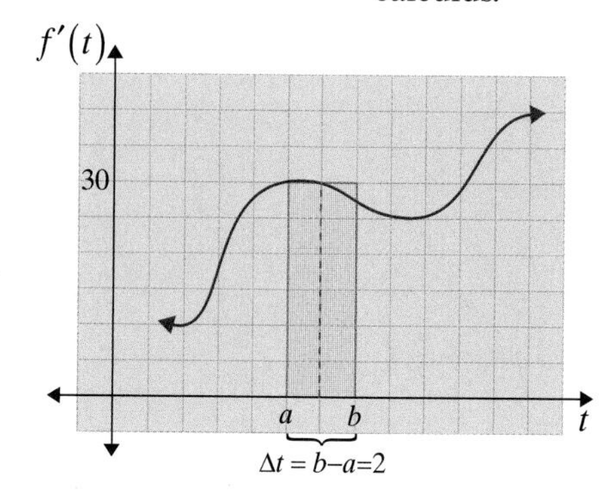

Figure 6.3.1

There is another very direct way to approach areas. We now consider the problem from another point of view. Let us consider what the area might represent. Suppose $f'(t)$ is the velocity of an object at time t and the graph of $f'(t)$ is that in Figure 6.3.1. We are interested in the area in Quadrant I under the curve from $t=a$ to $t=b$. Imagine a rectangle with base Δt and with height $f'(\hat{t})$ where $\hat{t}$ is some t-value in the interval Δt. The area of this rectangle is $f'(\hat{t})\Delta t$. Suppose $f'(\hat{t}) = 30$ meters/second and that $\Delta t = 2$ seconds. The area is then

$$(30 \text{ meters/second})(2 \text{ seconds}) = 60 \text{ meters}.$$

That is, the numerical value of the rectangular area represents a close approximation to the distance traveled during the 2 seconds.

The exact area represents the exact distance traveled. The distance function $f(t)$ gives the distance from the origin at time t, and the number $f(b) - f(a)$ is the net distance traveled from time $t = a$ to $t = b$. This shows that, from the numerical point of view, this net distance is exactly the numerical value of the area under the graph of $f'(t)$.

Having discovered what the area represents, we have also discovered a way to calculate it! Just use $f(x)$. This is part of what is called the **Fundamental Theorem of Calculus**, namely that the area under any "nice" curve, say $g(x)$, from $x = a$ to $x = b$, can be calculated using any antiderivative $G(x)$ of $g(x)$. The area will be $G(b) - G(a)$.

Determining which curves this applies to and proving the details in a formal way was not worked out satisfactorily until in the early 1800's – after 250 years of using calculus! Like everyone before us, we jump ahead and use the theorem, and then consider its proof later.

Example 1: Using the Fundamental Theorem of Calculus

Use the Fundamental Theorem of Calculus to determine the area under the graph of $f(x)=30-20x+x^5$ and over the x-axis interval $[1, 2]$.

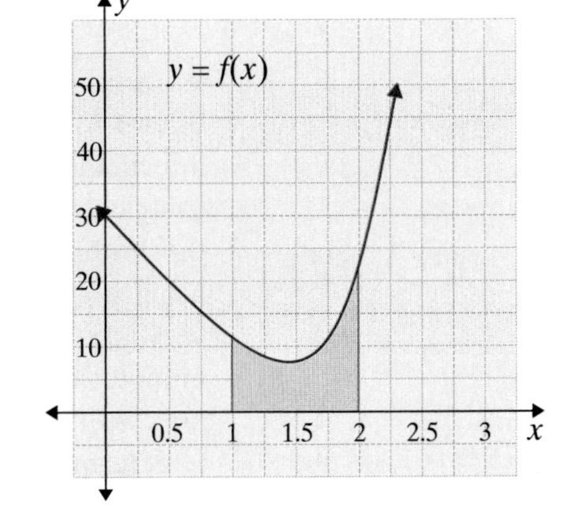

Solution: First, we must find the antiderivative, $F(x)$.

$$F(x)=30x-10x^2+\frac{1}{6}x^6$$

Note:

$$\frac{d}{dx}F(x)=\frac{d}{dx}\left(30x-10x^2+\frac{1}{6}x^6\right)$$

$$=30-20x+x^5$$

$$=f(x)$$

Using the antiderivative, solve for $F(b) - F(a)$. Use $a = 1$ and $b = 2$.

$$F(2)-F(1)=\left[30(2)-10(2)^2+\frac{(2)^6}{6}\right]-\left[30(1)-10(1)^2+\frac{(1)^6}{6}\right]$$

$$\approx[60-40+10.667]-[20.167]$$

$$=10.5$$

The area is 10.5 square units.

Calculations of other areas, the use of graphing utilities, and different applications of the Fundamental Theorem are deferred to the next section.

In the next subsection we digress to present an introduction to the modern approach to integral calculus using the Fundamental Theorem of Calculus. An overall view of this approach is shown in Figure 6.3.2.

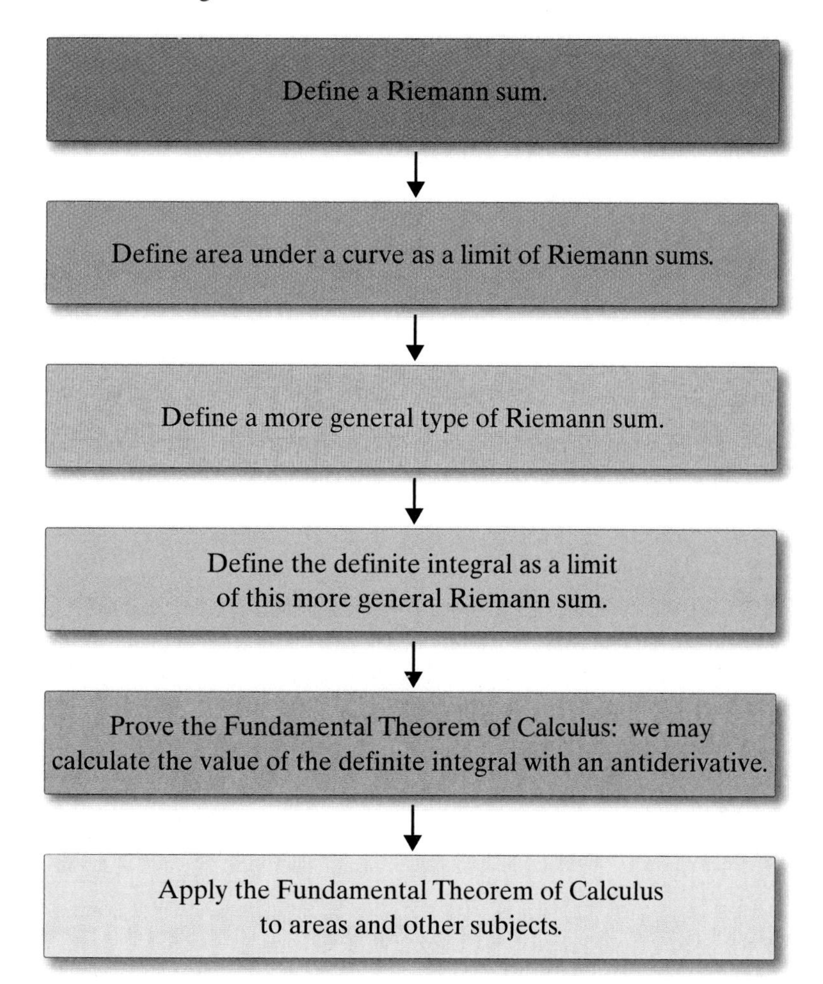

Define a Riemann sum.
Define area under a curve as a limit of Riemann sums.
Define a more general type of Riemann sum.
Define the definite integral as a limit of this more general Riemann sum.
Prove the Fundamental Theorem of Calculus: we may calculate the value of the definite integral with an antiderivative.
Apply the Fundamental Theorem of Calculus to areas and other subjects.

Figure 6.3.2

We return to the problem of using rectangles to estimate the area between a curve and the x-axis when the curve is above the x-axis and x is restricted to some closed interval $[a, b]$. This approach to the theoretical development of definite integrals is credited to Bernhard Riemann (1826 – 1866).

Riemann Sums

Suppose that $y = f(x)$ is a nonnegative continuous function on the interval $[a, b]$ and our objective is to estimate the area under the curve (i.e., the area between the curve and the x-axis). (See Figure 6.3.3.)

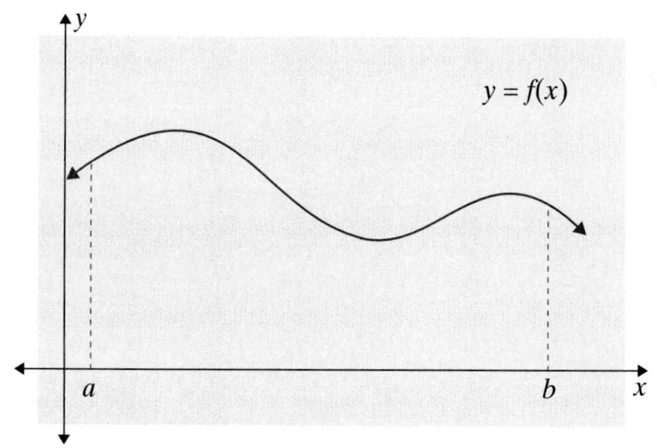

Figure 6.3.3

We estimate the area by summing the areas of adjacent rectangles as illustrated in the following discussion. In this step-by-step development, we use only five rectangles for simplicity.

Step 1: Partition the interval $[a, b]$ into $n = 5$ subintervals, each of width $\Delta x = \dfrac{b-a}{n} = \dfrac{b-a}{5}$ with end points $a = x_0$, x_1, x_2, x_3, x_4, and $x_5 = b$. (See Figure 6.3.4.)

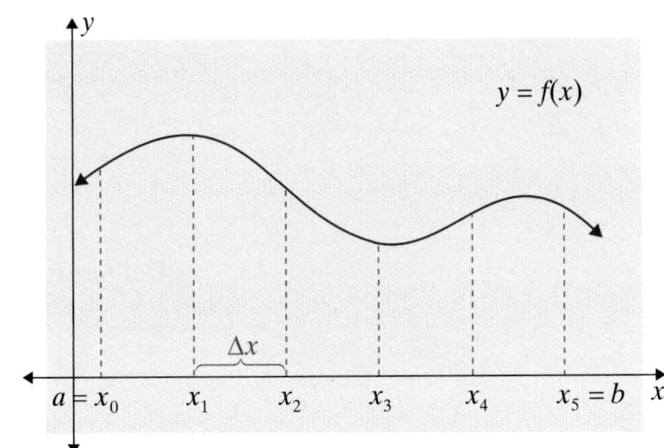

Figure 6.3.4

Step 2: Choose one value for x in each subinterval. In Figure 6.3.5 the chosen values are labeled c_1, c_2, c_3, c_4, and c_5.

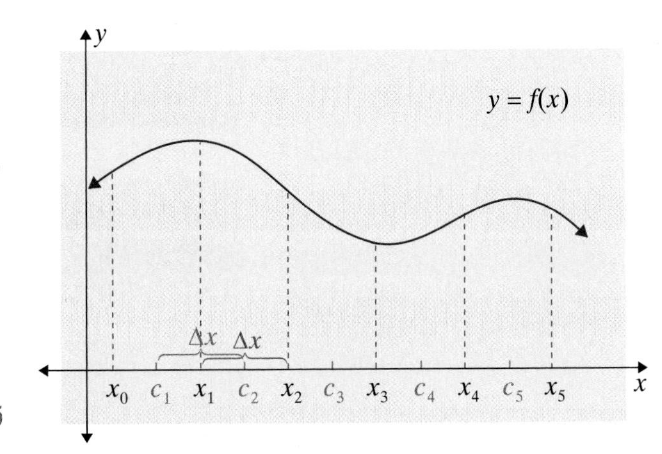

Figure 6.3.5

Step 3: Find the five corresponding y-values: $f(c_1), f(c_2), f(c_3), f(c_4)$, and $f(c_5)$. (See Figure 6.3.6.)

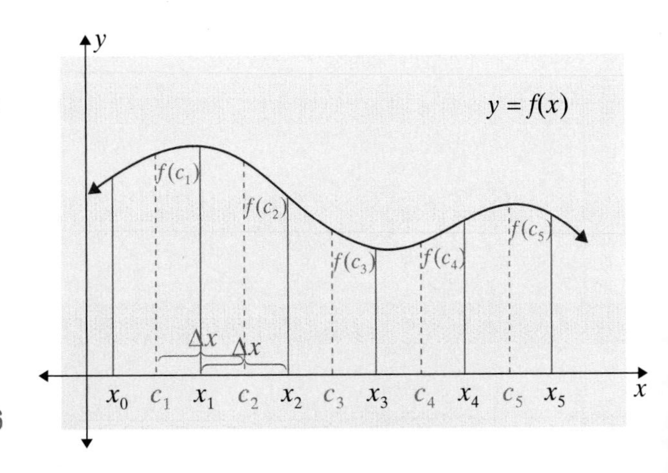

Figure 6.3.6

Step 4: Now, as illustrated in Figure 6.3.7, we treat each of these y-values as the height of a rectangle of width Δx. We multiply each y-value by Δx to find the areas of the rectangles and then form the sum of the products. This sum is called a **Riemann sum**, and it is an approximation of the area between the curve and the x-axis.

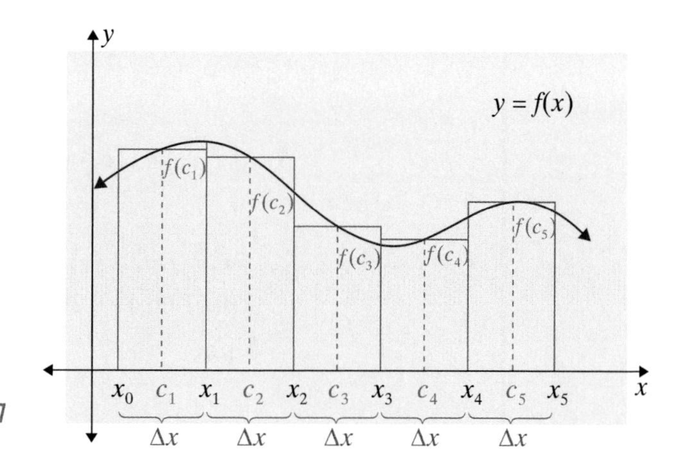

Figure 6.3.7

Here area $= f(c_1)\Delta x + f(c_2)\Delta x + \ldots + f(c_5)\Delta x$. The use of $n = 5$ subintervals in the previous discussion was purely arbitrary. Any finite number of subintervals could have been used, as we indicate in the following statement of the general form of a Riemann sum.

General Form of a Riemann Sum

*The **general form of a Riemann sum** for a function $y = f(x)$ continuous on $[a, b]$ is*

$$\boldsymbol{S_n = \left[f(c_1) + f(c_2) + \ldots + f(c_n)\right]\Delta x},$$

where $\Delta x = \dfrac{b-a}{n}$, n is the number of subintervals of $[a, b]$ and each of the numbers $c_1, c_2, \ldots, c_n$ represents one x-value from each subinterval.

Example 2: Riemann Sum

Use a Riemann sum to estimate the area between $f(x) = \dfrac{1}{x}$ and the x-axis on the interval $[a, b] = [1, 5]$ with $n = 4$.

Solution: $\Delta x = \dfrac{b-a}{n} = \dfrac{5-1}{4} = \dfrac{\not{4}}{\not{4}} = 1$

First, we find Δx by substituting the given values $a = 1, b = 5$, and $n = 4$.

continued on next page ...

In this example we use the midpoint of each subinterval for the values of c_1, c_2, c_3, and c_4.

Δx

$c_1 = 1.5$ $c_2 = 2.5$ $c_3 = 3.5$ $c_4 = 4.5$

$a = x_0 = 1$ $x_1 = 2$ $x_2 = 3$ $x_3 = 4$ $b = x_4 = 5$

Δx

$$
\begin{aligned}
S_4 &= \left[f(c_1)+f(c_2)+f(c_3)+f(c_4)\right]\Delta x \\
&= \left(\frac{1}{1.5}+\frac{1}{2.5}+\frac{1}{3.5}+\frac{1}{4.5}\right)\cdot 1 \\
&\approx (0.67+0.4+0.29+0.22) \\
&= 1.58
\end{aligned}
$$

Substitute the appropriate values into the general form of the Riemann sum.

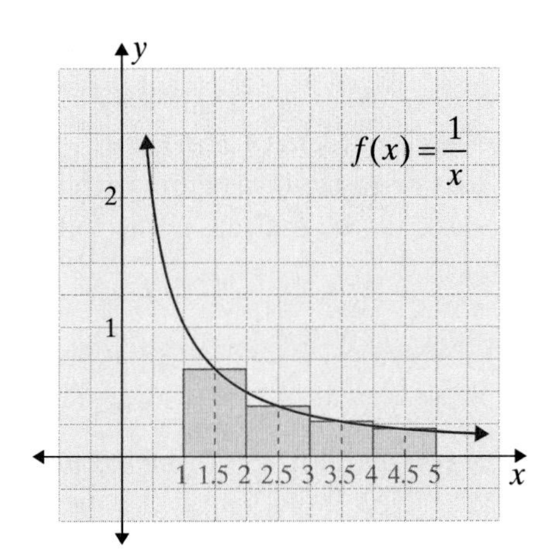

Example 3: Riemann Sum

Use a Riemann sum to estimate the area between $y = x^2 + 1$ and the x-axis on the interval $[a, b] = [-1, 2]$ with $n = 6$.

Solution: $\Delta x = \frac{2-(-1)}{6} = \frac{3}{6} = 0.5$

First, we find Δx by substituting the given values $a = -1$, $b = 2$, and $n = 6$.

In this example we use the left endpoint of each subinterval for the values of c_1, c_2, c_3, c_4, c_5, and c_6.

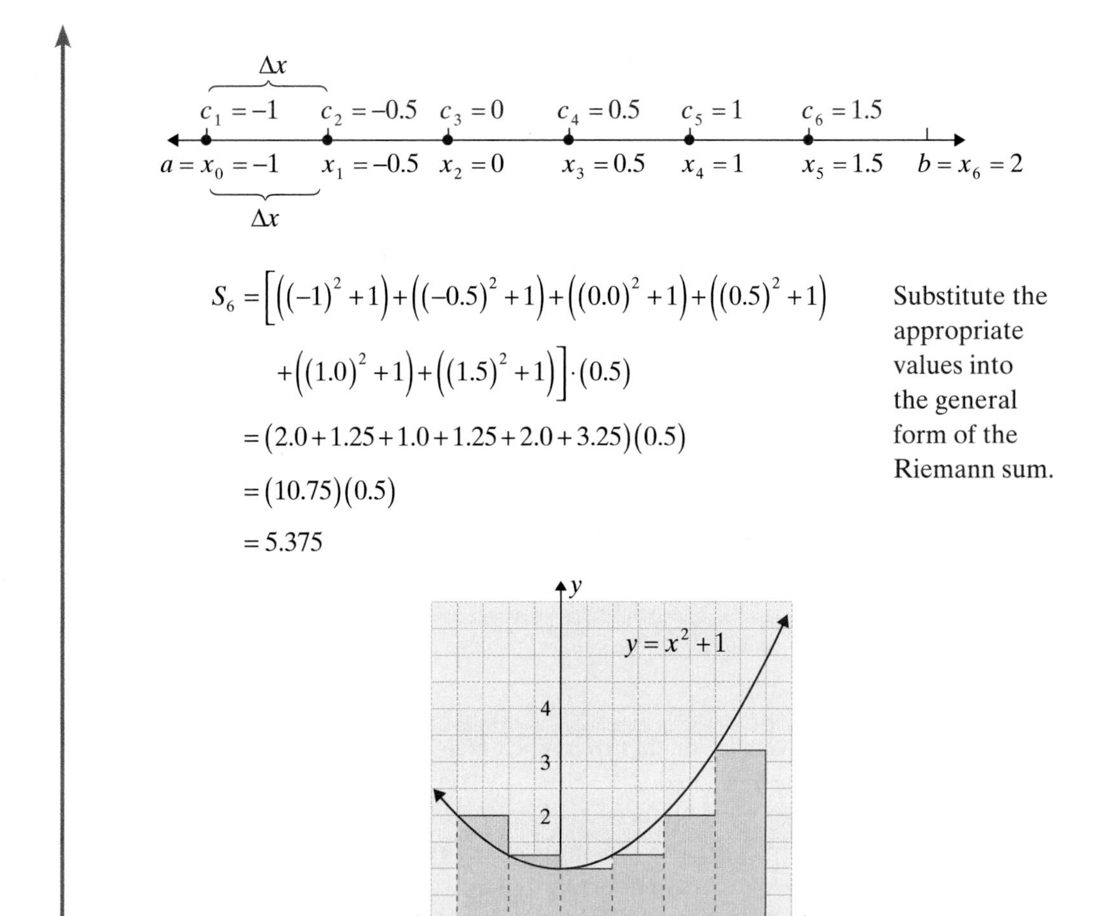

$$S_6 = \Big[\big((-1)^2+1\big)+\big((-0.5)^2+1\big)+\big((0.0)^2+1\big)+\big((0.5)^2+1\big)$$
$$+\big((1.0)^2+1\big)+\big((1.5)^2+1\big)\Big]\cdot(0.5)$$
$$=(2.0+1.25+1.0+1.25+2.0+3.25)(0.5)$$
$$=(10.75)(0.5)$$
$$=5.375$$

Substitute the appropriate values into the general form of the Riemann sum.

In a Riemann sum the more rectangles involved (and the smaller the value for Δx), the closer the sum is to the area under the curve. (See Figure 6.3.8.)

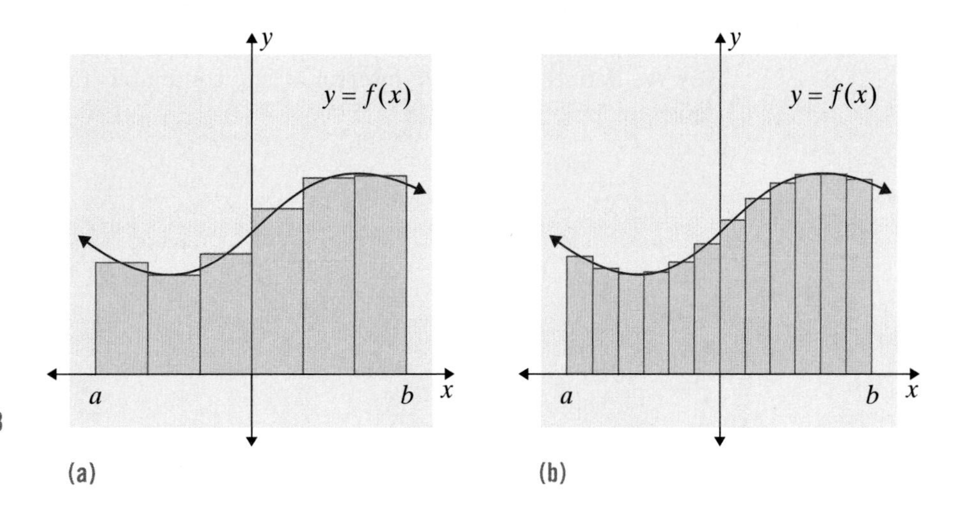

Figure 6.3.8

Programmable graphing calculators (like the TI-84 Plus) will calculate various types of Riemann sums directly using a menu. We do not include such problems in this text.

The Definite Integral

Our discussion of Riemann sums forms the basis for the following definition.

Area Under a Curve Defined

If a function $y = f(x)$ is nonnegative and continuous on the interval $[a, b]$ then the ***area under the curve*** *is defined to be*

$$\mathbf{A} = \lim_{n \to +\infty} \mathbf{S}_n,$$

where S_n is the general form of a Riemann sum for the function f.

We have made two implicit restrictions in our discussions of Riemann sums that should be explained.

1. Since our initial objective was to find the area under a curve, the discussion was restricted to nonnegative functions (i.e., functions whose graphs lie on or above the x-axis.) In general, Riemann sums can be defined for any continuous function. For those functions that are negative on some intervals, some of the corresponding rectangles may be below the x-axis. We will discuss this situation in detail in Section 6.4.

2. Also, for convenience, we required each subinterval to be the same width. Again, Riemann sums may be defined when the subintervals are of varying widths.

Now we define the **definite integral** of any continuous function in terms of Riemann sums.

The Definite Integral

If a function $y = f(x)$ is continuous on the interval $[a, b]$, then the ***definite integral*** *of f from a to b, symbolized as $\int_a^b f(x)dx$, is defined to be*

$$\int_a^b f(x)dx = \lim_{n \to +\infty} S_n,$$

where S_n is the general form of a Riemann sum for the function f.

The number a is called the ***lower limit*** *of integration and the number b is called the* ***upper limit*** *of integration.*

The two previous definitions lead to the following statement connecting area and the definite integral of nonnegative function.

Formula for Area Under a Curve

For a function $y = f(x)$, nonnegative and continuous on $[a, b]$, the area under the curve is given by

$$A = \int_a^b f(x)dx.$$

One way we have of finding the value indicated by a definite integral of a function is to find a general Riemann sum S_n for the function and then take the limit of S_n as $n \to +\infty$. This approach takes time and can involve techniques and formulas not covered in this course. However, it is essential in many applied areas of physics and engineering to construct a Riemann sum expression from theoretical considerations and only then consider the best way to calculate its value. We do exactly this later in this section when calculating the average value of a function. But now, we turn directly to the Fundamental Theorem of Calculus to provide the necessary tools for evaluating definite integrals. The full proof of the theorem, which involves Riemann sums, is beyond the scope of this text and therefore is not included.

The Fundamental Theorem of Calculus

If a function $y = f(x)$ is continuous on the interval $[a, b]$ and $F(x)$ is any antiderivative of $f(x)$, then

$$\int_a^b f(x)dx = \left[F(x)\right]_a^b = F(b) - F(a).$$

NOTES

We will use the handy notational convenience (the use of the labeled bracket notation) to indicate $F(b) - F(a)$ as shown on the previous page and as illustrated below:

$$F(b)-F(a)=F(x)\Big]_a^b$$

or even

$$\int_a^b f(x)\,dx=\left[\int f(x)\,dx\right]_a^b.$$

In some books other similar notation is used such as the following:

$$F(b)-F(a)=\left[F(x)\right]_a^b=F(x)\Big]_a^b=F(x)\Big|_a^b=\left[F(x)\right]_{x=a}^{x=b}.$$

As the Fundamental Theorem implies, the constant of integration, C, can be omitted when definite integrals are evaluated. Consider the following analysis.

Suppose that $F(x)$ is any antiderivative of $f(x)$. Then

$$\begin{aligned}\int_a^b f(x)\,dx &= F(x)+C\Big]_a^b\\ &=\left[F(b)+C\right]-\left[F(a)+C\right]\\ &=F(b)-F(a)+C-C\\ &=F(b)-F(a).\end{aligned}$$

Thus the constant of integration is subtracted out when a definite integral is evaluated.

Example 4: Definite Integrals

a. Evaluate $\int_1^2 2x\,dx$.

Solution: $\int_1^2 2x\,dx = x^2\Big]_1^2$ — Determine the antiderivative $F(x)$.

$=(2)^2-(1)^2$ — Use the Fundamental Theorem of Calculus with $a=1$ and $b=2$.

$=4-1$

$=3$

b. Find the value of $\int_1^4 \frac{1}{\sqrt{x}}dx$.

Solution: $\int_1^4 \frac{1}{\sqrt{x}}dx = \int_1^4 x^{-\frac{1}{2}}dx = \frac{2}{1}x^{\frac{1}{2}}\Big]_1^4 = 2\sqrt{x}\Big]_1^4$ Determine the antiderivative $F(x)$.

$= 2\sqrt{4} - 2\sqrt{1}$ Use the Fundamental Theorem of Calculus with $a = 1$ and $b = 4$.

$= 4 - 2$

$= 2$

c. Evaluate $\int_0^1 e^x dx$.

Solution: $\int_0^1 e^x dx = e^x\Big]_0^1$ Determine the antiderivate $F(x)$.

$= e^1 - e^0$ Use the Fundamental Theorem of Calculus with $a = 0$ and $b = 1$.

$= e - 1$

The properties of definite integrals listed here follow directly from the Fundamental Theorem of Calculus and provide the basis for expansion of the techniques used to evaluate definite integrals.

Properties of Definite Integrals

1. $\int_a^a f(x)dx = 0$

2. $\int_a^b kf(x)dx = k\int_a^b f(x)dx$

3. $\int_a^b [f(x) \pm g(x)]dx = \int_a^b f(x)dx \pm \int_a^b g(x)dx$

Example 5: Using Properties of Definite Integrals

a. Evaluate $\int_1^2 \left(x^2 + \frac{1}{x} - 3e^x\right)dx$.

Solution: $\int_1^2 \left(x^2 + \frac{1}{x} - 3e^x\right)dx = \frac{x^3}{3} + \ln|x| - 3e^x\Big]_1^2$ Determine the antiderivative $F(x)$.

continued on next page ...

$$= \left(\frac{(2)^3}{3} + \ln|2| - 3e^2\right) - \left(\frac{(1)^3}{3} + \ln|1| - 3e^1\right)$$

Use the Fundamental Theorem of Calculus with $a = 1$ and $b = 2$.

$$= \frac{8}{3} + \ln 2 - 3e^2 - \frac{1}{3} - 0 + 3e$$

$$= \frac{7}{3} + \ln 2 - 3e^2 + 3e$$

b. Find the value of $\int_0^1 x\sqrt{x^2+1}\,dx$.

Solution: Use the u-substitution technique with $u = x^2 + 1$ and $du = 2x\,dx$ to find the antiderivative. Then replace u with $x^2 + 1$ before evaluating the definite integral.

$$\int x\sqrt{x^2+1}\,dx = \frac{1}{2}\int \left(x^2+1\right)^{\frac{1}{2}} 2x\,dx$$

Use u-substitution to find the antiderivative.

$$= \frac{1}{2}\int u^{\frac{1}{2}}\,du$$

$$= \frac{1}{2}\cdot\frac{2}{3}u^{\frac{3}{2}} + C$$

$$= \frac{1}{3}\left(x^2+1\right)^{\frac{3}{2}} + C$$

$$\int_0^1 x\sqrt{x^2+1}\,dx = \frac{1}{3}\left(x^2+1\right)^{\frac{3}{2}} \Big]_0^1$$

Evaluate the definite integral.

$$= \left[\frac{1}{3}\left((1)^2+1\right)^{\frac{3}{2}}\right] - \left[\frac{1}{3}\left((0)^2+1\right)^{\frac{3}{2}}\right]$$

$$= \left(\frac{1}{3}\cdot 2^{\frac{3}{2}}\right) - \left(\frac{1}{3}\right)$$

$$= \frac{1}{3}\cdot 2\sqrt{2} - \frac{1}{3}$$

$$= \frac{1}{3}\left(2\sqrt{2} - 1\right)$$

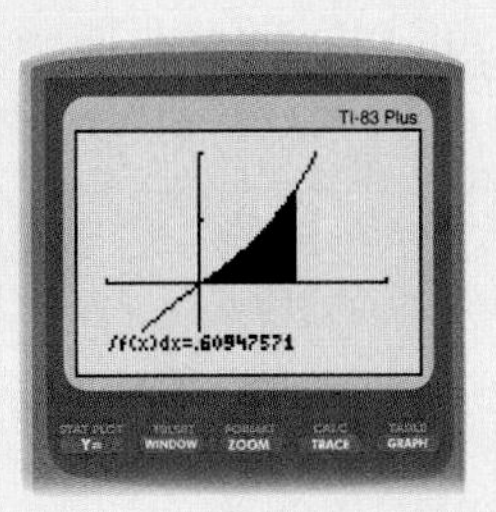

A graphing utility such as the TI-84 Plus can calculate an approximation to a definite integral. For example, to determine the area in Example 5**b**., $\int_0^1 x\sqrt{x^2+1}dx$:

1. Graph the function in a window which includes the endpoints a and b.

2. With the graph on screen, press 2ND TRACE, and select menu item 7.

3. When prompted for a lower bound, since $a = 0$, type Ø and ENTER. When prompted for an upper bound, since $b = 1$, type 1 and ENTER.

4. The area will be shaded and the approximate value of the definite integral, *.60947571,* appears at the bottom of the screen.

Average Value

Suppose that the last three cars you bought cost \$7500, \$8200, and \$10,400. To find the average cost, you would add these numbers and divide by 3:

$$\frac{7500+8200+10,400}{3}=\frac{26,100}{3}=\$8700.$$

To find the average value of a continuous function $y = f(x)$ on an interval $[a, b]$ is somewhat more difficult because $f(x)$ has an infinite number of values on $[a, b]$.

For example, let's numerically estimate an average y-value for $y=8x-x^2$ considered over the x-axis interval $[0, 8]$. (See Figure 6.3.9.)

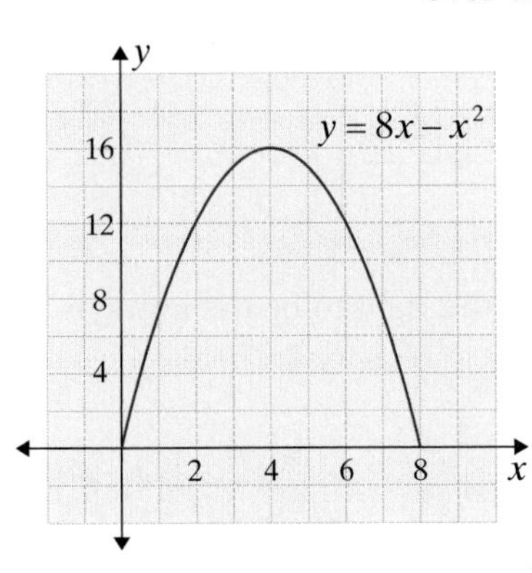

Figure 6.3.9

We could average the highest y-value, 16, with the lowest y-value, 0, and get 8. If we average $f(0)$ and $f(8)$, we get 0. Perhaps, it is better if we pick a few evenly spaced x-values and average the corresponding y-values. Choosing four values, we get

$$\frac{f(0)+f(2)+f(4)+f(6)}{4}=\frac{0+12+16+12}{4}=\frac{40}{4}=10$$

or we might get

$$\frac{f(1)+f(3)+f(5)+f(7)}{4}=\frac{7+15+15+7}{4}=\frac{44}{4}=11.$$

Evidently, the average y-value is somewhere near 10 or 11.

In general, one may estimate an average value of $f(x)$ on the interval $[a, b]$ by taking n evenly spaced x-values and computing the following expression:

$$\frac{f(x_1)+f(x_2)+...+f(x_n)}{n}.$$

Presumably, we get a better estimate as n gets bigger (i.e., if we use more points). If we could get a convenient limit expression, which could be seen to converge, and is easy to calculate, then we could always estimate an average y-value. Actually, the expression on the previous page is very suggestive of a Riemann sum since a Riemann sum involves the sum of y values. We proceed by analyzing a general Riemann sum; and, not surprisingly, we find a relationship between the concept of average value and the definite integral.

Step 1: Form the general Riemann sum S_n for the function f.

$$S_n = \left[f(c_1)+f(c_2)+\ldots+f(c_n)\right]\Delta x$$

Step 2: Substitute $\Delta x = \dfrac{b-a}{n}$.

$$S_n = \left[f(c_1)+f(c_2)+\ldots+f(c_n)\right]\frac{b-a}{n}$$
$$=(b-a)\left[\frac{f(c_1)+f(c_2)+\ldots+f(c_n)}{n}\right]$$

Observe that the factor in brackets is exactly what we need.

Step 3: We know that $\lim\limits_{n\to+\infty} S_n = \int_a^b f(x)\,dx$.

So

$$\int_a^b f(x)\,dx = (b-a)\cdot \lim_{n\to+\infty}\left[\frac{f(c_1)+f(c_2)+\ldots+f(c_n)}{n}\right],$$

or

$$\frac{1}{b-a}\int_a^b f(x)\,dx = \lim_{n\to+\infty}\left[\frac{f(c_1)+f(c_2)+\ldots+f(c_n)}{n}\right].$$

Thus the very limit we wanted exists! We define the limit on the right to be the **average value (AV)** of f on the interval $[a, b]$.

Average Value

*For a function $y = f(x)$, continuous on the interval $[a, b]$, the **average value** is*

$$AV = \frac{1}{b-a}\int_a^b f(x)\,dx.$$

Geometrically, the average value for a nonnegative continuous function is the height of a rectangle with base $(b - a)$ that has area exactly equal to the area under the curve. (See Figure 6.3.10.)

Figure 6.3.10

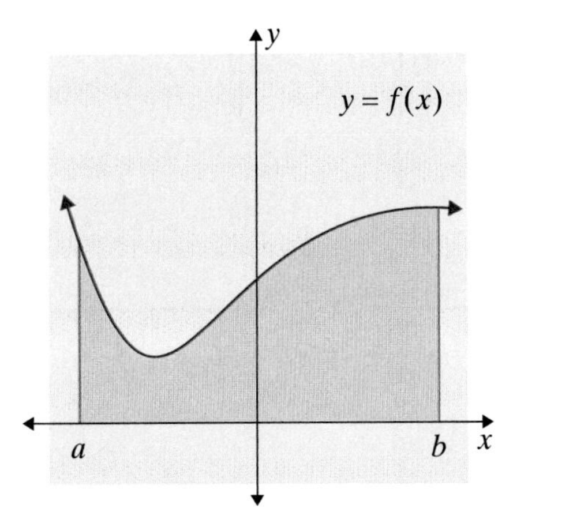

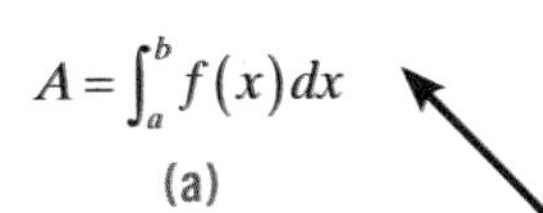

(a)

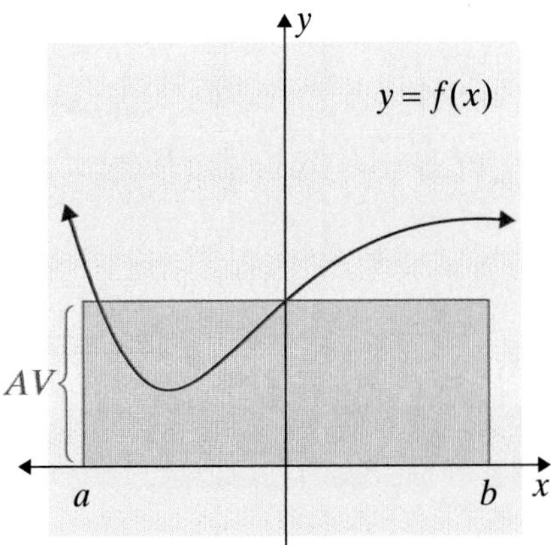

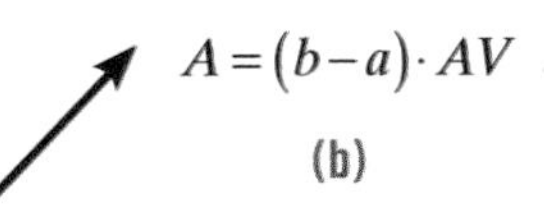

(b)

The shaded area exactly equals the rectangular area.

Example 6: Average Value

a. Find the average value of $f(x) = 8x - x^2$ on the interval [0, 8].

Solution: $AV = \frac{1}{8-0} \cdot \int_0^8 (8x - x^2)\,dx$ Substitute $a = 0, b = 8$, and the given $f(x)$ into the formula for average value.

$= \frac{1}{8} \cdot \left[\frac{8x^2}{2} - \frac{x^3}{3}\right]_0^8$ Solve.

$= \frac{1}{8} \cdot \left[\frac{8(8)^2}{2} - \frac{(8)^3}{3}\right] - \frac{1}{8} \cdot \left[\frac{8(0)^2}{2} - \frac{(0)^3}{3}\right]$

$= 32 - \frac{64}{3} - 0$

$\approx 10.\overline{6}$

Note that this is between 10 and 11 as estimated earlier.

b. Find the average value of $f(x) = x^2 + 2$ on the interval [0, 3].

Solution: $AV = \frac{1}{3-0} \int_0^3 (x^2 + 2)\,dx$ Substitute $a = 0, b = 3$, and the given $f(x)$ into the formula for average value.

$= \frac{1}{3}\left(\frac{x^3}{3} + 2x\right)\Bigg]_0^3$ Solve.

continued on next page ...

$$= \frac{1}{3}\left[\left(\frac{(3)^3}{3} + 2 \cdot 3\right) - \left(\frac{(0)^3}{3} + 2 \cdot 0\right)\right]$$

$$= \frac{1}{3}(9 + 6)$$

$$= 5$$

Example 7: Average Sales

A new electronics company sells $y = \frac{1}{2}t^2 + t$ DVD players (in hundreds) per month, where t is the number of months the company has been in business. Find the average sales per month during the company's first 6 months in business.

Solution: $AV = \frac{1}{6-0}\int_0^6 \left(\frac{1}{2}t^2 + t\right)dt$ Substitute $a = 0, b = 6,$ and the given $f(x)$ into the formula for average value.

$$= \frac{1}{6}\left(\frac{t^3}{6} + \frac{t^2}{2}\right)\Bigg]_0^6$$ Solve.

$$= \frac{1}{6}\left[\left(\frac{(6)^3}{6} + \frac{(6)^2}{2}\right) - \left(\frac{(0)^3}{6} + \frac{(0)^2}{2}\right)\right]$$

$$= \frac{1}{6}(36 + 18 - 0)$$

$$= \frac{1}{\cancel{6}}(\cancel{54})_9$$

$$= 9$$

The company averaged sales of 900 DVD players per month during its first 6 months in business.

6.3 Exercises

1. Find the area of the shaded region in the figure below.

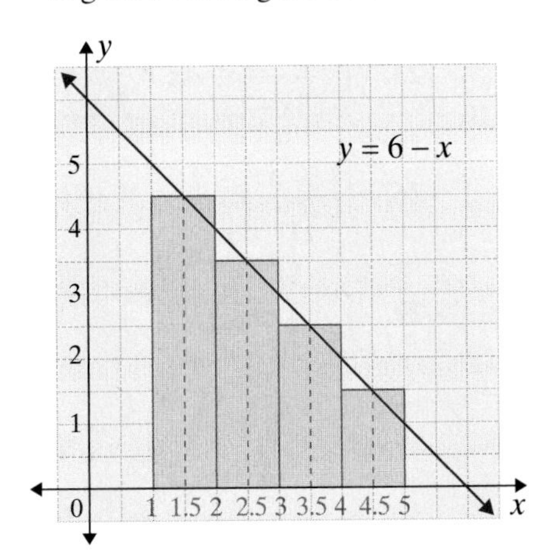

2. Find the area of the shaded region in the figure below.

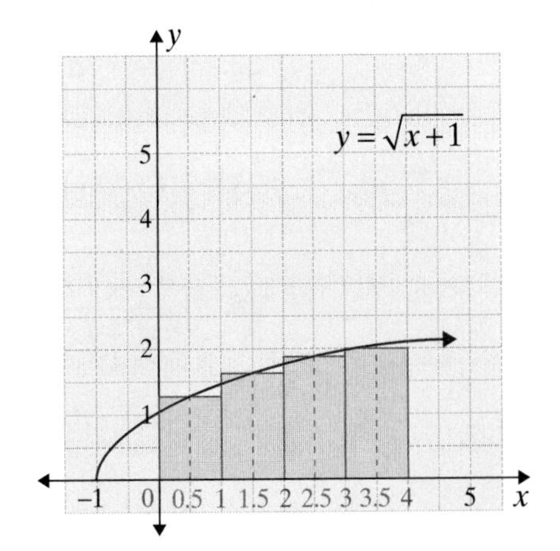

In Exercises 3 – 6, find the Riemann sum S_n for the given function, interval, and value of n.

3. $f(x)=x^2$; $[a, b]=[0, 4]$; $n=4$; $c_1=0.5$, $c_2=1.5$, $c_3=2.5$, $c_4=3.5$.

4. $f(x)=9-x^2$; $[a, b]=[-3, 2]$; $n=5$; $c_1=-2.5$, $c_2=-1.5$, $c_3=-0.5$, $c_4=0.5$, $c_5=1.5$.

5. $f(x)=\sqrt{2-x}$; $[a, b]=[-2, 2]$; $n=4$. Use the midpoint of each subinterval for the value of each c_k.

6. $f(x)=\dfrac{1}{x+2}$; $[a, b]=[-1, 3]$; $n=4$. Use the midpoint of each subinterval for the value of each c_k.

In Exercises 7 – 38, evaluate each definite integral.

7. $\int_{-1}^{0} 5x^2 dx$

8. $\int_{1}^{4} \frac{1}{x} dx$

9. $\int_{0}^{3} (1+e^x) dx$

10. $\int_{0}^{2} 6e^x dx$

11. $\int_{3}^{5} \frac{1}{x-2} dx$

12. $\int_{1}^{9} \sqrt{x}\, dx$

13. $\int_{-4}^{-2} \frac{1}{x^2} dx$

14. $\int_{2}^{4} (7x+2) dx$

15. $\int_{-1}^{3} (4x+1) dx$

16. $\int_{-2}^{1} \frac{4}{x+3} dx$ **17.** $\int_{-1}^{3} e^{x+1} dx$ **18.** $\int_{2}^{3} (x^2+2x-4) dx$

19. $\int_{2}^{4} x(x^2-3) dx$ **20.** $\int_{1}^{8} \left(1+\frac{1}{\sqrt[3]{x}}\right) dx$ **21.** $\int_{0}^{3} \frac{1}{3x+1} dx$

22. $\int_{1}^{3} 2e^{-1.5x} dx$ **23.** $\int_{3}^{5} \frac{1}{(3x+1)^2} dx$ **24.** $\int_{-1}^{0} \sqrt{3x+4}\, dx$

25. $\int_{2}^{3} (3-2x)^4 dx$ **26.** $\int_{0}^{2} \frac{x}{\sqrt[3]{x^2+4}} dx$ **27.** $\int_{2}^{6} \frac{3x}{x^2-3} dx$

28. $\int_{3}^{5} \frac{x+2}{x^2+4x+3} dx$ **29.** $\int_{0}^{1} e^x(e^x+1) dx$ **30.** $\int_{0}^{5} xe^{-0.24x^2} dx$

31. $\int_{1}^{3} xe^{x^2-1} dx$ **32.** $\int_{1}^{2} (x-1)(2x^2-4x+1)^2 dx$

33. $\int_{6}^{7} \frac{x-3}{\sqrt{x^2-6x+4}} dx$ **34.** $\int_{1}^{8} \left(2x^{\frac{2}{3}}-x^{-2}\right) dx$

35. $\int_{0}^{1} \frac{e^x}{e^x+1} dx$ **36.** $\int_{1}^{4} \frac{\ln x}{x} dx$

37. $\int_{1}^{3} \frac{1+\ln x}{x} dx$ **38.** $\int_{2}^{4} \frac{1}{x^2} e^{\frac{1}{x}} dx$

For Exercises 39 – 44, find the average value of the function on the given interval.

39. $f(x)=x^2+6;\ [1, 4]$ **40.** $f(x)=4x^2-3x+1;\ [-1, 3]$

41. $f(x)=\sqrt{x+1};\ [3, 8]$ **42.** $f(x)=\sqrt[3]{2x+1};\ [0, 13]$

43. $f(x)=2e^{-0.25x};\ [0, 4]$ **44.** $f(x)=1+e^{-0.4x};\ [0, 5]$

45. Pollution. The level of pollution in San Felipe Bay, due to an oil spill, is estimated to be $f(t)=\frac{1800t}{\sqrt{t^2+11}}$ parts per million, where t is the time in days since the spill occurred. Find the average level of pollution during the first 5 days after the spill occurred.

46. Bacterial population. It is estimated that the number of bacteria present in a culture t hours after bacteria are introduced to the culture is given by $N(t)=\frac{8000}{\sqrt{8-0.5t}}$. Find the average number of bacteria present during the first 8 hours.

47. Average production. The daily production level for a product is given by $N(t) = 240 - 240e^{-0.2t}$ units, where t is the time in hours after production begins. Find the average production during the first 4 hours.

48. Average marginal profit. The marginal profit from the production and the sale of x barbecue grills is given by $P'(x) = 52 - 0.8x$ dollars per grill. Find the average marginal profit for the first 40 grills produced and sold.

Hawkes Learning Systems: Essential Calculus

The Fundamental Theorem of Calculus	p. 491 - 498	Ex. 1 - 6
The Definite Integeral	p. 498 - 506	Ex. 7 - 48

6.4 Area (with Applications)

Objectives

After completing this section, you will be able to:

1. *Use definite integrals to find area when a function has negative values.*
2. *Apply what you have learned about definite integrals to marginal analysis.*

In Section 6.3, we discussed the fact that when a function is nonnegative, the definite integral of the function represents the area between the graph of the function and the x-axis on the interval $[a, b]$. In this section, we will continue that discussion and relate definite integrals to area when a function has negative values.

Area Bounded by a Curve and the x-Axis

A Riemann sum can involve rectangles both above and below the x-axis, as illustrated in Figure 6.4.1.

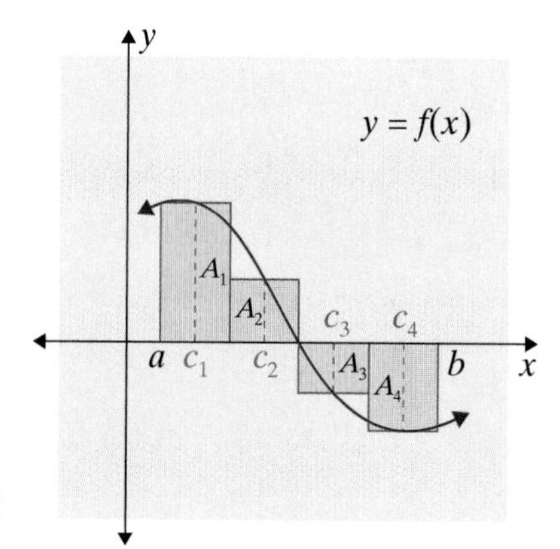

Figure 6.4.1

In Figure 6.4.1

$$S_4 = f(c_1)\Delta x + f(c_2)\Delta x + f(c_3)\Delta x + f(c_4)\Delta x$$
$$= A_1 + A_2 - A_3 - A_4.$$

Since $f(c_3)$ is negative, $f(c_3)\Delta x$ is **not** an area. It is instead the negative of an area, $f(c_3)\Delta x = -A_3$ (sometimes called a "signed area"). Similarly, $f(c_4)\Delta x = -A_4$. Thus, in this figure, the Riemann sum is not a total area; instead, it is the difference between areas. This concept leads to the following related statement about computing areas.

The Integral as Area

For $y = f(x)$, a continuous function on the interval $[a, b]$, the integral $\int_a^b f(x)\,dx$ represents:

1. *The total area bounded by the curve and the x-axis from $x = a$ to $x = b$ if $f(x)$ is nonnegative for all x in $[a, b]$.*
2. *The difference between the areas above the x-axis and below the x-axis that are bounded by the curve and the x-axis from $x = a$ to $x = b$ if $f(x)$ is negative for any x in $[a, b]$.*

Example 1: Interpreting the Integral

a. Evaluate $\int_0^2 (x+1)\,dx$ and interpret the integral geometrically.

Solution: $\int_0^2 (x+1)\,dx = \left.\frac{x^2}{2} + x\right]_0^2$

$$= \left(\frac{(2)^2}{2} + 2\right) - \left(\frac{(0)^2}{2} + 0\right)$$

$$= 4$$

Geometrically, 4 is the area between the curve $y = x + 1$ and the x-axis on the interval $[0, 2]$.

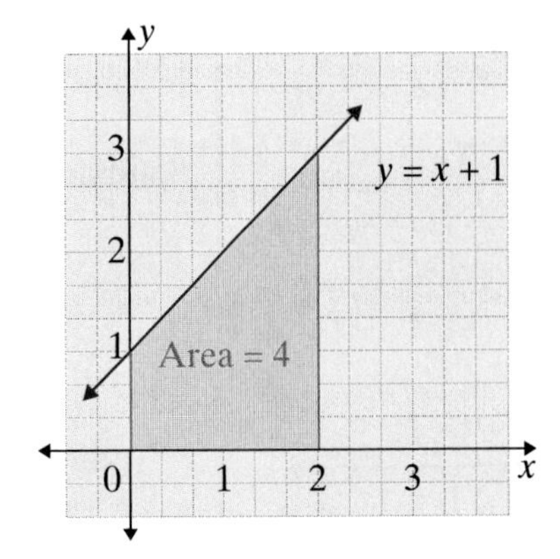

b. Evaluate $\int_0^2 (1-x)\,dx$ and interpret the integral geometrically.

Solution: $\int_0^2 (1-x)\,dx = \left.x - \frac{x^2}{2}\right]_0^2$

$$= \left(2 - \frac{(2)^2}{2}\right) - \left(0 - \frac{(0)^2}{2}\right)$$

$$= 0$$

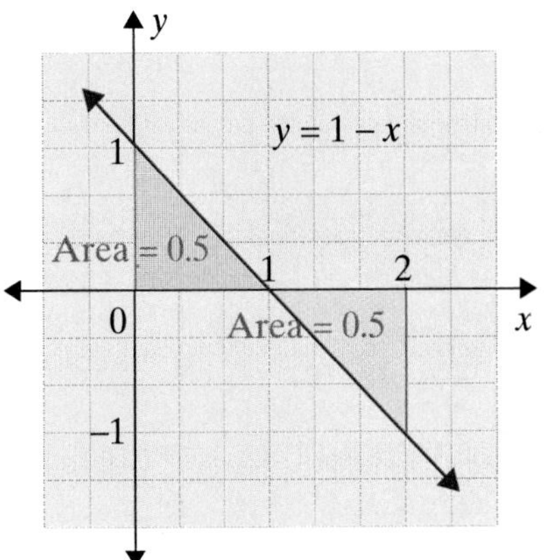

Geometrically, the 0 means the area above the x-axis and the area below the x-axis on the given interval are exactly equal.

In Example 1**b**. the triangular region above the x-axis has the same area (0.5) as the triangular region below the x-axis. Thus, if we wanted to find the total of the areas, we could, because of the symmetry in the function, integrate the function from $x = 0$ to $x = 1$ and double the result.

$$\begin{aligned}\text{Total area} &= 2\int_0^1 (1-x)\,dx \\ &= 2\left(x - \frac{x^2}{2}\right)\Bigg]_0^1 \\ &= 2\left(1-\frac{1}{2}\right) = 2\left(\frac{1}{2}\right) = 1\end{aligned}$$

Example 2 shows a case where the regions above and below the x-axis do not have the same area. In this case we must separate the integral into two parts to find the total area.

Example 2: Total Area

Find the total area bounded by the x-axis and the curve $y = 4 - x^2$ on the interval $[-2, 3]$.

Solution: The curve $y = 4 - x^2$ is a parabola that crosses the x-axis at $x = -2$ and $x = 2$. We integrate the function from $x = -2$ to $x = 2$ to find A_1. Then, to find A_2, we take the additive inverse (negative) of the integral from $x = 2$ to $x = 3$.

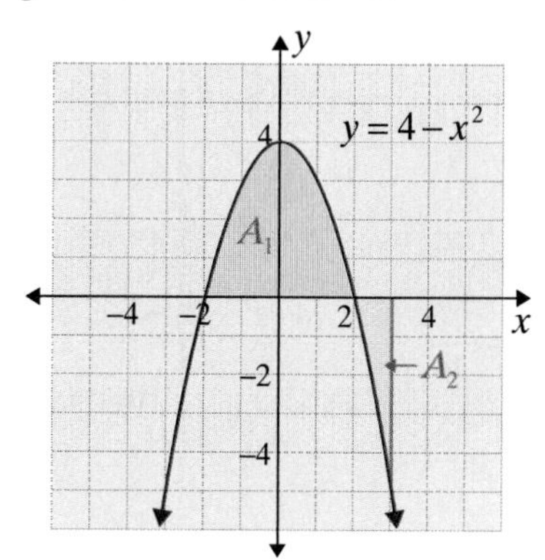

$$\begin{aligned}A_1 &= \int_{-2}^{2} (4-x^2)\,dx \\ &= 4x - \frac{x^3}{3}\Bigg]_{-2}^{2} \\ &= \left[4(2) - \frac{(2)^3}{3}\right] - \left[4(-2) - \frac{(-2)^3}{3}\right] \\ &= \left(8 - \frac{8}{3}\right) - \left(-8 + \frac{8}{3}\right) \\ &= 8 - \frac{8}{3} + 8 - \frac{8}{3} \\ &= 16 - \frac{16}{3} \\ &= \frac{32}{3}\end{aligned}$$

$$A_2 = -\int_2^3 (4 - x^2)\,dx$$
$$= -\left(4x - \frac{x^3}{3}\right)\Big]_2^3$$
$$= -\left[\left(4(3) - \frac{(3)^3}{3}\right) - \left(4(2) - \frac{(2)^3}{3}\right)\right]$$
$$= -\left[(12 - 9) - \left(8 - \frac{8}{3}\right)\right]$$
$$= -\left(12 - 9 - 8 + \frac{8}{3}\right)$$
$$= -\left(-5 + \frac{8}{3}\right)$$
$$= -\left(-\frac{15}{3} + \frac{8}{3}\right)$$
$$= -\left(-\frac{7}{3}\right)$$
$$= \frac{7}{3}$$

Note: Adding a minus sign will not apply to other examples in this book. In this case the goal was to find the area bounded by the curve and the x-axis.

$$\text{Total Area} = A_1 + A_2 = \frac{32}{3} + \frac{7}{3} = \frac{39}{3} = 13$$

Note that $\int_{-2}^3 (4 - x^2)\,dx = A_1 - A_2 = \frac{32}{3} - \frac{7}{3} = \frac{25}{3}$, which is not the total area.

NOTES

Graphs, as shown in Examples 1 and 2, are valuable visual aids in understanding definite integrals. A sketch of the graph of the function being integrated should be made whenever possible.

The following two properties of the definite integral can be added to the list of three properties stated in Section 6.3.

Additional Properties of the Definite Integral

$$\int_a^b f(x)\,dx = -\int_b^a f(x)\,dx$$

$$\int_a^b f(x)\,dx = \int_a^c f(x)\,dx + \int_c^b f(x)\,dx \quad \text{where } c \text{ is any point with } a \le c \le b.$$

The latter of the previous properties is particularly useful if a function is defined in pieces and we want to find the area bounded by the curve and the x-axis. (See Figure 6.4.2.)

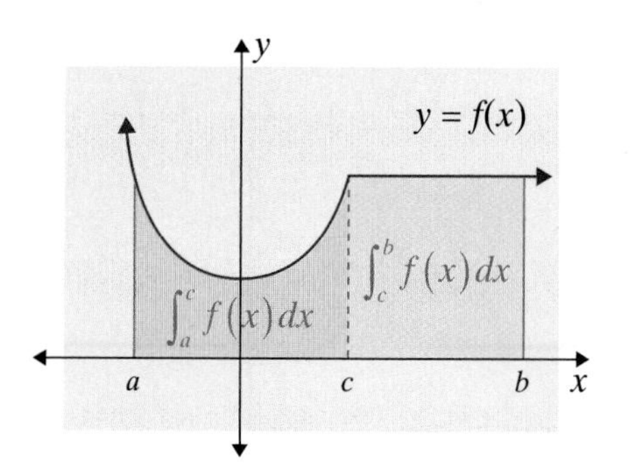

Figure 6.4.2

Example 3: Bounded Area

Find the area of the region bounded by the function

$$f(x)=\begin{cases} x+2 & \text{if } x\le 1 \\ 5-2x & \text{if } x>1 \end{cases}$$

and the x-axis from $x=-1$ to $x=2$.

Solution: The graph consists of two line segments.

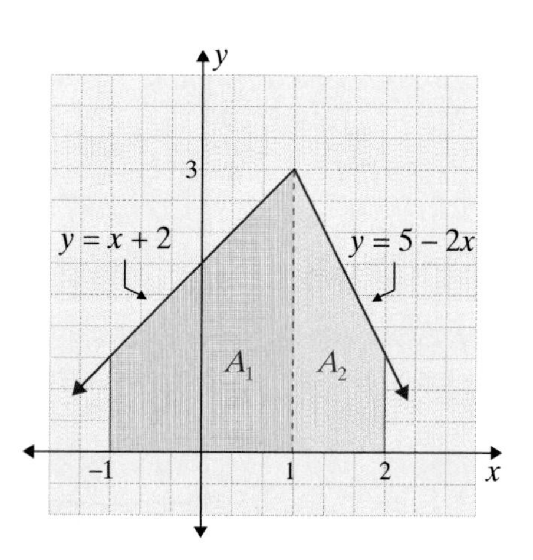

$$\begin{aligned}
\int_{-1}^{2} f(x)\,dx &= \int_{-1}^{1}(x+2)\,dx+\int_{1}^{2}(5-2x)\,dx \\
&= \left(\frac{x^2}{2}+2x\right)\Bigg]_{-1}^{1}+\left(5x-x^2\right)\Big]_{1}^{2} \\
&= \left[\left(\frac{(1)^2}{2}+2\right)-\left(\frac{(-1)^2}{2}-2\right)\right]+\left[\left(5(2)-4\right)-\left(5(1)-1\right)\right] \\
&= \left(\frac{1}{2}+2-\frac{1}{2}+2\right)+(10-4-5+1) \\
&= 4+2 \\
&= 6
\end{aligned}$$

Marginal Analysis

We know that marginal cost is the derivative $C'(x)$, marginal revenue is the derivative $R'(x)$, and marginal profit is the derivative $P'(x)$. Therefore, the integrals of each of these functions will yield the antiderivatives, which are, respectively, the cost function $C(x)$, the revenue function $R(x)$, and the profit function $P(x)$. The term "marginal" was discussed in Section 6.1 when we introduced the antiderivative. Now we will show how the definite integral can be used to measure changes in cost, revenue, and profit. The integrals do not take into account fixed costs such as rent. Such costs must be treated separately and are ignored here.

Suppose that a marginal profit function $P'(x) = 20 - 0.05x$ is known, where x is the number of items sold and the marginal profit is in dollars per item. Since P' represents the change in profit per item, we note that the profit is decreasing by 5 cents per item. The area under the curve corresponding to P' on an interval $[0, a]$ represents the profit when a items are produced and sold. (See Figure 6.4.3.) These ideas are illustrated in Example 4.

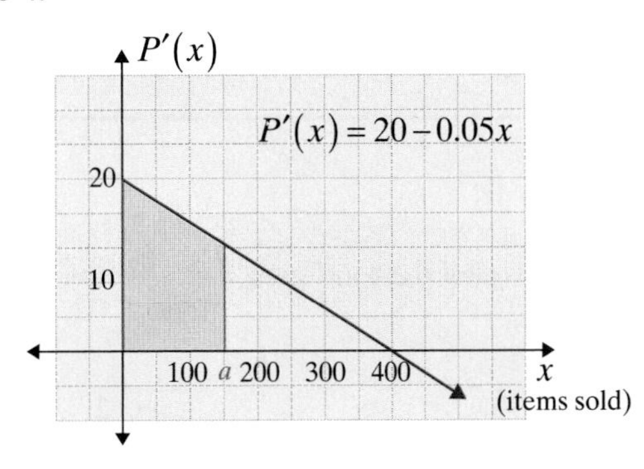

Figure 6.4.3

The figure shows that if more than 400 items are sold, the company's profits will decline (the area will be below the x-axis). Note that at $x = 100$ (say), marginal profit is decreasing, but profit (measured by the area) is increasing at $x = 100$. If marginal profit is decreasing, it might still be worthwhile to continue producing and selling the product.

Example 4: Marginal Analysis

A picture frame maker knows that his profit (in dollars) is changing at a rate given by the function $P'(x) = 20 - 0.05x$, where x is the number of frames he makes and sells.

a. Find the profit from selling the first 200 frames.

b. Find the change in his profit when sales increase from 200 to 400 frames.

Solutions: a. Integrate the marginal profit function from $x = 0$ to $x = 200$.

$$\int_0^{200} (20 - 0.05x)\,dx = 20x - 0.025x^2 \Big]_0^{200}$$

$$= \left[20(200) - 0.025(200)^2\right] - \left[20(0) - 0.025(0)^2\right]$$

$$= (4000 - 1000) - 0$$

$$= 3000$$

His profit is \$3000 on the first 200 frames if we ignore any fixed costs.

continued on next page ...

b. Integrate from $x = 200$ to $x = 400$.

$$\int_{200}^{400} (20 - 0.05x)\,dx = 20x - 0.025x^2 \Big]_{200}^{400}$$
$$= \left[20(400) - 0.025(400)^2\right] - \left[20(200) - 0.025(200)^2\right]$$
$$= (8000 - 4000) - (4000 - 1000)$$
$$= 1000$$

His profit changes by \$1000 when sales increase from 200 to 400 frames.

Note that from parts **a.** and **b.** we see that the profit on sales for the first 200 frames ($x = 0$ to $x = 200$) is greater than for the second 200 frames ($x = 200$ to $x = 400$). This result is quite reasonable because the marginal profit, $P'(x) = 20 - 0.05x$, is decreasing by 5 percent per frame. In this problem, the fixed costs are relevant. Suppose the fixed costs are \$1000. Then $P(x) = 20x - 0.025x^2 - 1000$ dollars, and $P(200) = \$2000$, which is the **net profit**. The integral from $x = 0$ to $x = 200$ gives the increase in profits, $P(200) - P(0) = 2000 - (-1000) = 3000$.

6.4 Exercises

For Exercises 1 – 18, find the total area bounded by the x-axis and the curve $y = f(x)$ on the indicated interval.

1. $f(x) = 3x + 1,\ [0, 5]$

2. $f(x) = 7 - 2x,\ [-1, 3]$

3. $f(x) = x^2 + 1,\ [-2, 2]$

4. $f(x) = 0.5x^2 + 2,\ [1, 4]$

5. $f(x) = x^3 + 2,\ [-1, 1]$

6. $f(x) = 2x^3 - 1,\ [1, 2]$

7. $f(x) = x^2 + x + 1,\ [-1, 3]$

8. $f(x) = x^2 + 2x - 3,\ [1, 3]$

9. $f(x) = \dfrac{4}{x+1},\ [0, 3]$

10. $f(x) = \dfrac{3}{2x+1},\ [0, 2]$

11. $f(x) = 3e^{0.6x},\ [0, 5]$

12. $f(x) = 1 + e^{-0.3x},\ [0, 4]$

13. $f(x) = x^2 - 2x - 8,\ [2, 5]$

14. $f(x) = x^2 + 3x - 4,\ [0, 4]$

15. $f(x) = \begin{cases} 2 - x & \text{if } -1 \le x \le 2 \\ x^2 - 4 & \text{if } 2 \le x \le 3 \end{cases}$

16. $f(x) = \begin{cases} x^2 & \text{if } -2 \le x \le 1 \\ 2x - 1 & \text{if } 1 \le x \le 2 \end{cases}$

17. $f(x) = \begin{cases} x + 2 & \text{if } -2 \le x \le 0 \\ \sqrt{x+4} & \text{if } 0 \le x \le 5 \end{cases}$

18. $f(x) = \begin{cases} 1 - 2x & \text{if } -2 \le x \le 0 \\ e^{2x} & \text{if } 0 \le x \le 1.5 \end{cases}$

In Exercises 19 and 20, explain the meaning of the shaded region in the graphs.

19.

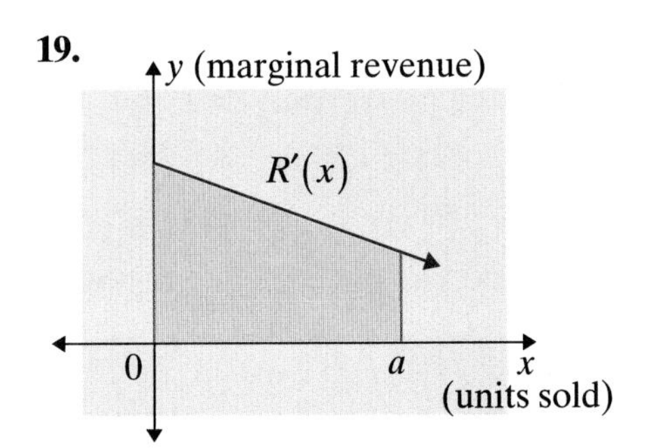

20.

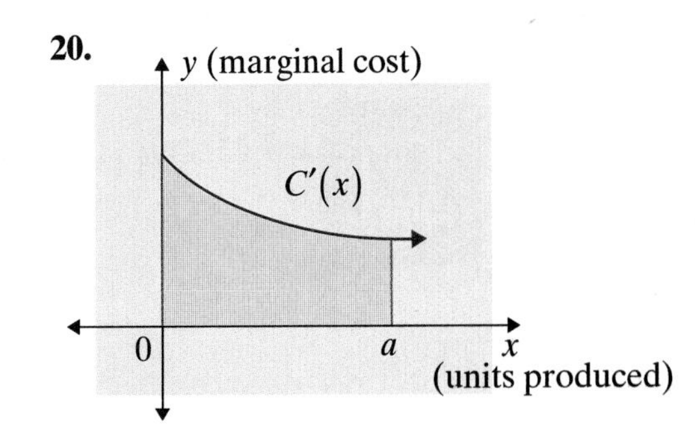

21. **Profit.** The marginal profit for a certain style of sports jacket is given by $P'(x) = 56 - 0.8x$ dollars per jacket, where x is the number of jackets produced and sold weekly. Find the profit for the first 50 jackets that are produced and sold. (Ignore any fixed costs.)

22. **Profit.** The marginal profit of an important product is given by $P'(x) = 10 - 0.015e^{0.6x}$ dollars per item, where x is the number of items produced and sold. Find the profit for the first 8 items. (Ignore any fixed costs.)

23. **Cost.** The marginal cost of a product is given by $15 + \frac{4}{\sqrt{x}}$ dollars per unit, where x is the number of units produced. The current level of production is 100 units weekly. If the level of production is increased to 169 units weekly, find the increase in the total costs.

24. **Revenue.** The marginal revenue from the sale of x bottles of a wine is given by $8.4 - 0.3\sqrt{x}$ dollars per bottle. Find the increase in total revenue if the number of bottles sold is increased from 225 to 350.

25. **Wildlife management.** The manager of a wildlife preserve has started a management program to control the population of the preserve's bison herd. It is estimated that the population will continue to grow according to the function $N'(t) = 15 - 6t^{\frac{1}{2}}$ bison per year, where t is the number of years after implementation of the plan and $0 \le t \le 5$. Find the increase in the population during the first 4 years of the program.

26. **Bacterial population.** It is estimated that t hours after some particular bacteria are introduced into a culture, the population will be increasing at a rate of

$$P'(t) = \frac{1200}{(12 - 0.5t)^{\frac{1}{2}}}$$

bacteria per hour. Find the increase in the population during the first 6 hours.

Hawkes Learning Systems: Essential Calculus

Area (with Applications)

6.5 Area Between Two Curves (with Applications)

Objectives

After completing this section, you will be able to:

1. *Find the area bounded by two curves.*
2. *Apply the concept of bounded area to business and economic situations.*

In Section 6.4 we studied the area between a curve and the x-axis. The x-axis itself is the line (or curve) corresponding to the equation $y = 0$. Thus the area between a curve and the x-axis can be thought of as the area between two curves. In this section we will expand on this idea by developing general techniques for finding the area bounded by two curves and by discussing some interesting applications.

Area Between Two Curves

Suppose that $y = f(x)$ and $y = g(x)$ are two continuous functions and $f(x) \geq g(x)$ on some interval $[a, b]$. Then the area of the region between the two curves can be found by the following definite integral. (See Figure 6.5.1.)

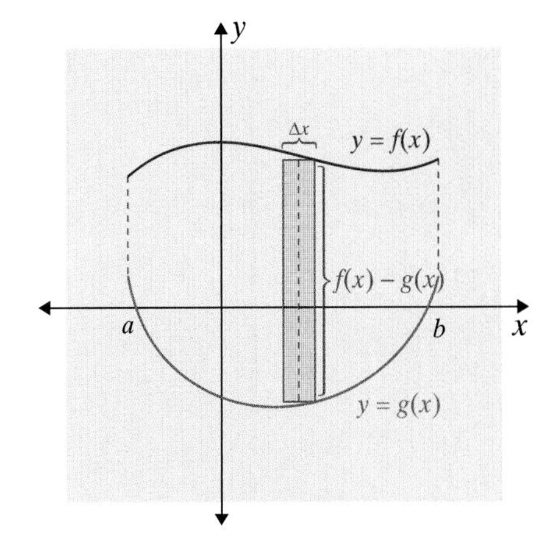

Figure 6.5.1

The area of the rectangle is $\left[f(x)-g(x)\right]\Delta x$ where x is a point in the interval Δx. The integral $A=\int_a^b\left[f(x)-g(x)\right]dx$ represents the limit of a Riemann sum of such rectangles.

Area Between Two Curves

If f and g are two continuous functions and $f(x) \geq g(x)$ on the interval $[a, b]$, then the area between the two curves on this interval is

$$\boldsymbol{A=\int_a^b\left[f(x)-g(x)\right]dx.}$$

Since $f(x) \geq g(x)$, we are assured that $[f(x) - g(x)]$ is nonnegative. Thus whether or not $f(x)$ and $g(x)$ are themselves positive or negative is not of concern when we are finding the area between the curves. The graph, though, is still a valuable tool, as we shall see in Examples 1 and 2.

Example 1: Finding the Area Between Two Curves

Find the area enclosed by the curves $y = x^2 - 1$ and $y = 1 - x$.

Solution: Sketch the graphs of both functions to help determine which function is larger. Now set the two y-values equal to each other to find the points of intersection. These x-values will be the limits of integration.

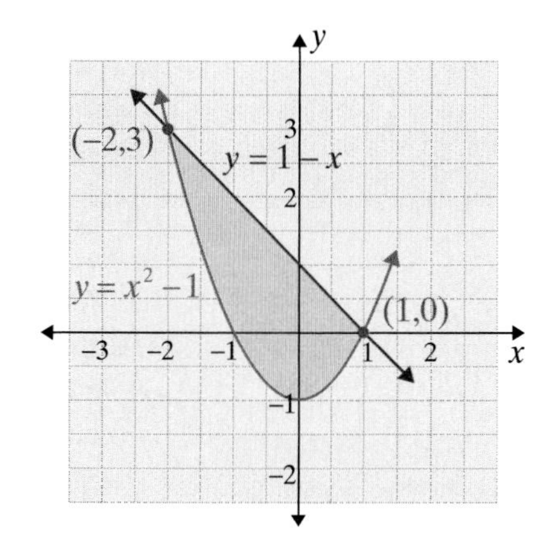

$$x^2 - 1 = 1 - x$$

$$x^2 + x - 2 = 0$$

$$(x+2)(x-1) = 0$$

$$x = -2, 1$$

Thus

$$\begin{aligned}
\text{area} &= \int_{-2}^{1} \left[(1-x) - (x^2 - 1)\right] dx \\
&= \int_{-2}^{1} (2 - x - x^2)\, dx \\
&= 2x - \frac{x^2}{2} - \frac{x^3}{3}\Big]_{-2}^{1} \\
&= \left(2(1) - \frac{(1)^2}{2} - \frac{(1)^3}{3}\right) - \left(2(-2) - \frac{(-2)^2}{2} - \frac{(-2)^3}{3}\right) \\
&= \left(2 - \frac{1}{2} - \frac{1}{3}\right) - \left(-4 - \frac{4}{2} - \frac{-8}{3}\right) \\
&= 2 - \frac{1}{2} - \frac{1}{3} + 4 + 2 - \frac{8}{3} \\
&= \frac{9}{2} \\
&= 4.5
\end{aligned}$$

Example 2: Finding the Area Between Two Curves

Find the area between the curves $y = x^3$ and $y = x^2 + 1$ on the interval $[-1, 1]$.

Solution: Sketch the graphs of both functions to help determine whether or not the curves intersect on the interval $[-1, 1]$.

These curves do not intersect on the interval $[-1, 1]$ and $y = x^2 + 1$ is larger on the entire interval.

Therefore,

$$\begin{aligned}
\text{area} &= \int_{-1}^{1}\left[\left(x^2+1\right)-\left(x^3\right)\right]dx \\
&= \int_{-1}^{1}\left(x^2+1-x^3\right)dx \\
&= \frac{x^3}{3}+x-\frac{x^4}{4}\Bigg]_{-1}^{1} \\
&= \left(\frac{(1)^3}{3}+1-\frac{(1)^4}{4}\right)-\left(\frac{(-1)^3}{3}+(-1)-\frac{(-1)^4}{4}\right) \\
&= \left(\frac{1}{3}+1-\frac{1}{4}\right)-\left(-\frac{1}{3}-1-\frac{1}{4}\right) \\
&= \frac{1}{3}+1-\frac{1}{4}+\frac{1}{3}+1+\frac{1}{4} \\
&= \frac{8}{3}
\end{aligned}$$

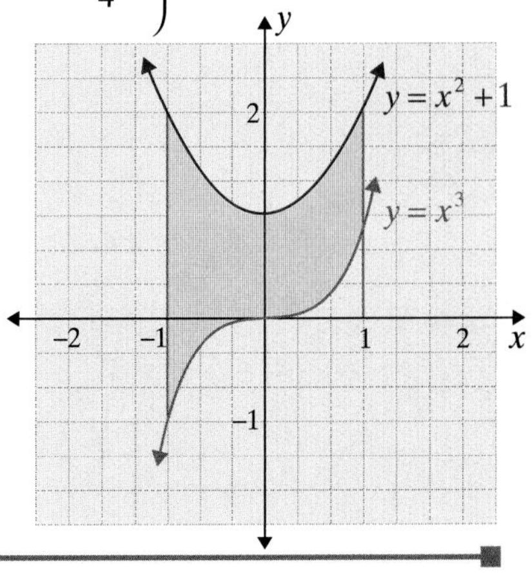

y
$y = x^2 + 1$
$y = x^3$
x
−2 −1 1 2
2
−1

Now we will show how the definite integral, along with the concept of area between two curves, can be applied to two topics from the realm of business and economics: consumers' and producers' surplus and Lorenz curves.

Consumers' Surplus and Producers' Surplus

For consumers, the demand curve $p = D(x)$ represents the price they are willing to pay per item when x items are available in the market place. For producers, the supply curve $p = S(x)$ represents the price per item at which they are willing to produce and sell x items. The equilibrium point (x_E, p_E) is the point where the two curves intersect (see Figure 6.5.2). These ideas were first discussed in Section 1.8.

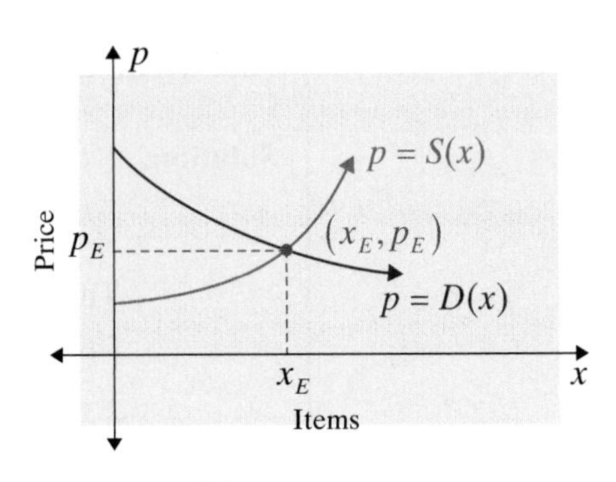

Figure 6.5.2

Some consumers are happy to pay the equilibrium price p_E because, as Figure 6.5.2 shows, they would have been willing to pay a higher price if there had been fewer than x_E items on the market. Thus these people have saved money by buying at a lower price. This savings, called the **consumers' surplus (CS)**, is the difference between the total value to the consumers and the actual total cost to the consumers, as illustrated in Figure 6.5.3.

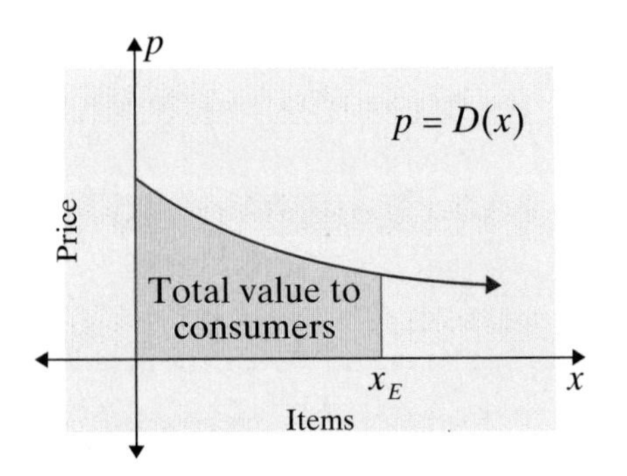

$\int_0^{x_E} D(x)dx =$ total value to consumers

(a)

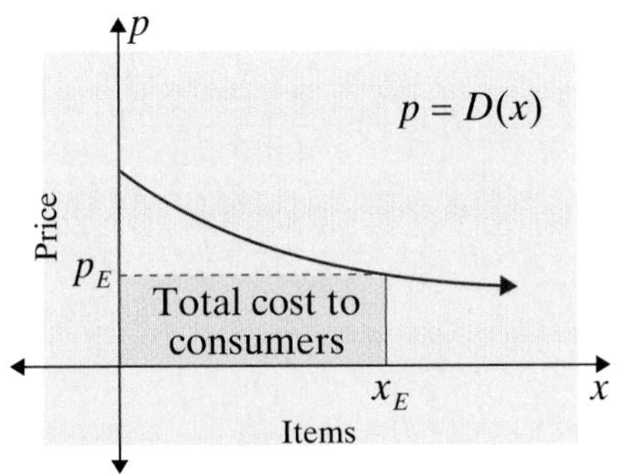

$x_E \cdot p_E =$ total cost to consumers

(b)

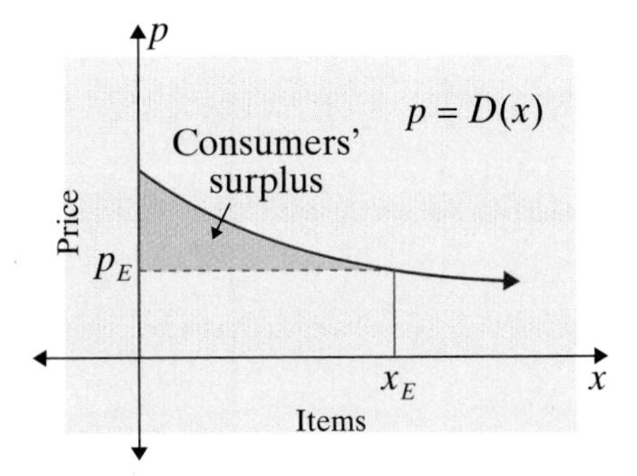

$\text{CS} = \int_0^{x_E} D(x)dx - x_E \cdot p_E$

(c)

Figure 6.5.3

Consumers' Surplus

*The **consumers' surplus** is defined to be*

$$\mathbf{CS} = \int_0^{x_E} D(x)\,dx - x_E \cdot p_E,$$

where $p = D(x)$ is the demand curve and (x_E, p_E) is the equilibrium point.

Some producers are pleased to sell at price p_E. They would have been willing to sell fewer items at a lower price. Thus the consumers have spent more than these producers anticipated. This extra income, called the **producers' surplus (PS)**, is the difference between the total cost to the consumers and the total value to the producers, as illustrated in Figure 6.5.4.

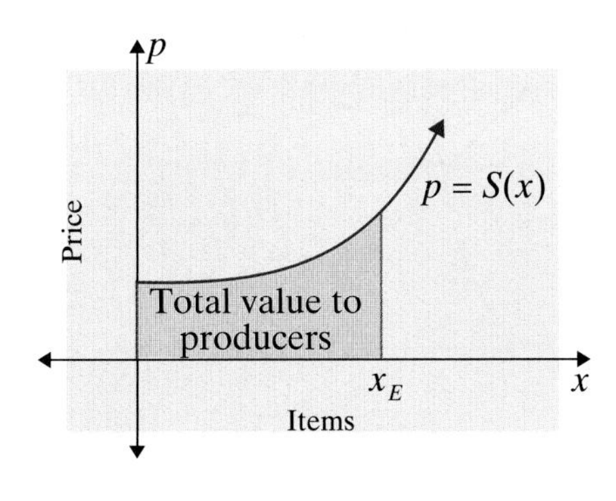

$\int_0^{x_E} S(x)\,dx$ = total value to producers

(a)

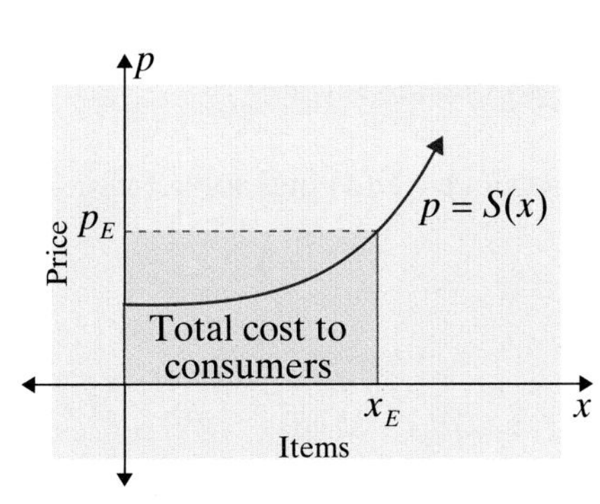

$x_E \cdot p_E$ = total cost to consumers

(b)

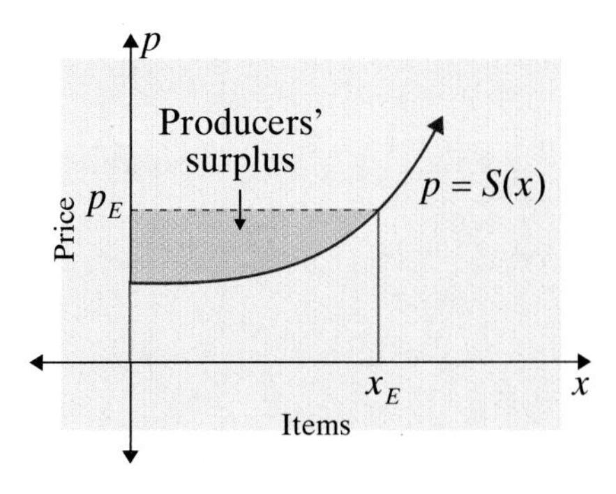

$\text{PS} = x_E \cdot p_E - \int_0^{x_E} S(x)\,dx$

(c)

Figure 6.5.4

Producers' Surplus

*The **producers' surplus** is defined to be*

$$\mathbf{PS} = x_E \cdot p_E - \int_0^{x_E} S(x)\,dx,$$

where $p = S(x)$ is the supply curve and (x_E, p_E) is the equilibrium point.

Both consumers' surplus and producers' surplus are illustrated in Figure 6.5.5.

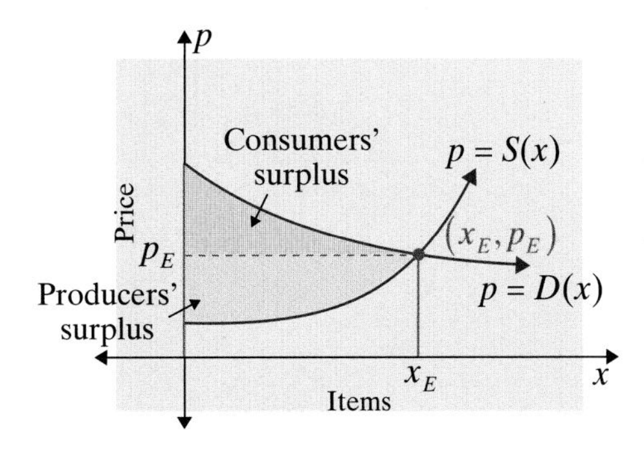

Figure 6.5.5

Example 3: Determining Surpluses

Suppose, for a certain new brand of car stereo, the demand function is $D(x) = 6 - x$ and the supply function is $S(x) = x^2$.

a. Find the equilibrium point.

b. Find the consumers' surplus.

c. Find the producers' surplus.

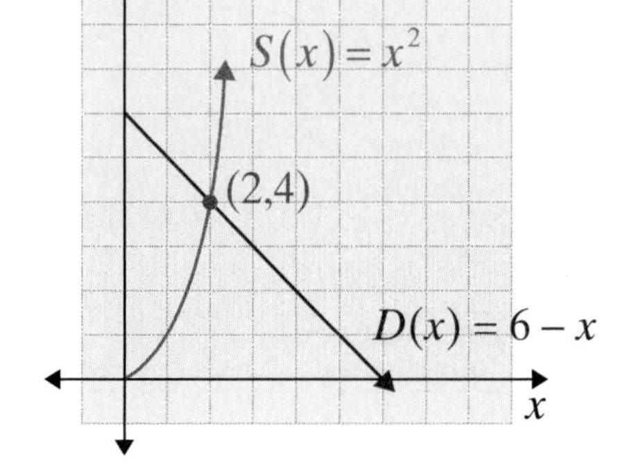

Solutions: a. Set $S(x) = D(x)$ and solve for x.

$$x^2 = 6 - x$$
$$x^2 + x - 6 = 0$$
$$(x+3)(x-2) = 0$$
$$\cancel{x = -3} \text{ or } x = 2 \qquad x \text{ cannot be negative.}$$

Therefore, $x_E = 2$ and $p_E = 6 - 2 = 4$.
The equilibrium point is $(2, 4)$.

b. $\text{CS} = \int_0^2 (6-x)\,dx - (2 \cdot 4)$ Use the formula for consumers' surplus.

$$= 6x - \frac{x^2}{2}\Big]_0^2 - 8$$
$$= \left(6(2) - \frac{(2)^2}{2}\right) - \left(6(0) - \frac{(0)^2}{2}\right) - 8$$
$$= 12 - \frac{4}{2} - 8$$
$$= \$2$$

c. $\text{PS} = (2 \cdot 4) - \int_0^2 x^2\,dx$ Use the formula for producers' surplus.

$$= 8 - \left(\frac{x^3}{3}\Big]_0^2\right)$$
$$= 8 - \left[\left(\frac{(2)^3}{3}\right) - \left(\frac{(0)^3}{3}\right)\right]$$
$$= 8 - \frac{8}{3}$$
$$= \frac{16}{3} \approx \$5.33$$

Lorenz Curves

Economists and sociologists use curves known as **Lorenz curves** to study the distribution of income within a society or country. In the graph of a Lorenz curve (see Figure 6.5.6), the x-axis shows the cumulative percentage of a country's families, starting from the lowest income families. The y-axis shows the cumulative percentage of the country's total income. Thus, for every number x in the interval $[0, 1]$, $f(x)$ is defined to be the percentage of the country's total income earned by the lower $100x$ percent of the families.

For example, in Figure 6.5.6, the point (0.3, 0.1) indicates that the bottom 30 percent of the families earn 10 percent of the country's income. Similarly, the point (0.7, 0.5) indicates that the bottom 70 percent earn 50 percent of the income.

In this graph, percent is represented in decimal form.

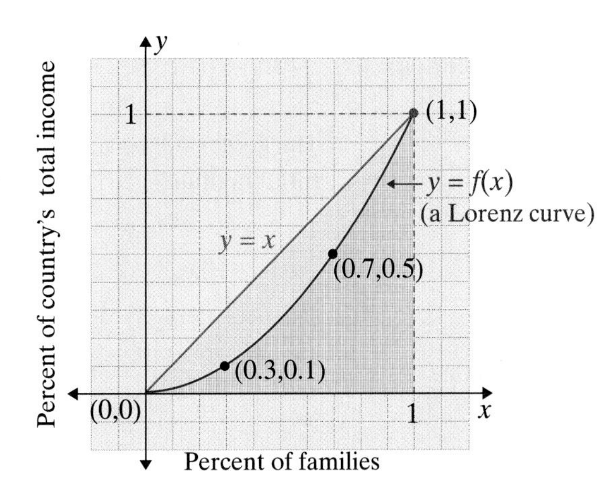

Figure 6.5.6

If every family had the same income, then the distribution of income would be represented by $y = x$, the **line of complete equality**. The area between the line $y = x$ and the Lorenz curve $y = f(x)$ is used to measure how much the distribution of income differs from complete equality. The ratio of the area between $y = x$ and $y = f(x)$ to the area under the line $y = x$ is called the **coefficient of inequality**. Since the square with upper right corner at $(1, 1)$ in Figure 6.5.6 has area 1, we see that the area under the line $y = x$ is $\frac{1}{2}$. Thus

$$\text{Coefficient of inequality} = \frac{\text{area between curves}}{\frac{1}{2}} = 2 \cdot (\text{area between curves}).$$

Coefficient of Inequality for a Lorenz Curve

If $y = f(x)$ represents a Lorenz curve, then

$$\textbf{\textit{Coefficient of inequality}} = 2\int_0^1 \left[x - f(x)\right] dx.$$

A coefficient of 0 corresponds to a distribution along the line of complete equality. A coefficient of 1 corresponds to a maximum of "inequality" in which $f(x) = 0$.

Any function that describes a Lorenz curve must have the following properties:

1. The domain is [0, 1].
2. The range is [0, 1].
3. $f(0) = 0$ and $f(1) = 1$.
4. $f(x) \le x$ for all x in the interval [0, 1].

Example 4: Using a Lorenz Curve

Suppose that $f(x) = 0.6x^2 + 0.4x$ represents a Lorenz curve for some country.

a. What percent of the country's total income is earned by the lower 50 percent of the families in this country?

b. Find the coefficient of inequality.

Solutions: a. $f(0.5) = 0.6(0.5)^2 + 0.4(0.5) = 0.35$ Since we want the lower 50% of families, set $x = 0.5$.

The lower 50 percent of the families earn 35 percent of the country's total income.

b.
$$2\int_0^1 \left[x - \left(0.6x^2 + 0.4x\right)\right]dx = 2\int_0^1 \left[\left(0.6x - 0.6x^2\right)\right]dx$$

Note: combine terms in the integrand before finding the antiderivative.

$$= 2\left(0.6\frac{x^2}{2} - 0.6\frac{x^3}{3}\right)\Bigg]_0^1$$

$$= 2\left[\left(0.6\frac{(1)^2}{2} - 0.6\frac{(1)^3}{3}\right) - \left(0.6\frac{(0)^2}{2} - 0.6\frac{(0)^3}{3}\right)\right]$$

$$= 2(0.3 - 0.2)$$

$$= 0.2$$

The coefficient of inequality is 0.2.

6.5 Exercises

In Exercises 1 – 16, find the area of the region bounded by the graphs of the given equations.

1. $y = x^2,\ y = x - 1,\ x = -1,\ x = 4$

2. $y = x^2 + 2,\ y = x,\ x = 2,\ x = 5$

3. $y = x^3,\ y = x^2,\ x = 0,\ x = 1$

4. $y = x^2 + 1,\ y = 1 - 2x,\ x = 0,\ x = 3$

5. $y = \sqrt{2x + 1},\ y = 3x + 2,\ x = 0,\ x = 2$

6. $y = e^{x-1},\ y = x,\ x = 1,\ x = 4$

7. $y = e^{-x},\ y = x+1,\ x = 0,\ x = 3$

8. $y = x^2+1,\ y = e^{-0.2x},\ x = 0,\ x = 4$

9. $y = \dfrac{1}{x},\ y = \dfrac{5}{2} - x,\ x = \dfrac{1}{2},\ x = 2$

10. $y = \dfrac{1}{x+1},\ y = e^{0.7x},\ x = 0,\ x = 2$

11. $y = x+1,\ y = x^2 + x$

12. $y = x^2+1,\ y = 6-x$

13. $y = \sqrt{x},\ y = x^2$

14. $y = x^2 - 6x,\ y = -x^2$

15. $y = x^2 - 2x - 3,\ y = 2x+2$

16. $y = x^2 + 5x - 1,\ y = 2 - x^2$

For Exercises 17 – 21, determine the area pictured (check each answer using a graphing utility if possible). In each case you must determine the limits of integration if necessary.

17.

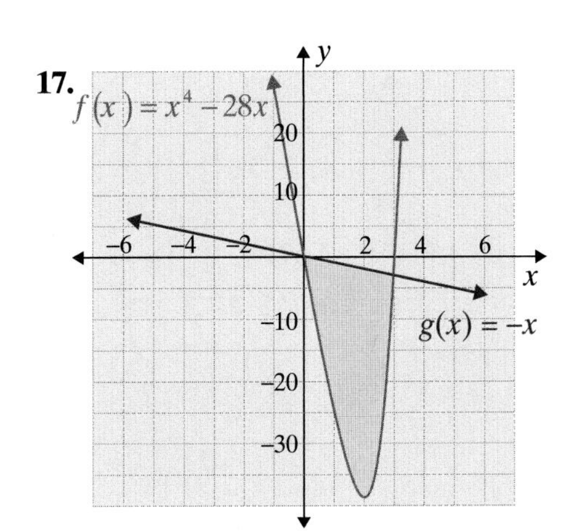

18.

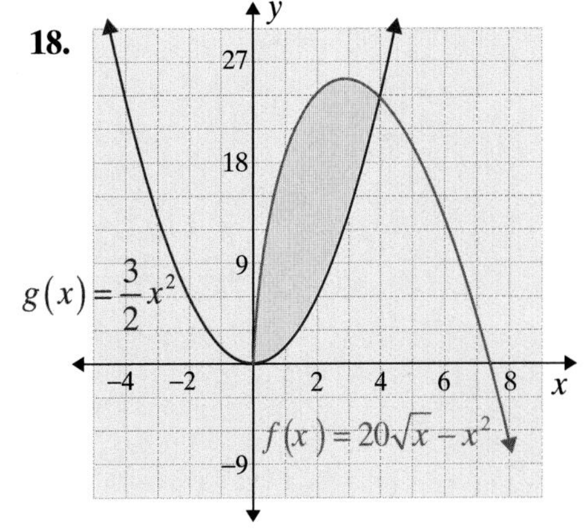

19.

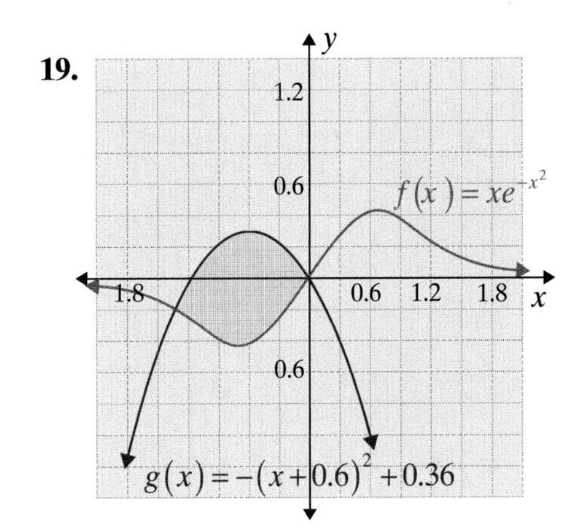

20.

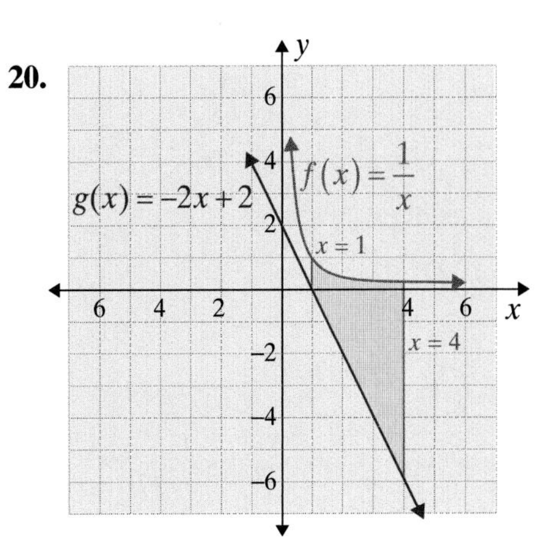

21.

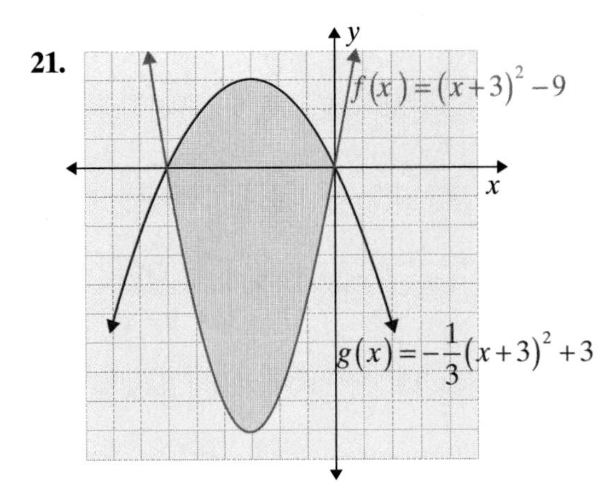

For exercises 22 – 25 use a graphing utility to determine the area.

22.

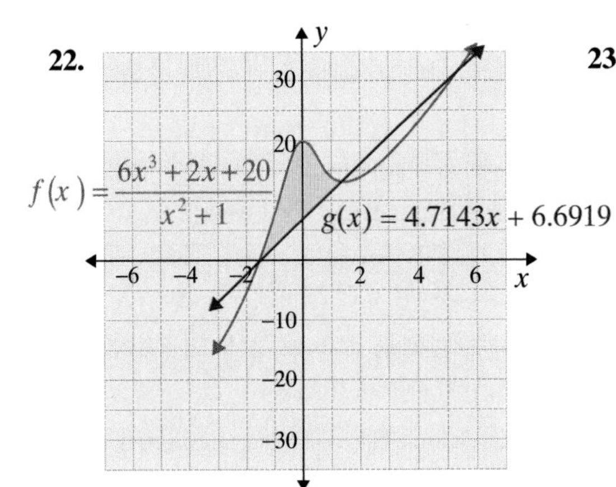

23.

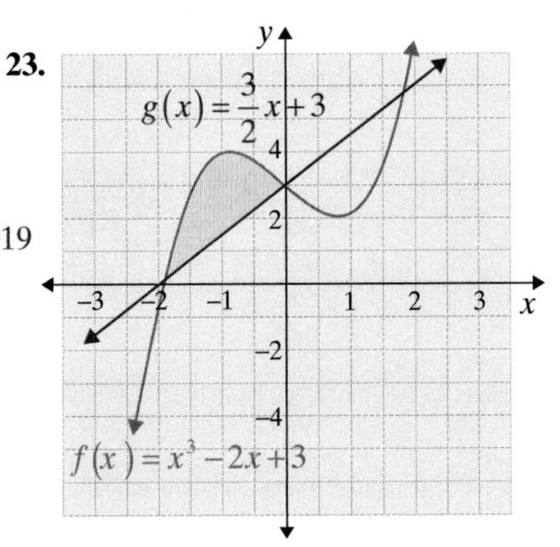

24.

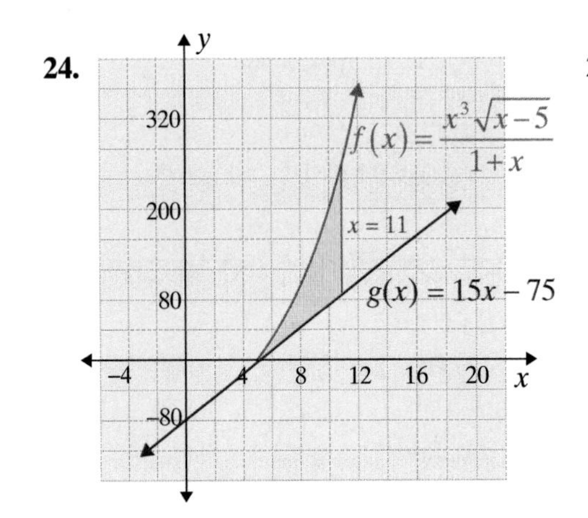

25.

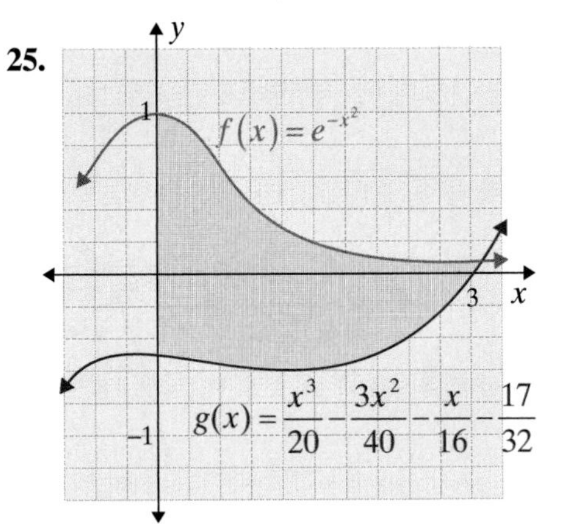

26. **Consumers' surplus.** The demand function for a particular product is given by the function $D(x)=24-0.6x-0.03x^2$. If $x_E=10$ units, find the consumers' surplus.

27. **Consumers' surplus.** Find the consumers' surplus for a product if the demand function is given by $D(x)=\dfrac{800}{x+4}$ and $x_E=4$ units.

28. **Producers' surplus.** Find the producers' surplus for a product if the supply function is given by $S(x)=9e^{0.4x}$ and $x_E=5$ units.

29. **Producers' surplus.** The supply function for a product is given by the function $S(x)=\sqrt{16+1.5x}$. If $x_E=6$ units, find the producers' surplus.

30. **Consumers' and producers' surplus.** The demand curve for a product is given by $D(x) = 18 - 3x$ and the corresponding supply curve is $S(x) = 3x + 6$. Find the consumers' surplus and the producers' surplus.

31. **Consumers' and producers' surplus.** The demand curve for a product is given by $D(x) = 125 - 15x$ and the corresponding supply curve is $S(x) = 50 + 10x$. Find the equilibrium point, the CS, and the PS.

For each of the demand and supply functions in Exercises 32 – 35, find **a.** *the equilibrium point,* **b.** *the consumers' surplus, and* **c.** *the producers' surplus.*

32. $D(x)=18-0.4x,\ S(x)=3+0.1x,\ 0\le x\le 40$

33. $D(x)=24-0.2x,\ S(x)=10+0.5x,\ 0\le x\le 100$

34. $D(x)=1000-30x,\ S(x)=200+0.5x^2,\ 0\le x\le 30$

35. $D(x)=66-5\sqrt{x},\ S(x)=16+x,\ 0\le x\le 120$

36. **Lorenz curve.** The income distribution of a small country is estimated by the Lorenz curve $f(x)=\dfrac{13}{18}x^2+\dfrac{5}{18}x$.
    a. What percentage of the country's total income is earned by the lower 80 percent of its families? Round to the nearest percentage point.
    b. Find the coefficient of inequality.

37. **Lorenz curve.** The Lorenz curve for estimating the income distribution of a country is given by $f(x)=\dfrac{7}{16}x^2+\dfrac{9}{16}x$.
    a. What percentage of the country's total income is earned by the lower 70 percent of its families? Round to the nearest percentage point.
    b. Find the coefficient of inequality.

38. Lorenz curve. A study shows that the income distribution of farmers in a certain state is estimated by $f(x)=0.47x^3+0.24x^2+0.29x$.

a. What percentage of the state's farming income is earned by the lower 60 percent of the state's farmers? Round to the nearest percentage point.

b. Find the coefficient of inequality.

39. Lorenz curve. In a certain state the income distribution for the lumber and the logging industry is estimated by $f(x)=1.16x^3-0.82x^2+0.66x$.

a. What percentage of the state's lumber and logging income is earned by the lower 50 percent of the companies? Round to the nearest percentage point.

b. Find the coefficient of inequality.

40. Lorenz curve. The income distribution for a certain country in 1996 was estimated by the function $f(x)=0.34x+0.66x^2$. In 2000 the income distribution was estimated by the function $f(x)=0.3x+0.72x^2-0.02x^3$.

a. Find the coefficient of inequality for each of the years.

b. Which year had a more equitable income distribution?

41. Lorenz curve. The income distribution for country A is estimated by the function $f(x)=0.24x+0.72x^2+0.04x^3$. The income distribution for country B is estimated by the function $f(x)=0.28x+0.69x^2+0.03x^3$.

a. Find the coefficient of inequality for each of the two countries.

b. Which country has a more equitable income distribution?

Hawkes Learning Systems: Essential Calculus

Area Between Two Curves (with Applications)

6.6 Differential Equations

Objectives

After completing this section, you will be able to:

1. *Verify that certain functions are solutions to given differential equations.*
2. *Solve differential equations by a technique known as separating variables.*
3. *Solve application problems using differential equations.*

A **differential equation** is an equation that involves differentials or derivatives. In Section 5.5 our discussion of exponential growth and decline and continuously compounded interest involved one type of differential equation,

$$\frac{dA}{dt} = kA,$$

and the general solution,

$$A = ce^{kt}.$$

Other examples of differential equations are

1. $\frac{dy}{dx} = -\frac{x}{y}$, **2.** $\frac{d^2y}{dt^2} + 2\frac{dy}{dt} = 0$,

3. $y' - 2y = 4x$, and **4.** $y'' - y' - 6y = 0$.

Applications of differential equations appear in such fields as engineering, physics, business, economics, biology, and psychology whenever motion or measurement of rates of change are involved. In general, the techniques and integrals used in solving differential equations are not elementary in nature. In fact, entire courses are devoted to the study of differential equations. Our presentation is limited to a few important types of problems.

Verification of Solutions

The order of a differential equation is the order of the highest derivative that appears in the equation. In the examples shown previously, equations (1) and (3) are the first-order differential equations and equations (2) and (4) are second-order differential equations.

Solution of a Differential Equation

A function $y = f(x)$ is a ***solution of a differential equation*** *if the function and its appropriate derivatives satisfy the equation.*

Solutions of differential equations are classified as

1. **general solutions** if arbitrary constants are involved,
2. **particular solutions** if no arbitrary constants are involved, or
3. the **trivial solution** $y = 0$.

For example, $y = ce^{3t}$ is a general solution of the equation $y' - 3y = 0$, and $y = 4e^{3t}$ is a particular solution of the same equation.

The following examples illustrate how to verify that a given function is a solution of a differential equation.

Example 1: Solutions of Differential Equations

Verify that the function $y = cx^3 - 4$ is a general solution of the differential equation $xy' - 3y = 12$.

Solution: Step 1: Find y'.

$$y = cx^3 - 4$$
$$y' = 3cx^2$$

Step 2: Substitute for y and y'.

$$\begin{aligned} xy' - 3y &= x\left(3cx^2\right) - 3\left(cx^3 - 4\right) \\ &= 3cx^3 - 3cx^3 + 12 \\ &= 12 \end{aligned}$$

Therefore, since the function contains an arbitrary constant, c, and is a solution, it is a general solution.

Example 2: Solutions of Differential Equations

Verify that the function $y = e^{3x} + e^{-2x}$ satisfies the second-order differential equation $y'' - y' - 6y = 0$ and is, therefore, a particular solution.

Solution: Step 1: Find y' and y''.

$$y = e^{3x} + e^{-2x}$$

$$y' = 3e^{3x} - 2e^{-2x}$$

$$y'' = 9e^{3x} + 4e^{-2x}$$

Step 2: Substitute for y, y', and y''.

$$\begin{aligned} y'' - y' - 6y &= \left(9e^{3x} + 4e^{-2x}\right) - \left(3e^{3x} - 2e^{-2x}\right) - 6\left(e^{3x} + e^{-2x}\right) \\ &= 9e^{3x} + 4e^{-2x} - 3e^{3x} + 2e^{-2x} - 6e^{3x} - 6e^{-2x} \\ &= (9 - 3 - 6)e^{3x} + (4 + 2 - 6)e^{-2x} \\ &= 0 \end{aligned}$$

Therefore, the given function is a solution, and since it contains no arbitrary constraints, it is a particular solution.

Separation of Variables

Sometimes a first-order differential equation can be written in the form

$$g(y)y' = f(x).$$

In such cases we can treat $y' = \frac{dy}{dx}$ as a ratio of two differentials dy and dx and rewrite the equation in the form

$$g(y)dy = f(x)dx.$$

This equation is called a **separable equation** because the variables x and y and their corresponding differentials are separated and appear on opposite sides of the equation. If $g(y)$ and $f(x)$ are continuous functions, the general solution can be found by integrating both sides:

$$\int g(y)dy = \int f(x)dx.$$

Only one constant of integration is necessary and is added after this step. This technique of solving first-order differential equations is called the method of **separation of variables** and is essentially a three-step process.

To Solve a First-Order Separable Differential Equation

Step 1: *Write y' as $\frac{dy}{dx}$.*

Step 2: *Separate variables by writing dx with all terms involving x on one side of the equation and dy with all terms involving y on the other side.*

Step 3: *Integrate both sides of the new equation.*

The method of solving differential equations by separating variables is illustrated in Examples 3 and 4.

Example 3: Using the Method of Separating Variables

Solve the differential equation $y' = \frac{y}{x-1}$ where $x > 1$ and $y > 0$.

Solution: Step 1: $\frac{dy}{dx} = \frac{y}{x-1}$ — Write y' as $\frac{dy}{dx}$.

Step 2: $\frac{dy}{y} = \frac{dx}{x-1}$ — Separate the variables.

Step 3: $\int \frac{1}{y} dy = \int \frac{1}{x-1} dx$ — Integrate both sides of the equation.

$$\ln y = \ln(x-1) + C$$

$$\ln y = \ln(x-1) + \ln k$$

$$\ln y = \ln k(x-1)$$

$$y = k(x-1)$$

We rewrite the constant C as the constant $\ln k$ to help simplify the expression.

Example 4: Using the Method of Separating Variables

Solve the differential equation $y' = 2xy$, where $y > 0$.

Solution: Step 1: $\frac{dy}{dx} = 2xy$ — Write y' as $\frac{dy}{dx}$.

Step 2: $\frac{dy}{y} = 2x\,dx$ — Separate the variables.

Step 3: $\int \frac{1}{y}\,dy = \int 2x\,dx$ — Integrate both sides of the equation.

$$\ln y = x^2 + C$$

$$y = e^{x^2+C}$$

$$y = e^C e^{x^2}$$

$$y = ke^{x^2}$$

Solve for y by rewriting the equation in exponential form. e^C is a positive constant and is represented by the single letter k.

The following example is called an **initial-value problem** because an initial condition is given on x and y that allows the evaluation of the arbitrary constant C in the solution of the differential equation. Thus an initial-value problem results in a particular solution.

Example 5: Initial-Value Problem

Solve the initial-value problem $y' = x$ with initial condition $f(0) = 1$. (That is, $y = 1$ when $x = 0$.)

Solution:

$$\frac{dy}{dx} = x \qquad \text{Step 1}$$

$$dy = x\,dx \qquad \text{Step 2}$$

$$\int dy = \int x\,dx \qquad \text{Step 3}$$

$$y = \frac{1}{2}x^2 + C \qquad \text{This is the general solution.}$$

Since $f(0) = 1$, we have

$$1 = \frac{1}{2}\cdot 0^2 + C$$

$$1 = C$$

Therefore, the particular solution is $y = \frac{1}{2}x^2 + 1$.

Applications

We have discussed elasticity of demand (see Section 5.5) as represented by the formula,

$$E = -\frac{1}{x}\cdot\frac{D(x)}{D'(x)},$$

where $p = D(x)$ is a demand function and $D'(x) = \frac{dp}{dx}$. If E is a known constant value, then the formula can be treated as a differential equation that has the demand function as its solution.

Example 6: Elasticity of Demand

Find the demand function $p = D(x)$ given that the elasticity of demand $E = 3$ for all positive x.

Solution:

$$3 = -\frac{1}{x} \cdot \frac{p}{\frac{dp}{dx}}$$ Use the formula for E.

$$\frac{1}{p}dp = -\frac{1}{3x}dx$$ Separate the variables.

$$\int \frac{1}{p}dp = -\frac{1}{3}\int \frac{1}{x}dx$$ Integrate both sides of the equation.

$$\ln p = -\frac{1}{3}\ln x + C$$ Since $x > 0$, no absolute value signs are needed for $\ln x$.

$$\ln p = -\frac{1}{3}\ln x + \ln k$$ We rewrite the constant C as the constant $\ln k$ to help simplify the expression.

$$\ln p = \ln\left(k \cdot x^{-\frac{1}{3}}\right)$$

$$p = k \cdot x^{-\frac{1}{3}}$$

$$p = \frac{k}{\sqrt[3]{x}}$$

Another interesting application involves an equation known as the **logistic equation**. This equation (and its corresponding curve, known as a **logistic curve** or ***S* curve**) represents a type of modified exponential growth. The rate of growth is slow at first, increases to a maximum rate, and then tapers off as the population approaches some upper limit (or upper bound). (See Figure 6.6.1.)

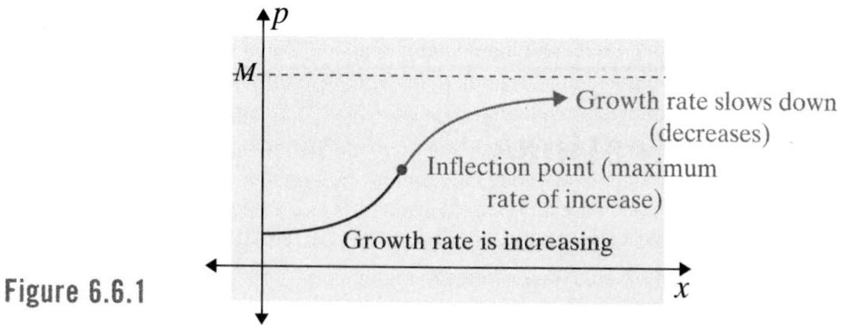

Figure 6.6.1

The graph of a logistic curve

We know from Section 5.5 that the exponential growth of a population P can be described by the differential equation

$$\frac{dP}{dt} = kP.$$

Prolonged exponential growth is unrealistic in many applications. In 1844, the Belgian mathematician P. F. Verhulst introduced an inhibiting factor proportional to $-P^2$ resulting in the **logistic equation**

$$\frac{dP}{dt} = k\left(MP - P^2\right).$$

We can solve this equation by separating variables in the form

$$\frac{dP}{P(M-P)} = k\,dt$$

and using the table of solutions (Table 6.6.1) given at the end of this section. The solution (where $P \neq M$ and $P \neq 0$) is

$$P(t) = \frac{M}{1+Ce^{-Mkt}}.$$

The constant M is the limiting value (or upper bound) of P as $t \to +\infty$.

Example 7: Population of Bacteria

At the beginning of an experiment, a culture of 100 bacteria is growing in a medium that will allow a maximum of 10,000 bacteria to survive. If the population is 200 after 5 hours, use the logistic equation to represent the number of bacteria present t hours after the experiment has begun.

Solution: We know that the solution function has the form

$$P(t) = \frac{10{,}000}{1+Ce^{-10{,}000kt}}.$$

We also know that $P(0) = 100$ and $P(5) = 200$. Substituting $t = 0$ and $P = 100$ gives

$$100 = \frac{10{,}000}{1+Ce^{0}}$$

$$100 = \frac{10{,}000}{1+C}$$

$$C = 99.$$

Substituting $t = 5$ and $P = 200$ gives

$$200 = \frac{10{,}000}{1+99e^{-10{,}000k(5)}}$$

continued on next page ...

$$200 = \frac{10,000}{1+99e^{-50,000k}}$$

$$1+99e^{-50,000k} = 50$$

$$e^{-50,000k} = \frac{49}{99}$$

$$-50,000k = \ln\left(\frac{49}{99}\right)$$

$$-50,000k \approx -0.7033$$

$$k \approx 0.000014$$

Thus the final form for $P(t)$ is $P(t) = \dfrac{10,000}{1+99e^{-0.14t}}$.

Table 6.6.1 contains the general solutions of differential equations that arise in three basic growth applications.

Table of General Solutions for Growth Applications

Application	Differential Equation	General Solution
Unbounded growth	$\frac{dy}{dt} = ky$	$y = Ce^{kt}$
Bounded growth	$\frac{dy}{dt} = k(M-y)$	$y = M + Ce^{-kt}$
Logistic curve	$\frac{dy}{dt} = ky(M-y)$	$y = \frac{M}{1+Ce^{-Mkt}}$

Table 6.6.1

6.6 Exercises

In Exercises 1 – 12, verify that the differential equation has the given function as a particular solution.

1. $\frac{dy}{dx} = 6,\ y = 6x - 1$

2. $\frac{dy}{dx} = 3x - 2,\ y = \frac{3}{2}x^2 - 2x - 4$

3. $\frac{dy}{dx} = 3 + y,\ y = e^x - 3$

4. $x\frac{dy}{dx} - x + 2 = 0,\ y = x - 2\ln x + 7$

5. $\frac{dy}{dx} = y^{\frac{3}{2}},\ y = \frac{4}{(5-x)^2}$

6. $\frac{dy}{dx} = -0.5y,\ y = 3e^{-0.5x}$

7. $2\frac{dy}{dx} + 3y = 1,\ y = \frac{1}{3} - 2e^{-1.5x}$

8. $x\frac{dy}{dx} = xy + y,\ y = 4xe^{x}$

9. $2x^2y'' - xy' - 2y = 4 - 15x,\ y = 3x^2 + 5x - 2$

10. $x^2y'' + xy' - y + \ln x = 0,\ y = x + \ln x$

11. $x^2y'' - xy' + y = 0,\ y = x\ln x$

12. $y'' - 2y' + y = 0,\ y = e^{x}(x+2)$

In Exercises 13 – 20, find the solutions of each separable differential equation.

13. $\frac{dy}{dx} = 3x + \frac{1}{x}$

14. $\frac{dy}{dx} = -3xy$

15. $\frac{dy}{dx} = \frac{2y-1}{x+1}$

16. $x\frac{dy}{dx} = \frac{x^2+1}{y^2}$

17. $\frac{dy}{dx} = -0.4y$

18. $\frac{dy}{dx} = -2(26-y)$

19. $\frac{dy}{dx} = (1-5x)y^2$

20. $\frac{dy}{dx} = (x^2+1)e^{-y}$

In Exercises 21 – 30, solve each initial-value problem or obtain a general solution as indicated. (Refer to the Table of Solutions in the text if necessary.)

21. $\frac{dy}{dx} = 0.6y(20-y)$

22. $\frac{dy}{dx} = 0.3y(50-y)$

23. $\frac{dy}{dx} = 0.25y,\ y = 5$ when $x = 0$

24. $\frac{dy}{dx} = -3x,\ y = 10$ when $x = 2$

25. $x\frac{dy}{dx} = y + 1,\ y = 14$ when $x = 3$

26. $\frac{dy}{dx} = -2xy,\ y = 18$ when $x = 0$

27. $\frac{dy}{dx} = -6x^2y^2,\ y = 2$ when $x = 1$

28. $\frac{dy}{dx} = \frac{xy}{x^2+1},\ y = 7$ when $x = 0$

29. $\frac{dy}{dx} = 8x + 2xy,\ y = 10$ when $x = 0$

30. $\frac{dy}{dx} = 0.3(80-y),\ y = 60$ when $x = 0$

31. $\frac{dy}{dx} = 0.8y(40-y), y = 30$ when $x = 0$

32. $\frac{dy}{dx} = 0.04y(60-y),\ y = 10$ when $x = 0$

33. Elasticity of demand. The elasticity of demand for a product is given by $E = 1.5$. Find the demand function $p = D(x)$ if $D(8) = 24$.

34. Elasticity of demand. The elasticity of demand for a product is given by $E = 2$. Find the demand function $p = D(x)$ if $D(25) = 30$.

35. Elasticity of demand. The elasticity of demand for a product is given by $E = \dfrac{2(120 - x)}{x}$. Find the demand function $p = D(x)$ if $D(20) = 180$.

36. Elasticity of demand. The elasticity of demand for a product is given by $E = \dfrac{60 - 0.4x}{0.2x}$. Find the demand function $p = D(x)$ if $D(70) = 28$.

37. Resale value. The resale or salvage value V of a machine decreases at a rate proportional to its value. Thus $\dfrac{dV}{dt} = -kV$, where t is the machine's age in years and k is its rate of decrease in value.

a. Find the expression for the value when the machine is t years old if the original value was \$24,000 and the rate of decrease is 6 percent.

b. Find the value of the machine when it is 7 years old.

38. Drug concentration. The amount A of a drug remaining in a body t hours after an injection decreases at a rate proportional to the amount present. This suggests the differential equation $\dfrac{dA}{dt} = -kA$. The amount of a certain drug decreases at a rate of 3 percent per hour. Find the amount of the drug remaining in the body 4 hours after an injection of 20 cc of the drug.

39. Newton's law of cooling. Newton's law of cooling states that the rate at which the temperature T of an object changes is proportional to the difference between the temperature of the object and the temperature of the surrounding medium. That is $\dfrac{dT}{dt} = -k(T - M)$, where k is the constant of proportionality, t is time, and M is the constant temperature of the medium.

a. Solve the differential equation for $T(t)$.

b. Find $T(5)$, if $T(0) = 78°$, $M = 26°$, and $k = 0.3$.

40. Newton's law of cooling. The temperature of a roast was 160° when it was removed from an oven and placed in a room with constant temperature of 76°. After 10 minutes, the temperature of the roast was 152°. Find the temperature 20 minutes after the roast was removed from the oven. (See Exercises 39.)

41. **Spread of a rumor.** In a small community with a population of 2800, a rumor about the mayor was started. The rate at which the rumor spread was approximated by $\frac{dN}{dt} = 0.0003N(2800 - N)$ people per day, where N is the number of people who have heard the rumor t days after the rumor was started.
   a. Write an equation for $N(t)$, assuming that 20 people have heard the rumor at $t = 0$.
   b. How many days will it take for 1500 people to hear the rumor? Round to the nearest day.

42. **Spread of a disease.** The population of seals on an island is about 600. Biologists estimate that there are 12 seals with a very infectious disease. The disease will spread at a rate $\frac{dN}{dt} = 0.006N(600 - N)$ seals per day, where N is the number of infected seals t days after the discovery of the disease.
   a. Write a function $N(t)$ for the number of seals infected t days after the discovery of the disease.
   b. At what time t will 300 seals will be infected? Round to the nearest day.

Hawkes Learning Systems: Essential Calculus

Differential Equations

Chapter 6 Index of Key Ideas and Terms

Section 6.1 The Indefinite Integral

Antiderivative pages 466 - 468

If F and f are two functions and $F'(x) = f(x)$ for all x in the domain of f, then F is an antiderivative of f. We also say that $F(x)$ is an antiderivative of $f(x)$.

Antiderivatives Differ by a Constant

If G and F are both antiderivatives of the same function f, then $G(x) = F(x) + C$, where C is a constant.

Integration pages 468 - 475

The Indefinite Integral

If $F(x)$ is any antiderivative of $f(x)$, then the indefinite integral of $f(x)$, symbolized by $\int f(x)\,dx$, is defined to be

$$\int f(x)\,dx = F(x) + C,$$

where C is an arbitrary constant. $f(x)$ is called the integrand, dx is called the differential, and C is called the constant of integration.

Formulas of Integration

I. $\int k\,dx = kx + C$

II. $\int x^r\,dx = \dfrac{1}{r+1} \cdot x^{r+1} + C,$ where $r \neq -1$.

III. $\int \dfrac{1}{x}\,dx = \int x^{-1}\,dx = \ln|x| + C$

IV. $\int e^x\,dx = e^x + C$

V. $\int kf(x)\,dx = k\int f(x)\,dx$ **Constant Multiple Rule**

VI. $\int [f(x) \pm g(x)]\,dx = \int f(x)\,dx \pm \int g(x)\,dx$ **Sum and Difference Rule**

Distance, Velocity, and Acceleration pages 475 - 477

Section 6.2 Integration by Substitution

Formulas of Integration with Function Notation page 482

I. $\int kg'(x)\,dx = kg(x) + C$

II. $\int [g(x)]^r g'(x)\,dx = \frac{1}{r+1}[g(x)]^{r+1} + C,$ where $r \neq -1$.

III. $\int \frac{g'(x)\,dx}{g(x)} = \int \frac{1}{g(x)} g'(x)\,dx = \int [g(x)]^{-1} g'(x)\,dx = \ln|g(x)| + C$

IV. $\int e^{g(x)} g'(x)\,dx = e^{g(x)} + C$

Formulas of Integration with *u*-Substitution pages 482 - 487

For $u = g(x)$ and $du = g'(x)\,dx$

I. $\int k\,du = ku + C$

II. $\int u^r\,du = \frac{1}{r+1}u^{r+1} + C,$ where $r \neq -1$.

III. $\int \frac{du}{u} = \int \frac{1}{u}\,du = \int u^{-1}\,du = \ln|u| + C$

IV. $\int e^u\,du = e^u + C$

Section 6.3 The Fundamental Theorem of Calculus and the Definite Integral

The Fundamental Theorem of Calculus pages 491 - 493

If a function $y = f(x)$ is continuous on the interval $[a, b]$ and $F(x)$ is any antiderivative of $f(x)$, then

$$\int_a^b f(x)\,dx = [F(x)]_a^b = F(b) - F(a).$$

Riemann Sums pages 493 - 498

The general form of a Riemann sum for a function $y = f(x)$ continuous on $[a, b]$ is

$$S_n = [f(c_1) + f(c_2) + \ldots + f(c_n)]\Delta x,$$

where $\Delta x = \frac{b-a}{n}$, n is the number of subintervals of $[a, b]$ and each of the numbers $c_1, c_2, \ldots, c_n$ represents one x-value from each subinterval.

continued on next page ...

Section 6.3 The Fundamental Theorem of Calculus and the Definite Integral (continued)

The Definite Integral pages 498 - 503

If a function $y = f(x)$ is continuous on the interval $[a, b]$, then the definite integral of f from a to b, symbolized as $\int_a^b f(x)dx$, is defined to be

$$\int_a^b f(x)dx = \lim_{n\to+\infty} S_n,$$

where S_n is the general form of a Riemann sum for the function f. The number a is called the lower limit of integration and the number b is called the upper limit of integration.

Area Under a Curve Defined

If a function $y = f(x)$ is nonnegative and continuous on the interval $[a, b]$ then the area under the curve is defined to be

$$A = \lim_{n\to+\infty} S_n,$$

where S_n is the general form of a Riemann sum for the function f.

Formula for Area Under a Curve

For a function $y = f(x)$, nonnegative and continuous on $[a, b]$, the area under the curve is given by

$$A = \int_a^b f(x)dx.$$

Properties of Definite Integrals

1. $\int_a^a f(x)dx = 0$
2. $\int_a^b kf(x)dx = k\int_a^b f(x)dx$
3. $\int_a^b [f(x) \pm g(x)]dx = \int_a^b f(x)dx \pm \int_a^b g(x)dx$

Average Value pages 503 - 506

For a function $y = f(x)$, continuous on the interval $[a, b]$, the average value is

$$AV = \frac{1}{b-a}\int_a^b f(x)dx.$$

Section 6.4 Area (with Applications)

Area Bounded by a Curve and the x-Axis

For $y = f(x)$, a continuous function on the interval $[a, b]$, the integral $\int_a^b f(x)dx$ represents:

1. The total area bounded by the curve and the x-axis from $x = a$ to $x = b$ if $f(x)$ is nonnegative for all x in $[a, b]$.
2. The difference between the areas above the x-axis and below the x-axis that are bounded by the curve and the x-axis from $x = a$ to $x = b$ if $f(x)$ is negative for any x in $[a, b]$.
3. $\int_a^b f(x)dx = -\int_b^a f(x)dx$
4. $\int_a^b f(x)dx = \int_a^c f(x)dx + \int_c^b f(x)dx$

Marginal Analysis

Section 6.5 Area Between Two Curves (with Applications)

Area Between Two Curves

If f and g are two continuous functions and $f(x) \geq g(x)$ on the interval $[a, b]$, then the area between the two curves on this interval is

$$A = \int_a^b [f(x) - g(x)]dx.$$

Consumers' Surplus

The consumers' surplus is defined to be

$$\text{CS} = \int_0^{x_E} D(x)dx - x_E \cdot p_E,$$

where $p = D(x)$ is the demand curve and (x_E, p_E) is the equilibrium point.

Producers' Surplus

The producers' surplus is defined to be

$$\text{PS} = x_E \cdot p_E - \int_0^{x_E} S(x)dx,$$

where $p = S(x)$ is the supply curve and (x_E, p_E) is the equilibrium point.

Lorenz Curves

In the graph of a Lorenz curve, the x-axis shows the cumulative percentage of a country's families, starting from the lowest income families. The y-axis shows the cumulative percentage of the country's total income. If every family had the same income, then the distribution of income would be represented by $y = x$, the **line of complete equality**. The ratio of the area between $y = x$ and $y = f(x)$ to the area under the line $y = x$ is called the **coefficient of inequality**

$$\left(2\int_0^1 [x - f(x)]dx\right).$$

Section 6.6 Differential Equations

Verification of Solutions pages 531 - 533

A function $y = f(x)$ is a **solution of a differential equation** if the function and its appropriate derivatives satisfy the equation.

Solutions of differential equations are classified as:

1. **general solutions** if arbitrary constants are involved,
2. **particular solutions** if no arbitrary constants are involved, or
3. the **trivial solution** $y = 0$.

Separation of Variables pages 533 - 535

A **separable equation** is an equation in which the variables x and y and their corresponding differentials are separated and appear on opposite sides of the equation. If $g(y)$ and $f(x)$ are continuous functions, the general solution can then be found by integrating both sides.

To Solve a First-Order Separable Differential Equation

Step 1: Write y' as $\dfrac{dy}{dx}$.

Step 2: Separate variables by writing dx with all terms involving x on one side of the equation and dy with all terms involving y on the other side.

Step 3: Integrate both sides of the new equation.

Growth Applications pages 535 - 538

Table of General Solutions for Growth Applications

Application	Differential Equation	General Solution
Unbounded growth	$\dfrac{dy}{dt} = ky$	$y = Ce^{kt}$
Bounded growth	$\dfrac{dy}{dt} = k(M - y)$	$y = M + Ce^{-kt}$
Logistic curve	$\dfrac{dy}{dt} = ky(M - y)$	$y = \dfrac{M}{1 + Ce^{-Mkt}}$

Chapter 6 Review

For a review of the topics and problems from Chapter 6, look at the following lessons from *Hawkes Learning Systems: Essential Calculus.*

The Indefinite Integral
Integration by Substitution
The Fundamental Theorem of Calculus
The Definite Integral
Area (with Applications)
Area Between Two Curves (with Applications)
Differential Equations

Chapter 6 Test

In Exercises 1 and 2, show that G(x) is an antiderivative for g(x) by differentiating G(x).

1. $g(x) = 8 - 6x;\ G(x) = 8x - 3x^2$

2. $g(x) = -\frac{6}{x^2} - \frac{2}{x};\ G(x) = \frac{6}{x} - 2\ln x + 11$

Find the indefinite integral in Exercises 3 – 6.

3. $\int (2x + 12x^5)\,dx$

4. $\int \sqrt{x}\,(3x^2 - 4x + 6)\,dx$

5. $\int \frac{2 + 3x + x^2}{x}\,dx$

6. $\int \frac{x^2 + 3x - 5}{\sqrt{x}}\,dx$

Use the technique of substitution to evaluate the integrals in Exercises 7 – 8.

7. $\int x^3 e^{x^4}\,dx$

8. $\int 12x^2 (x^3 + 4)^3\,dx$

9. Find the Riemann Sum S_n for the given function, interval, and value of n: $f(x) = 2 + x^2$, $n = 4$, $[0, 4]$. Use the midpoint of each sub-interval to get the values of $f(x)$ which are needed.

In Exercises 10 – 11, evaluate each definite integral.

10. $\int_1^3 \left(2 + \frac{5}{x}\right) dx$

11. $\int_2^4 (2x + e^{2x})\,dx$

12. What is the average y-value for $f(x) = 4 - x^2$ on the x-interval $[0, 2]$? Draw a sketch of the graph.

13. Determine the area bounded by the x-axis, the y-axis, the x-interval $[0, 3]$, and $y = 1 + x + e^{-x}$.

14. Integrate $\int \frac{e^{2x}}{4e^{2x} + 1}\,dx.$ **(Hint:** Let $u = 4e^{2x} + 1$.)

15. Graph $y = 3 + 2x$ and $y = x^2$ on the same coordinate axis. Shade the area in Quadrant I ($x \geq 0$ and $y \geq 0$) which is between the curves. Calculate this area.

16. Verify whether or not $y = f(x) = e^x + e^{2x}$ is a solution to $\frac{dy}{dx} = y + e^{2x}$.

17. Solve the initial value problem: $x^3 \frac{dy}{dx} = y^2$, $y = 1$ when $x = 1$.

18. **Insect population.** For a biology laboratory experiment, 200 mosquitos are placed in a container that can hold a maximum population of 3000 mosquitos. The population P grows at a rate given by $\frac{dP}{dt} = 0.02(3000 - P)$ mosquitos per day t days after the experiment begins. Find the population 10 days after the experiment is begun.

Cumulative Review

1. Answer each part based on the table at the right.
   a. What number do the x-values seem to be approaching?
   b. "The y-values are nearing 1.038." Is this statement true or false?
   c. Write an appropriate limit statement or conclusion based on the table if y refers to $f(x)$ and $f(x)=\dfrac{2x^3}{6x^2-x+1}$.
   d. Determine the exact limit (express your answer as a common fraction reduced to lowest terms).

x	y
2.5	0.868
2.8	0.970
2.9	1.004
2.99	1.035
2.999	1.038
2.9999	1.038

2. For each of the following quantities, determine whether or not it is (or could be) the value of a derivative at a point. If the quantity is not a derivative, explain why not. If it is a derivative, give, or briefly describe, a possible corresponding function f and identify what x represents.
   a. 35 calories per mouthful.
   b. 6 matches per meet.
   c. 3 radios per store.
   d. 1 milligram per mole.

3. Sketch the graph of a continuous function which satisfies each of the following conditions.

$$g(-2)=4,\quad g(0)=0,$$
$$g'(-2)=0,\quad g'(0)=0,$$
$$g''(-2)=-4,\; g''(0)=4$$

4. Use a graphing calculator to plot $y=3^x$ in the window $[-1,3]$ by $[-3,3]$. Determine the slope at the point (0, 1) and get the equation of the tangent line.

5. Suppose $G(0)=53$ and $G'(0)=-1$, where G denotes the population of gorillas in the wilds of Myst Mountains, Zimolayeo, and t is the time in years since 2000.
   a. Interpret $G(0)=53$.
   b. Interpret $G'(0)=-1$.

6. A manufaturer determines that the revenue from selling x units of his product is given by $R(x)=20x$. Suppose the average cost is given by $f(x)=\dfrac{300}{x}+10$.
   a. Determine the demand function $D(x)$.
   b. Find the cost function $C(x)$.
   c. Determine the marginal revenue and marginal cost function.
   d. Does profit have a maximum?

7. Suppose $f(t)$ is the average temperature in degrees Fahrenheit in New York City on a typical July day, where t is the time in hours after midnight ($t = 0$).

a. Is $f'(6)$ positive or negative?

b. What is the meaning of $f'(12) = 2$?

In Exercises 8 – 13, calculate the derivative $\frac{df}{dx}$ for each function f(x).

8. $f(x) = \dfrac{2x + 3x^2 - x^4}{x}$

9. $f(x) = \dfrac{12 - 10x + 15x^3}{x^2}$

10. $f(x) = x^2\left(4 - \sqrt{x} + x\right)$

11. $f(x) = \left(1 + x^2\right)\left(2 - x\right)$

12. $f(x) = \left(x^2 + x - 1\right)\sqrt{x}$

13. $f(x) = \dfrac{2x - 3 - x^{\frac{1}{2}}}{x^{\frac{1}{2}}}$

In Exercises 14 – 16, find the derivative of the given function.

14. $H(s) = \sqrt{1 + s^2}$

15. $g(t) = \sqrt{\dfrac{4 + 7t}{5 + 12t}}$

16. $g(t) = \dfrac{\sqrt{t + 100}}{\sqrt{t} + 100}$

Use the f(x) given in Exercise 17 for Exercises 18 – 22. Solve Exercises 17 – 22 with the aid of a graphing calculator.

17. Sketch the graph of $y = f(x) = \dfrac{16x + 5}{3x^2 + x - 100}$. The graph of $f(x)$ has two vertical asymptotes, $x = a$ and $x = b$. Determine a (negative) and b (positive). Is there a horizontal asymptote?

18. $\lim\limits_{x \to a^-} \left(f(x)\right)$

19. $\lim\limits_{x \to b^+} \left(f(x)\right)$

20. $\lim\limits_{x \to a^+} \left(f(x)\right)$

21. $\lim\limits_{x \to b^-} \left(f(x)\right)$

22. Which of the following limits exist:

a. $\lim\limits_{x \to a} \left(f(x)\right)$ **b.** $\lim\limits_{x \to b} \left(f(x)\right)$ **c.** $\lim\limits_{x \to \infty} \left(f(x)\right)$

23. Find the equation of the line tangent to $f(x) = 14 - 6x + x^2$ at the point where $x = 6$. Show all details (check your answer with a graphing utility).

In Exercises 24 – 26, use the following scenario:

The West Company makes leather belts for rodeo riders, cowboys, and tourists which it distributes through a chain of country and western novelty stores in several western states. Its total monthly cost function is given by $C(x) = 0.06x^2 + 3.5x + 800$. The West Company charges the novelty stores \$28 per belt.

24. Compute and interpret the meaning of $C(250)$ and $C'(250)$.

25. Determine the revenue function $R(x)$ and the marginal revenue $R'(x)$. What value is $R(250)$ and what value is $R'(250)$?

26. The profit function is defined as $P(x) = R(x) - C(x)$. Find the profit if 250 belts are sold in one month. Calculate and interpret the number $P'(250)$.

27. Population Growth. Two population models, $F(t)$ and $G(t)$, were given in Section 2.4 (Exercises 31 and 32) where t is the number of years since 2000. $F(t)$ is based on Total Fertility Rate (TFR) of 1.5 and $G(t)$ is based on a TFR of 2.0. (***Source:*** Population Reference Bureau, "Transition's in World Population," *Population Bulletin*, Vol. 59, No. 1, 36, (March 2004).)

$$F(t) = -\frac{2.1}{10^6}t^4 + \frac{1.6}{10^4}t^3 - 0.00457t^2 + 0.0994t + 6.03$$

$$G(t) = -\frac{2.1}{10^6}t^4 + \frac{1.66}{10^4}t^3 - 0.0045t^2 + 0.113t + 6.03$$

Let us define population spread $S(t)$ by $S(t) = G(t) - F(t)$. Use a calculator if necessary to answer the following questions.

a. Write a formula for $S(t)$.
b. Sketch a graph of $S(t)$.
c. Is $S(t)$ increasing for the entire range of $[0, 50]$?
d. Determine a formula for marginal spread.
e. Determine the marginal spread in 2030.
f. Determine the values $S(50)$ and $S'(50)$.

In Exercises 28 and 29, find the first, second, third, and fourth derivatives for the given function.

28. $f(x) = 1 - \frac{x^2}{2} + \frac{x^4}{24} - \frac{x^6}{720}$

29. $f(x) = 1 + x + \frac{x^2}{2} + \frac{x^3}{6} + \frac{x^4}{24} + \frac{x^5}{120}$

In Exercises 30 and 31, determine $f'(x)$ and $f''(x)$ for the functions given. Then determine the intervals for which $f(x)$ is increasing and decreasing as well as concave up and concave down. Finally, use this information to graph $f(x)$.

30. $f(x) = 6x^3 - 9x^2$

31. $f(x) = 3x^4 - 44x^3$

32. The value of vests manufactured by Castle Armored Vest Company decreases with time t in months according to the equation $V(t) = 450 - \frac{60t}{t+2}$.

a. Find $V(0)$ and $V(100)$.

b. Find $V'(100)$ and interpret this number.

c. Find $\lim_{t\to\infty} V(t)$. There is an interpretation for this number. What is it?

For Exercises 33 – 36 determine all extreme points and inflection points in the x-interval given.

33. $f(x) = x^6 + 6x^4$; $[-5, 5]$

34. $f(x) = x^6 - 6x^4$; $[-5, 5]$

35. $f(x) = 2x^5 - 135x^2$; $[-4, 4]$

36. $f(x) = x^6 - 192x$; $[-1, 3]$

37. The flu season at Castle College starts when 25 students have reported cases of the flu. After t additional days, the population, $P(t)$, of students who have reported the flu is modeled by $P(t) = \frac{2500}{1+99e^{-0.1t}}$.

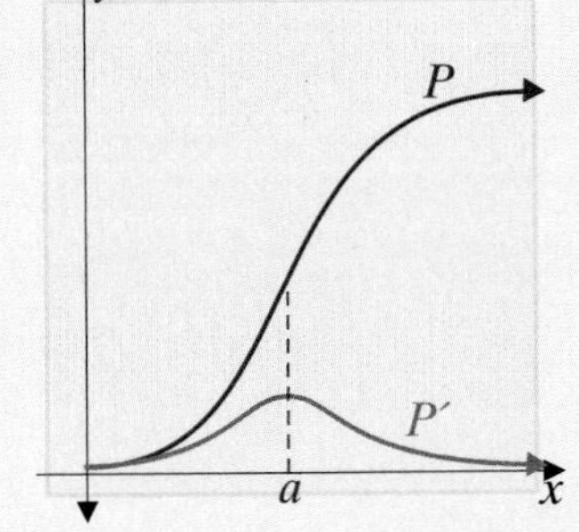

a. Using the graphs of P and P', determine the x-coordinate of the inflection point of P.

b. Interpret the meaning of $P'(a)$.

38. Suppose the volume in cubic feet of a large timber tree is given by $V = \frac{20000}{1+20000t^{-\frac{5}{2}}}$, where t is the age in years.

a. Determine $V(49)$ and $V(100)$.

b. Use a graphing utility to locate any max-min points or inflection points. Sketch the graph.

39. An old stone wall makes two legs of a right angle, one 40 feet long and the other 20 feet long. A contractor is told to add 220 feet of new stone wall to complete a rectangular enclosure. How should he complete the fence so as to maximize the enclosed area? Determine the maximum enclosed area he may obtain.

40. A front window on a new home is designed as a rectangle with a semicircle on the top. If the window is designed to let in a maximum amount of light, then the architect must fix the perimeter and then determine the radius r and height h so as to maximize the area. Determine r and h, if the total perimeter of the window is set at 600 inches.

41. A child rolls a hoop down a hilly street with the distance in feet traveled given by $S(t) = 4t + t^2$ where t is the time in seconds.

a. How far has the hoop traveled in 3 seconds?

b. What was the speed after 3 seconds?

c. At what rate was the speed changing at $t = 3$?

42. An investor plans to manufacture rectangular box containers where bottom and top are rectangular, x feet by $3x$ feet. The box must contain 18 cubic feet. The top and bottom will cost \$2 per square foot, and the four sides will cost \$3 per square foot. What should the height h be so as to minimize costs?

43. What is the point on $f(x)=10-x^2$ nearest to the line l ($y = -x + 14$)? (**Hint:** At the nearest point, (a, b), the tangent line will be parallel to l.)

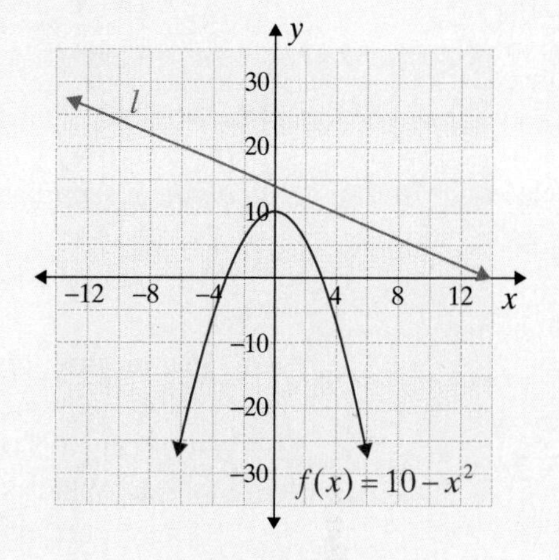

44. Locate the point or points on $f(x)=x^3$ so that the tangent line is parallel to $y = 12x - 18$.

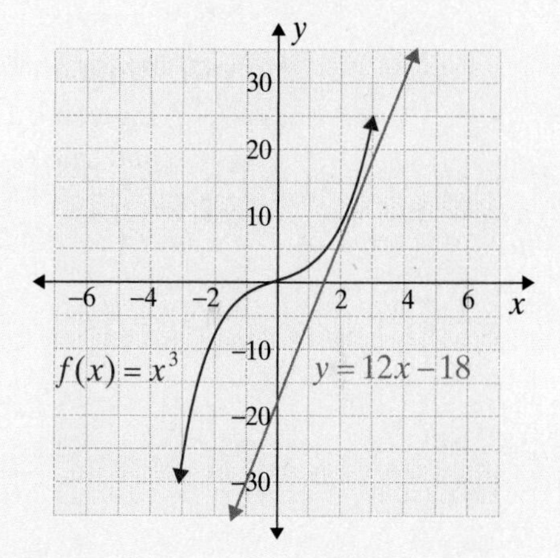

45. Suppose several functions are related as follows:

$$f(w)=3w^2+2, \quad w(t)=2t-1, \quad \text{and} \quad t(x)=1+2x-x^2.$$

a. Determine $\dfrac{df}{dx}$.

b. What number is $\dfrac{df}{dx}$ when $x = 1$?

In Exercises 46 – 49, find the indefinite integral.

46. $\int\left(x^2+4x-1\right)dx$

47. $\int\left(y^3+7y^2-2y\right)dy$

48. $\int\left(e^x-2\sqrt{x}\right)dx$

49. $\int\left(\dfrac{2}{\sqrt{x}}+\dfrac{1}{x}+2\right)dx$

In Exercises 50 – 53, evaluate each definite integral.

50. $\int_{-1}^{2} 5x^3 dx$

51. $\int_{1}^{3} \frac{5}{x} dx$

52. $\int_{-1}^{1} \frac{e^x}{4e^x + 1} dx$

53. $\int_{1}^{3} \frac{x+2}{x^2 + 4x - 3} dx$

In Exercises 54 and 55, find the total area bounded by the x-axis and the curve $y = f(x)$ on the given interval.

54. $f(x) = x + \sqrt{x}$; $[1, 4]$

55. $f(x) = \frac{1}{x^2} + e^{0.5x}$; $[1, 4]$

56. Income distribution. A recent study shows that the income distribution of commercial fishermen in a western coastal state is estimated by $f(x) = 0.34x^3 + 0.26x^2 + 0.40x$.

a. What percentage of the state's income from fishing is earned by the lower 50 percent of the state's fisherman?

b. Find the coefficient of inequality.

57. Profit. The marginal profit for the production and the sale of x units of a product is given by $P'(x) = 60 - 1.2x$ dollars per unit. Find the profit from the production and sale of the first 30 units.

CHAPTER 7

Additional Integration Topics

Did You Know?

Best known for his advances in non-Euclidean geometry and for posing the greatest unsolved problem in mathematics, Bernhard Riemann (1826 – 1866) was born the son of a Lutheran minister. Although he initially planned to study theology at the University of Göttingen, Riemann's interest in mathematics led him to study under the great mathematician C.F. Gauss (1777 – 1855).

While many forms of integrals exist, it was Riemann who developed the present form that is taught in this text as well as nearly every elementary calculus course. His form is the easiest to understand and can be used in nearly every setting. An interval is divided into equal parts and an estimate of the function taken for each part to produce an approximation of the area beneath the curve. This process, used by graphing calculators to approximate definite integrals, is called a Riemann sum. This is often the best approach to solving real world problems in engineering, economics, and finance, where an accurate approximation is needed.

By applying another concept from calculus, the limit, we are able to arrive at the exact value of a definite integral. This is accomplished by allowing the width of the partition to approach zero, and the Riemann integral is the result. Riemann is credited with establishing the idea of integrals as sums. Before his work in this area, integrals were simply thought of as antiderivatives, the reverse operation of taking the derivative.

Published in 1859, the Riemann Hypothesis for the Zeta Function is likely the single most pursued conjecture in mathematics. This function involves taking an integral over the field of complex numbers, and it has been linked with such diverse topics as the study of prime numbers and elliptic curves. Hundreds of mathematicians have spent nearly 150 years trying to prove or disprove his conjecture. In fact, a one million dollar prize has been set for a proof verifying the result.

Riemann also lent his genius to the field of non-Euclidean geometry. A branch of this field, Riemannian geometry was actually named in his honor. His studies in this area would later become the basis of Albert Einstein's (1879 – 1955) theory of relativity.

A calculator can provide an area or an evaluation of a definite integral numerically. In many applications, however, one wants to have a formula (function) which relates the variables directly (a table of 23 formulas is presented in Section 7.3). Moreover, students in business, liberal arts, and science want to know how certain apparently very complicated formulas could possibly have been obtained. One of the most powerful integration techniques in calculus was developed by both Newton and Leibniz, and it turns out to be based on a very simple idea. Today this technique is called integration by parts, or the parts method (Section 7.1). It enables us to integrate functions that involve products of different types of expressions such as $\int xe^x dx$. In Section 7.2 we apply this technique to business and economics applications, including annuities and income streams.

In Section 7.3 improper integrals, such as $\int_1^\infty xe^{-x} dx$, are defined and evaluated. These integrals involve $+\infty$ and $-\infty$ as the upper and lower limits of integration. The values of such integrals, if they exist, depend on the limit concept developed in Chapter 2. We further explore the idea of improper integrals in Section 7.5 by showing how the typical bell-shaped curve is related to such integrals. The last topic is that of computation of volumes in which a region in the xy-plane is revolved about the x-axis to generate a three-dimensional volume. The analysis uses Riemann sums in a manner similar to the development of the definite integral in Chapter 6, and the necessary integrations are carried out using all the techniques of the chapter.

7.1 Integration by Parts

Objectives

After completing this section, you will be able to:

Use the technique of integration by parts to evaluate integrals.

In Section 6.1 we introduced four basic formulas and two rules for integration. Then, in Section 6.2, we expanded the applications of these same four formulas by using the technique of substitution. In this section we will develop another substitution technique that will allow us to evaluate an even wider variety of integrals. This new technique, called **integration by parts**, is based on the Product Rule for Differentiation (see Section 3.1):

$$\frac{d}{dx}\left[f(x)\cdot g(x)\right]=f(x)\cdot g'(x)+g(x)\cdot f'(x).$$

If we substitute

$$u=f(x),\ v=g(x),\ \frac{du}{dx}=f'(x),\ \text{and}\ \frac{dv}{dx}=g'(x),$$

the formula can be written as

$$\frac{d}{dx}[u \cdot v] = u \cdot \frac{dv}{dx} + v \cdot \frac{du}{dx}.$$

Now, thinking in terms of differentials, we multiply both sides of the equation by dx and get

$$d[u \cdot v] = u \cdot dv + v \cdot du.$$

Integrating both sides of the equation gives

$$\int d[u \cdot v] = \int u \cdot dv + \int v \cdot du$$

and

$$u \cdot v = \int u \cdot dv + \int v \cdot du.$$

Solving for $\int u \cdot dv$ gives the desired form,

$$\int u \cdot dv = u \cdot v - \int v \cdot du.$$

This equation is known as the formula for **integration by parts** or just as the **parts formula**. The given integral is $\int u \cdot dv$, and the two parts are u and dv. The objective is to choose these two parts in such a way that the resulting integral on the right, $\int v \cdot du$, is easier to evaluate than the original integral.

Formula for Integration by Parts

$$\int \boldsymbol{u \cdot dv = u \cdot v} - \int \boldsymbol{v \cdot du}$$

To see how this formula works, consider the integral

$$\int xe^x \, dx$$

and let

$$u = x \quad \text{and} \quad dv = e^x \, dx.$$

Then

$$du = dx \quad \text{and} \quad v = \int dv = \int e^x \, dx = e^x.$$

NOTES The constant of integration C is not an explicit part of the formula. It will appear later when the last integration is performed.

Now, by using the formula for integration by parts, we obtain

$$\int u \cdot dv = u \cdot v - \int v \cdot du$$

$$\int x \cdot e^x \, dx = x \cdot e^x - \int e^x \, dx$$

This last integral, $\int e^x dx$, is one we are familiar with.

$$= xe^x - e^x + C.$$

Integration by parts can involve somewhat of a trial-and-error approach in the choice of u and dv. Suppose, for example, that we had chosen the parts as follows:

$$u = e^x \quad \text{and} \quad dv = x\,dx.$$

Then

$$du = e^x \, dx \quad \text{and} \quad v = \int dv = \int x\,dx = \frac{1}{2}x^2.$$

Then integrating by parts gives

$$\int xe^x \, dx = \frac{1}{2}x^2 e^x - \int \frac{1}{2}x^2 e^x \, dx$$

and the last integral on the right is more difficult to evaluate than the original integral. In such a case, we simply start over and make new choices for u and dv.

As an organizational aid, we will use the following box-type format when we integrate by parts.

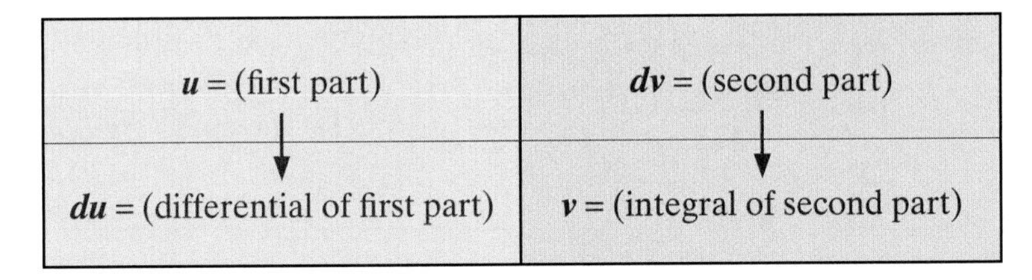

u = (first part)
dv = (second part)
du = (differential of first part)
v = (integral of second part)

The outline emphasizes that we have separated the original problem into two parts, one of which (the "u") needs to be differentiated (usually very easily) and the other of which (the "dv") needs to be integrated (also easy, if we have chosen well). The following integrals are evaluated with the technique of integration by parts.

Example 1: Integration by Parts

Find $\int \ln x \, dx$.

Solution: The differential dx will always be part of the differential dv. In this case, $dv = \ln x \, dx$ is not appropriate since the integration of $\ln x \, dx$ is the original problem. Thus we choose $dv = dx$ (a VERY easy integration) and $u = \ln x$.

$u = \ln x$	$dv = dx$
$du = \frac{1}{x}dx$	$v = \int dv = \int dx = x$

Now substituting in the parts for our formula gives

$$\int u\,dv = uv - \int v\,du$$
$$\int \ln x\,dx = (\ln x)x - \int x\frac{1}{x}dx$$
$$= x\ln x - x + C.$$

As a check, we differentiate:

$$\frac{d}{dx}(x\ln x - x + C) = x\cdot\frac{d}{dx}\ln x + \ln x\cdot\frac{d}{dx}x - 1 + 0$$
$$= x\cdot\frac{1}{x} + \ln x(1) - 1 + 0$$
$$= 1 + \ln x - 1 = \ln x.$$

This shows the solution is correct.

Example 2: Integration by Parts

Find $\int x\ln x\,dx$.

Solution: Since we differentiate the choice for u, we could let $u = x$ (with $dv = \ln x\,dx$) or $u = \ln x$ (with $dv = x\,dx$). However, we integrate the choice for dv, so that we do **not** choose $dv = \ln x\,dx$ because we do not have a standard way to integrate this choice of dv directly (in theory we could use Example 1, but this is not as easy as the choice we will make).

Thus

$u = \ln x$	$dv = x\,dx$
$du = \frac{1}{x}dx$	$v = \int dv = \int x\,dx = \frac{1}{2}x^2$

Therefore,

$$\int u\,dv = uv - \int v\,du$$
$$\int \ln x\cdot x\,dx = (\ln x)\frac{1}{2}x^2 - \int \frac{1}{2}x^2\cdot\frac{1}{x}dx$$

Rewrite $\int x\ln x\,dx$ as $\int \ln x\cdot x\,dx$

continued on next page ...

$$= \ln x \cdot \frac{1}{2}x^2 - \int \frac{1}{2}x^2 \cdot \frac{1}{x}\,dx$$

$$= \frac{1}{2}x^2 \ln x - \frac{1}{2}\int x\,dx$$

$$= \frac{1}{2}x^2 \ln x - \frac{1}{2}\cdot\frac{1}{2}x^2 + C$$

$$= \frac{1}{2}x^2 \ln x - \frac{1}{4}x^2 + C$$

The cancellation which occurs when we multiply v times du tells us that the choice for u and dv was excellent!

Example 3: Integration by Parts

Find $\int x^2 e^{-x}dx$.

Solution: In this problem we apply integration by parts **twice**.

$u = x^2$	$dv = e^{-x}\,dx$
$du = 2x\,dx$	$v = \int e^{-x}dx = -e^{-x}$

Therefore,

$$\int u\,dv = uv - \int v\,du$$

$$\int x^2 e^{-x}dx = x^2 \cdot -e^{-x} - \int -e^{-x} \cdot 2x\,dx$$

$$= -x^2 e^{-x} + \int 2xe^{-x}dx.$$

Now integrate $\int 2xe^{-x}dx$ by parts.

$u = 2x$	$dv = e^{-x}\,dx$
$du = 2\,dx$	$v = \int e^{-x}dx = -e^{-x}$

Therefore,

$$\int u\,dv = uv - \int v\,du$$

$$\int 2xe^{-x}dx = 2x \cdot -e^{-x} - \int -e^{-x} \cdot 2\,dx$$

$$= -2xe^{-x} + 2\int e^{-x}dx$$

$$= -2xe^{-x} - 2e^{-x} + C$$

Putting all the results together, we have

$$\int x^2 e^{-x}dx = -x^2 e^{-x} - 2xe^{-x} - 2e^{-x} + C.$$

Example 4: Integration by Parts

Evaluate the definite integral $\int_1^5 x\sqrt{x-1}\,dx$.

Solution:

$u = x$	$dv = \sqrt{x-1}\,dx$
$du = dx$	$v = \int_1^5 (x-1)^{\frac{1}{2}}\,dx = \frac{2}{3}(x-1)^{\frac{3}{2}}\Big]_1^5$

Therefore,

$$\int u\,dv = uv - \int v\,du$$

$$\int_1^5 x\sqrt{x-1}\,dx = x \cdot \frac{2}{3}(x-1)^{\frac{3}{2}}\Big]_1^5 - \int_1^5 \frac{2}{3}(x-1)^{\frac{3}{2}}\,dx$$

$$= \frac{2}{3}x(x-1)^{\frac{3}{2}} - \frac{2}{3}\cdot\frac{2}{5}(x-1)^{\frac{5}{2}}\Big]_1^5$$

$$= \frac{2}{3}x(x-1)^{\frac{3}{2}} - \frac{4}{15}(x-1)^{\frac{5}{2}}\Big]_1^5$$

$$= \left[\frac{10}{3}(4)^{\frac{3}{2}} - \frac{4}{15}(4)^{\frac{5}{2}}\right] - (0)$$

$$= \frac{80}{3} - \frac{128}{15}$$

$$= \frac{400}{15} - \frac{128}{15}$$

$$= \frac{272}{15}$$

7.1 Exercises

In Exercises 1 – 16, use the technique of integration by parts to evaluate the integrals.

1. $\int xe^{2x}dx$ **2.** $\int 3xe^{-x}dx$ **3.** $\int 2ye^{0.5y}dy$

4. $\int 5te^{0.4t}dt$ **5.** $\int \ln t\,dt$ **6.** $\int y^2\ln y\,dy$

7. $\int x^3\ln 5x\,dx$ **8.** $\int 8x\ln 3x\,dx$ **9.** $\int x\sqrt{x+2}\,dx$

10. $\int x\sqrt{x-3}\,dx$ **11.** $\int x(x+4)^{-2}dx$ **12.** $\int x(x-1)^{-3}dx$

13. $\int \frac{t}{2e^{0.6t}}dt$ **14.** $\int y^2e^{3y}dy$ **15.** $\int \sqrt{x}\ln 7x\,dx$

16. $\int 3x(x-6)^{-\frac{2}{3}}dx$

In Exercises 17 – 22, use the technique of integration by parts to evaluate each definite integral. Round your answer to the nearest hundredth.

17. $\int_0^2 xe^{-2x}dx$ **18.** $\int_0^3 (x+1)e^{-0.5x}dx$ **19.** $\int_0^1 (x+2)e^{-4x}dx$

20. $\int_0^4 (1-2x)e^{1.2x}dx$ **21.** $\int_{-2}^3 \frac{x}{\sqrt{6+x}}dx$ **22.** $\int_0^4 x\sqrt{1+2x}\,dx$

In each of Exercises 23 – 30, identify the u and dv which would solve the integral by using integration by parts. Use the TI-84 Plus to complete the exercises. Round your answer to the nearest hundredth.

23. $\int_0^1 4x(3x+1)^5dx$ **24.** $\int_1^2 \frac{x}{\sqrt{2x+5}}dx$ **25.** $\int_{-1}^2 (x+1)(x+2)^{\frac{3}{2}}dx$

26. $\int_1^4 \sqrt{x}\ln x\,dx$ **27.** $\int_1^5 x^2\ln x\,dx$ **28.** $\int_1^3 \frac{\ln t}{t^2}dt$

29. $\int_0^6 \ln(x+1)dx$ **30.** $\int_1^2 (2x+1)\ln x\,dx$

In Exercises 31 – 40, use the technique of substituting or integration by parts to evaluate the integrals.

31. $\int 5te^{-2t}dt$ **32.** $\int 5te^{-2t^2}dt$ **33.** $\int \sqrt{3x}\ln x\,dx$

34. $\int \frac{\ln x}{x}dx$ **35.** $\int 3x(2x^2-1)^{\frac{3}{2}}dx$ **36.** $\int 3x(2x-1)^{\frac{3}{2}}dx$

37. $\int \frac{(\ln x)^2}{x}\,dx$ **38.** $\int x \ln x^2\,dx$ **39.** $\int \frac{e^x}{1-e^x}\,dx$

40. $\int \frac{x}{\sqrt{5x^2-3}}\,dx$

41. Demand for a natural resource. The demand for a natural resource t years from now will be increasing at a rate of $te^{0.01t}$ million units per year. If the current demand is 80 million units, write a function for the demand t years from now.

42. Revenue. The marginal revenue for x units of a product is given by $R'(x) = (200 - 30x)e^{-0.15x}$ dollars per unit. Find the revenue function $R(x)$ if $R(0) = 0$.

43. Revenue. The marginal revenue for x units of a product is given by $R'(x) = 18 - 0.4\ln x$ dollars per unit, where $x \geq 1$. Find the revenue function if $R(1) = \$18.40$.

44. Resale Value. The value of a machine depreciates at a rate of $-200t(t+1)^{-2}$ dollars per year, where t is the age (in years) of the machine. If the original cost of the machine is \$540, find a function for the value of the machine when it is t years old.

Hawkes Learning Systems: Essential Calculus

Integration by Parts

7.2 Annuities and Income Streams

Objectives

After completing this section, you will be able to:

1. *Determine the value of a given annuity.*
2. *Calculate present and future values of an income stream.*

We are, in general, familiar with periodic payments and receipts of money such as rent payments, loan payments, and interest on savings accounts. In some large businesses, such as hotel and restaurant chains, banks, and department stores, money appears to flow continuously. We will find that even though this flow is not exactly continuous, we can represent the flow of money with a continuous function and obtain practical results. In this section we will discuss two topics related to the flow of money: annuities and income streams.

Annuities

An **annuity** consists of a series of equal payments made at equal time intervals. The **amount** (or **future value**) of an annuity is its total value, that is, the sum of the payments and interest earned on each payment. In the case of an **ordinary annuity**, the payments are made at the end of each time period. For an **annuity due**, the payments are made at the beginning of each time period. Mortgage payments and rent payments are examples of annuities due.

Suppose that you deposit \$100 into a savings account at the end of each month for 6 months. Each \$100 earns interest over a different period of time. (In fact, the last \$100 does not earn any interest during the 6 months.) If interest on each deposit is compounded continuously at 12 percent, then the amount of the annuity at the end of the sixth month is the sum

$$\begin{aligned}
S_6 &= 100e^{0.01(5)} + 100e^{0.01(4)} + 100e^{0.01(3)} + 100e^{0.01(2)} + 100e^{0.01(1)} + 100e^{0.01(0)} \\
&= 100\left(e^{0.01(5)} + e^{0.01(4)} + e^{0.01(3)} + e^{0.01(2)} + e^{0.01(1)} + e^{0.01(0)}\right) \\
&= 100(1.0513 + 1.0408 + 1.0305 + 1.0202 + 1.0101 + 1.00) \\
&= 100(6.1529) \\
&= \$615.29.
\end{aligned}$$

Note that since 12 percent $= 0.12$ is an annual rate of interest, $\frac{0.12}{12} = 0.01$ is the monthly rate of interest.

The process is illustrated in Figure 7.2.1.

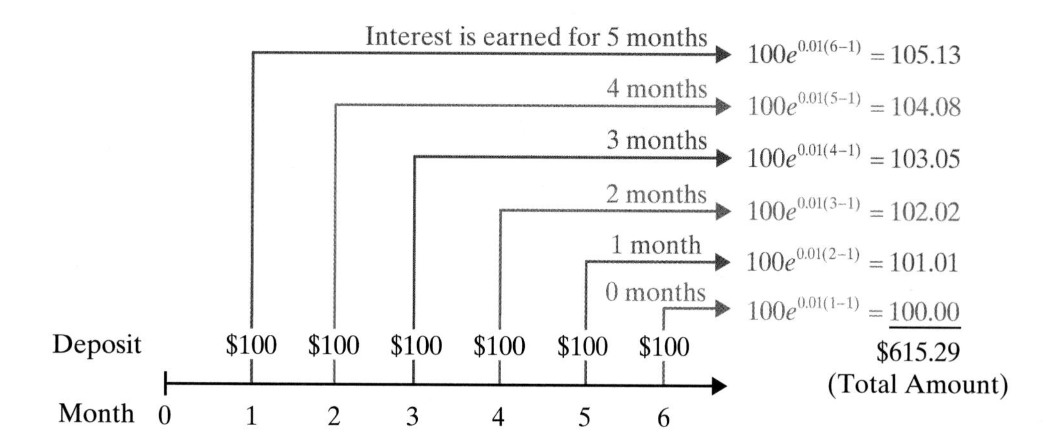

Figure 7.2.1

The sum S_6 is a Riemann sum with $\Delta t = 1$ month. Thus the integral

$$\int_0^6 100e^{0.01(6-t)}dt$$

gives a good approximation of the amount of the annuity at the end of the 6 months.

Specifically,

$$\begin{aligned}\int_0^6 100e^{0.01(6-t)}dt &= -\frac{100}{0.01}e^{0.01(6-t)}\Big]_0^6 \\ &= -10{,}000\left(e^{0.01(6-6)} - e^{0.01(6-0)}\right) \\ &= -10{,}000(1 - 1.061836547) \\ &= \$618.37.\end{aligned}$$

Future Value of an Annuity

*The **amount** (or **future value**) **of an annuity** at the end of N time periods is approximated by the integral*

$$\int_0^N Pe^{r(N-t)}dt = \frac{P}{r}\left(e^{rN} - 1\right),$$

where P is the number of dollars invested each time period, and r is the interest rate (as a decimal) per time period.

For an **annual annuity**, r is the annual rate of interest and N is the number of years that payments are made. The preceding formula gives a good approximation whether the annuity is an ordinary annuity or an annuity due.

Example 1: Future Value of an Annuity

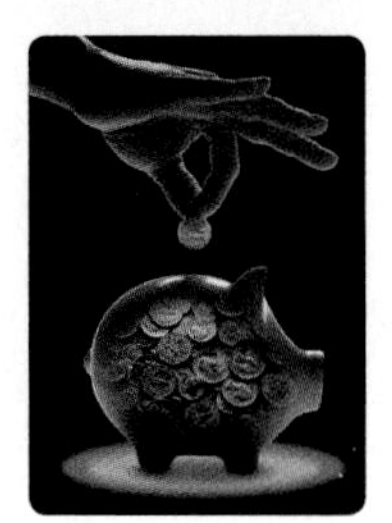

Suppose that the parents of a child set up an annuity account paying 10 percent compounded continuously for the child's college education. They deposit \$500 each year for 20 years. What will be the approximate amount of the annuity in 20 years?

Solution: Since the deposits are made yearly, $r = 0.10$, $N = 20$, and $P = 500$.

$$\begin{aligned}\int_0^{20} 500e^{0.10(20-t)}dt &= -\frac{500}{0.10}\left(e^{0.10(20-t)}\right)\Big]_0^{20} \\ &= -\frac{500}{0.10}\left(1-e^{0.10(20)}\right) \\ &= -5000\left(1-e^2\right) \\ &= -5000(1-7.38906) \\ &= 31,945.30\end{aligned}$$

The value of the annuity will be approximately \$31,945.30 in 20 years.

Income Streams

Suppose that money is not paid in equal amounts at regular time intervals, but instead, income flows almost continuously, as does the **income stream** from a video arcade game or from airline ticket sales. If $R(t)$ represents the rate of flow of revenue, we can use a definite integral to find the amount of income over a period of time.

Future Value of an Income Stream

The ***amount*** *(or* ***future value****)* ***of an income stream*** *at the end of T years is given by the integral*

$$A = \int_0^T R(t)e^{r(T-t)}dt,$$

where $R(t)$ is the rate of flow of revenue at time t, r is the annual interest rate (as a decimal), and interest is compounded continuously.

Example 2: Future Value of an Income Stream

A travel agency expects income from airline ticket sales to increase at a continuous rate represented by $R(t) = 210 + 0.1t$ in hundreds of dollars per year over the next 3 years. With interest at 5 percent compounded continuously, what income does the agency expect from airline ticket sales over the next 3 years?

Solution: Using the formula $\int_0^T R(t)e^{r(T-t)}dt$ with $R(t) = 210 + 0.1t$, $T = 3$, and $r = 0.05$, we have

$$A = \int_0^3 (210 + 0.1t)e^{0.05(3-t)}dt.$$

Integrating by parts, we obtain the following results.

$u = 210 + 0.1t$	$dv = e^{0.05(3-t)}dt$
$du = 0.1\,dt$	$v = \int e^{0.05(3-t)}dt = -\frac{1}{0.05}e^{0.05(3-t)} = -20e^{0.05(3-t)}$

$$A = (210+0.1t)\left(-20e^{0.05(3-t)}\right)\Big]_0^3 - \int_0^3 \left(-20e^{0.05(3-t)}\right)(0.1\,dt)$$ Using the parts formula, $uv - \int v\,du$.

$$= (210+0.1t)\left(-20e^{0.05(3-t)}\right)\Big]_0^3 + 2\int_0^3 e^{0.05(3-t)}dt$$

$$= (210+0.1t)\left(-20e^{0.05(3-t)}\right)\Big]_0^3 - 2\left(\frac{1}{0.05}e^{0.05(3-t)}\right)\Big]_0^3$$

$$= 210.3(-20)(e^0) - 210(-20)(e^{0.15}) - 2(20)(e^0) + 2(20)(e^{0.15})$$

$$\approx -4206 - 210(-23.23668485) - 40 + 2(23.23668485)$$

$$= 210(23.23668485) + 2(23.23668485) - 4246$$

$$= 212(23.23668485) - 4246$$

$$\approx 4926 - 4246$$

$$= 680$$ The units are hundreds of dollars.

The agency expects approximately $68,000 from airline ticket sales over the next 3 years.

From Chapter 5 we know that if P dollars are invested with interest compounded continuously at a rate r for t years, then the future value A is given by the formula

$$A = Pe^{rt}.$$

If this formula is solved for P, we have

$$P = Ae^{-rt}.$$

We call P the **present value of A**. That is, P is the amount to be invested now so that the amount A will be accumulated if interest is compounded continuously at a rate r for t years. Reasoning in a similar manner, we can relate the concept of the future value of an income stream to the **present value of an income stream**.

Present Value of an Income Stream

*The **present value of an income stream** over T years is given by the integral*

$$PV = \int_0^T R(t)e^{-rt}dt,$$

where $R(t)$ is the rate of flow of revenue at time t, r is the annual interest rate (as a decimal), and interest is compounded continuously.

Example 3: Present Value of an Income Stream

A new department store is expected to generate income at a continuous rate described by the function $R(t) = 50{,}000 + 2000t$ dollars per year over the next 5 years. Find the present value of the store if the current interest rate is 10 percent compounded continuously.

Solution: Using the formula $\int_0^T R(t)e^{-rt}dt$ with $R(t) = 50{,}000 + 2000t$, $T = 5$, and $r = 0.10$, we have

$$PV = \int_0^5 (50{,}000 + 2000t)e^{-0.10t}\,dt.$$

Integrating by parts gives the following results.

$u = 50{,}000 + 2000t$	$dv = e^{-0.10t}dt$
$du = 2000\,dt$	$v = \int e^{-0.10t}dt = -\frac{1}{0.10}e^{-0.10t} = -10e^{-0.10t}$

$$PV = (50{,}000+2000t)(-10e^{-0.10t})\Big]_0^5 - \int_0^5 (-10e^{-0.10t})(2000\,dt)$$ Using the parts formula, $uv - \int v\,du$.

$$= (50{,}000+2000t)(-10e^{-0.10t})\Big]_0^5 + 20{,}000\int_0^5 e^{-0.10t}\,dt$$

$$= (50{,}000+2000t)(-10e^{-0.10t})\Big]_0^5 + (-200{,}000e^{-0.10t})\Big]_0^5$$

$$= 60{,}000(-10e^{-0.5}) + 500{,}000 - 200{,}000(e^{-0.5}) + 200{,}000$$

$$= 700{,}000 - 800{,}000e^{-0.5}$$

$$\approx 700{,}000 - 800{,}000(0.6065307)$$

$$= 700{,}000 - 485{,}225$$

$$= 214{,}775$$

The present value of the new store is \$214,775.

7.2 Exercises

1. **Annuity.** Estimate the amount of an annuity if \$1000 is deposited annually for 10 years at a rate of 8 percent compounded continuously.

2. **Annuity.** \$6000 is invested in an account each year for 8 years. Find the approximate balance at the end of the 8 years if the account pays interest at a rate of 7 percent compounded continuously.

3. **Annuity.** Christine has decided to invest \$2000 each year into an IRA account that pays interest at the rate of 9 percent compounded continuously. Find the amount in the account at the end of 15 years.

4. **Annuity.** Bob and Ann plan to deposit \$4000 per year into their retirement account. If the account pays interest at a rate of 8.4 percent compounded continuously, approximately how much will be in their account after 12 years?

5. **Annuity.** Bryan plans to deposit \$1200 each year into an annuity. If the account pays interest at a rate of 7.5 percent compounded continuously, find the approximate balance of his account after 10 years.

6. **Income Stream.** Find the value of an income stream after 7 years if the rate of flow is estimated to be \$200,000 annually and the income is invested at a rate of 8 percent compounded continuously.

7. **Income stream.** The owner of a local convenience store estimates that the store will generate an annual income of \$340,000 for the next 4 years. If the rate of interest is 9 percent compounded continuously, find the value of the income stream.

8. **Income stream.** A real estate investment is expected to generate an income flow of \$12,000 annually for the next 6 years. Find the amount of the income stream if the interest rate is 7.8 percent compounded continuously.

9. **Income stream.** Find the value of an income stream after five years if $R(t) = 3600e^{0.02t}$ is the rate of flow of revenue and the income is deposited at a rate of 7 percent compounded continuously.

10. **Income stream.** A certain investment has a continuous flow of money at a rate of $R(t) = 7200e^{0.01t}$. Find the value of this flow after 4 years if the interest rate is 8.2 percent compounded continuously.

11. **Income stream.** Find the value of an income stream if $R(t) = 50 + 0.2t$ is the rate of flow of revenue reinvested at 6 percent compounded continuously for 8 years.

12. **Income stream.** Find the value of an income stream if $R(t) = 80 + 1.2t$ is the rate of flow of revenue reinvested at 6.4 percent compounded continuously over the next 6 years.

13. **Income stream.** The profit from a number of soft drink machines is estimated to be at the rate of $R(t) = 15 + 0.8t$ thousand dollars per year. If the profits are deposited into an account paying 6.5 percent compounded continuously, find the amount of the income stream after 7 years.

14. **Income stream.** It is estimated that a computer will save accounting fees at a small company at a rate of $R(t) = 4 + 0.6t$ thousand dollars per year. If the savings are reinvested at 5 percent compounded continuously, find the amount of the income stream after 4 years.

15. **Income stream.** Find the present value of an income stream with $R(t) = 60 - 0.4t$, $r = 8$ percent, and $T = 20$.

16. **Income stream.** Find the present value of an income stream with $R(t) = 150 - t$, $r = 12$ percent, and $T = 10$.

17. **Income stream.** The rate of flow of an income stream is estimated by $R(t) = 6000e^{0.015t}$ for the next 4 years. Find the present value of this flow if the interest rate is 6 percent compounded continuously.

18. Income stream. The rate of flow of an income stream for the next 6 years is estimated by $R(t) = 10{,}000e^{-0.01t}$. Find the present value of this flow if the interest rate is 8.5 percent compounded continuously.

19. Income stream. Sandy estimates that the profits from his ice cream store will be $R(t) = 24 + 3.6t$ thousand dollars per year for the next 5 years. Find the present value of the store if the current interest rate is 10 percent compounded continuously.

20. Income stream. Elco Grain Company expects their profits to be $R(t) = 30 + 12e^{0.02t}$ thousand dollars per year for the next 4 years. If the current interest rate is 8 percent compounded continuously, find the present value of the company.

21. In Figure 7.2.1, replace the column information with $100e^{0.01(5)}$, $100e^{0.01(4)}$, …, $100e^{0.01(0)}$. This suggests that the future value of an annuity could be given by $\int_0^N Pe^{rt}\,dt$. Does this give the same result as the formula in the box just below? Explain why or why not.

Hawkes Learning Systems: Essential Calculus

Annuities and Income Streams

7.3 Tables of Integrals

Objectives

After completing this section, you will be able to:

Use a table of integrals to evaluate a variety of integrals.

Extensive tables of integrals contain hundreds of integral formulas listed according to the form that fits the type of function in the integrand. To use such a table to evaluate an integral, we simply look through the table until we find a formula in which the integrand matches the form of the expression we are to integrate. The answer is given by the formula. Table 7.3.1 contains a selection of formulas that will be used in this text. In general, the formulas in these tables are categorized according to the types of expressions contained in the integrands.

In Table 7.3.1 we have listed elementary forms of integrals (all of which we have seen previously in this chapter), forms involving the expressions $ax + b$, $\sqrt{ax+b}$, $(ax+b)(cx+d)$, and $x^2 - a^2$, as well as forms involving exponential and logarithmic expressions. Thus, to evaluate an integral such as $\int \frac{1}{x(3x+4)}dx$, we first note that the integrand contains an expression of the form $ax + b$ with $a = 3$ and $b = 4$. Then we find this category in Table 7.3.1 and determine that Formula 10 is the correct form (see Example 1).

If an integrand contains an expression of more than one type, the appropriate integration formula may be difficult to find. In fact, some tables may not contain the correct formula for a specific problem. Before using a formula, we must be sure that the integrand in the problem precisely matches the integrand in the formula.

Table 7.3.1

Elementary Forms	
1. $\int k\,dx = kx + C$	
2. $\int x^r dx = \frac{1}{r+1}x^{r+1} + C$	$(r \neq -1)$
3. $\int e^x dx = e^x + C$	

continued on next page ...

4. $\int \frac{1}{x}\,dx = \ln|x| + C$

5. $\int kf(x)\,dx = k\int f(x)\,dx$

6. $\int [f(x) \pm g(x)]\,dx = \int f(x)\,dx \pm \int g(x)\,dx$

7. $\int u\,dv = uv - \int v\,du$

Forms Involving $(ax + b)$

8. $\int \frac{1}{ax+b}\,dx = \frac{1}{a}\ln|ax+b| + C$

9. $\int \frac{1}{(ax+b)^2}\,dx = -\frac{1}{a(ax+b)} + C$

10. $\int \frac{1}{x(ax+b)}\,dx = \frac{1}{b}\ln\left|\frac{x}{ax+b}\right| + C$

11. $\int \frac{x}{ax+b}\,dx = \frac{1}{a^2}(ax - b\ln|ax+b|) + C$

Forms Involving $(ax + b)(cx + d)$

12. $\int \frac{1}{(ax+b)(cx+d)}\,dx = \frac{1}{ad-bc}\ln\left|\frac{ax+b}{cx+d}\right| + C$

13. $\int \frac{x}{(ax+b)(cx+d)}\,dx = \frac{1}{ad-bc}\left(\frac{d}{c}\ln|cx+d| - \frac{b}{a}\ln|ax+b|\right) + C$

Forms Involving $\sqrt{ax+b}$

14. $\int \sqrt{ax+b}\,dx = \frac{2}{3a}(ax+b)^{\frac{3}{2}} + C$

15. $\int x\sqrt{ax+b}\,dx = \frac{2(3ax-2b)}{15a^2}(ax+b)^{\frac{3}{2}} + C$

continued on next page ...

Forms Involving $(x^2 - a^2)$	
16. $\int \frac{1}{x^2 - a^2}\,dx = \frac{1}{2a}\ln\left|\frac{x-a}{x+a}\right| + C$	$(x^2 > a^2)$
Forms Involving $\sqrt{x^2 \pm a^2}$	
17. $\int \sqrt{x^2 \pm a^2}\,dx = \frac{x}{2}\sqrt{x^2 \pm a^2} \pm \frac{a^2}{2}\ln\left|x + \sqrt{x^2 \pm a^2}\right| + C$	
18. $\int \frac{1}{\sqrt{x^2 \pm a^2}}\,dx = \ln\left|x + \sqrt{x^2 \pm a^2}\right| + C$	
Forms Involving e^{kx}	
19. $\int e^{kx}\,dx = \frac{1}{k}e^{kx} + C$	
20. $\int x^n e^{kx}\,dx = \frac{x^n e^{kx}}{k} - \frac{n}{k}\int x^{n-1}e^{kx}\,dx$	
21. $\int \frac{1}{a + be^{kx}}\,dx = \frac{x}{a} - \frac{1}{ak}\ln\left|a + be^{kx}\right| + C$	
Forms Involving $\ln x$	
22. $\int \ln x\,dx = x\ln x - x + C$	
23. $\int x^n \ln x\,dx = \frac{x^{n+1}\ln x}{n+1} - \frac{x^{n+1}}{(n+1)^2} + C$	

Comments about Integral Tables

1. In this text, as well as in general, the letters in the beginning of the alphabet, such as a, b, c, and d, represent constants. The letter d is used to emphasize certain patterns or symmetry even though it also appears as part of the differential dx. The two appearances of d should not cause confusion. The letter k is often used to represent a constant in an exponential function.
2. The differential might be in the numerator, so the integral $\int \frac{1}{ax+b}\,dx$ might look like $\int \frac{dx}{ax+b}$.
3. The natural logarithm might be written as $\log x$ instead of $\ln x$.
4. The constant C might be omitted.

The user of the table should be aware of the notation used and its implications.

The following examples illustrate the use of the formulas in Table 7.3.1.

Example 1: Using the Integral Table

Find $\int \frac{1}{x(3x+4)} dx$.

Solution: Using Formula 10 with $a = 3$ and $b = 4$, we have

$$\int \frac{1}{x(3x+4)} dx = \frac{1}{4} \ln \left| \frac{x}{3x+4} \right| + C.$$

Example 2: Using the Integral Table

Evaluate $\int \frac{9}{x^2 - 25} dx$.

Solution: $\int \frac{9}{x^2-25} dx = 9 \int \frac{1}{x^2-25} dx$ — Factor out 9 by Formula 5 with $k = 9$.

$= 9 \int \frac{1}{x^2 - 5^2} dx$ — Rewrite $x^2 - 25$ as $x^2 - 5^2$ and find an integral formula that contains the form $x^2 - a^2$.

$= 9 \cdot \frac{1}{2(5)} \ln \left| \frac{x-5}{x+5} \right| + C$ — Apply Formula 16 with $a = 5$.

$= \frac{9}{10} \ln \left| \frac{x-5}{x+5} \right| + C$ — Simplify.

In Example 3 we use a **reduction formula**, a formula that yields another simpler integral. In some cases, this simpler integral will result in yet another integral to be evaluated. This process continues until no integrals remain. Reduction formulas are said to be **iterative** in nature because of the repeated pattern of the steps involved.

Example 3: Using a Reduction Formula

Integrate $\int x^2 e^{-3x} dx$.

Solution: We will use Formula 20 twice and then use Formula 19.

$\int x^2 e^{-3x} dx = \frac{x^2 e^{-3x}}{-3} - \frac{2}{-3} \int x^{2-1} e^{-3x} dx$ — Use Formula 20 with $n = 2$ and $k = -3$.

continued on next page ...

$$= -\frac{1}{3}x^2e^{-3x} + \frac{2}{3}\left(\frac{x^1e^{-3x}}{-3} - \frac{1}{-3}\int x^{1-1}e^{-3x}dx\right)$$ Use Formula 20 again on the rightmost part of the expression with $n = 1$ and $k = -3$.

$$= -\frac{1}{3}x^2e^{-3x} - \frac{2}{9}xe^{-3x} + \frac{2}{9}\int e^{-3x}dx$$ Simplify.

$$= -\frac{1}{3}x^2e^{-3x} - \frac{2}{9}xe^{-3x} + \frac{2}{9}\cdot\frac{e^{-3x}}{-3} + C$$ Use Formula 19 with $k = -3$.

$$= -\frac{1}{3}x^2e^{-3x} - \frac{2}{9}xe^{-3x} - \frac{2}{27}e^{-3x} + C$$ Simplify.

In Example 4 we show how an integrand can be rewritten algebraically as a sum so that each part of the sum fits a formula in the table.

Example 4: Rewriting the Integrand

Integrate $\int \frac{x+1}{x^2+5x+6}dx$.

Solution: $$\int \frac{x+1}{x^2+5x+6}dx = \int \frac{x+1}{(x+2)(x+3)}dx$$ Factor the denominator.

$$= \int \frac{x}{(x+2)(x+3)}dx + \int \frac{1}{(x+2)(x+3)}dx$$ By Formula 6.

Now we apply Formula 13 to the left integral and Formula 12 to the right integral.

$$\int \frac{x}{(x+2)(x+3)}dx = \frac{1}{3-2}\left(\frac{3}{1}\ln|x+3| - \frac{2}{1}\ln|x+2|\right) + C_1$$ Left integral.

$$= 3\ln|x+3| - 2\ln|x+2| + C_1$$

$$\int \frac{1}{(x+2)(x+3)}dx = \frac{1}{3-2}\ln\left|\frac{x+2}{x+3}\right| + C_2$$ Right integral.

$$= \ln\left|\frac{x+2}{x+3}\right| + C_2$$

$$= \ln|x+2| - \ln|x+3| + C_2$$

Now combining the parts gives the result.

$$\int \frac{x+1}{x^2+5x+6}dx = 3\ln|x+3| - 2\ln|x+2| + C_1 + \ln|x+2| - \ln|x+3| + C_2$$

$$= 2\ln|x+3| - \ln|x+2| + C \qquad \text{Where } C = C_1 + C_2.$$

7.3 Exercises

Use Table 7.3.1 to find the following integrals.

1. $\int \frac{1}{4x+3}dx$ **2.** $\int \sqrt{9x+2}\,dx$ **3.** $\int e^{-0.15x}dx$

4. $\int \ln x\,dx$ **5.** $\int \frac{1}{(2x-5)^2}dx$ **6.** $\int \frac{x}{x+6}dx$

7. $\int x\sqrt{3x-4}dx$ **8.** $\int \frac{x}{(2x+1)(x-2)}dx$ **9.** $\int \sqrt{x^2+36}\,dx$

10. $\int x^4 \ln x\,dx$ **11.** $\int \frac{1}{x^2-16}dx$ **12.** $\int \frac{1}{x(4x-3)}dx$

13. $\int \frac{1}{(x+8)(5x-1)}dx$ **14.** $\int \frac{1}{2+e^{3x}}dx$ **15.** $\int 7x^5 \ln x\,dx$

16. $\int x^4 e^{-2x}\,dx$ **17.** $\int \frac{1}{8-5e^{-0.7x}}dx$ **18.** $\int \frac{1}{(0.3x+2)^2}dx$

19. $\int \frac{2}{x(3x-1)}dx$ **20.** $\int \frac{4}{x^2-8}dx$ **21.** $\int 14\sqrt{6x-5}\,dx$

22. $\int \frac{13}{(4x-1)(2x+3)}dx$ **23.** $\int \frac{8}{x(0.4x+1)}dx$ **24.** $\int 2x\sqrt{3x-4}\,dx$

25. $\int x^3 e^{1.5x}\,dx$ **26.** $\int \sqrt{x^2+9}\,dx$ **27.** $\int \frac{x}{(x-7)(5x+2)}dx$

28. $\int \sqrt{x^2-15}\,dx$ **29.** $\int \frac{1}{\sqrt{x^2-12}}dx$ **30.** $\int \frac{6}{24-9e^{3.1x}}dx$

Hawkes Learning Systems: Essential Calculus

Tables of Integrals

7.4 Improper Integrals

Objectives

After completing this section, you will be able to:

1. *Evaluate improper integrals by hand and by using a graphing calculator.*
2. *Determine whether a given improper integral is convergent or divergent.*

There are many practical situations where an integral on an unbounded interval has meaning. An integral on an unbounded interval is called an **improper integral**. In Section 7.5 we will see how improper integrals are related to probability distributions which have applications in statistics, economics, and the behavioral and social sciences.

Improper Integral

The integral $\int_a^{+\infty} f(x)\,dx$ *is called an* ***improper integral****. This integral is defined as the following limit:*

$$\int_a^{+\infty} f(x)\,dx = \lim_{b\to+\infty}\int_a^b f(x)\,dx.$$

If $\lim_{b\to+\infty}\int_a^b f(x)\,dx$ *exists, then the improper integral is said to be* ***convergent****.*

If $\lim_{b\to+\infty}\int_a^b f(x)\,dx$ *does not exist, then the improper integral is said to be* ***divergent****.*

It is a curiosity that an area can be finite yet have an infinite boundary. It is like having a backyard fence infinitely long but only a finite yard.

Example 1: Evaluating an Improper Integral

Evaluate $\int_1^{+\infty} \frac{1}{x^2}\,dx$.

Solution: First, evaluate the integral from $x = 1$ to $x = b$.

$$\int_1^b \frac{1}{x^2}\,dx = \int_1^b x^{-2}\,dx$$

$$= \frac{x^{-1}}{-1}\Big]_1^b = -\frac{1}{x}\Big]_1^b$$

$$= -\frac{1}{b} + 1$$

continued on next page ...

Next, find the limit as $b \to +\infty$.

$$\lim_{b \to +\infty}\left(-\frac{1}{b}+1\right)=0+1=1 \qquad \text{Note that } \lim_{b \to +\infty}\frac{1}{b}=0.$$

Therefore,

$$\int_1^{+\infty}\frac{1}{x^2}\,dx = 1.$$

We say that the integral **converges** to 1.

The result from Example 1 can be interpreted as the area under the curve $f(x)=\dfrac{1}{x^2}$ on the interval $[1, +\infty)$. As illustrated in Figure 7.4.1, the larger b gets, the closer the value of the integral $\int_1^b \frac{1}{x^2}\,dx$ gets to 1.

Figure 7.4.1

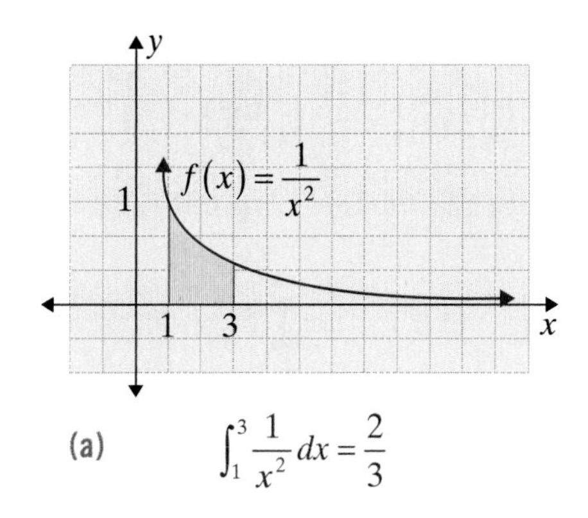

(a) $\int_1^3 \frac{1}{x^2}\,dx = \frac{2}{3}$

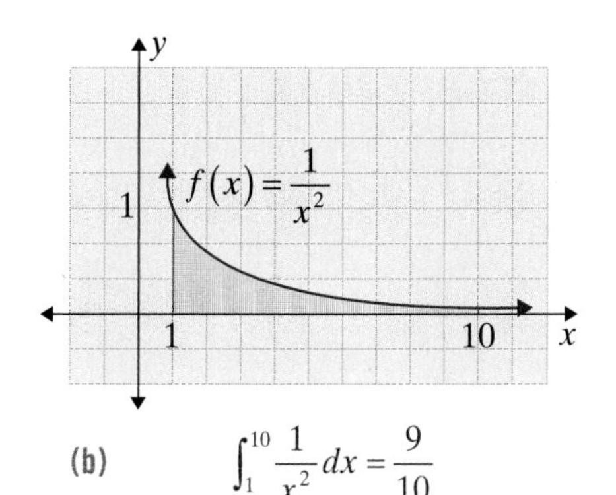

(b) $\int_1^{10} \frac{1}{x^2}\,dx = \frac{9}{10}$

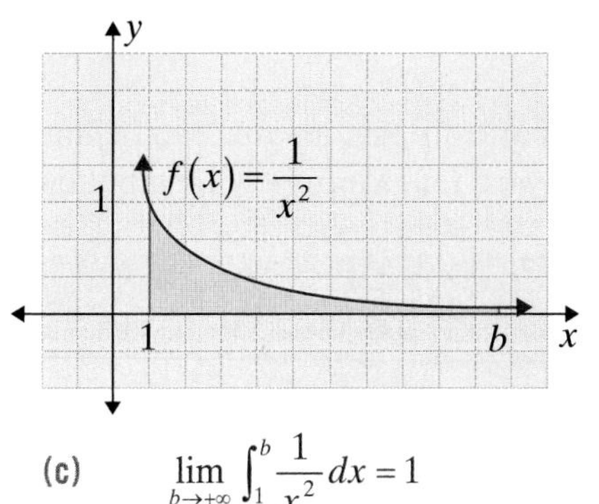

(c) $\lim_{b \to +\infty}\int_1^b \frac{1}{x^2}\,dx = 1$

In evaluating improper integrals, we must be able to find limits as $b \to +\infty$. These ideas were discussed in Section 2.4, and we list the following results for use in the examples and exercises that follow.

$$\lim_{b \to +\infty}\frac{1}{b}=0, \quad \lim_{b \to +\infty}\frac{1}{e^b}=0, \quad \lim_{b \to +\infty}\frac{b}{e^b}=0, \quad \lim_{b \to +\infty}(\ln b)=+\infty, \quad \text{and} \quad \lim_{b \to +\infty}b^{\frac{1}{n}}=+\infty \quad (\text{for } n>0).$$

Example 2: Determining Integral Convergence

Determine whether the improper integral $\int_0^{+\infty} e^{-3x}\,dx$ is convergent or divergent. Evaluate it if it is convergent.

Solution: First, evaluate the integral from 0 to b.

$$\int_0^b e^{-3x}\,dx = -\frac{1}{3}e^{-3x}\bigg]_0^b = -\frac{1}{3e^{3b}} + \frac{1}{3}$$

Next, find the limit as $b \to +\infty$.

$$\lim_{b\to+\infty}\left(-\frac{1}{3e^{3b}} + \frac{1}{3}\right) = 0 + \frac{1}{3} = \frac{1}{3}$$

This integral converges to $\frac{1}{3}$.

Example 3: Determining Integral Convergence

Determine whether the improper integral $\int_1^{+\infty} \frac{1}{x}\,dx$ is convergent or divergent. Evaluate it if it is convergent.

Solution: First, evaluate the integral from 1 to b.

$$\int_1^b \frac{1}{x}\,dx = \ln|x|\Big]_1^b = \ln b - \ln 1 = \ln b$$

Next, find the limit as $b \to +\infty$.

$$\lim_{b\to+\infty} \ln b = +\infty$$

The integral is divergent.

Example 3 illustrates the importance of the limit as well as the unusual and sometimes unexpected nature of improper integrals. In contrast to Example 1, where the area under the curve $f(x) = \frac{1}{x^2}$ equals 1, the area under the curve $f(x) = \frac{1}{x}$ is infinite. The curves seem similar on an intuitive basis but are quite different in terms of improper integrals and limits. (See Figure 7.4.2.)

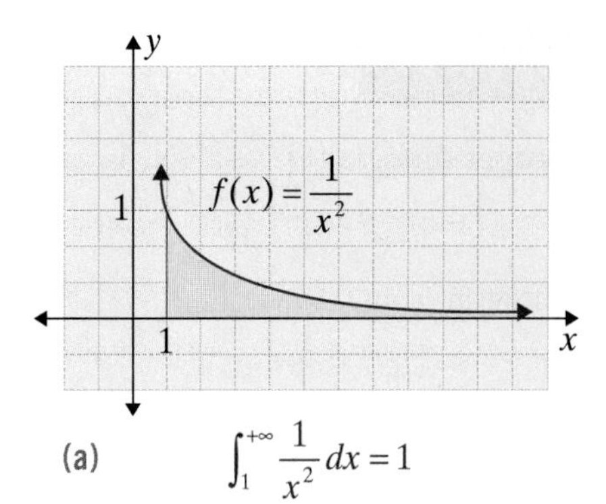

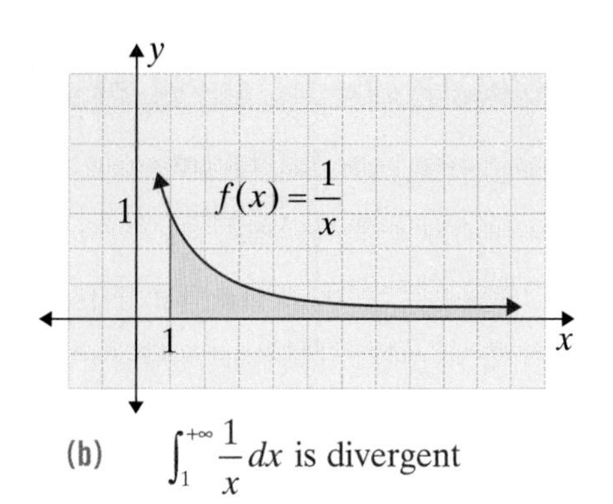

Figure 7.4.2 (a) $\int_1^{+\infty} \frac{1}{x^2}\,dx = 1$ (b) $\int_1^{+\infty} \frac{1}{x}\,dx$ is divergent

Example 4: Determining Integral Convergence

Find the value of the integral $\int_4^{+\infty} \frac{1}{\sqrt{x}}\,dx$ if it is convergent.

Solution: First, evaluate the integral from 4 to b.

$$\int_4^b \frac{1}{\sqrt{x}}\,dx = \int_4^b x^{-\frac{1}{2}}\,dx = 2x^{\frac{1}{2}}\Big]_4^b$$

$$= 2b^{\frac{1}{2}} - 4$$

Next, find the limit as $b \to +\infty$.

$$\lim_{b\to+\infty}\left(2b^{\frac{1}{2}} - 4\right) = +\infty$$

The integral is divergent.

Example 5: Determining Integral Convergence

Find the value of the integral $\int_1^{+\infty} \frac{1}{x^{\frac{3}{2}}}\,dx$ if it is convergent.

Solution: First, evaluate the integral from 1 to b.

$$\int_1^b \frac{1}{x^{\frac{3}{2}}}\,dx = \int_1^b x^{-\frac{3}{2}}\,dx = -2x^{-\frac{1}{2}}\Big]_1^b$$

$$= -2b^{-\frac{1}{2}} + 2$$

$$= -\frac{2}{b^{\frac{1}{2}}} + 2$$

Next, find the limit as $b \to +\infty$.

$$\lim_{b \to +\infty} \left(-\frac{2}{b^{\frac{1}{2}}} + 2 \right) = 0 + 2 = 2$$

The integral converges to 2.

There are other types of improper integrals. We consider one more in this text. We consider the case of an area bounded by an asymptote.

Example 6: Area Bounded by an Asymptote

Find the value of $\int_0^1 \frac{1}{x^2}\,dx$.

Solution: Since the integrand is not defined at the lower limit, we must use a limit to evaluate the integral.

$$\begin{aligned}
\int_0^1 \frac{1}{x^2}\,dx &= \lim_{a \to 0^+} \int_a^1 \frac{1}{x^2}\,dx \\
&= \lim_{a \to 0^+} \left[-x^{-1} \right]_a^1 \\
&= \lim_{a \to 0^+} \left[-1 - \left(-a^{-1} \right) \right] \\
&= \lim_{a \to 0^+} \left[-1 + \frac{1}{a} \right] \\
&= -1 + \lim_{a \to 0^+} \frac{1}{a}
\end{aligned}$$

Since this last limit does not exist, the integral diverges.

Improper Integrals on a TI-84 Plus

Once a convergent improper integral has been set up, it is possible to use the TI-84 Plus or other graphing utility to do the calculating. There are several methods to do this–the simplest is to use the previous method described. For example, let us integrate $\int_1^\infty 2xe^{-3x}\,dx$. The following steps are recommended:

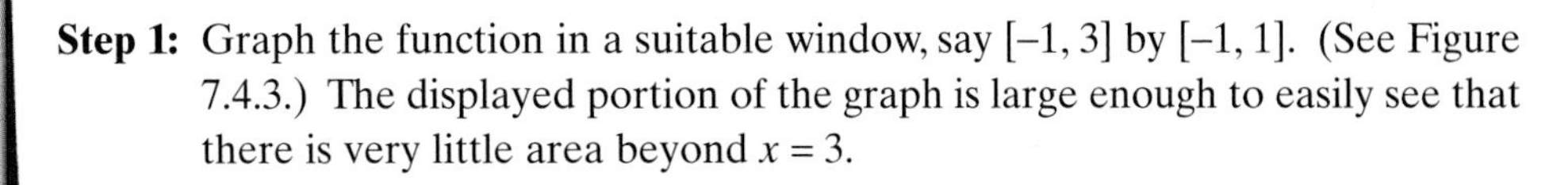

Step 1: Graph the function in a suitable window, say $[-1, 3]$ by $[-1, 1]$. (See Figure 7.4.3.) The displayed portion of the graph is large enough to easily see that there is very little area beyond $x = 3$.

Figure 7.4.3

continued on next page ...

Step 2: Press 2ND TRACE and select item `7:∫f(x)dx` (this is the integration symbol). See Figure 7.4.4. At the prompt type the lower limit `1`, and enter. At the next prompt, type the upper limit `3`. The area is shaded and the decimal answer `0.04398093` appears at the screen bottom (see Figure 7.4.5). This is an approximation to the actual value, but when using this method, the upper and lower limits must be in the range of x-values plotted on screen.

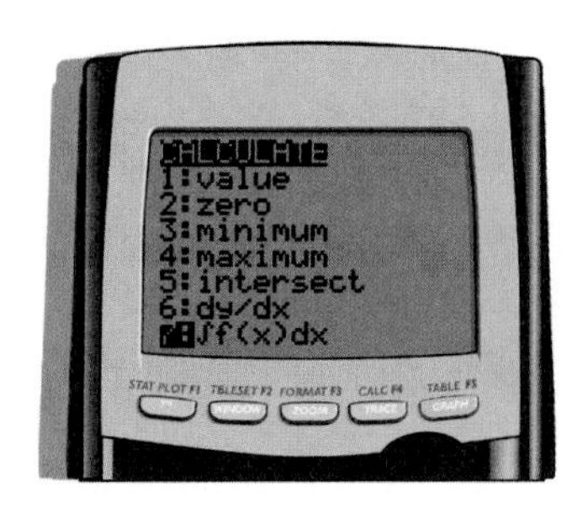

Figure 7.4.4

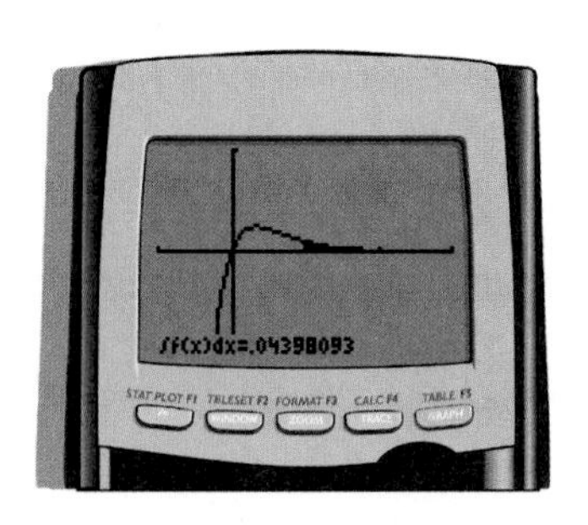

Figure 7.4.5

An Alternate Method

Step 1: Graph the function as in Step 1 of the previous method.

Step 2: Go to the MATH menu and select item `9: fnInt(` (this is the function integral symbol). See Figure 7.4.6. This selection now requires that you type the function, the variable of integration x, the lower limit, the upper limit, close the parentheses, and ENTER. The four items within the parentheses must be separated by commas:

```
fnInt(2Xe^(-3X),X,1,100).
```

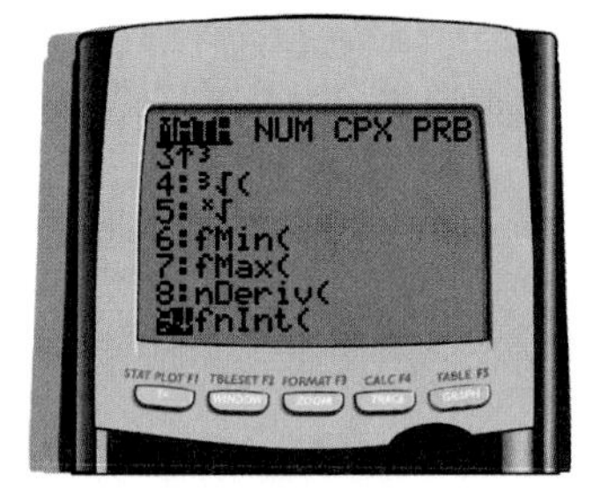

Figure 7.4.6

(**Note:** You may use any number for the upper limit; for this function 100 works well.)

The calculator will return `0.0442551719`, a more accurate answer than the previous result (Figure 7.4.7).

If the formula has been typed into Y_1 position, say, it is not necessary to retype the function. It can be accessed using the `VARS` button.

Figure 7.4.7

If the Function is Already in the Y_1 Position on the Calculator

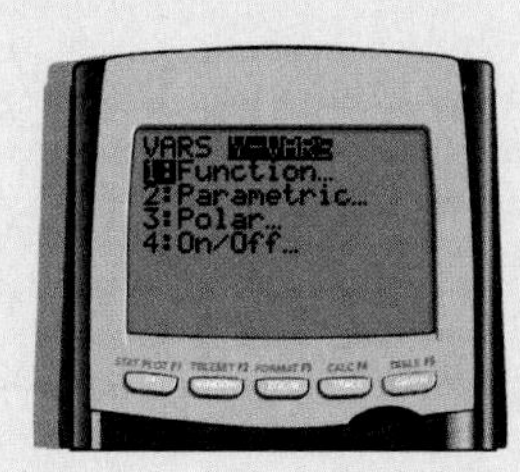

It is not necessary to retype the function in Step 2 of the alternate method if the function is already stored in the `Y1` position.

1. Graph the function as in Step 1 of the previous method.

2. Go the MATH menu and select item `9:fnInt`. Next, select VARS and, in the menu, arrow right one step to `Y-VARS` and press ENTER. This selects the list of functions. `Y1` is already selected so just press ENTER again. Now type a comma, X,T,θ,n `, 1, 100)` as before and press ENTER. This lets one use the `fnInt` feature without having to retype the formula.

7.4 Exercises

In Exercises 1 – 10, find the limit if it exists.

1. $\lim_{b\to+\infty} \frac{1}{b}$ **2.** $\lim_{b\to+\infty} \frac{1}{\sqrt[3]{b}}$ **3.** $\lim_{b\to+\infty} \frac{\sqrt{b}}{20}$

4. $\lim_{b\to+\infty} e^{0.1b}$ **5.** $\lim_{b\to+\infty} e^{-4b}$ **6.** $\lim_{b\to+\infty} (-12\ln b)$

7. $\lim_{b\to+\infty} \left(2+\frac{9}{\sqrt{3b+1}}\right)$ **8.** $\lim_{b\to+\infty} \left(5+e^{-2b}\right)$ **9.** $\lim_{b\to+\infty} 7b^4e^{-b}$

10. $\lim_{b\to+\infty} (7b+2)^{-\frac{2}{3}}$

In Exercises 11 – 34, determine whether the improper integrals are convergent or divergent, and evaluate those which are convergent. Use your graphing calculator or other graphing utility for those marked with a ✦.

11. $\int_2^{+\infty} \frac{4}{x^3}dx$ **12.** $\int_1^{+\infty} \frac{1}{\sqrt[3]{x}}dx$ **✦13.** $\int_8^{+\infty} x^{-\frac{2}{3}}dx$

14. $\int_4^{+\infty} 5x^{-\frac{3}{2}}dx$ **15.** $\int_{20}^{+\infty} 3e^{-x}dx$ **✦16.** $\int_4^{+\infty} e^{-2x}dx$

17. $\int_2^{+\infty} e^{-\frac{x}{3}}dx$ **18.** $\int_2^{+\infty} 4e^{-0.5x}dx$ **✦19.** $\int_2^{+\infty} e^{1.5x}dx$

20. $\int_{-1}^{+\infty} \frac{1}{80}e^{0.16x}dx$ **21.** $\int_0^{+\infty} \frac{1}{(x+3)^2}dx$ **✦22.** $\int_0^{+\infty} \frac{4}{\sqrt{3x+1}}dx$

23. $\int_{-1}^{+\infty} \frac{2}{\sqrt[3]{2x+3}}dx$ **24.** $\int_2^{+\infty} (3x+2)^{-\frac{4}{3}}dx$ **✦25.** $\int_0^{+\infty} \frac{5}{x+1}dx$

26. $\int_0^{+\infty} (5x+4)^{-\frac{3}{2}}dx$ **27.** $\int_0^{+\infty} x^2e^{-x^3}dx$ **✦28.** $\int_0^{+\infty} -4xe^{x^2}dx$

29. $\int_1^{+\infty} xe^{1-x^2}dx$ **30.** $\int_0^{+\infty} 7x^2e^{-x^2}dx$ **31.** $\int_2^{+\infty} \frac{1}{x(\ln x)^3}dx$

32. $\int_e^{+\infty} \frac{1}{x\ln x}dx$ **✦33.** $\int_0^{+\infty} xe^{-x}dx$ **✦34.** $\int_0^{+\infty} xe^{-0.2x}dx$

In Exercises 35 – 38, find the area, if exists, of the region under the curve y = f(x) on the given interval of the x-axis.

35. $f(x)=\frac{4}{x^2},\ x\geq 2$ **36.** $f(x)=3e^{-x},\ x\geq 0$

37. $f(x)=\frac{3}{x},\ x\geq 6$ **38.** $f(x)=2e^{0.8x},\ x\geq 0$

39. The integral $\int_1^{\infty} \frac{1}{x^p}\,dx$ converges if and only if (choose all that apply):

a. $0 < p < 1$.
b. $p \neq 1$.
c. p is an integer greater than or equal to 2.
d. $p > 1$.
e. p is positive.
f. none of the above.

40. Integrate $\int_1^{\infty} 2xe^{-3x}\,dx$ by evaluating the limits and compare your answer to the calculator values obtained at the end of Section 7.4.

Hawkes Learning Systems: Essential Calculus

Improper Integrals

7.5 Probability

Objectives

After completing this section, you will be able to:

1. *Calculate the probability of an event occurring given a function.*
2. *Determine the expected value of a random variable.*

Probability

In the study of probability and statistics, an **experiment** is defined to be an activity that yields a set of data. The data can be either **qualitative** (related to some descriptive characteristic such as color) or **quantitative** (resulting from counting or measuring). Quantitative data can be either **discrete** (resulting from a count, such as the number of cars in a parking lot) or **continuous** (resulting from a measure, such as age, weight, or height). Any one particular result of an experiment is called an **outcome**, and the set of all possible outcomes is called the **sample space *S***. For example, if an experiment consists of rolling a single six-sided die, then the sample space consists of the six possible outcomes, namely, $S = \{1, 2, 3, 4, 5, 6\}$.

If an experiment consists of tossing a fair coin, there are two possible outcomes, heads (H) or tails (T), and the sample space is $S = \{H, T\}$. If x represents a possible outcome and $P(x)$ represents the probability of x, then, since the coin is fair, $P(x = H) = \frac{1}{2}$ and $P(x = T) = \frac{1}{2}$. In this case x is called a **discrete random variable**.

A table or graph that represents all the possible outcomes and their probability is called a **probability distribution**. If each outcome is equally likely to occur (as in a fair coin toss or the roll of a fair die), then the distribution is called a **uniform probability distribution**. (See Figure 7.5.1.)

Figure 7.5.1

Two Uniform Probability Distributions

x	$p(x)$
H	$\frac{1}{2}$
T	$\frac{1}{2}$

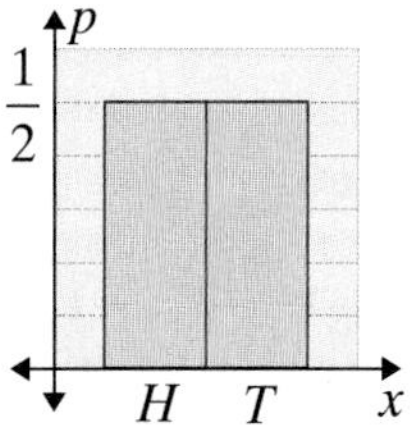

Experiment – Fair Coin Toss
(a)

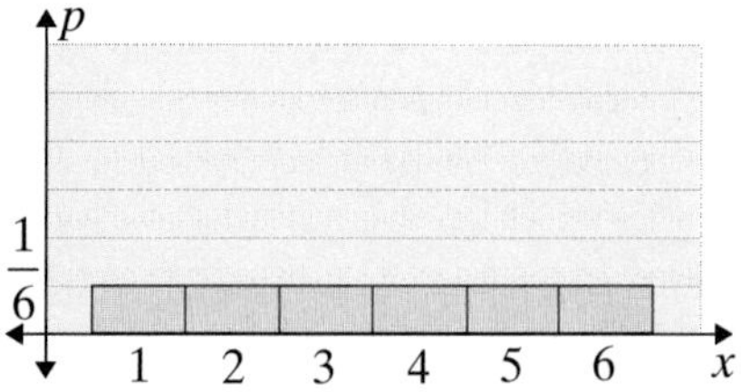

Experiment – Rolling a Single Fair Die
(b)

x	$p(x)$
1	$\frac{1}{6}$
2	$\frac{1}{6}$
3	$\frac{1}{6}$
4	$\frac{1}{6}$
5	$\frac{1}{6}$
6	$\frac{1}{6}$

Of course, not all probability distributions are uniform. Suppose that a box contains 10 chips: 3 green (G), 2 blue (B), and 5 yellow (Y), and the experiment is to draw one chip and record its color. If each chip is equally likely to be drawn, then the event of drawing a particular color has probability

$$P(x) = \frac{n(x)}{n(S)},$$

where the random variable x is the color of chips, $n(x)$ is the number of chips of that color, and $n(S)$ is the number of chips in the sample space. The probability distribution for this experiment is shown in Figure 7.5.2.

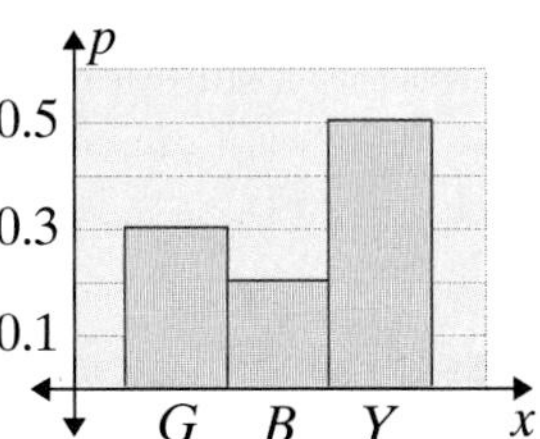

x	p(x)
G	$\frac{3}{10} = 0.3$
B	$\frac{2}{10} = 0.2$
Y	$\frac{5}{10} = 0.5$

Figure 7.5.2

Probability Distribution

*Every **probability distribution** with n outcomes for a discrete random variable x must have the following two characteristics:*

1. *The probability of each outcome is between* 0 *and* 1:
$$\mathbf{0 \le P(x) \le 1.}$$
2. *The sum of the probabilities of the outcomes is* 1:
$$\mathbf{P(x_1) + P(x_2) + \ldots + P(x_n) = 1.}$$

Probability Density Function

Because of calculus, we can also consider the case of continuous random variables, where a **continuous random variable** x can take on any value in an interval of real numbers. Experiments that involve any type of measurement allow for real number outcomes. Thus an experiment that involves measuring heights or weights, timing races, measuring reaction time to a drug, or measuring the life of a light bulb allows for a continuous random variable defined on some interval of real numbers.

In order to find the probability that a continuous random variable x is in a particular interval, we need the corresponding **probability density function** for the experiment.

Probability Density Function

If x is a continuous random variable distributed on the interval [a, b], then f is a ***probability density function*** *for x if*

1. $f(x) \geq 0$ *for all x in* $[a, b]$, *and*

2. $\int_a^b f(x)\,dx = 1$.

If $[c, d]$ is a subinterval of $[a, b]$, then the probability that x is in $[c, d]$ is

$$P(c \leq x \leq d) = \int_c^d f(x)\,dx.$$

Example 1: Probability Density Function

Show that $f(x) = \frac{3}{10}(3x - x^2)$ is a probability density function on the interval $[0, 2]$.

Solution: First, we note that $f(x) \geq 0$ for all x in $[0, 2]$. Then we evaluate the integral.

$$\begin{aligned} \int_0^2 \frac{3}{10}(3x - x^2)\,dx &= \frac{3}{10}\int_0^2 (3x - x^2)\,dx \\ &= \frac{3}{10}\left(\frac{3x^2}{2} - \frac{x^3}{3}\right)\Bigg]_0^2 \\ &= \frac{3}{10}\left(6 - \frac{8}{3}\right) \\ &= \frac{\cancel{3}}{\cancel{10}}\left(\frac{\cancel{10}}{\cancel{3}}\right) \\ &= 1 \end{aligned}$$

Thus f is a probability density function on the interval $[0, 2]$.

Example 2: Probability Density Function

The function $f(x)=\frac{1}{12}x$ is a probability function on the interval $[1, 5]$.

a. Find the probability that x is between 1 and 2.
b. Show that the probability at $x = 3$ is 0.

Solutions: a.

$$P(1 \le x \le 2)=\int_1^2 \frac{1}{12}x\,dx$$

Evaluate the integral of the function on the interval $[1, 2]$.

$$=\frac{1}{12}\cdot\frac{x^2}{2}\Big]_1^2=\frac{x^2}{24}\Big]_1^2$$

$$=\frac{4}{24}-\frac{1}{24}$$

$$=\frac{1}{8}$$

b.

$$P(x=3)=\int_3^3 \frac{1}{12}x\,dx=0$$

Evaluate the integral of the function at $x = 3$.

Part **b.** in Example 2 illustrates the fact that, **for a continuous random variable, the probability that x will have any one specific value is 0.**

Exponential Distributions

Measures such as the waiting time between landings at an airport, the lifetime of a product, or scores on an IQ test are continuous random variables distributed on unbounded intervals. Exponential distributions describe continuous random variables of this type.

Exponential Distribution

A continuous random variable x is ***exponentially distributed*** *on the interval $[a, b]$ if it has a probability density function of the form*

$$f(x)=c_1e^{-c_2x}$$

and

$$\int_a^b f(x)\,dx=1.$$

With $c_1 = c_2 = k$, since $\int_0^{\infty} ke^{-kx}\,dx = 1$, $f(x) = ke^{-kx}$ is exponentially distributed on $[0, \infty)$.

Example 3: Exponential Distributions

The Overseas Telephone Company determines that the length of time (in minutes) for a particular telephone call has the probability density function $f(t) = 0.01e^{-0.01t}$.

a. Find the probability that a phone call will last 30 minutes or less.

b. Find the probability that a phone call will last at least 1 hour.

Solutions: a.

$$P(0 \le t \le 30) = \int_0^{30} 0.01e^{-0.01t}\,dt$$
$$= -e^{-0.01t}\Big]_0^{30}$$
$$= -e^{-0.3} + 1$$
$$\approx 0.26$$

b.

$$P(t \ge 60) = \int_{60}^{+\infty} 0.01e^{-0.01t}\,dt$$

1 hour = 60 minutes.

$$= \lim_{b \to +\infty} \int_{60}^{b} 0.01e^{-0.01t}\,dt$$
$$= \lim_{b \to +\infty} -e^{-0.01t}\Big]_{60}^{b}$$
$$= \lim_{b \to +\infty} \left(-e^{-0.01b} + e^{-0.6}\right)$$
$$= e^{-0.6}$$
$$\approx 0.55$$

Standard Normal Distribution

*The **standard normal distribution** is represented by the probability density function*

$$f(x) = \frac{1}{\sqrt{2\pi}} e^{-\frac{x^2}{2}}$$

on the interval $(-\infty, +\infty)$.

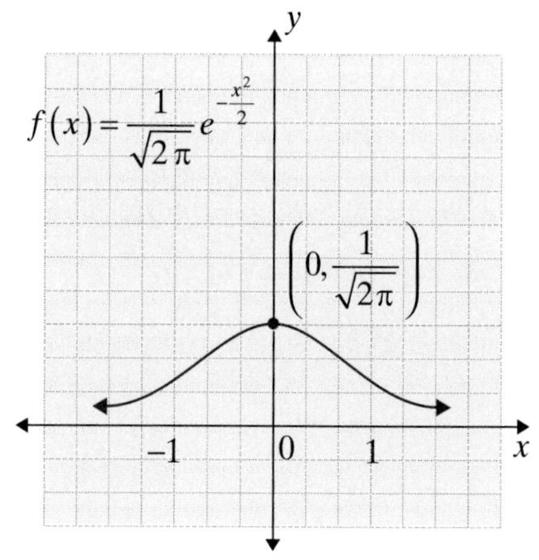

Figure 7.5.4

The graph of the standard normal distribution, known as the **bell-shaped curve**, is shown in Figure 7.5.4.

Techniques developed in higher-level calculus courses show that

$$\int_{-\infty}^{+\infty} \frac{1}{\sqrt{2\pi}} e^{-\frac{x^2}{2}} dx = 1.$$

This distribution has been studied and used extensively in the study of statistics, and tables are available that give the probabilities associated with the standard normal distribution.

Expected Value

The **expected value** of a random variable is also called the **mean value** or **average value** of the variable. In the context of a density function $F(x)$, the area above an interval $[x_1, x_2]$ measures the probability that x is in that interval.

Consider the function $F(x) = \frac{3x^2}{64}$ on the interval $[0, 4]$. Let us determine the probability that when a random x is selected it is "near" $x = 1.75$. For instance, the probability that a randomly selected x is within 0.25 of 1.75, that is, the probability $P[(1.75 - 0.25) \le x \le (1.75 + 0.25)] \to P[1.5 \le x \le 2]$ is the area $I(1.75) = \int_{1.5}^{2} F(x)\,dx \approx 0.072$. This area assigns an "expectation" $E(1.75)$ to $x = 1.75$, and we calculate the total expectation for $x = 1.75$ by multiplying x by its assigned area:

$$E(1.75) = 1.75 \cdot I(1.75) = 1.75(0.072) = 0.126.$$

We can do this for a sequence of x values (0.25, 0.75, 1.25, 1.75, 2.25, …, 3.75) over the domain $[a, b] = [0, 4]$. We will approximate the areas first by using a rectangle centered on each of the x-values. If the rectangles have the same base $\Delta x = 0.5$, then the area of a rectangle is $F(x) \cdot \Delta x$ and the total expectation is approximately

$$E(x) = \sum_{i=1}^{8} x_i F(x_i) \cdot \Delta x = 0.25 \cdot F(0.25) \cdot 0.5 + \ldots + 3.75 \cdot F(3.75) \cdot 0.5.$$

This approximation is a Riemann sum (see Chapter 6) whose exact value is given by an integral.

Expected Value

Suppose the function f defined on the interval [a, b] is the probability density function associated with a continuous random variable x. Then the ***expected value of x*** *is*

$$\boldsymbol{E(x) = \int_a^b x \cdot f(x)\,dx.}$$

E(x) is often denoted with the Greek letter mu (μ) and is also called the mean value of x.

Example 4: Calculating Expected Value

For the function $F(x) = \dfrac{3x^2}{64}$, **a.** verify that $F(x)$ is a density function over the interval $[0, 4]$ and **b.** determine the expected value of x.

Solutions: a. We must see if $\int_0^4 \frac{3x^2}{64}\,dx$ is equal to 1.

$$\begin{aligned} \int_0^4 \left(\frac{3x^2}{64}\right) dx &= \left(\frac{3}{64}\right)\left(\frac{x^3}{3}\right)\Bigg]_0^4 \\ &= \frac{x^3}{64}\Bigg]_0^4 \\ &= \frac{(4)^3}{64} - \frac{(0)^3}{64} \\ &= 1 \end{aligned}$$

Therefore, $F(x)$ is a density function over the interval $[0, 4]$.

b. $E(x) = \int_0^4 x\left(\frac{3}{64}\right)x^2 \cdot dx$

$$= \int_0^4 \left(\frac{3}{64}\right)x^3 \cdot dx$$

$$= \left(\frac{3}{64}\right)\left(\frac{x^4}{4}\right)\Big]_0^4$$

$$= \left(\frac{3}{64} \cdot \frac{(4)^4}{4}\right) - \left(\frac{3}{64} \cdot \frac{(0)}{4}\right)$$

$$= 3$$

Example 5: Calculating Expected Value

For the situation in Example 3, determine the average length of an overseas phone call.

Solution: We must evaluate $\int t\left(0.01e^{-0.01t}\right)dt.$ We use integration by parts or Formula 20, from Table 7.3.1 in Section 7.3, to get the expression $te^{-0.01t} - \frac{1}{0.01}e^{-0.01t} = \frac{t}{e^{0.01t}} - \frac{100}{e^{0.01t}}.$

Using limits, $\mu = \lim_{b \to +\infty}\left(\frac{b}{e^{0.01b}} - \frac{100}{e^{0.01b}}\right) - (0 - 100) = 100$ since the limit expression goes to zero.

Note that for an exponential distribution the constant k actually gives the mean value. That is, $\mu = \frac{1}{k}.$

7.5 Exercises

In Exercises 1 – 12, show that each function is a probability density function on the given interval. If the function is not a probability density function on the given interval, explain why.

1. $f(x) = \frac{3}{16}\sqrt{x},\ [0, 4]$

2. $f(x) = \frac{4}{45}\sqrt[3]{x},\ [1, 8]$

3. $f(x) = \frac{4}{15}\left(\frac{1}{2}x + 1\right),\ [-1, 2]$

4. $f(x) = \frac{1}{4}(x + 1),\ [0, 2]$

5. $f(x)=\frac{3}{68}\left(x-\sqrt{x}\right),\ [1, 9]$

6. $f(x)=\frac{2}{13}\left(x+\frac{2}{x}\right),\ [2, 4]$

7. $f(x)=\frac{1}{2x},\ \left[1, e^2\right]$

8. $f(x)=\frac{1}{3x},\ \left[1, e^3\right]$

9. $f(x)=\frac{3}{32}\left(4-x^2\right),\ [-2, 2]$

10. $f(x)=6\left(\sqrt{x}-x\right),\ [0, 1]$

11. $f(x)=2e^{-2x},\ [0, +\infty)$

12. $f(x)=\frac{1}{10}e^{-0.1x},\ [0, +\infty)$

In Exercises 13 – 20, **a.** *determine the value of k such that the function is a probability density function on the given interval and* **b.** *determine the average or expected value of x.*

13. $f(x)=k(3-x),\ [0, 3]$

14. $f(x)=k(5-2x),\ [-1, 2]$

15. $f(x)=\frac{k}{\sqrt{x}},\ [1, 4]$

16. $f(x)=ke^{2x},\ [0, 1]$

17. $f(x)=ke^{-0.25x},\ [0, +\infty)$

18. $f(x)=ke^{-0.5x},\ [2, +\infty)$

19. $f(x)=\frac{k}{x^3},\ [1, 4]$

20. $f(x)=\frac{k}{(x+1)^2},\ [0, 7]$

21. An experiment has the probability density function $f(x)=\frac{2}{9}\left(3x-x^2\right)$, where $0\le x\le 3$. Find the probability that x is between 1 and 2.

22. The probability density function for an experiment is $f(x)=\frac{1}{3x}$, where $1\le x\le e^3$. Find the probability that x is between 1 and 10.

23. The exponential probability density function for an experiment is given by $f(x)=0.4e^{-0.4x}$, where $x\ge 0$. What is the probability that $x\ge 10$?

24. The outcomes of an experiment are distributed exponentially according to $f(x)=0.7e^{-0.7x}$, where $x\ge 0$. What is the probability that $10\le x\le 20$?

25. **Waiting time.** The average waiting time in minutes for a shuttle at the airport parking lot has the probability density function $f(t)=\frac{1}{20}$, where $0\le t\le 20$. What is the probability that you will wait at least 12 minutes? What is the average waiting time?

26. **Travel time.** The length of time t (in minutes) it requires Frank to get to work ranges from 25 to 40 minutes. The probability density function is $f(t)=\frac{1}{15}$, where $25 \le t \le 40$. If Frank allots himself 36 minutes to get to work, what is the probability that he will be late? If he leaves his house at 8:00 A.M., what is his average arrival time?

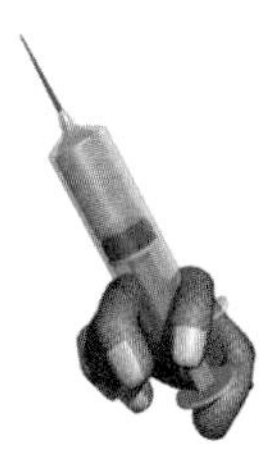

27. **Reaction time.** The length of time t (in seconds) that it takes the body to react to the injection of a particular drug has the probability density function $f(t)=\frac{81}{40t^3}$, where $1 \le t \le 9$. What is the probability density that it will take at least 6 seconds for the body to react to the drug? What is the average waiting time for the body to react?

28. **Waiting time.** The length of time t (in minutes) that you and your friends will wait to be seated at a popular restaurant has the probability density function $f(t)=\frac{26}{25(t+1)^2}$, where $0 \le t \le 25$. What is the probability that you will wait between 5 and 10 minutes?

29. **Reliability.** The manager of Electronic City guarantees a certain brand of television set for 2 years. The probability that a television set will last t years without service is given by $f(t)=0.15e^{-0.15t}$. What is the probability that a television will need service before it is 2 years old?

30. **Traffic spacing.** On the freeway during rush hour, the distance x (in feet) between your car and the car behind you has the probability density function $f(x)=0.02e^{-0.02x}$, where $x \ge 0$. What is the probability that the car behind you is within 30 feet of you?

Hawkes Learning Systems: Essential Calculus

Probability

7.6 Volume

Objectives

After completing this section, you will be able to:

Find the volume of a solid of revolution.

Solid of Revolution

In Section 6.4 we used definite integration to find the area of the region between the graph of a continuous function and the x-axis. If such a region is revolved about the x-axis, it sweeps through a **solid of revolution**. The objective of this section is to develop a technique for finding the volume of a solid of revolution. (See Figure 7.6.1.)

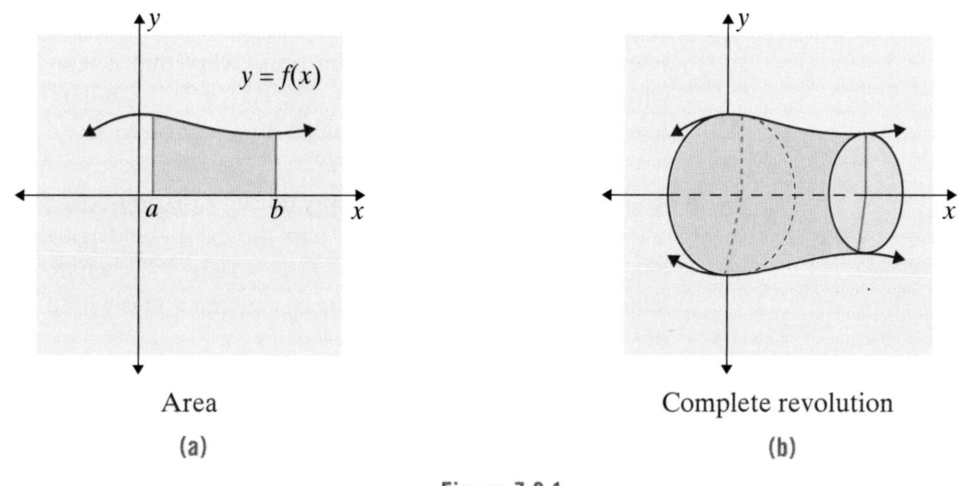

Figure 7.6.1

From geometry, we know that the area of a circle with radius r is given by $A = \pi r^2$ and that the volume of a cylinder with a radius r and height h is given by $V = \pi r^2 h$. (See Figure 7.6.2.)

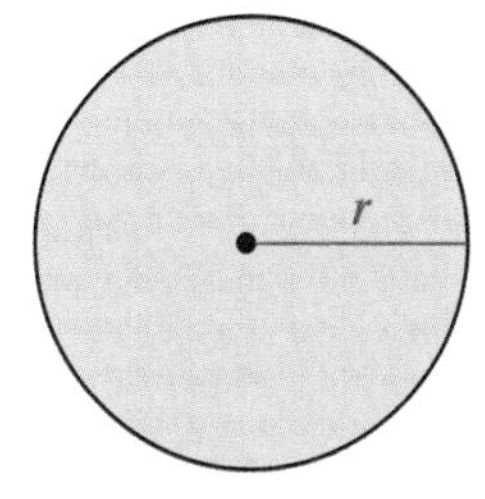

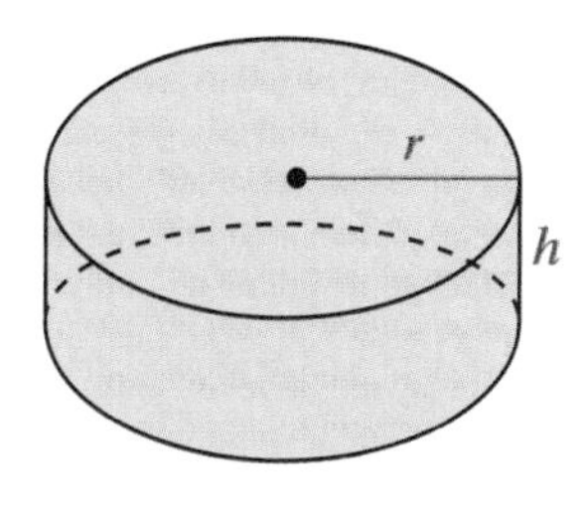

Figure 7.6.2 Area of a circle $A = \pi r^2$ Volume of a cylinder $V = \pi r^2 h$

For a solid of revolution, each cross section is a circle and each circle has a radius $f(x)$ that depends on x. The area of such a circular cross section is $A = \pi\left[f(x)\right]^2$. (See Figure 7.6.3(a).)

In Figure 7.6.3(b) the cylindrical cross section with radius $f(x)$ and height Δx is shown to have volume $V = \pi\left[f(x)\right]^2 \Delta x$.

Figure 7.6.3

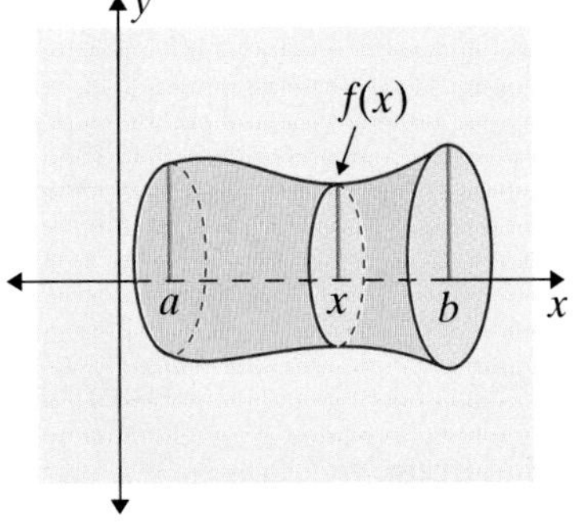

(a) Area of a circular cross section

$$A = \pi\left[f(x)\right]^2$$

(b) Volume of a cylindrical cross section

$$V = \pi\left[f(x)\right]^2 \Delta x$$

Now we partition the interval $[a, b]$ into n subintervals, each of width Δx, and choose a value in each subinterval: $c_1, c_2, \ldots, c_n$. The Riemann sum of the volumes of the cylinders with radii $f(c_1), f(c_2), \ldots, f(c_n)$ and height Δx is given by

$$S_n = \pi\left[f(c_1)\right]^2 \Delta x + \pi\left[f(c_2)\right]^2 \Delta x + \ldots + \pi\left[f(c_n)\right]^2 \Delta x.$$

The limit of this sum as $n \to +\infty$ is defined to be the definite integral that is the **volume of the solid of revolution.**

Volume of a Solid of Revolution

If $y = f(x)$ is a nonnegative continuous function on the interval $[a, b]$, then the volume of the solid formed by revolving the region bounded by the graph of the function and the x-axis $(a \le x \le b)$ about the x-axis is given by

$$V = \int_a^b \pi\left[f(x)\right]^2 dx.$$

Example 1: Finding the Volume of a Solid of Revolution

Find the volume of the solid of revolution generated by revolving the region under the curve $y = \sqrt{x}$ from $x = 0$ to $x = 4$ about the axis.

Solution: $V = \int_0^4 \pi [f(x)]^2 \, dx$

$$= \int_0^4 \pi \left(\sqrt{x}\right)^2 dx$$

$$= \pi \int_0^4 x \, dx$$

$$= \pi \left(\frac{x^2}{2}\right)\Bigg]_0^4$$

$$= 8\pi$$

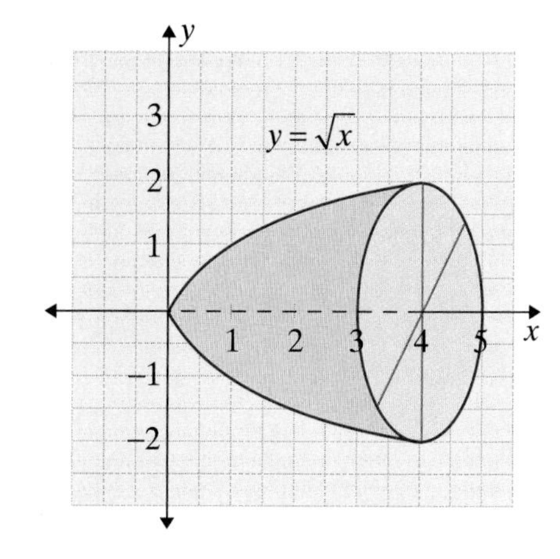

The volume is 8π cubic units.

Example 2: Finding the Volume of a Solid of Revolution

If the region under the line $y = \frac{r}{h}x$ from $x = 0$ to $x = h$ is revolved about the x-axis, a circular cone is formed. Find the volume of this cone.

Solution: $V = \int_0^h \pi [f(x)]^2 dx$

$$= \pi \int_0^h \left(\frac{r}{h}x\right)^2 dx$$

$$= \pi \frac{r^2}{h^2} \int_0^h x^2 dx$$

$$= \pi \frac{r^2}{h^2} \left(\frac{x^3}{3}\right)\Bigg]_0^h$$

$$= \pi \frac{r^2}{h^2} \left(\frac{h^3}{3}\right)$$

$$= \frac{1}{3}\pi r^2 h$$

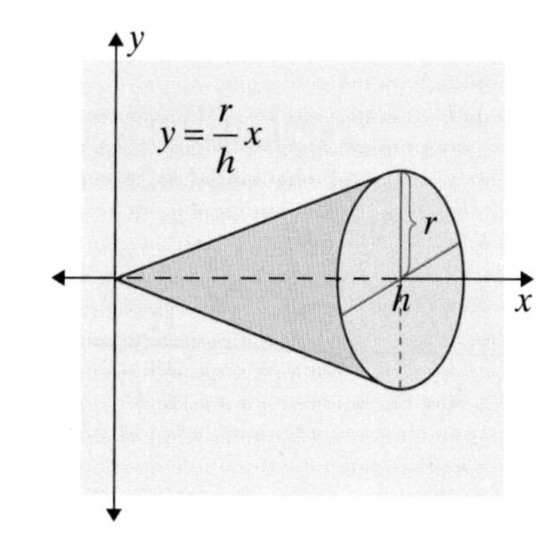

Thus the volume is given by the standard formula $\frac{1}{3}\pi r^2 h$.

Example 3: Finding the Volume of a Solid of Revolution

The region bounded by the parabola $y = 9 - x^2$ and the x-axis is revolved about the x-axis. Find the volume of the solid of revolution that is generated.

Solution: Set $y = 0$ to find the points where the parabola intersects the x-axis.

$$9 - x^2 = 0$$
$$x^2 = 9$$
$$x = \pm 3$$

$$\begin{aligned}
V &= \int_{-3}^{3} \pi\left[f(x)\right]^2 dx \\
&= \pi\int_{-3}^{3} \left(9 - x^2\right)^2 dx \\
&= \pi\int_{-3}^{3} \left(81 - 18x^2 + x^4\right) dx \\
&= \pi\left(81x - \frac{18x^3}{3} + \frac{x^5}{5}\right)\Bigg]_{-3}^{3} \\
&= \pi\left[\left(243 - 162 + \frac{243}{5}\right) - \left(-243 + 162 - \frac{243}{5}\right)\right] \\
&= \pi\left(486 - 324 + \frac{486}{5}\right) \\
&= \frac{1296\pi}{5}
\end{aligned}$$

The volume is $\frac{1296\pi}{5}$ cubic units.

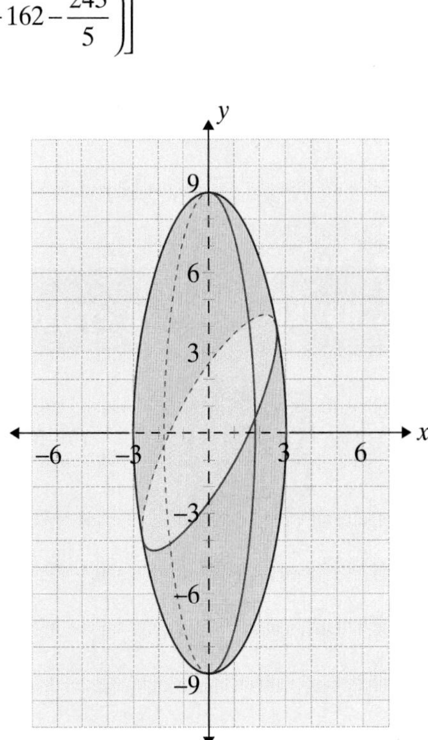

We have considered only solids of revolution in which a region is rotated about the x-axis. The formula we used would need some adjustment if the region were rotated about some other horizontal line, since the radius of revolution would be represented by some expression other than $f(x)$. Such solids of revolution will not be considered in this text.

7.6 Exercises

Find the volume of the solid generated when the regions bounded by the graphs of the given equations and the x-axis are rotated about the x-axis.

1. $y = x,\ x = 0,\ x = 2$

2. $y = 3x,\ x = 1,\ x = 3$

3. $y = 2\sqrt{x},\ x = 0,\ x = 4$

4. $y = \sqrt[3]{x},\ x = 0,\ x = 8$

5. $y = e^{x},\ x = -1,\ x = 2$

6. $y = e^{-x},\ x = -1,\ x = 2$

7. $y = 1 - x^2,\ x = -1,\ x = 1$

8. $y = 4 - x^2,\ x = -2,\ x = 2$

9. $y = \left(16 - x^2\right)^{\frac{1}{2}},\ x = 0,\ x = 4$

10. $y = \sqrt{3 - x^2},\ x = 0,\ x = 1$

11. $y = \frac{4}{x},\ x = 1,\ x = 3$

12. $y = \frac{2}{x},\ x = 1,\ x = 2$

13. $y = \frac{1}{\sqrt{x}},\ x = 1,\ x = 6$

14. $y = \frac{2}{\sqrt{x}},\ x = 1,\ x = 5$

15. $y = x + \sqrt{x},\ x = 1,\ x = 4$

16. $y = \sqrt{x} - x,\ x = 0,\ x = 1$

Hawkes Learning Systems: Essential Calculus

Volume

Chapter 7 Index of Key Ideas and Terms

Section 7.1 Integration by Parts

Formula for Integration by Parts page 556 - 558

$$\int u \cdot dv = u \cdot v - \int v \cdot du$$

Using the Box-Type Format for an Aid

u = (first part)	dv = (second part)
du = (differential of first part)	v = (integral of second part)

Section 7.2 Annuities and Income Streams

Annuities pages 564 - 566

An **annuity** consists of equal payments made at equal time intervals.

In the case of an **ordinary annuity**, the payments are made at the end of each time period.

For an **annuity due**, the payments are made at the beginning of each time period.

Future Value of an Annuity

The amount (or future value) of an annuity at the end of N time periods is approximated by the integral

$$\int_0^N Pe^{r(N-t)}dt = \frac{P}{r}\left(e^{rN} - 1\right),$$

where P is the number of dollars invested each time period, and r is the interest rate (as a decimal) per time period.

Income Streams pages 566 - 569

Future Value of an Income Stream

The amount (or future value) of an income stream at the end of T years is given by the integral

$$A = \int_0^T R(t)e^{r(T-t)}dt,$$

where $R(t)$ is the rate of flow of revenue at time t, r is the annual interest rate (as a decimal), and interest is compounded continuously.

continued on next page ...

Section 7.2 Annuities and Income Streams (continued)

Present Value of an Income Stream

The present value of an income stream over T years is given by the integral

$$PV = \int_0^T R(t)e^{-rt}\,dt,$$

where $R(t)$ is the rate of flow of revenue at time t, r is the annual interest rate (as a decimal), and interest is compounded continuously.

Section 7.3 Tables of Integrals

Table of Integrals pages 572 - 574

Elementary Forms
Forms Involving $(ax+b)$
Forms Involving $(ax+b)(cx+d)$
Forms Involving $\sqrt{ax+b}$
Forms Involving (x^2-a^2)
Forms Involving $\sqrt{x^2 \pm a^2}$
Forms Involving e^{kx}
Forms Involving $\ln x$

Section 7.4 Improper Integrals

Improper Integral page 579

The integral $\int_0^{+\infty} f(x)\,dx$ is called an improper integral. This integral is defined as the following limit:

$$\int_0^{+\infty} f(x)\,dx = \lim_{b\to+\infty}\int_a^b f(x)\,dx.$$

If $\lim_{b\to+\infty}\int_a^b f(x)\,dx$ exists, then the improper integral is said to be **convergent**.

If $\lim_{b\to+\infty}\int_a^b f(x)\,dx$ does not exist, then the improper integral is said to be **divergent**.

Improper Integrals on a TI-84 Plus pages 583 - 585

Section 7.5 Probability

Probability pages 588 - 589

Probability Distribution

Every probability distribution with n outcomes for a discrete random variable x must have the following two characteristics:

1. The probability of each outcome is between 0 and 1:
$$0 \le P(x) \le 1.$$
2. The sum of the probabilities of the outcomes is 1:
$$P(x_1) + P(x_2) + \ldots + P(x_n) = 1.$$

Probability Density Function pages 589 - 591

If x is a continuous random variable distributed on the interval $[a, b]$, then f is a probability density function for x if

1. $f(x) \ge 0$ for all x in $[a, b]$, and
2. $\int_a^b f(x)\,dx = 1.$

Note: For a continuous random variable, the probability that x will have any one specific value is 0.

Exponential Distributions pages 591 - 593

A continuous random variable x is exponentially distributed if it has a probability density function of the form

$$f(x) = ke^{-kx} \text{ and } \int_0^\infty ke^{-kx}\,dx = 1.$$

Standard Normal Distribution

The standard normal distribution is represented by the probability density function

$$f(x) = \frac{1}{\sqrt{2\pi}} e^{-\frac{x^2}{2}}$$

on the interval $(-\infty, +\infty)$.

Expected Value pages 593 - 595

Suppose the function f defined on the interval $[a, b]$ is the probability density function associated with a continuous random variable x. Then the expected value of x is

$$E(x) = \int_a^b x \cdot f(x)\,dx.$$

$E(x)$ is often denoted with the Greek letter *mu* (μ) and is also called the mean value of x.

Section 7.6 Volume

Solid of Revolution pages 598 - 602

Volume of a Solid of Revolution

If $y = f(x)$ is a nonnegative continuous function on the interval $[a, b]$, then the volume of the solid formed by revolving the region bounded by the graph of the function and the x-axis $(a \le x \le b)$ about the x-axis is given by

$$V = \int_a^b \pi[f(x)]^2\, dx.$$

Chapter 7 Review

For a review of the topics and problems from Chapter 7, look at the following lessons from *Hawkes Learning Systems: Essential Calculus.*

Integration by Parts
Annuities and Income Streams
Tables of Integrals
Improper Integrals
Probability
Volume

Chapter 7 Test

In Exercises 1 – 8, use the technique of integration by parts to evaluate the integral.

1. $\int xe^{1.6x}\,dx$

2. $\int (x+5)e^{3x}\,dx$

3. $\int x\sqrt{3x+1}\,dx$

4. $\int x(2x+1)^{-3}\,dx$

5. $\int (x-3)(x+6)^4\,dx$

6. $\int \sqrt[3]{x}\ln 6x\,dx$

7. $\int x^{-3}\ln x\,dx$

8. $\int (x+2)\ln 7x\,dx$

In Exercises 9 – 14, use the technique of integration by parts to evaluate each definite integral.

9. $\int_1^2 4xe^{-x}\,dx$

10. $\int_0^3 (x-1)e^{-0.5x}\,dx$

11. $\int_0^5 x\sqrt{3x+1}\,dx$

12. $\int_{-1}^1 (x-4)(x+3)^3\,dx$

13. $\int_1^8 \sqrt[3]{x}\ln x\,dx$

14. $\int_1^2 (2x-1)\ln 3x\,dx$

In Exercises 15 – 18, use Table 7.3.1 to find the integrals.

15. $\int \sqrt{x^2+121}\,dx$

16. $\int \frac{x}{4x+3}\,dx$

17. $\int \frac{1}{6x(40-x)}\,dx$

18. $\int \frac{1}{2+5e^{0.2x}}\,dx$

In Exercises 19 – 21, determine whether the improper integrals are convergent or divergent, and then evaluate those that are convergent.

19. $\int_0^\infty \frac{1}{3+7x}\,dx$

20. $\int_0^\infty -\frac{2}{e^x}\,dx$

21. $\int_1^\infty \frac{5}{(2x-1)^3}\,dx$

22. Show that each of the functions is a probability density function on the given interval. Use your calculator.

a. $f(x)=\frac{2}{45}\left(2x+\frac{3}{2}x^2\right)$; $[0, 3]$

b. $f(x)=\frac{10}{49}\left(x^{\frac{3}{2}}-x\right)$; $[1, 4]$

23. Using your calculator, determine the number a so that the function $f(x) = axe^{-0.5x}$ is a probability density function on the interval $[0, +\infty)$.

In Exercises 24 – 27, find the volumes of the solid generated when the region bounded by the given graphs is rotated about the x-axis.

24. $f(x) = x^3 + 1;\ x = 0,\ x = 2$

25. $f(x) = 6 - x^2;\ x = -1,\ x = 2$

26. $f(x) = 1 + e^{-x};\ x = 0,\ x = 3$

27. $f(x) = \sqrt{4 - x};\ x = 0,\ x = 4$

28. Annuity. Seven thousand dollars is invested in an annuity account each year for 4 years. Find the approximate balance in the account after 4 years if the rate of interest is 6.8 percent compounded continuously.

29. Income stream. The rate of income from a number of vending machines is estimated by $f(t) = 80e^{0.015t}$ thousand dollars per year. If the receipts are reinvested into an account paying 5.5 percent compounded continuously, find the amount of the income stream after 5 years.

30. Income stream. Find the value of an income stream in 6 years if the rate of flow of income at time t is estimated by the function $R(t) = 2000 + 800t$ dollars per year and is reinvested at 8 percent compounded continuously.

31. Present value. An income stream is expected to generate revenues at a rate given by $R(t) = 18 + 2.4t$ thousand dollars per year for the next 6 years. Find the present value of the stream if the interest rate is 8.5 percent compounded continuously.

32. Probability. At a local restaurant, the proportion of orders that are filled within 15 minutes at dinner time is given by the probability density function $f(x) = \frac{12}{5}(2x^2 - x^3)$, where $0 \le x \le 1$. What is the probability that between 50 and 60 percent of the orders will be filled in 15 minutes?

33. Probability. The average waiting time in minutes at a local bank between noon and 1 P.M. has the probability density function $f(t) = \frac{.481}{t+1}$ for $0 \le t \le 7$. What is the average waiting time for customers at that time of day?

34. Probability. The length of time t in minutes it takes for a response to the new pain killer Melochtabs has a probability density function $f(t) = \frac{t+3}{1980}$. Assuming $0 \le t \le 60$, what is the mean response time to a dose of Melochtabs?

35. Probability. The probability that a AAA battery will last some number of months between 0 and 48 has a density function $f(t) = \frac{e^{0.0003t^2} - 1}{13.79}$. What is the mean life-time of such a battery?

Cumulative Review

In Exercises 1 – 3, you are **a.** *to determine a certain limit and then* **b.** *determine a function f(x) and a value x = a so the limit in part* **a.** *is $f'(a)$.*

1. $\lim\limits_{h\to 0}\dfrac{(5+h)^2-25}{h}$ **2.** $\lim\limits_{h\to 0}\dfrac{\sqrt{4+h}-2}{h}$ **3.** $\lim\limits_{h\to 0}\dfrac{e^{2+h}-e^2}{h}$

For Exercises 4 – 6, a one-year contract for the use of a certain cell phone costs $15 per month plus $0.15 per minute with no other charges.

4. What is the company's monthly revenue function in terms of the number of minutes, x, of monthly use? What is the firm's marginal revenue?

5. The firm assumes that a phone costs $6.00 per month and if it pays two cents a minute for customer calls, what is the firm's cost per phone function $C(x)$ for x minutes used by a customer? What is its marginal cost function?

6. What is the company's marginal profit function?

In Exercises 7 – 12, determine a formula for $f'(x)$.

7. $f(x)=x^{.5}-3x^{-1}+12x^{.25}$ **8.** $f(x)=(2x+1)^5$ **9.** $f(x)=2x\sqrt{3x+1}$

10. $f(x)=2e^x(5x-1)$ **11.** $f(x)=\dfrac{3x+\ln x}{4x-1}$ **12.** $f(x)=2^x+\log(x)$

For Exercises 13 – 15, it is estimated that t years from 2000, the level of population of robins in Florida, Georgia, and South Carolina is given by $P(x)=\dfrac{3}{1+2e^{-x}}$ where $P(x)$ is in millions.

13. Determine the marginal population function $P'(t)$.

14. Determine the inflection point (x, y). Interpret the meaning of this value.

15. Is there a limiting value for P? If so, what is it?

In Exercises 16 and 17, for which intervals is f(x) increasing and/or decreasing?

16. $f(x)=2xe^{-x}+2$ **17.** $f(x)=(x-3)(x+6)^2$

18. Calculate $\int_1^4 (x+1)^3\,dx$. **19.** Calculate $f'(2)$ for $f(x)=2x\ln(x+1)$.

20. Given $f(x)=1+2x$ and $g(x)=22-x$,

a. Write a single integral which will give the area in Quadrant I between $f(x)$ and $g(x)$.

b. Calculate this area.

For problems 21 – 24, integrate by parts or u-substitution.

21. $\int 6x\sqrt{2x+1}\,dx$ **22.** $\int 3ye^{6y}dy$

23. $\int \sqrt{2x+1}\ln(2x+1)dx$ **24.** $\int \frac{(\ln x)^3\,dx}{x}$

25. What is the eventual sum accumulated by an income stream of $1200 per year for 15 years assuming continuous compounding at an annual rate of 5%?

26. A hotdog stand profits by $90 per day (or $32,850 per year) over salaries and expenses. Assuming continuous compounding at an annual rate of 4%, what will be the accumulated profits in 6 years?

In Exercises 27 – 29, use a table of integrals formulas for the indefinite integrals.

27. $\int \frac{1}{\sqrt{x^2+3^2}}dx$ **28.** $\int \frac{1}{x^2-9}dx$

29. $\int \frac{1}{x^2-5x+6}dx$ [Hint: $x^2-5x+6=(x-3)(x-2)$]

In Exercises 30 – 32, determine the improper definite integrals.

30. $\int_1^{+\infty} 2xe^{-x}dx$ **32.** $\int_0^{+\infty} (x+1)e^{-2x}dx$

31. $\int_1^{+\infty} e^{-x}\ln x dx$ (Hint: Use a calculator.)

33. Find a sequence of x-values which demonstrate that $\lim_{b\to\infty}(\ln(\ln b))=\infty$.

34. Determine a value for k so that $f(x) = k(3x + 1)$ is a probability density function on the interval [0, 5].

35. Sketch $y=3x+1$ in the window [–1, 6] by [–2, 20]. What is the average value of y over the interval [0, 5]?

36. A distribution for a random variable x is $f(x)=1.04(x-5)e^{5-x}$ for $5\le x\le 10$. What is the probability that $5\le x\le 7$?

In Exercises 37 – 40, determine the area of the solid of revolution generated by revolving the region under the curve specified from x = a to x = b about the x-axis.

37. $f(x)=(5-x)e^{.5x}$; $(a, b)=(0, 2)$ **38.** $f(x)=3+\ln x$; $(a, b)=(1, 4)$

39. $f(x)=\sqrt{9-x^2}$; $(a, b)=(0, 3)$ **40.** $f(x)=2+x+x^2$; $(a, b)=(0, 1)$

Multivariable Calculus

CHAPTER 8

Did You Know?

Born the son of working class Germans, Carl Friedrich Gauss (1777 – 1855) quickly showed the genius he would apply to mathematics as well as many other scientific and practical fields. He is reported to have discovered a complicated arithmetic error in his father's records at the age of three. Later, his elementary school class was given the time-consuming task of totaling the first one hundred integers by hand. Gauss instantaneously derived the formula for the sum of the first n integers to arrive at the correct answer, 5050.

Gauss spent his time in college preparatory school studying a wide range of topics including geometry, number theory, and astronomy. He likely invented the principle of least squares as an analytical tool in number theory. This was between 1792 and 1795, and his invention of this method was never published. In 1801 Italian astronomer Giuseppe Piazzi (1746 – 1826) discovered the asteroid *Ceres.* He subsequently was unable to relocate his discovery and many astronomers of the day joined in the chase. Despite having very little data with which to work, Gauss plotted the asteroid's future location with extreme accuracy. Although his feat was amazing, the young mathematician never disclosed his methods. Nearly ten years later, Adrien-Marie Legendre (1752 – 1833) rediscovered and published the method of least squares.

The method of least squares can be used to perform regression on collections of points resembling many different functions. Gauss used least squares approximation on an elliptic model to plot *Ceres'* location, but this text will concentrate on linear regression with a least squares fit since it is the method commonly used in science and industry.

Gauss' contributions can be seen in nearly every branch of advanced mathematics; however, he also had insight that can be appreciated by students at a beginning level. You were likely taught the Fundamental Theorem of Algebra and the Fundamental Theorem of Arithmetic in a previous course. The Fundamental Theorem of Algebra says that a polynomial of nth degree will have n complex roots up to multiplicity. Gauss gave the first proof of this theorem for his doctoral dissertation. He made a monumental discovery in number theory at the age of 24. This theory, the Fundamental Theorem of Arithmetic, says that every integer greater than one can be uniquely written as a product of primes. Indeed, the work of this mathematician was too far-reaching to describe him as a geometer, number theorist, analyst, or astronomer; for this reason, Carl Friedrich Gauss is known simply as the "Prince of Mathematics."

In Chapter 8 we discuss calculus as it relates to functions of more than one variable. For example, the function $C(x, y) = 12x^2 + 3y^2 + 100$ might represent the cost of producing two products, and our objective might be to find the minimum cost of production subject to certain manufacturing restrictions on the numbers of units of each product.

The chapter begins with graphing single points and planes in three dimensions. We discuss the curves determined when a plane intersects a surface and the slopes of the tangent lines to these curves. These slopes are found by a process called partial differentiation. The rules and corresponding notation for partial differentiation are defined in Section 8.2. With partial derivatives we develop two methods for finding the local extrema of functions of two variables whose graphs are "smooth" surfaces. The first method is called the *D*-Test and is explained in Section 8.3. The method of Lagrange multipliers, discussed in Section 8.4, involves local extrema where the variables are subject to certain constraints.

In statistics a scatter diagram is the graph of points that represent data relating two variables. In Section 8.5, with calculus and the method of least squares, we show how to find the line of best fit for the data. Such lines can be helpful in predicting future data.

In Section 8.6 we develop the techniques of double integration for functions of two variables and show how these double integrals can be used to find volumes of solids.

8.1 Functions of Several Variables

Objectives

After completing this section, you will be able to:

1. *Evaluate values for functions of two or more variables.*
2. *Graph a function of two variables in a 3D coordinate system.*

Up to this point we have studied functions of one variable. For example, the cost function

$$C(x) = x^2 + 3x + 1000$$

depends only on the single variable *x*, where *x* represents the number of items produced. Similarly, the distance function

$$s(t) = 180t - 16t^2$$

depends only on the single variable *t*, where *t* represents time. With functions of one variable such as these, we have discussed concepts related to differentiation such as maxima and minima, developed integration techniques, worked with a wide variety of applications, and learned detailed graphing techniques.

Now we want to consider several of these same ideas as they relate to functions that have more than one variable. For example, suppose that a company produces two products, x units of one and y units of the other. Then the cost function depends on both x and y, and we might write

$$C(x, y) = 20,000 + 10x + 30y,$$

where \$20,000 represents the fixed costs.

If another company produces three products, the cost function might be in the form

$$C(x, y, z) = 500 + x + e^{0.2y} + 10.6z^2,$$

where x, y, and z represent the numbers produced of each of the three types of items. Both cost functions are **functions of several variables**.

In the current discussion we will deal primarily with functions of two variables, although the basic ideas can easily be expanded to include functions of more than two variables.

Domain and Range of $f(x, y)$

If D is a set of ordered pairs of real numbers and for each ordered pair (x, y) in D there corresponds a unique real number f(x, y), then f is called a ***function of x and y****. The set D is the* ***domain*** *of the function. The set of all the values of f(x, y) is called the* ***range*** *of the function.*

We sometimes write

$$z = f(x, y)$$

and call z the **dependent variable** and x and y the **independent variables**.

Example 1: Evaluating $f(x, y)$

For $f(x, y) = 3x + y^2$, find the following:

a. $f(2, 3)$.

b. $f(2, \sqrt{2})$.

c. $f(0, 0)$.

Solutions: a. $f(2, 3) = 3 \cdot 2 + 3^2 = 15$ Substitute $x = 2$ and $y = 3$.

b. $f(2, \sqrt{2}) = 3 \cdot 2 + (\sqrt{2})^2 = 8$ Substitute $x = 2$ and $y = \sqrt{2}$.

c. $f(0, 0) = 3 \cdot 0 + 0^2 = 0$ Substitute $x = 0$ and $y = 0$.

Example 2: Evaluating $f(x, y)$

For $f(x, y) = e^{\sqrt{x}} + \ln y$, **a.** find the domain and **b.** find $f(0, 1)$.

Solutions: a. For $\sqrt{x}$ to be defined, we must have $x \geq 0$.
For $\ln y$ to be defined, we must have $y > 0$.
So the domain is $\{(x, y) | x \geq 0 \text{ and } y > 0\}$.

b. $f(0, 1) = e^{\sqrt{0}} + \ln 1 = e^0 + 0 = 1$ Substitute $x = 0$ and $y = 1$.

Example 3: Evaluating Revenue Influenced by Two Variables

A pharmacy sells two brands of aspirin. Brand A sells for \$1.25 per bottle and Brand B sells for \$1.50 per bottle.

a. What is the revenue function for aspirin?

b. What is the revenue for aspirin if 100 bottles of Brand A and 150 bottles of Brand B are sold?

Solutions: a. Let x = the number of bottles of Brand A sold, and
y = the number of bottles of Brand B sold.
Then the revenue function is

$$R(x, y) = 1.25x + 1.50y.$$

b.

$$\begin{aligned} R(100, 150) &= 1.25(100) + 1.50(150) \\ &= 125 + 225 \\ &= \$350 \end{aligned}$$

Example 4: Characteristics of a Box

Suppose that a box has a square base and an open top.

a. Write a function of two variables that represents the volume of the box.

b. Write a function of two variables that represents the surface area of the box.

c. Determine the surface area if $x = 4$ and $y = 5$.

Solutions: Since the base is square, the length and width of the box are equal, as illustrated in the previous page.
Let x = length of the box,
x = width of the box, and
y = height of the box.

y
x
x

a. The formula for the volume of a rectangular solid is $V = l \cdot w \cdot h$, where l = length, w = width, and h = height. Thus the function that represents the volume of this box is

$$V(x, y) = x \cdot x \cdot y = x^2 y.$$

b. The surface area consists of the sum of the areas of the bottom surface, the front and back faces, and the left and the right faces. For this box, we have

x^2 = area of bottom surface and
xy = area of each of the other four faces.

Thus the total surface area is represented by the function

$$S(x, y) = x^2 + 4xy.$$

c. Evaluate this function in two variables at $x = 4$ and $y = 5$. Then

$$S(4, 5) = 4^2 + 4(4)(5) = 16 + 80 = 96.$$

Cobb-Douglas Production Formula

Economists use a formula called the **Cobb-Douglas Production Formula** to model the production levels of a company (or a country). The formula is

$$P(x, y) = kx^a y^{1-a},$$

where P is the total units produced, x is a measure of labor units, y is a measure of capital invested, and k is a constant that varies from product to product. Notice that the sum of the exponents on x and y is 1. That is, $a + (1 - a) = 1$.

Example 5: Using the Cobb-Douglas Production Formula

Suppose that the function $P(x, y) = 500x^{0.3}y^{0.7}$ represents the number of units produced by a company with x units of labor and y units of capital.

a. How many units of a product will be manufactured if 300 units of labor and 50 units of capital are used?

b. How many units will be produced if twice the number of units of labor and capital are used?

Solutions: a. $P(300, 50) = 500(300)^{0.3}(50)^{0.7}$

$\approx 42{,}794$ units produced — Fractional units are not counted.

continued on next page ...

b. If the number of units of labor and capital are both doubled, then

$$x = 2 \cdot 300 = 600 \text{ and } y = 2 \cdot 50 = 100.$$

So

$$P(600, 100) = 500(600)^{0.3}(100)^{0.7} \quad \text{Substitute } x = 600 \text{ and } y = 100.$$
$$\approx 85{,}588 \text{ units produced.}$$

Thus we see that production is doubled if both labor and capital are doubled.

Graphs in Three Dimensions

Graphs of functions of two variables require a three-dimensional coordinate system. We will use the Cartesian system with three axes, x-axis, y-axis, and z-axis, each perpendicular to the other two. Points are represented by **ordered triples** in the form (x, y, z). Figure 8.1.1 illustrates the location of the point $(1, 2, 3)$.

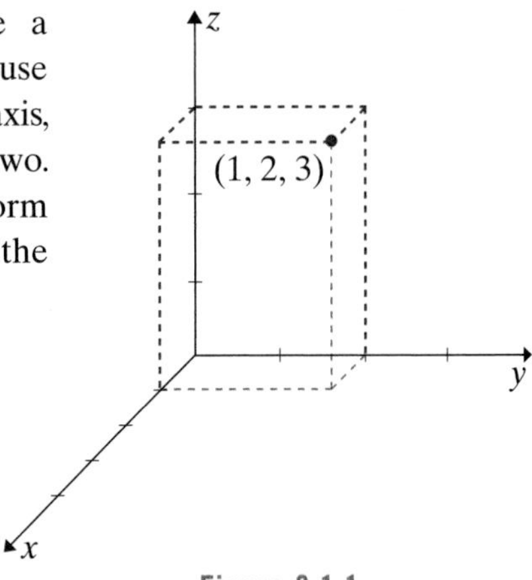

Figure 8.1.1

Intuitively, picture a corner of your classroom where two walls and the floor come together. The x-axis is where the floor meets one wall, the y-axis is where the floor meets the other wall, and the z-axis is where the two walls meet. If several students stand, their feet are in the xy-plane, and the tops of their heads can be considered to be points in the space of the room represented by ordered triples depending on how far they are standing from each wall and their height. (See Figure 8.1.2.)

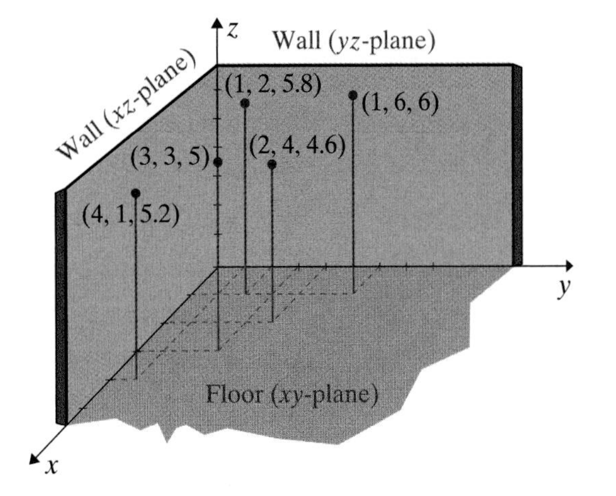

Figure 8.1.2

Example 6: Graphing 3D Points

Draw a three-dimensional coordinate system; then graph and label the following points:

$$A(1, -2, 3), \quad B(0, 0, 4), \quad C(0, 3, 2), \quad \text{and} \quad D(2, 2, -1).$$

Solution:

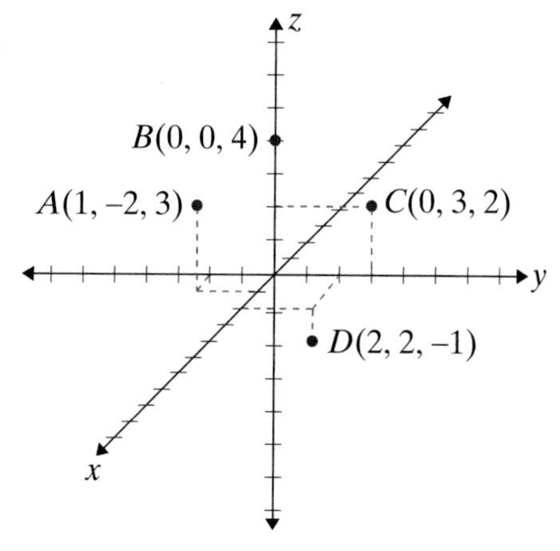

A point in space with coordinates (a, b, c) can be thought of as the intersection of three planes, each plane perpendicular to the other two. The equations of these planes are $x = a$ (where y and z can have all real values), $y = b$ (where x and z can have all real values), and $z = c$ (where x and y can have all real values). Portions of the graphs of these planes are illustrated in Figure 8.1.3.

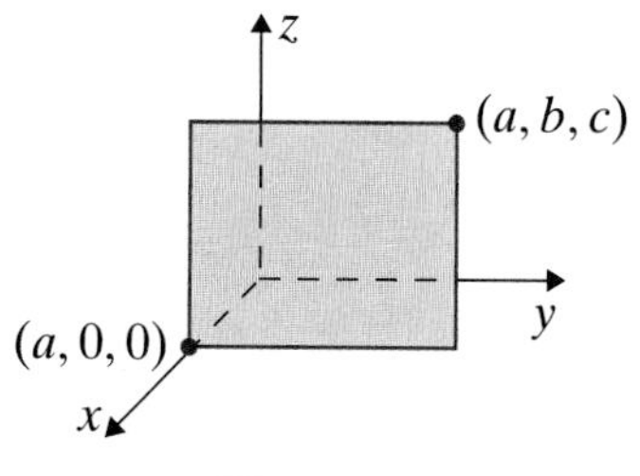

Plane $x = a$
(parallel to the yz-plane)
(a)

z
(a, b, c)
(0, b, 0)
y
x

Plane $y = b$
(parallel to the xz-plane)
(b)

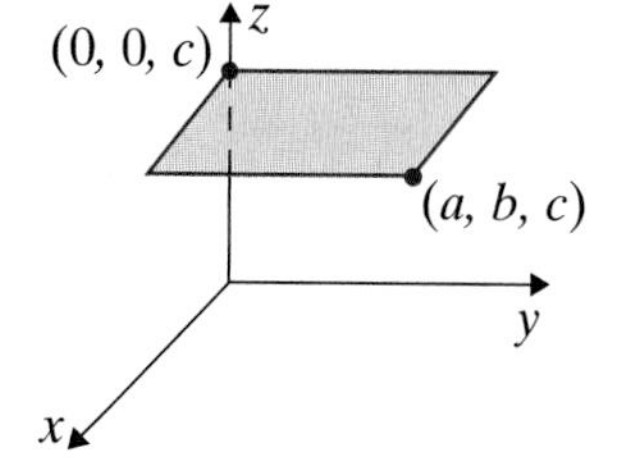

Plane $z = c$
(parallel to the xy-plane)
(c)

Figure 8.1.3

Example 7: Graphing 3D Planes

Set up a three-dimensional coordinate system to graph each of the following planes:

a. $x = 5$

b. $y = -2.$

Solutions: a.

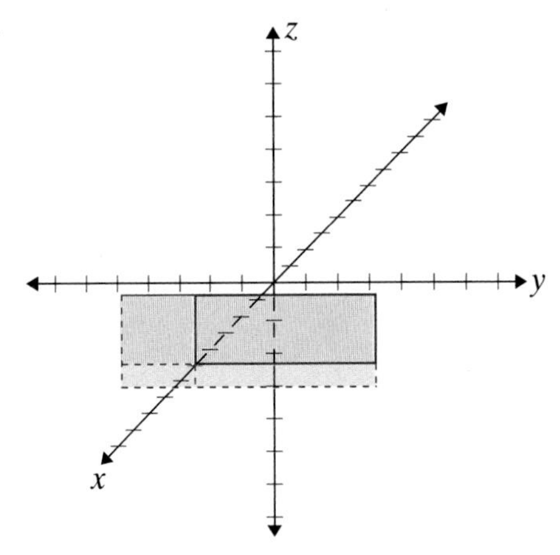

Note: $x = 5$ is a plane parallel to the yz-plane.

b.

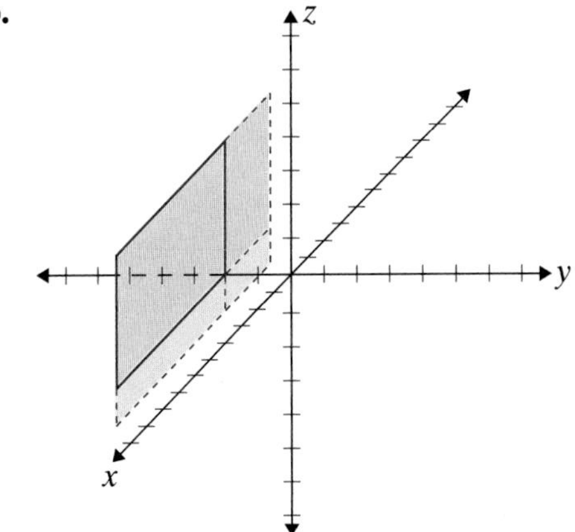

Note: $y = -2$ is a plane parallel to the xz-plane.

In each graph, the contrasting color corresponds to negative z-values

Note: In each case, only a portion of the plane is graphed. You might have chosen to sketch a different portion. For your sketch to have a proper perspective, the lines that are shown should be drawn parallel to the coordinate axes.

In general, the graph of a function of two variables is a **surface** and can be difficult to draw. In this text, drawing such surfaces is not part of the exercises; however, the graphs of several surfaces and their corresponding equations will be presented as aids in understanding certain basic concepts such as local maxima, local minima, and partial derivatives. Portions of four such surfaces are illustrated in Figure 8.1.4.

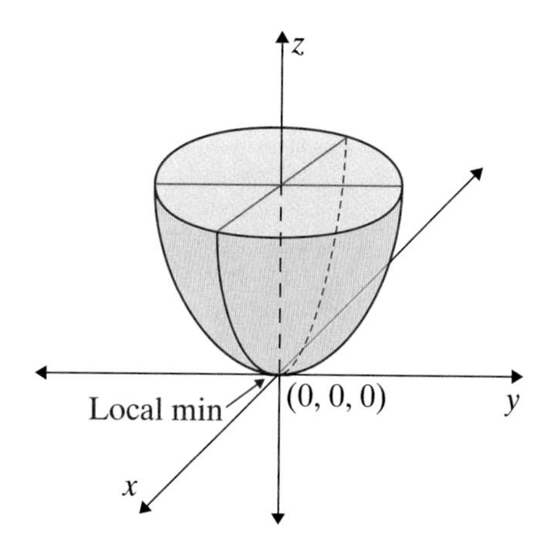

Circular Paraboloid

$z = x^2 + y^2$

(a)

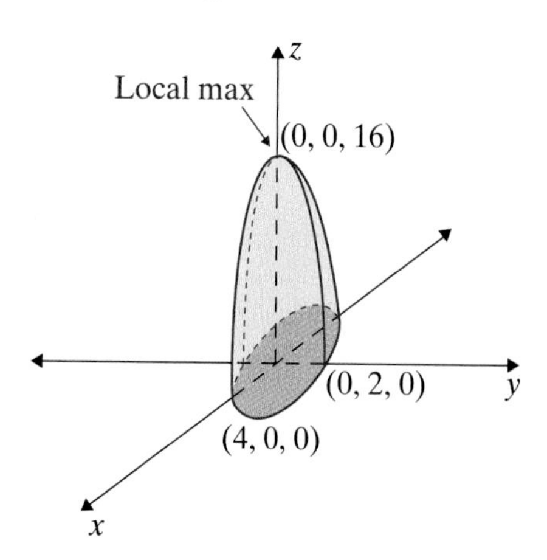

Elliptical Paraboloid

$z = 16 - x^2 - 4y^2$

(b)

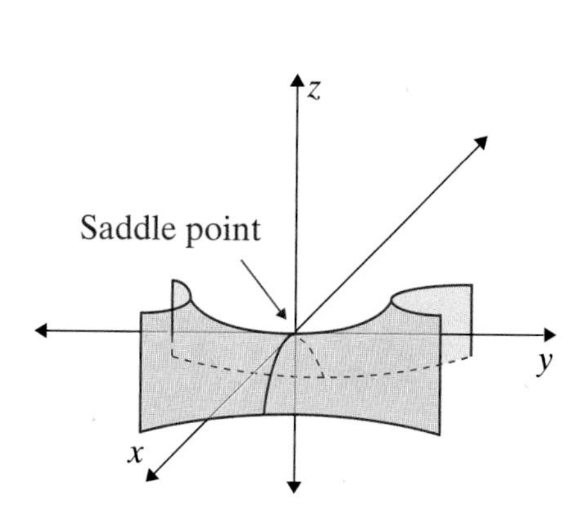

Hyperbolic Paraboloid

$z = y^2 - x^2$

(c)

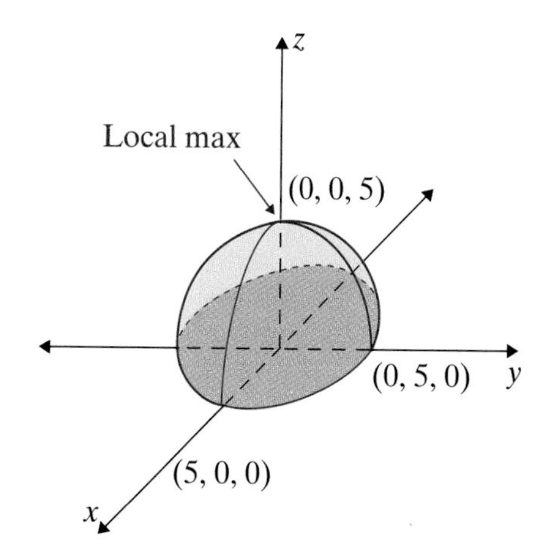

Hemisphere

$z = \sqrt{25 - x^2 - y^2}$

(d)

Figure 8.1.4

8.1 Exercises

In Exercises 1 – 15, find the indicated function values, if possible.

1. $f(x, y) = 12x - 3y + xy$
  a. $f(1, 3)$
  b. $f(0, 4)$

2. $f(x, y) = 7xy - 11x + 9y$
  a. $f(-2, 1)$
  b. $f(3, -2)$

3. $f(x, y) = 4x^2 - 3xy + y^2$
  a. $f(2, 5)$
  b. $f(0, 3)$

4. $g(x, y) = 2xy + 5x^2y + y^3$
  a. $g(-2, 2)$
  b. $g(-1, -1)$

5. $g(x, y) = \dfrac{4x + y}{x - y}$
  a. $g(3, -1)$
  b. $g(2, 2)$

6. $f(x, y) = \dfrac{2x - y}{x^2 + y}$
  a. $f(6, 2)$
  b. $f(3, -9)$

7. $g(x, y) = \dfrac{8x^2y}{\sqrt{2x + y}}$
  a. $g(1, -4)$
  b. $g(-2, 6)$

8. $f(x, y) = \dfrac{5x^2 + y^3}{\sqrt{4x + y^2}}$
  a. $f(-1, 3)$
  b. $f(3, 2)$

9. $g(x, y) = 3xe^{x+y}$
  a. $g(2, 1)$
  b. $g(-1, 0)$

10. $g(x, y) = ye^{4x} + 2xy$
  a. $g(1, 5)$
  b. $g(-1, 3)$

11. $f(x, y) = x \ln xy + y \ln x$
  a. $f(1, e)$
  b. $f(e^2, 1)$

12. $f(x, y) = 2x^3y + \ln y$
  a. $f(-2, e)$
  b. $f(3, 1)$

13. $A(P, r, t) = Pe^{rt}$
  a. $A(1000, 0.08, 4)$
  b. $A(500, 0.06, 5)$

14. $A(P, r, t) = P\left(1 + \dfrac{r}{4}\right)^{4t}$
  a. $A(1500, 0.08, 6)$
  b. $A(800, 0.06, 10)$

15. $S(l, w, h) = 2lw + 2lh + 2wh$
  a. $S(18, 15, 9)$
  b. $S(14, 9, 11)$

In Exercises 16 – 19, draw a three-dimensional coordinate system, and graph and label the given points.

16. $A(0, 0, 3);\ B(2, -1, 0);\ C(4, 2, 1);\ D(3, -2, -2)$

17. $A(0, 1, 0);\ B(3, 0, -4);\ C(1, 1, -3);\ D(-2, 3, 0)$

18. $A(-3, 0, 2)$; $B(0, 0, 2)$; $C(2, -2, 2)$; $D(0, -2, 2)$

19. $A(-3, -2, 1)$; $B(4, 0, -1)$; $C(0, 0, 4)$; $D(-1, 4, -1)$

In Exercises 20 – 25, draw a three-dimensional coordinate system, and graph the given planes.

20. $x = 0$

21. $y = 0$

22. $z = 0$

23. $x = 3$

24. $y = 2$

25. $z = 4$

26. Stock yield. The yield of a stock is given by the function $Y(d, p) = \frac{d}{p}$, where d is the dividends per share of stock and p is the price per share. Find the yield of a stock that sells for $5.88 if the dividend is $1.00.

27. Intelligence quotient. The intelligence quotient (IQ) of a person is determined by $f(M, C) = 100\frac{M}{C}$, where M is the mental age (determined by tests) and C is the actual or chronological age. Find the IQ of a child who is 13 years old and has a mental age of 15.4 years. (Round to the nearest integer.)

28. Cobb-Douglas production. The number of units of a product that are manufactured by a company is given by $f(L, K) = 300L^{0.4}K^{0.6}$, where L is the units of labor and K is the units of capital.

a. How many units of a product will be manufactured by utilizing 30 units of labor and 24 units of capital? (Round to the nearest unit.)

b. How many units will be produced if the number of labor and capital are doubled? (Round to the nearest unit.)

29. Cost. A company manufactures two lawnmower models, standard and self-propelled. The cost of producing each standard mower is $80, and the cost of producing each self-propelled mower is $140. If the fixed costs are $5200, the total cost function is given by $C(x, y) = 5200 + 80x + 140y$, where x is the number of standard and y is the number of self-propelled mowers.

a. Find $C(30, 20)$.

b. Find $C(36, 25)$.

30. Cost. The cost function for producing two models of a product is found to be $C(x, y) = 850 + 32x + 20y$, where x is the number of model A and y is the number of model B. The cost for model A is $32, the cost for model B is $20, and the fixed costs are $850 per week.

a. Find $C(40, 24)$.

b. Find $C(60, 38)$.

31. **Cost.** The cost of producing the standard model of a video recorder is \$160. The cost of producing the deluxe model is \$220.
   a. If a company has weekly fixed costs of \$1360, find the cost function $C(x, y)$, where x is the number of standard models and y is the number of deluxe models.
   b. Find $C(15, 12)$.

32. **Cost.** A company makes two grades of paint, grade I, guaranteed for 5 years, and grade II, guaranteed for 10 years. A gallon of grade I costs \$3.20 to make, while a gallon of grade II costs \$3.90 to make. The weekly fixed costs are \$4500.
   a. Find the cost function $C(x, y)$ for making x gallons of grade I and y gallons of grade II.
   b. What is the cost of making 200 gallons of grade I and 140 gallons of grade II?

33. **Revenue.** A grocery store sells two brands of a product, the "house" brand and a "name" brand. The manager estimates that if she sells the "house" brand for x dollars and the "name" brand for y dollars, she will be able to sell $64 - 20x + 18y$ units of the "house" brand and $52 + 16x - 22y$ units of the "name" brand.
   a. Find the revenue function $R(x, y)$.
   b. What is the revenue if she sells the "house" brand for \$4.00 and the "name" brand for \$4.50?

34. **Revenue.** A pharmacy sells two cold remedies, one a generic remedy and the other a "name" brand. The store manager has determined that he can sell $26 - 6x + 8y$ bottles of the generic remedy and $22 + 5x - 9y$ bottles of the "name" brand if the price is x dollars per bottle and y dollars per bottle, respectively.
   a. Find the revenue function $R(x, y)$.
   b. What is the revenue if the generic remedy is priced at \$2.20 per bottle and the "name" brand is priced at \$3.00 per bottle?

35. **Volume and surface area.** A rectangular box has no top and one partition (see diagram). 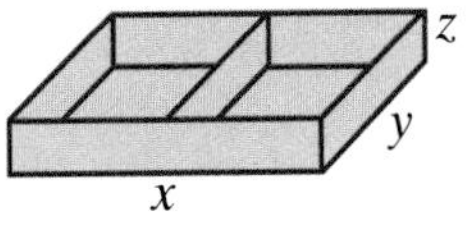
   a. Write a function of three variables for the number of cubic units in the volume of the box.
   b. Write a function of three variables for the number of square units of material needed to construct the box.

36. **Volume and surface area.** A rectangular box has no top and two intersecting partitions (see diagram). 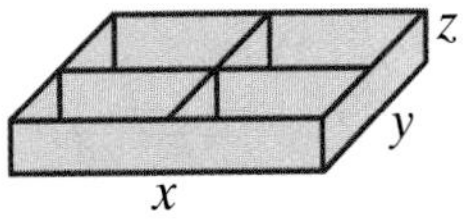
   a. Write a function of three variables for the number of cubic units in the volume of the box.
   b. Write a function of three variables for the number of square units of material needed to construct the box.

37. Compound interest. A deposit of \$1000 is made into a savings account earning r percent interest compounded quarterly. The amount $A(r, t)$ after t years is given by $A(r,t) = 1000\left(1+\frac{r}{4}\right)^{4t}$. Use this function of two variables to complete the following table.

	Number of Years (t)		
Rate (r)	3	5	10
0.06			
0.08			
0.10			

38. Interest compounded continuously. A deposit of \$1000 is made into a savings account earning r percent interest compounded continuously. The amount $A(r, t)$ after t years is given by $A(r, t) = 1000e^{rt}$. Use this function of two variables to complete the following table.

	Number of Years (t)		
Rate (r)	5	8	12
0.080			
0.085			
0.100			

Hawkes Learning Systems: Essential Calculus

Functions of Several Variables

8.2 Partial Derivatives

Objectives

After completing this section, you will be able to:

1. *Find partial derivatives of functions of two and three variables.*
2. *Find second-order partial derivatives.*

For a function of one variable, $y = f(x)$, the first derivative $\frac{dy}{dx} = f'(x)$, if it exists, can be interpreted as the instantaneous rate of change of y with respect to x. Geometrically, $f'(x)$ is the slope of a line tangent to the graph of the function, as shown in Figure 8.2.1.

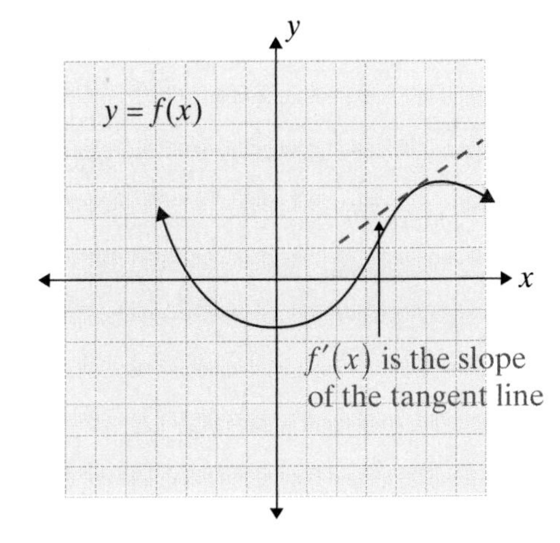

Figure 8.2.1

Now we will consider the same ideas for a function of two variables, $z = f(x, y)$. We want to find the instantaneous rate of change of z with respect to each variable, one at a time. That is, we find a derivative with respect to x by temporarily treating y as a constant and a derivative with respect to y by temporarily treating x as a constant. Derivatives of this type (if they exist) are called **partial derivatives**.

Partial Derivatives

Let $z = f(x, y)$.

a. *The **first partial derivative of f with respect to x** (if it exists) is*

$$\frac{\partial f}{\partial x} = \lim_{h \to 0} \frac{f(x+h, y) - f(x, y)}{h}.$$

b. *The **first partial derivative of f with respect to y** (if it exists) is*

$$\frac{\partial f}{\partial y} = \lim_{k \to 0} \frac{f(x, y+k) - f(x, y)}{k}.$$

c. $\frac{\partial f}{\partial x} = \frac{dz}{dx}$ *and* $\frac{\partial f}{\partial y} = \frac{dz}{dy}$ *in our notation.*

Notice that in the definition of $\frac{\partial f}{\partial x}$, x is changed by h and y is unchanged (treated as a constant):

x is increased by h — y is unchanged

$$\frac{\partial f}{\partial x} = \lim_{h \to 0} \frac{f(x+h, y) - f(x, y)}{h}.$$

Similarly, in the definition of $\frac{\partial f}{\partial y}$, y is changed by k and x is unchanged (treated as a constant):

x is unchanged — y is increased by k

$$\frac{\partial f}{\partial y} = \lim_{k \to 0} \frac{f(x, y+k) - f(x, y)}{k}.$$

The two symbols $\frac{\partial}{\partial x}$ and $\frac{\partial}{\partial y}$ are called **operators**. These operators are similar to $\frac{d}{dx}$ but are used to indicate that the operation of partial differentiation is to be performed relative to the variable in the denominator with other variables held constant.

Geometrically, if $\frac{\partial f}{\partial x}$ or $\frac{dz}{dx}$ exists, it can be interpreted as the slope of a line tangent to the surface represented by the function $z = f(x, y)$. This line is in a plane where y is constant (a plane parallel to the xz-plane) and is also tangent to the curve formed by the intersection of the surface and the plane. Similarly, $\frac{\partial f}{\partial y}$ can be interpreted as the slope of a line tangent to the surface in a plane where x is constant (a plane parallel to the yz-plane). Both situations are illustrated in Figure 8.2.2.

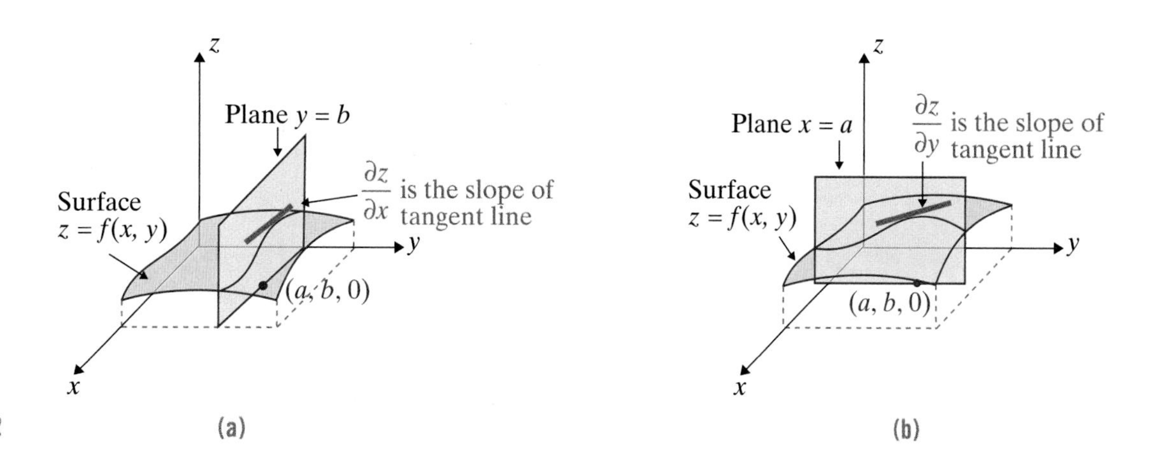

Figure 8.2.2

In effect, to find $\frac{\partial f}{\partial x}$, we simply treat y and any function of y as any other constant and follow the rules of differentiation with one variable. For example, each of the expressions

$$\ln y, \ -5y, \ y^4, \ \text{and} \ e^y$$

are all treated as constants when finding $\frac{\partial f}{\partial x}$.

Similarly, x and any function of x are treated as constants when finding $\frac{\partial f}{\partial y}$.

The following notations are also used to indicate partial derivatives:

$$\frac{\partial f}{\partial x} = \frac{\partial z}{\partial x} = f_x(x, y) = f_x$$

and

$$\frac{\partial f}{\partial y} = \frac{\partial z}{\partial y} = f_y(x, y) = f_y.$$

The value of the partial derivative of $z = f(x,y)$ with respect to x at the point $(a, b, f(a,b))$ can be denoted by any of the following forms:

$$\left.\frac{\partial f}{\partial x}\right|_{(a,\,b)}, \quad \left.\frac{\partial z}{\partial x}\right|_{(a,\,b)}, \quad \text{and} \quad f_x(a, b).$$

Similar notations denote the value of the partial derivative with respect to y at that point.

Study the following examples carefully so that you become accustomed to treating y (or x, as the case may be) as a constant.

Example 1: Finding a Partial Derivative

For the function $f(x, y) = 4x^2 - 3xy + 5y^2$, find **a.** $\frac{\partial f}{\partial x}$ and **b.** $\frac{\partial f}{\partial y}$.

Solutions: a. Treating y as a constant, we obtain

$$\frac{\partial f}{\partial x} = 8x - 3y.$$

Note that in the expression $-3xy$, we treat $-3y$ as a constant coefficient of x. Also, the expression $5y^2$ is treated as a constant.

b. Treating x as a constant, we obtain

$$\frac{\partial f}{\partial y} = -3x + 10y.$$

In this case $4x^2$ is treated as a constant and $-3x$ is treated as a constant coefficient of y.

Example 2: Finding a Partial Derivative

For the function $f(x, y) = e^{xy} - \ln x + y^3$, find **a.** f_x and **b.** f_y.

Solutions: a. To differentiate e^{xy} with respect to x, we use the Chain Rule and differentiate the exponent xy by treating y as a constant. Thus we have

$$f_x = e^{xy} \cdot y - \frac{1}{x} = ye^{xy} - \frac{1}{x}.$$

b. To find f_y, we treat x as a constant. (This means that $\ln x$ is treated as a constant, too.)

$$f_y = x \cdot e^{xy} + 3y^2$$

Example 3: Finding a Partial Derivative

For the function $f(x, y) = xe^{xy} + y^2$, find **a.** $f_x(1, 2)$ and **b.** $f_y(1, 2)$.

Solutions: a. To find $f_x(x, y)$, we treat y^2 as a constant. However, we treat xe^{xy} as a product of two functions of x. The two functions are x and e^{xy}.

$$\begin{aligned} f_x(x, y) &= x \cdot \frac{\partial}{\partial x}\left(e^{xy}\right) + e^{xy}\frac{\partial}{\partial x}(x) + \frac{\partial}{\partial x}\left(y^2\right) \\ &= x \cdot e^{xy} \cdot y + e^{xy} \cdot 1 + 0 \\ &= xye^{xy} + e^{xy} \end{aligned}$$

Now we evaluate $f_x(1, 2)$.

$$\begin{aligned} f_x(1, 2) &= 1 \cdot 2 \cdot e^{1 \cdot 2} + e^{1 \cdot 2} \\ &= 2e^2 + e^2 \\ &= 3e^2 \end{aligned}$$

continued on next page ...

b. To find $f_y(x, y)$, we do **not** use the Product Rule to differentiate xe^{xy} with respect to y, since x is treated as a constant.

$$\begin{aligned} f_y(x, y) &= x \cdot e^{xy} \cdot x + 2y \\ &= x^2 e^{xy} + 2y \end{aligned}$$

Now we evaluate $f_y(1, 2)$.

$$\begin{aligned} f_y(1, 2) &= 1^2 \cdot e^{1 \cdot 2} + 2 \cdot 2 \\ &= e^2 + 4 \end{aligned}$$

Example 4: Using the Partial Derivative

Given the function $z = x^2 + xy + y^2 - 7x - 8y + 1$, find all the ordered pairs (x, y) where both $\frac{\partial z}{\partial x} = 0$ and $\frac{\partial z}{\partial y} = 0$.

Solution: First, we find the partial derivatives.

$$\frac{\partial z}{\partial x} = 2x + y - 7 \text{ and } \frac{\partial z}{\partial y} = x + 2y - 8$$

We want to solve the following system of equations.

$$\begin{cases} 2x + y - 7 = 0 \\ x + 2y - 8 = 0 \end{cases}$$

Multiply each term in the first equation by –2, add to eliminate y, and then solve the resulting equation for x.

$-4x - 2y = -14$	Multiply each term by –2.
$x + 2y = 8$	
$-3x = -6$	Add the equations to eliminate y.
$x = 2$	Solve for x.
$2 + 2y - 8 = 0$	Substitute $x = 2$ in one of the original equations and solve for y.
$2y = 6$	
$y = 3$	

Thus both partial derivatives are 0 at (2, 3).

In Section 8.1 we introduced the Cobb-Douglas Production Formula $P(x, y) = kx^a y^{1-a}$, where x is a measure of units of labor and y is a measure of units of capital invested. The partial derivative $\frac{\partial P}{\partial x}$ is called the **marginal productivity of labor**, and the partial derivative $\frac{\partial P}{\partial y}$ is called the **marginal productivity of capital**. Thus, when $x = x_1$ and $y = y_1$,

$\left.\frac{\partial P}{\partial x}\right|_{(x_1, y_1)}$ is the approximate increase in production for one unit of increase in labor, and

$\left.\frac{\partial P}{\partial y}\right|_{(x_1, y_1)}$ is the approximate increase in production for one unit of increase in capital.

Example 5: Using the Cobb-Douglas Production Formula

Suppose that the production function $P(x, y) = 2000x^{0.5}y^{0.5}$ is known. Determine the marginal productivity of labor and the marginal productivity of capital when 16 units of labor and 144 units of capital are used.

Solution: $$\frac{\partial P}{\partial x} = 2000(0.5)x^{-0.5}y^{0.5} = \frac{1000y^{0.5}}{x^{0.5}}$$

$$\frac{\partial P}{\partial y} = 2000(0.5)x^{0.5}y^{-0.5} = \frac{1000x^{0.5}}{y^{0.5}}$$

Substituting $x = 16$ and $y = 144$, we have

$$\left.\frac{\partial P}{\partial x}\right|_{(16, 144)} = \frac{1000(144)^{0.5}}{(16)^{0.5}} = \frac{1000(12)}{4} = 3000 \text{ units,}$$

and $$\left.\frac{\partial P}{\partial y}\right|_{(16, 144)} = \frac{1000(16)^{0.5}}{(144)^{0.5}} = \frac{1000(4)}{12} = 333 \text{ units.}$$

Only whole units are counted.

Thus we see that adding one unit of labor will increase production by about 3000 units and adding one unit of capital will increase production by about 333 units.

Second-Order Partial Derivatives

For a function of two variables, $z = f(x, y)$, each of the first partial derivatives $f_x(x, y)$ and $f_y(x, y)$ is also a function of two variables. This means that each of the functions f_x and f_y has two partial derivatives, provided that they exist. These four partial derivatives of partial derivatives are called **second-order partial derivatives** (or simply **second partial derivatives**) and can be denoted by the following forms:

$$\frac{\partial}{\partial x}\left(\frac{\partial z}{\partial x}\right) = \frac{\partial^2 z}{\partial x^2} = \frac{\partial^2 f}{\partial x^2} = f_{xx},$$

$$\frac{\partial}{\partial y}\left(\frac{\partial z}{\partial x}\right) = \frac{\partial^2 z}{\partial y \partial x} = \frac{\partial^2 f}{\partial y \partial x} = f_{xy},$$

$$\frac{\partial}{\partial y}\left(\frac{\partial z}{\partial y}\right) = \frac{\partial^2 z}{\partial y^2} = \frac{\partial^2 f}{\partial y^2} = f_{yy}, \quad \text{and}$$

$$\frac{\partial}{\partial x}\left(\frac{\partial z}{\partial y}\right) = \frac{\partial^2 z}{\partial x \partial y} = \frac{\partial^2 f}{\partial x \partial y} = f_{yx}.$$

The notation indicates the specific order in which the partial differentiation is to take place. For example, f_{xy} indicates that the partial differentiation is to be first with respect to x and then with respect to y. When the notation ∂ is used, this same order of partial differentiation is shown with the positions of x and y reversed.

$$f_{xy} = \frac{\partial^2 f}{\partial y \partial x}$$

Left-to-right order   Right-to-left order

In Section 8.3 we will show how second-order partial derivatives for a function of two variables can be used to determine whether a local extremum is a local maximum or a local minimum.

Example 6: Finding Second Partial Derivatives

a. Find all four second partial derivatives of $f(x, y) = \ln(x^2 + 4y)$.

b. Find $f_{xx}\left(2, \frac{1}{2}\right)$.

Solutions: a. We must find the first partial derivatives f_x and f_y before we can find the second partial derivatives.

$$f_x = \frac{1}{x^2 + 4y}(2x) = \frac{2x}{x^2 + 4y}$$

$$f_y = \frac{1}{x^2 + 4y}(4) = \frac{4}{x^2 + 4y}$$

Now we can find the second partial derivatives.

$$f_{xx} = \frac{(x^2+4y)\cdot 2 - 2x(2x)}{(x^2+4y)^2} = \frac{-2x^2+8y}{(x^2+4y)^2}$$

$$f_{xy} = \frac{(x^2+4y)\cdot 0 - 2x\cdot 4}{(x^2+4y)^2} = \frac{-8x}{(x^2+4y)^2}$$

$$f_{yx} = \frac{(x^2+4y)\cdot 0 - 4\cdot 2x}{(x^2+4y)^2} = \frac{-8x}{(x^2+4y)^2}$$

$$f_{yy} = \frac{(x^2+4y)\cdot 0 - 4\cdot 4}{(x^2+4y)^2} = \frac{-16}{(x^2+4y)^2}$$

b. $$f_{xx}\left(2,\frac{1}{2}\right) = \frac{-2\cdot 2^2+8\left(\frac{1}{2}\right)}{\left[2^2+4\left(\frac{1}{2}\right)\right]^2} = \frac{-8+4}{(4+2)^2} = \frac{-4}{36} = -\frac{1}{9}$$

Partial Derivatives of Functions of Three Variables

The concept of partial derivatives is easily extended to include functions of three or more variables. If, for example, $w = f(x, y, z)$, then there are three first partial derivatives:

$$\frac{\partial w}{\partial x} = f_x(x, y, z),$$ found by treating y and z as constants,

$$\frac{\partial w}{\partial y} = f_y(x, y, z),$$ found by treating x and z as constants, and

$$\frac{\partial w}{\partial z} = f_z(x, y, z),$$ found by treating x and y as constants.

In Section 8.4 we use partial derivatives of functions of three variables to determine the maximum or minimum value of a function of two variables with restrictions on the variables.

Example 7: Finding Partial Derivatives with 3 Variables

For $w = 2ye^{xy} + z^2$, find

a. $\frac{\partial w}{\partial x}$ **b.** $\frac{\partial w}{\partial y}$ **c.** $\frac{\partial w}{\partial z}$

Solutions: a. Treat y and z as constants.

$$\frac{\partial w}{\partial x} = 2y \cdot e^{xy} \cdot y + 0 = 2y^2 e^{xy}$$

b. Treat x and z as constants. Here $2ye^{xy}$ is a product of two functions of y, and we use the Product Rule to differentiate.

$$\frac{\partial w}{\partial y} = 2y \cdot e^{xy} \cdot x + e^{xy} \cdot 2 + 0 = 2e^{xy}(xy + 1)$$

c. Treat x and y as constants. In this case, the entire expression $2ye^{xy}$ is treated as a constant.

$$\frac{\partial w}{\partial z} = 0 + 2z = 2z$$

8.2 Exercises

Find $\frac{\partial f}{\partial x}$ *and* $\frac{\partial f}{\partial y}$ *for each of the functions in Exercises 1 – 28.*

1. $f(x,y) = 4x + 7y - 10$

2. $f(x,y) = 11x - 19y + 2$

3. $f(x,y) = 2x^2 + 5y^2$

4. $f(x,y) = 5x^3 - 6y^4$

5. $f(x,y) = x^2y + 4xy^3 + 6$

6. $f(x,y) = x^3 + 5x^2y^3 - 95x^2y^3 - 9$

7. $f(x,y) = y\sqrt{25 + x^2}$

8. $f(x,y) = x^3\sqrt{y^2 + 5}$

9. $f(x,y) = \sqrt{49 - x^2 - y^2}$

10. $f(x,y) = \sqrt{16 + 2x^2 + y^2}$

11. $f(x,y) = 4e^{x-y}$

12. $f(x,y) = 7e^{xy}$

13. $f(x,y) = x\ln y$

14. $f(x,y) = \ln(x^2 + y^2)$

15. $f(x,y) = \ln(x^2 + 3xy)$

16. $f(x,y) = \frac{y}{\ln x}$

17. $f(x,y) = \frac{2x^2}{y^2 + 1}$

18. $f(x,y) = \frac{x^2 + 3}{5y - 9}$

19. $f(x,y)=\dfrac{x^2}{xy+3}$

20. $f(x,y)=\dfrac{x+y}{4x-y}$

21. $f(x,y)=x^3e^y+ye^{x^2}$

22. $f(x,y)=xe^{y^3}-y^2e^x$

23. $f(x,y)=x\sqrt{xy+2}$

24. $f(x,y)=y\sqrt{x^2+y^3}$

25. $f(x,y)=x^4e^{xy}$

26. $f(x,y)=y^3e^{-xy}$

27. $f(x,y)=y^5\ln\left(x^2-5y^2\right)$

28. $f(x,y)=3x\ln xy^4$

In Exercises 29 – 32, find $\dfrac{\partial f}{\partial x}, \dfrac{\partial f}{\partial y},$ *and* $\dfrac{\partial f}{\partial z}$.

29. $f(x,y,z)=xy+2xz+9yz$

30. $f(x,y,z)=3x^2y+2xyz+7xz^2$

31. $f(x,y,z)=\left(8x^2+5y^2-2z^2\right)^2$

32. $f(x,y,z)=\sqrt{x^2+2y^2+4z^2}$

In Exercises 33 – 44, find all second-order partial derivatives $f_{xx}, f_{yy}, f_{xy},$ *and* f_{yx}.

33. $f(x,y)=3xy+x^2y^3-19$

34. $f(x,y)=5x^3y-3x^3y^2-13$

35. $f(x,y)=x^4y^{\frac{2}{3}}$

36. $f(x,y)=(xy)^{\frac{3}{4}}$

37. $f(xy)=xe^{2y}$

38. $f(x,y)=\dfrac{e^{x^3}}{y^4}$

39. $f(x,y)=(4x-3y)^{\frac{5}{3}}$

40. $f(x,y)=\sqrt{7x^3+y^2}$

41. $f(x,y)=\dfrac{3x+1}{5y+3}$

42. $f(x,y)=\dfrac{6y^2-5}{2x+7}$

43. $f(x,y)=\dfrac{2xy}{x-y}$

44. $f(x,y)=\dfrac{x-y}{xy}$

In Exercises 45 – 48, find $\dfrac{\partial S}{\partial m}$ *and* $\dfrac{\partial S}{\partial b}$.

45. $S(m,b)=(2m+b-9)^2+(4m+b-13)^2+(5m+b-18)^2$

46. $S(m,b)=(8m+b-17)^2+(9m+b-23)^2+(10m+b-28)^2$

47. $S(m,b)=(6m+b-40)^2+(8m+b-49)^2+(9m+b-55)^2+(10m+b-62)^2$

48. $S(m,b)=(12m+b-81)^2+(13m+b-88)^2+(14m+b-96)^2+(15m+b-101)^2$

In Exercises 49 – 52, find $\frac{\partial F}{\partial x}, \frac{\partial F}{\partial y},$ *and* $\frac{\partial F}{\partial \lambda}$. *($\lambda$ is the Greek letter lambda.)*

49. $F(x,y,\lambda) = 8x + 15xy - 2y^2 + \lambda(x + y - 60)$

50. $F(x,y,\lambda) = 3x^2 + 12y^2 + \lambda(x + 2y - 84)$

51. $F(x,y,\lambda) = 5x^2 + 3xy - 10y^2 + \lambda(14x + 17y - 49)$

52. $F(x,y,\lambda) = 7x^2 - 2xy + 9y^2 + \lambda(8x + 15y - 120)$

53. Marginal productivity. The number of units of a product that are manufactured by a company is given by $f(L,K) = 80L^{\frac{2}{3}}K^{\frac{1}{3}}$, where L is the units of labor and K is the units of capital. Find the marginal productivity of labor and the marginal productivity of capital if the company is currently utilizing 27 units of labor and 64 units of capital.

54. Marginal productivity. The productivity of a company is approximated by $f(L,K) = 20L^{\frac{2}{5}}K^{\frac{3}{5}}$, where L is the units of labor and K is the units of capital. Find the marginal productivity of labor and the marginal productivity of capital if the company is currently utilizing 32 units of labor and 32 units of capital.

55. Marginal cost. A company manufactures two items, product A and product B. The cost of producing x units of A and y units of B is $C(x,y) = 3000 + 7x + 5.8y + 0.03x^2 - xy + 0.02y^2$.
a. Find the marginal cost with respect to x.
b. Find the marginal cost with respect to y.

56. Marginal profit. The profit from the sale of two products is given by the function $P(x,y) = 88x + 54y - 0.02x^2 - 0.015y^2 - 68$, where x is the number of units of product A sold, and y is the number of units of product B sold.
a. Find the marginal profit with respect to x.
b. Find the marginal profit with respect to y.

57. Marginal profit. A company produces two models of a product. The cost function is given by $C(x,y) = x^2 - 2xy + 2y^2 + 4x + 3y + 8$ and the revenue function is given by $R(x,y) = 20x + 15y$, where x is the number of units of model A and y is the number of units of model B produced and sold.
a. Find the profit function.
b. Find $P_x(20,14)$ and $P_y(20,14)$ and interpret the results.

58. **Marginal profit.** A firm produces and sells x units of product A and y units of product B. Its revenue function is given by $R(x,y)=80x+100y$ and its cost function is given by $C(x,y)=x^2+1.5y^2-xy+1500$. Find $P_x(50,25)$ and $P_y(50,25)$ and interpret the results.

59. **Marginal profit.** A marketing manager of a department store has determined that revenue is related to the number of units of television advertising x and the number of units of newspaper advertising y by the function $R(x,y)=500(20x+5y+20xy-x^2)$. Each unit of television advertising costs \$5000 and each unit of newspaper advertising costs \$2500.

a. Find the marginal profit with respect to x.

b. Find the marginal profit with respect to y.

60. **Marginal profit.** An electronics firm manufactures and sells two models of automobile radios. The standard model of the radio sells for \$300, and the deluxe model of the radio sells for \$400. The total cost function is $C(x,y)=90{,}000+0.05x^2+0.1y^2+0.125xy$, where x is the number of standard models and y is the number of deluxe models.

a. Find the marginal profit with respect to x.

b. Find the marginal profit with respect to y.

Hawkes Learning Systems: Essential Calculus

Partial Derivatives

8.3 Local Extrema for Functions of Two Variables

Objectives

After completing this section, you will be able to:

1. *Find critical points for functions of two variables.*
2. *Apply the D-test to classify critical points for functions of two variables.*

In Chapter 3, we discussed differentiation of functions of one variable and how the first derivative can be used to help locate and test local extrema. In Chapter 4, we also tested local extrema with the Second-Derivative Test. In this section, we will define local maxima and local minima for functions of two variables and discuss a test involving second-order partial derivatives for determining the nature of these local extrema.

We will find that the conditions and concepts related to local maxima and local minima are quite similar for functions of one variable and functions of two variables. For functions of one variable, we dealt with intervals on real number lines: open intervals (endpoints not included) and closed intervals (endpoints included). For functions of two variables, we deal with regions in planes: **open regions** (boundary points not included) and **closed regions** (boundary points included).

Local Extrema for a Function $z = f(x, y)$

Suppose that $z = f(x, y)$ is a function defined on a region in the xy-plane and (a, b) is a point in that region.

1. *If there is an open region R containing (a, b) such that*

$$f(a,b) \geq f(x,y)$$

for all (x, y) in R, then $f(a, b)$ is called a ***local maximum*** *of f.*

2. *If there is an open region R containing (a, b) such that*

$$f(a,b) \leq f(x,y)$$

for all (x, y) in R, then $f(a, b)$ is called a ***local minimum*** *of f.*

As with functions of one variable, local extrema for functions of two variables may or may not be absolute extrema. In this text we will not be concerned with determining absolute extrema for functions of two variables. Several local extrema are illustrated in Figure 8.3.1.

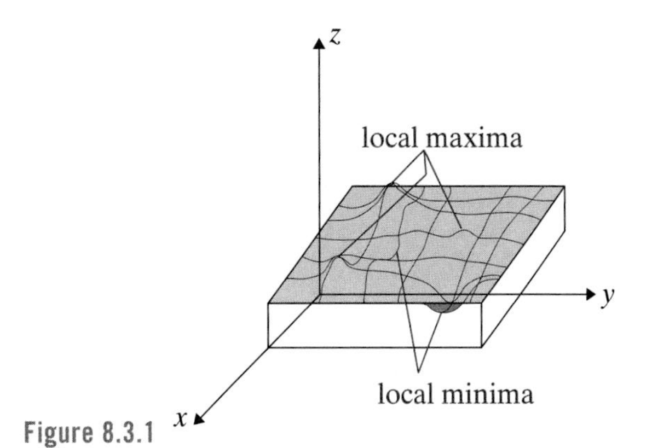

Figure 8.3.1

Recall that $y = f(x)$, a function of one variable, $x = c$ is a critical value if $f'(c) = 0$ or $f'(c)$ does not exist. Similarly, for a function $z = f(x, y)$, a point (a, b) in the domain of f is a **critical point** if $f_x(a,b) = 0$ and $f_y(a,b) = 0$.

In this text we will not discuss points where the partial derivatives f_x and f_y of a function are undefined. Such points also can be classified as critical points.

If $f(a, b)$ is a local minimum (or a local maximum), then (a, b) is a critical point. However, a critical point does not guarantee a local extremum. In some cases, a critical point gives a **saddle point** where there is neither a local minimum nor a local maximum.

A saddle point for a function of two variables is analogous to a point of inflection for a function of one variable.

Example 1: Locating Critical Points

Locate the critical points for the function $f(x,y) = 4 + (x-3)^2 + (y+5)^2$.

Solution: We find the first partial derivatives, set them equal to zero, and solve the resulting system of equations.

$$f_x(x,y) = 2(x-3) \text{ and } f_y(x,y) = 2(y+5).$$

The solution of the system

$$\begin{cases} 2(x-3) = 0 \\ 2(y+5) = 0 \end{cases}$$

is the point (3, –5). Thus the only critical point is (3, –5). Therefore, if f has a local minimum or a local maximum, then it must occur at (3, –5).

To determine whether or not a critical point is a local minimum, a local maximum, or a saddle point, we can use the following test, known as the **Second Partials Test** (or the ***D*-Test**).

Second Partials Test (or *D*-test)

Suppose that a function $z = f(x, y)$ and the first partial derivatives and the second partial derivatives are all defined in an open region R and that ***(a, b) is a critical point*** *in R such that*

$$f_x(a,b)=0 \text{ and } f_y(a,b)=0.$$

Define the quantity D as follows:

$$D = f_{xx}(a,b)\cdot f_{yy}(a,b)-\left[f_{xy}(a,b)\right]^2.$$

Case 1: *If $D > 0$ and $f_{xx}(a,b)>0$, then $f(a, b)$ is a* ***local minimum.***

Case 2: *If $D > 0$ and $f_{xx}(a,b)<0$, then $f(a, b)$ is a* ***local maximum.***

Case 3: *If $D < 0$, then $(a, b, f(a, b))$ is a* ***saddle point.***

Case 4: *If $D = 0$, then this test gives no information.*

Example 2: Applying the *D*-test

Find all the local minima, local maxima, and saddle points for the function $f(x,y)=x^2-xy+y^2-9x+5$.

Solution: Find the first partial derivatives f_x and f_y.

$$f_x(x,y)=2x-y-9 \text{ and } f_y(x,y)=-x+2y$$

Now, to find any critical points, solve the following system.

$$\begin{cases}2x-y-9=0\\-x+2y=0\end{cases}$$

$$2x-y=9$$

$$\underline{-2x+4y=0}$$ Multiply the second equation by 2 and add the two equations.

$$3y=9$$

$$y=3$$

$$-x+2\cdot 3=0$$ Substitute $y=3$ and solve for x.

$$x=6$$

The only critical point is $(6, 3)$.
The second partials are

$f_{xx}(x,y)=2$, $f_{yy}(x,y)=2$, and $f_{xy}(x,y)=-1$.

In this case each of the second partials is a constant and will have that constant value at (6, 3). Thus

$$D = f_{xx}(6,3) f_{yy}(6,3) - [f_{xy}(6,3)]^2$$
$$= 2 \cdot 2 - (-1)^2 = 4 - 1 = 3 > 0.$$

Since $D > 0$, we check the sign of $f_{xx}(6,3)$ to determine whether $(6,3)$ yields a local minimum or a local maximum.

$$f_{xx}(6,3) = 2 > 0$$

Therefore, by Case 1 in the *D*-Test, the critical point (6, 3) yields a local minimum value. This value is

$$f(6,3) = 6^2 - 6 \cdot 3 + 3^2 - 9 \cdot 6 + 5 = -22.$$

Example 3: Locating and Classifying Critical Points

Locate and classify all critical points of the function

$$g(x,y) = \frac{1}{3}x^3 - \frac{1}{3}y^3 + xy.$$

Solution: The first partial derivatives are

$$g_x = x^2 + y \text{ and } g_y = -y^2 + x.$$

Now solve the following system where $g_x = 0$ and $g_y = 0$.

$$\begin{cases} x^2 + y = 0 \\ -y^2 + x = 0 \end{cases}$$

Solving the second equation for *x* gives $x = y^2$, and substituting y^2 for *x* in the first equation gives

$$(y^2)^2 + y = 0$$
$$y^4 + y = 0$$
$$y(y^3 + 1) = 0$$
$$y = 0 \text{ or } y^3 = -1$$
$$y = -1$$
$$x = 0^2 \qquad x = (-1)^2 = 1$$

Thus there are two critical points: (0, 0) and (1, –1).
Now find the second partials and calculate *D* for each critical point.

$$g_{xx} = 2x, g_{yy} = -2y, \text{ and } g_{xy} = 1$$

continued on next page ...

For (0, 0),

$$g_{xx}(0,0)=0,\ g_{yy}(0,0)=0, \text{ and } g_{xy}(0,0)=1.$$

$$D=0\cdot 0-(1)^2=-1<0.$$

Therefore, by Case 3 of the *D*-Test, the point $(0,0,g(0,0))=(0,0,0)$ is a saddle point.
For (1, –1),

$$g_{xx}(1,-1)=2,\ g_{yy}(1,-1)=2, \text{ and } g_{xy}(1,-1)=1.$$

$$D=2\cdot 2-(1)^2=4-1=3>0.$$

Since $D>0$ and $g_{xx}(1,-1)=2>0$, by Case 1 of the *D*-Test, the critical point (1, –1) yields a local minimum. The minimum value is

$$g(1,-1)=\frac{1}{3}(1)^3-\frac{1}{3}(-1)^3+1(-1)$$

$$=\frac{1}{3}+\frac{1}{3}-1=-\frac{1}{3}.$$

Example 4: Optimizing an Equation in Two Variables

A company produces and sells two styles of umbrellas. One style sells for $20 each and the other sells for $25 each. The company has determined that if x thousand of the first style and y thousand of the second style are produced, then the total cost in thousands of dollars is given by the function

$$C(x,y)=3x^2-3xy+\frac{3}{2}y^2+32x-29y+70.$$

How many of each style of umbrella should the company produce and sell in order to maximize profit?

Solution: Since x thousand umbrellas sell for $20 each and y thousand umbrellas sell for $25 each, the revenue function (in thousands of dollars) is given by

$$R(x,y)=20x+25y.$$

Thus the profit function is

$$P(x,y)=R(x,y)-C(x,y)$$

$$=20x+25y-\left(3x^2-3xy+\frac{3}{2}y^2+32x-29y+70\right)$$

$$=-3x^2+3xy-\frac{3}{2}y^2-12x+54y-70.$$

The first partial derivatives are

$$P_x = -6x + 3y - 12 \text{ and } P_y = 3x - 3y + 54.$$

Now solve the following system of equations.

$$\begin{cases} -6x + 3y - 12 = 0 \\ 3x - 3y + 54 = 0 \end{cases} \qquad \begin{array}{r} -6x + 3y = 12 \\ 3x - 3y = -54 \\ \hline -3x \quad = -42 \\ x = 14 \end{array}$$

Substitute $x = 14$ and solve for y.

$$3(14) - 3y = -54$$
$$-3y = -96$$
$$y = 32$$

The company will make the maximum profit if it produces and sells 14,000 of the first style of umbrella and 32,000 of the second style. [The student can verify (with the D-Test) that the profit is indeed a maximum at (14, 32).]

8.3 Exercises

For each of the Exercises 1 – 24, find all local maxima, local minima, and saddle points.

1. $f(x,y) = x^2 + y^2 - 6x + 2y - 4$

2. $f(x,y) = x^2 + 2y^2 + 8x - 4y + 2$

3. $f(x,y) = x^2 - y^2 - 10x + 2y + 9$

4. $f(x,y) = 12x + 8y - x^2 - y^2 - 7$

5. $f(x,y) = 5x + 8y - x^2 - y^2 + 11$

6. $f(x,y) = y^2 - x^2 + 6x - y - 10$

7. $f(x,y) = x^2 + xy - y^2 - 5y - 8$

8. $f(x,y) = x^2 - 2xy + 4y^2 - 6y + 3$

9. $f(x,y) = 2x^2 - 3xy + 3y^2 + 5x - 13$

10. $f(x,y) = 10x - 2x^2 + 2xy - 3y^2 + 5$

11. $f(x,y) = x^2 - xy + y^2 - 2x - 2y + 1$

12. $f(x,y) = 3x^2 - 2xy + y^2 - 16x + 4y + 14$

13. $f(x,y) = -x^2 + xy - y^2 + 4x - 5y + 6$

14. $f(x,y) = x^2 + 3y^2 + 5xy + 4x - 3y + 15$

15. $f(x,y) = x^3 + y^2 - 3x - 6y + 11$

16. $f(x,y) = x^3 - 3x^2 - 2y^2 - 9x + 8y + 7$

17. $f(x,y) = x^2 - y^3 + 9y^2 - 4x - 15y - 14$

18. $f(x,y) = 2x^2 + y^3 - 3y^2 - 12x - 24y + 21$

19. $f(x,y) = x^3 - 3x^2y + 12y$

20. $f(x,y) = 9x - xy^2 + 2y^3$

21. $f(x,y) = 2x^2 - x^2y + y^2$

22. $f(x,y) = x^2 - 2xy^2 + 4y^2$

23. $f(x,y) = xy + \frac{2}{x} + \frac{4}{y}$

24. $f(x,y) = xy + \frac{9}{x} + \frac{3}{y}$

25. Profit. An automobile agency sells two models of a car. The manager has determined that the annual profit is estimated by $P(x,y) = -0.1x^2 - 0.2y^2 + 6x + 10y - 160$ in thousands of dollars. Find the number of each model that should be sold to maximize profit.

26. Sales. The owner of a small business advertises in the newspaper and on radio. He has found that the number of units that he sells is approximated by $N(x,y) = -0.5x^2 - y^2 + 8x + 12y + 240$, where x (in thousands of dollars) is the amount spent on newspaper advertising and y (in thousands of dollars) is the amount spent on radio advertising. How much should he spend on each to maximize the number of units sold?

27. Profit. A firm produces two kinds of magazine racks, one selling for \$50 and the other for \$45. The total cost of producing x of the \$50 racks and y of the \$45 racks is given by $C(x,y) = 0.15x^2 + 0.1y^2 - 10x - 3y + 4760$ dollars. Find the number of each kind that should be produced and sold to maximize the profit.

28. Profit. A company makes two types of work gloves, leather and cloth. The leather gloves sell for \$5.80 and the cloth gloves for \$1.60. The total cost function is $C(x,y) = 0.25x^2 + 0.03y^2 + 1.3x + 0.4y + 14$ in thousands of dollars, where x (in thousand pairs) is the number of leather gloves and y (in thousand pairs) is the number of cloth gloves. How many gloves of each type should be produced and sold to maximize profits?

29. Revenue. A department store sells two types of alarm clocks, a quartz type and a keywind type. The store manager has determined that he can sell $23 - 6x + 8y$ of the quartz type and $26 + 5x - 9y$ of the keywind type if the price is x dollars for the quartz and y dollars for the keywind. Find the price of each that will yield maximum revenue.

30. Revenue. A grocery store sells two brands of a product, a "name" brand and a "house" brand. The manager estimates that if she sells the "name" brand for x dollars per unit and the "house" brand for y dollars per unit, she will be able to sell $62 - 20x + 18y$ units of the "name" brand and $53 + 16x - 22y$ units of the "house" brand. Find the price of each that will yield maximum revenue.

31. Construction. A rectangular box is to be constructed without a top and one partition (see drawing). The volume of the box must be 162 in.^{3} Find the dimensions that will minimize the material required to construct the box.

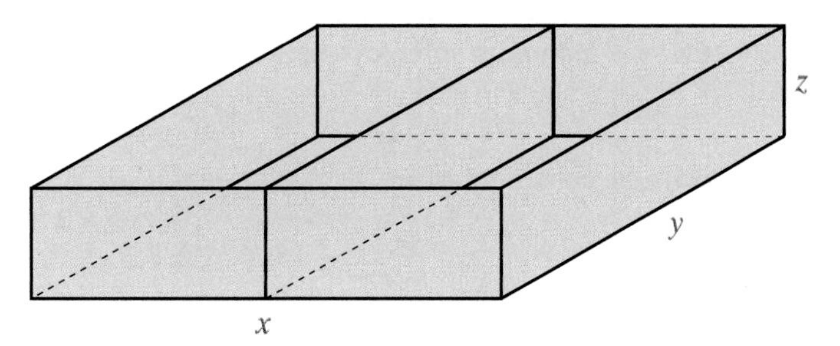

32. Packaging. The Postal Service has a limit of 108 inches on the combined length and girth of a rectangular package to be sent by parcel post. Find the dimensions of the package of maximum volume that can be sent (see drawing).

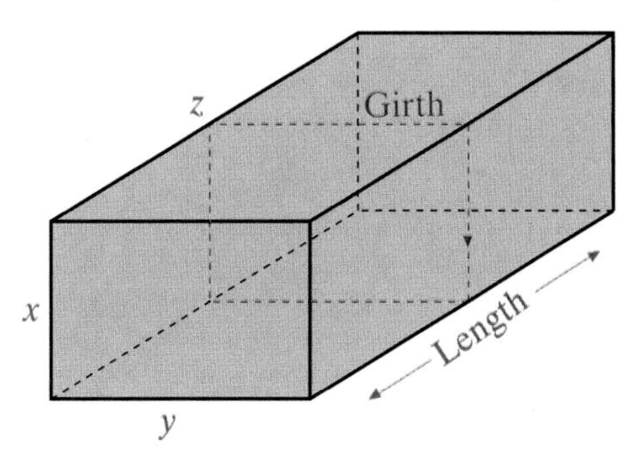

Hawkes Learning Systems: Essential Calculus

Local Extrema for Functions of Two Variables

8.4 Lagrange Multipliers

Objectives

After completing this section, you will be able to:

Apply the method of Lagrange multipliers to solve constrained optimization problems.

In this section we will consider problems of locating the maxima and minima of functions of two variables in which there are certain restrictions on the variables. The restrictions are also called **side conditions** or **constraints**. The problems themselves are known as **constrained optimization problems**. For example, suppose that we want to find the minimum value of the function

$$f(x, y) = x^2 + y^2 + 2$$

with the constraint that

$$x + y - 3 = 0.$$

The situation is illustrated geometrically in Figure 8.4.1

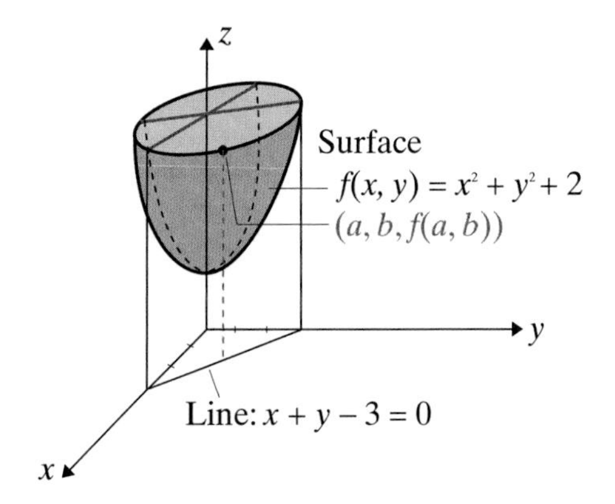

Figure 8.4.1

We want to find the point (a, b) where $f(a, b)$ is a minimum under the constraint that $a + b - 3 = 0$. One general method for finding maxima or minima of functions when there is a constraint on the variables is the method of **Lagrange multipliers**, named after the French mathematician Joseph Louis Lagrange (1736-1813). In this method the Greek letter lambda (λ) is used to represent a variable called the **Lagrange multiplier**.

The technique is based on treating λ as an independent variable and forming the **Lagrange function**

$$F(x,y,\lambda) = f(x,y) + \lambda \cdot g(x,y),$$

where $z = f(x, y)$ is the object function and $g(x, y) = 0$ is the constraint.

The Method of Lagrange Multipliers

To maximize or minimize a function $z = f(x, y)$ subject to the constraint $g(x, y) = 0$,

Step 1: *Form the Lagrange function*

$$F(x,y,\lambda) = f(x,y) + \lambda \cdot g(x,y).$$

Step 2: *Find each of the partial derivatives F_x, F_y, and F_λ, provided all partials exist.*

Step 3: *Solve the system of three equations*

$$\begin{cases} F_x(x,y,\lambda) = 0 \\ F_y(x,y,\lambda) = 0 \\ F_\lambda(x,y,\lambda) = 0 \end{cases}$$

Step 4: *If f has a local maximum or local minimum subject to the constraint $g(x, y) = 0$, then the corresponding x and y values will be found among the solutions to the system in Step 3.*

To determine whether the solution (a, b) found by using the method of Lagrange multipliers gives a maximum or a minimum, we can compare the value $f(a, b)$ with the values $f(x, y)$ where the points (x, y) are near (a, b) and satisfy the constraint $g(x, y) = 0$. For the problems in this text, the values found in the solutions to the system in Step 3 do give the maximum value (or minimum value) as requested in the problems.

Example 1: Applying the Method of Lagrange Multipliers

Use the method of Lagrange multipliers to find the minimum value of the function $f(x,y) = x^2 + y^2 + 2$ with the constraint $x + y - 3 = 0$.

Solution: **Step 1:** Form the Lagrange function.

$$F(x,y,\lambda) = x^2 + y^2 + 2 + \lambda(x + y - 3)$$

Step 2: Find each of the partial derivatives.

$$F_x(x,y,\lambda) = 2x + \lambda \qquad (1)$$

$$F_y(x,y,\lambda) = 2y + \lambda \qquad (2)$$

$$F_\lambda(x,y,\lambda) = x + y - 3. \qquad (3)$$

continued on next page ...

Note: The constraint function $g(x, y)$ will always be equal to F_λ because it is treated as a constant coefficient of λ in the Lagrange function.

Step 3: Solve the following system of equations by isolating λ in equations (1) and (2) and solving the resulting equations.

$$2x+\lambda=0 \qquad (1)$$
$$\lambda=-2x$$
$$2y+\lambda=0 \qquad (2)$$
$$\lambda=-2y$$

Thus $-2x = -2y$ and $x = y$. Substituting y for x in equation (3) gives

$$x+x-3=0$$
$$2x=3$$
$$x=1.5$$

Since $x = y$, we have $y = 1.5$.

Step 4: From the graph in Figure 8.4.1, we can see that

$$f(1.5, 1.5)=(1.5)^2+(1.5)^2+2=6.50$$

is a minimum. In order to verify this result algebraically, we choose the point (1.4, 1.6), which is close to (1.5, 1.5) and satisfies the constraint $x + y - 3 = 0$, and find that $f(1.4, 1.6) = 6.52 > 6.50$. The D-Test can also be used to verify this.

Example 2: Applying the Method of Lagrange Multipliers

Use the method of Lagrange multipliers to find the maximum value of the function $f(x, y) = xy$ under the restriction $g(x,y)=x^2+y^2-200=0$, where $x > 0$ and $y > 0$.

Solution: **Step 1:** Form the Lagrange function.

$$F(x, y, \lambda)=xy+\lambda\left(x^2+y^2-200\right)$$

Step 2: Find each of the partial derivatives.

$$F_x=y+\lambda\cdot 2x \ (1),\ F_y=x+\lambda\cdot 2y \ (2), \text{ and } F_\lambda=x^2+y^2-200 \ (3)$$

Step 3: Solve the following system of equations by isolating λ in equations (1) and (2) and solving the resulting equations.

$$y + \lambda \cdot 2x = 0 \qquad (1)$$

$$\lambda = \frac{-y}{2x}$$

$$x + \lambda \cdot 2y = 0 \qquad (2)$$

$$\lambda = \frac{-x}{2y}$$

$$x^2 + y^2 - 200 = 0 \qquad (3)$$

We have $\frac{-y}{2x} = \frac{-x}{2y}$, which gives $-2x^2 = -2y^2$ and $x^2 = y^2$. Substituting x^2 for y^2 in equation (3), we have

$$x^2 + x^2 - 200 = 0$$
$$2x^2 = 200$$
$$x^2 = 100$$
$$x = \pm 10.$$

Since $y^2 = x^2$, $y^2 = 100$, and $y = \pm 10$. Thus there are four points that satisfy the system: $(10, 10)$, $(10, -10)$, $(-10, 10)$, and $(-10, -10)$. However, among these points, only the point $(10, 10)$ satisfies the additional conditions $x > 0$ and $y > 0$.

Step 4: $f(10, 10) = (10)(10) = 100$ is a maximum for $f(x, y) = xy$ under the given constraints. As a check that 100 is indeed a maximum, we find that

$$f(9.88, 10.12) = (9.88)(10.12) = 99.9856 < 100.$$

Note: Points close to (10, 10) have irrational coordinates and we have chosen 9.88 and 10.12 as rounded off values that nearly satisfy the constraints.

Example 3: Optimization Using the Cobb–Douglas Production Formula

Suppose that the Cobb-Douglas Production Formula for a particular product is $f(x, y) = 100x^{0.6}y^{0.4}$, where x represents units of labor at a cost of \$200 per unit and y represents units of capital at a cost of \$300 per unit. If the company's budget allows \$120,000 for labor and capital, then the constraint is represented by the equation $200x + 300y = 120{,}000$. Find the maximum level of production for this product.

continued on next page ...

Solution: **Step 1:** Form the Lagrange function.

$$F(x, y, \lambda) = 100x^{0.6}y^{0.4} + \lambda(200x + 300y - 120{,}000)$$

Step 2: Find each of the partial derivatives.

$$F_x = 60x^{-0.4}y^{0.4} + 200\lambda$$

$$F_y = 40x^{0.6}y^{-0.6} + 300\lambda$$

$$F_\lambda = 200x + 300y - 120{,}000$$

Step 3: Solve the following system of equations by isolating λ in equations (1) and (2) and solving the resulting equations.

$$60x^{-0.4}y^{0.4} + 200\lambda = 0 \qquad (1)$$

$$40x^{0.6}y^{-0.6} + 300\lambda = 0 \qquad (2)$$

From equations (1) and (2) we have

$$\frac{-60x^{-0.4}y^{0.4}}{200} = \lambda = \frac{-40x^{0.6}y^{-0.6}}{300}.$$

Multiply both sides of the equation by $-600x^{0.4}y^{0.6}$ to get

$$180y = 80x$$

$$y = \frac{4}{9}x.$$

Substitute for y in the partial derivative F_λ.

$$200x + 300\left(\frac{4}{9}x\right) - 120{,}000 = 0$$

$$3000x = 1{,}080{,}000$$

$$x = 360$$

$$y = \frac{4}{9}(360) = 160$$

Step 4: To maximize production, the company should invest 360 units of labor and 160 units of capital. This investment will result in a production level of approximately

$$f(360, 160) = 100(360)^{0.6}(160)^{0.4}$$

$$\approx 100(34.181)(7.615) \approx 26{,}029 \text{ units.}$$

(The student should verify that 26,029 is indeed a maximum.)

Comment Concerning λ

In economics, the absolute value of the Lagrange multiplier λ used in maximizing a production function is called the **marginal productivity of money**. That is, $|\lambda|$ indicates the increase in production that could be expected for an increased investment of one dollar. In this text we will not be concerned with the marginal productivity of money or any other interpretations of λ. This is why we have not found the values of λ in the examples.

8.4 Exercises

In Exercises 1 – 8, use the method of Lagrange multipliers to find the minimum value of f subject to the given constraint.

1. $f(x,y)=x^2+y^2$, subject to $x + y - 4 = 0$.

2. $f(x,y)=4x^2+3y^2$, subject to $x + y - 7 = 0$.

3. $f(x,y)=5x^2+4y^2-2x$, subject to $x - y - 2 = 0$.

4. $f(x,y)=2x^2+y^2-18x$, subject to $3x - y - 8 = 0$.

5. $f(x,y)=6x^2+5y^2-xy$, subject to $2x + y = 24$.

6. $f(x,y)=2x^2+3y^2-3xy$, subject to $x + y = 16$.

7. $f(x,y)=x^3+y^3$, subject to $x + y = 8$.

8. $f(x,y)=x^3-y^3$, subject to $x - y = 10$.

In Exercises 9 – 16, use the method of Lagrange multipliers to find the maximum value of f subject to the given constraint.

9. $f(x,y)=2x^2-5y^2$, subject to $x - y = 3$.

10. $f(x,y)=5y^2-8x^2$, subject to $x + y = 6$.

11. $f(x,y)=8xy-3x^2$, subject to $x + 2y = 14$.

12. $f(x,y)=6x^2-5xy$, subject to $2x - y = 8$.

13. $f(x,y)=x^2+y^2+4xy$, subject to $3x + 4y = 23$.

14. $f(x,y)=x^2-4y^2+84xy$, subject to $5x + 2y = 18$.

15. $f(x,y)=15x^{0.4}y^{0.6}$, subject to $10x + 8y = 200$.

16. $f(x,y)=8x^{\frac{1}{2}}y^{\frac{1}{2}}$, subject to $6x + 15y = 450$.

In Exercises 17 – 20, use the method of Lagrange multipliers to find the maximum and minimum values of f subject to the given constraint.

17. $f(x, y) = 4xy$, subject to $x^2 + 4y^2 = 72$.

18. $f(x, y) = 5xy$, subject to $9x^2 + y^2 = 162$.

19. $f(x,y)=x^3+4y^3$, subject to $x + y = 6$.

20. $f(x,y)=3x^3+y^3$, subject to $3x + y = 8$.

21. Cost. A company has a plant in Los Angeles and a plant in Oklahoma City. The firm is committed to produce a total of 40 units of a product each week. The cost function is given by $C(x,y)=0.3x^2+0.2y^2+20x+7y+200$, where x is the number of units produced in Los Angeles and y is the number of units produced in Oklahoma City. How many units should be produced in each plant to minimize the total weekly costs?

22. Profit. A department store sells two styles of a jacket, lined and unlined. During the month of January, the management expects to sell exactly 250 jackets. The profit function is given by $P(x,y)=-0.3x^2-0.4y^2-0.3xy+80x+65y-1000$, where x is the number of lined jackets sold and y is the number of unlined jackets sold. How many of each type should be sold to maximize the profit?

23. Production. The production function for a certain product is given by $f(x,y)=75x^{0.3}y^{0.7}$, where x is the number of units of labor and y is the number of units of capital. Each unit of labor costs \$300, and each unit of capital costs \$200. If the company's budget allows a total of \$20,000 for labor and capital, find the maximum level of production.

24. Production. The management of a company has determined that x units of labor and y units of capital are required to produce $f(x,y)=130x^{0.4}y^{0.6}$ units of a product. Each unit of labor costs \$450, and each unit of capital costs \$360. Find the maximum number of units that can be produced if a total of \$90,000 is available for labor and capital.

25. Sales. A sales representative for a textbook publishing company estimates her monthly sales for March to be $S(x, y) = 30x + 18y - 1.2x^2 - 0.6y^2$ in thousands of dollars, where x and y represent the number of days spent in each of the two metropolitan areas that comprise her sales territory. If she plans to work 20 days during the month, how many days should she spend in each area to maximize her sales?

26. Revenue. The marketing manager of a department store has determined that revenue, in dollars, is related to the number of units of television advertising x and the number of units of newspaper advertising y by the function $R(x, y) = 500(20x + 5y + 6xy - x^2)$. Each unit of television advertising costs \$3000, and each unit of newspaper advertising costs \$1500. If the advertising budget is \$30,000, find the maximum revenue.

27. Construction. A farmer wants to build a rectangular pen and then divide it with two interior fences (see drawing). The total area enclosed is to be 2484 ft². The exterior fence costs \$18 per foot, and the interior fence costs \$16.50 per foot. Find the dimensions of the pen that will minimize the cost of fencing.

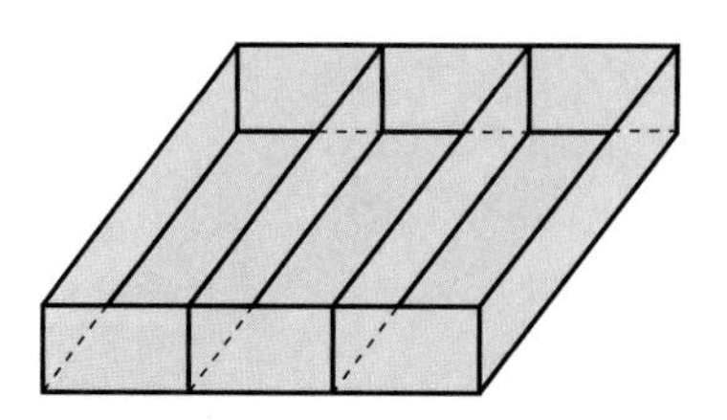

28. Shipping. A container manufacturer is asked to design a closed rectangular shipping crate with a square base. The volume of the crate is 36 ft³. The material for the top costs \$1 per square foot, the material for the sides costs \$0.90 per square foot, and the material for the bottom costs \$1.40 per square foot. Find the dimensions that will minimize the total cost of building the crate.

Hawkes Learning Systems: Essential Calculus

Lagrange Multipliers

8.5 The Method of Least Squares

Objectives

After completing this section, you will be able to:

Apply the method of least squares to obtain a line of best fit.

Suppose that an automobile company is running a new television commercial in five cities with approximately the same population. The company tests the success of the commercial by looking at two variables: x (the number of times the commercial is run on TV in each city) and y (the sales in each city). Sales are in hundreds of cars. The results are shown in the table and scatter diagram in Figure 8.5.1.

For a sample of five cities:

Number of TV Commercials x	Sales of Cars (in hundreds) y
4	3
5	5
12	5
16	8
18	7

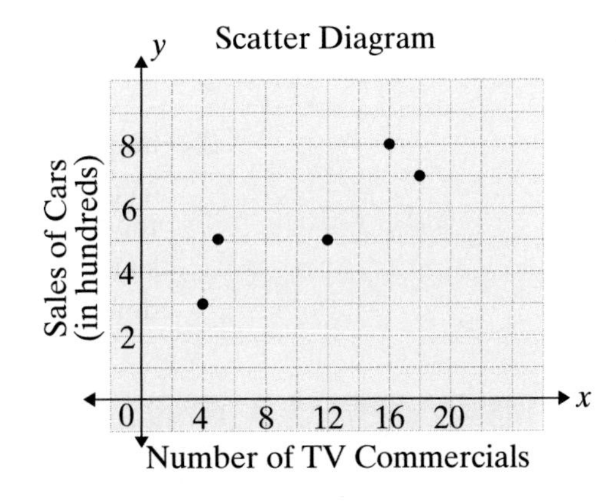

Figure 8.5.1

The scatter diagram indicates that there is a positive correlation between the two variables x and y. That is, there seems to be a tendency for an increase in sales whenever the number of commercials shown is increased.

With this data, the company would like to know what sales to expect in one of these cities if it uses this commercial a particular number of times. For example, if $x = 10$, what would be the expected value of y (the sales in that city)?

One solution to this problem is to find a **curve of best fit**. In this text we will be concerned only with **regression lines** (or **lines of best fit**). An approximate position for the regression line for the data given in Figure 8.5.1 is shown in Figure 8.5.2. Also indicated in Figure 8.5.2 are the vertical distances between the data points and the regression line: d_1, d_2, d_3, d_4 and d_5.

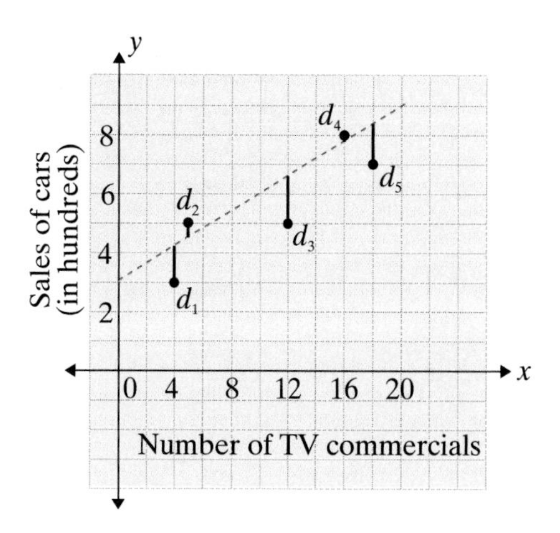

Figure 8.5.2

The dashed line indicates the approximate position of the **regression line** (or **line of best fit**) for the given data.

One algebraic method for finding a regression line is called the **Method of Least Squares**. In this method, the regression line is represented by an equation in the form $y = mx + b$, and the values of the vertical distances between the given points and this line are squared. The values of m and b in the linear equation are determined by minimizing the sum of these squares. With reference to Figure 8.5.2, we want to find the minimum value of the sum

$$\sum_{k=1}^{5} d_k^2 = d_1^2 + d_2^2 + d_3^2 + d_4^2 + d_5^2.$$

NOTES The use of the Greek capital letter sigma (Σ) to indicate summations is a standard notation in mathematics.

For the data given here:

$$d_1 = (4m+b)-3 \quad \text{since } x_1 = 4 \text{ and } y_1 = 3,$$

$$d_2 = (5m+b)-5 \quad \text{since } x_2 = 5 \text{ and } y_2 = 5,$$

$$d_3 = (12m+b)-5 \quad \text{since } x_3 = 12 \text{ and } y_3 = 5,$$

$$d_4 = (16m+b)-8 \quad \text{since } x_4 = 16 \text{ and } y_4 = 8,$$

$$d_5 = (18m+b)-7 \quad \text{since } x_5 = 18 \text{ and } y_5 = 7.$$

We see that the value for each d_k depends on the two variables m and b. Thus the sum of the squares is also a function of m and b:

$$f(m,b) = \sum_{k=1}^{5} d_k^2 = (4m+b-3)^2 + (5m+b-5)^2 + (12m+b-5)^2 + (16m+b-8)^2 + (18m+b-7)^2.$$

Our objective is to find the values of m and b that minimize $f(m, b)$. We use the D-Test discussed in Section 8.2.

$$\begin{aligned} f_m &= 2(4m+b-3)\cdot 4+2(5m+b-5)\cdot 5+2(12m+b-5)\cdot 12 \\ &\quad +2(16m+b-8)\cdot 16+2(18m+b-7)\cdot 18 \\ &= 1530m+110b-702 \\ f_b &= 2(4m+b-3)+2(5m+b-5)+2(12m+b-5) \\ &\quad +2(16m+b-8)+2(18m+b-7) \\ &= 110m+10b-56 \end{aligned}$$

The solutions to the system $f_m = 0$ and $f_b = 0$ are $m = 0.26875$ and $b = 2.64375$.

For our purposes, the rounded-off values $m = 0.27$ and $b = 2.64$ are sufficiently accurate. For the D-Test, we have

$$f_{mm} = 1530,\ f_{bb} = 10, \text{ and } f_{mb} = 110.$$

Thus $D = (1530)(10) - (110)^2 = 3200 > 0$ with $f_{mm} = 1530 > 0$, and the values $m = 0.27$ and $b = 2.64$ do give a minimum value for the sum of the squares represented by the function $f(m, b)$.

Therefore, the regression line is

$$y = 0.27x + 2.64;$$

and for $x = 10$, the company can expect to sell

$$y = 0.27(10) + 2.64 = 5.34 \text{ (about 534 cars).}$$

For any significant amount of data, finding the representation of the function $f(m, b)$, calculating the partials, and solving the simultaneous equations can be a prohibitive exercise. General formulas for finding the values of m and b for the regression line have been determined. These formulas make use of the following notations where n is the number of pairs of data items:

$$\bar{x} = \frac{\sum_{k=1}^{n} x}{n}$$ The average (or mean) of the x values.

$$\bar{y} = \frac{\sum_{k=1}^{n} y}{n}$$ The average (or mean) of the y values.

For simplicity, the subscript notation on Σ has been omitted in the following formulas.

The Least-Squares Regression Line

For a set of data points $(x_1, y_1), (x_2, y_2), \ldots, (x_n, y_n)$, *the regression line is* $y = mx + b$ *where (omitting indices in the summations):*

1. $m = \dfrac{n \cdot \sum(xy) - \sum x \sum y}{n \cdot \sum x^2 - \left(\sum x\right)^2},$

2. $b = \overline{y} - m\overline{x},$

and

3. $\overline{y} = \dfrac{1}{n}\sum y$ *and* $\overline{x} = \dfrac{1}{n}\sum x.$

Example 1: Using Linear Regression on Tabular Data

Use the formulas to find the regression line for the data given in the table in Figure 8.5.1.

Solution: The data are exhibited and calculations are made in table form.

	x	y	xy	x^2
	4	3	12	16
	5	5	25	25
	12	5	60	144
	16	8	128	256
	18	7	126	324
Sums	**55**	**28**	**351**	**765**

We find that

$$\overline{x} = \frac{55}{5} = 11 \text{ and } \overline{y} = \frac{28}{5} = 5.6.$$

Thus, from Formula 1,

$$m = \frac{5(351) - (55)(28)}{5(765) - (55)^2} = \frac{1755 - 1540}{3825 - 3025} = \frac{215}{800} = 0.27(\text{rounded off}),$$

and, from Formula 2,

$$b = \overline{y} - m\overline{x} = 5.6 - (0.27)(11) = 5.6 - 2.97 = 2.63.$$

continued on next page ...

The line of regression is

$$y = 0.27x + 2.63$$

which is in agreement (except for slight round-off errors) with the previous calculations.

Example 2: Forecasting Grade Point Averages with Linear Regression

The following table shows the high school grade-point averages (HS-GPA) and the college grade-point averages (C-GPA) after 1 year of college for 10 students.

a. Find the regression line for the data given in the table.
b. What would be your best prediction for the C-GPA of a high school student with an HS-GPA of 2.5?
c. Graph the data and the regression line on the same set of axes.

HS-GPA(x)	2.0	2.0	2.2	2.2	2.7	3.2	3.2	3.3	3.5	3.7
C-GPA(y)	1.5	1.8	2.0	1.5	2.0	2.8	3.0	3.5	3.5	3.4

Solutions: a. Set up a table to calculate the values needed for Formulas 1 and 2.

	x	y	xy	x^2
	2.0	1.5	3.00	4.00
	2.0	1.8	3.60	4.00
	2.2	2.0	4.40	4.84
	2.2	1.5	3.30	4.84
	2.7	2.0	5.40	7.29
	3.2	2.8	8.96	10.24
	3.2	3.0	9.60	10.24
	3.3	3.5	11.55	10.89
	3.5	3.5	12.25	12.25
	3.7	3.4	12.58	13.69
Sums	**28.0**	**25.0**	**74.64**	**82.28**

Now

$$\bar{x} = \frac{28}{10} = 2.8 \text{ and } \bar{y} = \frac{25}{10} = 2.5.$$

From Formula 1, we have

$$m = \frac{10(74.64) - (28.0)(25.0)}{10(82.28) - (28.0)^2} = \frac{46.4}{38.8} = 1.20 \quad (\text{rounded off}).$$

From Formula 2, we have

$$b = 2.5 - 1.20(2.8) = -0.86.$$

Thus the line of regression is $y = 1.20x - 0.86$.

b. The best prediction for HS-GPA of 2.5 is found by substituting $x = 2.5$ in the equation for the line of regression and solving for y:

$$y = 1.20(2.5) - 0.86 = 2.14.$$

c.

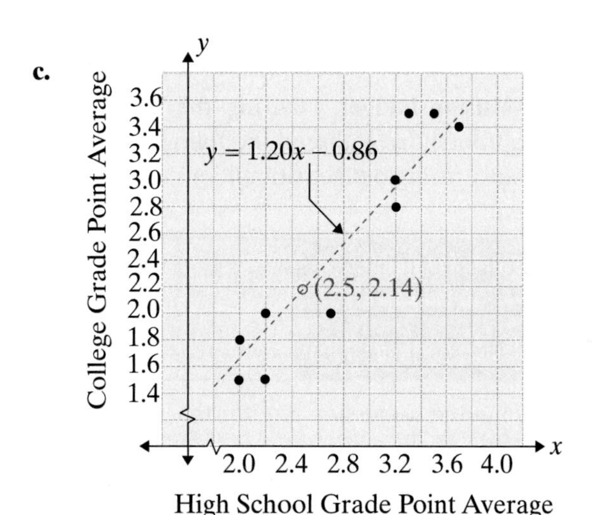

On a subjective basis, a look at a carefully detailed scatter diagram can help determine whether the curve of the best fit is a straight line or some other curve such as a parabola. To deal with a nonlinear relationship between two variables, we would need to modify the model from linear $(y = mx + b)$ to quadratic $(y = ax^2 + bx + c)$ or some higher-degree function and minimize the sum of the squares by using partial derivatives as before. A nonlinear regression curve is illustrated in Figure 8.5.3. We will not analyze nonlinear regression curves in this text.

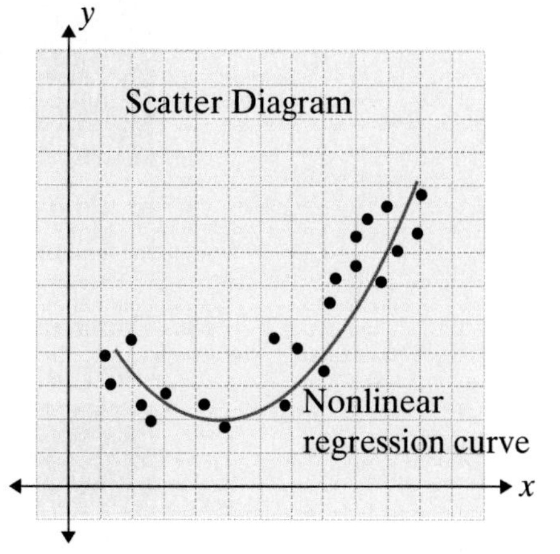

Figure 8.5.3

8.5 Exercises

*In Exercises 1 – 4, **a.** find the equation of the regression line for the given points, and **b.** draw the scatter diagram and graph the regression line.*

1. $(0, 3), (1, 5), (2, 7), (3, 8), (5, 9), (6, 9)$
2. $(1, 10), (2, 8), (3, 7), (4, 6), (5, 5), (6, 5), (7, 4)$
3. $(1, 9.6), (2, 8.7), (3, 7.7), (4, 6.1), (5, 5.0)$
4. $(1, 5.2), (2, 6.4), (3, 8.1), (4, 9.2), (5, 10.6)$

In Exercises 5 – 10, find the equation of the regression line for the given points.

5. $(10, 6.5), (20, 5.8), (30, 5.6), (40, 3.1), (50, 1.8)$
6. $(1, 0.2), (2, 0.4), (3, 0.3), (4, 0.6), (5, 0.6)$
7. $(1, 236), (2, 248), (3, 270), (4, 285), (5, 291)$
8. $(1, 0.45), (3, 0.71), (4, 0.82), (5, 0.94)$
9. $(0.6, 4.8), (0.8, 5.0), (1.0, 4.8), (1.2, 5.2), (1.4, 5.8)$
10. $(3.2, 0.10), (4.1, 0.15), (4.8, 0.20), (5.1, 0.23), (6.0, 0.29)$

11. Advertising budget. During the last 5 years, the advertising manager for a corporation has gathered the following data that shows the relationship between the advertising budget (in millions of dollars) and the total sales (in thousands of units).

Advertising Budget (x) (in millions)	\$4.5	\$6.5	\$3.5	\$3.2	\$2.6
Sales (y) (in thousands)	37	46	42	32	29

a. Find the regression line for the data.
b. Estimate the sales if \$4 million is budgeted for advertising.

12. Price. Records at a company for the last 5 years show the following relationship between the units sold (in thousands) and the price of a product.

Price (p)	\$8.80	\$8.00	\$7.50	\$6.90	\$6.20
Quantity sold (x) (in thousands)	3.8	5.2	7.3	8.0	9.6

a. Find the regression line for the price in terms of units.
b. Estimate the price that should be charged in order to sell 10,000 units.

13. Construction. The following data shows the amount spent on office building construction (in thousands) for a particular county during a 6-month period.

Month	Apr	May	June	July	Aug	Sept
Amount (in thousands)	\$24	\$19	\$30	\$49	\$68	\$69

a. Find the regression line for the data.
b. Estimate the amount spent in October.

14. Revenue. The annual revenue (in millions of dollars) for a corporation is given in the following table.

Year	1999	2000	2001	2002	2003	2004
Revenue (in millions)	\$66	\$82	\$127	\$201	\$310	\$392

a. Find the line of regression for the data.
b. Estimate the revenue for 2005.

15. Livestock futures. The price of livestock futures is the estimated market price of livestock on the delivery date (end of the indicated month). The cattle futures (in cents per pound) for the months February through July are as follows.

Month	Feb	Mar	Apr	May	June	July
Price (¢ per pound)	79.10	76.02	71.80	71.45	71.45	72.50

a. Find the line of regression for the data.
b. Estimate the price for August.

16. Tourism. The total number of foreign tourists visiting the United States, as reported by the U.S. Travel and Tourism Administration, is shown in the following table.

Year (x)	2000	2001	2002	2003	2004
Tourists (y) (in millions)	25.7	26.3	29.7	34.2	38.3

a. Find the regression line.
b. Estimate the number of foreign tourists that will visit the United States during 2005.

Hawkes Learning Systems: Essential Calculus

The Method of Least Squares

8.6 Double Integrals

Objectives

After completing this section, you will be able to:

Evaluate double integrals on regions of type I, II, and III.

In Chapter 6 we developed the concept of a definite integral of a continuous function of one variable. A definite integral was defined on a closed interval $[a, b]$ to be the limit of a Riemann Sum. The closed interval was partitioned into subintervals of width $\Delta x = \dfrac{b-a}{n}$, where n was the number of subintervals. In an analogous manner, we define the **double integral** of a continuous function of two variables, $z = f(x, y)$, on a closed region R in the xy-plane where R is partitioned into n rectangular subregions each with area $\Delta A = \Delta x\, \Delta y$. In each of these subregions, a point (x_k, y_k) is chosen and the n products $f(x_k, y_k)\Delta x \Delta y$ are added to form a Riemann Sum for the function of two variables. The limit of this sum as $n \to \infty$ is called a **double integral** (or an **iterated integral**) of the function $z = f(x, y)$ on the region R.

Definition of Double Integral

For a function $z = f(x, y)$ continuous on a region R in the xy-plane

$$\iint_R f(x,y)\,dA = \iint_R f(x,y)\,dx\,dy = \lim_{n\to\infty} \sum_{k=1}^{n} f(x_k, y_k)\Delta x \Delta y.$$

For a continuous function of one variable, $y = f(x) \geq 0$, the definite integral can be interpreted geometrically as the area between the graph of the function and the x-axis on the closed interval $[a, b]$. Similarly, if a continuous function $z = f(x, y) \geq 0$ for all (x, y) in the region R, then each of the products $f(x_k, y_k)\Delta x \Delta y$ in the Riemann sum can be interpreted as the volume of a rectangular solid. As $n \to \infty$ and the rectangular regions $\Delta x \Delta y$ shrink in size, the sum of the volumes of these rectangular solids approaches the volume of the solid bounded above by f and below by the xy-plane with a vertical boundary passing through the boundary of the region R. (See Figure 8.6.1 and 8.6.2.)

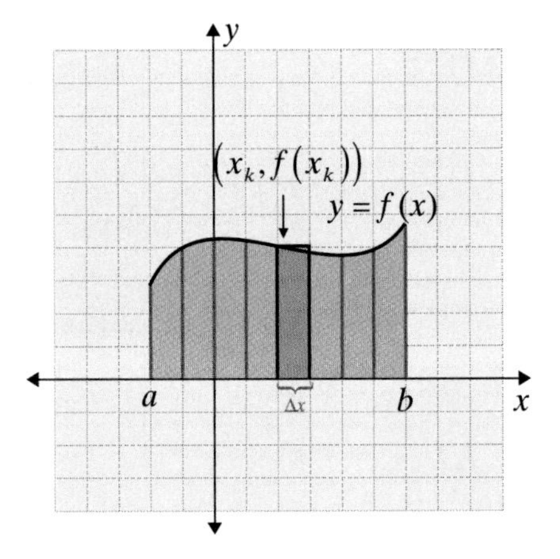

Figure 8.6.1

$$\int_a^b f(x)\,dx = \lim_{n\to\infty} S_n$$

$$= \lim_{n\to\infty} \sum_{k=1}^{n} f(x_k)\Delta x$$

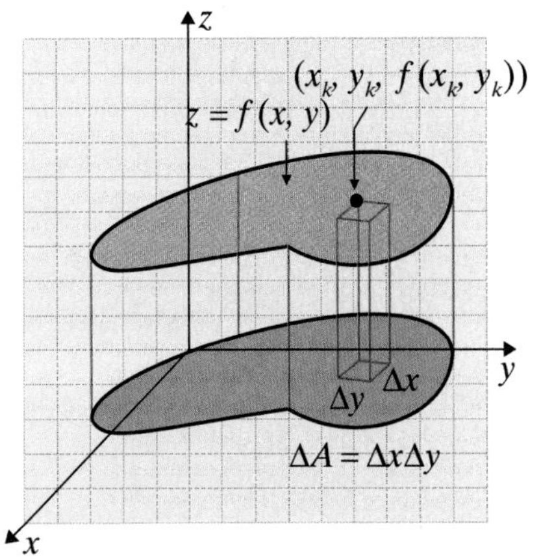

Figure 8.6.2

$$\iint_R f(x,y)\,dA = \iint_R f(x,y)\,dx\,dy$$

$$= \lim_{n\to\infty} \sum_{k=1}^{n} f(x_k, y_k)\,\Delta x \Delta y$$

Thus we define the volume of a solid as depicted in Figure 8.6.2 to be the **double integral** of f on R.

Value of the Double Integral

If $z = f(x, y)$ is a continuous function on a region R in the xy-plane and $f(x,y) \geq 0$ on R, then the volume of the solid bounded above by the graph of f and below by R is the ***value of the double integral*** *of f on R:*

$$V = \iint_R f(x,y)\,dx\,dy.$$

The actual evaluation of a double integral, whether the interpretation is geometric or not, depends a great deal on the nature of the region R in the xy-plane. In this text we will discuss three types of regions, which we designate **Type I**, **Type II**, and **Type III** for reference purposes only. A general form of each of these types is shown in Figure 8.6.3 and Figure 8.6.4.

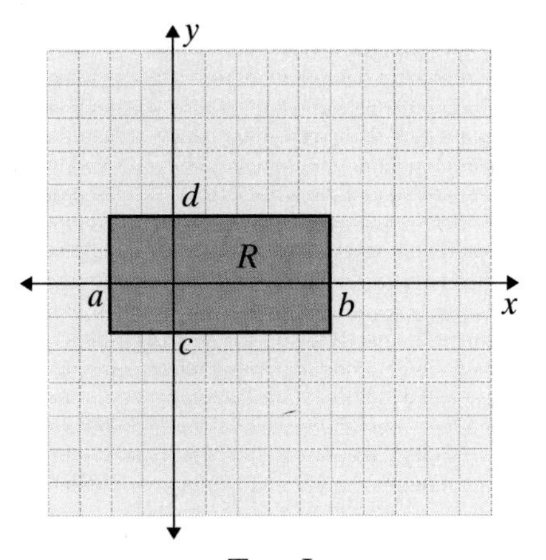

Figure 8.6.3

Type I
Region R is a rectangle:
$R = \{(x, y) \mid a \leq x \leq b \text{ and } c \leq y \leq d\}$

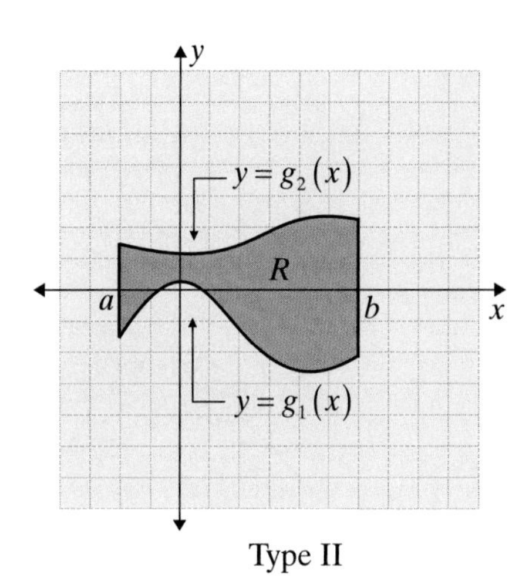

Figure 8.6.4

Type II

$x = h_1(y)$ d $x = h_2(y)$ R a x c

Type III

Region R is bound left and right by two constant values of x and above and below by two continuous functions of x:

$R = \{(x, y) | a \le x \le b \text{ and } g_1(x) \le y \le g_2(x)\}$

Region R is bound left and right by two continuous functions of y and above and below by two constant values of y:

$R = \{(x, y) | h_1(y) \le x \le h_2(y) \text{ and } c \le y \le d\}$

The following theorem, stated here without proof, shows how the evaluation of a double integral depends on the region R and its type.

Theorem for Evaluating Double Integrals

If $z = f(x, y)$ is a continuous function on a closed region R, and

Case 1: *R is of Type I, then*

$$\iint_R f(x, y)\,dxdy = \int_c^d \left(\int_a^b f(x, y)\,dx\right)dy = \int_a^b \left(\int_c^d f(x, y)\,dy\right)dx.$$

Case 2: *R is of Type II, then*

$$\iint_R f(x, y)\,dxdy = \int_a^b \left(\int_{g_1(x)}^{g_2(x)} f(x, y)\,dy\right)dx.$$

Case 3: *R is of Type III, then*

$$\iint_R f(x, y)\,dxdy = \int_c^d \left(\int_{h_1(y)}^{h_2(y)} f(x, y)\,dx\right)dy.$$

Note: *Only in Case 1 have we shown that the order of integration may be reversed. There are problems of the types in Case 2 and Case 3 in which the order of integration may be reversed; however, we will not discuss reversing the order of integration for such problems in this text.*

In each case, the evaluation of a double integral is accomplished in two steps:

Step 1: The "inside" integration is performed with respect to the variable indicated by the inside differential (dy or dx), and the second variable is treated as a constant. Then the integral is evaluated by using the limits on the inside integral sign and substituting for the first variable. The result is a function of the second variable.

Step 2: The resulting function from Step 1 is integrated with respect to the second variable and then evaluated by using the limits on the outside integral sign.

Example 1: Evaluating a Double Integral on a Type I Region

Evaluate the double integral $\int_1^3 \int_1^e \frac{y^2}{x}\, dx dy$.

Solution: Since R is of Type I (a rectangular region), we may integrate in either order and obtain the same result. Both integrations are shown here to clarify the idea of reversing the order of integration.

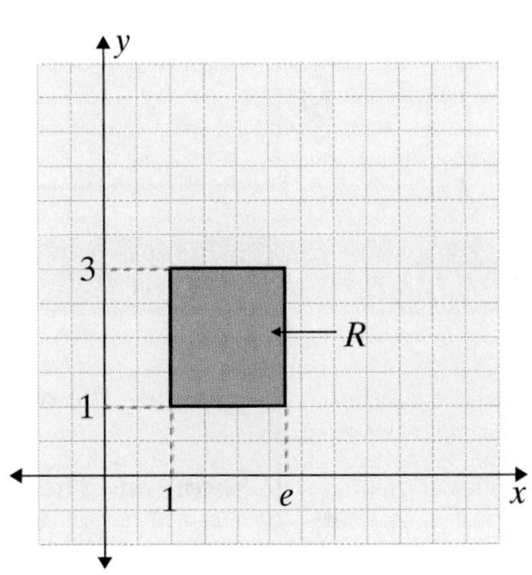

1.
$$
\begin{aligned}
\int_1^3 \int_1^e \frac{y^2}{x}\, dxdy &= \int_1^3 \left(\int_1^e \frac{y^2}{x}\, dx \right) dy \\
&= \int_1^3 \left(y^2 \ln|x| \Big]_1^e \right) dy \\
&= \int_1^3 \left(y^2 \ln|e| - y^2 \ln|1| \right) dy \\
&= \int_1^3 \left(y^2 - 0 \right) dy \\
&= \frac{y^3}{3} \Big]_1^3 \\
&= \frac{3^3}{3} - \frac{1^3}{3} \\
&= \frac{27}{3} - \frac{1}{3} \\
&= \frac{26}{3}
\end{aligned}
$$

Integrate first with respect to x and evaluate at $x = e$ and $x = 1$. The variable y is treated as constant.

Now integrate with respect to y and evaluate.

continued on next page ...

2. $\int_1^3 \int_1^e \frac{y^2}{x} dxdy = \int_1^e \left(\int_1^3 \frac{y^2}{x} dy \right) dx$

Integrate first with respect to y and evaluate at $y = 3$ and $y = 1$. The variable x is treated as a constant.

$= \int_1^e \left(\frac{1}{x} \cdot \frac{y^3}{3} \Big]_1^3 \right) dx$

$= \int_1^e \left(\frac{1}{x} \cdot \frac{3^3}{3} - \frac{1}{x} \cdot \frac{1^3}{3} \right) dx$

$= \int_1^e \left(\frac{27}{3} - \frac{1}{3} \right) \frac{1}{x} dx$

$= \frac{26}{3} \int_1^e \frac{1}{x} dx$

Now integrate with respect to x and evaluate.

$= \frac{26}{3} (\ln|x|) \Big]_1^e = \frac{26}{3} (\ln|e| - \ln|1|)$

$= \frac{26}{3}(1-0) = \frac{26}{3}$

Example 2: Evaluating a Double Integral on a Type II Region

Evaluate the double integral $\int_0^1 \int_{x^2}^{x+1} 2x^2 y dydx$.

Solution: The region R is of Type II and is illustrated in the figure at the right.

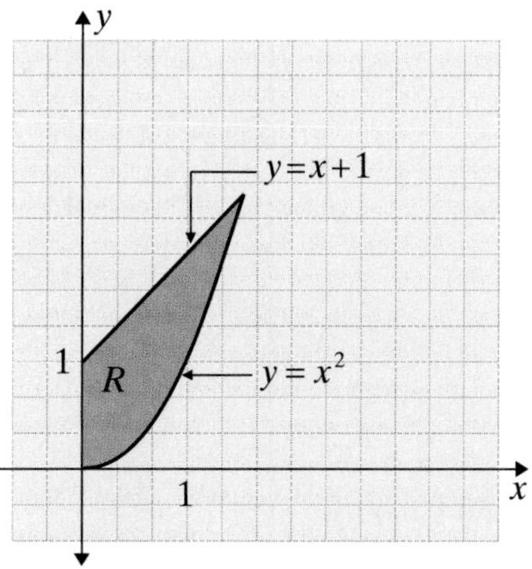

$\int_0^1 \int_{x^2}^{x+1} 2x^2 y dydx = \int_0^1 \left(\int_{x^2}^{x+1} 2x^2 y dy \right) dx$

Integrate first with respect to y and evaluate at $y = x + 1$ and $y = x^2$.

$= \int_0^1 \left(2x^2 \frac{y^2}{2} \Big]_{x^2}^{x+1} \right) dx$

$= \int_0^1 \left(x^2 y^2 \Big]_{x^2}^{x+1} \right) dx$

$= \int_0^1 \left[x^2 (x+1)^2 - x^2 \left(x^2\right)^2 \right] dx$

$= \int_0^1 [x^2(x^2+2x+1) - x^6] dx$

$$= \int_0^1 \left(x^4 + 2x^3 + x^2 - x^6\right)dx = \frac{x^5}{5} + \frac{x^4}{2} + \frac{x^3}{3} - \frac{x^7}{7}\Bigg]_0^1$$

$$= \left(\frac{1}{5} + \frac{1}{2} + \frac{1}{3} - \frac{1}{7}\right) - (0) = \frac{42}{210} + \frac{105}{210} + \frac{70}{210} - \frac{30}{210}$$

$$= \frac{187}{210} = 0.89\,(\text{to the nearest hundredth})$$

Example 3: Evaluating a Double Integral on a Type III Region

Evaluate the double integral $\int_0^1 \int_0^{\frac{y}{3}} e^{3x+y}\,dx\,dy.$

Solution: $\int_0^1 \int_0^{\frac{y}{3}} e^{3x+y}\,dx\,dy = \int_0^1 \left(\int_0^{\frac{y}{3}} e^{3x+y}\,dx \right) dy$

Integrate first with respect to x and evaluate at $x = \frac{y}{3}$ and $x = 0$.

$$= \int_0^1 \left(\frac{1}{3} e^{3x+y} \Bigg]_0^{\frac{y}{3}} \right) dy$$

$$= \int_0^1 \left(\frac{1}{3} e^{3\left(\frac{y}{3}\right)+y} - \frac{1}{3} e^{3\cdot 0+y} \right) dy$$

$$= \frac{1}{3} \int_0^1 \left(e^{2y} - e^{y}\right) dy = \frac{1}{3}\left(\frac{1}{2}e^{2y} - e^{y}\right)\Bigg]_0^1$$

$$= \frac{1}{3}\left[\left(\frac{1}{2}e^{2} - e^{1}\right) - \left(\frac{1}{2}e^{0} - e^{0}\right)\right]$$

$$= \frac{1}{6}e^2 - \frac{1}{3}e + \frac{1}{6}$$

Example 4: Finding the Volume of a Geometric Solid

Find the volume of the solid bounded above by the graph of the function $z = 10 - x - y$ and below by the triangular region R in the xy-plane with vertices at $(0, 0, 0)$, $(4, 0, 0)$, and $(0, 8, 0)$.

Solution: The surface $z = 10 - x - y$ is a plane, and the region R is a triangle in the xy-plane with the given points as vertices. The volume to be found is illustrated in the figure on the next page.

continued on next page ...

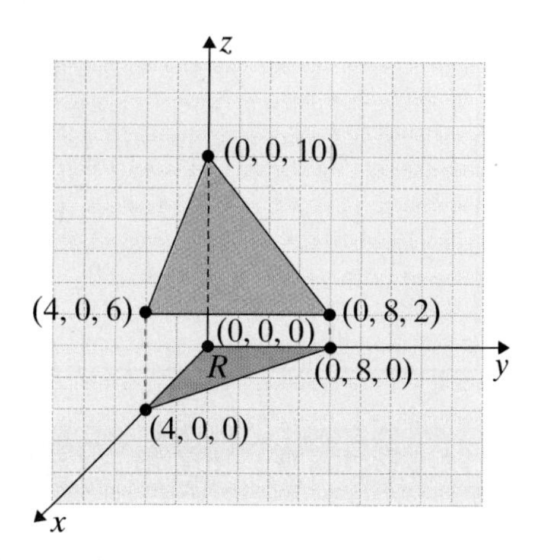

To set the limits of integration, we need to represent the region R in terms of x and y. We can do this by determining the equation of the line through the two points (4, 0) and (0, 8) on an xy-coordinate system and then setting constant restrictions on one of the variables so that the triangular region is described.

The slope of the line is

$$m = \frac{0-8}{4-0} = -2,$$

and the equation of the line is

$$y - 8 = -2(x-0)$$

$$y = -2x + 8.$$

Thus R can be described by the following restrictions on x and y:

$0 \le x \le 4$ and $0 \le y \le -2x+8$.

The described volume is now found by using double integration.

$$\int_0^4 \left(\int_0^{-2x+8} (10 - x - y)\, dy \right) dx = \int_0^4 \left(10y - xy - \frac{y^2}{2} \Big]_0^{-2x+8} \right) dx$$

$$= \int_0^4 \left(10(-2x+8) - x(-2x+8) - \frac{(-2x+8)^2}{2} \right) dx$$

$$= \int_0^4 (-12x + 48)\, dx = -6x^2 + 48x \Big]_0^4$$

$$= (-96 + 192) - 0 = 96$$

Thus the volume of the solid described is 96 cubic units.

8.6 Exercises

In Exercises 1 – 10, evaluate the given double integral.

1. $\int_0^1\int_1^2 (x+1)\,dydx$

2. $\int_0^2\int_1^3 (4-x)\,dydx$

3. $\int_{-1}^2\int_1^4 (3x+2y)\,dydx$

4. $\int_{-2}^2\int_3^4 (2x-y)\,dydx$

5. $\int_1^2\int_2^3 \frac{2y}{x}\,dydx$

6. $\int_1^3\int_{-1}^2 \frac{3y}{2x}\,dydx$

7. $\int_1^3\int_{-2}^2 \left(x^2+3y^2-1\right)dxdy$

8. $\int_2^3\int_{-1}^1 \left(2x^2+y^2-x\right)dxdy$

9. $\int_0^2\int_0^1 e^{x+y}\,dxdy$

10. $\int_0^1\int_{-1}^2 ye^{xy}\,dxdy$

In Exercises 11 – 16, evaluate the double integral on the given rectangular region.

11. $\iint_R \left(x-y^2\right)dA \quad R: 0\le x\le 2 \text{ and } 0\le y\le 1$

12. $\iint_R (xy+x)\,dA \quad R: 0\le x\le 3 \text{ and } 0\le y\le 3$

13. $\iint_R y\sqrt{x+1}\,dA \quad R: 0\le x\le 3 \text{ and } 1\le y\le 5$

14. $\iint_R x^2\sqrt{3+y}\,dA \quad R: -1\le x\le 4 \text{ and } 1\le y\le 6$

15. $\iint_R e^{x+2y}\,dA \quad R: 0\le x\le 3 \text{ and } 0\le y\le 4$

16. $\iint_R e^{2x+y}\,dA \quad R: 0\le x\le 2 \text{ and } 0\le y\le 1$

In Exercises 17 – 24, evaluate the double integral.

17. $\int_0^2\int_0^{3x} xy^2\,dydx$

18. $\int_0^1\int_{2x}^{x^2} x^2y^2\,dydx$

19. $\int_1^4\int_0^{x^2} \sqrt{\frac{y}{x}}\,dydx$

20. $\int_1^4\int_x^{x^2} \sqrt{\frac{x}{y}}\,dydx$

21. $\int_1^3\int_1^{e^y} \frac{y}{x}\,dxdy$

22. $\int_0^2\int_0^y e^{y^2}\,dxdy$

23. $\int_0^4\int_0^y \sqrt{9+y^2}\,dxdy$

24. $\int_0^2\int_0^{4-y^2} y\sqrt{x}\,dxdy$

In Exercises 25 – 30, evaluate the double integral on the given region.

25. $\iint_R 2xy\,dA$ $\quad R: 0 \le x \le 1$ and $x^2 \le y \le \sqrt{x}$

26. $\iint_R 3xy^2\,dA$ $\quad R: 0 \le x \le 1$ and $x^3 \le y \le \sqrt[3]{x}$

27. $\iint_R \left(x^2 - y\right)dA$ $\quad R: 1 \le x \le 2$ and $x \le y \le x^2$

28. $\iint_R (3-2x-2y)\,dA$ $\quad R: 0 \le x \le 1$ and $0 \le y \le (2-x)$

29. $\iint_R e^y\,dA$ $\quad R: 0 \le x \le 2$ and $x \le y \le 3x$

30. $\iint_R e^y\,dA$ $\quad R: 0 \le x \le 1$ and $0 \le y \le 2x$

31. $\iint_R (x+y)\,dA$

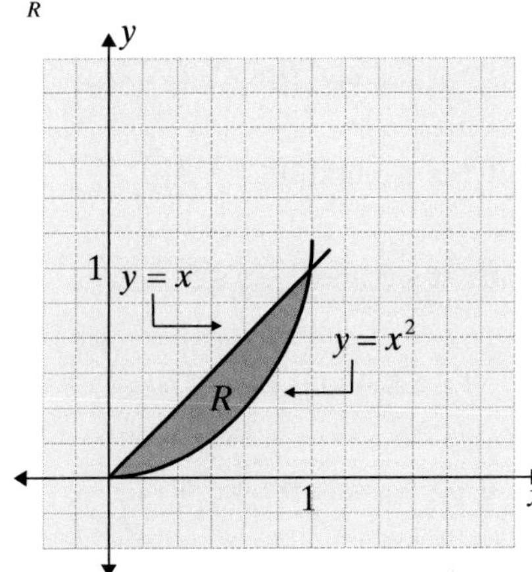

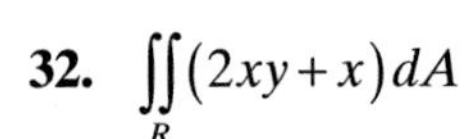

32. $\iint_R (2xy+x)\,dA$

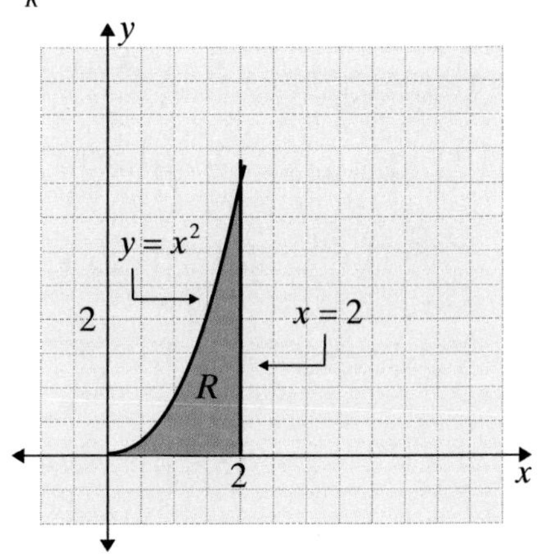

33. $\iint_R (3-2xy)\,dA$

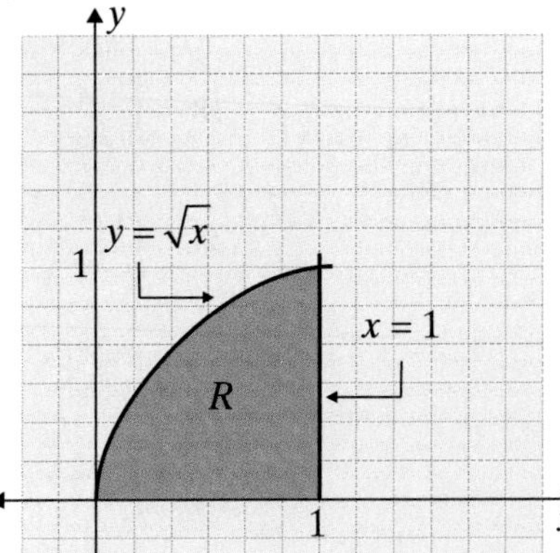

34. $\iint_R \sqrt{4-x^2}\,dA$

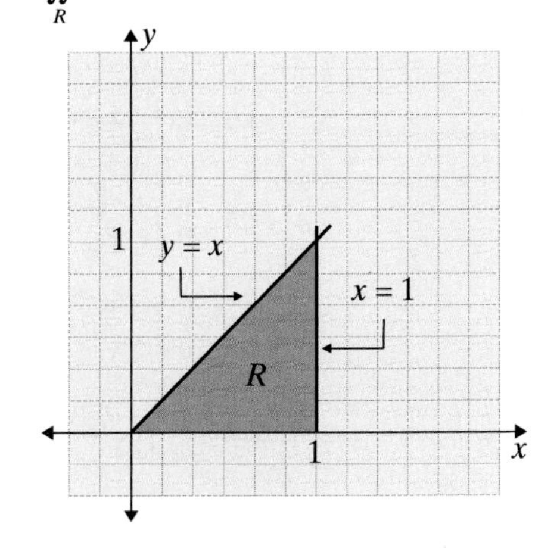

35. Find the volume of the solid bounded above by the graph of $f(x, y) = 8 - x^2 - y^2$ and below by the rectangle $R: -1 \le x \le 2$ and $0 \le y \le 2$.

36. Find the volume of the solid bounded above by the graph of $f(x, y) = 2 + x^2 + y^2$ and below by the rectangle $R: 0 \le x \le 1$ and $0 \le y \le 3$.

37. Find the volume of the solid bounded above by the graph of $f(x, y) = 8 - 4x - 2y$ and below by the triangle with vertices (0, 0, 0), (2, 0, 0), and (0, 4, 0).

38. Find the volume of the solid bounded above by the graph of $f(x, y) = 3 + x + 2y$ and below by the triangle with vertices (0, 0, 0), (0, 2, 0), and (2, 0, 0).

39. Find the volume of the solid bounded above by the graph of $f(x, y) = 2xy$ and below by the region bounded by $y = \sqrt{x}$, $y = 0$, and $x = 1$.

40. Find the volume of the solid bounded above by the graph of $f(x, y) = 4x^2 y$ and below by the region bounded by $y = x^2$, $y = 0$, and $x = 1$.

Hawkes Learning Systems: Essential Calculus

Double Integrals

Chapter 8 Index of Key Ideas and Terms

Section 8.1 Functions of Several Variables

For a Function of Two Variables $z = f(x, y)$ pages 612 - 619

domain, range, dependent variable, independent variables, Cobb-Douglas Production Formula, ordered triples in the form (x, y, z), points in space, surfaces

Section 8.2 Partial Derivatives

Partial Derivatives pages 624 - 629

Given a function $z = f(x, y)$,

$$\frac{\partial f}{\partial x} = \lim_{h \to 0} \frac{f(x+h, y) - f(x, y)}{h} \quad \text{and} \quad \frac{\partial f}{\partial y} = \lim_{k \to 0} \frac{f(x, y+k) - f(x, y)}{k}$$

Notation for Partial Derivatives page 624

$$\frac{\partial f}{\partial x} = \frac{\partial z}{\partial x} = f_x$$

$$\frac{\partial f}{\partial y} = \frac{\partial z}{\partial y} = f_y$$

These partial derivatives are also called **first partials** or **first-order partial derivatives.**

Notation for Second-Order Partial Derivatives pages 630 - 631

$$\frac{\partial}{\partial x}\left(\frac{\partial z}{\partial x}\right) = \frac{\partial^2 z}{\partial x^2} = \frac{\partial^2 f}{\partial x^2} = f_{xx}$$

$$\frac{\partial}{\partial y}\left(\frac{\partial z}{\partial y}\right) = \frac{\partial^2 z}{\partial y^2} = \frac{\partial^2 f}{\partial y^2} = f_{yy}$$

$$\frac{\partial}{\partial y}\left(\frac{\partial z}{\partial x}\right) = \frac{\partial^2 z}{\partial y \partial x} = \frac{\partial^2 f}{\partial y \partial x} = f_{xy}$$

$$\frac{\partial}{\partial x}\left(\frac{\partial z}{\partial y}\right) = \frac{\partial^2 z}{\partial x \partial y} = \frac{\partial^2 f}{\partial x \partial y} = f_{yx}$$

Partial Derivatives for Functions of Three Variables pages 631 - 632

Given a function $w = f(x, y, z)$

$$\frac{\partial w}{\partial x} = f_x(x, y, z), \quad \frac{\partial w}{\partial y} = f_y(x, y, z), \quad \text{and} \quad \frac{\partial w}{\partial z} = f_z(x, y, z).$$

Section 8.3 Local Extrema for Functions of Two Variables

Local Extrema for a Function $z = f(x, y)$ Defined on a Region page 636

If (a, b) is an open region R and for all (x, y) in R

1. $f(a,b) \geq f(x,y)$, then $f(a, b)$ is a **local maximum**.
2. $f(a,b) \leq f(x,y)$, then $f(a, b)$ is a **local minimum**.

Critical Point, Saddle Point page 637

Second Partials Test (or *D*-Test) pages 638 - 641

Suppose that a function $z = f(x, y)$ and the first partial derivatives and the second partial derivatives are all defined in an open region R and that **(a, b) is a critical point** in R such that

$$\boldsymbol{f_x(a,b) = 0 \text{ and } f_y(a,b) = 0.}$$

Define the quantity D as follows:

$$\boldsymbol{D = f_{xx}(a,b) \cdot f_{yy}(a,b) - \left[f_{xy}(a,b)\right]^2.}$$

Case 1: If $D > 0$ and $f_{xx}(a,b) > 0$, then $f(a, b)$ is a **local minimum**.

Case 2: If $D > 0$ and $f_{xx}(a,b) < 0$, then $f(a, b)$ is a **local maximum**.

Case 3: If $D < 0$, then $(a, b, f(a, b))$ is a **saddle point**.

Case 4: If $D = 0$, then this test gives no information.

Section 8.4 Lagrange Multipliers

The Method of Lagrange Multipliers pages 644 - 649

To maximize or minimize a function $z = f(x, y)$ subject to the constraint $g(x, y) = 0$,

Step 1: Form the **Lagrange function**

$$F(x,y,\lambda) = f(x,y) + \lambda \cdot g(x,y).$$

Step 2: Find each of the partial derivatives F_x, F_y, and F_λ, provided all partials exist.

Step 3: Solve the system of three equations

$$\begin{cases} F_x(x,y,\lambda) = 0 \\ F_y(x,y,\lambda) = 0 \\ F_\lambda(x,y,\lambda) = 0 \end{cases}$$

Step 4: If f has a local maximum or a local minimum subject to the constraint $g(x, y) = 0$, then the corresponding x and y values will be found among the solutions to the system in Step 3.

Section 8.5 The Method of Least Squares

Least Squares Regression Line pages 652 - 657

For a set of data points $(x_1, y_1), (x_2, y_2), ..., (x_n, y_n)$, the regression line is $y = mx + b$ where (omitting indices in the summations):

1. $m = \dfrac{n \cdot \sum(xy) - \sum x \sum y}{n \cdot \sum x^2 - \left(\sum x\right)^2}$,

2. $b = \bar{y} - m\bar{x}$,

and

3. $\bar{y} = \dfrac{1}{n}\sum y$ and $\bar{x} = \dfrac{1}{n}\sum x$.

Section 8.6 Double Integrals

Double integrals are symbolized as follows: pages 660 - 666

$$\iint_R f(x,y)\,dxdy \text{ or } \int_R \left(\int f(x,y)\,dx\right)dy \text{ or } \int_R \left(\int f(x,y)\,dy\right)dx \text{ or } \iint f(x,y)\,dy\,dx$$

The evaluation of a double integral is accomplished in two steps:

Step 1: The "inside" integration is performed with respect to the variable indicated by the inside differential (*dy* or *dx*), and the second variable is treated as a constant. Then the integral is evaluated by using the limits on the inside integral sign and substituting for the first variable. The result is a function of the second variable.

Step 2: The resulting function from Step 1 is integrated with respect to the second variable and then evaluated by using the limits on the outside integral sign.

For double integrals, we consider three types of regions:

Type I Region: Region R is a rectangle:

$$R = \{(x,y) \mid a \le x \le b \text{ and } c \le y \le d\}$$

Type II Region: Region R is bound left and right by two constant values of x and above and below by two continuous functions of x:

$$R = \{(x,y) \mid a \le x \le b \text{ and } g_1(x) \le y \le g_2(x)\}$$

Type III Region: Region R is bound left and right by two continuous functions of y and above and below by two constant values of y:

$$R = \{(x,y) \mid h_1(y) \le x \le h_2(y) \text{ and } c \le y \le d\}$$

Chapter 8 Review

For a review of the topics and problems from Chapter 8, look at the following lessons from *Hawkes Learning Systems: Essential Calculus.*

Functions of Several Variables
Partial Derivatives
Local Extrema for Functions of Two Variables
Lagrange Multipliers
The Method of Least Squares
Double Integrals

Chapter 8 Test

In Exercises 1 – 4, find the indicated function values, if possible.

1. $f(x, y) = 4x^2 + 9xy - 3y^2$, $f(-1, 2)$, $f(3, 5)$

2. $f(x, y) = 5xy - 3y^2 + 2\ln|x|$, $f(0, 1)$, $f(-2, 4)$

3. $f(x, y) = x^2\sqrt{1+y}$, $f(-1, 3)$, $f(4, 5)$

4. $f(x, y, z) = 6x^2 ye^{2yz}$, $f(1, 1, 0)$, $f(5, 1, -1)$

In Exercises 5 – 10, find $\frac{\partial f}{\partial x}$ *and* $\frac{\partial f}{\partial y}$.

5. $f(x,y) = 5x^2 - \frac{4x}{y} + 2y^2$

6. $f(x,y) = 8e^{\frac{x}{y}}$

7. $f(x,y) = \frac{3}{2}x^2 \ln(y^2+1)$

8. $f(x,y) = \frac{3x}{x-5y}$

9. $f(x,y) = \sqrt{6 - x^2y^2}$

10. $f(x,y) = xe^{xy^2}$

In Exercises 11 and 12, find $\frac{\partial f}{\partial x}$, $\frac{\partial f}{\partial y}$, *and* $\frac{\partial f}{\partial \lambda}$.

11. $f(x,y,\lambda) = x^3y + y + \lambda(9x - 2y - 6)$

12. $f(x,y,\lambda) = 7xy + \lambda(x^2 + y^2 - 16)$

In Exercises 13 and 14, find $f_{xx}(x, y)$, $f_{xy}(x, y)$, $f_{yx}(x, y)$, *and* $f_{yy}(x, y)$.

13. $f(x,y) = x^5 - 2y^2 + \frac{2}{3}y^3$

14. $f(x,y) = xy^2 + y\ln x - e^{2y}$

In Exercises 15 – 20, find all local maxima, local minima, and saddle points.

15. $f(x,y) = 31 + x - 2xy - 2y^2$

16. $f(x,y) = x^2 - xy + y^2 + 9x - 6y + 3$

17. $f(x,y) = \frac{1}{3}x^3 + xy - 4x - \frac{1}{6}y^2$

18. $f(x,y) = 2xy + x^2 + \frac{1}{3}y^3 + 11$

19. $f(x,y) = x^3 - 4.5x^2 - 30x + 3y^2 - 6y$

20. $f(x,y) = x^2 - 3y^2 + 4xy - 16x - 4y + 80$

In Exercises 21 – 24, use the method of Lagrange multipliers to find the maximum and minimum values of f subject to the given constraint.

21. $f(x,y)=2x^2+y^2-xy$, subject to $x+y=24$

22. $f(x,y)=10xy-3x^2-y^2$, subject to $2x+5y=1790$

23. $f(x,y)=60x^{0.2}y^{0.8}$, subject to $9x+15y=450$

24. $f(x,y)=7xy$, subject to $4x^2+y^2=128$

In Exercises 25 and 26, find the equation of the regression line for the given points.

25. (0, 15), (2, 13), (4, 12), (6, 11), (8, 8)
26. (1, 3.6), (2, 4.1), (3, 4.5), (4, 5.2), (5, 5.9)

In Exercises 27 – 32, evaluate the double integral.

27. $\int_0^1\int_1^3(3-x-2y)\,dydx$

28. $\int_1^e\int_{-1}^2\frac{2y}{x}\,dydx$

29. $\int_0^3\int_3^5 y\sqrt{y^2+16}\,dxdy$

30. $\int_0^3\int_1^{x^2}(2y-6x)\,dydx$

31. $\int_0^1\int_0^x e^{x^2+1}\,dydx$

32. $\int_{-2}^1\int_y^{2-y}\left(y^2+2x\right)dxdy$

In Exercises 33 – 38, evaluate the double integral on the given region.

33. $\iint_R\left(x^2+y^2\right)dA \quad R: -2\le x\le 0 \text{ and } -1\le y\le 2$

34. $\iint_R x\sqrt{1-x^2}\,dA \quad R: 0\le x\le 1 \text{ and } 2\le y\le 3$

35. $\iint_R e^{3x+y}\,dA \quad R: 0\le x\le 1 \text{ and } 0\le y\le 3$

36. $\iint_R xy^2\,dA \quad R: 1\le x\le 2 \text{ and } \sqrt[3]{x}\le y\le x^2$

37. $\iint_R 2xy\,dA \quad R: 0\le x\le 3 \text{ and } 0\le y\le \sqrt{9-x^2}$

38. $\iint_R\left(2xy-x^2\right)dA \quad R: 0\le x\le 3 \text{ and } x\le y\le 6-x$

39. Volume and surface area. A rectangular box has no top and two partitions (see drawing).

a. Write a formula of three variables for the volume of the box.
b. Write a function of three variables for the surface area of the box.

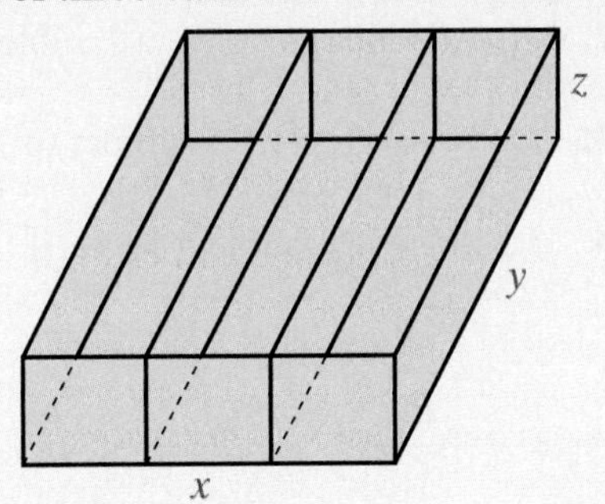

40. Marginal productivity. The productivity of a certain company is given by $f(L,K)=80L^{0.4}K^{0.6}$ units where L is the units of labor and K is the units of capital. If the company is currently utilizing 40 units of labor and 30 units of capital, find the marginal productivity of labor and the marginal productivity of capital and interpret the results.

41. Cost. A firm produces its product on two assembly lines. The total production cost is given by $C(x,y)=3x^2+4.5y^2-xy,$ where x is the number of units produced on one line and y is the number of units produced on the other. If the total output of the two lines must be 255 units, find the number of units that should be produced on each line to minimize total production cost.

42. Sales. The foreign sales of an automobile manufacturer are given in the following table:

Year	2000	2001	2002	2003	2004	2005
Sales (in million units)	2.01	2.02	2.10	2.05	2.14	2.30

a. Write the equation for the regression line.
b. Estimate the foreign sales for 2006.

43. Volume. Find the volume of the solid bounded above by the graph of $f(x,y)=9-x-y$ and below by the triangle with vertices $(0,0,0)$, $(0,1,0)$, and $(1,0,0)$.

Cumulative Review

1. If the distance in feet of a particle from an origin is given by $s = 3 + 3t - t^2$, where t is time in seconds, then determine:

a. the velocity at $t = 3$ seconds,

b. the acceleration at $t = 4$ seconds.

For Exercises 2 to 4, determine the indefinite integral indicated.

2. $\int (3e^{2x} - \frac{5}{x} + 6x + 3x^{-4})dx$

3. $\int \frac{2 + 6x - 5x^4}{x} dx$

4. The marginal profit in dollars from selling small green flags at Notre Dame football games is $P'(x) = 2 - .002x$, where x is the quantity of flags. What is the profit from selling 40 flags? (Assume there are no fixed costs.)

5. The marginal revenue from selling baskets of oranges at a certain roadside stand is modeled by $R'(x) = 6 - .004x$. What revenue is obtained from the sales of 1500 baskets of oranges?

6. What is the average y-value for $f(x) = \sqrt{x}$ over the interval [25, 36]?

7. What is the cost of producing 150 roll-top desks if the marginal cost function is $C'(x) = .003x^2 + .05x + 135$?

8. Form a Riemann Sum for $f(x) = 2xe^{-x}$ on the interval [0, 5] with $n = 10$ terms. Use the right-hand endpoint of Δx_i of each subinterval as t_i. Do not evaluate the Riemann Sum.

9. What is the area between $y_1 = 2x + 1$ and $y_2 = x^2 + 1$?

10. Solve the differential equation $y' = 6xy$, given that $y = 1$ when $x = 0$.

11. Using your calculator, sketch the logistic curve $f(x) = \frac{20}{1 + 19e^{-4t}}$ over the interval [−1, 5]. Add the derivative, $f'(x)$, to your sketch. Find the inflection point graphically with your calculator and mark it on your sketch.

12. Suppose the demand function for a new video game is $D(x) = 42 - 3x$ and the supply function is $S(x) = x^2 + 2$.

a. Find the equilibrium point.

b. Find the consumers' surplus.

c. Find the producers' surplus.

13. Solve the differential equation $y' = 3x^2 y$ given $y = 2e$ if $x = 1$.

14. Use the Table of Integrals in Chapter 7 to determine the indefinite integral $\int \frac{120}{x^2 - 36} dx$.

15. Determine the area of the region between $y = 12e^{-3x}$ and the x-axis for $x \geq 0$.

16. Determine the volume, if finite, of the solid of revolution determined by the region in problem 15 when the region is revolved about the x-axis.

17. Determine a value of k so that $f(x) = k(3xe^{-x})$ is a probability density function over the interval [0, 9].

18. Given that $f(x) = \frac{2}{9}x - \frac{2}{3}$ is a probability density function for a random variable X where $3 \leq x \leq 6$. What is the probability that a randomly selected x is in the range $3 \leq x \leq 5$?

19. What is the expected value of x, given that $f(x) = .25x + .25$ is a probability density function for x over the range [0, 2]?

20. Suppose the region bounded by the cubic $y = 16(x - x^3)$ and the x-axis is rotated about the x-axis (x is in [0, 1]). What is the volume of the corresponding solid of revolution?

For Exercises 21 – 24, evaluate the double integral given.

21. $\int_0^1 \int_1^2 (2x + 1) dy dx$

22. $\int_1^3 \int_0^1 (3x - 3y) dy dx$

23. $\int_0^2 \int_0^1 xe^{x^2} dx dy$

24. $\int_1^2 \int_1^2 \ln(xy) dx dy$

In Exercises 25 – 26, integrate as indicated.

25. $\int_0^1 \int_0^x (2x + 2y) dy dx$

26. $\int_0^2 \int_x^{2x} (e^y) dy dx$

For the Exercises 27 – 34, determine the partials f_x, f_y, f_{xx}, f_{yy}, f_{xy}, f_{yx}.

27. $f(x) = 3 + 2x + 3y$

28. $f(x) = 2x - 3y + 3x^2 - 11y^4$

29. $f(x, y) = e^{x+y}$

30. $f(x, y) = e^{x^2+y^2}$

31. $f(x, y) = \ln(3 + x + y)$

32. $f(x, y) = 3x^2y^2$

33. $f(x, y) = \dfrac{x+y}{x-y}$

34. $f(x, y) = xe^{-y}$

For the Exercises 35–37, find all local maxima, local minima and saddle points.

35. $f(x, y) = x^2 - xy + y^2 - 9x + 10$

36. $f(x, y) = 6x^2 - 6xy + 3y^2 + 64x - 58y + 100$

37. $f(x, y) = 2x^3 + 2y^3 - 6x - 12y + 30$

38. A rectangular box with a top is to be constructed with a partition. The volume is to be 300 cubic centimeters. Find the dimensions that will minimize the material used to make the box. (See problem 31, section 8.3.)

The Trigonometric Functions

CHAPTER 9

Did You Know?

This chapter extends our work with calculus to include the class of functions called the **trigonometric functions**. These functions are used in connection with a broad array of applications in astronomy, navigation, surveying, engineering, and medicine. The invention of logarithms by John Napier (1550 – 1617) helped to simplify the study of complicated trigonometric tables used in navigation and established trigonometry as a branch of mathematics in its own right.

The beating of a heart occurs periodically. The motion of tides is periodic. The rise and fall of daily temperatures, the phases of the moon, the motion of the Earth around the sun, the movement of a pendulum, and countless other phenomena are periodic. The continuous functions in mathematics (and calculus) which are used to describe periodic outcomes are the six trigonometric functions and their (partial) inverses. These functions are introduced to students in geometry in high school and again in trigonometry or precalculus. In these previous courses, the functions are first used to describe triangular relationships, and then numerous formulas are developed relating the formulas to one another or to specific angles.

It has been said that in calculus there are no triangles. Where are the triangles when one records the beating of a heart on an oscilloscope? In calculus, one uses the trigonometric functions to describe phenomena and as an aid in evaluating integrals of the sort $\int_0^1 \frac{1}{1+x^2}\,dx = \frac{\pi}{4}$. A TI-84 Plus will give the area as 0.78539816, but most of us will not recognize this number written as a decimal or have any idea of the connections that relate fractions of the number π to certain areas. The student who has only seen trigonometric functions as a way to calculate ratios corresponding to similar right triangles is in for a treat. For one example, $\frac{d}{dx}\sin x = \cos x$ and another, $\frac{d}{dx}\arctan x = \frac{1}{1+x^2}$.

Our goal is to introduce the calculus of the trigonometric functions in the most natural way, and we therefore only briefly present a review of both the degree and radian measures of angles. Since radians are abstract real numbers with no unit of measure, we may define the six trigonometric functions in Section 9.1 with reference to the unit circle and then graph the related curves in terms of ordered pairs (x, y) of real numbers, as triangle-free numbers. In Sections 9.2 and 9.3 we develop the formulas for differentiation and integration with the trigonometric functions, with emphasis on the sine, cosine, and tangent functions. We close the chapter in Section 9.4 with a discussion of the inverse sine, inverse cosine, and inverse tangent functions and related derivatives and integrals.

The approach in one's course depends on the goals of the school and/or the instructor. A brief approach emphasizing calculators in this chapter is quite satisfactory for the rest of the book and, in fact, is sufficient for all but practicing scientists. A sufficient selection of problems and examples is provided for a variety of choice and preference.

9.1 The Trigonometric Functions

Objectives

After completing this section, you will be able to:

1. *Sketch and evaluate trigonometric functions.*
2. *Verify basic trigonometric identities.*

Definitions of the Trigonometric Functions

The graph that corresponds to the equation $x^2 + y^2 = 1$ is a circle of radius 1 with center at the origin. This circle is called the **unit circle**. If $P(x, y)$ is any point on the unit circle, then the radius OP forms an angle of radian measure θ (where $0 \leq \theta \leq 2\pi$) with the positive x-axis as the initial side of the angle. (**Note:** We consider θ to be the name of the angle as well as its measure in radians.) The **sine** of angle θ (written $\sin \theta$) is defined to be the y-coordinate of P. The **cosine** of angle θ (written $\cos \theta$) is defined to be the x-coordinate of P. (See Figure 9.1.1.)

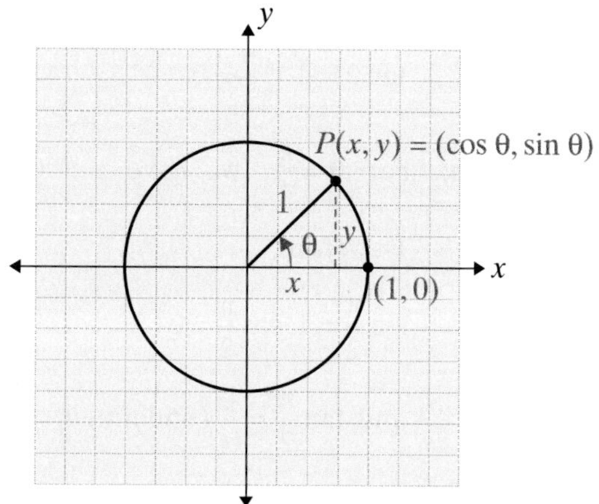

Figure 9.1.1 $x^2 + y^2 = 1$ is the unit circle. For $P(x,y)$ on the unit circle, $x = \cos\theta$ and $y = \sin\theta$.

There are four other trigonometric functions of angle θ: the **tangent** (or tan θ), the **cotangent** (or cot θ), the **secant** (or sec θ), and the **cosecant** (or csc θ). The definitions of all six functions are given in Table 9.1.1, which follows.

Table 9.1.1 The Trigonometric Functions

If $P(x, y)$ is a point on the unit circle and θ is the corresponding central angle, then

$$\sin\theta = y \quad \text{and} \quad \cos\theta = x$$

$$\tan\theta = \frac{y}{x} = \frac{\sin\theta}{\cos\theta} \qquad (x \neq 0)$$

$$\cot\theta = \frac{x}{y} = \frac{\cos\theta}{\sin\theta} \qquad (y \neq 0)$$

$$\sec\theta = \frac{1}{x} = \frac{1}{\cos\theta} \qquad (x \neq 0)$$

$$\csc\theta = \frac{1}{y} = \frac{1}{\sin\theta} \qquad (y \neq 0)$$

Note that if $x = 0$, then tan θ and sec θ are **undefined**. Also, if $y = 0$, then cot θ and csc θ are **undefined**.

Example 1: Trigonometric Functions

Use the given figure and the definitions to find sin θ, cos θ, and tan θ for each of the following values of θ.

a. 0 **b.** $\frac{\pi}{2}$ **c.** $\frac{\pi}{4}$

continued on next page ...

Solutions: a. For $\theta = 0$, we have $x = 1$ and $y = 0$. Therefore, $\sin 0 = y = 0$, $\cos 0 = x = 1$, and

$$\tan 0 = \frac{y}{x} = \frac{0}{1} = 0.$$

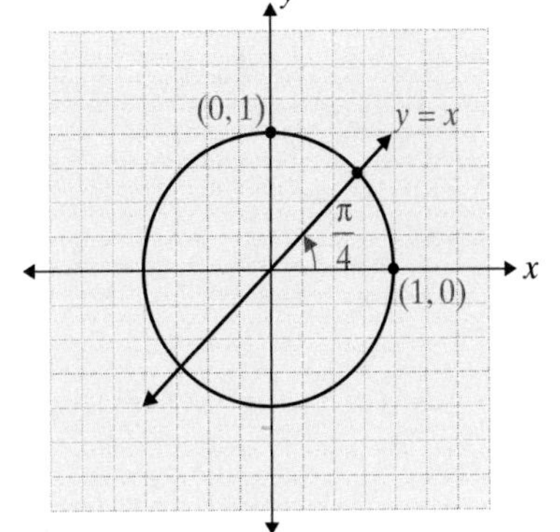

b. For $\theta = \frac{\pi}{2}$, we have $x = 0$ and $y = 1$.

Therefore,

$$\sin\frac{\pi}{2} = 1, \cos\frac{\pi}{2} = 0$$

and $\tan\frac{\pi}{2} = \frac{1}{0}$ (undefined).

c. For $\theta = \frac{\pi}{4}$, we locate the point of intersection in the first quadrant of the line $y = x$ and the unit circle.

$$x^2 + y^2 = 1 \quad \text{This defines the unit circle.}$$

$$x^2 + x^2 = 1 \quad \text{Next, substitute } x \text{ for } y.$$

$$2x^2 = 1$$

$$x^2 = \frac{1}{2}$$

$$y = x = \frac{1}{\sqrt{2}} \quad \text{This is in the first quadrant.}$$

Therefore, $\sin\frac{\pi}{4} = \cos\frac{\pi}{4} = \frac{1}{\sqrt{2}}$ and $\tan\frac{\pi}{4} = \frac{y}{x} = 1$.

Degree Measure of Angles

Suppose 360 points are evenly spaced counter-clockwise around a circle, say $a_1, a_2, \ldots, a_{360}$ as in Figure 9.1.2. Then an angle with its vertex at the circle's center and whose two rays intersect the circle at points a_i and a_j $(j > i)$ is said to have degree measure of $(j - i)^\circ$. Thus 90° corresponds to a right angle.

Since the unit circle has a circumference of 2π units, one may convert degree measure to radians or radians to degrees.

Figure 9.1.2

Table 9.1.2 contains the values for sine, cosine, and tangent of several commonly used angles. These angles can be found by using algebraic and geometric methods similar to those used in Example 1.

Table 9.1.2

θ in radians	0	$\frac{\pi}{6}$	$\frac{\pi}{4}$	$\frac{\pi}{3}$	$\frac{\pi}{2}$	$\frac{2\pi}{3}$	$\frac{3\pi}{4}$	$\frac{5\pi}{6}$	π	$\frac{3\pi}{2}$	2π
sin θ	0	$\frac{1}{2}$	$\frac{1}{\sqrt{2}}$	$\frac{\sqrt{3}}{2}$	1	$\frac{\sqrt{3}}{2}$	$\frac{1}{\sqrt{2}}$	$\frac{1}{2}$	0	−1	0
cos θ	1	$\frac{\sqrt{3}}{2}$	$\frac{1}{\sqrt{2}}$	$\frac{1}{2}$	0	$-\frac{1}{2}$	$-\frac{1}{\sqrt{2}}$	$-\frac{\sqrt{3}}{2}$	−1	0	1
tan θ	0	$\frac{1}{\sqrt{3}}$	1	$\sqrt{3}$	–	$-\sqrt{3}$	−1	$-\frac{1}{\sqrt{3}}$	0	–	0

Example 2: Angle Conversion

Convert between radians and degrees as requested:

a. convert 30° to radian measure.

b. convert 1.7 radians to degree measure.

Solutions:

a. 30° corresponds to $30°\left(\frac{\pi \text{ radians}}{180°}\right) = \frac{\pi}{6}$ radians. Use $x° = x\left(\frac{\pi}{180}\right)$ rad.

b. 1.7 radians corresponds to $1.7\left(\frac{180°}{\pi}\right) \approx 97.4°$. Use $x \text{ rad.} = x\left(\frac{180}{\pi}\right)^{\circ}$.

Graphing Trigonometric Functions

Again, since 2π radians corresponds to 360° (one revolution), if any radius is rotated 2π radians, then it will return to its original position. This means that the values of the sine and cosine functions also will remain the same for the new angle as they were for the original angle. Thus, geometrically, an angle of θ corresponds to an angle of θ+2π, and

$$\sin(\theta+2\pi) = \sin\theta \text{ and } \cos(\theta+2\pi) = \cos\theta.$$

This is illustrated in Figure 9.1.3. More generally, since 2π is the smallest positive number for which these equations are true, we say that the sine and cosine functions are **periodic with period 2π**, and we have the following results where k is any integer:

$$\sin(\theta+2k\pi) = \sin\theta \text{ and } \cos(\theta+2k\pi) = \cos\theta.$$

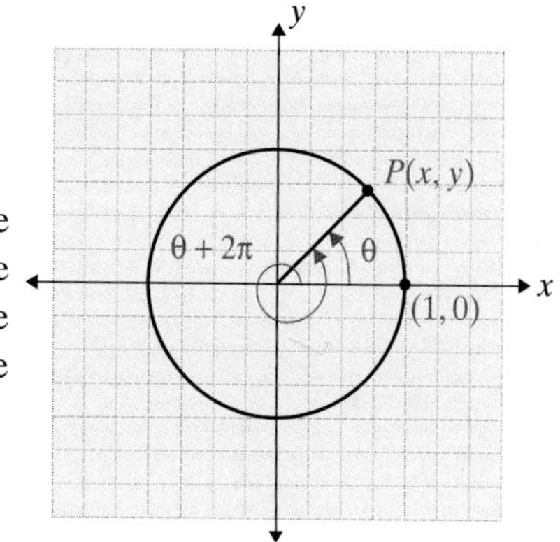

Figure 9.1.3

Also, as illustrated in Figure 9.1.4, we have the following relationships:

$$\sin(-\theta) = -\sin\theta \quad \text{and} \quad \cos(-\theta) = \cos\theta.$$

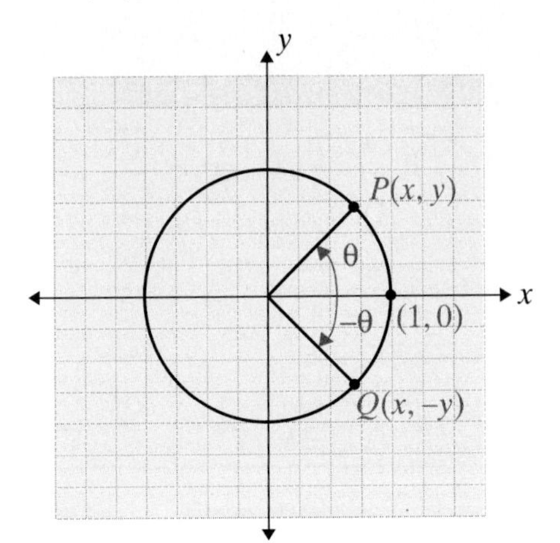

Figure 9.1.4

Example 3: Trigonometric Values

Use the fact that the sine and cosine functions have period 2π and Table 9.1.2 to find the following functional values.

a. $\sin\left(\frac{5\pi}{2}\right)$ **b.** $\cos(3\pi)$ **c.** $\sin\left(-\frac{9\pi}{4}\right)$

Solutions: a. $\sin\left(\frac{5\pi}{2}\right) = \sin\left(\frac{\pi}{2} + 2\pi\right) = \sin\left(\frac{\pi}{2}\right) = 1$

b. $\cos(3\pi) = \cos(\pi + 2\pi) = \cos(\pi) = -1$

c. $\sin\left(-\frac{9\pi}{4}\right) = \sin\left(-\frac{\pi}{4} - 2\pi\right) = \sin\left(-\frac{\pi}{4}\right) = -\sin\left(\frac{\pi}{4}\right) = -\frac{1}{\sqrt{2}}$

NOTES

We have used θ to represent angles and angle measure and x and y to represent coordinates on the unit circle, and we will continue to do so whenever it is convenient and the context is clear. However, for the purposes of graphing trigonometric functions and developing the related calculus concepts in Sections 9.2 through 9.4, we will return to the standard notation of using x for values of the domain and y for the values of the range. Thus the trigonometric functions for sine, cosine, and tangent can be represented in the forms

$$y = \sin x, \quad y = \cos x, \quad \text{and} \quad y = \tan x.$$

The graphs of these functions can be sketched by using values from Table 9.1.2 and a calculator. (See Figures 9.1.5, 9.1.6, and 9.1.7) Also, both the functions $y = \sin x$ and $y = \cos x$ have period 2π and are **continuous for all real numbers**. This means that the pattern of each graph on the interval $[0, 2\pi]$ is continuously repeated on the entire interval of real numbers $(-\infty, +\infty)$.

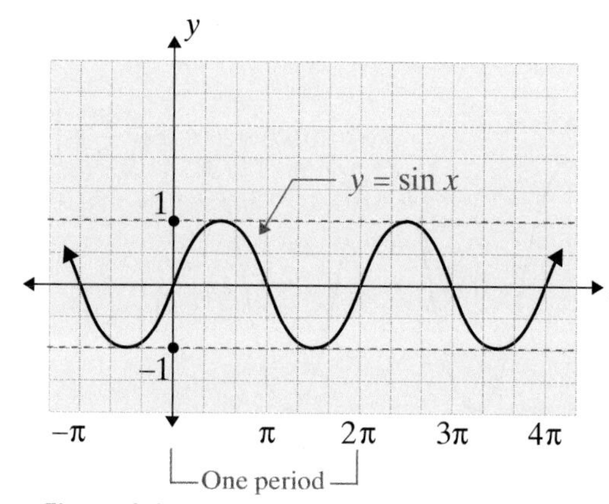

Figure 9.1.5

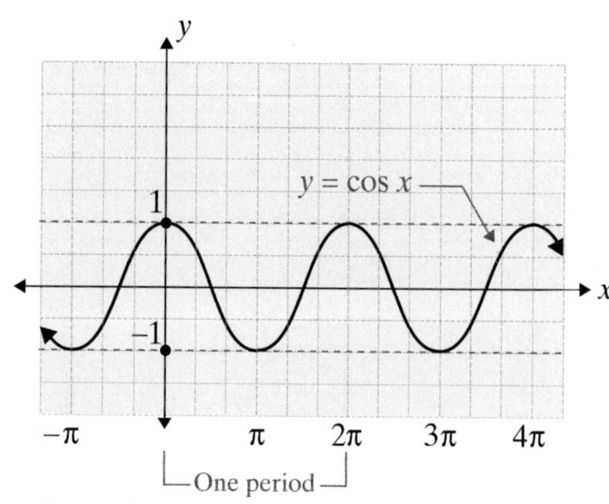

Figure 9.1.6

From the basic definition of tan x, we can write $y = \tan x = \dfrac{\sin x}{\cos x}$. This means that whenever $\cos x = 0$, $\tan x$ is undefined. This occurs for $x = \dfrac{\pi}{2} + k\pi$ (where k is an integer), and the lines represented by the equations $x = \dfrac{\pi}{2} + k\pi$ are vertical asymptotes. A general function $f(x)$ is called **periodic of period** T if $f(x) = f(x + T)$ for every x, and if T is the smallest positive number for which this is true.

From Figure 9.1.7, we observe that the tangent function is periodic with period $T = \pi$.

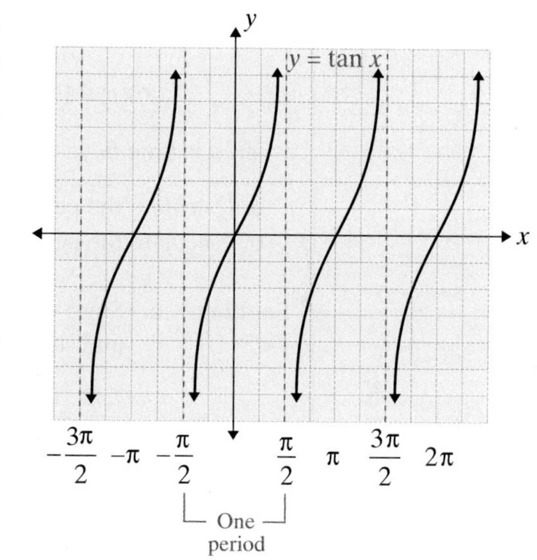

Figure 9.1.7

Example 4: The Predator-Prey Model

Ranchers in a certain area note that the populations of coyotes (predators) and rabbits (prey) interact with each other and are cyclic over a period of 3 years. As the population of rabbits increases, the population of coyotes also increases until the prey begins to decrease. The predator population continues to increase to the point where there is not enough food for the coyotes and their population begins to decrease. Then the decreasing population of coyotes allows an increase in the rabbit population, and the cycle begins again.

continued on next page ...

The coyote-rabbit population interaction might be represented by the following two functions and their graphs:

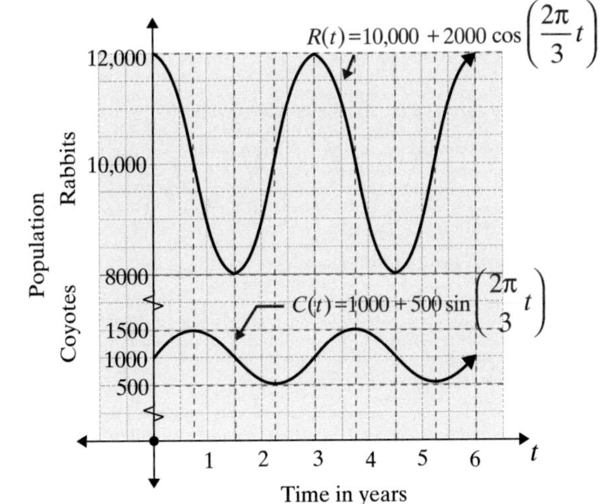

$$C(t) = 1000 + 500\sin\left(\frac{2\pi}{3}t\right) \text{ and } R(t) = 10{,}000 + 2000\cos\left(\frac{2\pi}{3}t\right)$$

where t is time in years.

a. Calculate the period of the function $C(t)$ algebraically.
b. Calculate the period of the function $R(t)$ graphically.

Solutions: **a.** Since the sine function has a period of 2π, we set $\frac{2\pi t}{3} = 0$ and $\frac{2\pi t}{3} = 2\pi$ and solve for t in each case.

$$\left.\begin{aligned} \frac{2\pi t}{3} &= 0 \quad \Rightarrow t = 0 \\ \frac{2\pi t}{3} &= 2\pi \Rightarrow t = 3 \end{aligned}\right\}$$

The period is the distance between the two times: $t_2 - t_1 = 3 - 0 = 3$.

b. Since two corresponding y-values on the graph occur for the two minimums $(x_1, y_1) = (1.5, 8000)$ and $(x_2, y_2) = (4.5, 8000)$, the period is $x_2 - x_1 = 4.5 - 1.5 = 3$. Note, one complete cycle (or period) occurs between x_1 and x_2.

Some Basic Trigonometric Identities

An **identity** is an equation that is true for every value of the variable (or variables) for which both sides of the equation are defined. By substituting the values $x = \cos\theta$ and $y = \sin\theta$ into the equation for the unit circle, we have one of the basic **trigonometric identities** used throughout mathematics.

$$x^2 + y^2 = 1$$

$$(\cos\theta)^2 + (\sin\theta)^2 = 1$$

We generally write $(\cos\theta)^2$ as $\cos^2\theta$, and $(\sin\theta)^2$ is usually written as $\sin^2\theta$. Then the previous identity is rewritten as

$$\cos^2\theta+\sin^2\theta=1.$$

Other basic trigonometric identities are listed in Table 9.1.3.

Table 9.1.3 Basic Trigonometric Identities

1. $\cos^2\theta+\sin^2\theta=1$	Pythagorean identity
2. $\sin(-\theta)=-\sin\theta$ **3.** $\cos(-\theta)=\cos\theta$	Even/Odd identities
4. $\sin(\theta\pm\phi)=\sin\theta\cos\phi\pm\cos\theta\sin\phi$ **5.** $\cos(\theta\pm\phi)=\cos\theta\cos\phi\mp\sin\theta\sin\phi$	Sum and difference identities
6. $\cos^2\frac{\theta}{2}=\frac{1+\cos\theta}{2}$ **7.** $\sin^2\frac{\theta}{2}=\frac{1-\cos\theta}{2}$	Half-angle identities
8. $\sin 2\theta=2\sin\theta\cos\theta$ **9.** $\cos 2\theta=\cos^2\theta-\sin^2\theta$ $=2\cos^2\theta-1$ $=1-2\sin^2\theta$	Double-angle identities
10. $\sin\theta=\cos\left(\frac{\pi}{2}-\theta\right)$ **11.** $\cos\theta=\sin\left(\frac{\pi}{2}-\theta\right)$	Cofunction identities

Example 5: Trigonometric Identities

Verify the trigonometric identity $1+\tan^2\theta=\sec^2\theta$.

Solution: Divide each term in Identity 1 in Table 9.1.3 by $\cos^2\theta$, and then refer to the definitions in Table 9.1.1 for $\tan\theta=\frac{\sin\theta}{\cos\theta}$ and $\sec\theta=\frac{1}{\cos\theta}$.

$$\cos^2\theta+\sin^2\theta=1 \qquad \text{By Identity 1, Table 9.1.3.}$$

$$\frac{\cos^2\theta}{\cos^2\theta}+\frac{\sin^2\theta}{\cos^2\theta}=\frac{1}{\cos^2\theta} \qquad \text{Divide by } \cos^2\theta.$$

$$1+\tan^2\theta=\sec^2\theta \qquad \text{Use the definitions in Table 9.1.1.}$$

Right Triangle Trigonometry

If θ is in standard position in quadrant I, then the trigonometric functions can be expressed in terms of ratios of the sides of a right triangle such as triangle AOB as illustrated in Figure 9.1.8. Since the triangles POQ and AOB are similar, their corresponding sides are proportional. The definitions for $\sin\theta$, $\cos\theta$, and $\tan\theta$ are given in terms of the lengths of the sides of triangle AOB in Figure 9.1.8. and again in Table 9.1.4 in terms of the lengths of the sides of any right triangle.

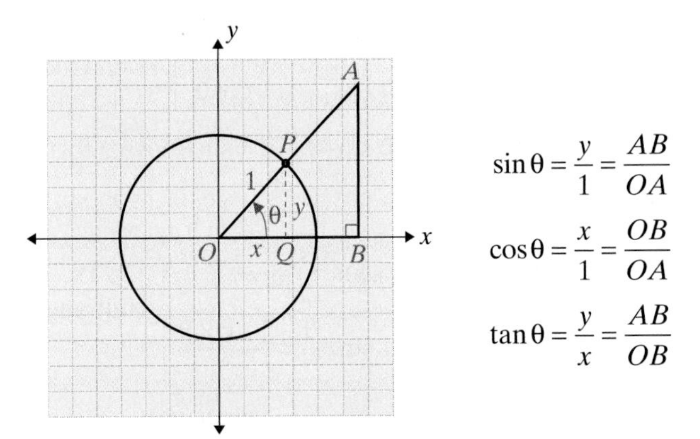

Figure 9.1.8

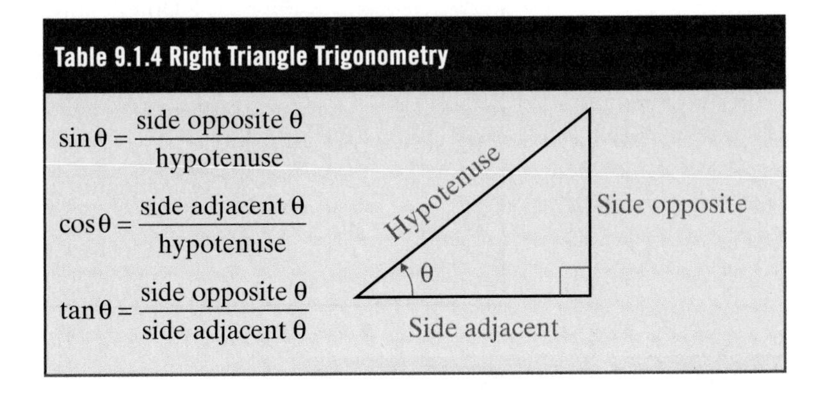

Table 9.1.4 Right Triangle Trigonometry

$\sin\theta = \dfrac{\text{side opposite } \theta}{\text{hypotenuse}}$	
$\cos\theta = \dfrac{\text{side adjacent } \theta}{\text{hypotenuse}}$	
$\tan\theta = \dfrac{\text{side opposite } \theta}{\text{side adjacent } \theta}$	

Example 6: Right Triangle Trigonometry

Find the lengths of C and A in the right triangle shown in the figure.

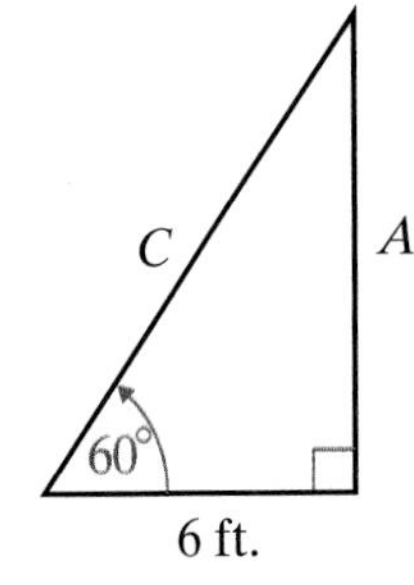

Solutions: a. $\cos 60^\circ = \dfrac{6}{C}$, so

$$C = \frac{6}{\cos 60^\circ} = \frac{6}{0.5} = 12 \text{ ft.}$$

b. $\tan 60^\circ = \dfrac{A}{6}$, so

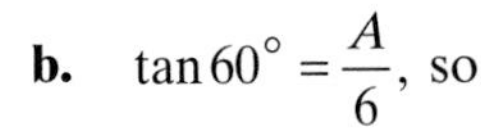

$$A = 6\tan 60^\circ = 6\left(\sqrt{3}\right) \approx 6(1.73) = 10.38 \text{ ft.}$$

9.1 Exercises

In Exercises 1 – 12, find the exact value of each of the given trigonometric functions.

1. $\cos\frac{\pi}{3}$ **2.** $\tan\frac{\pi}{4}$ **3.** $\sin\frac{3\pi}{4}$

4. $\cos\frac{5\pi}{6}$ **5.** $\tan\frac{13\pi}{6}$ **6.** $\sin\frac{8\pi}{3}$

7. $\cot\frac{\pi}{2}$ **8.** $\sec\frac{9\pi}{4}$ **9.** $\csc\left(-\frac{\pi}{6}\right)$

10. $\tan\left(-\frac{2\pi}{3}\right)$ **11.** $\cos\left(-\frac{5\pi}{4}\right)$ **12.** $\sin\left(-\frac{\pi}{2}\right)$

In Exercises 13 – 20, use a calculator to find the value of each of the given trigonometric functions, correct to four decimal places.

13. $\tan 64.3°$ **14.** $\cos 102.6°$ **15.** $\sin 246.1°$

16. $\sin(-53.2°)$ **17.** $\cos 2.31$ **18.** $\tan 0.891$

19. $\cos(-1.32)$ **20.** $\sin(-3.69)$

In Exercises 21 – 24, sketch the graph of each of the given functions over the interval $[-\pi, 2\pi]$.

21. $y = 3\cos x$ **22.** $y = 4\sin x$ **23.** $y = \sin 2x$

24. $y = \cos 4x$

In Exercises 25 – 28, verify each of the given identities.

25. $\csc^2\theta - 1 = \cot^2\theta$

26. $\sin 2\theta = 2\sin\theta\cos\theta \quad (\text{Hint}: 2\theta = \theta + \theta)$

27. $1 - \frac{\cos\theta}{\sec\theta} = \sin^2\theta$

28. $\frac{\cos\theta}{\sec\theta} + \frac{\sin\theta}{\csc\theta} = 1$

In Exercises 29 and 30, find the values of the six trigonometric functions of θ.

29.

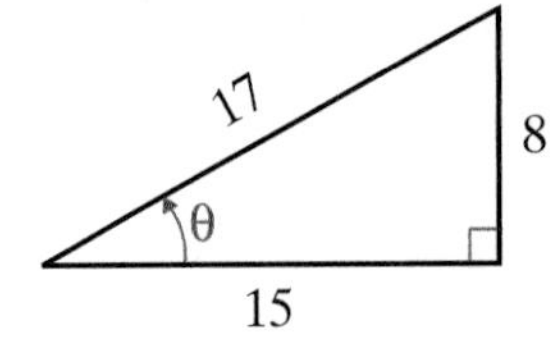

30.

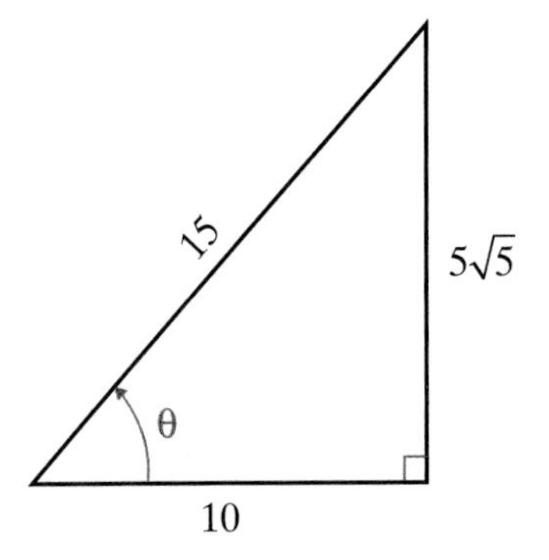

31. Aviation. An airplane that is approaching an airport is descending at an angle of 3.2°. Find the decrease in altitude if the plane travels a horizontal distance of 40 miles. Round your answer to the nearest thousandth.

32. Safety. In order to use a ladder safely, the angle that the ladder forms with the ground should not exceed 70°. If you have a ladder that is 16 ft long, what is the maximum height on the wall of a building that the ladder will safely reach? Round your answer to the nearest thousandth.

33. Sales. A plumbing contractor sells and installs both heating and air-conditioning units. The number of units he expects to sell each month is given by $N(t) = 28 + 12\cos\left(\frac{\pi}{3}(t-1)\right)$, where $t = 1$ represents the month of January.

a. Find $N(1)$, $N(4)$, $N(7)$, $N(10)$, and $N(13)$.
b. How many units does the contractor expect to sell in September?

34. Predator-prey model. A predator-prey model (see Example 4) for the relationship between a population of coyotes and a population of rabbits is given by $C(t) = 1000 + 500\sin\frac{\pi}{3}t$ for the number of coyotes, and the function $R(t) = 10{,}000 + 2000\cos\frac{2\pi}{9}t$ for the number of rabbits.

a. Find the period, T_1, of $C(t)$.
b. Find the period, T_2, of $R(t)$.

Hawkes Learning Systems: Essential Calculus

The Trigonometric Functions

9.2 Derivatives of the Trigonometric Functions

Objectives

After completing this section, you will be able to:

Compute derivatives of trigonometric functions.

In this section, we will first develop the formula for the derivative of the sine function by using the definition of a derivative from Section 2.3. Then we will use this result, some trigonometric identities, and the rules of differentiation to find the formulas for the derivatives of the remaining five trigonometric functions. These formulas automatically supply us with integral formulas that will be discussed in Section 9.3.

The development of the formula for the derivative of the function $y = \sin x$ involves the trigonometric identity

$$\sin(\theta + \phi) = \sin\theta\cos\phi + \cos\theta\sin\phi \qquad \text{(See Table 9.1.3, Identity 4.)}$$

and the two limits

$$\lim_{h \to 0} \frac{\sin h}{h} = 1 \qquad \text{and} \qquad \lim_{h \to 0} \frac{\cos h - 1}{h} = 0\,.$$

The values of these two limits are suggested from the results of the calculations in Table 9.2.1. The table shows values of h approaching 0 from the right. (Note: h is measured in radians.) A calculator can be used to check that the results are the same as h approaches 0 from the left.

Table 9.2.1

h	$\frac{\sin h}{h}$	$\frac{\cos h - 1}{h}$
0.5	0.958851077	−0.244834876
0.1	0.998334166	−0.049958347
0.01	0.999983333	−0.004999958
0.001	0.999999833	−0.000500000
0.0001	0.999999998	−0.000050000
↓	↓	↓
0	1	0

For a function $y = f(x)$, the definition of the derivative from Section 2.3 is

$$f'(x) = \lim_{h\to 0}\frac{f(x+h)-f(x)}{h}$$

provided the limit exists. We apply this definition to $f(x) = \sin x$.

$$\begin{aligned}
f'(x) &= \lim_{h\to 0}\frac{\sin(x+h)-\sin x}{h} \\
&= \lim_{h\to 0}\frac{\sin x\cos h+\cos x\sin h-\sin x}{h} && \text{Rewrite } \sin(x+h) \text{ using the identity.} \\
&= \lim_{h\to 0}\frac{(\sin x\cos h-\sin x)+\cos x\sin h}{h} && \text{Rearrange the terms.} \\
&= \lim_{h\to 0}\frac{\sin x(\cos h-1)+\cos x\sin h}{h} && \text{Factor out } \sin x. \\
&= \sin x\cdot\left(\lim_{h\to 0}\frac{\cos h-1}{h}\right)+\cos x\cdot\left(\lim_{h\to 0}\frac{\sin h}{h}\right) && \text{The limit of a sum is the sum of the limits.} \\
&= \sin x\cdot(0)+\cos x\cdot 1 && \text{From Table 9.2.1.} \\
&= \cos x
\end{aligned}$$

Thus we have the result stated in Theorem 9.2.1a. Theorem 9.2.1b follows immediately from the Chain Rule and Theorem 9.2.1a.

Theorem 9.2.1

a. *If* $y = \sin x$, *then*

$$\frac{dy}{dx} = \frac{d}{dx}[\sin x] = \cos x.$$

b. *If* $y = \sin[g(x)]$, *then*

$$\frac{dy}{dx} = \cos[g(x)]g'(x)$$

provided $g'(x)$ *exists.*

The formula for the derivative of the function $y = \cos x$ follows from Theorem 9.2.1b and the following two identities from Table 9.1.3:

$$\cos x = \sin\left(\frac{\pi}{2}-x\right) \quad \text{and} \quad \sin x = \cos\left(\frac{\pi}{2}-x\right).$$

For $f(x)=\cos x$,

$$f'(x)=\frac{d}{dx}[\cos x]=\frac{d}{dx}\left[\sin\left(\frac{\pi}{2}-x\right)\right]$$ From Theorem 9.2.1b.

$$=\cos\left(\frac{\pi}{2}-x\right)\cdot(-1)$$

$$=\sin x\cdot(-1)=-\sin x.$$ From Identity 10, Table 9.1.3.

Theorem 9.2.2

a. *If* $y = \cos x$, *then*

$$\frac{dy}{dx}=\frac{d}{dx}[\cos x]=-\sin x.$$

b. *If* $y = \cos[g(x)]$, *then the Chain Rule gives*

$$\frac{dy}{dx}=-\sin\left[g(x)\right]g'(x)$$

provided $g'(x)$ *exists.*

Example 1: Trigonometric Derivatives

Find $f'(x)$ for each of the following functions.

a. $f(x)=\sin^2 6x$ **b.** $f(x) = x\cos x$ **c.** $f(x)=\sin\left(e^x\right)$

Solutions: a. For $f(x)=\sin^2 6x$, use the Chain Rule.

$$f'(x)=2\cdot\sin 6x\cdot\cos 6x\cdot(6)=12\sin 6x\cos 6x$$

Recall that $\sin^2 6x$ means $[\sin(6x)]^2$.

b. For $f(x) = x\cos x$, use the Product Rule.

$$f'(x)=x(-\sin x)+\cos x\cdot(1)=-x\sin x+\cos x$$

c. For $f(x)=\sin\left(e^x\right)$, use the Chain Rule (or Theorem 9.2.1).

$$f'(x)=\cos\left(e^x\right)\cdot e^x=e^x\cos\left(e^x\right)$$

The formula for the derivative of the function $y = \tan x$ can be developed by using the Quotient Rule and the identity $\tan x = \dfrac{\sin x}{\cos x}$ as follows.

$$\frac{d}{dx}[\tan x] = \frac{d}{dx}\left[\frac{\sin x}{\cos x}\right] = \frac{\cos x \frac{d}{dx}[\sin x] - \sin x \frac{d}{dx}[\cos x]}{\cos^2 x}$$

$$= \frac{\cos x(\cos x) - \sin x(-\sin x)}{\cos^2 x}$$

$$= \frac{\cos^2 x + \sin^2 x}{\cos^2 x} = \frac{1}{\cos^2 x} = \sec^2 x$$

Theorem 9.2.3

a. *If* $y = \tan x$, *then*

$$\frac{dy}{dx} = \frac{d}{dx}[\tan x] = \sec^2 x.$$

b. *If* $y = \tan[g(x)]$, *then*

$$\frac{dy}{dx} = \sec^2[g(x)]g'(x)$$

provided $g'(x)$ *exists.*

The formulas for the derivative of cot x, sec x, and csc x can all be derived from basic identities and the Quotient Rule or the General Power Rule. The results are listed in Table 9.2.2.

Table 9.2.2 Derivatives of Trigonometric Functions

$$\frac{d}{dx}[\sin x] = \cos x$$

$$\frac{d}{dx}[\cos x] = -\sin x$$

$$\frac{d}{dx}[\tan x] = \sec^2 x$$

$$\frac{d}{dx}[\cot x] = -\csc^2 x$$

$$\frac{d}{dx}[\sec x] = \sec x \tan x$$

$$\frac{d}{dx}[\csc x] = -\csc x \cot x$$

Example 2: Trigonometric Derivatives

Find the equation of the line tangent to the graph of $y = \tan x$ at the point where $x = \frac{\pi}{4}$.

Solution: If $x = \frac{\pi}{4}$, then $y = \tan\left(\frac{\pi}{4}\right) = 1$. Use the derivative to find m in the formula $y - y_1 = m(x - x_1)$.

$$\frac{dy}{dx} = \frac{d}{dx}[\tan x] = \sec^2 x$$

$$\left.\frac{dy}{dx}\right|_{x=\frac{\pi}{4}} = \sec^2\left(\frac{\pi}{4}\right) = \left(\sqrt{2}\right)^2 = 2$$

$$\sec\left(\frac{\pi}{4}\right) = \frac{1}{\cos\left(\frac{\pi}{4}\right)} = \frac{1}{\frac{1}{2}\sqrt{2}} = \frac{2}{\sqrt{2}}$$

$$= \frac{2}{\sqrt{2}} \cdot \frac{\sqrt{2}}{\sqrt{2}} = \frac{2\sqrt{2}}{2} = \sqrt{2}$$

This gives $m = 2$, $x_1 = \frac{\pi}{4}$, and $y_1 = 1$. The equation of the tangent line is

$$y - 1 = 2\left(x - \frac{\pi}{4}\right) \text{ or}$$

$$y = 2x - \frac{\pi}{2} + 1 \approx 2x - 0.57.$$

Example 3: Marginal Cost

A pharmaceutical company finds that the cost (in dollars) of producing a particular drug is given by the function

$$C(x) = 38x - 36x\cos\left(\frac{\pi}{12}x\right)$$

where x is in thousands of units. Find the marginal cost when 24,000 units are produced.

Solution:

$$C'(x) = 38 - 36x\left[-\sin\left(\frac{\pi}{12}x\right)\cdot\frac{\pi}{12}\right] + \cos\left(\frac{\pi}{12}x\right)\cdot(-36)$$

$$= 38 + 3\pi x\sin\left(\frac{\pi}{12}x\right) - 36\cos\left(\frac{\pi}{12}x\right)$$

$$C'(24) = 38 + 3\pi\cdot 24\cdot\sin(2\pi) - 36\cos(2\pi)$$

$$= 38 + 72\pi(0) - 36(1)$$

$$= 38 + 0 - 36 = 2$$

The marginal cost is \$2 per unit when 24,000 units are produced.

Example 4: A Population of Geese

The number of geese that live near a lake in Canada is known to vary (approximately) according to the function

$$N(t) = 200 + 200\cos\left(\frac{\pi}{6}t\right)$$

where t is the number of months after May 1 each year.

a. About how many geese are in the area at the end of August?

b. At what rate is the population of geese changing at that time?

Solutions: a. At the end of August, $t = 4$. Therefore, the approximate number of geese in the area is

$$N(4) = 200 + 200\cos\left(\frac{\pi}{6}\cdot 4\right)$$
$$= 200 + 200\cos\left(\frac{2\pi}{3}\right)$$
$$= 200 + 200\left(-\frac{1}{2}\right) = 200 - 100 = 100$$

Approximately 100 geese are in the area at the end of August.

b. Find the derivative and evaluate the derivative for $t = 4$.

$$N'(t) = 200\left[-\sin\left(\frac{\pi}{6}t\right)\right]\cdot\frac{\pi}{6}$$
$$= -\frac{100\pi}{3}\sin\left(\frac{\pi}{6}t\right)$$
$$N'(4) = -\frac{100\pi}{3}\sin\left(\frac{\pi}{6}\cdot 4\right)$$
$$= -\frac{100\pi}{3}\sin\left(\frac{2\pi}{3}\right) = -\frac{100\pi}{3}\left(\frac{\sqrt{3}}{2}\right)$$
$$= \frac{50\pi}{\sqrt{3}} \approx -91$$

At the end of August, the population is decreasing by about 91 geese per month.

9.2 Exercises

In Exercises 1 – 28, find the derivative for each function.

1. $f(x) = 6\cos 4x$

2. $f(x) = 2\tan 3x$

3. $f(x) = -3\sin^3 x$

4. $f(x) = 5\cos^4 x$

5. $y = \ln x + \tan 2x$

6. $y = e^x - \sin 5x$

7. $y = x^2 \cos 8x$

8. $y = 2x^3 \tan 6x$

9. $f(x) = \sin x \cos x$

10. $f(x) = e^x \cos x$

11. $y = \dfrac{\sin x}{x}$

12. $y = \dfrac{\tan 2x}{x}$

13. $y = \tan e^{3x}$

14. $y = \sin(\ln x)$

15. $f(x) = \sqrt{\cos 5x}$

16. $f(x) = \sqrt{\sin 4x}$

17. $y = \sec 4x$

18. $y = \cot \dfrac{1}{x}$

19. $y = \tan^2 x$

20. $y = \sec^2 x$

21. $y = \sin^2 x + \cos^2 x$

22. $y = \sec^2 x - \tan^2 x$

23. $y = 3e^{\cos x}$

24. $y = -6e^{\sin x}$

25. $f(x) = \ln(\sin x)$

26. $f(x) = \ln(\tan x)$

27. $y = \dfrac{\sin x}{1 + \cos x}$

28. $y = \dfrac{1 + \sin x}{\cos x}$

In Exercises 29 – 34, find $\dfrac{\partial f}{\partial x}$ *and* $\dfrac{\partial f}{\partial y}$ *for each function.*

29. $f(x, y) = 4x + \sin xy$

30. $f(x, y) = 3\cos xy + x^2 y$

31. $f(x, y) = e^{xy} + y \sin x$

32. $f(x, y) = \cos xy + x \ln y$

33. $f(x, y) = 5x^3 \cos(3x + y)$

34. $f(x, y) = e^{-x} \sin(x + y)$

In Exercises 35 – 37, verify each of the formulas.

35. $\dfrac{d}{dx}[\cot x] = -\csc^2 x$

36. $\dfrac{d}{dx}[\sec x] = \sec x \tan x$

37. $\dfrac{d}{dx}[\csc x] = -\csc x \cot x$

38. Find an equation for the line tangent to the graph of $y = 3\cos\frac{\pi}{2}x$ at the point where $x = 4$.

39. Find an equation for the line tangent to the graph of $y = 2\sin\frac{\pi}{3}x$ at the point where $x = 2$.

40. **Air quality.** On a typical summer day in southern California, the level of pollutants in the air can be estimated by $L(x) = 45.5 - 10.5\cos 0.39t$ PSI (Pollution Standard Index), where t is the number of hours after 6:00 A.M. and $1 \le t \le 12$. Round your answers to the nearest thousandths.

a. Find the level of pollutants at 2:00 P.M.

b. Find the rate of change in the level of pollutants at 12:00 noon.

41. **Marginal revenue.** The Mammoth Firewood Company cuts and sells firewood in a mountain resort. The number of cords of wood sold each month is estimated by $N(t) = 100 - 60\sin\frac{\pi}{6}t$, where t is the time in months (t = 1 corresponds to January). Firewood sells for \$140 per cord.

a. Find the revenue for May.

b. Find the marginal revenue for September.

c. Interpret the results in part **b.**

42. **Population**. The fox population in a midwestern state is estimated by $P(t) = 2400 + 400\sin\frac{\pi t}{18}$, where t is in months.

a. Find the number of foxes when $t = 15$.

b. Find the rate of change in the fox population when $t = 24$.

c. Interpret the results in part **b.**

Hawkes Learning Systems: Essential Calculus

Derivatives of the Trigonometric Functions

9.3 Integration of the Trigonometric Functions

Objectives

After completing this section, you will be able to:

1. *Find the antiderivative for a trigonometric function.*
2. *Calculate definite integrals for trigonometric functions.*

Each of the formulas for differentiation of trigonometric functions discussed in Section 9.2 automatically yields a corresponding formula for integration. Three of these formulas, which will be used in this section, are listed in Table 9.3.1.

With the formulas in Table 9.3.1, the various rules of integration, the basic integral formulas for polynomial, logarithmic, and exponential functions, and the techniques of u-substitution and integration by parts, we can integrate an interesting variety of functions. The examples that follow show some of the possibilities.

Table 9.3.1 Integration Formulas for Trigonometric Functions
1. **a.** $\int \cos x\,dx = \sin x + C$
b. $\int \cos[g(x)] \cdot g'(x)\,dx = \sin[g(x)] + C$
2. **a.** $\int \sin x\,dx = -\cos x + C$
b. $\int \sin[g(x)] \cdot g'(x)\,dx = -\cos[g(x)] + C$
3. **a.** $\int \sec^2 x\,dx = \tan x + C$
b. $\int \sec^2[g(x)] \cdot g'(x)\,dx = \tan[g(x)] + C$

Example 1: Antiderivatives of Trigonometric Functions

Evaluate each integral.

a. $\int \cos 5x\,dx$

b. $\int \sin^3 x \cos x\,dx$

continued on next page ...

Solutions: a. Use u-substitution with $u = 5x$ and $du = 5\,dx$.

$$\begin{aligned}\int \cos 5x\,dx &= \frac{1}{5}\int \cos 5x \cdot 5\,dx \\ &= \frac{1}{5}\int \cos u\,du \\ &= \frac{1}{5}\sin u + C \\ &= \frac{1}{5}\sin 5x + C\end{aligned}$$

b. Use u-substitution with $u = \sin x$ and $du = \cos x\,dx$.

$$\begin{aligned}\int \sin^3 x \cos x dx &= \int u^3 du \\ &= \frac{1}{4}u^4 + C \\ &= \frac{1}{4}\sin^4 x + C\end{aligned}$$

Example 2: Definite Integrals of Trigonometric Functions

Evaluate each integral.

a. $\int_0^{\frac{\pi}{12}} \sec^2 3x\,dx$ **b.** $\int_0^{\frac{\pi}{2}} x \sin x\,dx$

Solutions: a. Use u-substitution with $u = 3x$ and $du = 3\,dx$

$$\begin{aligned}\int \sec^2 3x\,dx &= \frac{1}{3}\int \sec^2 3x \cdot 3\,dx \\ &= \frac{1}{3}\int \sec^2 u\,du \\ &= \frac{1}{3}\tan u + C \\ &= \frac{1}{3}\tan 3x + C\end{aligned}$$

The value of the definite integral is

$$\begin{aligned}\int_0^{\frac{\pi}{12}} \sec^2 3x\,dx &= \frac{1}{3}\tan 3x\Big]_0^{\frac{\pi}{12}} \\ &= \frac{1}{3}\tan\left(3 \cdot \frac{\pi}{12}\right) - \frac{1}{3}\tan(3 \cdot 0)\end{aligned}$$

$$= \frac{1}{3}\tan\left(\frac{\pi}{4}\right) - \frac{1}{3}\tan(0)$$

$$= \frac{1}{3}(1) - 0 = \frac{1}{3}$$

b. Use integration by parts as follows (see also Section 7.1):

$u = x$	$dv = \sin x\,dx$
$du = dx$	$v = \int \sin x\,dx = -\cos x$

Thus

$$\int_0^{\frac{\pi}{2}} x \sin x\,dx = x(-\cos x)\Big]_0^{\frac{\pi}{2}} - \int_0^{\frac{\pi}{2}} -\cos x\,dx$$

$$= -x\cos x\Big]_0^{\frac{\pi}{2}} + \int_0^{\frac{\pi}{2}} \cos x\,dx$$

$$= -x\cos x + \sin x\Big]_0^{\frac{\pi}{2}}$$

$$= \left(-\frac{\pi}{2}\cdot\cos\frac{\pi}{2} + \sin\frac{\pi}{2}\right) - (-0\cdot\cos 0 + \sin 0)$$

$$= \left(-\frac{\pi}{2}\cdot 0 + 1\right) - (0)$$

$$= 1$$

Example 3: Area Between Curves

Find the area between the curves $y = \cos x$ and $y = \sin x$ on the interval $\left[0, \frac{\pi}{4}\right]$.

Solution: As illustrated in the figure at the right, $y = \cos x$ is greater than or equal to $y = \sin x$ on the interval $\left[0, \frac{\pi}{4}\right]$.

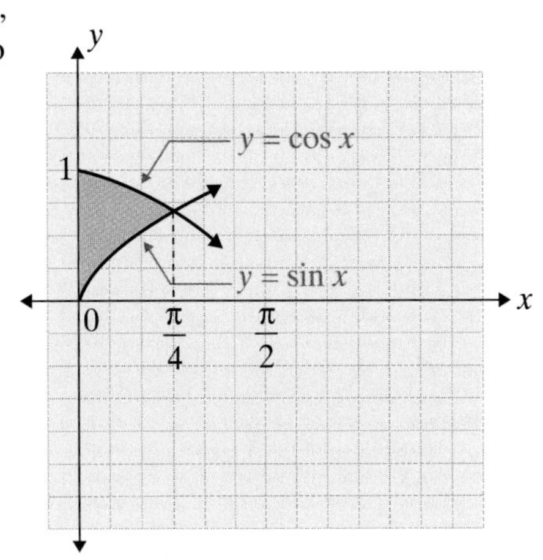

continued on next page ...

To find the area between the two curves, we integrate as follows:

$$\int_0^{\frac{\pi}{4}} (\cos x - \sin x)\,dx = \int_0^{\frac{\pi}{4}} \cos x\,dx - \int_0^{\frac{\pi}{4}} \sin x\,dx = \sin x + \cos x\Big]_0^{\frac{\pi}{4}}$$

$$= \left[\sin\left(\frac{\pi}{4}\right) + \cos\left(\frac{\pi}{4}\right)\right] - \left[\sin(0) + \cos(0)\right]$$

$$= \left(\frac{1}{\sqrt{2}} + \frac{1}{\sqrt{2}}\right) - (0+1)$$

$$= \frac{2}{\sqrt{2}} - 1 = \sqrt{2} - 1 \approx 0.4142$$

Now, to develop the formula for the integral of tan x, rewrite tan x by using the identity

$$\tan x = \frac{\sin x}{\cos x}$$

and the method of u-substitution with

$$u = \cos x \quad \text{and} \quad du = -\sin x\,dx.$$

With these substitutions, we have

$$\int \tan x dx = \int \frac{\sin x}{\cos x} dx$$

$$= -\int \frac{-\sin x}{\cos x} dx$$

$$= -\int \frac{1}{u} du \qquad \text{Use } u = \cos x \text{ and } du = -\sin x.$$

$$= -\ln|u| + C$$

$$= -\ln|\cos x| + C$$

Integration Formula for the Tangent Function

a. $\int \tan x dx = -\ln|\cos x| + C$

b. $\int \tan[g(x)] \cdot g'(x) dx = -\ln|\cos[g(x)]| + C$

Example 4: Integrating the Tangent

Find $\int x^2 \tan x^3 dx$.

Solution: Use u-substitution with $u = x^3$ and $du = 3x^2 dx$. Then

$$\begin{aligned}\int x^2 \tan x^3 dx &= \frac{1}{3}\int \tan x^3 \cdot 3x^2 dx \\ &= \frac{1}{3}\int \tan u\, du \\ &= -\frac{1}{3}\ln|\cos u| + C \\ &= -\frac{1}{3}\ln\left|\cos x^3\right| + C\end{aligned}$$

Example 5: Population Rate of Change

Suppose that the rate of change of the deer population in a region of the Rocky Mountains is approximately

$$N'(t) = 100\cos\left(\frac{\pi}{12}t\right)$$

where t is the number of months after January 1 each year. What is the change in the deer population from January 1 to July 1?

Solution: On January 1, $t = 0$, and on July 1, $t = 6$. The change in the deer population is approximated by the definite integral

$$\int_0^6 100\cos\left(\frac{\pi}{12}t\right)dt.$$

Use u-substitution with $u = \frac{\pi}{12}t$ and $du = \frac{\pi}{12}dt$. Then

$$\begin{aligned}\int 100\cos\left(\frac{\pi}{12}t\right)dt &= 100\cdot\frac{12}{\pi}\int \cos\left(\frac{\pi}{12}t\right)\cdot\frac{\pi}{12}dt \\ &= \frac{1200}{\pi}\int \cos u\, du \\ &= \frac{1200}{\pi}\sin u + C \\ &= \frac{1200}{\pi}\sin\left(\frac{\pi}{12}t\right) + C\end{aligned}$$

continued on next page ...

Therefore, the value of the definite integral is

$$\begin{aligned}\int_0^6 100\cos\left(\frac{\pi}{12}t\right)dt &= \frac{1200}{\pi}\sin\left(\frac{\pi}{12}t\right)\Bigg]_0^6\\ &= \frac{1200}{\pi}\sin\left(\frac{\pi}{12}\cdot 6\right)-\frac{1200}{\pi}\sin\left(\frac{\pi}{12}\cdot 0\right)\\ &= \frac{1200}{\pi}\left[\sin\left(\frac{\pi}{2}\right)-\sin(0)\right]\\ &= \frac{1200}{\pi}(1-0)\\ &\approx 382\end{aligned}$$

By July 1, the deer population has grown by about 382 deer.

Example 6: Estimating Revenue

The revenue (in hundreds of dollars) from the ski section of a sporting goods store is estimated to be

$$R(t) = 60 + 10\pi\cos\left(\frac{\pi}{12}t\right)$$

where t is the number of months after January 1 each year. Find the total estimated revenue for one year.

Solution: The total estimated revenue is the area under the curve as shown in the figure to the right. This area is the value of the definite integral.

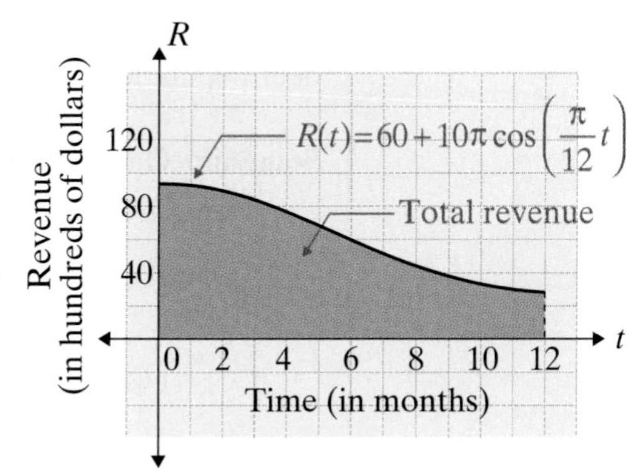

$$\begin{aligned}\int_0^{12}\left[60+10\pi\cos\left(\frac{\pi}{12}t\right)\right]dt &= 60t+10\pi\cdot\frac{12}{\pi}\cdot\sin\left(\frac{\pi}{12}t\right)\Bigg]_0^{12}\\ &= \left[60(12)+120\sin(\pi)\right]-0\\ &= 720\end{aligned}$$

The total estimated revenue from the ski department is \$72,000.

9.3 Exercises

In Exercises 1 – 30, evaluate the given integral.

1. $\int \cos 4x\,dx$

2. $\int \sin \frac{1}{3}x\,dx$

3. $\int \sec^2 4x\,dx$

4. $\int \tan \pi x\,dx$

5. $\int \tan(7x+1)\,dx$

6. $\int \sec^2(2x+5)\,dx$

7. $\int x \sin x^2\,dx$

8. $\int x^2 \cos x^3\,dx$

9. $\int_0^{\frac{\pi}{3}} \sin 2x\,dx$

10. $\int_{\frac{\pi}{6}}^{\frac{\pi}{3}} \cos 3x\,dx$

11. $\int_{-\frac{\pi}{2}}^{\frac{\pi}{2}} (\cos x + \sin x)\,dx$

12. $\int_0^{\frac{\pi}{4}} 2\sin(\pi - x)\,dx$

13. $\int_0^{\frac{\pi}{6}} \cos x \sin x\,dx$

14. $\int_0^{\frac{\pi}{4}} \sec^2 x \tan x\,dx$

15. $\int e^x \cos e^x\,dx$

16. $\int e^x \sec^2 e^x\,dx$

17. $\int \sin x e^{\cos x}\,dx$

18. $\int \cos x e^{\sin x}\,dx$

19. $\int \frac{\tan \sqrt{x}}{\sqrt{x}}\,dx$

20. $\int \frac{\sin \sqrt{x}}{\sqrt{x}}\,dx$

21. $\int \frac{\sin(\ln x)}{x}\,dx$

22. $\int \frac{\cos(\ln x)}{x}\,dx$

23. $\int_0^{\frac{\pi}{6}} \frac{\cos x}{1+\sin x}\,dx$

24. $\int_{\frac{\pi}{4}}^{\frac{\pi}{3}} \frac{\sin 2x}{1-\cos 2x}\,dx$

25. $\int_{\frac{\pi}{3}}^{\frac{\pi}{2}} \frac{\sin 3x}{(1-\cos 3x)^2}\,dx$

26. $\int_0^{\frac{\pi}{2}} \sqrt{1+\sin x}\cos x\,dx$

27. $\int \cot x\,dx \left(\text{Hint} : \cot x = \frac{\cos x}{\sin x} \right)$

28. $\int \tan^2 x\,dx \left(\text{Hint} : \tan^2 x = \sec^2 x - 1 \right)$

29. $\int \sin^2 x\,dx \left(\text{Hint} : \sin^2 x = \frac{1}{2}(1-\cos 2x) \right)$

30. $\int \cos^2 x\,dx \left(\text{Hint} : \cos^2 x = \frac{1}{2}(1+\cos 2x) \right)$

In Exercises 31 – 34, integrate by parts.

31. $\int x\cos 2x\,dx$

32. $\int x\sin 5x\,dx$

33. $\int x\sec^2 x\,dx$

34. $\int 3x\sec^2 2x\,dx$

35. Find the area of the region bounded by the x-axis and $y = 3\sin\frac{x}{2}$ on the interval $[0, \pi]$.

36. Find the area of the region bounded by the x-axis and $y = 5\cos\frac{x}{3}$ on the interval $\left[0, \frac{\pi}{2}\right]$.

37. Find the area of the region bounded by $y = 2\cos x$ and $y = \sin 2x$ on the interval $\left[0, \frac{\pi}{4}\right]$.

38. Find the area of the region bounded by $y = \sin x$ and $y = \tan x$ on the interval $\left[\frac{\pi}{4}, \frac{\pi}{3}\right]$.

39. Profit. Wes operates a boat rental concession at a fishing resort from mid-April to mid-September. His marginal profit is approximately $P'(t) = \frac{200\pi}{3}\sin\left[\frac{\pi}{12}(t+4)\right]$ dollars per week, where t is the number of weeks after mid-April and $0 \le t \le 20$. Find the weekly profit function P, if $P(0) = \$1300$.

40. Average population. The population of a farming community during harvest season is estimated by $P(t) = 2600 + 180\sin\frac{\pi}{12}t$, where t is the number of weeks after the start of harvest and $0 \le t \le 12$. Find the average weekly population of the community from $t = 4$ to $t = 8$. (See Section 6.3)

41. Volume. Find the volume of the solid generated when the region bounded by the graph of $y = 2\sin x$, $x = 0$ and $x = \pi$ is rotated about the x-axis. (See Section 7.6)

Hawkes Learning Systems: Essential Calculus

Integration of the Trigonometric Functions

9.4 Inverse Trigonometric Functions

Objectives

After completing this section, you will be able to:

1. *Sketch and evaluate basic inverse trigonometric functions.*
2. *Find the derivative of basic inverse trigonometric functions.*
3. *Find the definite and indefinite integrals of basic inverse trigonometric functions.*

If a function f is one-to-one, then the corresponding inverse function (symbolized f^{-1}) can be found by exchanging x and y and solving for y. For example, if

$$f(x) = y = e^x$$

then exchanging x and y gives

$$x = e^y$$

and solving the new equation for y gives the inverse function (in this case, the natural logarithm function):

$$y = \ln x \qquad \text{or} \qquad f^{-1}(x) = \ln x.$$

The trigonometric function $y = \sin x$ is not one-to-one; however, by restricting the values of x (the domain) to the interval $\left[-\frac{\pi}{2}, \frac{\pi}{2}\right]$, we can define the **inverse sine function** (symbolized $\sin^{-1}$) with the y values (the range) restricted to the interval $\left[-\frac{\pi}{2}, \frac{\pi}{2}\right]$. (See Figure 9.4.1)

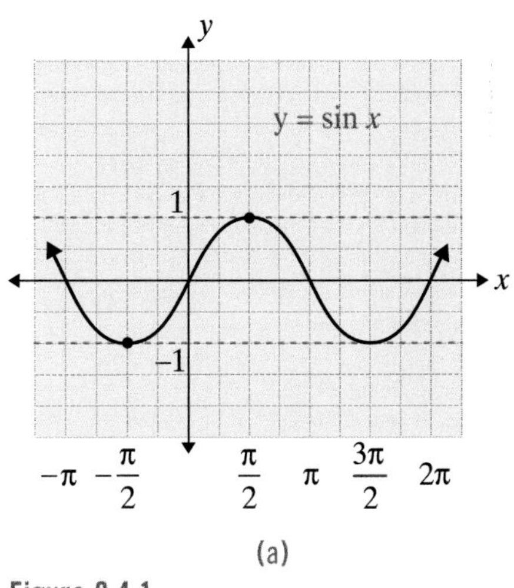

(a)

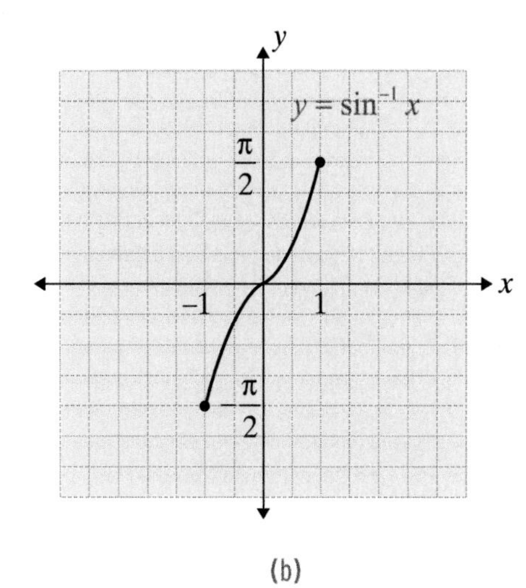

(b)

Figure 9.4.1

In the function $y = \sin^{-1} x$, y can be thought of as an angle (or the radian measure of an angle), and for this reason, the notation $y = \arcsin x$ ("angle whose sine is x") is also used to indicate the inverse sine function. Sometimes this notation helps in understanding the inverse trigonometric relationships. For example, if $y = \sin^{-1}\left(\frac{1}{2}\right)$, then y is the angle with $\frac{1}{2}$ as its sine. That is, $\sin y = \frac{1}{2}$, and given the restriction that y must be in the interval $\left[-\frac{\pi}{2}, \frac{\pi}{2}\right]$, we have $y = \frac{\pi}{6}$.

Note: If we did not have this restriction, then the result would be an infinite number of possible values for y, namely, $\frac{\pi}{6} + 2k\pi$ and $\frac{5\pi}{6} + 2k\pi$, where k is an integer.

Example 1: Inverse Trigonometric Functions

Find each of the indicated values of y.

a. $y = \sin^{-1}(1)$

b. $y = \sin^{-1}\left(-\frac{1}{\sqrt{2}}\right)$

Solutions: a. For $y = \sin^{-1}(1)$, we have $\sin y = 1$. Since y must be in the interval $\left[-\frac{\pi}{2}, \frac{\pi}{2}\right]$, $y = \frac{\pi}{2}$. That is, $\sin^{-1}(1) = \frac{\pi}{2}$.

b. For $y = \sin^{-1}\left(-\frac{1}{\sqrt{2}}\right)$, we have $\sin y = -\frac{1}{\sqrt{2}}$. Since y must be in the interval $\left[-\frac{\pi}{2}, \frac{\pi}{2}\right]$, $y = -\frac{\pi}{4}$. That is, $\sin^{-1}\left(-\frac{1}{\sqrt{2}}\right) = -\frac{\pi}{4}$.

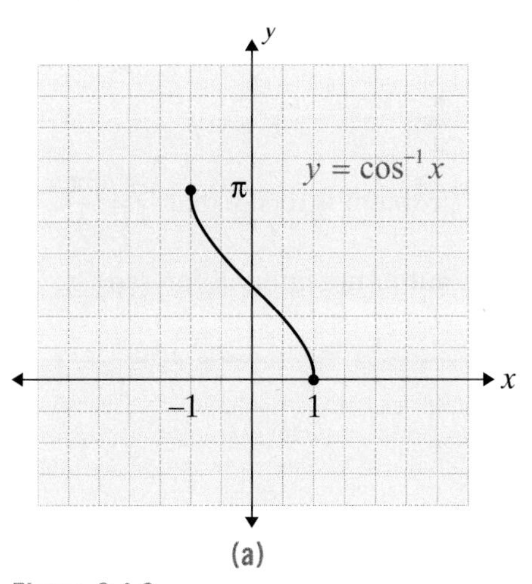

(a)

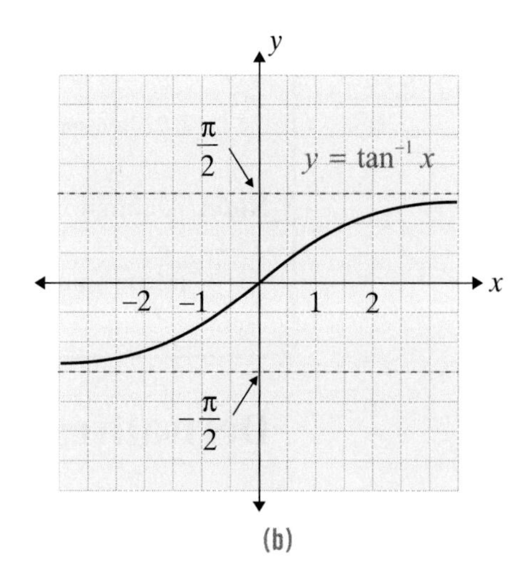

(b)

Figure 9.4.2

The three most commonly used inverse trigonometric functions, their domain, and their restricted ranges are presented in Table 9.4.1. The corresponding graphs of $y = \cos^{-1} x$ and $y = \tan^{-1} x$ are illustrated at the in Figure 9.4.2(a) and 9.4.2(b).

Table 9.4.1 Inverse Trigonometric Functions

Function	Domain (x values)	Range (y values)
$y = \sin^{-1} x$	$[-1, 1]$	$\left[-\frac{\pi}{2}, \frac{\pi}{2}\right]$
$y = \cos^{-1} x$	$[-1, 1]$	$[0, \pi]$
$y = \tan^{-1} x$	$(-\infty, +\infty)$	$\left(-\frac{\pi}{2}, \frac{\pi}{2}\right)$

Example 2: Inverse Trigonometric Functions

Find each of the indicated values of y.

a. $y = \cos^{-1}\left(\frac{1}{2}\right)$ **b.** $y = \tan^{-1}(-1)$

Solutions: a. For $y = \cos^{-1}\left(\frac{1}{2}\right)$, we have $\cos y = \frac{1}{2}$. Therefore, since y must be in the interval $[0, \pi]$, $y = \frac{\pi}{3}$.

continued on next page ...

b. For $y = \tan^{-1}(-1)$, we have $\tan y = -1$ with y in the interval $\left(-\frac{\pi}{2}, \frac{\pi}{2}\right)$.

Therefore, $y = -\frac{\pi}{4}$.

Note: A common error in this problem is to say that $y = \frac{3\pi}{4}$. While $\tan\left(\frac{3\pi}{4}\right) = -1$ is a true statement, $\frac{3\pi}{4}$ is **not** in the interval for $\tan^{-1} x$.

Derivatives

To find the derivative of the inverse trigonometric function $y = \sin^{-1} x$, we use the definition $\sin y = x$ and differentiate implicitly as follows: For $y = \sin^{-1} x$, we have $\sin y = x$, and differentiating implicitly gives

$$\cos y \cdot \frac{dy}{dx} = 1.$$

Solving for $\frac{dy}{dx}$, we have

$$\frac{dy}{dx} = \frac{1}{\cos y}.$$

Now, from the identity $\sin^2 y + \cos^2 y = 1$, $\cos y = \sqrt{1 - \sin^2 y}$.
(The positive square root is used because $\cos y$ is nonnegative in the interval $\left[-\frac{\pi}{2}, \frac{\pi}{2}\right]$.)

Replacing $\sin y$ with x gives the standard result:

$$\frac{dy}{dx} = \frac{1}{\cos y} = \frac{1}{\sqrt{1 - \sin^2 y}} = \frac{1}{\sqrt{1 - x^2}}.$$

The formulas for the derivatives of $y = \cos^{-1} x$ and $y = \tan^{-1} x$ can be developed in a similar manner and are listed in Table 9.4.2. An interesting and helpful observation is that these formulas are algebraic expressions and not trigonometric expressions, as might be expected.

A reference triangle is helpful. The acute angle is y (radians) and the sides are arranged to satisfy the Pythagorean Theorem and the condition that $\sin y = x$. From the triangle's geometry, $\cos y = \frac{\text{adjacent}}{\text{hypotenuse}} = \sqrt{1 - x^2}$.

1

x

y

$\sqrt{1 - x^2}$

Table 9.4.2 Differentiation Formulas for Inverse Trigonometric Functions

1. **a.** If $y = \sin^{-1} x$, then

$$\frac{dy}{dx} = \frac{1}{\sqrt{1-x^2}} \text{ for } x \text{ in } (-1, 1).$$

b. If $y = \sin^{-1}\left[g(x)\right]$, then

$$\frac{dy}{dx} = \frac{1}{\sqrt{1-\left[g(x)\right]^2}} \cdot g'(x) \text{ for } g(x) \text{ in} (-1, 1)$$

provided $g'(x)$ exists.

2. **a.** If $y = \cos^{-1} x$, then

$$\frac{dy}{dx} = \frac{-1}{\sqrt{1-x^2}} \text{ for } x \text{ in } (-1, 1).$$

b. If $y = \cos^{-1}\left[g(x)\right]$, then

$$\frac{dy}{dx} = \frac{-1}{\sqrt{1-\left[g(x)\right]^2}} \cdot g'(x) \text{ for } g(x) \text{ in} (-1, 1)$$

provided $g'(x)$ exists.

3. **a.** If $y = \tan^{-1} x$, then

$$\frac{dy}{dx} = \frac{1}{1+x^2} \text{ for } x \text{ in} (-\infty, +\infty).$$

b. If $y = \tan^{-1}\left[g(x)\right]$, then

$$\frac{dy}{dx} = \frac{1}{1+\left[g(x)\right]^2} \cdot g'(x)$$

provided $g'(x)$ exists.

Example 3: Differentiating Inverse Trigonometric Functions

Find $\frac{dy}{dx}$ for each of the following functions.

a. $y = \sin^{-1}\left(e^x\right)$

b. $y = \tan^{-1}\left(x^2\right)$

continued on next page ...

Solutions: a. Use Formula 1(b) from Table 9.4.2 as well as the Chain Rule with $g(x)=e^x$.

$$\frac{dy}{dx}=\frac{1}{\sqrt{1-(e^x)^2}}\cdot e^x=\frac{e^x}{\sqrt{1-e^{2x}}}$$

b. Use Formula 3(b) from Table 9.4.2 as well as the Chain Rule with $g(x)=x^2$.

$$\frac{dy}{dx}=\frac{1}{1+(x^2)^2}\cdot 2x=\frac{2x}{1+x^4}$$

Example 4: Differentiating Inverse Trigonometric Functions

Find $f'(2)$ for $f(x)=\tan^{-1}(2x)$.

Solution: From Formula 3(b) from Table 9.4.2,

$$f'(x)=\frac{1}{1+(2x)^2}\cdot 2=\frac{2}{1+4x^2}.$$

Thus

$$f'(2)=\frac{2}{1+4\cdot 2^2}=\frac{2}{17}.$$

Integrals

Each of the formulas for differentiation in Table 9.4.2 leads immediately to a corresponding formula for integration listed in Table 9.4.3.

Table 9.4.3 Integration Formulas Yielding Inverse Trigonometric Functions

1. a. $\int\frac{1}{\sqrt{1-x^2}}dx=\sin^{-1}x+C$

b. $\int\frac{1}{\sqrt{1-[g(x)]^2}}\cdot g'(x)dx=\sin^{-1}[g(x)]+C$

2. a. $\int\frac{-1}{\sqrt{1-x^2}}dx=\cos^{-1}x+C$

b. $\int\frac{-1}{\sqrt{1-[g(x)]^2}}\cdot g'(x)dx=\cos^{-1}[g(x)]+C$

continued on next page ...

3. a. $\int \frac{1}{1+x^2}dx = \tan^{-1}x + C$

b. $\int \frac{1}{1+[g(x)]^2}\cdot g'(x)dx = \tan^{-1}[g(x)] + C$

Example 5: Integrating Inverse Trigonometric Functions

Evaluate the following integrals.

a. $\int \frac{1}{\sqrt{1-9x^2}}dx$

b. $\int_{-1}^{1} \frac{1}{1+x^2}dx$

Solutions: a. Use u-substitution with $u = 3x$ and $du = 3\,dx$ along with Formula 1(a) from Table 9.4.3.

$$\begin{aligned}\int \frac{1}{\sqrt{1-9x^2}}dx &= \frac{1}{3}\int \frac{1}{\sqrt{1-9x^2}}\cdot 3dx \\ &= \frac{1}{3}\int \frac{1}{\sqrt{1-u^2}}du \\ &= \frac{1}{3}\sin^{-1}u + C \\ &= \frac{1}{3}\sin^{-1}(3x) + C\end{aligned}$$

b. Use Formula 3(a) from Table 9.4.3 and the definition of $\tan^{-1}x$.

$$\begin{aligned}\int_{-1}^{1} \frac{1}{1+x^2}dx &= \tan^{-1}x\Big]_{-1}^{1} \\ &= \tan^{-1}(1) - \tan^{-1}(-1) \\ &= \frac{\pi}{4} - \left(-\frac{\pi}{4}\right) = \frac{\pi}{2}\end{aligned}$$

9.4 Exercises

In Exercises 1 – 4, find the value in degrees.

1. $\sin^{-1}\left(\frac{\sqrt{3}}{2}\right)$

2. $\cos^{-1}\left(-\frac{1}{2}\right)$

3. $\tan^{-1}\left(-\sqrt{3}\right)$

4. $\sin^{-1}\left(\frac{1}{\sqrt{2}}\right)$

In Exercises 5 – 8, find the value in radian measure.

5. $\cos^{-1}\left(-\frac{1}{\sqrt{2}}\right)$

6. $\tan^{-1}0$

7. $\tan^{-1}\left(\frac{1}{\sqrt{3}}\right)$

8. $\sin^{-1}\left(\frac{1}{2}\right)$

In Exercises 9 – 14, use a calculator to find the value in radian measure, correct to the nearest hundredth.

9. $\sin^{-1}(-0.5913)$

10. $\tan^{-1}(1.5763)$

11. $\cos^{-1}(0.8136)$

12. $\cos^{-1}(-0.3876)$

13. $\tan^{-1}(-0.8752)$

14. $\sin^{-1}(0.6971)$

In Exercises 15 – 24, find the derivative of each function.

15. $y=\cos^{-1}x^2$

16. $y=\tan^{-1}\frac{1}{x}$

17. $f(x)=\sin^{-1}\sqrt{x}$

18. $f(x)=\cos^{-1}(\ln|x|)$

19. $f(x)=\tan^{-1}e^x$

20. $f(x)=\sin^{-1}e^{2x}$

21. $y=\cos^{-1}(2x+1)$

22. $y=\tan^{-1}\sqrt{2x}$

23. $y=\sin^{-1}(1-3x)$

24. $y=\cos^{-1}\frac{1}{x^2}$

In Exercises 25 – 34, evaluate each integral.

25. $\int\frac{1}{1+4x^2}dx$

26. $\int\frac{1}{\sqrt{1-16x^2}}dx$

27. $\int\frac{e^x}{1+e^{2x}}dx$

28. $\int\frac{x}{\sqrt{1-0.25x^4}}dx$

29. $\int\frac{1}{\sqrt{x}\left(\sqrt{1-x}\right)}dx$ $\left(\text{Hint: Let } u=\sqrt{x}.\right)$

30. $\int\frac{1}{x\left[1+(\ln 2x)^2\right]}dx$ $(\text{Hint: Let } u=\ln 2x.)$

31. $\int_0^{\frac{1}{4}}\frac{1}{\sqrt{1-4x^2}}dx$

32. $\int_0^{\frac{\sqrt{2}}{2}}\frac{x}{\sqrt{1-x^4}}dx$

33. $\int_{-\frac{\pi}{2}}^{\frac{\pi}{2}}\frac{\cos x}{1+\sin^2 x}dx$ $(\text{Hint: Let } u=\sin x.)$

34. $\int_1^3\frac{4}{\sqrt{x}(1+x)}dx$ $\left(\text{Hint: Let } u=\sqrt{x}.\right)$

35. **Worker efficiency.** A manufacturer estimates the time it takes for a new employee to produce the xth item is given by $T(x) = 90 - 25\tan^{-1} 0.2x$ minutes.
   a. Find $T'(6)$.
   b. Interpret the results in part (a).

36. **Election campaign.** A candidate for the board of trustees for a college plans to do an intensive door-to-door campaign. She predicts that the number of votes gained from this activity will be $N(x) = 900\sin^{-1} 0.007x$, where x is the percent of the homes in the district visited by her workers and $0 < x \leq 100$.
   a. Find $N'(60)$.
   b. Interpret the results in part (a).

37. **Drug acceptance.** The percent of doctors who are prescribing a new "miracle drug" is changing at a rate given by $P(t) = \dfrac{16.5}{1+0.09t^2}$ percent per month, where t is the number of months after the drug is made available. What percent of the doctors are prescribing the drug at the end of 6 months?

38. **Volume.** Find the volume of the solid of revolution determined when the region bounded by $y = \dfrac{1}{\sqrt{1+x^2}}$, $x = 0$, and $x = 1$ is rotated about the x-axis.

Hawkes Learning Systems: Essential Calculus

Inverse Trigonometric Functions

Chapter 9 Index of Key Ideas and Terms

Section 9.1 The Trigonometric Functions

Unit Circle pages 682 - 683

The graph that corresponds to the equation $x^2 + y^2 = 1$ is a circle of radius 1 with center at the origin.

The Trigonometric Functions pages 683 - 688

If $P(x, y)$ is a point on the unit circle and θ is the corresponding central angle, then

$$\sin\theta = y \quad \text{and} \quad \cos\theta = x$$

$$\tan\theta = \frac{y}{x} = \frac{\sin\theta}{\cos\theta} \qquad (x \neq 0)$$

$$\cot\theta = \frac{x}{y} = \frac{\cos\theta}{\sin\theta} \qquad (y \neq 0)$$

$$\sec\theta = \frac{1}{x} = \frac{1}{\cos\theta} \qquad (x \neq 0)$$

$$\csc\theta = \frac{1}{y} = \frac{1}{\sin\theta} \qquad (y \neq 0)$$

The sine and cosine functions are **periodic** with **period 2π** and

$$\sin(\theta + 2k\pi) = \sin\theta \qquad \text{and} \qquad \cos(\theta + 2k\pi) = \cos\theta$$

where k is any integer.

The tangent function is **periodic** with **period** π and

$$\tan(\theta + k\pi) = \tan\theta \qquad \text{for } \theta \neq \frac{\pi}{2} + k\pi$$

where k is any integer.

Both the functions $y = \sin x$ and $y = \cos x$ are continuous for all real numbers.

Basic Trigonometric Identities pages 688 - 689

1. $\cos^2\theta + \sin^2\theta = 1$ Pythagorean identity

2. $\sin(-\theta) = -\sin\theta$

3. $\cos(-\theta) = \cos\theta$

(2–3: Even/odd identities)

4. $\sin(\theta \pm \phi) = \sin\theta\cos\phi \pm \cos\theta\sin\phi$

5. $\cos(\theta \pm \phi) = \cos\theta\cos\phi \mp \sin\theta\sin\phi$

(4–5: Sum and difference identities)

6. $\cos^2\frac{\theta}{2} = \frac{1+\cos\theta}{2}$

7. $\sin^2\frac{\theta}{2} = \frac{1-\cos\theta}{2}$

(6–7: Half-angle identities)

continued on next page ...

Section 9.1 The Trigonometric Functions (continued)

8. $\sin 2\theta = 2\sin\theta\cos\theta$

9. $\cos 2\theta = \cos^2\theta - \sin^2\theta$

$= 2\cos^2\theta - 1$

$= 1 - 2\sin^2\theta$

(8–9: Double-angle identities)

10. $\sin\theta = \cos\left(\dfrac{\pi}{2} - \theta\right)$

11. $\cos\theta = \sin\left(\dfrac{\pi}{2} - \theta\right)$

(10–11: Complementary-angle identities)

Right-Triangle Trigonometry page 690

$$\sin\theta = \frac{\text{side opposite } \theta}{\text{hypotenuse}}$$

$$\cos\theta = \frac{\text{side adjacent } \theta}{\text{hypotenuse}}$$

$$\tan\theta = \frac{\text{side opposite } \theta}{\text{side adjacent } \theta}$$

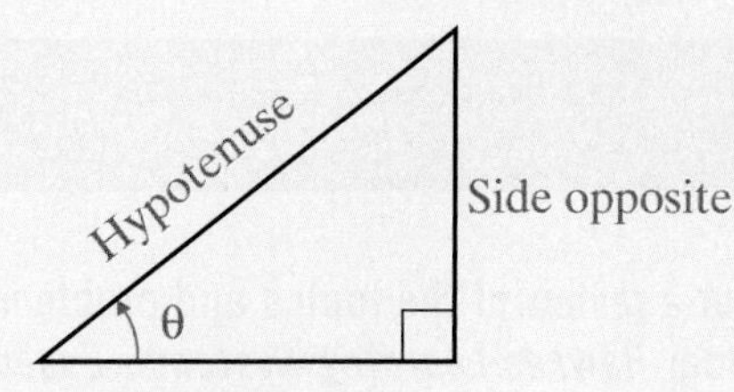

Section 9.2 Derivatives of the Trigonometric Functions

Formulas You Should Know pages 694 - 698

$$\frac{d}{dx}[\sin x] = \cos x \qquad \frac{d}{dx}[\cot x] = -\csc^2 x$$

$$\frac{d}{dx}[\cos x] = -\sin x \qquad \frac{d}{dx}[\sec x] = \sec x \tan x$$

$$\frac{d}{dx}[\tan x] = \sec^2 x \qquad \frac{d}{dx}[\csc x] = -\csc x \cot x$$

Section 9.3 Integration of the Trigonometric Functions

Formulas You Should Know pages 701 - 706

$$\int \cos x\,dx = \sin x + C$$

$$\int \sin x\,dx = -\cos x + C$$

$$\int \sec^2 x\,dx = \tan x + C$$

$$\int \tan x\,dx = -\ln|\cos x| + C$$

Section 9.4 Inverse Trigonometric Functions

Formulas You Should Know pages 713 - 715

$$\frac{d}{dx}\left[\sin^{-1} x\right]=\frac{1}{\sqrt{1-x^2}}, \text{ for } x \text{ in } (-1,1) \quad \text{and} \quad \int \frac{1}{\sqrt{1-x^2}}\,dx=\sin^{-1} x+C$$

$$\frac{d}{dx}\left[\cos^{-1} x\right]=\frac{-1}{\sqrt{1-x^2}}, \text{ for } x \text{ in } (-1,1) \quad \text{and} \quad \int \frac{-1}{\sqrt{1-x^2}}\,dx=\cos^{-1} x+C$$

$$\frac{d}{dx}\left[\tan^{-1} x\right]=\frac{1}{1+x^2}, \text{ for } x \text{ in } (-\infty,+\infty) \quad \text{and} \quad \int \frac{1}{1+x^2}\,dx=\tan^{-1} x+C$$

Chapter 9 Review

For a review of the topics and problems from Chapter 9, look at the following lessons from *Hawkes Learning Systems: Essential Calculus.*

The Trigonometric Functions
Derivatives of the Trigonometric Functions
Integration of the Trigonometric Functions
Inverse Trigonometric Functions

Chapter 9 Test

In Exercises 1 – 4, find the exact value of each of the trigonometric functions.

1. $\cos\left(-\frac{4\pi}{3}\right)$

2. $\tan\left(\frac{5\pi}{6}\right)$

3. $\csc\left(\frac{\pi}{2}\right)$

4. $\sec\left(\frac{5\pi}{4}\right)$

In Exercises 5 – 8, use a calculator to find the value of each of the given trigonometric functions, correct to four decimal places.

5. $\sin 131.2°$

6. $\cot 59.6°$

7. $\sec 2.31$

8. $\tan(-1.22)$

In Exercises 9 and 10, sketch the graph of each of the given functions over the interval $[-2\pi, 2\pi]$.

9. $y = 2\cos x$

10. $y = 3\sin 2x$

In Exercises 11 and 12, verify each of the given identities.

11. $1 + \frac{\tan\theta}{\cot\theta} = \sec^2\theta$

12. $\cos 2\theta = \cos^2\theta - \sin^2\theta$

$(\text{Hint}: 2\theta = \theta + \theta)$

13. Find the values of the six trigonometric functions of θ.

16

9

θ

14. Height of a tree. A tree casts a shadow that is 17 ft. long. If the sun is at 42° above the horizon, find the height of the tree.

15. Predator-prey model. In a certain environment, the number of predators is estimated by $P(t) = 600 + 180\sin\frac{\pi}{12}t$. The number of prey is estimated by $p(t) = 3400 + 420\cos\frac{\pi}{12}t$.

a. Find $P(2)$, $P(4)$, $P(6)$, and $P(12)$.

b. Find $p(2)$, $p(4)$, $p(6)$, and $p(12)$.

In Exercises 16 – 22, find the derivative for each function.

16. $y = 4 \sin \pi x$

17. $y = 2e^{\tan x}$

18. $f(x) = 5x^3 \cos 4x$

19. $f(x) = \csc^2 3x$

20. $f(x) = \ln(\cos x)$

21. $y = \sqrt{1 + \cos 2x}$

22. $y = \dfrac{1 - \cos x}{\sin x}$

In Exercises 23 and 24, find $\dfrac{\partial f}{\partial x}$ *and* $\dfrac{\partial f}{\partial y}$ *for each function.*

23. $f(x, y) = \sin xy + xe^y$

24. $f(x, y) = x \tan y + y \ln xy$

25. Find an equation for the line tangent to the graph of $y = \sin \dfrac{\pi}{6} x$ at the point where $x = 4$.

26. Hours of daylight. The number of hours of daylight in a certain community can be estimated by $D(t) = 12 + 1.36 \sin \dfrac{2\pi}{365} t$, where t is in days and $t = 0$ corresponds to March 20.

a. Estimate the number of hours of daylight on June 20.

b. Find $D'(274)$.

c. Interpret the results in part (b).

In Exercises 27 – 32, evaluate the given integral.

27. $\int x \sin(x^2 - 1)\, dx$

28. $\int \dfrac{\cos x}{(2 + \sin x)^{\frac{3}{2}}} dx$

29. $\int_0^{\frac{\pi}{12}} \sec^2(3x) e^{\tan 3x} dx$

30. $\int_0^{\frac{\pi}{3}} \dfrac{\sin x}{\cos^2 x} dx$

31. $\int \sin x \sqrt{\cos x + 1}\, dx$

32. $\int \dfrac{\sec^2 x}{1 + \tan x} dx$

In Exercises 33 and 34, integrate by parts.

33. $\int 4x \sin \dfrac{x}{2} dx$

34. $\int x \sec^2 4x dx$

35. Area. Find the area of the region bounded by the graph of $y = x + \sin x$ and the x-axis on the interval $\left[0, \dfrac{\pi}{3}\right]$.

36. Population growth. The size of a population of deer is dependent on the plant life and other elements in the environment. The rate of change of the size of a certain deer population is given by $P'(t) = 60\pi \sin \frac{\pi}{12} t$, where t is the number of months after the population is counted. Find the change in the population from $t = 2$ to $t = 6$.

In Exercises 37 and 38, find the value in degrees.

37. $\sin^{-1}\left(-\frac{1}{\sqrt{2}}\right)$

38. $\cos^{-1}(1)$

In Exercises 39 and 40, find the value in radian measure.

39. $\tan^{-1}(-1)$

40. $\sin^{-1}\left(\frac{1}{2}\right)$

In Exercises 41 and 42, use a calculator to find the value in radian measure, correct to the nearest hundredth.

41. $\cos^{-1}(0.4892)$

42. $\tan^{-1}(-1.1361)$

In Exercises 43 – 46, find the derivative of each function.

43. $f(x) = \sin^{-1} 4x$

44. $f(x) = \tan^{-1}(\ln|x|)$

45. $y = \tan^{-1}(\sin x)$

46. $y = \cos^{-1} e^{4x}$

In Exercises 47 – 50, evaluate each integral.

47. $\int \frac{2x}{\sqrt{1-x^4}} dx$

48. $\int \frac{e^{2x}}{\sqrt{1-e^{4x}}} dx$

49. $\int \frac{\sin x}{1+\cos^2 x} dx$

50. $\int \frac{5}{\sqrt{x}(1+4x)} dx$

Cumulative Review

In Exercises 1 – 6, determine $\frac{df}{dx}$ *for the function* $f(x)$.

1. $f(x) = 2x \sin x$

2. $f(x) = \frac{3+\cos x}{5+2x}$

3. $f(x) = \sqrt{5+\sin x}$

4. $f(x) = \left(200+8x-3x^2\right)^8$

5. $f(x) = \frac{1}{3+x-x^2}$

6. $f(x) = e^x \sin 2x$

7. Using the fact that $\frac{d}{dx}\left(e^{3x}\cos x\right)=e^{3x}\left(3\cos x-\sin x\right)$, evaluate $\int_0^1 e^{3x}\left(3\cos x-\sin x\right)dx$.

8. What is the total area between $f(x)=e^x$ and $g(x)=1-x$ for $0\le x\le 1$? Sketch this region.

9. Using calculus and a calculator, determine the closest point on the graph of $f(x)=\ln x$ to the point (3, 5)? Sketch these functions on the same axis.

Exercises 10 – 13 relate to the function $f(x)=1+\sin 2x$.

10. If $f(x)$ is a periodic function, determine its period T.

11. Determine the smallest positive x_0 so that the point (x_0, y_0) is a local maximum.

12. Determine the smallest positive x_1 so that the point (x_1, y_1) is a point of inflection.

13. Evaluate the integral $\int_{x_0}^{x_1}(1+\sin 2x)\,dx$ where x_0 and x_1 are the numbers mentioned in Exercises 11 and 12. Sketch the region represented by the definite integral.

14. **a.** Suppose $y=ax^3+bx^2+cx+d$ (and $a\neq 0$). Determine an x-value, say x_0, so that $(x_0,\ y_0)$ is an inflection point.
b. Argue that your answer to part (a) proves that a cubic polynomial always has an inflection point.

15. Suppose the marginal cost of producing hand woven reed baskets is $c'(x)=4x+0.01x^2$. If the fixed cost is \$10, what is the total cost of producing 6 baskets?

16. Use differentials to compute a rational number which approximates $\sqrt{82}$.

17. Suppose a bacterial culture doubles every 1.5 hours. How long will it take to triple?

18. Suppose a particle is shot upward with an initial speed of 64 ft./sec. from an initial height of 40 feet.
a. At what time *t*, in seconds, will it reach its maximum height?
b. What is the maximum height?

19. Suppose that, for a certain type of earphones for a portable phone, the demand function is $30-2x=D(x)$ and the supply function is $S(x)=0.5x^2$.
a. Find the equilibrium point.
b. Find the consumers' surplus.
c. Find the producer's surplus.

20. Use a Table of Integrals to compute $\int_0^1 \frac{2x+6}{0.5x+1}dx$. Check your answer with a graphing calculator.

Exercises 21 – 23 correspond to $f(x)=k\left(3x^2+2\right)$.

21. Determine k so that $f(x)$ is a probability density function over the interval [0,4].

22. Using k from Exercise 21, what is the average value of x over the range of [0, 4]?

23. Using k from Exercise 21, what is the probability that a randomly selected x-value lies in the range $0\le x\le 2$?

24. What is the average value of $y=\sin 2x$ over the range $0\le x\le \frac{\pi}{4}$?

25. Determine $\int \sin^3 x\cos x\,dx$. Hint: use u-substitution with $u=\sin x$.

26. Determine $\int \cos^4 x\sin x\,dx$. Hint: use u-substitution with $u=\cos x$.

27. Determine $\int \cos^2 x\,dx$. Hint: substitute using the half-angle identity $\cos^2 x=\frac{1}{2}(1+\cos 2x)$.

28. Determine $\int \sin^2 x\,dx$. Hint: substitute using the half-angle identity $\sin^2 x=\frac{1}{2}(1-\cos 2x)$.

29. Determine both first partial derivatives f_x and f_y for $f(x,y)=e^x\sin 2xy$.

30. Integrate $\int \frac{6}{4^2+9x^2}dx$. Hint: factor $\frac{6}{16}\int \frac{dx}{1+\frac{9}{16}x^2}$, and substitute $u=\frac{3}{4}x$.

Sequences, Taylor Polynomials, and Power Series

CHAPTER 10

Did You Know?

In this chapter we present a series of special topics in applied calculus. The emphasis is on understanding rather than technical expertise, and the role of clarifying examples is emphasized.

We begin with a puzzle from medieval mathematics. The logicians at Merton College at Oxford in the early 1300's formulated rules for calculations in what today we would call applied science. One problem widely circulated was to calculate the average velocity of a point moving under constant acceleration ($y'' = k$). Their success with this problem led to a famous puzzle:

The Merton College Problem is stated as follows: *find the average velocity of a point which moves at a constant velocity throughout half of a certain time interval but moves the next quarter interval with twice the velocity, the next 8th time interval at three times the velocity, and so on "to infinity."*

We will solve this problem in two ways, one using a standard technique and the other solution using a geometric approach given by Nichole Oresme (1323 - 1382). Oresme invented the technique of using areas for quantities not actually areas themselves, an idea central to calculus today. The problem is equivalent to this: determine the limiting value of the sum of terms

$$\frac{1}{2}+\frac{2}{4}+\frac{3}{8}+\frac{4}{16}+\ldots+\frac{n}{2^n}+\ldots$$

if it is finite. To deal with such problems and many others, we begin with infinite sequences of numbers and then proceed to additive expressions with infinitely many terms.

Often, a student's first exposure to the notion of infinity occurs in a calculus course. Limits, derivatives, antiderivatives, and integrals are concepts making use of this notion. Chapter 10 introduces infinite lists of numbers which we call sequences. When progressions such as these are summed, the result is called a series. While series and sequences are truly the building blocks for all of mathematics, an in-depth study has been reserved thus far in order to emphasize applications. Chapter 10 also includes an extremely useful tool for approximation, Taylor polynomials. Next, Euler's formula, one of the truly beautiful theorems in mathematics, is discussed along with several applications. The final topic included is a method for solving different equations based on the student's newly gained knowledge of series.

10.1 Infinite Sequences

Objectives

After completing this section, you will be able to:

1. *Understand infinite sequences and calculate the corresponding limits.*
2. *Understand both convergent and divergent series and calculate sums when they exist.*

Sequences

A sequence is an ordered set of numbers. There may be finitely or infinitely many members in a sequence, but the infinite case is the principal aim in a calculus course. We usually represent the terms of a sequence with a subscripted variable. For example, the infinite sequence $\{1, 3, 5, 7, ...\}$ of odd positive integers might be represented by $\{a_1, a_2, a_3, a_4, ...\}$ and we mean

$$a_1 = 1, a_2 = 3, a_3 = 5, a_4 = 7, \text{ and so on.}$$

The sequences of numbers we consider in this book will always have a pattern associated with them or will be generated by some function or formula. There is a function which generates the odd integers. It is

$$f(n) = 2n - 1, \text{ where } n \text{ is an integer.}$$

Using the function $f(n)$ is convenient and we say it gives *the formula for the* n^{th} *term* of the sequence.

We could equivalently write $a_n = 2n - 1$. This emphasizes the placement of the index variable in the subscript, and this is the common notation. We also use the notation

$$\{a_n\}_{n=1}^{\infty} \text{ or } \{a_n\}_1^{\infty} \text{ or } \{a_n\} \text{ to denote a sequence.}$$

We usually use the set of positive integers beginning with 1 for the domain of any function f which specifies a sequence, but this is not necessary, as Example 1 shows.

Example 1: Sequences

Determine a formula for the nth term of the sequence $\{b_n\}_{n=0}^{\infty} = \{1, 3, 5, 7, ...\}$.

Solution: The sequence $\{b_n\}_{n=0}^{\infty}$ where $b_n = 2n + 1$ gives the sequence $\{1, 3, 5, 7, ...\}$. Here $b_0 = 2(0) + 1 = 0 + 1 = 1$, $b_1 = 2(1) + 1 = 2 + 1 = 3$, and so on.

Example 2: Expressing a Sequence

Express the sequence $\{4, 6, 8, 10, ...\}$ as $\{a_n\}_{n=1}^{\infty}$ and as $\{b_n\}_{n=3}^{\infty}$.

Solution: The sequence is the set of even positive integers beginning with 4. We modify the function $f(n) = 2n$, and note $a_n = 2n + 2$ and $b_n = 2n - 2$ give the same sequence.
$a_1 = 2(1) + 2 = 4$, $a_2 = 2(2) + 2 = 6$, and so on.
$b_3 = 2(3) - 2 = 6 - 2 = 4$, $b_4 = 2(4) - 2 = 8 - 2 = 6$, and so on.

Example 3: Terms of a Sequence

Write the first five terms of each of the following sequences.

a. $\{2^n\}_{n=1}^{\infty}$

b. $\{a_n\}_{n=1}^{\infty} = \left\{\frac{1}{10^n}\right\}_{n=1}^{\infty}$

c. $\{(-1)^n(2n-1)\}_{n=1}^{\infty}$

d. $a_n = \frac{1}{n}$

e. $\{n \cdot (-1)^{n+1}\}_{n=1}^{\infty}$

Solutions: a. $\{2^1, 2^2, 2^3, 2^4, 2^5 ...\}$ or $\{2, 4, 8, 16, 32, ...\}$.

b. $\left\{\frac{1}{10}, \frac{1}{100}, \frac{1}{1000}, \frac{1}{10{,}000}, \frac{1}{100{,}000}, ...\right\}$.

c. $\{-1, 3, -5, 7, -9, ...\}$. When the terms alternate in sign, the sequence is called an alternating sequence. It is possible to factor the minus sign. For example, $\{2^n(-1)^n\}_{n=1}^{\infty} = \{(-2)^n\}_{n=1}^{\infty}$.

d. $\left\{\frac{1}{1}, \frac{1}{2}, \frac{1}{3}, \frac{1}{4}, \frac{1}{5}, ...\right\}$. When the domain is not specified, it is assumed that $n = 1$ is the first case.

e. $\{1, -2, 3, -4, 5, ...\}$

Example 4: Find the Function

Find a function with domain N = {1, 2, 3, ...} for each of the following sequences.

a. $\left\{\frac{1}{2}, \frac{1}{-2}, \frac{1}{2}, \frac{1}{-2}, \ldots\right\}$

b. $\{0.1, 0.01, 0.001, 0.0001, \ldots\}$

c. $\{2, 6, 12, 20, 30, \ldots\}$

d. $\left\{\frac{1}{\sqrt{2}}, \frac{1}{\sqrt{3}}, \frac{1}{\sqrt{4}}, \ldots\right\}$

e. $[0.9, 0.99, 0.999, 0.9999, \ldots\}$

Solutions: a. Several solutions are possible. All are equally acceptable.

$$f(n) = \frac{1}{2(-1)^{n+1}} = \frac{1}{2}(-1)^{n+1} = \frac{(-1)^{n+1}}{2}$$

b. $f(n) = \frac{1}{10^n}$

c. $f(n) = n(n+1)$

d. $f(n) = \frac{1}{\sqrt{n+1}}$

e. $f(n) = \frac{10^n - 1}{10^n} = 1 - \frac{1}{10^n}$

Limit of a Sequence

For interesting sequences defined by a formula in terms of the index variable n, it is possible to observe that the terms may approach a limit just as the functions studied earlier had limits. In the previous example, in parts (b) and (d), it is easy to see that the terms are getting closer and closer to 0. We write $\lim_{n\to\infty} f(n) = 0$. In part (a) the terms are not all getting closer and closer to a particular number. There is no limiting value. In part (c), the terms increase without bound, that is they are said to approach ∞. In part (e), the terms are increasing but they are bounded just below 1. In fact, they approach 1 as a limit.

Limit of a Sequence

Let $\{f(n)\}_{n=1}^{\infty}$ be any sequence. If there is a number L such that $\lim_{n\to\infty}|f(n) - L| = 0$, then we say L is the ***limit*** *of the nth term of the sequence. If $f(x)$ is continuous on the interval $(1, \infty)$, then $\lim_{n\to\infty} f(n) = L$ if and only if $\lim_{x\to\infty} f(x) = L$.*

Example 5: Limit of a Sequence

Determine the limit of the nth term of each sequence given below.

a. $g_n = \dfrac{3n}{(4n+1)}$ **b.** $g_n = 4 - \dfrac{1}{(2^n)}$ **c.** $g_n = \left(\dfrac{4}{7}\right)^n$

d. $g_n = \dfrac{2^n + 3^n}{5^n}$ **e.** $\left\{1, -\dfrac{1}{2}, \dfrac{1}{4}, -\dfrac{1}{8}, \ldots\right\}$ **f.** $\dfrac{2n+3}{e^n}$

g. $\{3, 3.1, 3.14, 3.141, 3.1415, \ldots\}$

Solutions: a. By substituting large values for n, and using a calculator, one might infer the limit is 0.75. One can confirm this by analysis since

$$\frac{3n}{4n+1} = \frac{3n}{4n+1} \cdot \frac{\frac{1}{n}}{\frac{1}{n}} = \frac{3}{4+\frac{1}{n}}$$

Taking the limit of the right hand side shows:

$$\lim_{n\to\infty}\left(\frac{3}{4+\frac{1}{n}}\right) = \frac{3}{4+0} = \frac{3}{4}.$$

Since $\dfrac{1}{n}$ approaches 0 as n gets large

b. $\lim\limits_{n\to\infty}\left(4 - \dfrac{1}{2^n}\right) = 4 - 0 = 4.$

c. For any fraction $\dfrac{a}{b}$ with $0 < \dfrac{a}{b} < 1$, the powers of $\dfrac{a}{b}$ are smaller than $\dfrac{a}{b}$ and are decreasing towards zero. That is

$$0 < \ldots < \left(\frac{a}{b}\right)^3 < \left(\frac{a}{b}\right)^2 < \frac{a}{b}$$

Thus $\lim\limits_{n\to\infty}\left(\dfrac{4}{7}\right)^n = 0.$

d. In this case, $g_n = \dfrac{2^n + 3^n}{5^n} = \dfrac{2^n}{5^n} + \dfrac{3^n}{5^n} = \left(\dfrac{2}{5}\right)^n + \left(\dfrac{3}{5}\right)^n$. Thus

$$\begin{aligned}\lim_{n\to\infty} g_n &= \lim_{n\to\infty}\left[\left(\frac{2}{5}\right)^n + \left(\frac{3}{5}\right)^n\right] = \lim_{n\to\infty}\left(\frac{2}{5}\right)^n + \lim_{n\to\infty}\left(\frac{3}{5}\right)^n \\ &= 0 + 0 = 0.\end{aligned}$$

e. Although the terms alternate, they get smaller and smaller in absolute value and "the terms go to zero." This explains the necessity of the absolute value sign in the definition of sequence convergence.

continued on next page ...

f. $f(x) = \dfrac{2x+3}{e^x}$ is continuous everywhere. By the remark in the text-box, we need to determine the limit of $f(x)$ as $x \to \infty$. For this $f(x)$, it is easy to see the positive x-axis is an asymptote since $f(x)$ is positive (as $x \to \infty$.) yet $f(x)$ is decreasing. One way to see this is to use the quotient rule to get y'.

$$f'(x) = \frac{e^x(2) - (2x+3)e^x}{e^{2x}} = \frac{e^x(-2x-1)}{e^{2x}} = \frac{-2x-1}{e^x}$$

Since the numerator is negative (x is positive) and the denominator is positive, the fraction is negative. Therefore, $f(x)$ is decreasing and

$$\lim_{x\to\infty} \frac{2x+3}{e^x} = 0.$$

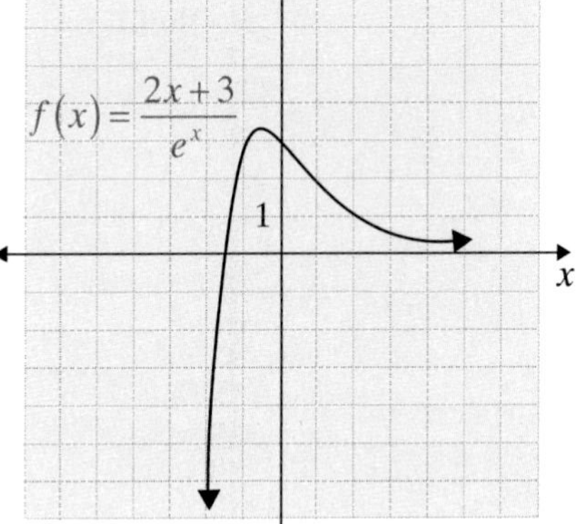

g. There is no determined pattern so the limit cannot be determined. However, the sequence of digits (so far) consists of the decimal digits of the number π.

Infinite Series

An **infinite series** is an expression involving *a sum of an infinite number of terms*. For example, if we add the terms in Example 3 part (b), we get

$$\frac{1}{10} + \frac{1}{100} + \frac{1}{1000} + \frac{1}{10000} + \ldots$$

This expression is usually first encountered early in academic life since it is equivalent to the number 0.1111... which is a decimal equivalent of $\frac{1}{9}$. A Riemann sum is another kind of series.

A **recursive sequence** is defined by specifying one or more initial terms and then expressing the nth term using previous terms.

Example 6: Fibonacci Sequence

Use a recursive presentation for the following sequence $\{1, 1, 2, 3, 5, 8, 13, 21, \ldots\}$.

Solution: Let $F_1 = 1$, $F_2 = 1$, and for $n \geq 3$, $F_n = F_{n-1} + F_{n-2}$. This sequence is called the **Fibonacci sequence** after Leonardo of Pisa (1170 - 1240) "of the family of Bonaccio."

Example 7: Define a Sequence

Use recursion to define the sequence $a_1 = \frac{1}{2}, a_2 = \frac{1}{2} + \frac{1}{2^2}, a_3 = \frac{1}{2} + \frac{1}{2^2} + \frac{1}{2^3}, \ldots$.

Solution: Let $a_1 = \frac{1}{2}$ and for $n > 1$, let $a_n = a_{n-1} + \frac{1}{2^n}$.

In calculus, sequences are studied because areas or other quantities may often be expressed or evaluated with limits of terms of sequences. Let us consider the sequence $\{a_n\}_{n=1}^{\infty} = \left\{\frac{1}{2^n}\right\}_{n=1}^{\infty}$ and represent each term using a segment of the same length on a number line. Although there are an infinite number of terms, we only need a finite interval to picture each segment. (See Figure 10.1.1.) It is clear in the picture that each segment of length a_n can be placed end-to-end and just cover the unit segment [0, 1] except for the right hand end-point. The segment a_1 covers the first half of the unit interval, the element a_2 covers half the remainder, the segment a_3 covers half the remainder, and so on. No matter how large a particular n we might examine or how many segments are drawn up to and including that a_n, there still remains a piece of the unit interval of length $\frac{1}{2^n}$ which is still uncovered. This geometric argument shows

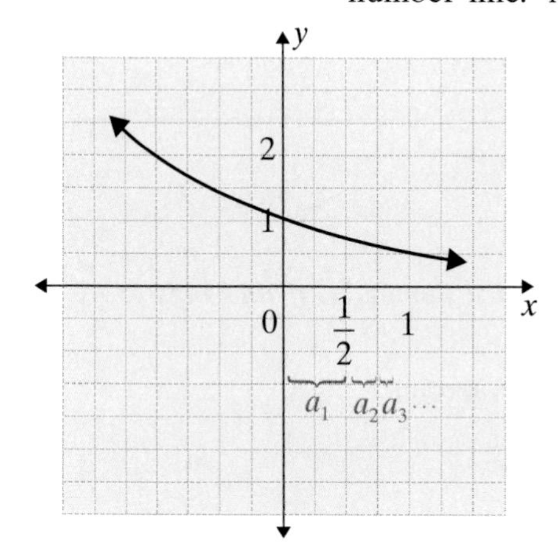

Figure 10.1.1

$$\sum_{n=1}^{\infty} a_n = a_1 + a_2 + a_3 + \ldots = \frac{1}{2} + \frac{1}{4} + \frac{1}{8} + \ldots \leq 1$$

That is, the sum of the terms in the expression cannot exceed 1. On the other hand, by choosing n large enough, we can find a finite sum of terms greater than any number less than 1. Clearly the limit is 1 and that is how we define such infinite expressions, by a limit.

Now let $\sum_{n=1}^{\infty} a_n = a_1 + a_2 + a_3 + \ldots$ denote an arbitrary series. We define a special sequence using the series called **the sequence of partial sums**.

$$S_1 = a_1,\ S_2 = a_1 + a_2,\ S_3 = a_1 + a_2 + a_3, \ldots \text{ and so on.}$$

Example 8: Sequence of Partial Sums

Given the series $\sum_{n=1}^{\infty} a_n = a_1 + a_2 + a_3 + \ldots$, define the sequence of partial sums recursively.

Solution: Let $S_1 = a_1$ and for $n > 1$, let $S_n = S_{n-1} + a_n$.

Example 9: Terms of a Partial Sum Sequence

Write out the first five partial sums of the sequence described by the Merton College Problem (see introduction).

Solution: $S_1 = \frac{1}{2}$

$$S_2 = \frac{1}{2} + \frac{2}{4} = \frac{2}{4} + \frac{2}{4} = 1$$

$$S_3 = \frac{1}{2} + \frac{2}{4} + \frac{3}{8} = \frac{4}{8} + \frac{4}{8} + \frac{3}{8} = \frac{11}{8}$$

$$S_4 = \frac{1}{2} + \frac{2}{4} + \frac{3}{8} + \frac{4}{16} = \frac{8}{16} + \frac{8}{16} + \frac{6}{16} + \frac{4}{16} = \frac{26}{16}$$

$$S_5 = \frac{1}{2} + \frac{2}{4} + \frac{3}{8} + \frac{4}{16} + \frac{5}{32} = \frac{26}{16} + \frac{5}{32} = \frac{52+5}{32} = \frac{57}{32}$$

Convergence Principle for Infinite Series

An infinite series $\sum_{n=1}^{\infty} a_n = a_1 + a_2 + a_3 + \ldots$ ***converges to a number L if and only if its corresponding sequence*** $\{S_n\}_{n=1}^{\infty}$ ***of partial sums converges to L.***

Example 10: Converging Series

Determine the number to which the Merton College series $\frac{1}{2} + \frac{2}{4} + \frac{3}{8} + \frac{4}{16} + \ldots + \frac{n}{2^n} + \ldots$ converges.

Solution: In a problem like this, one hopes to see a pattern which clarifies *whether* or not the sum converges and which also suggests the particular number to which the sum converges. This is not yet clear in the terms calculated in the previous example. However, the next two terms in the sequence of partial sums are $S_6 = \frac{120}{64}$ and $S_7 = \frac{247}{128}$.

All of these terms are below 2 and seem to be increasing towards 2. Perhaps 2 is the limit. This suggests seeing how close each term gets to 2. The results can be put into a table for analysis.

S_1	S_2	S_3	S_4	S_5	S_6	S_7	...	S_n
$2-\frac{3}{2}$	$2-1$	$2-\frac{5}{8}$	$2-\frac{6}{16}$	$2-\frac{7}{32}$	$2-\frac{8}{64}$	$2-\frac{9}{128}$	...	$2-\frac{(n+2)}{2^n}$

Assuming the guess for S_n is correct, we have

$$\sum_{n=1}^{\infty}\frac{n}{2^n}=\lim_{n\to\infty}\left(2-\frac{n+2}{2^n}\right)=2-\lim_{n\to\infty}\left(\frac{n+2}{2^n}\right)=2-0=2.$$

The last limit expression is very similar to the problem in Example 5(*f*).
In general, using small cases in order to determine an expression like that in the Table for S_n involves guesswork and inference. The proof that the format is correct usually involves a formal technique called mathematical induction which we do not discuss in this text. However, the main step in such a proof requires that we show the guess for the formula for S_n leads to a correct version for S_{n+1}. That is, one must show, given a formula for S_n, that $S_n + a_{n+1}$ leads to S_{n+1}. We illustrate for this example.

$$\begin{aligned}S_n+a_{n+1}&=2-\frac{n+2}{2^n}+\frac{n+1}{2^{n+1}}=2-\frac{2(n+2)}{2\cdot 2^n}+\frac{n+1}{2^{n+1}}\\&=2-\frac{2n+4}{2^{n+1}}-\frac{-n-1}{2^{n+1}}=2-\frac{2n+4-n-1}{2^{n+1}}\\&=2-\frac{(n+1)+2}{2^{n+1}}=S_{n+1}\end{aligned}$$

Example 11: Calculating an Infinite Series

Determine the sums of the following infinite series providing they converge.

a. $\sum_{n=1}^{\infty}\frac{1}{3^n}$ **b.** $\sum_{n=1}^{\infty}(-1)^n$ **c.** $\sum_{n=1}^{\infty}\frac{1}{n^2+n}$ **d.** $\sum_{n=1}^{\infty}\frac{1}{n+1}$

Solutions: a. The partial sums converge.

$$S_1=\frac{1}{3},\ S_2=\frac{1}{3}+\frac{1}{9}=\frac{3}{9}+\frac{1}{9}=\frac{4}{9},\ S_3=\frac{4}{9}+\frac{1}{27}=\frac{13}{27},$$

$$S_4=\frac{13}{27}+\frac{1}{3^4}=\frac{40}{81},\ S_5=\frac{40}{81}+\frac{1}{3^5}=\frac{121}{243}=\frac{\frac{(3^5-1)}{2}}{3^5}=\frac{1}{2}\cdot\left(1-\frac{1}{3^5}\right)$$

$$\lim_{n\to\infty}S_n=\lim_{n\to\infty}\frac{1}{2}\cdot\left(1-\frac{1}{3^n}\right)=\frac{1}{2}\cdot(1-0)=\frac{1}{2}$$

b. This series diverges. The sequence of partial sums does not converge.
$S_1=-1,\ S_2=0,\ S_3=-1,\ S_4=0,\ldots$ $\quad S_{2n}-S_{2n+1}=1$ for all n

c. The partial sums converge.

$$S_1=0.5=\frac{1}{2},S_2=0.666\ldots=\frac{2}{3},S_3=0.75=\frac{3}{4},S_4=0.8=\frac{4}{5},\text{ and}$$

$S_5=0.833\ldots=\frac{5}{6}$. So, apparently, $S_n=\frac{n}{n+1}$. We have $\lim_{n\to\infty}\left(\frac{n}{n+1}\right)=1$.

continued on next page ...

a_1	$\frac{1}{2}=0.5$
a_2	$\frac{1}{3}=0.\bar{3}$
a_3	$\frac{1}{4}=0.25$
a_4	$\frac{1}{5}=0.2$
a_5	$\frac{1}{6}=0.1\bar{6}$

d. This diverges. This is an important infinite series. It seems to converge since the sequence of partial sums is not too large and the terms seem to be getting closer together. Nevertheless, the terms S_n increase without bound ("go to infinity"). One can see this with a nice argument by partitioning the terms of the series in a new way.

Let $b_1 = \frac{1}{2}$ and

$$b_2 = \frac{1}{3}+\frac{1}{4} > \frac{1}{4}+\frac{1}{4} = 2\cdot\left(\frac{1}{4}\right) = \frac{1}{2}$$

$$b_3 = \frac{1}{5}+\frac{1}{6}+\frac{1}{7}+\frac{1}{8} > \frac{1}{8}+\frac{1}{8}+\frac{1}{8}+\frac{1}{8} = 4\cdot\left(\frac{1}{8}\right) = \frac{1}{2}$$

$$b_4 = \frac{1}{9}+\frac{1}{10}+\ldots+\frac{1}{16} > \frac{1}{16}+\frac{1}{16} = 8\cdot\left(\frac{1}{16}\right) = \frac{1}{2}$$

The nth term b_n will have 2^{n-1} fractions all of which will be greater than or equal to $\frac{1}{2^n}$. So b_n itself will be greater than $\frac{1}{2}$. That is,

$$b_n = \frac{1}{2^{n-1}}+\frac{1}{2^{n-1}+1}+\ldots+\frac{1}{2^n} > \frac{1}{2^n}+\frac{1}{2^n}+\ldots+\frac{1}{2^n} = 2^{n-1}\cdot\left(\frac{1}{2^n}\right) = \frac{1}{2}$$

But then $\sum_{n=1}^{\infty}\frac{1}{n+1} = \sum_{n=1}^{\infty} b_n > \frac{1}{2}+\frac{1}{2}+\ldots+\frac{1}{2}+\ldots \to \infty$.

The given series is greater than any given number because enough terms of size $\frac{1}{2}$ will surpass any given number.

Gottfried Wilhelm Von Leibniz (1646 - 1716) graduated from his university at 20, far too young to be hired as a professor or to practice law. He went to Paris instead and began to frequent intellectual circles. He only then began his study of mathematics in the scientific group there centered around Christian Huygens. He became well-known for his clever solution to Example 11(c) just above. He re-wrote each of the terms as a difference and thereby created the first "collapsing sum." That is

$$\sum_{n=1}^{\infty}\frac{1}{n^2+n} = \sum_{n=1}^{\infty}\frac{1}{n(n+1)} = \sum_{n=1}^{\infty}\left(\frac{1}{n}-\frac{1}{n+1}\right)$$

$$=\left(\frac{1}{1}-\frac{1}{2}\right)+\left(\frac{1}{2}-\frac{1}{3}\right)+\left(\frac{1}{3}-\frac{1}{4}\right)+\left(\frac{1}{4}-\frac{1}{5}\right)+\ldots+\left(\frac{1}{n}-\frac{1}{n+1}\right)+\ldots = 1.$$

If an infinite series is to converge, the individual terms must eventually all be very small. In fact, $\lim_{n\to\infty} a_n = 0$ is absolutely necessary if $\sum a_n$ is to converge. But this is not enough. Example 11(c) shows that not only must individual terms shrink to zero, they must do so faster than the terms in Example 11(d).

There are many important theorems which help students and professionals determine whether a series converges or not; however, they and their use are beyond the scope of this book.

Example 12: Limits

Show $\lim_{x\to\infty}(\ln x)$ does not exist.

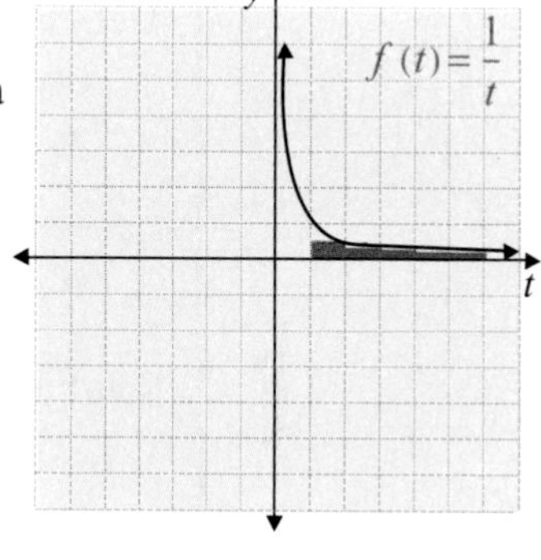

Solution: We use the Fundamental Theorem of Calculus, the results of Example 11**d.**, and a Riemann sum. The area under the graph is $\int_1^x \frac{1}{t}\,dt = \ln x$. The shaded area is less than the integral, and the shaded area is given by the Riemann sum

$$\sum_{n=1}^{\infty} f(t_n)\cdot \Delta t = \frac{1}{2}\cdot 1 + \frac{1}{3}\cdot 1 + \frac{1}{4}\cdot 1 + \ldots \to \infty.$$

Thus $\ln x$ increases without bound.

Geometric Series

One of the most important series in mathematics is called the **geometric series**. It has the form

$$a + ar + ar^2 + ar^3 + \ldots + ar^n + \ldots$$

Each term after the first is r times the preceding term. The number r is called the ratio of the series. We index the summation starting with $n = 0$, so $S_0 = a$, $S_1 = a + ar$, and S_n is the nth partial sum. Then

$$S_n = a + ar + ar^2 + \ldots + ar^n$$

$$rS_n = ar + ar^2 + \ldots + ar^n + ar^{n+1}$$

$$S_n - rS_n = a - ar^{n+1} \qquad \text{(the difference on the right "collapses")}$$

$$(1-r)S_n = a\left(1-r^{n+1}\right)$$

Thus $S_n = \frac{a}{1-r}\cdot\left(1-r^{n+1}\right)$. This is the sum of any finite number of terms. When the ratio satisfies $-1 < r < 1$, then $r^n \to 0$. In this case, by taking the limit as $n \to \infty$, we see

$$S = \lim_{n\to\infty} S_n = \frac{a}{1-r}\cdot \lim_{n\to\infty}\left(1-r^{n+1}\right) = \frac{a}{1-r}(1-0) = \frac{a}{1-r}.$$

Convergence of a Geometric Series

When $-1 < r < 1$, *the geometric series* $a + ar + ar^2 + ar^3 + \ldots = \sum_{n=0}^{\infty} ar^n$ *is finite and converges to* $\frac{a}{1-r}$.

Example 13: Geometric Series

Determine the sums of the series, if they converge.

a. $\sum_{n=0}^{\infty} \frac{3}{2^n}$

b. $2 - 1 + \frac{1}{2} - \frac{1}{4} + \frac{1}{8} - \frac{1}{16} + \ldots$

c. $\frac{1}{4} + \frac{1}{4^2} + \frac{1}{4^3} + \ldots$

Solutions: a. Here $a = 3$ and $r = \frac{1}{2}$. Thus $S = \frac{a}{1-r} = \frac{3}{1-0.5} = 6$.

b. Here $a = 2$ and $r = -\frac{1}{2}$. Thus $S = \frac{a}{1-r} = \frac{2}{1-\frac{-1}{2}} = \frac{2}{\frac{3}{2}} = \frac{4}{3}$.

c. The sequence does not start with $n = 0$ (using $a = 1$ and $r = \frac{1}{4}$). We use $a = \frac{1}{4}$ and $r = \frac{1}{4}$. Then $S = \frac{a}{1-r} = \frac{\frac{1}{4}}{1-\frac{1}{4}} = \frac{\frac{1}{4}}{\frac{3}{4}} = \frac{1}{3}$.

Alternatively, multiply by $\frac{1}{4}$ and get a second equation and subtract:

$$S = \frac{1}{4} + \frac{1}{4^2} + \frac{1}{4^3} + \ldots$$

$$\frac{1}{4} \cdot S = \frac{1}{4^2} + \frac{1}{4^3} + \frac{1}{4^4} + \ldots$$

$S - \frac{1}{4} \cdot S = \frac{1}{4}$, since the terms on the right collapse,

$\frac{3}{4} \cdot S = \frac{1}{4}$ and $S = \frac{1}{3}$.

Example 14: Merton College Puzzle

We now show Oresme's area solution to the Merton College puzzle (see Introduction). He used two decompositions of the area. (See Figure 10.1.2.) On the left, the rectangles have bases which are decreasing powers of 2 and heights which are the integers 1, 2, 3, ..., and the area is $1 \cdot \frac{1}{2} + 2 \cdot \frac{1}{4} + 3 \cdot \frac{1}{8} + \ldots$.

Recall, this is the original infinite series. On the other hand, *the same area* is decomposed into rectangles of height 1 and bases which are decreasing powers of 2, and there the area is the sum

$$1+\frac{1}{2}+\frac{1}{4}+\frac{1}{8}+\ldots=\frac{1}{1-\frac{1}{2}}=\frac{1}{\frac{1}{2}}=2.$$

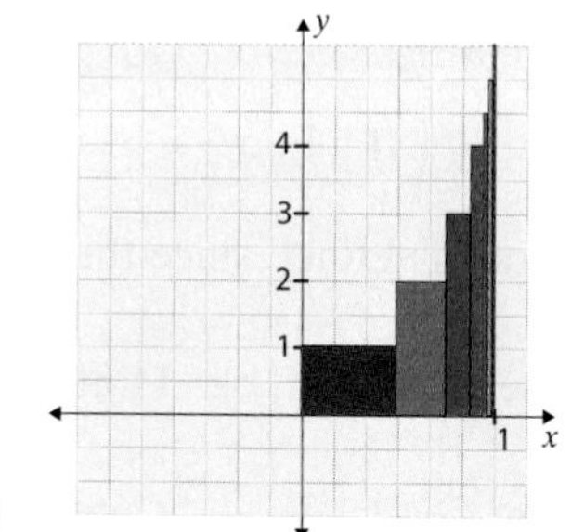

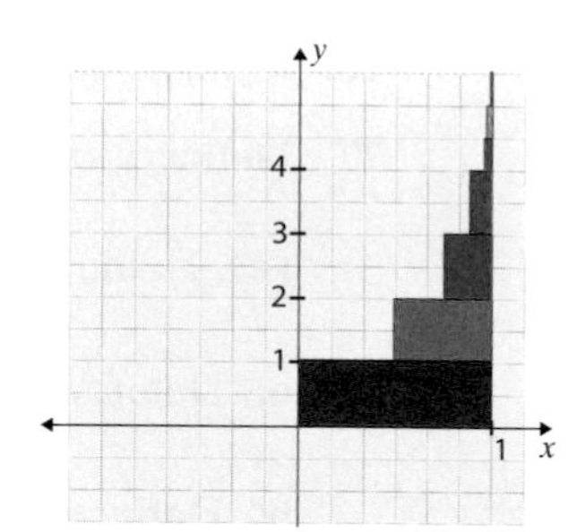

Figure 10.1.2

10.1 Exercises

In Exercises 1 – 6, write the first five terms of the sequence.

1. $a_n=\frac{(-1)^n}{n}$

2. $a_n=\frac{3}{e^n}$

3. $a_n=\sqrt{2n+1}$

4. $a_n=\frac{\ln n}{\ln(n+1)}$

5. $a_n=\frac{n^2-1}{n+1}$

6. $a_n=\frac{n^3+1}{n+1}$

In Exercises 7 – 12, write a general term for the sequence, if one exists, assuming a domain starting with n = 1.

7. $3, -5, 7, -9, \ldots$

8. $11, 14, 17, 20, \ldots$

9. $5, 10, 15, 20, 25, \ldots$

10. $3, -2, 1, 0, 1, -2, \ldots$

11. $2, 5, 10, 17, 26, 37, 50, \ldots$

12. $3, 8, 15, 24, 35, 48, \ldots$

In Exercises 13 – 18, compute the sums indicated.

13. $\sum_{n=1}^{5}(2n-1)^2$

14. $\sum_{n=1}^{5}n(-1)^n$

15. $\sum_{n=4}^{8}n-1$

16. $\sum_{i=1}^{4}3^i-2^i$

17. $\sum_{i=1}^{3}\frac{i^3-1}{i^2+i+1}$

18. $\sum_{i=1}^{5}\frac{i+1}{i^2+i}$

In Exercises 19 – 26, find the sum if the series converges.

19. $\sum_{j=0}^{\infty}\left(\frac{\sqrt{2}}{2}\right)^j$

20. $\sum_{n=0}^{\infty}\frac{2n}{(n+2)}$

21. $\sum_{n=0}^{\infty}(-1)^{n+1}\left(\frac{1}{2}\right)^n$

22. $\sum_{n=0}^{\infty}\left(\frac{e}{\pi}\right)^n$

23. $\sum_{j=0}^{\infty}\frac{2j+1}{10j+1}$

24. $1+e^{-1}+e^{-2}+e^{-3}+\ldots$

25. $4-0.4+0.04-0.004+\ldots$

26. $1-2+3-4+5-6+\ldots$

27. Define a function $f(x)=1+x+x^2+x^3+\ldots=\sum_{j=0}^{\infty}x^j$. For each x in the interval $(-1, 1)$, the function is defined since the series converges for each x.

a. Determine a series for $f'(x)$. Hint: Differentiate term-by-term.

b. Define $G(x)$ by $G(x)=\frac{1}{1-x}$. Determine $G'(x)$.

c. Determine $f'(a)$ and $G'(a)$ for a = 0.25, –0.25, –0.5, and 0.5. Discuss or explain your answers in a short essay or paragraph.

28. A tennis ball drops from a height of one foot above the ground. It falls and bounces back to $\frac{1}{4}$ of its previous position. It continues to fall and bounce back in the same proportion until it comes to rest. (Take downward as a negative direction.)

a. What was the net distance traveled by the ball?
b. What was the total (absolute value) of the distance traveled by the ball?

29. **a.** Write out the first 10 partial sums to evaluate the expression $\sum_{n=1}^{10}(2n-1)$.

b. Use the results of part (a) to determine a shortcut to evaluate $\sum_{n=1}^{100}(2n-1)$.

30. The XYZ Corporation, a small business, generates \$300,000 in operating cash during the year which it spends on equipment, supplies, and services. Assume half of this amount is then spent by the recipients of XYZ's business in the general economy on similar items with similar results. Assuming this pattern of economic growth continues, how much total revenue is spent in the economy from the original \$300,000?

31. Determine a rational number equal to the number (repeating decimal) given.

a. $\frac{4}{10}+\frac{3}{100}+\frac{4}{1000}+\frac{3}{10000}+\ldots=0.4343\ldots$

b. $\sum_{j=0}^{\infty}\frac{5}{10^{2j}}$

32. The marginal propensity to consume. In the summer of 2003, about 50 million tax payers received a \$1000 one time tax cut. Assume each person spent 90% and saved 10%. The proportion that each person will spend is called the marginal propensity to consume and the proportion saved is the marginal propensity to save. Assume the money spent is then spent by others in the same proportion, and so on. How much total spending is generated by the tax cut?

33. A patient takes a 10-milligram tablet of a blood pressure drug everyday, and 30% of each day's dose is still in the patient's system after 24 hours. After a few weeks, how much of the medicine is in the patient's body at any one time (to the nearest milligram)?

34. A patient in a VA Hospital takes a 5-milligram tablet of an experimental antidepressant drug daily. After 24 hours, a certain percentage is retained in the patient's system. It is determined that, after several weeks of trying the new medicine, the patient has 7.5 milligrams in his body. What percentage of a daily dose is retained for the next day (to the nearest percent)?

Hawkes Learning Systems: Essential Calculus

Infinite Sequences

10.2 Taylor Polynomials

Objectives

After completing this section, you will be able to:

1. *Linearly approximate a function at a point.*
2. *Calculate the Taylor polynomials of degree two and greater at a point.*

Linear Approximation

Accurate approximations have always been an important feature of mathematics. We know that -32 ft/sec^2 is only an approximation to the acceleration due to the Earth's gravity, that 3.14159 is an approximation to π, that 0.3333 is an approximation to $\frac{1}{3}$, and so on. Technically speaking,

$$\frac{1}{3}, \sqrt{2}, \pi, \text{ and } e \text{ are exact and}$$

$$0.333, 1.41, 3.14, \text{ and } 2.72 \text{ are approximations of these.}$$

Numerical approximations are generally simple to express and make computations easier. Except for certain exact integer calculations, the numbers calculated by calculators and computer software are always approximations. Since a calculator or computer program may generate excellent numerical approximations, the purpose of this section is to indicate how many of these calculations are actually made and to provide background for understanding and applying one of the most striking discoveries related to calculus – that many of the familiar numbers and functions have polynomial representations with infinitely many terms.

In calculus we develop techniques not only to approximate numbers but also to approximate functions. The main approximation developed is one already presented, namely the linear approximation to a given function $f(x)$ at a point $(a, f(a))$

Linear Approximation

The linear (tangent line) approximation to $f(x)$ at the point $(a, f(a))$ is given by

$$f(x) \approx P_1(x) = f(a) + f'(a) \cdot (x - a)$$

This formula, of course, comes from the slope-intercept form of a line, $y-y_1=m(x-x_1)$ where $x_1=a$, $y_1=f(a)$, $m=f'(a)$ and $y=P_1(x)$.

Example 1: Linear Approximation

Find the linear approximation $P_1(x)$ to $y=\sqrt{x}$ at the point (1, 1).

Solution: Since $f'(x)=\dfrac{1}{2\sqrt{x}}$, $f'(1)=0.5$. Thus

$$P_1(x)=f(1)+f'(1)(x-1)=1+0.5(x-1)=1-0.5+0.5x=0.5x+0.5.$$

It is important to point out that $P_1(x)$ is the only linear approximation to y which passes through (1, 1) and has the same slope at that point. In fact, a **quadratic approximation** to y can always be found which goes through a given point, has the same slope at that point, and also *has the same concavity at that point.* Let $f(x)$ be any function whose first and second derivatives exist at the point $(a, f(a))$. Let us denote this quadratic approximation by

$$P_2(x)=c_0+c_1x+c_2x^2$$

where c_0, c_1, and c_2 are constants to be determined. They will depend on the nature of the given function f and on the value $x = a$.
We want to insure

$$P_2(a)=f(a)$$

$$P_2'(a)=f'(a) \quad \text{and}$$

$$P_2''(x)=f''(a)$$

Let us take a = 0 in order to simplify and standardize the necessary calculations. Since $P_2(x)=c_0+c_1x+c_2x^2$, we have

$$P_2(0)=f(0)=c_0, \text{ and thus } c_0=f(0).$$

Since $P_2'(x)=c_1+2c_2x$, we have

$$P_2'(0)=c_1 \text{ and thus } c_1=f'(0).$$

Since $P_2''(x)=2c_2$, we have

$$P_2''(0)=2c_2 \text{ and thus } c_2=\frac{P_2''(0)}{2}=\frac{f''(0)}{2}.$$

This gives

$$P_2(x)=f(0)+f'(0)x+\frac{f''(0)}{2}\cdot x^2.$$

The Taylor Polynomial of Degree 2 Approximating $f(x)$ near $x = 0$

$$f(x) \approx P_2(x) = f(0) + f'(0)x + \frac{f''(0)}{2} \cdot x^2$$

Example 2: Taylor Polynomial of Degree 2

Determine the degree 2 Taylor polynomial approximating $y = e^x$ near $x = 0$.

Solution: $f(0) = f'(0) = f''(0) = e^0 = 1$. Thus $c_0 = 1, c_1 = 1$, and $c_2 = 1$.

So $e^x = 1 + x + \frac{1}{2}x^2$ near $x = 0$.

Example 3: Taylor Polynomial of Degree 2

Determine the degree 2 Taylor polynomial approximating $y = \sqrt{x}$ at the point $(0, 0)$.

Solution: The derivatives do not exist at $x = 0$, and there is no quadratic Taylor polynomial approximation at $x = 0$.

Higher Degree Polynomials

An extension to degree three of the calculation just above gives the same general formulas for c_0, c_1, and c_2. The new condition is that $f'''(0) = P_3'''(0) = 6c_3 = 3!c_3$.

Thus $c_3 = \frac{f'''(0)}{3!}$.

In general, for any degree k polynomial approximation,

$$c_k = \frac{f^{(k)}(0)}{k!}$$

where $f^{(k)}$ means the k-th derivative of f and $k!$ denotes the product $k(k-1)(k-2)...(3)(2)(1)$.

The Taylor Polynomial of Degree k Approximating $f(x)$ near $x = 0$

$$f(x) \approx f(0) + f'(0)x + \frac{1}{2!}f''(0)x^2 + \frac{1}{3!}f'''(0)x^3 + ... + \frac{1}{k!}f^{(k)}(x)x^k.$$

We call this a Taylor approximation centered at $x = 0$.

Example 4: Taylor Polynomial of Degree 7

Construct the Taylor polynomial of degree 7 for $y = \sin x$.

Solution:

$$f(x) = \sin x \qquad f(0) = 0$$
$$f'(x) = \cos x \qquad f'(0) = 1$$
$$f''(x) = -\sin x \qquad f''(0) = 0$$
$$f'''(x) = -\cos x \qquad f'''(0) = -1$$
$$f^{(4)}(x) = \sin x \qquad f^{(4)}(0) = 0$$
$$f^{(5)}(x) = \cos x \qquad f^{(5)}(0) = 1$$
$$f^{(6)}(x) = -\sin x \qquad f^{(6)}(0) = 0$$
$$f^{(7)}(x) = -\cos x \qquad f^{(7)}(0) = -1$$

Substituting these values into the Taylor formula we get

$$\sin x \approx P_7(x) = x - \frac{1}{3!}x^3 + \frac{1}{5!}x^5 - \frac{1}{7!}x^7$$

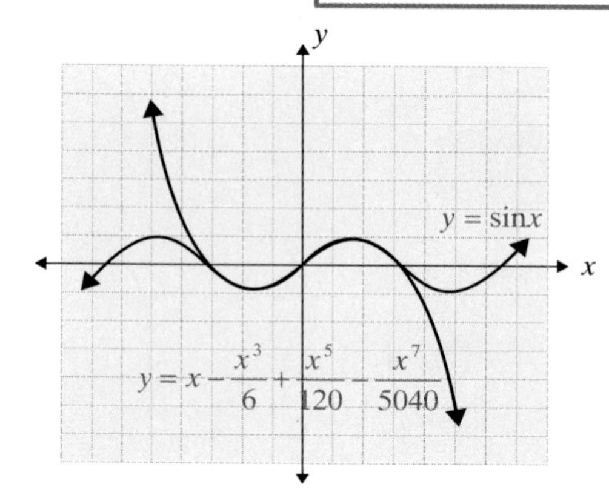

Figure 10.2.1

See Figure 10.2.1 for a comparison of $y = \sin x$ and $y = P_7(x)$. A numerical comparison can be made using any hand-held calculator. For example:

$$\sin(0.1) = 0.0998334166 \qquad P_7(0.1) = 0.09983342$$
$$\sin(0.5) = 0.4794255386 \qquad P_7(0.5) = 0.47942553$$
$$\sin\left(\frac{\pi}{3}\right) = 0.8660254038 \qquad P_7\left(\frac{\pi}{3}\right) = 0.86602127$$
$$= \frac{1}{2}\sqrt{3}$$

This is remarkable accuracy for so few terms. Observe, $P_7(x)$ diverges from $\sin x$ as x moves away from $x = 0$.

Example 5: Taylor Polynomial of Degree 20

Construct the Taylor polynomial for degree 20 for $y = e^x$ at $x = 0$.

Solution: All the derivatives are equal to 1 at $x = 0$. Thus

$$e^x \approx P_{20}(x) = 1 + x + \frac{1}{2!}x^2 + \frac{1}{3!}x^3 + \ldots + \frac{1}{20!}x^{20}.$$

Example 6: Taylor Polynomial of Degree k

Construct the Taylor polynomial of degree k for $y = e^{3x}$ near $x = 0$.

Solution: $f'(x) = 3e^{3x}$ and in general, $f^{(k)}(x) = 3^k e^{kx}$.

But $f^{(k)}(0) = 3^k e^{k\cdot 0} = 3^k \cdot 1 = 3^k$.

Thus the general polynomial is

$$e^{3x} \approx 1 + 3x + \frac{3^2x^2}{2!} + \frac{3^3x^3}{3!} + \frac{3^4x^4}{4!} + \ldots + \frac{3^kx^k}{k!}.$$

It may happen that we require a polynomial approximation but $x = 0$ is not a good choice either because we are interested in x values far removed from $x = 0$ or because the derivatives do not exist at $x = 0$. For example, if $y = \ln x$ or $y = \sqrt{x}$ then the derivatives are not defined at $x = 0$. We need a more general expansion.

The General Taylor Polynomial of Degree k centered at $x = a$

$$f(x) \approx P_k(x) = f(a) + f'(a)(x-a) + \frac{f''(a)(x-a)^2}{2!} + \frac{f'''(a)(x-a)^3}{3!} + \ldots + \frac{f^{(k)}(a)(x-a)^k}{k!}$$

Example 7: Taylor Polynomial of the Natural Log Function

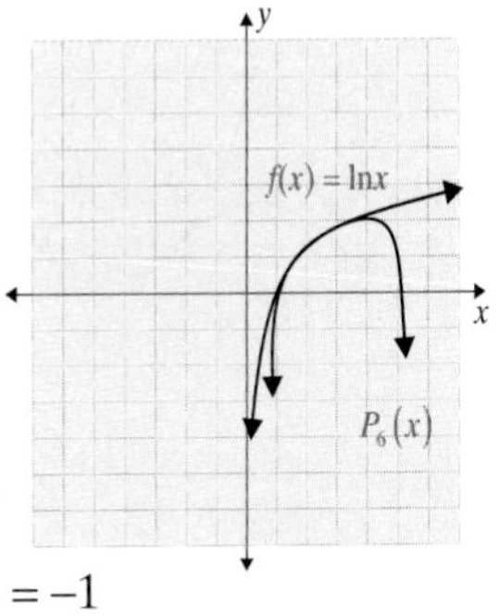

Construct the Taylor polynomial for $y = \ln x$ centered at $x = 1$.

Solution:

$f(x) = \ln x$ $\qquad$ $\ln(1) = 0$

$f'(x) = x^{-1}$ $\qquad$ $f'(1) = 1^{-1} = 1$

$f''(x) = -x^{-2}$ $\qquad$ $f''(1) = -(1)^{-2} = -1$

$f^{(3)}(x) = -(-2)x^{-3}$ $\qquad$ $f^{(3)}(1) = 2$

$f^{(k)}(x) = -(-2)(-3)\ldots(-k+1)x^k$ and $f^{(k)}(1) = (k-1)!(-1)^{k+1}$

The signs alternate and the derivative is negative when k is even.

Figure 10.2.2

The general polynomial is

$$\ln x \approx 0 + (x-1) - \frac{(x-1)^2}{2!} + \frac{2(x-1)^3}{3!} - \frac{3!(x-1)^4}{4!} + \ldots + \frac{(-1)^{k+1}(k-1)!(x-1)^k}{k!}$$

$$= (x-1) - \frac{(x-1)^2}{2} + \frac{(x-1)^3}{3} - \frac{(x-1)^4}{4} + \ldots + \frac{(-1)^{k+1}}{k}(x-1)^k$$

A comparison of ln x and $P_6(x)$ appears in Figure 10.2.2. Again, the general rule about accuracy is that near $x = a$, the approximation is dramatic for even small degree polynomials. Technical estimations in numerical analysis of accuracy consider how many terms are needed to estimate $f(x)$ for a Taylor polynomial centered at $x = a$. These considerations are beyond the scope of this book.

Example 8: Rational Approximation

Obtain a rational number approximation to $\sqrt[3]{8.5}$ using a Taylor polynomial of degree 3.

Solution: We use $f(x)=\sqrt[3]{x}$ centered at $a = 8$. Note $f(8) = 2$ is easy to obtain without a calculator. We use the derivatives

$$f'(x)=\frac{1}{3}x^{-\frac{2}{3}}=\frac{1}{3\cdot\sqrt[3]{x^2}}=\frac{1}{3\left(\sqrt[3]{x}\right)^2} \quad \text{and} \quad f'(8)=\frac{1}{3\left(\sqrt[3]{8}\right)^2}=\frac{1}{3(2)^2}=\frac{1}{12}$$

$$f''(x)=\frac{1}{3}\cdot\frac{-2}{3}\cdot x^{-\frac{5}{3}}=\frac{-2}{9}\cdot\frac{1}{\left(\sqrt[3]{x}\right)^5} \quad \text{and} \quad f''(8)=\frac{-2}{9}\cdot\frac{1}{2^5}=\frac{-1}{9\cdot 2^4}=\frac{-1}{144}$$

$$f'''(x)=\frac{-2}{9}\cdot\frac{-5}{3}\cdot x^{-\frac{8}{3}}=\frac{10}{27}\cdot\frac{1}{\left(\sqrt[3]{x}\right)^8} \quad \text{and} \quad f'''(8)=\frac{10}{27}\cdot\frac{1}{2^8}=\frac{5}{27\cdot 2^7}=\frac{5}{3456}.$$

Thus

$$\sqrt[3]{8.5}\approx P_3(8.5)=f(8)+f'(8)(8.5-8)+\frac{f''(8)(8.5-8)^2}{2!}+\frac{f'''(8)(8.5-8)^3}{3!}$$

$$=2+\frac{1}{12}\cdot\left(\frac{1}{2}\right)+\frac{\frac{-1}{144}\cdot\frac{1}{4}}{2}+\frac{\frac{5}{3456}\cdot\frac{1}{8}}{6}=2+\frac{1}{24}-\frac{1}{1152}+\frac{5}{165888}$$

$$=\frac{331776}{165888}+\frac{6912}{165888}-\frac{144}{165888}+\frac{5}{165888}=\frac{338549}{165888}$$

This fraction in decimal form is 2.040828752 which compares exactly for six digits to a calculator value of 2.040827551 for $\sqrt[3]{8.5}$.

10.2 Exercises

For Exercises 1 – 8, find the Taylor polynomial of degree n approximating the given function near x = 0. Using a graphing utility, sketch the given function and the Taylor approximation on the same coordinate system.

1. $y = \cos x, n = 6$

2. $y = \dfrac{1}{\sqrt{1+x}}, n = 3$

3. $y = \sqrt{x+4}, n = 3$

4. $y = \sqrt[3]{x+2}, n = 2$

5. $\ln(x+1), n = 3$

6. $y = x\ln(x+1), n = 3$

7. $y = e^{x+1}, n = 3$

8. $y = xe^{x}, n = 3$

For Exercises 9 – 16, find the Taylor polynomial of degree n near x = a for the given n and a.

9. $y = e^{x}, a = 1, n = 4$

10. $y = \sin x,\ a = \pi,\ n = 5$

11. $y = \cos x, a = \pi, n = 6$

12. $y = e^{3x}, a = 1, n = 4$

13. $y = \sqrt{x}, a = 1, n = 4$

14. $y = x^{\frac{1}{3}}, a = 1, n = 4$

15. $y = \dfrac{1}{1-x}, a = 0.5, n = 4$

16. $y = \dfrac{x}{1+x}, a = 0, n = 5$

17. Use the Taylor polynomial of degree 5 for $y = \sin x$, centered at $x = 0$, to estimate the area under the graph of $y = \sin x$ from $x = 0$ to $x = 3.14159$.

18. Use the Taylor polynomial of degree 6 for $y = \cos x$, centered at $x = 0$, to estimate the area from $x = 0$ to $x = 1.5708$.

19. Use the Taylor polynomial of degree 3 for $f(x) = \sqrt{x+1}$, centered at $x = 0$, to estimate a value for $\sqrt{2}$. Compare this to the value given by your calculator.

20. Construct the Taylor polynomial of degree 3 centered at $x = 2$, for the function $f(x) = \sqrt{x+1}$. Use this polynomial to estimate a value for $\sqrt{2}$. Compare this to the value given by your calculator.

21. Use the Taylor polynomial of degree 3 centered at $x = 0$, for $f(x) = \sqrt[4]{x+1}$ to estimate the value of $\sqrt[4]{0.5}$ (that is, calculate $P_3(-0.5)$). Compare this with the value obtained from your calculator.

22. Use a Taylor polynomial of degree 3 centered at $x = 1$ for $f(x) = \sqrt[4]{x}$ to estimate the value of $\sqrt[4]{0.5}$. Compare your value with the answer to Exercise 21.

23. Use a quadratic Taylor polynomial for an appropriate function $f(x)$ and an appropriate value of $x = a$ in order to estimate $\sqrt[3]{27.5}$.

24. Choose an appropriate function $y = f(x)$, a suitable value of $x = a$, and a corresponding Taylor polynomial of degree 4 in order to estimate $\sqrt{37}$.

25. Look up Brook Taylor in a history of mathematics book and write a one page report on his major accomplishments.

26. Determine the Taylor polynomial of degree 4 centered at $x = 0$ for the function $f(x) = 3 - 2x + 4x^2 - 2x^3$. Use your solution as a basis for a conjecture about the Taylor polynomial of a given polynomial function.

27. Differentiate the Taylor polynomial of degree 7 for $y = \sin t$ in order to get a polynomial approximation for $y = \cos t$. Compare your result with your answer to Exercise 1. Make a conjecture about derivatives of Taylor polynomial approximations to given functions.

28. **a.** Multiply the Taylor polynomial of degree 3 for $y = \sin x$ (centered at $x = 0$) by itself.

b. Is the result equal to the Taylor polynomial of degree 6 for $y = \sin^2 x$?

29. Use the identity $\sin^2 t = 1 - \cos^2 t$ and the answer to Exercise 28 **b.** in order to get a degree 6 polynomial representation for $y = \cos^2 t$.

Hawkes Learning Systems: Essential Calculus

Taylor Polynomials

10.3 Taylor Series, Infinite Expressions, and Their Applications

Objectives

After completing this section, you will be able to:

1. *Calculate infinite series based on Taylor polynomials and binomial expansions.*
2. *Utilize Euler's formula to solve equations in the complex plane.*
3. *Use a power series to find the solution for a differential equation.*

Taylor polynomials can be developed for any value of n. Could it be meaningful to write a Taylor polynomial with infinitely many terms? We have seen in Section 10.1 that expressions like

$$1+\frac{1}{2}+\frac{1}{2^2}+\frac{1}{2^3}+\ldots+\frac{1}{2^n}+\ldots$$

can be evaluated. If we write

$$f(x)=1+x+x^2+x^3+\ldots+x^n+\ldots$$

as a formula for a function f, it is clearly meaningful if the expression converges for a particular value of x. The infinite series above is just $f\left(\frac{1}{2}\right)$. Among the profound mathematical questions one might ask are these: Can infinite polynomial expressions (or other infinite expressions) properly represent the familiar functions of calculus? Can we use them to calculate slopes, convexity or areas as if they were finite polynomials?

Formal proofs and their complications are beyond the scope of this book. The theoretical results we need in brief are provided in this section of the text:

The Taylor Series Centered at $x = a$

If a function $f(x)$ is defined at $x = a$ and if all its derivatives exist at $x = a$, then the Taylor series

$$T(x)=f(a)+f'(a)(x-a)+\frac{f''(a)(x-a)^2}{2!}+\frac{f'''(a)(x-a)^3}{3!}+\ldots+\frac{f^{(k)}(a)(x-a)^k}{k!}+\ldots$$

converges for all x-values in some interval $(a - R, a + R)$ centered at $x = a$. Further, for all b in this interval, $T(b) = f(b)$. Thus, provided $x = b$ is inside this interval, any number $f(b)$ can be calculated with this series by using enough terms to guarantee the accuracy desired. The Taylor series may be differentiated or integrated term-by-term to obtain a derivative or antiderivative of $f(x)$. The number R is called the **radius of convergence.**

The determination of R for any particular function is not discussed. We write $R = \infty$ to mean that the Taylor series converges for all x.

The Binomial Series

The numbers in Pascal's triangle (See Figure 10.3.1.) are called **binomial coefficients.** The numbers occur as coefficients in the polynomial $(a + b)^n$. For example, with $n = 3$,

$$(a+b)^3 = 1a^3 + 3a^2b + 3ab^2 + 1b^3$$

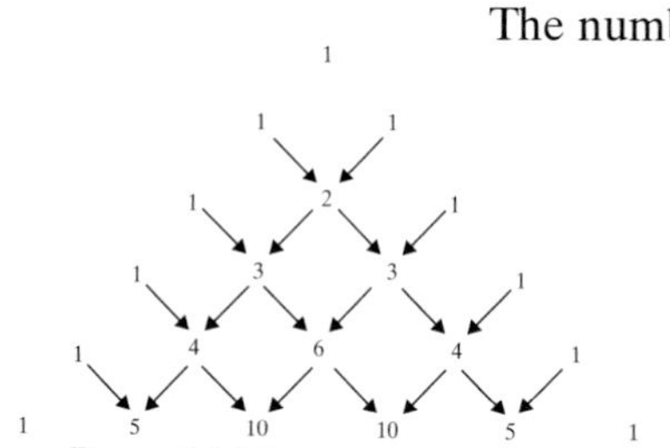

1
1 1
1 2 1
1 3 3 1
1 4 6 4 1
1 5 10 10 5 1

Figure 10.3.1

and the 3rd row in the triangle has the sequence 1-3-3-1. The triangle provides a convenient memory device for quickly generating the coefficients. In Pascal's triangle each number after the first row is the sum of the two numbers just above. The $(r + 1)^{\text{th}}$ term in row n is designated ${}_nC_r$ or $C(n, r)$. To get ${}_5C_3$ on a TI-84 Plus, type 5, then, MATH , and arrow right to PRB, select nCr, and ENTER , 3, ENTER . The calculator returns 10, which is the fourth number in row 5. That is ${}_5C_3 = 10$. There is also a formula

$${}_nC_r = \frac{n!}{(n-r)!r!} = \frac{n(n-1)(n-2)...(n-r+1)}{1\cdot 2\cdot 3\cdot ...\cdot r}.$$

For the purposes of this and other formulas, we define $0! = 1$. Thus, ${}_nC_0 = 1$ (and ${}_nC_0$ is the first element in row n). These numbers in the triangle occur in probability theory, statistics, and mathematical combinatorics, where they have many applications. For example ${}_nC_r$ also gives the number of different r-element size subsets which may be chosen from an n-element size set. For this reason ${}_nC_r$ is read as "n-choose-r".

Example 1: Binomial Series

How many 2-element subsets are there in a set of six elements?

Solution: ${}_6C_2 = \dfrac{6(6-1)}{1\cdot 2} = 15.$

By a mixture of examples, detailed calculation, and inference, Newton discovered as a teenager the extension of Pascal's triangle to the case where the exponent is not an integer.

The Binomial Series

$$(1+x)^p = 1 + px + \frac{p(p-1)}{2!}\cdot x^2 + \frac{p(p-1)(p-2)}{3!}\cdot x^3 + ...$$

The series converges provided $-1 < x < 1$ (the radius of convergence $R = 1$).

Example 2: Binomial Series

Use the binomial series when $p = \frac{1}{2}$ and get a series for $f(x) = \sqrt{1+x}$.

Solution: $$(1+x)^{\frac{1}{2}} = 1 + \frac{1}{2}x + \frac{\left(\frac{1}{2}\right)\left(\frac{1}{2}-1\right)}{2!}x^2 + \frac{\left(\frac{1}{2}\right)\left(\frac{1}{2}-1\right)\left(\frac{1}{2}-2\right)}{3!}x^3 + \ldots$$
$$= 1 + \frac{1}{2}x - \frac{1\cdot 3}{2!2^2}x^2 + \frac{1\cdot 3\cdot 5}{3!2^3}x^3 - \ldots$$

Example 3: Taylor Series

Find a Taylor series for $y = \frac{1}{1+x}$ centered at $a = 0$.

Solution: Use the binomial series with $p = -1$ since $\frac{1}{1+x} = (1+x)^{-1}$.

$$\frac{1}{1+x} = (1+x)^{-1} = 1 + (-1)x + \frac{(-1)(-2)x^2}{2!} + \frac{(-1)(-2)(-3)x^3}{3!} + \ldots$$
$$= 1 - x + x^2 - x^3 + \ldots$$
$$= \sum_{k=0}^{\infty} (-1)^k x^k$$

The solution in this example is recognizable as the geometric series for $r = -x$ (See Section 10.1). That is,

$$\frac{1}{1-r} = 1 + r + r^2 + r^3 + \ldots = 1 + (-x) + (-x)^2 + (-x)^3 + \ldots = 1 - x + x^2 - x^3 + \ldots$$

Example 4: Taylor Series

Obtain a Taylor series for $y = \frac{1}{1-x^2}$.

Solution: Use the geometric series with $r = x^2$.

$$\frac{1}{1-r} = 1 + r + r^2 + r^3 + \ldots = 1 + x^2 + (x^2)^2 + (x^2)^3 + \ldots$$
$$= 1 + x^2 + x^4 + x^6 + \ldots = \sum_{k=0}^{\infty} x^{2k}$$

Or use the binomial series with $p = -1$ and replacing x by $-x^2$:

$$\frac{1}{1-x^2} = \left(1+\left(-x^2\right)\right)^{-1}$$

$$= 1+(-1)\left(-x^2\right)+\frac{(-1)(-2)\left(-x^2\right)^2}{2!}+\frac{(-1)(-2)(-3)\left(-x^2\right)^3}{3!}+\ldots = \sum_{k=0}^{\infty} x^{2k}$$

Example 5: Taylor Series

Find a Taylor series for $y = e^{kx}$ where k is some constant.

Solution: $f'(x) = ke^{kx}$ and in general $f^{(n)}(x) = k^n e^{kx}$. Thus $f(0) = 1$, $f'(0) = k$, and in general $f^{(n)}(0) = k^n$. Now, substituting into the Taylor formula,

$$e^{kx} = 1+kx+\frac{k^2x^2}{2!}+\frac{k^3x^3}{3!}+\ldots = \sum_{n=0}^{\infty}\frac{(kx)^n}{n!}.$$

Example 6: Taylor Series

Find a function whose Taylor series coefficients are specified by the sequence given:

a. $(0, 1, 1, 1, 1, \ldots)$

b. $\left(1, 0, \frac{1}{2!}, \frac{1}{3!}, \frac{1}{4!}, \ldots\right)$

c. $(1, 2, 1, 0, 0, 0, \ldots)$

d. $(1, 0, 1, 0, 1, 0, \ldots)$

Solutions: a. The function is $f(x) = x+x^2+x^3+\ldots$. By adding and subtracting 1, this can be written as

$$f(x) = x+x^2+x^3+\ldots = -1+1+x+x^2+x^3+\ldots$$

$$= -1+\frac{1}{1-x} = \frac{-1+x+1}{1-x} = \frac{x}{1-x}$$

Another solution is to factor x and substitute:

$$f(x) = x+x^2+x^3+\ldots = x\left(1+x+x^2+\ldots\right) = x\cdot\frac{1}{1-x} = \frac{x}{1-x}.$$

b. The coefficients are almost the ones for e^x, except for the coefficient of x. This suggests the solution: $f(x) = e^x - x$.

c. The function is $f(x) = 1 + 2x + 1x^2 + 0x^3 + \ldots = (1 + x)^2$.

d. The function is $1+x^2+x^4+x^6+\ldots = \sum_{k=0}^{\infty} x^{2k}$, which is the geometric series with $r = x^2$. The solution is $y = \frac{1}{1-x^2}$.

Euler's Formula From Trigonometry

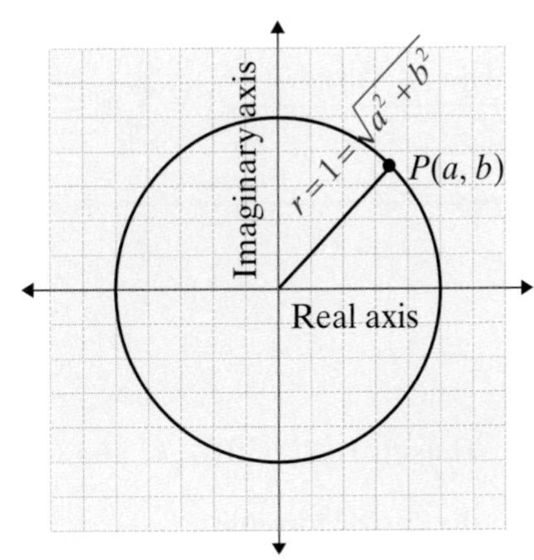

Figure 10.3.2

Figure 10.3.2 shows the unit circle in the complex plane. In the complex plane, every point z may be written as $z = a + bi$ for some real numbers a and b, and $i = \sqrt{-1}$. The algebra properties of i include

$$i^2 = -1,\ i^3 = i^2(i) = -i,\ i^4 = 1,\ \frac{1}{i} = i^3$$

and $i^{(4n+1)} = i,\ i^{(4n+2)} = i^2 = -1,\ i^{(4n+3)} = i^3,\ i^{4n} = 1$ for any integer n.

The length of a complex number is the distance from $a + bi$ to the origin, and this distance is denoted $|a+bi|$. Using the Pythagorean Theorem, we have

$$|a+bi| = \sqrt{a^2+b^2}\,.$$

Every point on the unit circle (radius 1) satisfies

$$|a+bi| = \sqrt{a^2+b^2} = 1.$$

If t denotes the arc length along the unit circle, measured counterclockwise from the point $1 + 0i$ (on the real axis), then a point on the unit circle can also be denoted $\cos t + i \sin t$. (See Figure 10.3.3.) Thus $z = a+bi = \cos t + i \sin t$. The expression $a + bi$ is called the rectangular form of z, and the expression $\cos t + i \sin t$ is called the polar form of z.

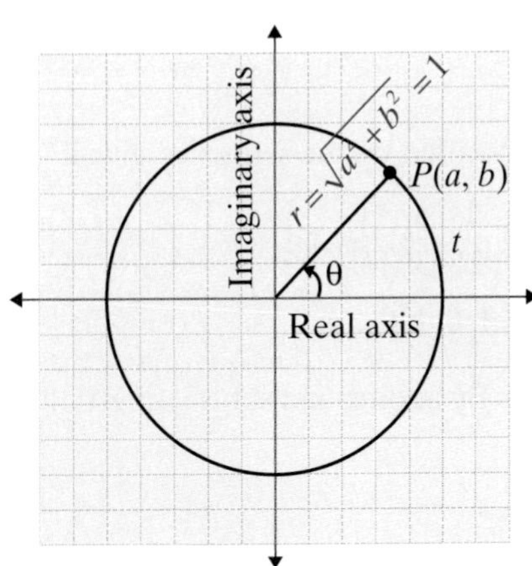

Figure 10.3.3

We now follow in the footsteps of Leonhard Euler (1707 - 1783) and use the Taylor series of the sine and cosine functions as well as the algebraic properties of i to get

$$z = \cos t + i \sin t = \left[1 - \frac{t^2}{2!} + \frac{t^4}{4!} - \frac{t^6}{6!} + \ldots\right] + i\left[t - \frac{t^3}{3!} + \frac{t^5}{5!} - \frac{t^7}{7!} + \ldots\right]$$

$$= \left[1 + \frac{i^2t^2}{2!} + \frac{i^4t^4}{4!} + \frac{i^6t^6}{6!} + \ldots\right] + \left[it + \frac{i^3t^3}{3!} + \frac{i^5t^5}{5!} + \frac{i^7t^7}{7!} + \ldots\right]$$

$$= 1 + it + \frac{i^2t^2}{2!} + \frac{i^3t^3}{3!} + \frac{i^4t^4}{4!} + \frac{i^5t^5}{5!} + \ldots = e^{it}$$

Euler's Formula

Thus we have ***Euler's Formula,***

$$e^{it} = \cos(t) + i\sin(t)$$

This formula and Euler's other work created a new subject called complex analysis, that is, calculus with complex variables. The development or further discussion of this subject is beyond the scope of this course. However, we will develop a famous identity by applying Euler's Formula. First, the point -1 is in the complex plane. Its rectangular form is $-1 + 0i$.

Since the distance along the unit circle from +1 to −1 is $t = \pi$ (half of the circumference of 2π), the polar form of −1 is $\cos \pi + i \sin \pi$. Thus

$$-1 = \cos \pi + i \sin \pi$$

and of course, $\cos \pi = -1$ and $i \sin \pi = i(0) = 0$. Now, using Euler's Formula, $-1 = e^{\pi i}$. Putting all these terms on the same side we get

$$1 + e^{\pi\sqrt{-1}} = 0$$

This amazing formula puts the five most important numbers in mathematics together in one expression. The formula $(\cos\theta + i\sin\theta)^2 = \cos(2\theta) + i\sin(2\theta)$ was well known in mathematical circles in the 1600's. With induction, the exponent can be replaced by any integer. Using Euler's Formula, this can be proved as the next example shows.

Example 7: Application of Euler's Formula

Prove the identity: for any real number t, $(\cos\theta + i\sin\theta)^t = \cos(t\theta) + i\sin(t\theta)$.

Solution: Using Euler's Formula twice, we have

$$(\cos\theta + i\sin\theta)^t = \left(e^{i\theta}\right)^t = e^{it\theta} = \cos(t\theta) + i\sin(t\theta).$$

Example 8: Solving an Equation

Solve the equation: $x^5 = 1$.

Solution: Since $|x| = 1$, we have $x = \cos\theta + i\sin\theta$ and any x satisfying the equation lies on the unit circle in the complex plane. Then

$$x^5 = (\cos\theta + i\sin\theta)^5 = \cos 5\theta + i\sin 5\theta = 1.$$

From this we see $5\theta = 2k\pi$ for some integer k. So $\theta = \frac{2k\pi}{5}$. We take $k = 0, 1, 2, 3, 4$ (since there are 5 solutions). We get

$$x = 1 = \cos 0 + i\sin 0 \qquad k = 0$$

$$x = \cos\left(\frac{2\pi}{5}\right) + i\sin\left(\frac{2\pi}{5}\right) \qquad k = 1$$

$$x = \cos\left(\frac{4\pi}{5}\right) + i\sin\left(\frac{4\pi}{5}\right) \qquad k = 2$$

$$x = \cos\left(\frac{6\pi}{5}\right) + i\sin\left(\frac{6\pi}{5}\right) \qquad k = 3$$

$$x = \cos\left(\frac{8\pi}{5}\right) + i\sin\left(\frac{8\pi}{5}\right) \qquad k = 4$$

Power Series Solutions to Differential Equations

A general expression like

$$A_0 + A_1x + A_2x^2 + A_3x^3 + \ldots = \sum_{n=0}^{\infty} A_n x^n$$

is called a **power series**. An approximate solution to a differential equation can often be found by assuming $y = \sum_{n=0}^{\infty} A_n x^n$ is a solution and using the methods of this subsection to determine as many coefficients as necessary for the accuracy needed.

In general, when a condition on the first derivative is given, one needs one data point (or "boundary value") to get a full solution. When a condition on the second derivative is given, one needs two data points given. These problems are also called boundary value problems.

Example 9: Power Series of Degree 5

Find a polynomial of degree at least five which represents the function $y = f(x)$ which satisfies the conditions given: $f'(x) = 2y$ and $f(0) = 1$.

Solution: Let $y = f(x) = A_0 + A_1x + A_2x^2 + A_3x^3 + \ldots = \sum_{n=0}^{\infty} A_n x^n$. Since $A_0 = f(0) = 1$, we have

$$f(x) = 1 + A_1x + A_2x^2 + A_3x^3 + A_4x^4 + A_5x^5 + \ldots, \text{ and}$$

$$f'(x) = A_1 + 2A_2x + 3A_3x^2 + 4A_4x^3 + 5A_5x^4 + \ldots$$

Also, using the given condition that $y' = 2y$,

$$f'(x) = 2y = 2 + 2A_1x + 2A_2x^2 + 2A_3x^3 + 2A_4x^4 + 2A_5x^5 + \ldots$$

We equate the two expressions for $f'(x)$, stopping with the degree 4 coefficient, and get

$$2 + 2A_1x + 2A_2x^2 + 2A_3x^3 + 2A_4x^4 = A_1 + 2A_2x + 3A_3x^2 + 4A_4x^3 + 5A_5x^4$$

$$(2 - A_1) + (2A_1 - 2A_2)x + (2A_2 - 3A_3)x^2 + (2A_3 - 4A_4)x^3 + (2A_4 - 5A_5)x^4 = 0$$

This equation is valid for all x in some interval about $x = 0$. This means that the coefficient of each power of x is zero. Thus

$2 - A_1 = 0$ which gives $A_1 = 2$

$2A_1 - 2A_2 = 0$ which gives $A_2 = 2$

$2A_2 - 3A_3 = 0$ which gives $A_3 = \frac{2A_2}{3} = \frac{4}{3}$

$2A_3 - 4A_4 = 0$ which gives $A_4 = \frac{2A_3}{4} = \frac{2}{3}$

$2A_4 - 5A_5 = 0$ which gives $A_5 = \frac{2A_4}{5} = \frac{4}{15}$

Therefore, $f(x) \approx P_5(x) = 1 + 2x + 2x^2 + \frac{4}{3}x^3 + \frac{2}{3}x^4 + \frac{4}{15}x^5.$

We may check that $f(0) = 1$ and

$$P_5'(x) = 2 + 4x + 4x^2 + \frac{8}{3}x^3 + \frac{4}{3}x^4 = 2\left(1 + 2x + 2x^2 + \frac{4}{3}x^3 + \frac{2}{3}x^4\right) = 2 \cdot P_5(x)$$

The conditions in the problem lead to the so-called *recursion* equation $2A_k - (k+1)A_{k+1} = 0,$ for all $k \geq 0.$

Since $A_0 = 1$ is given, all the coefficients can be found using the recursion equation.

Example 10: Power Series for a Polynomial

Determine a polynomial for a function which satisfies $\frac{dy}{dx} = 2xy$ subject to the condition that $f(0) = 1$.

Solution: Let $y = f(x) = A_0 + A_1x + A_2x^2 + A_3x^3 + \ldots = \sum_{n=0}^{\infty} A_n x^n$ and

$$\frac{dy}{dx} = A_1 + 2A_2x + 3A_3x^2 + 4A_4x^3 + 5A_5x^4 + \ldots$$

Since $\frac{dy}{dx} = 2xy,$ we get

$$2xy = 2A_0x + 2A_1x^2 + 2A_2x^3 + 2A_3x^4 + 2A_4x^5 + \ldots$$
$$= A_1 + 2A_2x + 3A_3x^2 + 4A_4x^3 + \ldots$$

Combining terms gives

$$A_1 + (2A_0 - 2A_2)x + (2A_1 - 3A_3)x^2 + (2A_2 - 4A_4)x^3 + (2A_3 - 5A_5)x^4 + \ldots = 0$$

This equation tells us that $A_1 = 0$ (substituting $x = 0$) and it gives the recursion formula

$$2A_k - (k+2)A_{k+2} = 0$$

or, more conveniently,

$$A_k = \frac{2A_{k-2}}{k} \text{ for } k \geq 2.$$

Now, using the recursion equation and the facts that $f(0) = 1 = A_0$, and $A_1 = 0$, we get

continued on next page ...

$A_1 = A_3 = A_5 = \ldots = A_{2k-1} = \ldots = 0$, and

$$A_2 = \frac{2A_0}{2} = \frac{2(1)}{2} = 1$$

$$A_4 = \frac{2A_2}{4} = \frac{2(1)}{4} = \frac{1}{2}$$

$$A_6 = \frac{2A_4}{6} = \frac{2\left(\frac{1}{2}\right)}{6} = \frac{1}{6}$$

$$A_8 = \frac{2\left(\frac{1}{6}\right)}{8} = \frac{1}{6}\cdot\frac{2}{8} = \frac{1}{2\cdot 3\cdot 4} = \frac{1}{4!}$$

$$A_{10} = \frac{2A_8}{10} = \frac{1}{5}\cdot A_8 = \frac{1}{5}\cdot\frac{1}{4!} = \frac{1}{5!}$$

Clearly, a nice pattern has developed.

$$y = 1 + 0x + 1x^2 + 0x^3 + \frac{1}{2!}x^4 + 0x^5 + \frac{1}{3!}x^6 + \ldots = \sum_{k=0}^{\infty} \frac{x^{2k}}{k!}$$

Example 11: Polynomial of Degree 6

Find a polynomial of degree six which represents the function satisfying $y' = y + \frac{1}{1+x}$ and which contains the point (0, 1). In other words, $f(0) = 1$.

Solution: We have two representations for $f'(x)$:

$$f'(x) = A_1 + 2A_2x + 3A_3x^2 + 4A_4x^3 + 5A_5x^4 + \ldots + nA_nx^{n-1} + \ldots \text{ and}$$

$$f'(x) = y + \frac{1}{1+x} = 1 + A_1x + A_2x^2 + A_3x^3 + \ldots + \frac{1}{1+x}$$

$$= 1 + A_1x + A_2x^2 + A_3x^3 + \ldots + 1 - x + x^2 - x^3 + x^4 - \ldots$$

$$= 2 + (A_1 - 1)x + (A_2 + 1)x^2 + (A_3 - 1)x^3 + (A_4 + 1)x^4 + (A_5 - 1)x^5 + \ldots$$

When we equate the corresponding coefficients in the two representations, we get

$A_1 = 2$

$2A_2 = A_1 - 1$ which gives $A_2 = \frac{1}{2}$

$3A_3 = A_2 + 1$ which gives $A_3 = \frac{1}{2}$

$4A_4 = A_3 - 1$ which gives $A_4 = -\frac{1}{8}$

$$5A_5 = A_4 + 1 \quad \text{which gives} \quad A_5 = \frac{7}{40}$$

$$6A_6 = A_5 - 1 \quad \text{which gives} \quad A_6 = -\frac{33}{240}$$

The recursion is given by

$$kA_k = A_{k-1} + (-1)^{k-1}.$$

We conclude $y \approx P_6(x) = 1 + 2x + \frac{1}{2}x^2 + \frac{1}{2}x^3 - \frac{1}{8}x^4 + \frac{7}{40}x^5 - \frac{33}{240}x^6$.

Example 12: Power Series

Solve $y'' = -y$ subject to the conditions $y(0) = 1$ and $y'(0) = 0$.

Solution: As before let $y = \sum_{n=0}^{\infty} A_n x^n$.

Then $y' = A_1 + 2A_2x + 3A_3x^2 + \ldots = \sum_{n=1}^{\infty} nA_n x^{n-1}$

and $y'' = 2A_2 + 2\cdot 3A_3x + 3\cdot 4A_4x^2 + 4\cdot 5A_5x^3 + \ldots = \sum_{n=2}^{\infty}(n-1)nA_n x^{n-2}$.

Since $y(0) = 1$, we get $A_0 = 1$. Since $y'(0) = 0$ we have $A_1 = 0$.
A second expression for y'' comes from the given condition:

$$y'' = -y = -\sum_{n=0}^{\infty} A_n x^n = 2A_2 + 2\cdot 3A_3x + 3\cdot 4A_4x^2 + \ldots.$$

$$-A_0 - A_1x - A_2x^2 - A_3x^3 - \ldots = 1\cdot 2A_2 + 2\cdot 3A_3x + 3\cdot 4A_4x^2 + \ldots$$

And as usual we equate the corresponding coefficients to get

$$-A_0 = 2A_2 \quad \text{which gives} \quad A_2 = -\frac{A_0}{2} = -\frac{1}{2}$$

$$-A_1 = 6A_3 \quad \text{which gives} \quad A_3 = -\frac{A_1}{6} = 0$$

$$-A_2 = 12A_4 \quad \text{which gives} \quad A_4 = -\frac{A_2}{12} = \frac{1}{24} = \frac{1}{4!}$$

$$-A_3 = 4(5)A_5 \quad \text{which gives} \quad A_5 = 0$$

$$-A_4 = 5(6)A_6 \quad \text{which gives} \quad A_6 = -\frac{1}{6!}$$

and so on.

We get $y = 1 + 0x - \frac{1}{2!}x^2 + 0x^3 + \frac{1}{4!}x^4 + 0x^5 - \frac{1}{6!}x^6 + \ldots = \sum_{n=0}^{\infty}(-1)^n \frac{x^{2n}}{(2n)!}$.

This is recognizable as the Taylor series for $y = \cos x$.

10.3 Exercises

1. Use the Taylor series for e^x and substitution to obtain a power series for $y = x^2e^x$.

2. Use the Taylor series for e^x and substitution to obtain a power series for $y = \dfrac{e^x - 1}{x}$.

3. Use the geometric series $\dfrac{1}{1-r} = 1 + r + r^2 + r^3 + ... + r^n + ...$ to obtain a power series for each of the following:

a. $\dfrac{1}{1+t}$ **b.** $\dfrac{1}{1+t^2}$ **c.** $\dfrac{t}{1+t}$ **d.** $\dfrac{t}{1+t^2}$

4. Since $\dfrac{d}{dt}\arctan t = \dfrac{1}{1+t^2}$, we have $\arctan x = \int_0^x \dfrac{1}{1+t^2}\,dt$.

a. Substitute your answer to 3**b.** into the integral and integrate term-by-term to get a power series expansion for the arctangent function.

b. The interval of convergence for the geometric series is $(-1, 1)$. This means the interval of convergence for the solution to 4**a.** is also expected to be $(-1, 1)$. However, in this case, the interval includes the endpoint $x = 1$. Since the angle whose tangent is 1 is the angle $\dfrac{\pi}{4}$, $\arctan(1) = \dfrac{\pi}{4}$. Thus $\pi = 4\arctan(1)$. Substitute $x = 1$ into your power series for 4(a) and get a numerical representation for π as an infinite series (due originally to Leibniz).

5. a. Integrate $\int_0^{\frac{1}{2}} \dfrac{1}{1-t}\,dt$ using a graphing utility.

b. Integrate $\int_0^{\frac{1}{2}} \dfrac{1}{1-t}\,dt$ exactly.

c. Integrate $\int_0^{\frac{1}{2}} \dfrac{1}{1-t}\,dt$ by replacing the integrand with a Taylor series and integrate term by term.

d. Re-write your answer to **c.** using summation notation and equate it to your answer to part **b.**

6. Replace the integral in Exercise 5 with $\int_0^1 \dfrac{1}{1+t}\,dt$ and repeat the four steps.

7. Find a Taylor series for each of the following:

a. $y = e^{-x}$ **b.** $y = e^{-x^2}$ **c.** $y = e^{7x}$ **d.** $y = \dfrac{x}{e^x}$

8. In engineering, the hyperbolic sine function, abbreviated sinh, is defined by $\sinh(t) = \dfrac{e^t - e^{-t}}{2}$.

a. Calculate the first four derivatives of sinh(t) and determine a degree 4 Taylor polynomial for $y = \sinh(t)$.

b. Combine the known Taylor series for e^t and e^{-t} and get a Taylor series for sinh(t). Compare your results with part(a).

9. In engineering, the hyperbolic cosine function, abbreviated cosh, is defined by $\cosh t = \dfrac{e^t + e^{-t}}{2}$.

a. Calculate the first four derivatives of cosh(t) and determine a degree 4 Taylor polynomial for $y = \cosh(t)$.

b. Combine the known Taylor series for e^t and e^{-t} and get a Taylor series for cosh(t). Compare your results with part **a.**

10. Determine a calculus relationship between the cosh t and sinh t and their derivatives. (See Exercises 8, 9.)

11. a. Compute the first four derivatives of $g(x) = \dfrac{1}{1-x^2}$

b. Use your answer to part (a) to get a degree four Taylor polynomial for g(x) centered at $x = 0$. Hint: as a check, we calculate that

$$g^{(4)}(x) = 384x^4\left(1-x^2\right)^{-5} + 288x^2\left(1-x^2\right)^{-4} + 24\left(1-x^2\right)^{-3}.$$

c. Use the geometric series for $g(x)$ with $r = x^2$ and compare results.

12. A function is called *even* if $f(x) = f(-x)$ for all values of x in its domain. For example $y = x^2 + 3$ is even. Using what you know about Taylor series, make a conjecture which allows one to tell at a glance which functions are even.

13. A function is called odd if it satisfies $f(x) = -f(-x)$ for all x in its domain. For example $y = x^3 - x$ is odd. Using what you know about Taylor series, make a conjecture which allows one to tell at a glance which functions are odd.

14. a. Find a power series representation for the function $y = \dfrac{1}{\sqrt{1-t}}$ by using the binomial theorem. Let $p = -\dfrac{1}{2}$ and replace x by $-t$.

b. Find a summation expression for your answer to part (a) so that your answer has the form $\dfrac{1}{\sqrt{1-t}} = \sum_0^\infty \dfrac{a_n}{b_n} x^n$.

15. Suppose $f(x) = \dfrac{1}{1-x}$.

a. Find $f'(x)$.

b. Replace $\dfrac{1}{(1-x)}$ by a power series expansion and differentiate term by term.

c. Equate your answers to **a.** and **b.** Express your result as

$$\frac{d}{dx}\left(\frac{1}{1-x}\right) = ?? = \sum_{n=?}^{\infty} a_n x^n.$$

16. a. Get a power series representation for $y = \dfrac{1}{\sqrt{1-u^2}}$. Hint: In Exercise 14, replace t by u^2 and write $\dfrac{1}{\sqrt{1-u^2}} = \sum_0^\infty \dfrac{a_n}{b_n} x^{2n}$.

b. Since $\dfrac{d}{du}\arcsin u = \dfrac{1}{\sqrt{1-u^2}}$, we can say $\arcsin x = \int_0^x \dfrac{1}{\sqrt{1-u^2}}\,du$. Thus one can substitute, integrate term by term, and get a power series representation for arcsin x. Carry out these details to represent arcsin x as a power series.

Hint: $\arcsin x = \displaystyle\int_0^x \frac{1}{\sqrt{1-u^2}}\,du = \sum_{n=?}^{\infty}\left(\frac{a_n}{b_n}\cdot\int_0^x u^{2n}\,du\right)$.

17. Evaluate the limit $\displaystyle\lim_{t\to 0}\left(\frac{\sin t}{t}\right)$ as follows: replace $\sin t$ by its Taylor series, simplify the quotient, and then calculate the limit as t goes to zero.

18. Calculate $\displaystyle\int_0^1 \frac{\sin t}{t}\,dt$ in two ways.

a. Use a graphing utility. (Remember to set your calculator to radian mode.)

b. Replace $\dfrac{\sin t}{t}$ by a degree six polynomial and then integrate term by term. Compare the answers.

19. Use long division of polynomials to confirm that $\dfrac{1}{1-r} = 1 + r + r^2 + r^3 + \ldots + r^n + \ldots$.

20. Multiply $(1-r)(1+r+r^2+r^3+\ldots+r^n+\ldots)$ and confirm the result of Exercise 19.

21. Find the functions whose Taylor series coefficients are given by the sequences below.

a. $(0, 0, 0, 1, 1, 1, 1, ..., 1, ...)$

b. $(1, 2, 3, 4, 5, ...)$

c. $(1, 1, 1, 0, 1, 1, ..., 1, ...)$

d. $(1, 6, 15, 20, 15, 6, 1, 0, 0, 0, ...)$

22. Find the functions whose Taylor series coefficients are given by the sequences below.

a. $\left(0, 1, 0, \frac{-1}{3!}, 0, \frac{1}{7!}, 0, \frac{-1}{9!}, 0, \frac{1}{11!}, 0, \frac{-1}{13!}, 0, ...\right)$

b. $(2, 2, 2, 2, ...)$

c. $\left(0, 1, \frac{1}{2}, \frac{1}{3}, \frac{1}{4}, ...\right)$

d. $(1, 5\ 10, 10, 5, 1, 0, 0, 0, ...)$

In problems 23 – 26, find a Taylor series polynomial of degree at least four which is a solution of the boundary value problem.

23. $f'(x) = y + 1,\ f(0) = 1.$

24. $f'(x) = (1 + x)y,\ f(0) = 1.$

25. $f''(x) = -f(x) + 1,\ f(0) = 1,\ f'(0) = 1.$

26. $f'(x) = y - x + 1,\ f(0) = 1.$

Hawkes Learning Systems: Essential Calculus

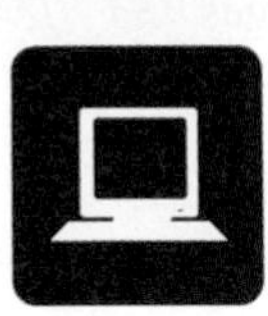

Taylor Series, Infinite Expressions, and Their Applications

Chapter 10 Index of Key Ideas and Terms

Section 10.1 Infinite Sequences

Sequences pages 730 - 732

Limit of a Sequence pages 730 - 732

Let $\{f(n)\}_{n=1}^{\infty}$ be any sequence. If there is a number L such that $\lim_{n\to\infty}|f(n)-L|=0$ then we say L is the **limit** of the nth term of the sequence. If $f(x)$ is continuous on the interval $(1, \infty)$, then $\lim_{n\to\infty} f(n)=L$ if and only if $\lim_{x\to\infty} f(x)=L$.

Infinite Series pages 732 - 737

An **infinite series** is an expression involving a sum of an infinite number of terms.

Recursive Sequence pages 732 - 737

A **recursive sequence** is defined by specifying one or more initial terms and then expressing the nth term using previous terms.

Fibonacci Sequence page 732

Let $F_1 = 1, F_2 = 1$, and for $n \geq 3, F_n = F_{n-1} + F_{n-2}$. This sequence is called the **Fibonacci sequence**.

Sequence of Partial Sums pages 733 - 734

Let $\sum_{n=1}^{\infty} a_n = a_1 + a_2 + a_3 + ...$ denote an arbitrary series. We define a special sequence using the series called the **sequence of partial sums**.

$$S_1 = a_1,\ S_2 = a_1 + a_2,\ S_3 = a_1 + a_2 + a_3, ... \text{ and so on.}$$

Convergence Principle for Infinite Series pages 734 - 737

An infinite series $\sum_{n=1}^{\infty} a_n = a_1 + a_2 + a_3 + ...$ converges to a number L if and only if its corresponding sequence $\{S_n\}_{n=1}^{\infty}$ of partial sums converges to L.

Geometric Series pages 737 - 739

A **geometric series** has the form $a + ar + ar^2 + ar^3 + ... + ar^n + ...$

Convergence of a Geometric Series pages 738 - 739

When $-1 < r < 1$, the geometric series $a + ar + ar^2 + ar^3 + ... = \sum_{n=0}^{\infty} ar^n$ is finite and converges to $\frac{a}{1-r}$.

Section 10.2 Taylor Polynomials

Linear Approximation pages 742 - 744

The **linear** (tangent line) **approximation** to $f(x)$ at the point $(a, f(a))$ is given by

$$f(x) \approx P_1(x) = f(a) + f'(a) \cdot (x-a)$$

Taylor Polynomial of Degree 2 Approximating $f(x)$ near $x = 0$ page 744

$$f(x) \approx P_2(x) = f(0) + f'(0)x + \frac{f''(0)}{2} \cdot x^2$$

Higher Degree Polynomials pages 744 - 747

Taylor Polynomial of Degree k Approximating $f(x)$ near $x = 0$ pages 744 - 746

$$f(x) \approx f(0) + f'(0)x + \frac{1}{2!}f''(0)x^2 + \frac{1}{3!}f'''(0)x^3 + \ldots + \frac{1}{k!}f^{(k)}(x)x^k.$$

We call this a Taylor approximation centered at $x = 0$

Taylor Polynomial of Degree k Approximating $f(x)$ near $x = a$ pages 746 - 747

$$f(x) \approx P_k(x) = f(a) + f'(a)(x-a) + \frac{f''(a)(x-a)^2}{2!} + \frac{f'''(a)(x-a)^3}{3!} + \ldots + \frac{f^{(k)}(a)(x-a)^k}{k!}$$

Section 10.3 Taylor Series, Infinite Expressions, and Their Applications

Taylor Series Centered at $x = a$ pages 750 - 751

Binomial Series pages 751 - 753

$$(1+x)^p = 1 + px + \frac{p(p-1)}{2!} \cdot x^2 + \frac{p(p-1)(p-2)}{3!} \cdot x^3 + \ldots$$

The series converges provided $-1 < x < 1$ (the radius of convergence $R = 1$).

Euler's Formula from Trigonometry pages 754 - 755

$$\boldsymbol{e^{it} = \cos(t) + i\sin(t)}$$

Power Series pages 756 - 759

A general expression like

$$A_0 + A_1x + A_2x^2 + A_3x^3 + \ldots = \sum_{n=0}^{\infty} A_n x^n$$

is called a **power series.**

Chapter 10 Test

For Exercises 1 – 4, write the first five terms of the sequence indicated.

1. $a_n = \dfrac{2n}{3}$

2. $a_n = 1 - \dfrac{1}{n}$

3. $a_n = a_{n-1} + 2^{n-1}, a_1 = 1$

4. $a_n = \sin\left(\dfrac{\pi n}{2}\right)$

For Exercises 5 – 8, determine the limit $\lim\limits_{n\to\infty} f(n)$.

5. $f(n) = 1 + \dfrac{3}{n}$

6. $f(n) = \dfrac{(-1)^n}{2}$

7. $f(n) = \dfrac{17n}{18n+1}$

8. $f(n) = \sum\limits_{i=0}^{n} \dfrac{1}{2^i}$

For Exercises 9 – 12, determine the indicated sum if it exists.

9. $\sum\limits_{i=0}^{\infty} \left(\dfrac{1}{7}\right)^i$

10. $\sum\limits_{i=0}^{\infty} (-1)^i \left(\dfrac{5}{2^i}\right)$

11. $3 - \dfrac{1}{3} + \dfrac{1}{9} - \dfrac{1}{27} + \ldots$

12. $\sum\limits_{i=0}^{\infty} \dfrac{\sin\left(\pi/4\right)}{2^i}$

13. Determine a rational number (common fraction) to which the sum $\dfrac{2}{10} - \dfrac{3}{100} + \dfrac{2}{1000} - \dfrac{3}{10{,}000} + \ldots$ converges.

14. Suppose that a patient takes 81 milligrams of aspirin per day and that 90% of the aspirin dissipates by the next day. After thirty days, how much aspirin is in the patient's body just after taking his usual dose?

15. Find a linear approximation to $y = e^x + \ln x$ near $x = 1$.

16. Find a Taylor polynomial of degree 3 for $f(x) = x^4 + x^3 + x^2 + x + 1$ near $x = 0$.

17. Find a rational approximation to $\sqrt{83}$ using a Taylor polynomial of degree 2.

18. Find a Taylor polynomial for $y = \dfrac{\sin x}{x}$ of degree 4.

19. Using a binomial series expansion, get a polynomial of degree 3 representation for $y = \sqrt{2+t}$. Hint: note $y = \sqrt{2\left(1+\dfrac{t}{2}\right)} = \sqrt{2}\left(1+\dfrac{t}{2}\right)^{\frac{1}{2}}$. Let $x = \dfrac{t}{2}$ in the binomial series.

20. Obtain a series representation for $y = \dfrac{1}{1-x^3}$.

21. Find seven (possibly complex) solutions to $x^7 = 1$.

22. Find a function whose Taylor series coefficients are given by the sequence $(3, 4, 5, 6, \ldots)$.

23. Solve the differential equation $\dfrac{dy}{dx} = xy$ subject to $f(0) = 1$ by using $f(x) = \sum\limits_{n=0}^{\infty} A_n x^n$. (Determine an expression for A_n.)

Cumulative Review

For Exercises 1 – 4, determine the derivative indicated.

1. $f(x) = \sin\left(2\pi x - \frac{\pi}{4}\right)$

2. $f(x) = e^{2x}\sin(3x)$

3. $f(x) = \sum_{k=0}^{\infty} \frac{(-1)^k x^k}{2^k}$

4. $f(x) = \frac{\tan x}{1+\sin x}$

5. What is the average y-value for $y = \frac{1}{1-x}$ on the interval $\left[-\frac{1}{2}, \frac{1}{2}\right]$?

6. Find a Taylor polynomial for $f(x) = x\ln x$ of degree 5 at $x = 1$.

7. Integrate by parts to determine $\int_0^1 x\ln x\,dx$. Note, the function is undefined at $x = 0$.

8. Evaluate $\int x\sin(x^2 - 2)\,dx$.

9. Determine $\frac{d}{dx}\left(\sin^{-1}(4x)\right)$.

10. Determine $\frac{d}{dx}\left(\tan^2 x\right)$.

Determine the integrals in Exercises 11 – 13.

11. $\int \frac{x}{x^2+1}\,dx$

12. $\int x\left(x^2+1\right)^2 dx$

13. $\int \frac{1}{x^2+1}\,dx$

14. $\int \frac{x}{\left(x^2+1\right)^3}\,dx$

15. Determine the area between $y = \tan x$ and $y = 2 - \frac{4x}{\pi}$ over the interval $\left[0, \frac{\pi}{4}\right]$. (The graphs intersect at $\left(\frac{\pi}{4}, 1\right)$.)

16. Evaluate $\int_0^{\frac{\pi}{2}} e^{\sin\theta}\cos\theta\,d\theta$.

17. Determine $\int_0^{\pi} e^{-x}\sin x\,dx$. Integrate by parts twice or use a calculator.

18. Verify that $y = \frac{2x^2}{1+x^2}$ satisfies $x^3\frac{dy}{dx} = y^2$.

19. The temperature of water in a small pond is estimated by $50 + 2t - .05t^2$ degrees, where t is the number of hours after 8 A.M. How fast is the temperature changing at noon?

20. In a biology lab experiment, the population P of mosquitos grows at the rate of $0.03(6000 - P) = \frac{dP}{dt}$ mosquitos per day t days after the experiment has begun. Suppose 300 mosquitos are put in the lab containment area initially and in 10 days there are 2000 mosquitos. Determine the particular solution for P in terms of t.

21. Determine the area between $f(x) = x + \pi$ and $g(x) = \sin x$ with $-\pi \le x \le 0$.

22. A roller coaster car travels parallel to a path shaped like $y = 60 + 60\sin\left(x + \frac{\pi}{2}\right)$. A math student wondered how high she was at the first point of greatest slope. Answer her question. Assume the loading of passengers occurs at $x = 0$.

23. During the three months of July, August, and September last year, the corn price in dollars per bushel offered to farmers varied according to the curve $y = 3 + 0.5\cos(0.1x)$. On what calendar day was the offered price the highest? (Take $x = 2.5$ to refer to July 3, for example.)

24. Determine the equation of the tangent line to $y = x\sin x$ at $x = \pi$.

25. Determine if the formula $\int 2e^x \sin x\,dx = e^x(\sin x - \cos x) + C$ is correct.

26. Using your calculator, determine a number k so that $f(x) = 2ke^x \sin x$ is a probability density function on the interval $[0, \pi]$.

27. What is the volume of the solid of revolution if the region bounded by $f(x) = \sqrt{\sin x}$, the x-axis, and the interval $[0, \pi]$ is revolved about the x-axis?

Answers

Chapter 1

Exercises 1.1, pages 8 - 9

1. $\left\{0,\sqrt{16}\right\}$ 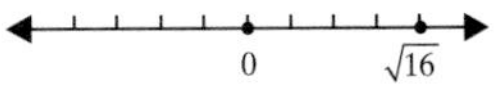

3. $\left\{\sqrt{16}\right\}$ 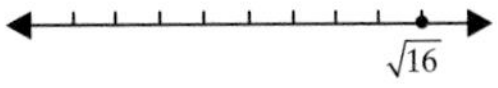

5. $\left\{-6,-\sqrt{4},-\frac{5}{4},-1.2,0,\frac{3}{8},\sqrt{16},5.2\right\}$ 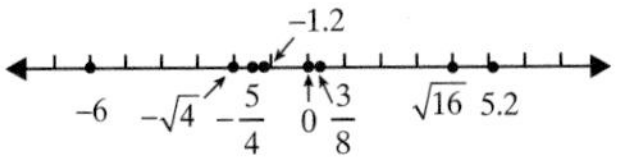

7. always **9.** sometimes **11.** sometimes **13.** 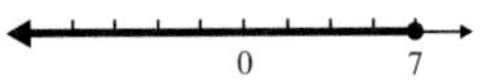

$(-\infty,7]$

15. 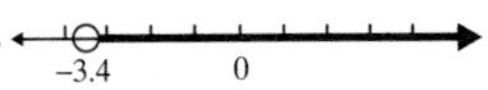

$(-3.4,\infty)$ **17.** 

$(-\infty,0]$ **19.** 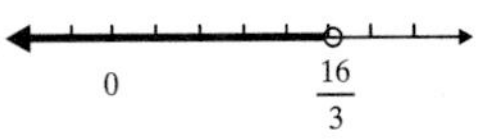

$\left(-\infty,\frac{16}{3}\right)$

21. 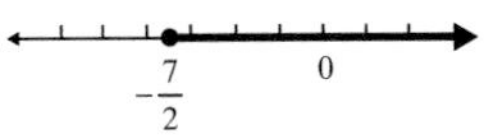

$\left[-\frac{7}{2},\infty\right)$ **23.** 1.4 **25.** $\frac{3}{8}$ **27.** $\sqrt{2}$ **29.** 12 **31.** −0.6 **33.** $x=\pm 8$ **35.** $x=\pm\frac{7}{16}$ **37.** $\varnothing$

39. $x=\pm 3\sqrt{2}$ **41.** $x\geq 0$ **43.** $x=1$; yes, it is rational.

Exercises 1.2, pages 12 - 13

1. 343 **3.** $\frac{1}{36}$ **5.** −16 **7.** −50 **9.** $\frac{1}{x^2}$ **11.** $\frac{5}{y}$ **13.** x^7 **15.** x **17.** $\frac{1}{x^7}$ **19.** $\frac{1}{a^3}$ **21.** y^2 **23.** x^5 **25.** $\frac{1}{x^3}$ **27.** x^3 **29.** b^4

31. x^6 **33.** a^8 **35.** y^{-1} **37.** x^{-5} **39.** x^{10} **41.** $\frac{1}{3}a^{-2}b^7$ **43.** $\frac{4}{3}x^6y^{-2}$ **45.** $\frac{27}{8}x^{-3}y^9$ **47.** $\frac{1}{36}s^2t^{-8}$ **49.** $\frac{1}{25}x^4y^{-2}$

Exercises 1.3, pages 20 - 21

1. 4 **3.** 8 **5.** $\frac{1}{125}$ **7.** −3 **9.** 4 **11.** $x^{\frac{5}{2}}$ **13.** $x^{-\frac{1}{6}}$ **15.** $a^{\frac{8}{5}}$ **17.** $x^{\frac{1}{3}}$ **19.** $x^{-\frac{7}{4}}$ **21.** $b^{\frac{3}{10}}$ **23.** $s^{\frac{7}{8}}$ **25.** $x^{\frac{1}{4}}$ **27.** $x^{\frac{4}{5}}$ **29.** $3x^2y^2$

31. $\frac{2t^{\frac{5}{4}}}{25s^{\frac{1}{6}}}$ **33.** $\sqrt{y}$ **35.** $\frac{1}{\sqrt[4]{x}}$ **37.** $\sqrt[5]{(x-2)^4}$ **39.** $\frac{2}{\sqrt{y^2+4}}$ **41.** $-\frac{2}{7\sqrt[4]{t^2-4}}$ **43.** $x^{\frac{1}{4}}$ **45.** $s^{-\frac{1}{4}}$ **47.** $(y+10)^{\frac{2}{3}}$

49. $(a^2+2)^{-\frac{1}{3}}$ **51.** $\frac{4}{9}x(x+5)^{-\frac{2}{3}}$ **53.** $4\sqrt{3}$ **55.** $\frac{3}{4}$ **57.** $5\sqrt{6x}$ **59.** $6t\sqrt{3t}$

Exercises 1.4, pages 29 - 30

1. $2x^2+x+5$ **3.** $5a^2+14a-2$ **5.** x^2+8x+2 **7.** $4x^2+3x+11$ **9.** $3a^2-8ab$ **11.** $-7x^2-5xy+5y^2$ **13.** $6x^4+10x^3-2x^2$

15. $12xy$ **17.** x^4, x^2y, y^2 **19.** $14x^2-9x-18$ **21.** $12x^2-25x+12$ **23.** $16x^2-24x+9$ **25.** $9x^2+12xy+4y^2$

27. $36x^2-y^2$ **29.** $2x^3+6x^2-8x$ **31.** x^3+27 **33.** x^3+8y^3 **35.** $-7x+9$ **37.** $7t+7$ **39.** $4x^2-x$ **41.** $3y, 4, 6$

43. $(s-7)(s+2)$ **45.** $(x+13)(x-2)$ **47.** $4b(b-4)(b+4)$ **49.** $3(3a-2)(3a+2)$ **51.** $(x-1)(x+1)(x^2-x+1)(x^2+x+1)$

53. $(5x^4-4)(5x^4+4)$ **55.** $2(t+2y)(t^2-2ty+4y^2)$ **57.** $(s+1)(s-1)(s^2+1)$ **59.** $100x(y+1)^2$ **61.** Answers will vary

Exercises 1.5, pages 37 - 39

1. $\frac{1}{3}$ **3.** 3 **5.** $\frac{-64}{85}$ **7.** No **9.** No **11.** $-7x+2y=7$ **13.** $y=\frac{2}{3}x+3$ **15.** $2x+5y=14$

17. $y=-\frac{1}{5}x+2;\ m=-\frac{1}{5};$ $b=2$ **19.** $y=-2x+\frac{11}{2};m=-2;$ $b=\frac{11}{2}$

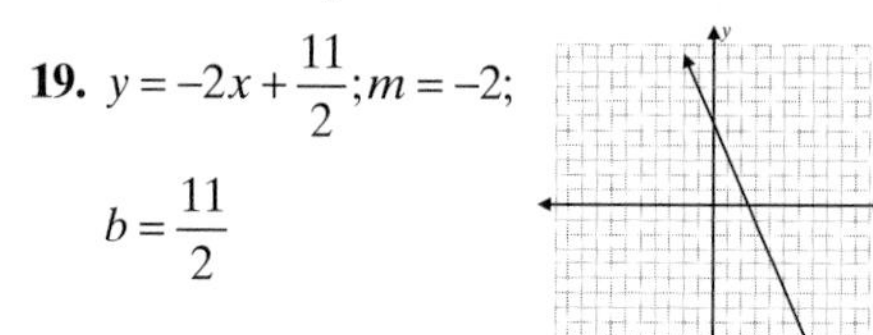

21. $y=\frac{5}{6}x-2;\ m=\frac{5}{6};$ $b=-2$

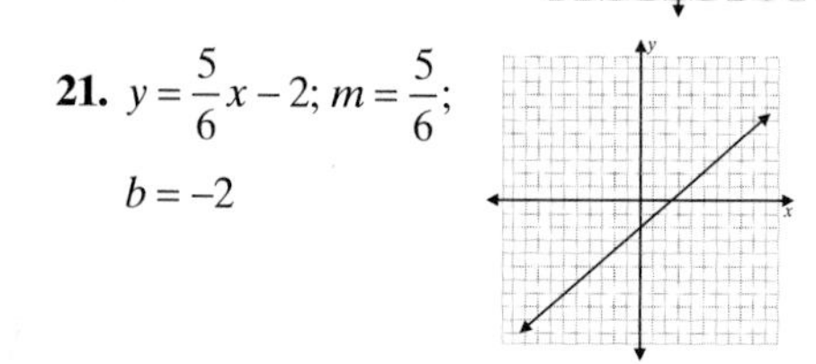

23. No **25.** No **27.** Neither; because the slopes of the lines are neither the same nor negative reciprocals of one another. **29.** Parallel; because the slopes of the lines are the same. **31.** Perpendicular; because the slopes of the lines are negative reciprocals of one another. **33.** Neither; because the slopes of the lines are neither the same nor negative reciprocals of one another.

35. Perpendicular; because the slopes of the lines are negative reciprocals of one another. **37.** $y=-\frac{1}{3}x+2$

39. $y=\frac{7}{5}x+\frac{17}{5}$ **41.** $y=-\frac{4}{3}x+7$ **43.** The line is $y=-1$;

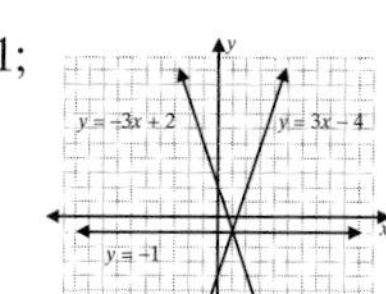

45. $b=\frac{d_1c_2-d_2c_1}{c_2-c_1}$

47. $b_1=0;\ b_2=60$

51. \$5/month

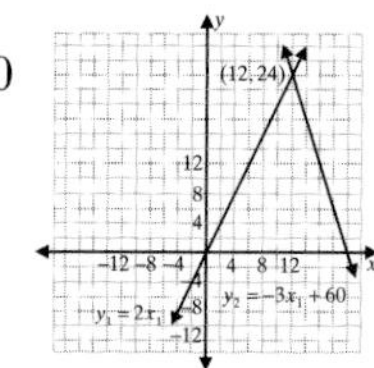

49. 100 hot dogs

Exercises 1.6, pages 53 - 57

1. a. 3 **b.** -11 **c.** $2a-5$ **d.** $2a-6$ **3. a.** 9 **b.** 4 **c.** a^2 **d.** a^2-2a+2 **5. a.** 4 **b.** -8 **c.** a^3+4a^2+2a **d.** a^3+a^2-3a+2

7. a. 35 **b.** $4a^2+16a+15$ **c.** $4x^2+8xh+4h^2-1$ **d.** 12 **9. a.** 2 **b.** $\sqrt{a+7}$ **c.** $\sqrt{x+h+5}$ **d.** $3-\sqrt{6}$ **11. a.** 3 **b.** $\frac{7}{9}$ **c.** $-\frac{20}{9}$

d. $-\frac{10}{9}$ **e.** (a) and (b) are points on the graph, (c) is the change in y, and (d) is the difference quotient, or slope. **13. a.** -5

b. -2 **c.** 0.25 **d.** 3 **15.** $x\neq 5,6$ **17.** $\mathbb{R}$ **19.** $\frac{1}{x}+x^2-2x$ **21.** $\frac{\sqrt{x}}{x}$ **23.** $x\sqrt{x}$ **25.** -1 **27.** $\sqrt{3}$ **29.** $\frac{1}{\sqrt{3}}=\frac{\sqrt{3}}{3}$

31. t^2-1 **33.** $\sqrt{2t-1}$ **35.** $\frac{1}{x^2}-\frac{2}{x}$ **37.** $x-2\sqrt{x}$ **39.** $\frac{1}{\sqrt{u}}=\frac{\sqrt{u}}{u}$ **41.** $5h$ **43.** h^2+2hx **45.** $3h^2+6hx$ **47.** $h^2+2hx+2h$

49. $4h^2+8hx-h$ **51.** $h^3+3h^2x+3hx^2$ **53.** $h^3+3h^2x+3hx^2$ **55.** $x\neq 2$ **57.** $x\neq -3, 4$ **59.** $\mathbb{R}$ **61.** $x\geq 0$

63. Is a function **65.** Is a function **67.** Not a function **69.** Not a function **71.** Is a function **73.** Not a function

Exercises 1.7, pages 66 - 67

1. $1+3\times x+x\wedge(1\div 2)$ **3.** $(3\times x)\div(x-1)\wedge(1\div 2)$ **5.** $(-3\times x+16)\wedge 4\div(3\times x+6)$ **7.** $(2\times x\wedge(2\div 3)+3\times x\wedge(5\div 3))\wedge 5$

9. The standard window displays a graph that is not informative. A window of $(-10, 10)$ by $(-14, 6)$ is better.
11. The standard window does not display the graph on the screen. A window of $(-10, 10)$ by $(-10{,}000, 210{,}000)$ is better.
13. In the standard window, the curve is split into two parts. A window of $(-10, 10)$ by $(-500, 500)$ is better.
15. In the standard window, the curve is split into two parts. A window of $(-10, 10)$ by $(-500, 500)$ is better.
17. The standard window does not closely display the interesting part of the graph. A window of $(-5, 5)$ by $(-5, 5)$ is better.
19. The standard window does not closely display the interesting part of the graph. A window of $(-5, 5)$ by $(-5, 5)$ is better.
21. In the standard window, the curve is split into two parts. A window of $(-10, 10)$ by $(-100, 100)$ is better.

23.

x	y
−2	48
−1	11
0	10
2	−28

Zeros in the given range: $x \approx 1.417$

25.

x	y
−2	−3.826
−1	undefined
0	0
2	−0.962

Zeros in the given range: $x = 0, 5$

27.

x	y
−2	990
−1	472
0	210
2	70

Zeros in the given range: $x = 3$

29.

x	y
−2	undefined
−1	undefined
0	1.732
2	5.469

Zeros in the given range: $\varnothing$

31. $y_{min} = 20$ at $x = 52$ **33.** $y_{min} = -21.98$ at $x = 2.38$ **35.** $y_{min} = -90.85$ at $x = 28.44$ **37.** $y_{min} = -236.50$ at $x = 6.17$
39. $y_{min} = -130783$ at $x = 2$

Exercises 1.8, pages 79 - 84

1. a **3.** c **5.** a **7.** a **9.** a **11. a.** $C(x) = 135 + 0.5x$ **b.** \$385 **13. a.** $P(x) = 4.5x - 135$ **b.** \$2115 **15.** $P(x) = 100(4.5x - 135)$

17. 0.75 atm **19.** ≈ 56.03 atm **21.** $x = 3$ **23.** $x = 30, 50$ **25.** $(4, 11)$ **27.** $(5, 30)$ **29. a.** $R(x) = 6.5x$

b. $C(x) = 1.1x + 378$ **c.** $P(x) = 5.4x - 378$ **d.** $x = 70$ pies **31. a.** $R(x) = 243x$ **b.** $C(x) = 73x + 5780$

c. $P(x) = 170x - 5780$ **d.** $x = 34$ sets of clubs **33. a.** \$0.15/pen **b.** $C(x) = 0.15x + 260$ **c.** \$260

35. a. $R(x) = 31x - 0.5x^2$ **b.** $P(x) = -0.5x^2 + 20x - 500$ **37. a.** $D(x) = -0.02x + 400$ **b.** $R(x) = -0.02x^2 + 400x$

c. $P(x) = -0.01x^2 + 150x - 3600$ **39. a.** $I = 0.0625P$ **b.** \$375 **41. a.** $p = \dfrac{340}{x}$ **b.** \$10

43. the least integer y such that $y \geq \dfrac{800}{x}$ **45.** $C(x) = \begin{cases} 0.65; \text{ for } 0 < x \leq 3 \\ 0.65 + 0.15(x-3); \\ \text{for } x > 3 \end{cases}$ **47.** $P(x) = 0.2575x$

49. $P(x) = -x^2 + 72x - 720$ **51.** $A(x) = 138x - x^2$ **53.** $P(x) = \dfrac{576}{x} + 2x$ **55.** $A(x) = 360x - \dfrac{3}{2}x^2$

Chapter 1 Test, pages 91 - 94

1. $\dfrac{3}{x}$ **2.** $12x^2$ **3.** x^4 **4.** $\dfrac{2}{5x^2}$ **5.** x^4 **6.** $-20x^6y^{-5}$ **7.** y^{-8} **8.** $4x^8y^8$ **9.** 27 **10.** $\dfrac{1}{8}$ **11.** $24x^{\frac{1}{6}}$ **12.** $x^{\frac{7}{4}}$ **13.** $\dfrac{3x}{y^2}$

14. $\dfrac{y^{\frac{1}{3}}}{x}$ **15.** $\sqrt[3]{x^8}$ **16.** $7\sqrt[5]{x^2y^3}$ **17.** $\dfrac{1}{\sqrt{3x+1}}$ **18.** $\sqrt[3]{(x+6)^2}$ **19.** $3x^{\frac{3}{4}}$ **20.** $-2x^{\frac{2}{3}}y^{\frac{1}{3}}$ **21.** $(4x-3)^{\frac{1}{5}}$ **22.** $\dfrac{6}{5(9-x^2)^{\frac{1}{2}}}$

23. $7x^2+10x-5$ **24.** $9x^2+3x-9$ **25.** $-x^3+5x^2-6$ **26.** $3x^3+4x^2-7x+3$ **27.** x^2-9x+5 **28.** $3x^2+x+6$

29. $49x^2-36$ **30.** $4x^2+36x+81$ **31.** $8x^3+125y^3$ **32.** $8x^4-18x^3-35x^2$ **33.** $4x+3y=23$ **34.** $5x-8y=-71$

35. $y=\frac{5}{4}x+13$ **36.** $y=-\frac{3}{10}x+15$ **37.** $y=\frac{5}{3}x+8$ **38.** $7x-2y=16$

39. $y=-\frac{2}{3}x+4$; $m=-\frac{2}{3}$; $b=4$

40. $y=\frac{5}{2}x-4$; $m=\frac{5}{2}$; $b=-4$

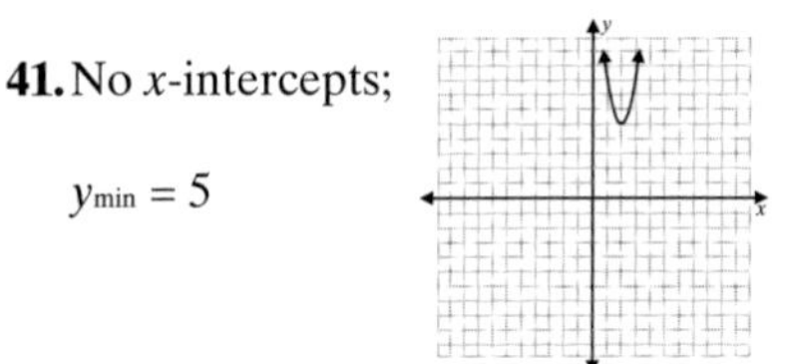

41. No x-intercepts; $y_{min}=5$

42. x-intercepts: $x=-1,-5$; $y_{min}=-4$

43. x-intercepts: $x=1,3$; $y_{max}=1$

44. x-intercepts: $x=-\frac{1}{2}, 2$; $y_{min}=-3.125$

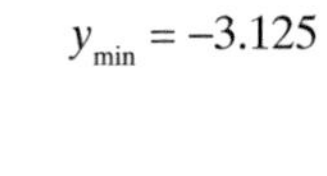

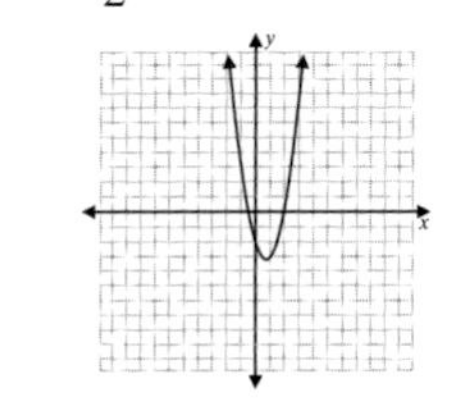

45. No x-intercepts; $y_{min}=1$

46. x-intercepts: $x=9$; $y_{max}=9.043$ **47. a.** 10 **b.** −25 **c.** $7a+3$ **d.** $7a-3$

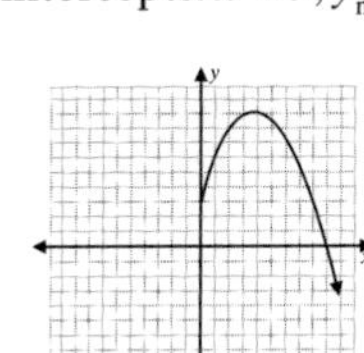

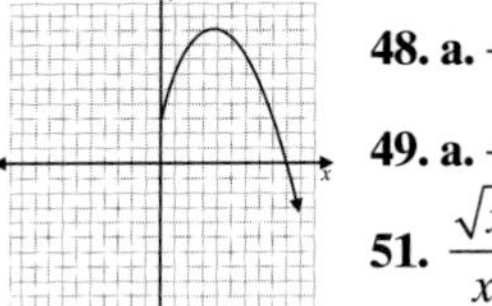

48. a. −3 **b.** $8a-2a^2-3$ **c.** $5-2x^2-4hx-2h^2$ **d.** −10

49. a. −3 **b.** −1 **c.** −1.75 **d.** −2 **50.** $5x-2+\frac{1}{x+4}$

51. $\frac{\sqrt{x+3}}{x+4}$ **52.** $5x^2+18x-8$ **53.** $5\sqrt{x+3}-2$

54. $\frac{1}{5x+2}$ **55.** $\sqrt{5x+1}$ **56.** $6h$ **57.** $4hx+2h^2-5h$ **58.** $3hx^2+3h^2x+h^3$ **59.** Is a function

60. Is a function **61.** $(66, 2145)$ **62.** $(18, 378)$ **63. a.** $R(x)=0.65x$ **b.** $C(x)=0.1x+220$ **c.** $P(x)=0.55x-220$

d. $x=400$ **64. a.** \$16/unit **b.** $C(x)=558+16x$ **c.** \$558 **65. a.** $D(x)=-\frac{1}{10}x+22$ **b.** $R(x)=-\frac{1}{10}x^2+22x$

66. $R(x)=18{,}000-1800x-200x^2$ **67.** $P(x)=-0.8x^2+42.5x$ **68.** $A(x)=55x-\frac{1}{2}x^2$

Chapter 2

Exercises 2.1, pages 110 - 120

1. lim = 14

x	y
6.5	13.5
6.9	13.9
6.99	13.99
6.999	13.999

3. lim = 0

x	y
3.5	0.25
3.1	0.01
3.01	0.0001
3.001	0.000001

5. lim = $1.\overline{3}$

x	y
−0.5	1.2785
−0.1	1.3311
−0.01	1.3333
−0.001	1.3333

7. lim = $2\sqrt{2}=2.828$

n	y
1.0	2.414
1.4	2.814
1.41	2.824
1.414	2.828

9. a. $a = 2$ **b.** lim = 0.25 **11. a.** $a = \pi$ **b.** No limit **13. a.** −2, −1.84, −1.36, −1.19, −1.0199, −1.001999 **b.** −1 **15. a.** 7, 8.76, 10.84, 11.41, 11.9401, 11.994001 **b.** 12 **17. a.** 3, $4.\overline{3}$, 11, 21, 201, 2001 **b.** $+\infty$ **19. a.** 0.2, 0.217, 0.238, 0.244, 0.249, 0.2499 **b.** 0.25 **21.** 3 **23.** 0 **25.** 2 **27.** $-\infty$ **29.** 2 **31.** 2 **33.** 0 **35.** $+\infty$ **37.** 1 **39.** 5 **41.** −17 **43.** 3 **45.** 0 **47.** −24 **49.** $-\infty$ **51.** $+\infty$ **53.** $\frac{1}{4}$ **55.** $-\frac{1}{4}$ **57. a.** 11 **b.** 11 **59. a.** 8 **b.** 9 **61. a.** 3 **b.** 3 **c.** 3 **d.** $+\infty$ **63. a.** 4 **b.** 4 **c.** 4 **d.** 3 **65. a.** 3 **b.** −1 **c.** does not exist **d.** 4 **67. a.** 3 **b.** 1 **c.** does not exist **d.** 3 **e.** 3 **f.** 3 **69. a.** −1 **b.** 1 **c.** $+\infty$ **d.** $+\infty$ **71. a.** 5 **b.** 2 **c.** does not exist **d.** 4 **73.** 12 **75.** 6 **77.** 32

79. a.

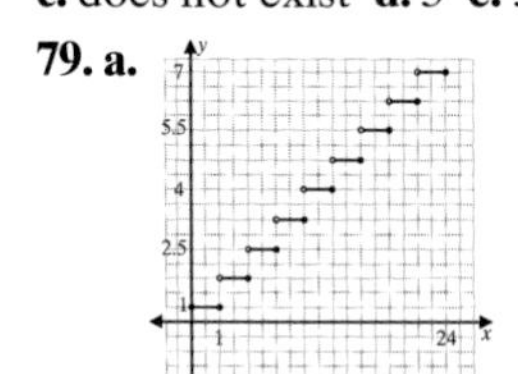

b. \$2.50 **c.** \$3.25 **d.** \$6.25 **e.** \$7.00 **81. a.** yes **b.** yes **c.** yes

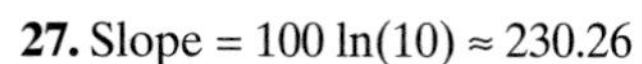

Exercises 2.2, pages 137 - 140

1. $f'(t)$ is the velocity of the car in ft./min at t minutes. **3.** $f'(n)$ is the rate of change of birds per tree. **5.** The rate of change of revenue per radio. **7.** The rate of change in bytes fed per second. **9.** The rate of change of GPA when the average SAT is s. **11.** −1 **13.** $f'(15) = 1$ **15.** 12 P.M. **17.** 3 **19.** 5.5 **21.** −0.343 **23.** 0.4545

25. Slope = 27

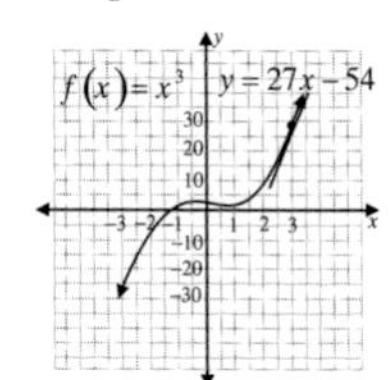

27. Slope = $100 \ln(10) \approx 230.26$

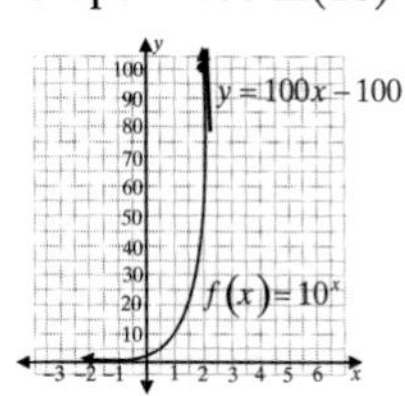

29. $f'(x) = 3x^2$

a	$f(a)$	$f'(a)$
−1	−1	3
0	0	0
1	1	3
2	8	12
3	27	27
4	64	48

31. $f(3) = 14$; 14 is the value of the function at the point $x = 3$.
$f'(3) = 7$; 7 is the slope of the tangent line at $x = 3$.

33. Slope at (1, 0) is 1.

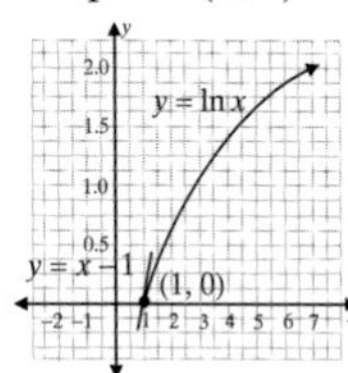

35. a. 5 **b.** positive **c.** Greater for an SUV.

37. a. $f(8) = 9$ means that after 8 days, 9 students on the campus have the flu.
b. $f'(8) = 3$ is the rate of change of y at the point (8, 9). 3 new cases are expected to be reported on the 9th day.

39. a. $f(0) = 72{,}000$ means that the cost of a 2200 ft.² house is \$72,000 in the year 2000.
b. $f'(x) = 3000$ means $f(x)$ is changing at the rate of \$3000 per year.
c. In the year 2003, a 2200 ft.² house cost \$81,000.

41. Window [−5, 8] by [−110, 800]

a.

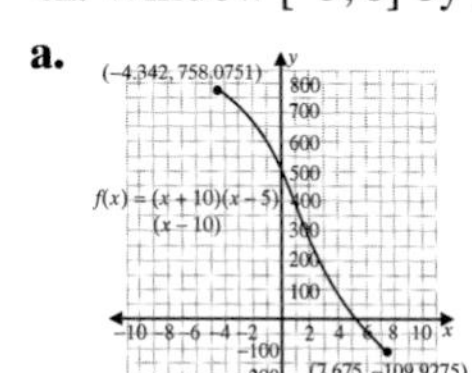

b. $f'(2) = -108$ **43.**

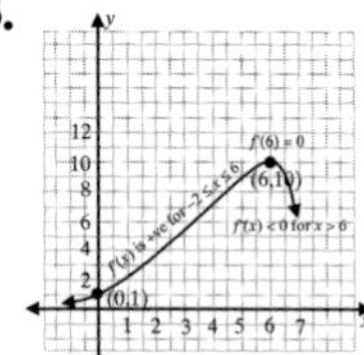

45. a. $F(x) = \frac{9}{5}x + 32$ **b.** $m = \frac{9}{5}$; $F'(x) = \frac{9}{5}$

47. a. $f(20) = 4.25$; it represents the average number of children per woman in 1990. **b.** $f'(20) = -0.11$; it represents a decrease in the average number of children per woman in the year 1990.

49. a. $f(30) = 16.57$; it represents the yearly birth rate per 1000 people in 1994. **b.** $f'(30) = -0.641$; it represents a decrease in the yearly birth rate per 1000 people in the year 1994.

Exercises 2.3, pages 156 - 160

1. 0 **3.** 7 **5.** $8x$ **7.** $-\frac{7}{x^2}$ **9.** $-\frac{3}{2x^4}$ **11.** $\frac{3}{2\sqrt{x}}$ **13.** $2.3t^{1.3}$ **15.** $2.4x^{-0.2}$ **17.** $-\frac{1}{2}u^{-\frac{3}{2}}$ **19.** $-\frac{15}{4}x^{-\frac{1}{4}}$ **21.** $3x^2-7$ **23.** $0.6x-4$

25. $3x^2-12x+5$ **27.** $3x^{\frac{1}{2}}+2x^{-\frac{1}{2}}$ **29.** $-t^{-\frac{3}{2}}+\frac{1}{2}t^{-\frac{1}{2}}+1$ **31.** $y=2x^2-x-3,\ y'=4x-1$

33. $f(v)=v^{\frac{7}{2}}+2v^{\frac{5}{2}}-v^{\frac{3}{2}},\ f'(v)=\frac{7}{2}v^{\frac{5}{2}}+5v^{\frac{3}{2}}-\frac{3}{2}v^{\frac{1}{2}}$ **35.** $f(x)=x^2+5x,\ f'(x)=2x+5$ **37.** $g(t)=\sqrt{t}-\frac{2}{\sqrt{t}},\ g'(t)=\frac{1}{2}t^{-\frac{1}{2}}+t^{-\frac{3}{2}}$

39. $g(x)=4\sqrt{x}+5-\frac{1}{\sqrt{x}},\ g'(t)=2x^{-\frac{1}{2}}+\frac{1}{2}x^{-\frac{3}{2}}$

41. a. $m = 5$
b. $y = 5x - 2$
c. $f(x)=x^3+2x-4$

43. a. $m = 4, m = 0, m = -2$
b. $y = 4x + 4, y = -4, y = -2x - 11$
c.

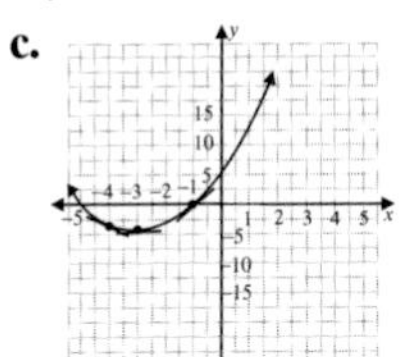

45. a. $m = 4, m = 0, m = -2$
b. $y = 4x + 12, y = 8, y = -2x + 9$

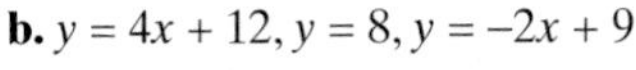

c.

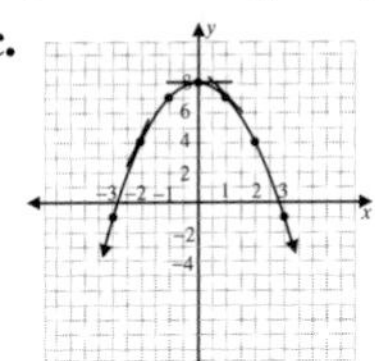

47. a. 192 ft./sec **b.** 64 ft./sec **c.** 6 sec **49. a.** 482 people/year **b.** 488 people/year **51. a.** 87.35 ft. **b.** 34.94 ft./sec **c.** –28 ft./sec **53. a.** 2 bacteria units **b.** 6.82 bacteria units **c.** 2.41 bacteria/hr. **d.** 2.60 bacteria/hr. **55. a.** 15 days **b.** –2.4, the number of students who have not heard the rumor is decreasing by 2.4 students per day. **c.** The number of students who have not heard the new rumor is always decreasing. **57. a.** 18.73. 18.73 is births per 1000 people in the year 1994. **b.** –0.277. The number of births per 1000 people is decreasing by 0.277 in 1994. **c.** 2005 **59. a.** 2.65 billion **b.** 0.04 billion per year. **c.** 8.02 billion **61. a.** $f(10) = 3.62$. This is the average number of births per woman in Thailand from 1970 to 1980. **b.** $f'(10) = -0.16$. The number of births per woman in Thailand was decreasing by 0.16 births per woman in the year 1980.

Exercises 2.4, pages 167 - 172

1. $\overline{C}(x)=\frac{220}{x}+0.4$ **3.** $D(x) = 3;\ R(x) = 3x$ **5.** $C(x) = 1.15x + 50$ **7.** $P(x) = 0.75x + 25$ **9. a.** $\overline{C}(x)=\frac{100}{x}+150$ **b.** $D(x) = 450$ **c.** $R(x) = 450x$ **d.** $P(x) = 300x - 100$ **11. a.** $C(10) = 2705, C(20) = 3060, C(30) = 3465$ **b.** $C' = 28 + 0.5x$ **c.** $C'(10)=33, C'(20)=38, C'(30)=43$ **d.** $\overline{C}(x)=\frac{2400}{x}+28+0.25x;\ \overline{C}'(x)=-2400x^{-2}+0.25$ **e.** $\overline{C}'(10)=-23.75;\ \overline{C}'(20)=-5.75;\ \overline{C}'(30)=-2.42$ **13. a.** $C(4) = 92, C(6) = 102, C(9) = 109.5$ **b.** $C'(x)=10-x$ **c.** $C'(4) = 6, C'(6) = 4, C'(9) = 1$ **d.** $\overline{C}(x)=\frac{60}{x}+10-0.5x$ **e.** $\overline{C}'(4)=-4.25;\ \overline{C}'(6)=-2.17;\ \overline{C}'(9)=-1.24$
15. a. $P(x)=-0.03x^2+60x-10{,}800$ **b.** $P(200) = \$0, P(400) = \$8400, P(600) = \$14{,}400$ **c.** $P'(x) = -0.06x + 60$ **d.** $P'(200) = 48, P'(400)=36, P'(600)=24$ **e.** $x=200, x=1800$ **17. a.** $P(x)=-0.2x^2+75x-2070$ **b.** $P(60)=\$1710, P(80)=\$2650, P(100)=\$3430$ **c.** $P'(x) = -0.4x + 75$ **d.** $P'(60) = 51, P'(80) = 43, P'(100) = 35$ **e.** $x = 30, x = 345$ **19. a.** $R(x)=46x+0.25x^2$ **b.** $P(x)=0.1x^2+40x-190$ **c.** $P(25) = \$872.50, P(30) = \$1100, P(40) = \$1570$ **d.** $P'(x) = 0.2x + 40$ **e.** $P'(25) = 45, P'(30) = 46, P'(40) = 48$ **21. a.** $R(x)=5.5x-0.0004x^2$ **b.** $P(x)=-0.0006x^2+4.5x-4650$ **c.** $P(3000) = 3450, P(3500) = 3750, P(4000) = 3750$ **d.** $P'(x) = -0.0012x + 4.5$ **e.** $P'(3000) = 0.9, P'(3500) = 0.3, P'(4000) = -0.3$ **23. a.** $C'(x)=-0.6-0.03x^{-0.4}$ **b.** $S'(x)=1.6+0.03x^{-0.4}$ **25. a.** $\overline{C}(x)=\frac{30}{x}+2+0.003x$ **b.** $\overline{C}'(x) = -\frac{30}{x^2}+0.003$ **c.** $x = 100$ **27.** Agree; Answers will vary.

29. a. $f(0) = 12{,}519.3$. This year, the value of the car is \$12,519.30. **b.** $f'(x) = -1391.10$. As the age of the car increases, the value of the car decreases at the rate of \$1391.10/year. **31. a.** $F'(t) = \frac{-8.4t^3}{10^6} + \frac{4.8t^2}{10^4} - 0.00914t + 0.0994$ **b.** $F(30) = 7.518$ billion **c.** $F'(30) = 0.0304$ billion. In the year 2030, the world population is increasing at the rate of 0.0304 billion/year. **33. a.** $H'(t) = \frac{-10.4t^3}{10^6} + \frac{5.61t^2}{10^4} - 0.00792t + 0.116$ **b.** $H(30) = 8.909$ billion **c.** $H'(30) = 0.1025$ billion. In the year 2030, the world population is increasing at the rate of 0.1025 billion/year.

Exercises 2.5, pages 187 - 194

1. 6 **3.** 10 **5.** $+\infty$ **7.** $\frac{6}{7}$ **9.** $\frac{1}{2}$ **11.** 0 **13.** $-\infty$ **15.** $\frac{3}{4}$ **17.** $-\infty$ **19.** 8 **21.** 3 **23.** -2 **25. a.** $\frac{1}{4}$ **b.** 6 **27.** 1 **29. a.** -3 **b.** -3 **c.** -3 **d.** -2 **31. a.** 3 **b.** 3 **c.** 3 **d.** 5 **33. a.** 2 **b.** 2 **c.** Does not exist. **d.** $f(x)$ is not continuous at $x = 3$ because it has a missing point discontinuity ($f(3)$ is undefined). **35. a.** -2 **b.** -2 **c.** -2 **d.** Yes, because the limit exists and the function exists at $x = 0$. i.e., $\lim_{x\to 0} f(x) = f(0)$ **37. a.** -1 **b.** $+\infty$ **c.** -1 **d.** $f(x)$ is not continuous at $x = 4$, as the limit doesn't exist at that point.
39. a. No, $f(x)$ is discontinuous at $x = -1$, because $\lim_{x\to -1} f(x) \neq f(-1)$. **b.** $f(x)$ is continuous at $x = 3$, as $\lim_{x\to 3} f(x) = f(3) = 7$.
41. At $x = -3$, $f(x)$ is discontinuous; it has a missing point discontinuity. **43.** At $x = 0$, $f(x)$ is discontinuous; it has a jump discontinuity. **45.** $f(x)$ is continuous at $x = 1$. **47.** $f(x)$ is discontinuous at $x = 2$. **49.** $f(x)$ is continuous at $x = 1$. **51.** $f(x)$ is discontinuous at $x = 0$. **53.** No points of discontinuity. **55.** $f(x)$ is discontinuous at $x = -3$; it has an infinite discontinuity.
57. $f(x)$ is discontinuous at $x = 3$; it has an infinite discontinuity. $f(x)$ is discontinuous at $x = -3$; it has an infinite discontinuity.
59. No points of discontinuity. **61.** $k = 2$ **63.** $k = 4$ **65. a.** \$11.88 **b.** \$11.88 **c.** \$11.88 **67.** \$73

69. a.

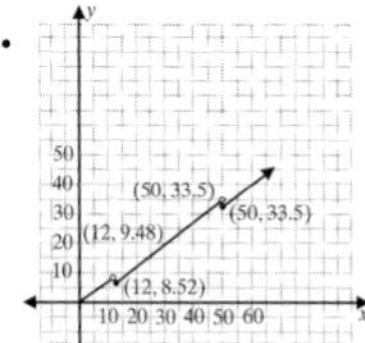

b. C is discontinuous at $x = 12$ and at $x = 50$. These are jump discontinuities.

71. a.

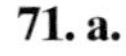

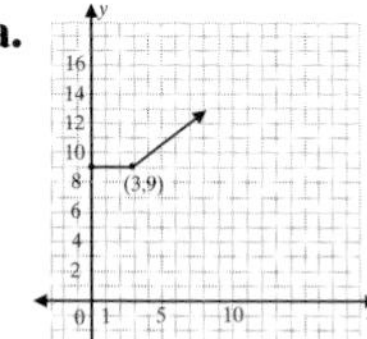

b. $C(x)$ is continuous for all $x > 0$.

Chapter 2 Test, pages 202 - 205

1. 0

x	y
2.5	−0.5
2.9	−0.1
2.99	−0.01
2.999	−0.001

2. 0

x	y
3.5	0.93
3.1	0.20
3.01	0.02
3.001	0.00

3. 27

x	y
3.5	31.75
3.1	27.91
3.01	27.09
3.001	27.01

4. $+\infty$

x	y
2.5	58.75
2.9	206.11
2.99	1826.91
2.999	18,027.99

5. $f'(x) = -30 + 32x$ **6.** $f'(x) = \frac{16x^2 - 2}{x^2}$ **7.** $f'(x) = 8x^{-\frac{1}{3}} - 15x^{-\frac{1}{2}} - 5x^{-6}$ **8.** $f'(x) = 64x^3 - 90x^2 + 4x$ **9.** $R(60) = 75$

10. Window used is [0, 100] by [0, 80]. **11.** $x = 30$

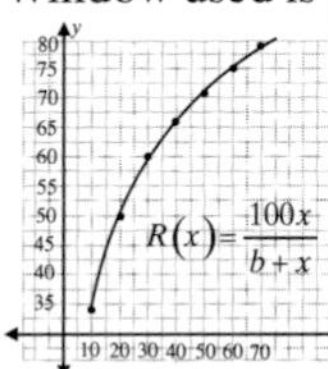

12. $R'(20)$ is the instantaneous rate of change in muscle force when the concentration of acetylcholine is 20.

13.

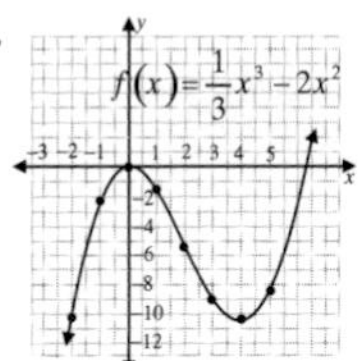

14. –8.33 **15.** 5

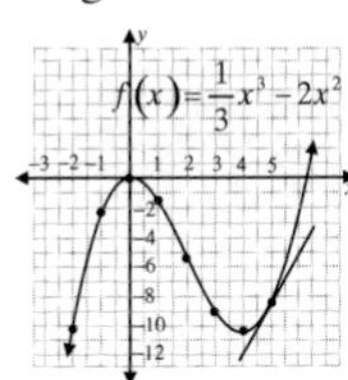

16. $y = -10.66$

17. $C(0) = 3000$. The fixed costs are \$3000. **18.** $C'(1000) = 1.20$. The rate of change in cost at $x = 1000$ is an increase of \$1.20 per additional mug made. **19.** $y = 1.20x + 2500$ **20.** $\frac{2+2h-\sqrt{4+h}}{h}$ **21.** 1.75 **22.** $2-\frac{1}{2\sqrt{x}}$ **23.** 1.75 **24.** $3x^2-3$

25. $y = 9x - 14$ **26.** $y = 4$ at $x = -1$ **27.** 4 meters **28** –8 meters/second **29.** $R(x) = 135x$ **30.** $P(x) = 131x - 8000 - 0.05x^2$

31. 2 **32.** 2 **33.** 3 **34. a.** 22,260,203 **b.** $P'(70) = 377.20$. The number of African Americans is increasing by 377.20 thousand per year in 1970.

Chapter 3

Exercises 3.1, pages 215 - 217

1. $f'(x) = -8x^3 + 9x^2 + 2x$ **3.** $f'(x) = \frac{15x^2+1}{2\sqrt{x}}$ **5.** $f'(x) = 8x$ **7.** $f'(x) = -\frac{1}{x^{\frac{3}{2}}}$ **9.** $f'(x) = 6 - 10x^4$ **11.** $5x^2(x^2+3)$

13. $\frac{3}{2}\left(\frac{4t+1}{\sqrt{t}}\right)$ **15.** $\frac{\sqrt{x}(5x+3)}{2}$ **17.** $-18u^2 + 20u - 9$ **19.** $\frac{1}{5t^2}(50t^3 + 5t^2 - 1)$ **21.** $18(x+6)^{-2}$ **23.** $-15(x-7)^{-2}$

25. $\frac{x^4+3x^2+10x}{(x^2+1)^2}$ **27.** $\frac{9-x}{2\sqrt{x}(x+9)^2}$ **29.** $\frac{3u^{\frac{3}{2}}+4u}{2(\sqrt{u}+1)^2}$ **31.** $\frac{3t^2-16t^{\frac{3}{2}}-3}{2\sqrt{t}(t^2+3)^2}$ **33.** $\frac{6x-15+10x^{\frac{4}{3}}-20x^{\frac{1}{3}}}{3(1+2\sqrt[3]{x})^2}$ **35.** $h'(2) = 1$

37. $h'(2) = \frac{10}{3}$ **39.** $h'(2) = 9$ **41.** $h'(2) = \frac{1}{3}$ **43.** $h'(2) = -\frac{7}{121}$ **45.** $y = 616.5x - 1766$ **47.** $y = -\frac{19}{25}x + \frac{2}{5}$ **49.** $y = -\frac{17}{20}x + \frac{37}{20}$

51. $(x, y) = (-2, 36)$ and $\left(\frac{8}{3}, -\frac{400}{27}\right)$ **53.** $(x, y) = (10, 100)$

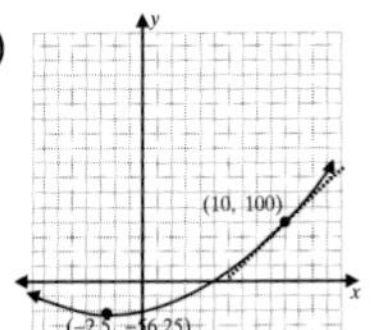

55. (–1.58, –4.74), (1.58, 4.74)

(1.58, 4.74)
(–1.58, –4.74)

57. \$1.15/item **59.** \$13.20/item **61.** 6100 people/year

Exercises 3.2, pages 226 - 229

1. $8(2x-5)^3$ **3.** $-12(1-4x)^2$ **5.** $-4x(x^2+4)^{-3}$ **7.** $-3(4t+3)(2t^2+3t)^{-4}$ **9.** $16x^3+60x^2-6x-70$ **11.** $\frac{3x^2}{2\sqrt{x^3+1}}$

13. $\frac{8x}{3(4x^2+1)^{\frac{2}{3}}}$ **15.** $-\frac{3x^2}{2(1-2x^3)^{\frac{3}{4}}}$ **17.** $5(13t^3+3)(t^3+3)^3$ **19.** $6x^2(x^2-8)^2(3x^2-8)$ **21.** $-\frac{x}{(x^2+6)^{\frac{3}{2}}}$ **23.** $\frac{8}{(t^2+8)^{\frac{3}{2}}}$

25. $-\frac{(10x+18)}{3x^3(2x+3)^{\frac{2}{3}}}$ **27.** $\frac{18x-13}{2(3x-4)^{\frac{1}{2}}}$ **29.** $\frac{26(3x-2)}{(5x+1)^3}$ **31.** $\frac{5t-17}{3(t-1)^{\frac{5}{3}}}$ **33.** $-9(2t+5)^2(t+1)^{-4}$ **35.** $\frac{5}{(4-2x)^{\frac{3}{2}}(x+3)^{\frac{1}{2}}}$

37. $-\frac{7}{2(3x-1)^{\frac{3}{2}}(x+2)^{\frac{1}{2}}}$ **39.** $(7-2x)^{-\frac{3}{2}}(-3x^2+13x+7)$ **41.** -48 **43.** $-\frac{3}{4}$ **45.** $\frac{135}{2}$ **47.** $-\frac{7}{8}$ **49.** 295 **51.** 64 **53.** 8

55. 64 **57.** $y = 36x - 28$ **59.** $y=\frac{13}{8}x+\frac{3}{2}$ **61. a.** $R(x)=20x\sqrt{280-4x}$ **b.** $5880 **c.** $40\sqrt{70-x}-\frac{20x}{\sqrt{70-x}}$ **d.** $220 per oven

63. a. 18,560 people **b.** 640 people/year **65.** 218 units per day **67. a.** $140.83/hour **b.** $109.48/hour
69. a. 2.5 ppm/thousand people **b.** 30 ppm/thousand people **71. a.** 1789 people **b.** $N'(5) = .05$ **c.** The rate of attendance is increasing at a rate of 50 people per week. **73. a.** $P(0) = 2.0$ million students; represents the number of students registered in calculus in 1995. **b.** $P'(t)=-\frac{1}{12}\left(1-\frac{t}{30}\right)^{1.5}$ **c.** $P'(10) = -0.045$. The rate of registration will be decreasing by 0.045 million students/year in 2005.

Exercises 3.3, pages 240 - 244

1. $\frac{dy}{dx}=-\frac{2x}{y}$ **3.** $\frac{dy}{dx}=-\frac{2x^2}{y^2}$ **5.** $\frac{dy}{dx}=-\frac{\sqrt{y}}{\sqrt{x}}$ **7.** $\frac{dy}{dx}=-\frac{2y}{x}$ **9.** $\frac{dy}{dx}=-\frac{(2x+y)}{(x+2y)}$ **11.** $\frac{dy}{dx}=\frac{2-8x-3y}{3x+2y}$

13. $\frac{dy}{dx}=\frac{x^2y-2x^2y^2-y^2}{x^3}$ **15.** $\frac{dy}{dx}=\frac{4x^{\frac{3}{2}}y^{\frac{1}{2}}+y}{8x^{\frac{1}{2}}y^{\frac{3}{2}}-x}$ **17.** $\frac{dy}{dx}=\frac{4x^3-2xy^2-y^3}{2x^2y+3xy^2}$ **19.** $\frac{dy}{dx}=\frac{x}{2-y}$ **21.** $\frac{dy}{dx}=\frac{x}{3y^2}$; $m=\frac{2}{3}$

23. $\frac{dy}{dx}=\frac{4+2xy}{2y-x^2}$; $m=\frac{12}{7}$ **25.** $\frac{dy}{dx}=\frac{3x^2+2y}{2(y-x)}$; $m=-\frac{7}{4}$ **27.** $\frac{dy}{dx}=\frac{x^2y-y^3-2x^2y^3}{2x^3}$; $m=1$

29. $\frac{dy}{dx}=\frac{20x^{\frac{3}{2}}y^{\frac{1}{2}}-y}{4x^{\frac{1}{2}}y^{\frac{1}{2}}+x}$; $m=\frac{76}{9}$ **31.** $\frac{dy}{dt}=\frac{x}{4y}\frac{dx}{dt}$ **33.** $\frac{dy}{dt}=\frac{2-3x^2}{10y}\frac{dx}{dt}$ **35.** $\frac{dy}{dt}=-\frac{y}{x}\frac{dx}{dt}$

37. $\frac{dy}{dt}=-\frac{(2x+y)}{x+2y}\frac{dx}{dt}$ **39.** $\frac{dy}{dt}=\frac{2(x+y)}{3y^2-2x}\frac{dx}{dt}$ **41.** $\frac{5}{4}$ speakers per receiver; the number of speakers sold in a week increases by 1.25 for each additional reciever sold. **43.** $-\frac{27}{64}$ units of capital/week **45.** $312/week. **47.** $-\frac{45}{8}$ ft./sec

49. 35.4 ft./sec **51. a.** 4.5 seconds **b.** $\frac{20x}{y}$ **c.** $10\sqrt{2}$ ft./sec **53. a.** $p = \$2$ **b.** $0.38/day

55. a. 800 bears **b.** $\frac{dx}{dp}=-\frac{x}{p-1}$; 228.571 **57. a.** (504, 672) **b.** $\frac{dy}{dx}=-\frac{x}{y}$ **c.** $\left.\frac{dy}{dx}\right|_{t=t_0}=-0.75$; at this point in time, for every 75 ft. the car travels vertically, it also travels 100 ft. to the negative horizontal. **d.** $\left.\frac{dy}{dt}\right|_{t=t_0}=37.5$ mph **e.** 60 mph
f. 1.2 minutes **59. a.** $S = 12$ **b.** $\frac{dp}{dn}=-\frac{p-p^2}{n(1-2p)}$; $-.00049$

Exercises 3.4, pages 261 - 267

1. a. $(2, +\infty)$ **b.** $(-\infty, 2)$ **3. a.** $(-\infty, 2)$ **b.** $(2, +\infty)$ **5. a.** $(-1, 2)$ **b.** $(-\infty, -1), (2, +\infty)$ **7. a.** $(-2, -1), (2, +\infty)$ **b.** $(-\infty, -2), (-1, 2)$
9. a. $(-\infty, -2), (0, 6)$ **b.** $(-2, 0), (6, +\infty)$

11. a. Inc.: $(0, +\infty)$;
Dec.: $(-\infty, 0)$

b.

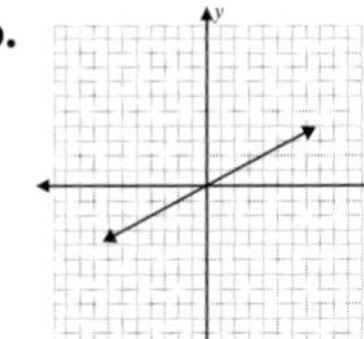

13. a. Inc.: $(-2, 2)$;
Dec.: $(-\infty, -2), (2, +\infty)$

b.

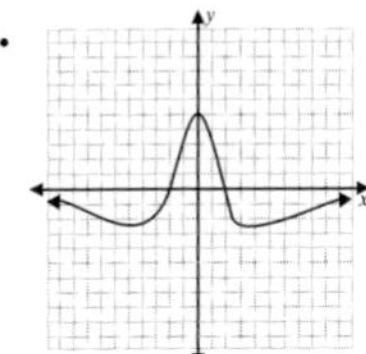

15. a. $x = -3$
b. Inc.: $(-3, +\infty)$; Dec.: $(-\infty, -3)$

c.

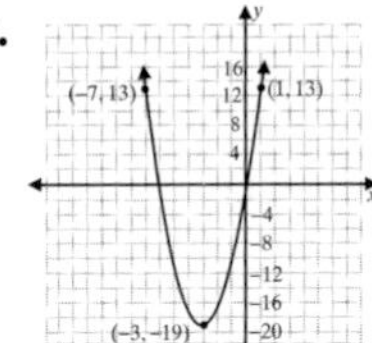

17. a. $x = \frac{7}{4}$

b. Inc.: $\left(-\infty, \frac{7}{4}\right)$;

Dec.: $\left(\frac{7}{4}, +\infty\right)$

c.

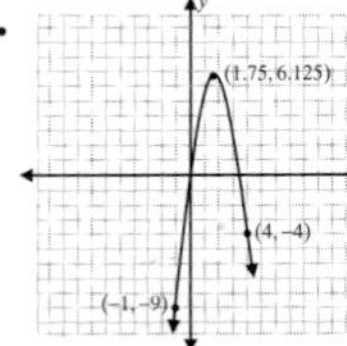

19. a. $x = \frac{2}{3}$

b. Inc.: $\left(\frac{2}{3}, +\infty\right)$;

Dec.: $\left(-\infty, \frac{2}{3}\right)$

c.

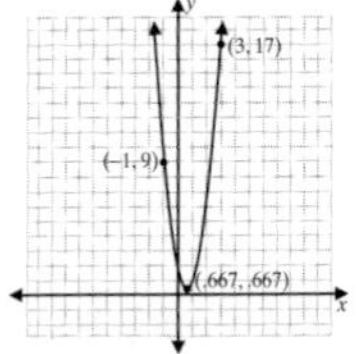

21. a. $x = \frac{2}{3}$

b. Inc.: $\left(\frac{2}{3}, +\infty\right)$;

Dec.: $\left(-\infty, \frac{2}{3}\right)$

c.

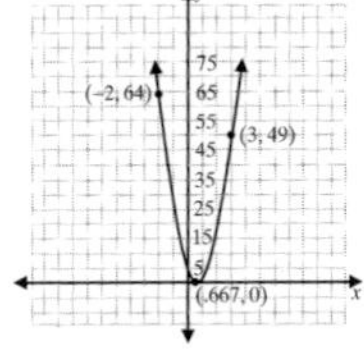

23. a. $x = 0$

b. Inc.: $(-\infty, +\infty)$;
Dec.: $\varnothing$

c.

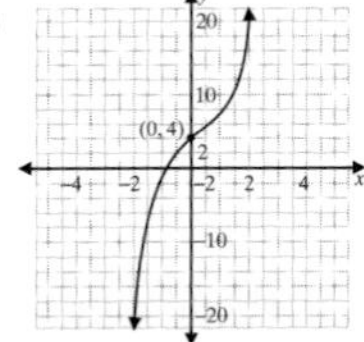

25. a. $x = 0, x = 4$

b. Inc.: $(-\infty, 0), (4, +\infty)$;

c. Dec.: $(0, 4)$

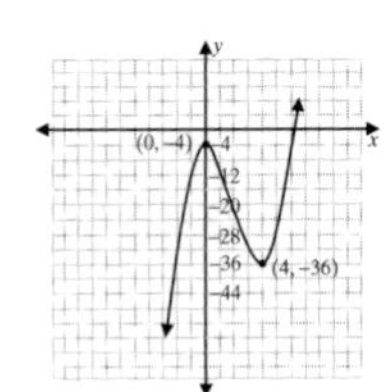

27. a. $x = -\frac{1}{3}, x = 1$

b. Inc.: $\left(-\infty, -\frac{1}{3}\right)$, $(1, +\infty)$;

c. Dec.: $\left(-\frac{1}{3}, 1\right)$

29. a. $x = -1, x = \frac{5}{3}$

b. Inc.: $(-\infty, -1)$, $\left(\frac{5}{3}, +\infty\right)$;

c. Dec.: $\left(-1, \frac{5}{3}\right)$

31. a. $x = 1, x = 3$

b. Inc.: $(-\infty, 1), (3, +\infty)$;

c. Dec.: $(1, 3)$

33. a. $x = -1, x = -4$
b. Inc.: $(-\infty, -4), (-1, +\infty)$;
Dec.: $(-4, -1)$

c.

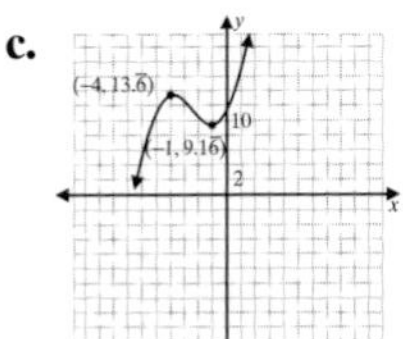

35. a. None **b.** Inc.: $\varnothing$; Dec.: $(-\infty, 2), (2, +\infty)$ **37. a.** None **b.** Inc.: $(-\infty, 0), (0, +\infty)$; Dec.: $\varnothing$
39. a. $x = -3$ **b.** Inc.: $(-\infty, -3), (0, +\infty)$; Dec.: $(-3, 0)$ **41. a.** None **b.** Inc.: $(-\infty, -3), (-3, +\infty)$; Dec.: $\varnothing$

43. a. $x = 4$ **b.** Inc.: $(4, +\infty)$; Dec.: $(-\infty, 0), (0, 4)$ **45.** $x = \frac{9}{2}$ **b.** local max: $f\left(\frac{9}{2}\right) = \frac{81}{4}$

47. a. $x = 5$ **b.** local min: $f(5) = -13$ **49. a.** $x = 3$ **b.** local max: $f(3) = \frac{5}{2}$

51. a. $x = -1, \; x = -\frac{1}{3}$ **b.** local max: $f(-1) = -2$; local min: $f\left(-\frac{1}{3}\right) = -\frac{58}{27}$

53. a. $x = -1, \; x = \frac{2}{3}$ **b.** local max: $f(-1) = 3$; local min: $f\left(\frac{2}{3}\right) = -\frac{44}{27}$ **55. a.** $x = 0, x = -2$ **b.** local max: $f(-2) = 0$; local min: $f(0) = -4$
57. a. No critical values. **b.** No local extrema. **59. a.** No critical values. **b.** No local extrema.

61. a. $x=-\frac{1}{5},\ x=\frac{1}{5}$ **b.** local max: $f\left(-\frac{1}{5}\right)=-10$; local min: $f\left(\frac{1}{5}\right)=10$ **63. a.** $x=-\frac{1}{3}$ **b.** local max: $f\left(-\frac{1}{3}\right)=-27$
65. Inc.: (0, 8000) Dec.: (8000, 10,32.5)Dec.: (62.5, 100) **69.** 4 years **71.** $\overline{C}(x)=\frac{250}{x}+45-0.2x\ \overline{C}'(x)=-0.2-\frac{250}{x^2}<0$

73. $W'(t)=\frac{1400t}{(t^2+20)^2}>0$ so $W(t)$ is increasing; $\lim\limits_{t\to+\infty} w(t)=75$ **75.** $F'(t)=\frac{64{,}000t}{(t^2+100)^2}>0$, so $F(t)$ is increasing;

$\lim\limits_{t\to+\infty} F(t)=350$ **77.** $F(33.8) = 54.4$ is a local max; $F(65.4) = 45.9$ is a local min. **79.** $F'(T)=\frac{1080T}{(T^2+180)^2}>0$, so $F(T)$ is

increasing; $\lim\limits_{T\to\infty} F(T)=4$ **81.** $F(30) = 20$ relative max.; $F(83.4) = 4.6$ relative min. The computer does not crash.

Exercises 3.5, pages 276 - 282

1. Abs. min.: (5, 3); Abs. max.: (3, 6) **3.** Abs. min.: (3, 2); Abs. max.: (–5, 8) **5.** Abs. min.: (4, 4); Abs. max.: (2, 10)
7. Abs. min.: (5, –1); Abs. max.: (3, 6) **9.** $f(4) = -16$ min; $f(0) = 0$ max **11.** $f(3) = 27$ min; $f(5) = 31$ max
13. $f(4) = 2$ min; $f(0) = 14$ max **15.** $f(3) = -19$ min; $f(-1) = 9$ max **17.** $f(2) = -16$ min; $f(-2) = f(4) = 16$ max
19. $f\left(\frac{1}{3}\right)=-\frac{1}{27}$ min; $f(2)=12$ max **21.** $f(-3) = -27$ min; $f(1) = 5$ max **23.** $f(2) = -30$ min; $f(4) = 22$ max
25. $f(0) = -4$ min; $f(8) = 0$ max **27.** $f(1)=-1$ min; $f\left(\frac{1}{8}\right)=1$ max **29.** $f(0)=2$ min; $f(2)=2\sqrt{2}$ max

31. $f(0)=-1$ min; $f(-2)=f(2)=\sqrt[3]{3}$ max **33.** $f(2)=4$ min; $f\left(\frac{1}{2}\right)=8.5$ max **35.** $f(1.5)=12$ min; $f(3)=15$ max
37. $f(2)=12$ min; $f\left(\frac{1}{2}\right)=32.25$ max **39.** 24 units **41.** 40 telescopes **43.** 120 refrigerators **45.** 1:00 P.M.
47. 1600 bacteria at time $t = 6$. **49.** $F\left(\frac{65}{3}\right)=7041.\overline{6}$ ft **51.** $t = 34.32$; 41.77%; yes **53.** 500 Newtons, $v = 2500$
55. $F(0.8)=666.\overline{6}$; yes **57.** max at (0, 75); min at (5, 10) **59.** at $t = 25.11$, 0.048 moles/sec., at $t = 62.01$; 0.025 moles/sec.
61. at $t = 20$ nanosec.; $E(20) = 15$. **63.** 31st day; 3.298 million.

Chapter 3 Test, pages 287 - 289

1. $60x^2+48x+25$ **2.** $-4x^3+12x^2-7$ **3.** $\frac{1}{2}x^{-\frac{1}{2}}(25x^2-6x+3)$ **4.** $\frac{1}{2}x^{\frac{1}{2}}(42x^2+3)$ **5.** $\frac{2x^3+15x^2+14}{(x+5)^2}$ **6.** $\frac{2x^2-6x+3}{(2x-3)^2}$

7. $-\frac{x+4}{2\sqrt{x}(x-4)^2}$ **8.** $-\frac{x+30}{2x^{\frac{5}{2}}}$ **9.** $\frac{21}{(4-7t)^4}$ **10.** $-\frac{5x}{(5x^2-6)^{\frac{3}{2}}}$ **11.** $\frac{x(5x-18)}{\sqrt{2x-9}}$ **12.** $-\frac{55x+24}{3x^5(5x+2)^{\frac{2}{3}}}$ **13.** $-\frac{68(2x+7)}{(4x-3)^3}$

14. $\frac{12x^2+x-18}{\sqrt{x^2-3}}$ **15.** $-\frac{2x-25}{2(x-7)^{\frac{3}{2}}}$ **16.** $\frac{2(4x^2-9x+72)}{3(x^2+6)^{\frac{4}{3}}}$ **17.** $-\frac{8(4x-1)^3(2x^2-x-4)}{(x^2+2)^5}$ **18.** $\frac{39}{2(5x+2)^{\frac{3}{2}}\sqrt{2x-7}}$

19. $(12x^4+36x^2+29)(4x)$; 308 **20.** $f'(x)=\frac{dy}{dx}=8x-9$ for $x\geq 1$; 15 **21.** $\frac{4x+1}{\sqrt{4x^2+2x-3}}; -\frac{7}{3}$ **22.** $\frac{6x\left(5-12\sqrt{2x^2-1}\right)}{\sqrt{2x^2-1}}$; 42

23. $-\frac{2x+3y}{3x+4y}; -\frac{2}{3}$ **24.** $\frac{5-6x^2}{3y^2}; -\frac{1}{12}$ **25.** $\frac{6\sqrt{y}-2y-4x\sqrt{y}}{x-2\sqrt{y}}; -8$ **26.** $\frac{y}{x}; \frac{3}{2}$ **27.** –0.32 dollars per week

28. a. 3.43 million dollars **b.** \$0.2016/yr. **29. a.** (–5, 2), (4, 7) **b.** (–7, –5), (2, 4) **30. a.** (–2, 1) **b.** (–∞, –2), (1, +∞)

31. a. $x=\frac{3}{2}$ **b.** Inc.: $\left(\frac{3}{2},+\infty\right)$; Dec.: $\left(-\infty,\frac{3}{2}\right)$ **32. a.** $x = -2, x = 2$ **b.** Inc.: (–∞, –2), (2, +∞); Dec.:(–2, 2)

33. a. $x = -1, x = 0, x = 1$ **b.** Inc.: (–1, 0), (1, +∞); Dec.: (–∞, –1), (0, 1) **34. a.** $x = -6$ **b.** Inc.: (–6, 0); Dec.: (–∞, –6), (0, +∞)

35. a. $x = 3$ **b.** Local max: $f(3) = 13$ Local min: ∅ **36. a.** $x=-\frac{5}{3}, x=1$ **b.** Local max: $f\left(-\frac{5}{3}\right)=\frac{229}{27}$; Local min: $f(1) = -1$

37. a. $x=0, x=-2, x=\frac{5}{4}$ **b.** Local max: $f(0)=3$; Local min: $f(-2)=-9$ and $f\left(\frac{5}{4}\right)=-\frac{107}{256}\approx -0.42$ **38. a.** $x=-\frac{5}{2},\ x=\frac{5}{2}$

b. Local max: $f\left(-\frac{5}{2}\right)=-20$; Local min: $f\left(\frac{5}{2}\right)=20$ **39. a.** Inc.: (–∞, 0) Dec.: (0, +∞) **b.**

40. Max: $f(2)=9$; Min: $f\left(\frac{3}{4}\right)=\frac{47}{8}$ **41.** Max: $f(1) = 5$; Min: $f(4) = -40$

42. Max: $f(2)=\sqrt{7}$; Min: $f(0)=\sqrt{3}$ **43.** Max: $f(-2) = -12$; Min: $f(-0.1) = -120.3$

44. $0 < x < 3000$ **45.** 250 shirts **46.** 4600 people

Exercises 3 Cumulative Review pages 289 - 293

1. $f'(x)=3-24x^3$ **2.** $f'(x)=-3x^{-\frac{1}{2}}+4x^{-\frac{2}{3}}$ **3.** $f'(x)=12x^2+\frac{1}{2\sqrt{x}}$ **4.** $f'(x)=3-\frac{3}{2}x^{-\frac{3}{2}}$ **5.** $f'(x)=\frac{1}{\sqrt{3+2x}}$

6. $f'(x)=\frac{17}{2\sqrt{5+17x}}$ **7.** $F'(u)=80(16u+1)^4$ **8.** $F'(u)=3(u^2+u-1)^2(2u+1)$ **9.** $g'(t)=45t^2+22t-28$

10. $g'(t)=31+42t-66t^2$ **11.** $F'(u)=-\frac{2}{(2u-1)^2}$ **12.** $F'(u)=\frac{6u^2-20u+8}{(u^2+u-3)^2}$ **13.** 2 **14.** 1 **15.** 3 **16.** 0 **17.** –3

18. a. It is not a derivative since it is not a rate of change.
b. It is a derivative and the possible function may be $f(x) = 19x$ where x represents number of gallons.
c. It is not a derivative since it is not a rate of change. **d.** It is not a derivative since it is not a rate of change.
e. It is a derivative and the possible function may be $f(x) = 9x$ where x represents number of hours.
f. It is a derivative and the possible function may be $f(x) = 9x$ where x represents number of man-hours.

19. $A_4 < A_3 \le A_5 < A_2 < A_1 < A_6$ **20. a.**

$h = \Delta x$	$\frac{\Delta y}{\Delta x}$
0.1	2.8220
0.01	2.8321
0.001	2.8332
0.0001	2.8333

b. $2.8\overline{3}$

21. a. In the year 2000 the population is 250 million. **b.** Population is increasing by 0.78 million / yr. **c.** 257.8 million
d. 0.78 million / yr.

22. a.

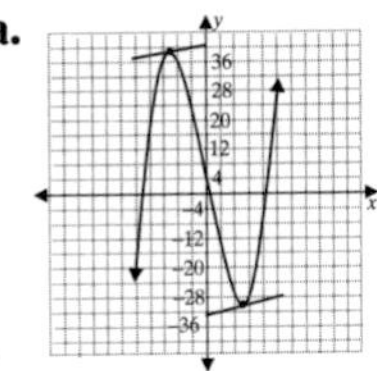

b. $(-2.65, 38.39), (2.65, -30.39)$

23. a. \$1 **b.** $C'(x)=0.01+0.006x$ **c.** $P(x)=-0.003x^2+0.99x-500$ **d.** $A(1000) = 3.51$
e. $C'(408.25)=A(408.25)=2.46$ If $A'(x) = 0$, the marginal cost equals the average cost.

24. a. \$0.40 / unit of production **b.** $\frac{dx}{dy}=40, \frac{dc}{dy}=16$ **c.** $\frac{dc}{dx}=\frac{dc}{dy}\cdot\frac{dy}{dx}$ **25.** $R'(x)=1800$ per item

26. $y = 2x + 14$;

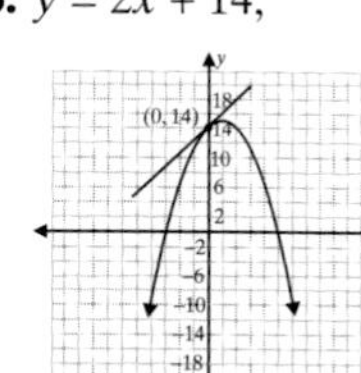

27. $m \approx 1.10$; $y = 1.10x + 1$;

28. $y = -11x - 11$

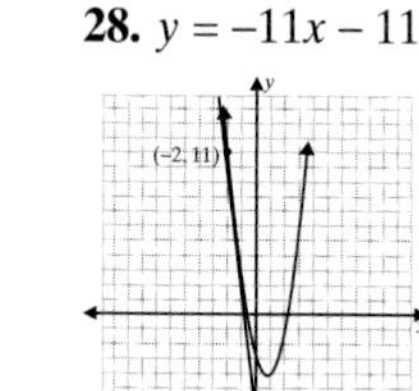

29. $f'(t) = 3 + \frac{450}{t^3}, 136.33$

30. a. In the year 2000 the population of gorillas is 53.

b. In the year 2000 the population is decreasing at the rate of 1 gorilla per year.

31. a.

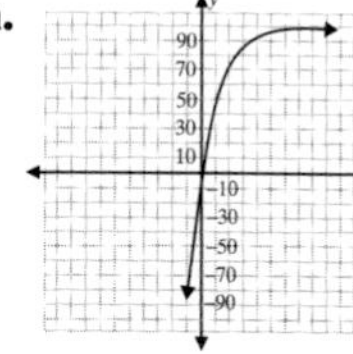

b. When the test is given 2 times, the percentage of students scoring 90% or better is 75%.

c. The percentage of students scoring 90% or better is increasing at the rate of 17.3% when the test has been given twice.

d. Answers will vary.

32. a. Inc.: $(-\infty, -3)$, $(3, +\infty)$; Dec.: $(-3, 3)$

b.

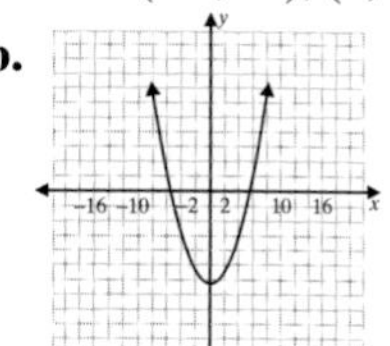

33. a. Inc.: $(0, +\infty)$; Dec.: $\varnothing$

b.

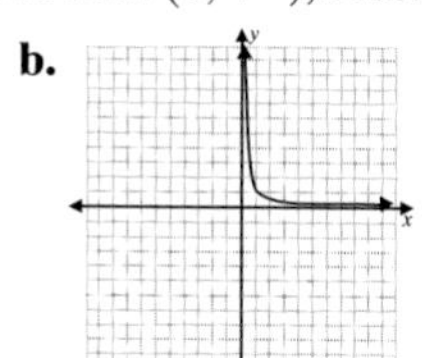

Chapter 4

Exercises 4.1, pages 318 - 323

1. At A: f' is positive; f'' is positive. At B: f' is positive; f'' is zero. At C: f' is zero; f'' is negative. At D: f' is negative; f'' is zero. At E: f' is negative; f'' is positive

3.

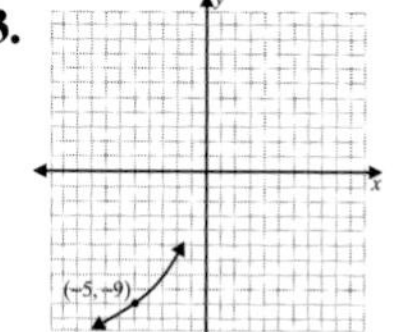
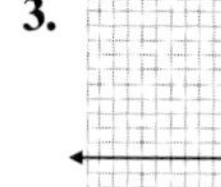

5.

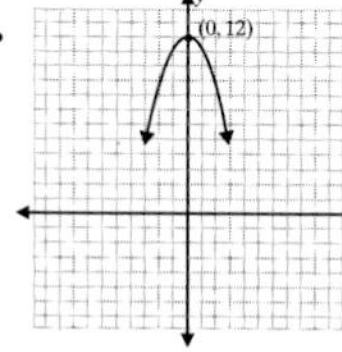

7. $f'(x) = 10x - 9$; $f''(x) = 10$; Max: $\varnothing$; Min: $(0.9, -2.05)$; I: $\varnothing$; Con. up: $\mathbb{R}$ Con. down: $\varnothing$ **9.** $f'(x) = 9x^2 + 6$; $f''(x) = 18x$; Max: $\varnothing$; Min: $\varnothing$; I: $(0, -8)$ Con. up: $(0, +\infty)$; Con. down: $(-\infty, 0)$

11. $f'(x) = 4x^3 - \frac{3}{2\sqrt{x}}$; $f''(x) = 12x^2 + \frac{3}{4}x^{-\frac{3}{2}}$; Max: $\varnothing$; Min: $(0.7556, -0.2818)$; I: $\varnothing$; Con. up: $(0, +\infty)$; Con. down: $\varnothing$

13. $f'(x) = 8x(2x^2 - 5)$; $f''(x) = 48x^2 - 40$; Max: $(0, 25)$; Min: $\left(-\frac{\sqrt{10}}{2}, 0\right), \left(\frac{\sqrt{10}}{2}, 0\right)$; I: $x = \pm\frac{\sqrt{30}}{6}$, $y = \frac{100}{9}$;

Con. up: $\left(-\infty, -\frac{\sqrt{30}}{6}\right), \left(\frac{\sqrt{30}}{6}, +\infty\right)$; Con. down: $\left(-\frac{\sqrt{30}}{6}, \frac{\sqrt{30}}{6}\right)$

15. $f'(x) = \frac{2}{3}x(x^2 + 9)^{-\frac{2}{3}}$; $f''(x) = \frac{2}{3}(x^2 + 9)^{-\frac{2}{3}} - \frac{8}{9}x^2(x^2 + 9)^{-\frac{5}{3}}$ Max: $\varnothing$; Min: $(0, \sqrt[3]{9})$; I: $x = \pm 3\sqrt{3}$; $y = \sqrt[3]{36}$

Con. up: $(-3\sqrt{3}, 3\sqrt{3})$; Con. down: $(-\infty, -3\sqrt{3}), (3\sqrt{3}, +\infty)$

17. $f'(x)=-\dfrac{2\left(x^2+x+4\right)}{\left(x^2-4\right)^2}; f''(x)=\dfrac{2\left(2x^3+3x^2+24x+4\right)}{\left(x^2-4\right)^3}$; Max: ∅; Min: ∅ I: (–0.170, –0.166);
Con. up: (–2, –0.170), (2, +∞); Con. down: (–∞,–2), (–0.170, 2)
19. $f''(x) = 6x - 2$; $f''(0) = -2$; $f''(1) = 4$; $f''(4) = 22$ **21.** $f''(x)=2-\dfrac{1}{2x^{\frac{3}{2}}}$; $f''(0)=$ does not exist; $f''(1)=\dfrac{3}{2}$; $f''(4)=\dfrac{31}{16}$

23. $f''(x)=-\dfrac{1}{(2x+1)^{\frac{3}{2}}}; f''(0)=-1; f''(1)=-\dfrac{1}{3\sqrt{3}}; f''(4)=-\dfrac{1}{27}$

25. $f''(x)=-\dfrac{12}{(x+4)^3}; f''(0)=-\dfrac{3}{16}; f''(1)=-\dfrac{12}{125}; f''(4)=-\dfrac{3}{128}$

27. POI: $(0, 0), (-0.267, -0.69), (0.267, 0.69)$

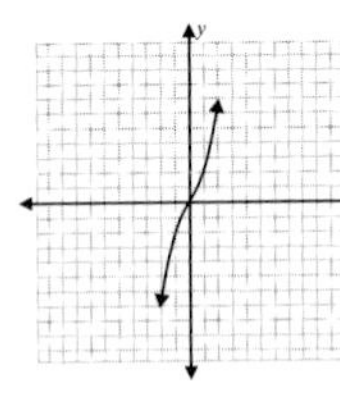

29. POI: (–2.2, 19.17)

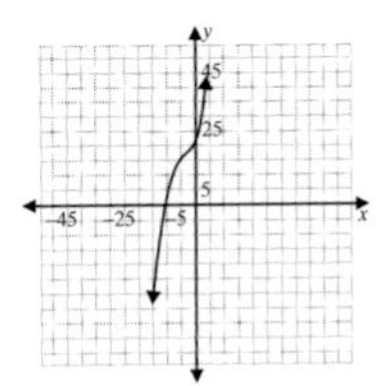

31.

33.

35. a. (–3, –1), (–1, 1) **b.** (–∞, –3), (1, ∞) **c.** (–3, 2), (1, 2)
37. a. (–∞, –3), (1, 5), (5, ∞) **b.** (–3, 1) **c.** (–3, 3), (1, 4)
39. a. (–∞, +∞) **b.** None **c.** None **41. a.** (–2, +∞) **b.** (–∞, –2) **c.** (–2, 6)
43. a. (0, +∞) **b.** (–∞, 0) **c.** (0, 2) **45. a.** (–3, +∞)**b.** (–∞, –3) **c.** (–3, 7)
47. a. $\left(-\infty, \dfrac{3}{5}\right)$ **b.** $\left(\dfrac{3}{5}, \infty\right)$ **c.** $\left(\dfrac{3}{5}, 0\right)$

49. a. $\left(-\sqrt{5}, 0\right), \left(\sqrt{5}, \infty\right)$ **b.** $\left(-\infty, -\sqrt{5}\right), \left(0, \sqrt{5}\right)$ **c.** (0,0) **51.** Local max: $f\left(\dfrac{7}{4}\right)=\dfrac{113}{8}$
53. Local max: $f(-4) = 22$; Local min: $f(0) = -10$ **55.** Local max: $f(-1) = 6$; Local min: $f(1) = 2$
57. Local max: $f(-3) = 8$; Local min: $f(1)=-\dfrac{8}{3}$ **59.** Local max: $f(0) = 3$; Local min: $f(-1) = 2$, $f(1) = 2$
61. Local min: $f\left(\dfrac{9}{2}\right)=-\dfrac{2059}{16}$ **63.** Local max: $f(-3) = -6$; Local min: $f(3) = 6$ **65.** $G(x) = x^4$ **67.** $J(x) = (x-4)^3$ **69.** x = \$6000
71. $C'(8)$ = \$4.80/unit **73. a.** $N(9)$ = 104.6 crimes **b.** $N'(4) = 8$ crimes/month **75. a.** Max where the bottle is skinniest; min where the bottle is widest **b.** – **c.** POI: yes, where the width begins increasing or decreasing

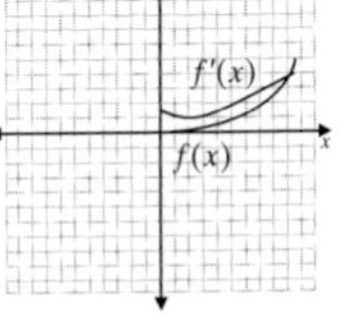

Exercises 4.2, pages 333 - 335

1.

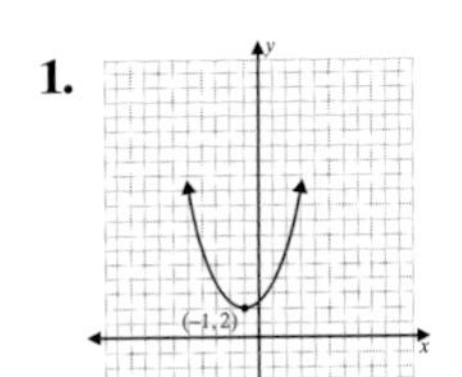

3.

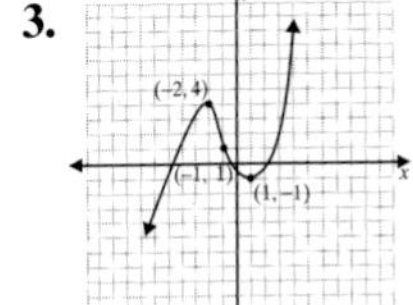

5.

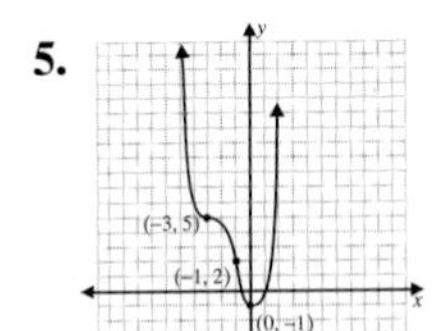

7.

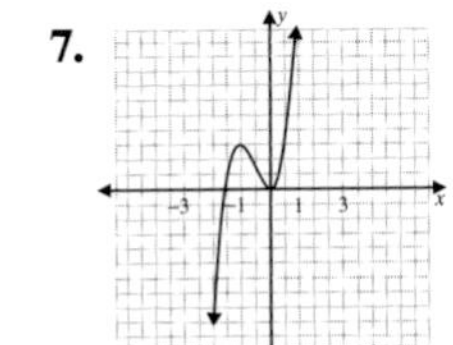

9.

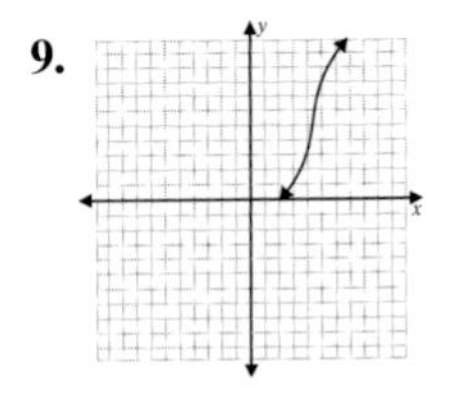

11. 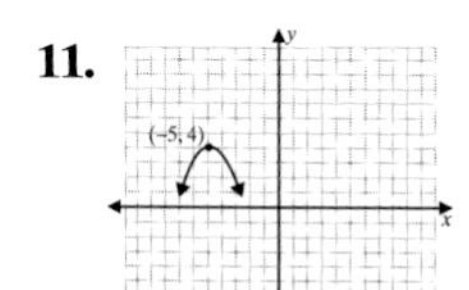

13. 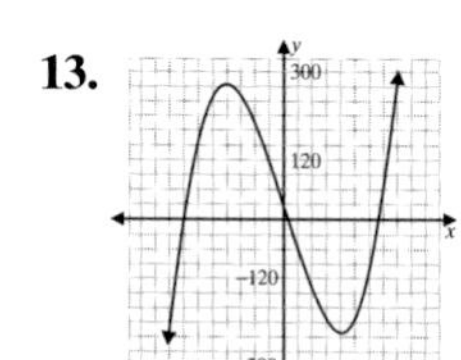

15. 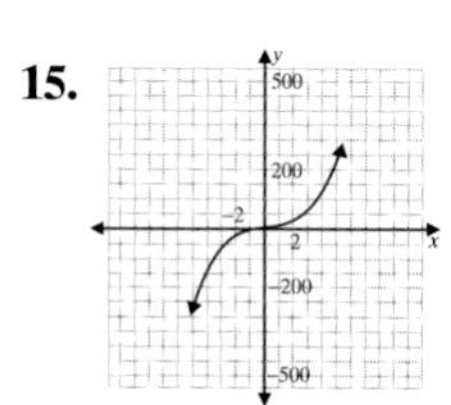

$y = \frac{4}{3}x^3 + 2x^2 + 5x + 2$ is implied.

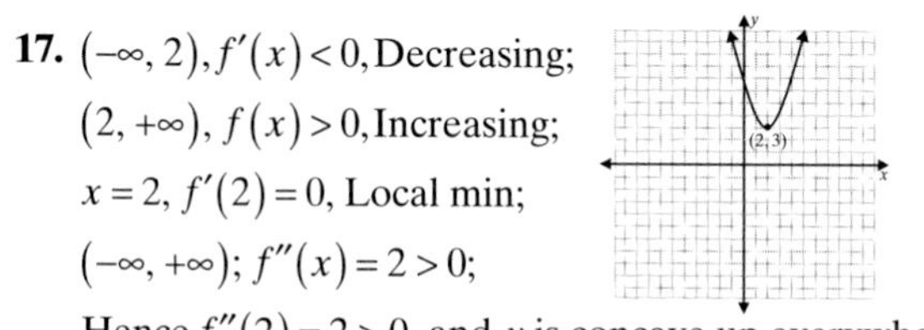

17. $(-\infty, 2), f'(x) < 0$, Decreasing;
$(2, +\infty), f(x) > 0$, Increasing;
$x = 2,\ f'(2) = 0$, Local min;
$(-\infty, +\infty);\ f''(x) = 2 > 0;$
Hence $f''(2) = 2 > 0$ and y is concave up everywhere.

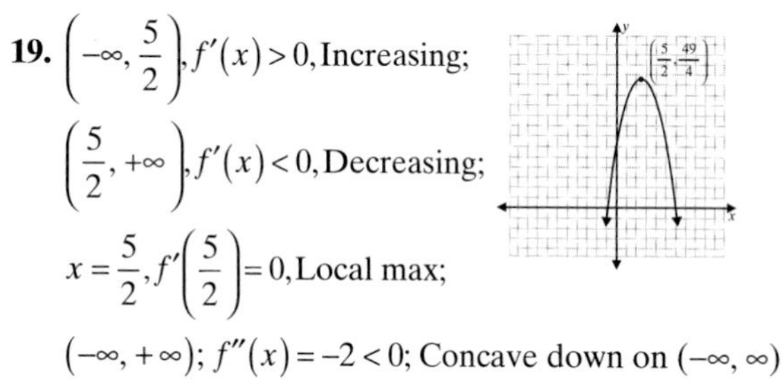

19. $\left(-\infty, \frac{5}{2}\right), f'(x) > 0$, Increasing;
$\left(\frac{5}{2}, +\infty\right), f'(x) < 0$, Decreasing;
$x = \frac{5}{2}, f'\left(\frac{5}{2}\right) = 0$, Local max;
$(-\infty, +\infty);\ f''(x) = -2 < 0$; Concave down on $(-\infty, \infty)$

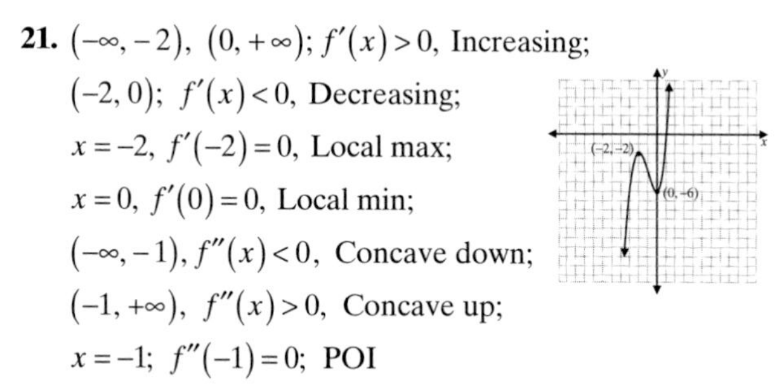

21. $(-\infty, -2),\ (0, +\infty);\ f'(x) > 0$, Increasing;
$(-2, 0);\ f'(x) < 0$, Decreasing;
$x = -2,\ f'(-2) = 0$, Local max;
$x = 0,\ f'(0) = 0$, Local min;
$(-\infty, -1), f''(x) < 0$, Concave down;
$(-1, +\infty),\ f''(x) > 0$, Concave up;
$x = -1;\ f''(-1) = 0$; POI

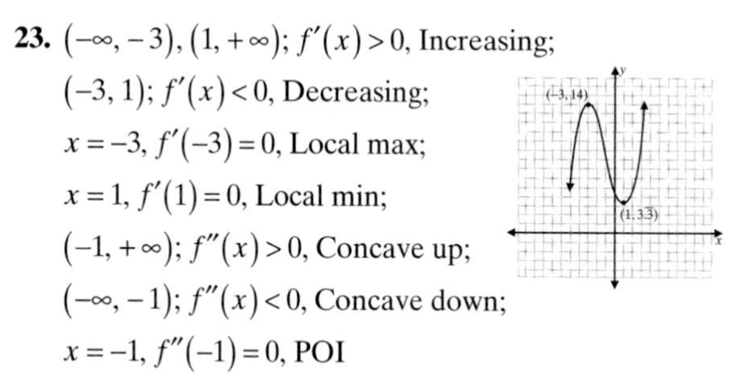

23. $(-\infty, -3), (1, +\infty);\ f'(x) > 0$, Increasing;
$(-3, 1);\ f'(x) < 0$, Decreasing;
$x = -3,\ f'(-3) = 0$, Local max;
$x = 1,\ f'(1) = 0$, Local min;
$(-1, +\infty);\ f''(x) > 0$, Concave up;
$(-\infty, -1);\ f''(x) < 0$, Concave down;
$x = -1,\ f''(-1) = 0$, POI

25. $(-1, 0), (1, +\infty);\ f'(x) > 0$, Increasing;
$(-\infty, -1), (0,1);\ f'(x) < 0$, Decreasing;
$x = 0,\ f'(0) = 0$, Local max;
$x = -1,\ f'(-1) = 0$, Local min;
$x = 1,\ f'(1) = 0$, Local min;
$\left(-\infty, -\frac{1}{\sqrt{3}}\right), \left(\frac{1}{\sqrt{3}}, +\infty\right);\ f''(x) > 0$, Concave up;
$\left(-\frac{1}{\sqrt{3}}, \frac{1}{\sqrt{3}}\right);\ f''(x) < 0$, Concave down;
$x = -\frac{1}{\sqrt{3}},\ f''\left(-\frac{1}{\sqrt{3}}\right) = 0$, POI;
$x = \frac{1}{\sqrt{3}},\ f''\left(\frac{1}{\sqrt{3}}\right) = 0$, POI

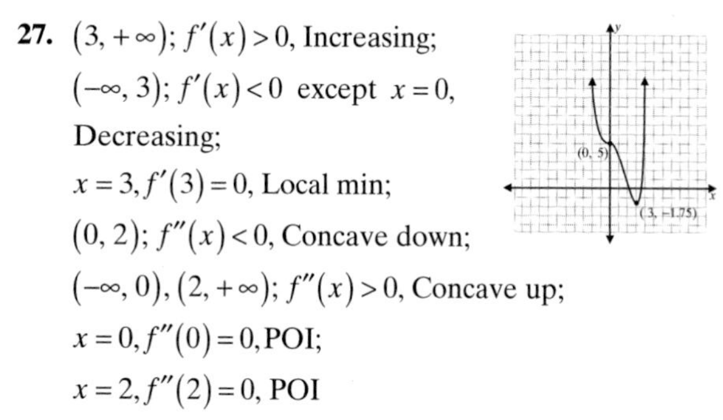

27. $(3, +\infty);\ f'(x) > 0$, Increasing;
$(-\infty, 3);\ f'(x) < 0$ except $x = 0$, Decreasing;
$x = 3, f'(3) = 0$, Local min;
$(0, 2);\ f''(x) < 0$, Concave down;
$(-\infty, 0), (2, +\infty);\ f''(x) > 0$, Concave up;
$x = 0, f''(0) = 0$, POI;
$x = 2, f''(2) = 0$, POI

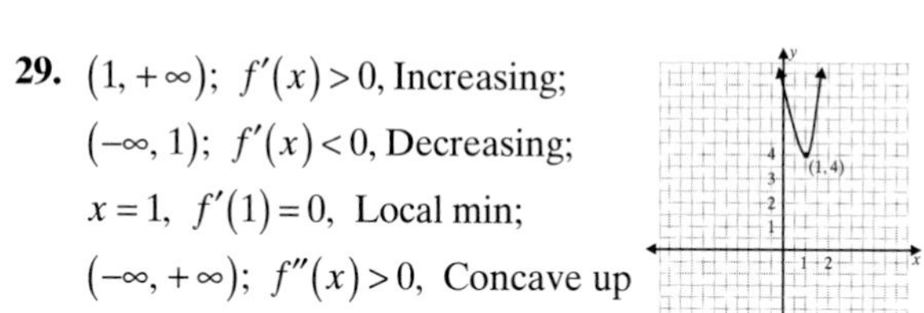

29. $(1, +\infty);\ f'(x) > 0$, Increasing;
$(-\infty, 1);\ f'(x) < 0$, Decreasing;
$x = 1,\ f'(1) = 0$, Local min;
$(-\infty, +\infty);\ f''(x) > 0$, Concave up

31. $\left(-\infty, -\frac{7}{3}\right)$, $(3, +\infty)$; $f'(x) > 0$, Increasing;
$\left(-\frac{7}{3}, 3\right)$; $f'(x) < 0$, Decreasing;
$x = -\frac{7}{3}$, $f'\left(-\frac{7}{3}\right) = 0$, Local max;
$x = 3$; $f'(3) = 0$, Local min;

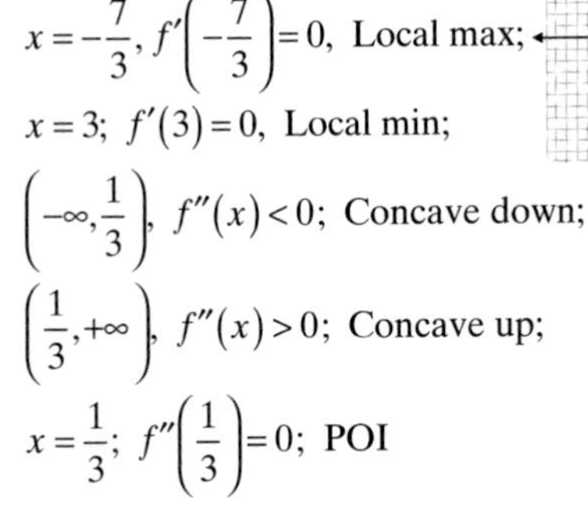

$\left(-\infty, \frac{1}{3}\right)$, $f''(x) < 0$; Concave down;
$\left(\frac{1}{3}, +\infty\right)$, $f''(x) > 0$; Concave up;
$x = \frac{1}{3}$; $f''\left(\frac{1}{3}\right) = 0$; POI

33. $(-\infty, 1.62)$, $f'(x) < 0$, Decreasing;
$(1.62, +\infty)$, $f'(x) > 0$ except $x = 8$, Increasing;
$x = 1.62$, $f'(1.62) = 0$, Local min;
$(-\infty, 3.75)$, $(8, +\infty)$; $f''(x) > 0$, Concave up;
$(3.75, 8)$; $f''(x) < 0$, Concave down;
$x = 3.75$, $f''(3.75) = 0$, POI;
$x = 8$, $f''(8) = 0$, POI

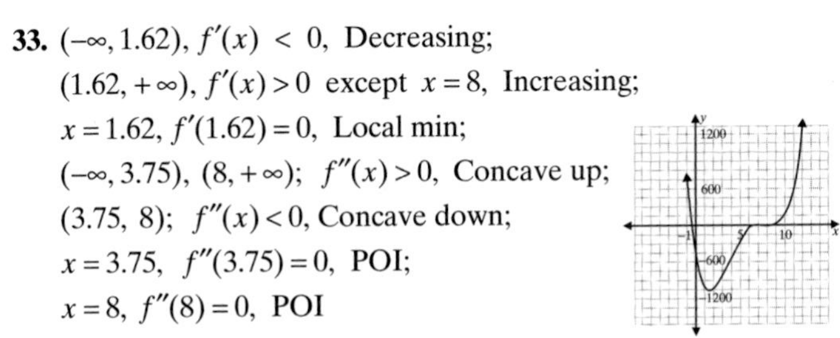

35. $(-\infty, -0.4)$; $f'(x) < 0$ except $x = -1.6$, Decreasing;
$(-0.4, +\infty)$, $f'(x) > 0$; Increasing;
$x = -0.4$, $f'(-0.4) = 0$, Local min;
$(-1.6, -0.8)$; $f''(x) < 0$; Concave down;
$(-\infty, -1.6)$, $(-0.8, +\infty)$; $f''(x) > 0$; Concave up;
$x = -1.6$, $f''(-1.6) = 0$; $x = -0.8$, $f''(-0.8) = 0$, POI

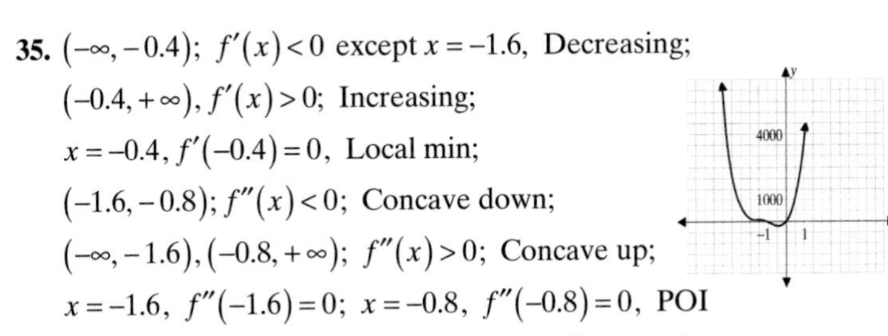

37. 3 **39.** 3 **41.** No **43. a.** At $t = 9$ minutes, E-readings are satisfactory and above $t = 91$, E-readings are low. **b.** The effective readings are high at $E(25) = 50$. **c.** If the first injection is given in the zero hour, the next injection should be given no more than 90 minutes later; perhaps 85 minutes is optimal, and every 80-85 minutes thereafter.

Exercises 4.3, pages 344 - 346

1. a. $x = 4$ **b.** $y = 0$ **c.** None **3. a.** $x = -8$ **b.** $y = 2$ **c.** None **5. a.** None **b.** $y = 0$ **c.** None **7. a.** $x = 1, x = -\frac{1}{3}$ **b.** $y = \frac{5}{3}$ **c.** None **9. a.** $x = 0$ **b.** None **c.** $y = x$ **11. a.** $x = -1$ **b.** None **c.** $y = x - 1$

13.

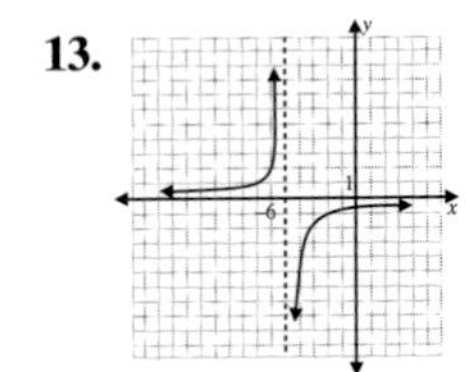

15.

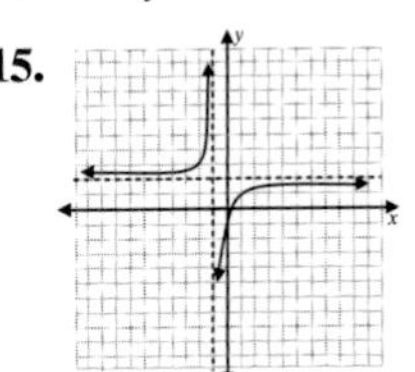

17.

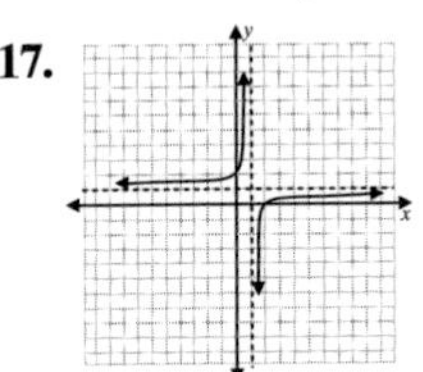

19.

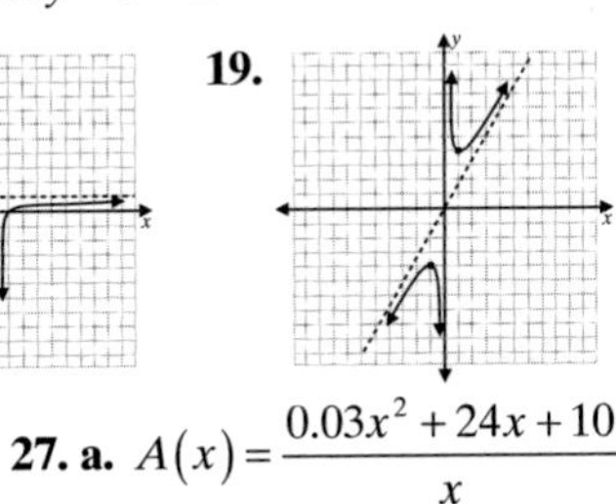

21.

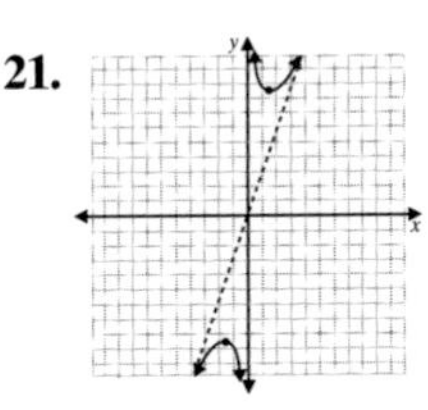

23.

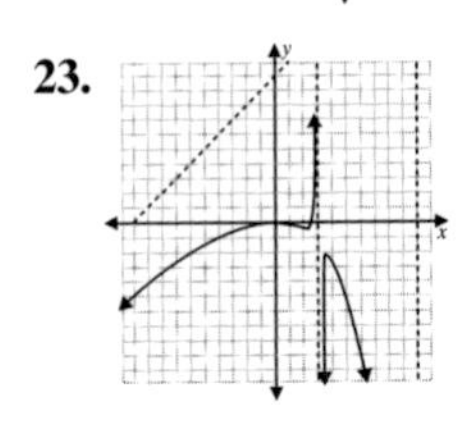

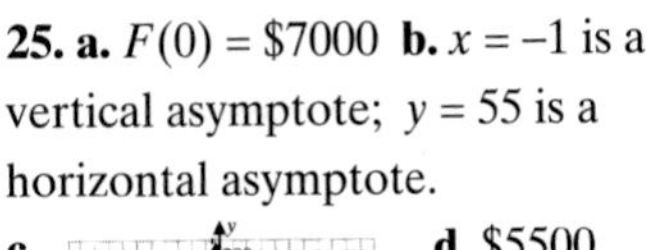

25. a. $F(0) = \$7000$ **b.** $x = -1$ is a vertical asymptote; $y = 55$ is a horizontal asymptote.
c.

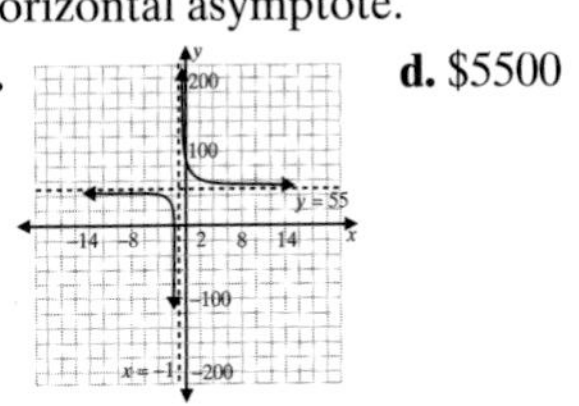

d. \$5500

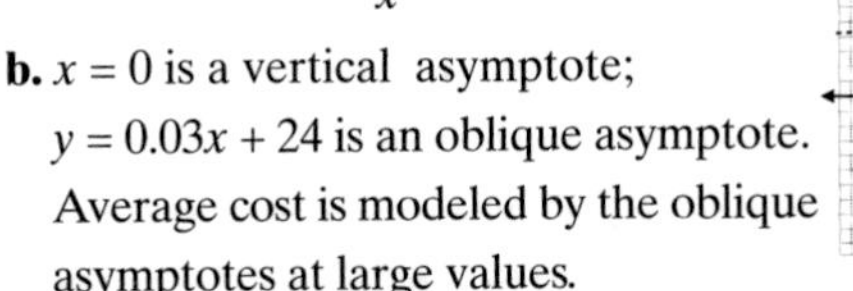

27. a. $A(x) = \dfrac{0.03x^2 + 24x + 10}{x}$
b. $x = 0$ is a vertical asymptote; $y = 0.03x + 24$ is an oblique asymptote. Average cost is modeled by the oblique asymptotes at large values.

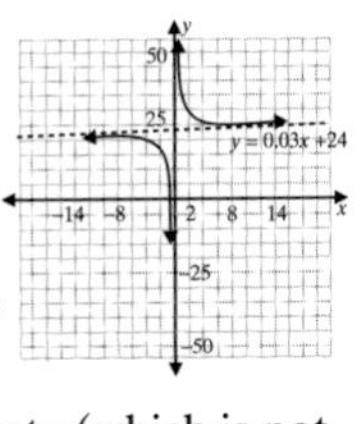

29. a. 32 dinars **b.** $x = -3$ is a vertical asymptote (which is not significant), $y = 20$ is horizontal asymptote, (Average cost always exceeds 20 dinars.) **c.** In the year 2007.
31. a. $y = 8$ is a horizontal asymptote. **b.** 4 million

Exercises 4.4, pages 354 - 357

1. 25 vacuum cleaners per order; 5 orders per year **3.** 30 copies per order; 3 orders per year **5.** 200 sweatshirts per order; 20 orders per year **7.** 120 snowmobiles per order; 8 orders per year **9.** \$24 **11.** 75 cents **13.** 5 salespeople **15.** \$6 **17.** 13 units at \$36 **19.** 1900 color televisions at \$326 **21. a.** The company is making \$12,000 **b.** The profit will rise \$2 per phone. **c.** The rate of profit increase is dropping. **23. a.** $A'(x)=\dfrac{xC'(x)-C(x)}{x^2}$ **b.** $C'(x)=\dfrac{C(x)}{x}$ **c.** Average costs are minimal when marginal cost equals $A(x)$. **25. a.** $C(x)=0.2x^2+3x$ **b.** $A'(x)=0.2$ **c.** $C'(x)=0.4x+3$

27. 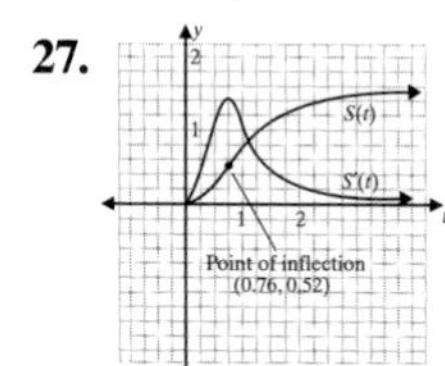

29. a. \$26,66.67 **b.** Value is increasing. **c.** Yes. We can use test values with V' to determine if V' is in fact increasing or not (this rate of change is V'').

Exercises 4.5, pages 365 - 369

1. 4 in. by 4 in. **3.** $\frac{5}{3}$ in. by $\frac{14}{3}$ in. by $\frac{35}{3}$ in. **5. a.** 12 in. by 12 in. by 6 in. **b.** 432 in.^{2} **7.** 2 ft. by 2 ft. by 2.5 ft. **9.** 1.82 ft. **11.** 18.5 ft. by 37 ft. **13.** 70 ft by 70 ft.; 4900 ft^2. **15.** 36 ft. by 69 ft. **17.** $\frac{65}{3}$ in. by $\frac{65}{3}$ in. by $\frac{130}{3}$ in.; $\frac{260}{9}$ in. by $\frac{260}{9}$ in. by $\frac{65}{3}$ in. yields a package of lesser volume. **19.** 3 miles from point A. **21.** 0.8 miles from point A. **23. a.** 67 ft./sec **b.** 129 ft. **25. a.** 324 ft. **b.** 8 sec **c.** 144 ft./sec

Exercises 4.6, pages 377 - 379

1. $dy=3x^2dx$ **3.** $du=8t(2t^2+1)dt$ **5.** $dA=2\pi r dr$ **7.** $dV=4\pi r^2 dr$ **9.** $dS=\left(8x-\frac{1350}{x^2}\right)dx$ **11.** $dC=\left(3+\frac{0.2}{\sqrt{x}}\right)dx$ **13.** $dP=(75-0.4x)dx$ **15. a.** 0.64 **b.** 0.6 **c.** 0.04 **17. a.** −14.03 **b.** −15.36 **c.** 1.33 **19. a.** −17.7 **b.** −16.8 **c.** −0.9 **21. a.** 0.148 **b.** 0.15 **c.** −0.002 **23.** $6+\frac{1}{12}$ **25.** $3-\frac{1}{27}$ **27.** $7+\frac{1}{10}$ **29.** $4-\frac{17}{480}$ **31.** \$20.40 **33.** \$42 **35.** −340 people **37.** 19.44 in.^{3} **39.** 0.013 ft. **41.** 2400π in.^{3}

Chapter 4 Test, pages 384 - 386

1. $12x^2+\frac{9}{2\sqrt{x}}$ **2.** $\frac{8(2x^2+21)}{9(2x^2+7)^{\frac{4}{3}}}$ **3.** $\frac{4}{(x+1)^3}$ **4.** $\frac{42x+24}{x^5}$ **5.** $f''(x)=6x-8$; $f''(-2)=-20$; $f''(0)=-8$; $f''(1)=-2$ **6.** $f''(x)=2-\frac{28}{(x-2)^3}$; $f''(-2)=\frac{39}{16}$; $f''(0)=\frac{11}{2}$; $f''(1)=30$ **7. a.** $(-\infty,-4),(0,+\infty)$ **b.** $(-4,0)$ **c.** $(-4,1),(0,2)$ **8. a.** $(-\infty,-6),(-1,+\infty)$ **b.** $(-6,-4),(-4,-1)$ **c.** $(-6,5),(-1,6)$ **9. a.** $(2,+\infty)$ **b.** $(-\infty,2)$ **c.** $(2,-6)$ **10. a.** $(-\infty,+\infty)$ **b.** None **c.** None **11. a.** $(-\infty,0),(\sqrt[3]{16},+\infty)$ **b.** $(0,\sqrt[3]{16})$ **c.** $(\sqrt[3]{16},0)$ **12. a.** $(-\infty,-\sqrt{3}),(\sqrt{3},+\infty)$ **b.** $(-\sqrt{3},\sqrt{3})$ **c.** $(-\sqrt{3},-13),(\sqrt{3},-13)$

13. Local min: $\left(\frac{3}{2}, -\frac{7}{4}\right)$ **14.** Local max: $\left(-1, \frac{17}{3}\right)$; Local min: $\left(7, -\frac{239}{3}\right)$ **15.** Local max: (–2, –12); Local min: (2, 12)

16. Local max: (0, 2); Local min: (–4, –126),(1, –1) **17.**

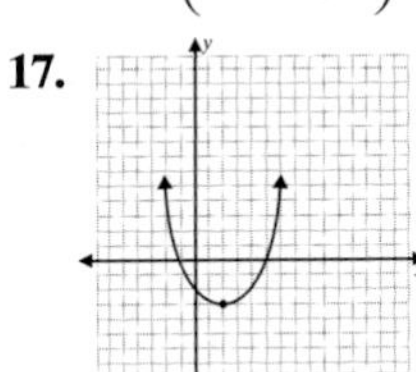

18. \$18 per unit

19. a. 808 people **b.** 12 people per month

20. $\left(-\infty, -\frac{5}{2}\right), f'(x) > 0$, Increasing;
$\left(-\frac{5}{2}, +\infty\right), f'(x) < 0$, Decreasing;
$x = -\frac{5}{2}, f'\left(-\frac{5}{2}\right) = 0$, Local max;
$(-\infty, +\infty), f''(x) < 0$, Concave down

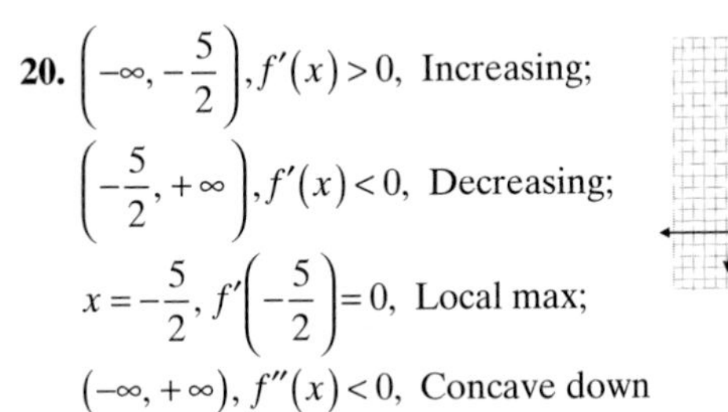

21. $(-\infty, -2), \left(\frac{1}{2}, +\infty\right), f'(x) > 0$, Increasing;
$\left(-2, \frac{1}{2}\right), f'(x) < 0$, Decreasing;
$x = -2, f'(-2) = 0$, Local max;
$x = \frac{1}{2}, f'\left(\frac{1}{2}\right) = 0$, Local min;
$\left(-\infty, -\frac{3}{4}\right), f''(x) < 0$, Concave down;
$\left(-\frac{3}{4}, +\infty\right), f''(x) > 0$, Concave up;
$x = -\frac{3}{4}, f''\left(-\frac{3}{4}\right) = 0$, POI

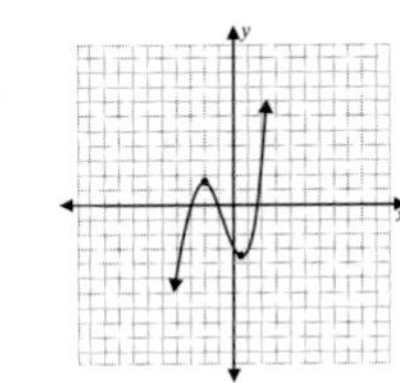

22. $(-\sqrt{3}, 0), (\sqrt{3}, +\infty), f'(x) > 0$, Increasing;
$(-\infty, -\sqrt{3}), (0, \sqrt{3}), f'(x) < 0$, Decreasing;
$x = -\sqrt{3}, f'(-\sqrt{3}) = 0$, Local min;
$x = 0, f'(0) = 0$, Local max;
$x = \sqrt{3}, f'(\sqrt{3}) = 0$, Local min;
$(-1, 1), f''(x) < 0$, Concave down;
$(-\infty, -1), (1, +\infty), f''(x) > 0$, Concave up;
$x = -1, f''(-1) = 0$, POI
$x = 1, f''(1) = 0$, POI

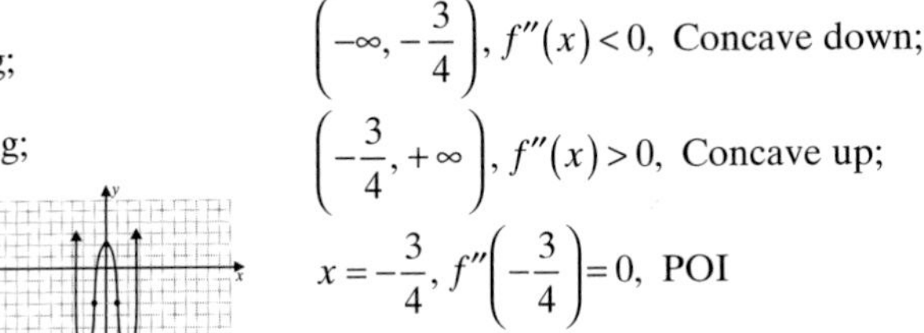

23. $(-\infty, -1), (0, +\infty), f'(x) > 0$, Increasing;
$(-1, 0), f'(x) < 0$, Decreasing;
$x = -1, f'(-1) = 0$, Local max
$(-\infty, 0), (0, +\infty), f''(x) < 0$, Concave down

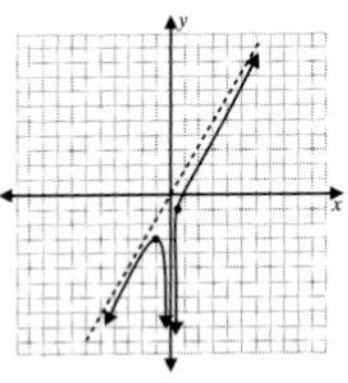

24. a. 35 **b.** 8 **c.** 27 **25. a** –0.095 **b.** –0.094 **c.** –0.001 **26.** $6\frac{13}{14} \approx 6.929$ **27.** $4\frac{1}{48} \approx 4.021$

28. a. $x = 4$ hours **b.** 0.15 percent **29.** 60 radios per order; 9 times per year **30.** \$360 **31. a.** 125 bikes **b.** \$262.50

32. 1.5 ft. by 1.5 ft. by 2 ft. **33.** 1200 ft.^{2} **34.** $\frac{1}{3}$ mile down the shore **35. a.** 484 ft. **b.** 11 sec

c. –176 ft./sec, or 176 ft./sec downward **36.** \$92 **37.** 32 bacteria **38.** 2.88π cm^3

Chapter 4 Cumulative Review, pages 387 - 390

1. $+\infty$ **2.** $-\infty$ **3.** 1 **4.** 1 **5.** 38

x	y
3.5	30.5
3.9	36.42
3.99	37.84
3.9999	37.98
3.9999	38.00

6. ∞

x	y
–0.5	3.71
–0.9	8.12
–0.99	22.60
–0.999	66.12
–0.9999	202.96

7. 6

x	y
2.5	5.50
2.9	5.90
2.99	5.99
2.999	6.00
2.9999	6.00

8. 9

x	y
4.5	9
4.1	9
4.01	9
4.001	9
4.0001	9

9. a. - c. Answers will vary. **10.** Instantaneous rate of change in oil consumed per week **11.** $m = 3$ **12.** $12x$ **13.** $18x^2 - 12x$ **14.** $\frac{7}{\sqrt{x}}$ **15. a.** yes; $f = \#$ of aces; $x = \#$ of decks **b.** yes; $f = \#$ of weekends; $x = \#$ of semesters **c.** yes; $f = \#$ calories; $x = \#$ servings **d.** yes; $f =$ dollars; $x = \#$ of weeks **16. a.** \$0.40/unit **b.** $\frac{dc}{dy}$ = change in cost per man-hr., $\frac{dx}{dy}$ = change in units/man-hr., $\frac{dc}{dx}$ = change in cost/unit production. **c.** $\frac{dc/dy}{dx/dy} = \frac{dc}{dx}$ **17.** \$1800

18. $f'(x) = -\frac{115}{2(5x+17)^{\frac{3}{2}}}$ **19.** $g'(t) = \frac{5}{(3t+1)^2}$ **20.** $g'(t) = \frac{1}{\sqrt{1+2t}} + \frac{3}{2\sqrt{3t+1}} + 5$ **21.** $g'(t) = \frac{25(9t+4)}{2\sqrt{t}}$

22. $H'(s) = \frac{4+16s}{\sqrt{1+8s+16s^2}}$ **23.** $F'(u) = \frac{3}{(3u-1)^{\frac{1}{2}}(3u+1)^{\frac{3}{2}}}$ **24.** $f'(t) = \frac{450}{t^3} + 3$; $f'(1.5) = 136\frac{1}{3} \approx 136.33$

25. Local max: (–7.5, 5); Local min: (5.1, –3.2) **26.** Local min: (0.076, –35.088); Local max: (–20.076, 2011.088) **27.** Local max: (0.0816, 3.305); Local min: (–0.0816, 3.295) **28.** None **29.** 6.675 ft./sec **30.** 11.268 ft./sec **31. a.** 2005 **b.** 1990 **c.** yes **32.** $f'(x) = 17 - 22x + 24x^2 - 8x^3$; $f''(x) = -22 + 48x - 24x^2$; $f'''(x) = 48 - 48x$; $f^{(4)}(x) = -48$

33. $f'(x) = -\frac{1}{x^2}$; $f''(x) = \frac{2}{x^3}$; $f'''(x) = -\frac{6}{x^4}$; $f^{(4)}(x) = \frac{24}{x^5}$

34. $f'(x) = -\frac{11}{(x-3)^2}$; $f''(x) = \frac{22}{(x-3)^3}$; $f'''(x) = -\frac{66}{(x-3)^4}$; $f^{(4)}(x) = \frac{264}{(x-3)^5}$

35. $f'(x) = \frac{3}{2}(5+3x)^{-\frac{1}{2}}$; $f''(x) = -\frac{9}{4}(5+3x)^{-\frac{3}{2}}$; $f'''(x) = \frac{81}{8}(5+3x)^{-\frac{5}{2}}$; $f^{(4)}(x) = -\frac{1215}{16}(5+3x)^{-\frac{7}{2}}$

36.

37.

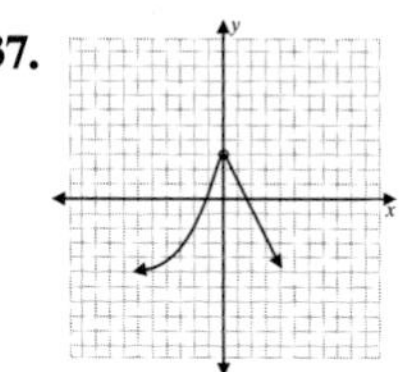

38. $(-\infty, 6)$; $f'(x) < 0$, Decreasing;
$(6, 8), (8, +\infty)$; $f'(x) > 0$, Increasing;
$x = 6$, $f'(6) = 0$, Local min;
$(-\infty, 8), (12, +\infty)$; $f''(x) > 0$, Concave up;
$(8, 12)$; $f''(x) < 0$, Concave down;
$x = 8$, $f''(8) = 0$; $x = 12$, $f''(12) = 0$, POI

39. $(-\infty, -12), (1.33, \infty)$; $f'(x) > 0$, Increasing; $(-12, 1.33)$; $f'(x) < 0$, Decreasing;
$x = 1.33, f'(1.33) = 0$, Local min; $x = -12, f'(-12) = 0$, Local max;
$(-\infty, -5.33)$; $f''(x) < 0$, Concave down; $(-5.33, +\infty)$; $f''(x) > 0$, Concave up;
$x = -5.33, f''(-5.33) = 0$, POI

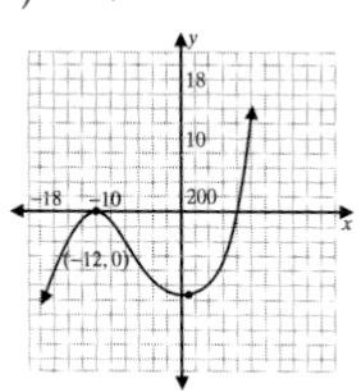

40. Vertical asymptote: $x = 0$ **41.** Vertical asymptote: $x = -10^{\frac{2}{3}} = \sqrt[3]{-100} = -\sqrt[3]{100}$; Horizontal asymptote: $y = 0$ **42. a.** –8.78; the average cost of a vest is decreasing by $8.78 when the company is producing 50 vests. **b.** There is no horizontal asymptote for $A(x)$. **43. a.** $W(0) = 5$; in 1900, 5% of the male workers over 45 were veterans. **b.** August, 1902 **c.** 39.38% **44. a.** 135 cm^3 **b.** $328.02

Chapter 5

Exercises 5.1, pages 402 - 404

1. 199.53 **3.** 0.61 **5.** 20.09 **7.** 0.40 **9. a.** 1 **b.** 3 **c.** 27 **11.**

13.

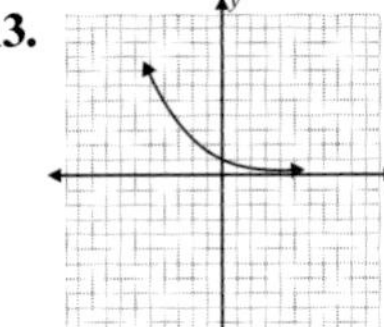

15.

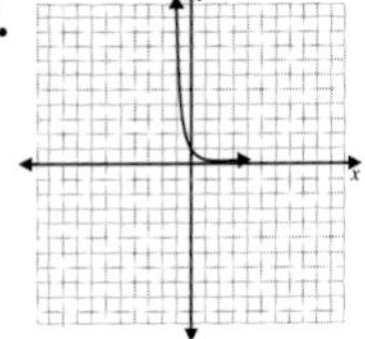

17.

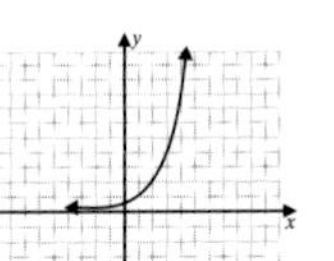

19.

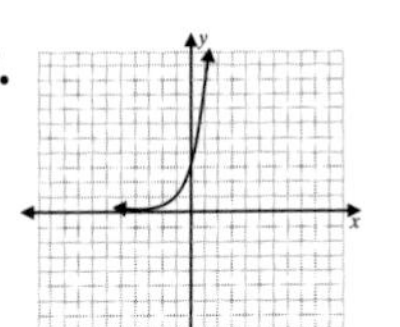

21.

23.

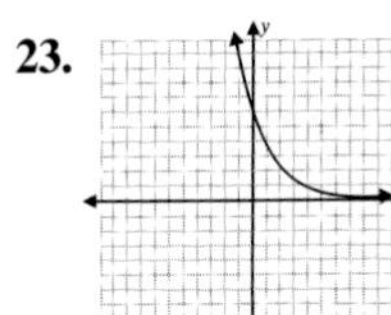

25.

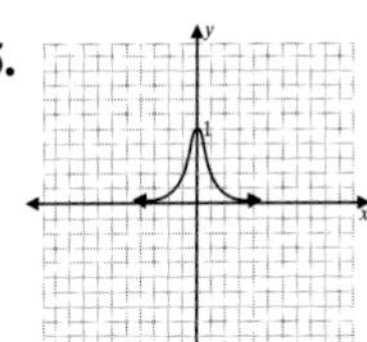

27. a. $6312.38 **b.** $6333.85 **c.** $6344.93 **d.** $6356.12 **e.** $6356.25 **29.** $8481.89 **31.** The 8 percent investment is better by $88.43. **33.** 0.82 mg. **35.** 225,594 people **37.** $3168.06 **39.** $15,059.71

Exercises 5.2, pages 412 - 414

1. $\ln(1.648721) \approx 0.5$ **3.** $\ln(0.122456) \approx -2.1$ **5.** $\ln(1.284025) \approx 0.25$ **7.** $e^{0.6} \approx 1.822119$ **9.** $e^{3.161247} \approx 23.6$ **11.** $e^{-2.667} \approx 0.069460$ **13.** 4.6 **15.** 3 **17.** $\frac{1}{3}$ **19.** $5x$ **21.** $2t-1$ **23.** $\ln(xy^3)$ **25.** $\ln\left(\frac{xz}{y}\right)$ **27.** $\ln(x(x+3))$ **29.** $\ln\left(\frac{x^2}{x+1}\right)$ **31.** $\ln(x-1)$ **33.** $\ln\frac{(x+3)(x+1)}{x^{\frac{1}{2}}}$ **35.** $3\ln x + 2\ln y$ **37.** $\frac{1}{3}(\ln x - \ln y)$ **39.** $\ln 15 + \ln x - \frac{1}{2}\ln y$ **41.** $\frac{1}{2}\ln(4x+1)$ **43.** $1 - 2\ln(x+2)$ **45.** $\ln(x+4) + \ln(x-4)$ **47.** $-\frac{1}{2} + \frac{\ln 13}{2} \approx 0.7825$ **49.** $-\frac{\ln 0.8}{0.6} \approx 0.3719$ **51.** $\frac{\ln 18}{\ln 5} \approx 1.7959$ **53.** $\frac{1}{3}\left(1 + \frac{\ln 8.6}{\ln 10}\right) \approx 0.6448$ **55.** 3 **57.** $\sqrt{2}$ **59.** $1 + e \approx 3.7183$ **61. a.** 7.56 years **b.** 7.31 years **c.** 7.25 years **63.** 8.66 years **65.** 3.98 years **67. a.** $36,725.89, $73,377.52, $80,670.38 **b.** $86,836.41 **c.** 17,456 books

Exercises 5.3, pages 423 - 425

1. $f'(x) = 2x - \frac{1}{x}$ **3.** $f'(x) = 1 + \ln x$ **5.** $\frac{dy}{dx} = \frac{1 - \ln x}{x^2}$ **7.** $\frac{dy}{dx} = \frac{3}{x}(\ln x)^2$ **9.** $f'(x) = \frac{5}{5x+3}$ **11.** $f'(x) = \frac{2x}{x^2+2}$

13. $f'(x)=\dfrac{2x}{2x^2-1}$ **15.** $\dfrac{dy}{dx}=\dfrac{8}{4x+3}$ **17.** $\dfrac{dy}{dx}=\dfrac{2x^{\frac{3}{2}}}{x^2+2}+\dfrac{\ln(x^2+2)}{2\sqrt{x}}$ **19.** $\dfrac{dy}{dx}=\dfrac{2x+2}{x(x^2+2x-1)}-\dfrac{\ln(x^2+2x-1)}{x^2}$

21. $f'(x)=\dfrac{9x^2+2x+9}{(3x+1)(x^2+3)}$ **23.** $f'(x)=\dfrac{8x^2-2x-8}{(2x-1)(x^2-2)}$ **25.** $\dfrac{dy}{dx}=-\dfrac{3}{(x+1)(x-2)}$ **27.** $f'(x)=\dfrac{2x^2+18x+10}{(x^2-5)(2x+9)}$

29. $f''(x)=\dfrac{2-\ln x}{x(\ln x)^3}$; $x=e^2\approx 7.39$; $(e^2, 3.7)$ **31. a.** $f''(x)=9+6\ln x$ **b.** $x=e^{-\frac{3}{2}}$ **c.** $\left(e^{-\frac{3}{2}},-0.224\right)$

33. a. $f''(x)=\dfrac{3\ln x}{x^2}[2-\ln x]$ **b.** $x=1, x=e^2\approx 7.39$ **c.** $(1,0), (e^2,8)$ **35.** $\dfrac{dy}{dx}=\dfrac{(2x-5)^2(8x^2-19x+5)}{\sqrt{x^2-2x}}$

37. $\dfrac{dy}{dx}=\dfrac{2x(11x^2-14)(x^2+2)^3}{3(x^2-1)^{\frac{4}{3}}}$ **39.** $\dfrac{dy}{dx}=\dfrac{(16x^2-46x-54)(2x+x^2)^2}{(4x-9)^3}$ **41.** $\dfrac{dy}{dx}=\dfrac{(21x^2-114x-248)(x+2)(3x+4)}{2(x-6)^{\frac{3}{2}}}$

43. Abs. max.: $(2, 1.31)$; Abs. min.: $(1, 1)$ **45.** Abs. max.: $(1.2, 6.58)$; Abs. min.: (e, e)

47. Abs. max.: $(e, 4.39)$; Abs. min.: $(\sqrt{1.5}, 0.89)$

49.

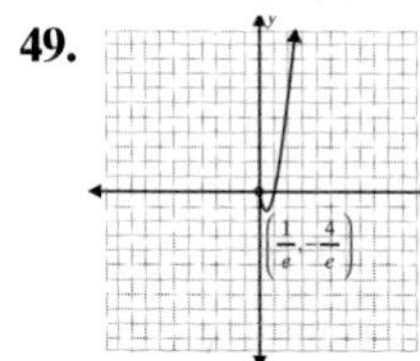

51.

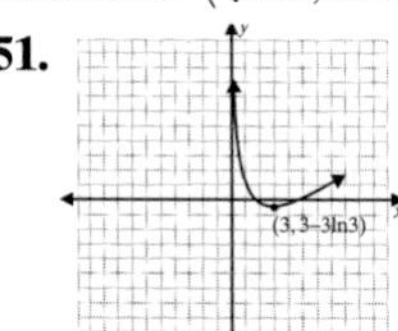

53.

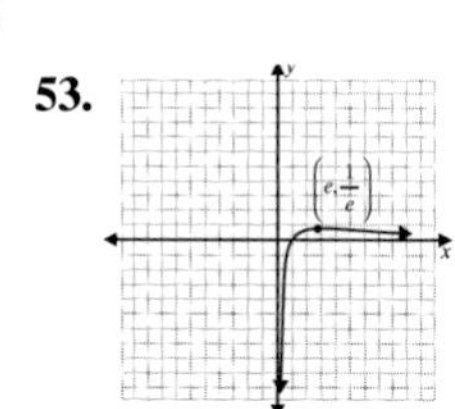

55. $96+\dfrac{8x}{4x^2+15}$ dollars per table

57. \$294.30 **59. a.** 54 flies, 61 flies, 0 flies **b.** 63 flies

Exercises 5.4, pages 432 - 435

1. $f'(x)=3e^x$; $f''(x)=3e^x$ **3.** $f'(x)=2x+5e^x$; $f''(x)=2+5e^x$ **5.** $f'(x)=e^x(x+1)$; $f''(x)=e^x(x+2)$

7. $f'(x)=e^x\left(\dfrac{1}{x}+\ln x\right)$; $f'(a)=e$ **9.** $f'(x)=-\dfrac{e^x}{(e^x-1)^2}$; $f'(a)=-0.181$ **11.** $f'(x)=8e^{4x}$; $f'(a)=0.147$

13. $f'(x)=6e^{2x+1}$; $f'(a)=6e\approx 16.310$ **15.** $f'(x)=-2xe^{1-x^2}$; increasing on $(-\infty, 0)$; decreasing on $(0, \infty)$

17. $f'(x)=4e^{2x}(e^{2x}-4)$; decreasing on $(-\infty, 0.693)$; increasing on $(0.693, \infty)$

19. $f'(x)=-0.1e^{-0.2x}(e^{-0.2x}+11)^{\frac{-1}{2}}$; always decreasing; horiz. asym.: $y=\sqrt{11}$

21. $f'(x)=\dfrac{e^x}{e^x+1}$; There is a diag. asym. since slope $\to 1$ as $x\to\infty$.

23. $f'(x)=2xe^{-x^2}\cdot\left[\dfrac{1}{x^2+2}-\ln(x^2+2)\right]$; Abs. max.: $(0, 0.693)$; infl. pts.: $(-1.039, 0.382)$ and $(1.039, 0.382)$

25. Abs. min.: $(-0.5, -0.18)$; Abs. max.: $(1, 7.39)$ **27.** Abs. min.: $(0, 0)$; Abs. max.: $(-1, e)$

29. Abs. min.: $(-2, 0.25)$; Abs. max.: $(0, 13.59)$ **31.** Abs. min.: $(1, 4.09)$; Abs. max.: $(3.5, 4.97)$

33. a. $x = 2.5$
b. $x = 5$
c. down: $(-\infty, 5)$; up: $(5, \infty)$
d.

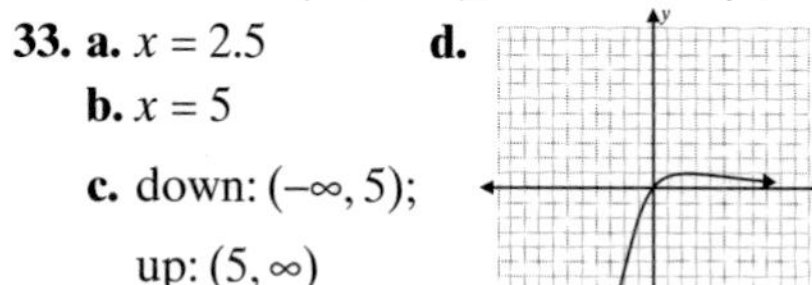

35. a. $x = 0, 2$
b. $x = 2\pm\sqrt{2}$
c. down: $(2-\sqrt{2}, 2+\sqrt{2})$; up: $(-\infty, 2-\sqrt{2})$, $(2+\sqrt{2}, \infty)$
d.

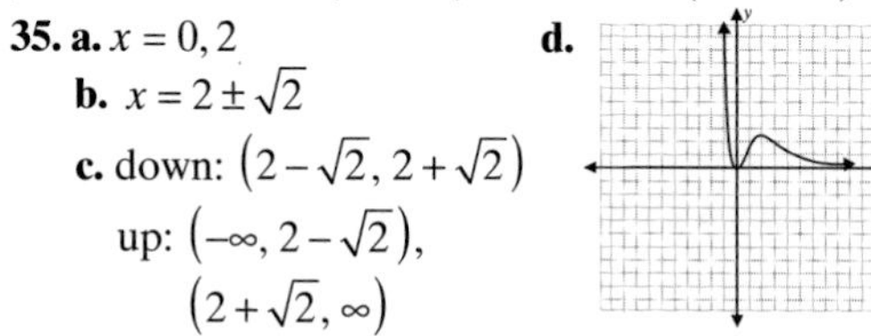

37. a. $x = 0$
b. None
c. up: $(-\infty, \infty)$
d.

39. a. 20 units **b.** \$51.50 each **41.** 10.432 thousand per day **43.** 209 bacteria/hr. **45. a.** \$772.73, \$5674.72 **b.** \$1131.72/yr., \$859.59/yr. **c.** $(3.8, 4250)$, The rifle's value begins to increase at a much slower rate and level off. **d.** 2004 **47. a.** June 1st **b.** Aug 1st **49. a.** 36.35% **b.** The percentage is decreasing by 3.81% each week. **c.** 30% of the information is retained. **51. a.** $y = -0.4x + 4.9$ **b.** Answers will vary.

Exercises 5.5, pages 450 - 454

1. a. $p(t)=8400e^{0.035t}$ **b.** 11,920 people **c.** 12 years, or in the year 2012 **3.** \$21.33 **5.** 9.5 hours **7.** \$300.44 **9.** 26.7 years **11.** 8700 years **13.** 11.8 lb/in.^{2} **15.** \$18,846 **17.** 0.777; 23.1 hours **19. a.** 800 fish **b.** 4303 fish **c.** 5.9 years **21.** 13.5 minutes **23.** \$11,880.94 **25. a.** $E(x)=\frac{28-x}{x}$ **b.** $x=14$ **27. a.** $E(x)=\frac{200-x}{x}$ **b.** $x=100$ **29. a.** $E(x)=\frac{5}{x}$ **b.** $x=5$ **31. a.** $E(x)=\frac{40}{x}$ **b.** $x=40$ **33. a.** $E(x)=\frac{2(150-x)}{x}$ **b.** $x=100$ **35. a.** $E(x)=\frac{2(54-x)}{x}$ **b.** $x=36$ **37. a.** $E(x)=\frac{2(18-\sqrt{x})}{\sqrt{x}}$ **b.** $x=144$ **39. a.** $E(x)=\frac{363-x^2}{2x^2}$ **b.** $x=11$ **41.** 24 sharpeners **43.** $x<121$ **45. a.** 19 videogames **b.** $E=-\frac{pe^{-\frac{p}{10}}}{10\left(1-e^{-\frac{p}{10}}\right)}$ At $p=10$, $E=-0.58<1$ and revenue is decreasing and demand is inelastic. **c.** \$189.64 **47. a.** $E=0.073<1$, the demand is inelastic. **b.** $R(x)=xe^{\frac{1800}{x}-10}$ **c.** R is decreasing at $p=20$ ($x\approx 137.99$); yes. **49. a.** $p=\$264.75$ **b.** $D'(x)=-3xe^{-\frac{x^2}{200}}$ **c.** $E=4$ **d.** $x=10$

Chapter 5 Test, pages 459 - 460

1. 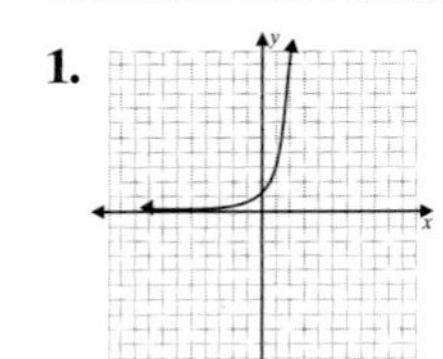 **2.** 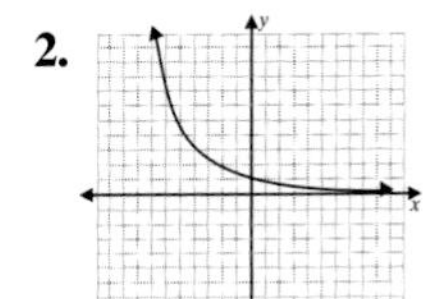 **3.** 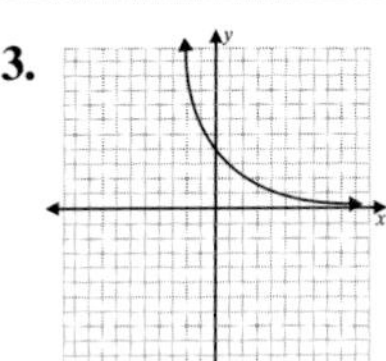 **4.** 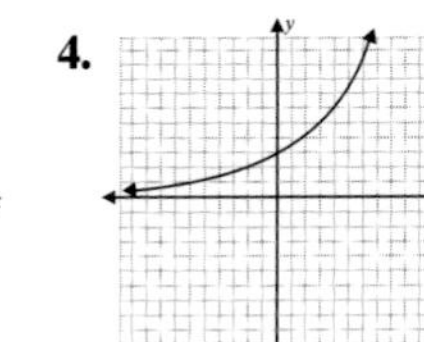

5. a. \$10,486.37 **b.** \$10,534.47 **c.** \$10,559.43 **d.** \$10,584.75 **e.** \$10,585.04 **6.** \$6313.97 **7.** $\ln 1.284=0.25$ **8.** $\ln 14.6=m$ **9.** $e^{0.6831}=1.98$ **10.** $P=e^{1.3741}$ **11.** $\ln(7x-2)-\ln(5)$ **12.** $\frac{1}{3}\ln(1-12x)$ **13.** $2\ln(x)-[\ln(y+1)+\ln(y-1)]$ **14.** $3[\ln x+2\ln y]$ **15.** $x\approx 11.224$ **16.** $x\approx -0.962$ **17.** $x\approx 0.248$ **18.** $x\approx 12.888$ **19.** $x=1$ **20.** $x=1+e^2\approx 8.389$ **21.** $f'(x)=3x^2-2e^{-2x}$ **22.** $f'(x)=2e^x(2x^2+4x-3)$ **23.** $\frac{dy}{dx}=15e^{3x}(e^{3x}-7)^4$ **24.** $\frac{dy}{dx}=e^{-x}\left[\frac{1}{x+2}-\ln(x+2)\right]$ **25.** $\frac{dy}{dx}=\frac{14x^2+36x+5}{(2x^2+5)(x+3)}$ **26.** $\frac{dy}{dx}=-\frac{12(2x+3)}{(6x+1)(2x-5)}$ **27.** Abs. min.: $(0, 200)$; Abs. max.: $(4, 200e)$ **28.** Abs. min.: $(2,-e^2)$; Abs. max.: $(4,e^4)$

29. 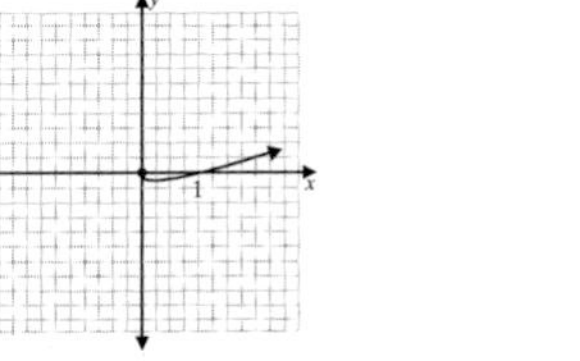

30. 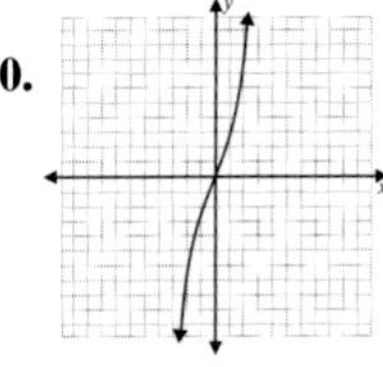

31. \$4661.99 **32.** 17.7 hours **33.** \$3821.75 **34. a.** $E(x)=\frac{4(108-1.5x)}{3x}$ **b.** $x=48$ **35. a.** $E(x)=\frac{45-2\sqrt{x}}{\sqrt{x}}$ **b.** $x=225$

Chapter 5 Cumulative Review, pages 461 - 464

1. $+\infty$ **2.** $-\infty$ **3.** 0 **4.** 0 **5.** $y=2x+14$

6. **b.** When the test has been given twice, 75 percent of the students scored 90 percent or better on the last test. **c.** When the test has been given twice, the percentage of students scoring 90 percent or better on the most recent test is increasing by 17.3 percent **d.** The rate of change is not constant on [2, 3].

7. $y'=-\frac{5}{2}x^{\frac{-3}{2}}+2x^{\frac{-4}{3}}$

8. a. $p = 1$ **b.** $P(x) = -500 + 0.99x - 0.003$ **c.** $C'(x) = 0.01 + 0.006x$ **d.** $P'(x) = 0.99 - 0.006x$ **e.** $A(1000) = 3.51$ **f.** True

9. a. $A(x) = 0.04x + 90 + \dfrac{1500}{x}$ **b.** $A(200) = \$105.50$ **c.** $A'(200) = \$0.0025$ **10.** $k = \dfrac{1}{3}\ln\left(\dfrac{3}{2}\right)$ **11.** $k = \dfrac{1}{4}e^{\frac{1}{4}}$

12. $f'(x) = -3 + 12x;\ f''(x) = 12$ **13.** $f'(x) = 122 - 88x - 192x^2;\ f''(x) = -88 - 384x$

14. $f'(x) = \dfrac{3}{2\sqrt{100+3x}};\ f''(x) = \dfrac{-9}{4(3x+100)^{\frac{3}{2}}}$ **15.** $f'(x) = \dfrac{3}{2}\sqrt{xe^x + x^2}\left(e^x + xe^x + 2x\right);\ f'(2) = 170.09$

16. $f'(x) = \dfrac{3}{4}\dfrac{\left(4e^{3x} + \sqrt{x}\right)}{\sqrt{x^{\frac{3}{2}} + 2e^{3x}}};\ f'\left(\dfrac{1}{2}\right) = 4.58$

17. $(-\infty, 2),\ f'(x) > 0$, increasing;
$(2, +\infty),\ f'(x) < 0$, decreasing;
$x = 2,\ f'(2) = 0$, local max;
$(-\infty, +\infty),\ f''(x) < 0$, concave down

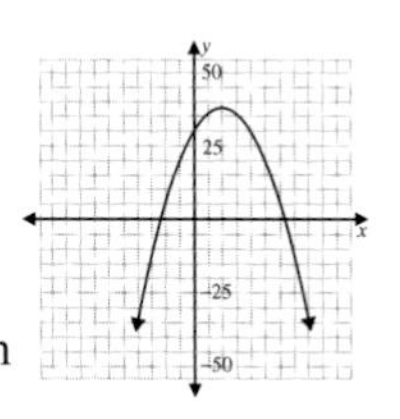

18. $\left(-\infty, -\sqrt{\dfrac{3}{2}}\right), f'(x) > 0$, increasing;
$\left(-\sqrt{\dfrac{3}{2}}, 0\right), f'(x) < 0$, decreasing;
$\left(0, \sqrt{\dfrac{3}{2}}\right), f'(x) < 0$, decreasing;
$\left(\sqrt{\dfrac{3}{2}}, \infty\right), f'(x) > 0$, increasing;
$f'\left(-\sqrt{\dfrac{3}{2}}\right) = 0$, local max; $f'\left(\sqrt{\dfrac{3}{2}}\right) = 0$, local min;
$(-\infty, 0), f''(x) < 0$, con. down; $(0, \infty), f''(x) > 0$, con. up

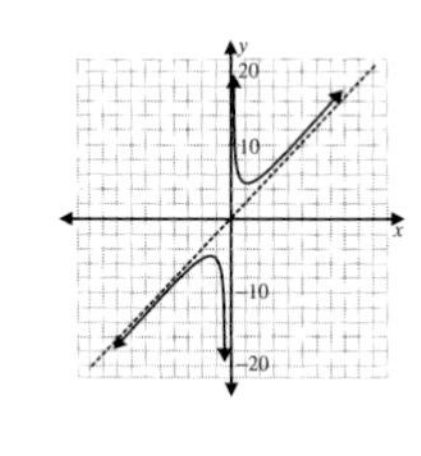

19. a. $x = 0$ is a vertical asymptote; $y = x$ is an oblique asymptote
b. $f'(x) = \dfrac{x^2 - 9}{x^2};\ f''(x) = \dfrac{18}{x^3}$ **c.**

20. infl. pt.: (0, 11); Concave down: $(-\infty, 0)$; Concave up: $(0, +\infty)$

21. a. $\left(\sqrt{\dfrac{1}{3}}, \dfrac{3}{4}\right), \left(-\sqrt{\dfrac{1}{3}}, \dfrac{3}{4}\right)$; **b.** Concave down: $\left(-\sqrt{\dfrac{1}{3}}, \sqrt{\dfrac{1}{3}}\right)$. Concave up: $\left(-\infty, -\sqrt{\dfrac{1}{3}}\right)$, and $\left(\sqrt{\dfrac{1}{3}}, \infty\right)$.

22. In the interval $(0, \infty)$, W is increasing. **23. a.** $\dfrac{dy}{dx} = 9x^2 - 144;\ \dfrac{d^2y}{dx^2} = 18x$ **b.** $x = \pm 4$ **c.** $(4, -384)$ min. $(-4, 384)$ max
d. $(0, 0)$ POI **24. a.** $\dfrac{dy}{dx} = 30 - 2x;\ \dfrac{d^2y}{dx^2} = -2$ **b.** $x = 15$ **c.** $(15, 365)$ max **d.** No inflection points

25.

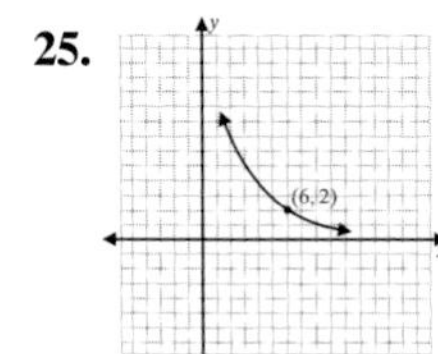

26.

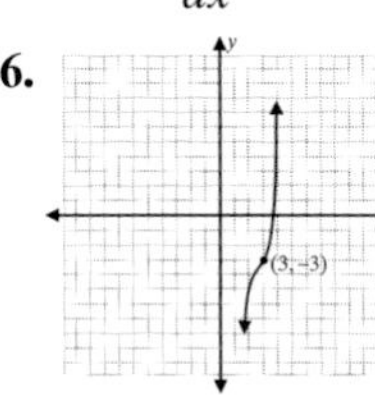

27. a. $C'(x) = 0.004x + 29$, $R'(x) = 45$ **b.** $P(x) = -0.002x^2 + 16x - 2400$
$P'(x) = -0.004x + 16$ **c.** $x = 4000$ **28.** \$5.10 **29.** $\dfrac{dy}{dx} = 5 - \dfrac{1}{x}$
30. $\dfrac{dy}{dx} = x(1 + 2\ln x)$ **31.** $f'(x) = \dfrac{x}{x^2 + 2}$

32. $f'(x) = 2x^{-3}\left[\dfrac{x}{2x+1} - \ln(2x+1)\right]$ **33.** $\dfrac{dy}{dx} = \dfrac{2\left(10x^2 + 62x + 25\right)(2x+1)}{(x+4)^2}$ **34.** $\dfrac{dy}{dx} = \dfrac{11x^2\left(12x^2 - 70x - 135\right)}{(2x-9)^2(4x+5)^{\frac{3}{2}}}$

35. Abs. min.: $(1, -0.386)$; Abs. max.: $(3, 0.227)$ **36.** Abs. min.: $(6, -15.501)$; Abs. max.: $(1, 1)$

37.

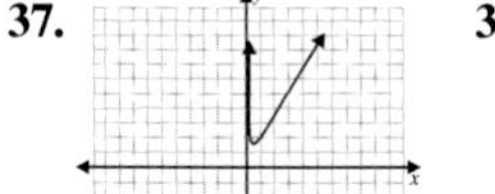

38.

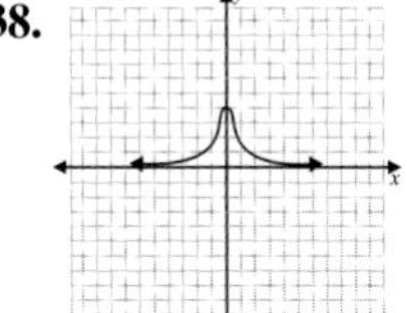

39. 3436 years **40.** 22 minutes; 12.2 minutes **41.** 187 units

42. a. $E(x) = \dfrac{63 - 0.9x}{0.9x}$ **b.** $x = 35$

43. a. $E(x) = \dfrac{1}{0.3x}$ **b.** $x = \dfrac{10}{3}$

Chapter 6

Exercises 6.1, pages 477 - 480

1. $F'(x)=4$ **3.** $F'(x)=6x+5$ **5.** $F'(x)=\frac{1}{x}+\frac{1}{x^2}-8xe^{x^2}$ **7.** $F'(x)=8x(x^2+3)^3$ **9.** $F'(x)=5e^x(e^x-4)^2$

11. $F'(x)=\frac{2x+5}{x^2+5x-3}$ **13.** $7x+C$ **15.** x^5+C **17.** $\frac{1}{3}x^3-3x+C$ **19.** $\frac{1}{3}t-e^t+C$ **21.** $\ln|y|+\frac{1}{4}y^4+C$

23. $\frac{4}{3}x^3+2\ln|x|-\frac{1}{x}+C$ **25.** $\frac{3}{2}x^{\frac{4}{3}}+\frac{10}{3}x^{\frac{3}{2}}+C$ **27.** $4e^y-\frac{1}{3}y^6-\frac{1}{5}y+C$ **29.** $3t^{\frac{2}{3}}-\frac{7}{2t^2}+C$ **31.** $\frac{2}{3}e^x-2x^{-\frac{1}{2}}-7\ln|x|+C$

33. $\frac{2}{5}x^5-\frac{1}{4}x^4+C$ **35.** $3t^3+6t^2+4t+C$ **37.** $\frac{2}{7}y^{\frac{7}{2}}+\frac{4}{5}y^{\frac{5}{2}}-\frac{2}{3}y^{\frac{3}{2}}+C$ **39.** $3x+5\ln|x|+\frac{4}{x}+C$ **41.** $4\ln|x|+2\sqrt{x}-3x+C$

43. $\frac{2}{3}x^{\frac{3}{2}}+6\ln|x|-2e^x+C$ **45. a.** $F(x)=\frac{1}{3}x^3-e^x+2$ **b.** $F(x)=2x^3+\frac{1}{2}x^2-10x$ **c.** $F(x)=20\sqrt{x}$ **d.** $F(x)=6e^x-2x-16$

47.

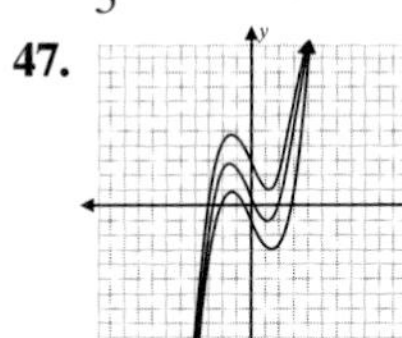

49. $C(x)=28x+0.025x^2+2400$ **51.** $R(x)=94x-0.03x^2$ **53. a.** $P(x)=24x-0.2x^2-400$ **b.** \$140

55. $P(t)=8600+120t-10t^{\frac{3}{2}}$ **57.** $L(x)=1+0.2x+0.001x^2$ **59. a.** $s(t)=-16t^2+96t+18$ **b.** 162 ft.

Exercises 6.2, pages 487 - 490

1. $\frac{1}{8}(x+4)^8+C$ **3.** $\frac{2}{3}(x^2-1)^{\frac{3}{2}}+C$ **5.** $\ln|t+2|+C$ **7.** $\ln|t^3+4|+C$ **9.** $e^{y+5}+C$ **11.** $e^{-0.2x}+C$ **13.** $\frac{1}{5}\ln|5x+3|+C$

15. $\frac{1}{2}\sqrt{4x-1}+C$ **17.** $\frac{1}{4}e^{2x^2}+C$ **19.** $-\frac{1}{6}(3x^2-1)^{-1}+C$ **21.** $\ln|t^2+t-4|+C$ **23.** $-\frac{5}{2}e^{-y^2}+C$ **25.** $(5+2x^2)^{\frac{3}{2}}+C$

27. $-4e^{\frac{1}{x}}+C$ **29.** $-\frac{1}{18}(1-3e^{2x})^3+C$ **31.** $\frac{1}{2}(\ln x)^2+C$ **33.** $\ln|\ln x|+C$ **35.** $\ln(e^x+e^{-x})+C$ **37. a.** $x+\ln|x-2|+C$

b. $x+2\ln|x+1|+C$ **39.** $x-2\ln|x+4|+C$ **41.** $f(x)=e^{5x^2}+10$ **43.** $f(x)=\frac{(2x+2)^6}{12}+10$ **45. a.** $V(t)=5000(25-1.8t)^{-\frac{1}{2}}$

b. \$1250 **47.** $p(x)=14-\frac{1}{2}\ln x$ **49.** $N(t)=6\sqrt{t^2+5t}$ bikes **51.** $s(t)=36t-\frac{60}{t+1}+4$ **53. a.** $v(t)=-8(2t+1)^3+14$

b. $s(t)=-(2t+1)^4+14t+\frac{1}{4}$ **55. a.** Average age of death per year **b.** $f(t)=30\ln|1+0.01t|+30$ **c.** 50.8 years

57. a. 127 students/yr. **b.** 2121 students **59. a.** $a(t)=30+\frac{12}{\sqrt{t}}$ **b.** $s(t)=200t+15t^2+16t^{\frac{3}{2}}+5000$

Exercises 6.3, pages 507 - 509

1. 12 **3.** 21 **5.** 5.38 **7.** $\frac{5}{3}$ **9.** $2+e^3\approx 22.09$ **11.** $\ln 3\approx 1.10$ **13.** $\frac{1}{4}$ **15.** 20 **17.** $e^4-1\approx 53.60$ **19.** 42 **21.** $\frac{1}{3}\ln 10\approx 0.77$

23. $\frac{1}{80}$ **25.** $\frac{121}{5}$ **27.** $\frac{3}{2}\ln 33\approx 5.24$ **29.** $\frac{1}{2}\left[(e+1)^2-4\right]\approx 4.91$ **31.** $\frac{1}{2}(e^8-1)\approx 1489.98$ **33.** 1.32 **35.** $\ln(e+1)-\ln 2\approx 0.62$

37. $\frac{1}{2}\left[(1+\ln 3)^2-1\right]\approx 1.7$ **39.** 13 **41.** $\frac{38}{15}$ **43.** $2(1-e^{-1})\approx 1.26$ **45.** $360(6-\sqrt{11})\approx 966$ ppm **47.** 74.80 units

Exercises 6.4, pages 516 - 517

1. $\frac{85}{2}$ **3.** $\frac{28}{3}$ **5.** 4 **7.** $\frac{52}{3}$ **9.** $4\ln 4 \approx 5.55$ **11.** $5\left(e^{3}-1\right) \approx 95.43$ **13.** $\frac{38}{3}$ **15.** $\frac{41}{6}$ **17.** $\frac{44}{3}$ **19.** Total revenue from the sale of the first *a* units of a product. **21.** \$1800 **23.** \$1059 **25.** 28 bison

Exercises 6.5, pages 526 - 530

1. $\frac{115}{6}$ **3.** $\frac{1}{12}$ **5.** $\frac{31-5\sqrt{5}}{3} \approx 6.61$ **7.** $6.5+e^{-3} \approx 6.55$ **9.** $\frac{15}{8}-\ln 2+\ln\frac{1}{2} \approx 0.489$ **11.** $\frac{4}{3}$ **13.** $\frac{1}{3}$ **15.** 36 **17.** $A = 72.9$ sq. units **19.** $A \approx 0.693$ sq. units **21.** $A = 48$ sq. units **23.** $A \approx 3.06$ sq. units **25.** $A \approx 2.42$ sq. units **27.** \$154.52 **29.** \$2.89 **31.** $(3,80)$, $CS = \frac{135}{2}$, $PS = 45$ **33. a.** (20, \$20) **b.** \$40 **c.** \$100 **35. a.** (25, \$41) **b.** \$208.33 **c.** \$312.50 **37. a.** 61 percent **b.** $\frac{7}{48}$ **39. a.** 27 percent **b.** $\frac{23}{75}$ **41. a.** *A*: 0.26; *B*: 0.245 **b.** *B*

Exercises 6.6, pages 538 - 541

1. $\frac{dy}{dx}=6$ **3.** $\frac{dy}{dx}=e^{x}=3+y$ **5.** $\frac{dy}{dx}=\frac{8}{(5-x)^{3}}=y^{\frac{3}{2}}$ **7.** $2\frac{dy}{dx}+3y=2\left(3e^{-1.5x}\right)+3\left(\frac{1}{3}-2e^{-1.5x}\right)=6e^{-1.5x}+1-6e^{-1.5x}=1$

9. $2x^{2}y''-xy'-2y=2x^{2}(6)-x(6x+5)-2\left(3x^{2}+5x-2\right)=4-15x$ **11.** $x^{2}y''-xy'+y=x^{2}\left(\frac{1}{x}\right)-x(1+\ln x)+x\ln x=0$

13. $y=\frac{3}{2}x^{2}+\ln|x|+C$ **15.** $y=\frac{C(x+1)^{2}+1}{2}$ **17.** $y=Ce^{-0.4x}$ **19.** $y=\frac{2}{5x^{2}-2x+C}$ **21.** $y=\frac{20}{1+Ce^{-12x}}$ **23.** $y=5e^{0.25x}$

25. $y=5x-1$ **27.** $y=\frac{2}{4x^{3}-3}$ **29.** $y=14e^{x^{2}}-4$ **31.** $y=\frac{120}{3+e^{-32x}}$ **33.** $D(x)=\frac{96}{x^{\frac{2}{3}}}$ **35.** $D(x)=18\sqrt{120-x}$

37. a. $V=24{,}000e^{-0.06t}$ **b.** \$15,769.12 **39. a.** $T(t)=Ce^{-kt}+M$ **b.** 37.6° **41. a.** $N(t)=\frac{2800}{1+139e^{-0.84t}}$ **b.** 6 days

Chapter 6 Test, page 548

1. $G'(x)=8-6x$ **2.** $G'(x)=-\frac{6}{x^{2}}-\frac{2}{x}$ **3.** $x^{2}+2x^{6}+C$ **4.** $\frac{6}{7}x^{\frac{7}{2}}-\frac{8}{5}x^{\frac{5}{2}}+4x^{\frac{3}{2}}+C$ **5.** $2\ln|x|+3x+\frac{x^{2}}{2}+C$

6. $\frac{2}{5}x^{\frac{5}{2}}+2x^{\frac{3}{2}}-10x^{\frac{1}{2}}+C$ **7.** $\frac{e^{x^{4}}}{4}+C$ **8.** $\left(x^{3}+4\right)^{4}+C$ **9.** 29 **10.** 9.49 **11.** 1475.18

12. $\frac{8}{3}$

13. 8.45 **14.** $\frac{1}{8}\ln\left|4e^{2x}+1\right|+C$ **15.** 9

16. $y'=e^{x}+2e^{2x}=e^{x}+e^{2x}+e^{2x}=y+e^{2x}$ The given function is a solution. **17.** $y=\frac{2x^{2}}{x^{2}+1}$ **18.** 708 mosquitos

Chapter 6 Cumulative Review, pages 549 - 554

1. a. 3.0 **b.** True **c.** $\lim_{x\to 3^-} f(x) = 1.038$ **d.** $\frac{27}{26}$

2. a. Yes, answers will vary. **b.** Yes, answers will vary. **c.** Yes, answers will vary. **d.** Yes, answers will vary. **3.**

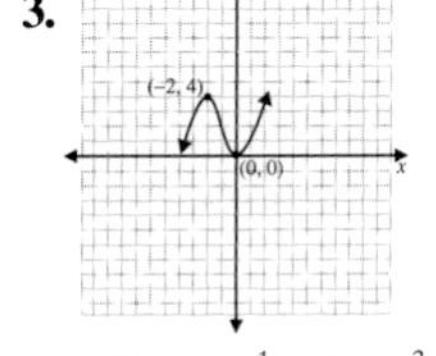

4. slope is ln 3; $y = x\ln 3 + 1$ **5. a.** Pop. in 2000 is 53. **b.** Pop. is decreasing by 1 gorilla per year.
6. a. $D(x) = 20$ **b.** $C(x) = 300 + 10x$ **c.** $R'(x) = 20;\ C'(x) = 10$ **d.** No **7. a.** $f'(6)$ is positive

b. At noon the average temperature is increasing by 2 degrees Fahrenheit per hour. **8.** $\frac{df}{dx} = 3(1 - x^2)$

9. $\frac{df}{dx} = \frac{15x^3 + 10x - 24}{x^3}$ **10.** $\frac{df}{dx} = 3x^2 - \frac{5}{2}x^{\frac{3}{2}} + 8x$ **11.** $\frac{df}{dx} = -3x^2 + 4x - 1$ **12.** $\frac{df}{dx} = \frac{5x^2 + 3x - 1}{2\sqrt{x}}$ **13.** $\frac{df}{dx} = x^{-\frac{1}{2}} + \frac{3}{2}x^{-\frac{3}{2}}$

14. $H'(s) = \frac{s}{\sqrt{s^2 + 1}}$ **15.** $g'(t) = \frac{-13}{2(12t + 5)^2\sqrt{\frac{7t + 4}{12t + 5}}}$ **16.** $g'(t) = \frac{50(-1 + \sqrt{t})}{(100 + \sqrt{t})^2(\sqrt{t})(\sqrt{100 + t})}$

17. $a = -5.943; b = 5.609;\ y = 0$ is the horizontal asymptote **18.** $-\infty$ **19.** $+\infty$ **20.** $+\infty$ **21.** $-\infty$
22. a. limit does not exist **b.** limit does not exist **c.** limit exists **23.** $y = 6x - 22$

24. $C(250) = 5425$; It costs \$5425 to make 250 belts. $C'(250) = 33.5$; The cost is increasing \$33.50/belt at $x = 250$.
25. $R(x) = 28x; R'(x) = 28; R(250) = \$7000; R'(250) = 28$ **26.** $P(250) = \$1575; P'(250) = -5.5$; Profit is decreasing \$5.50/belt at $x = 250$.

27. a. $S(t) = \frac{6x^3}{10^6} + \frac{7x^2}{10^5} + 0.136x$ **b.** **c.** yes **d.** $S'(t) = \frac{18x^2}{10^6} + \frac{14x}{10^5} + 0.136$ **e.** 0.1564

f. $S(50) = 7.725;\ S'(50) = 0.188$

28. $f'(x) = -x + \frac{x^3}{6} - \frac{x^5}{120};\ f''(x) = -1 + \frac{x^2}{2} - \frac{x^4}{24};\ f'''(x) = x - \frac{x^3}{6};\ f^{(4)}(x) = 1 - \frac{x^2}{2}$

29. $f'(x) = 1 + x + \frac{x^2}{2} + \frac{x^3}{6} + \frac{x^4}{24};\ f''(x) = 1 + x + \frac{x^2}{2} + \frac{x^3}{6};\ f'''(x) = 1 + x + \frac{x^2}{2};\ f^{(4)}(x) = 1 + x$

30. $(-\infty, 0), (1, +\infty); f'(x) > 0$; Increasing $(0, 1); f'(x) < 0$; Decreasing
$x = 0; f'(0) = 0$; Local max. $x = 1; f'(1) = 0$; Local min.
$\left(-\infty, \frac{1}{2}\right); f''(x) < 0$; Concave down. $\left(\frac{1}{2}, +\infty\right); f''(x) > 0$; Concave up; $x = \frac{1}{2};\ f''\left(\frac{1}{2}\right) = 0$; POI

31. $(-\infty, 0), (0, 11), f'(x) < 0$, Decreasing. $(11, +\infty), f'(x) > 0$, Increasing. $x = 11; f'(11) = 0$; Local min.
$(-\infty, 0), (7.\overline{3}, +\infty), f''(x) > 0$, Concave up. $(0, 7.\overline{3}), f''(x) < 0$, Concave down. $f''(7.\overline{3}) = f''(0) = 0; x = 0$ and $x = 7.\overline{3}$, POI.

32. a. 450, 391.18 **b.** –0.012; The value of a vest is decreasing when time is 100 months by \$0.012/month.
c. 390; \$390 is the value at which each vest levels off.
33. $x = 0$ is a local minimum
34. $x = \pm 2$ are local mins; $x = 0$ is a local max; (–1.55, –20.74) and (1.55, –20.74) are points of inflection.
35. $x = 3$ is a local min; $x = 0$ is a local max; (1.89, –433.96) is a point of inflection.

36. $x = 2$ is a local minimum. **37. a.** $x = 45.95$ **b.** $P'(a)$ is the maximum rate of change for $P(a)$.
38. a. 9132.5 ft.^{3}; 16,666.7 ft.^{3}

b. 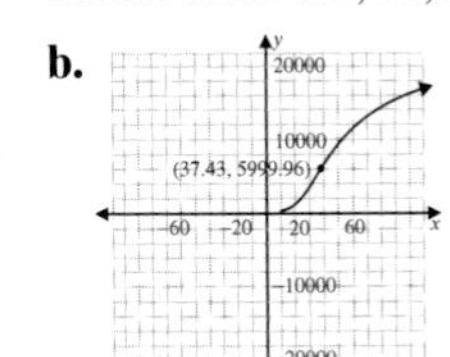

39. 70 ft. by 70 ft.; 4900 ft.^{2} **40.** $r = 84.015$ in.; $h = 84.015$ in. **41. a.** 21 ft. **b.** 10 ft per sec. **c.** 2 ft. per sec^2. **42.** 1.82 ft. **43.** $(0.5, 9.75)$ **44.** $(2, 8), (-2, -8)$

45. a. $\frac{df}{dx} = 24(1-x)(1+4x-2x^2)$ **b.** $\frac{df}{dx} = 0$ **46.** $\frac{1}{3}x^3 + 2x^2 - x + C$ **47.** $\frac{1}{4}y^4 + \frac{7}{3}y^3 - y^2 + C$

48. $e^x - \frac{4}{3}x^{\frac{3}{2}} + C$ **49.** $4x^{\frac{1}{2}} + \ln|x| + 2x + C$ **50.** $\frac{75}{4}$ **51.** $5\ln 3 \approx 5.49$ **52.** $\frac{1}{4}\left[\ln(4e+1) - \ln\left(\frac{4}{e}+1\right)\right] \approx 0.39$

53. $\frac{1}{2}(\ln 18 - \ln 2) = \ln 3 \approx 1.10$ **54.** $\frac{73}{6}$ **55.** $\frac{3}{4} + 2e^2 - 2e^{0.5} \approx 12.23$ **56. a.** 30.75 percent **b.** 0.257 **57.** \$1260

Chapter 7

Exercises 7.1, pages 562 - 563

1. $\frac{1}{4}e^{2x}(2x-1)+C$ **3.** $4e^{0.5y}(y-2)+C$ **5.** $t(\ln t - 1) + C$ **7.** $\frac{1}{16}x^4(4\ln 5x - 1)+C$ **9.** $\frac{2}{15}(x+2)^{\frac{3}{2}}(3x-4)+C$

11. $-x(x+4)^{-1} + \ln|x+4| + C$ **13.** $-\frac{5}{18}e^{-0.6t}(3t+5)+C$ **15.** $\frac{2}{9}x^{\frac{3}{2}}(3\ln 7x - 2)+C$ **17.** $\frac{1}{4} - \frac{5}{4}e^{-4} \approx 0.23$

19. $\frac{9}{16} - \frac{13}{16}e^{-4} \approx 0.55$ **21.** $\frac{2}{3}$ **23.** $u = 4x, dv = (3x+1)^5\,dx;\ \frac{5158}{7} \approx 736.86$ **25.** $u = x+1, dv = (x+2)^{\frac{3}{2}}\,dx;\ \frac{836}{35} \approx 23.89$

27. $u = \ln x, dv = x^2;\ \frac{125}{3}\ln 5 - \frac{124}{9} \approx 53.28$ **29.** $u = \ln(x+1), dv = dx;\ 7\ln 7 - 6 \approx 7.62$ **31.** $-\frac{5}{4}e^{-2t}(2t+1)+C$

33. $\frac{2\sqrt{3}}{9}x^{\frac{3}{2}}(3\ln x - 2)+C$ **35.** $\frac{3}{10}(2x^2-1)^{\frac{5}{2}}+C$ **37.** $\frac{1}{3}(\ln x)^3 + C$ **39.** $-\ln|1-e^x|+C$

41. $D(t) = 100te^{0.01t} - 10{,}000e^{0.01t} + 10{,}080$ million units **43.** $R(x) = 18.4x - 0.4x\ln x$ dollars

Exercises 7.2, pages 569 - 571

1. \$15,319.26 **3.** \$63,498.35 **5.** \$17,872.00 **7.** \$1,637,022.23 **9.** \$22,600.56 **11.** \$520.96 **13.** \$155,907.16 **15.** \$568.89
17. \$21,963.97 **19.** \$126,906.09 **21.** No, it gives the approximate value.

Exercises 7.3, page 577

1. $\frac{1}{4}\ln|4x+3|+C$ **3.** $-\frac{20}{3}e^{-0.15x}+C$ **5.** $-\frac{1}{2(2x-5)}+C$ **7.** $\frac{2(9x+8)}{135}\cdot(3x-4)^{\frac{3}{2}}+C$

9. $\frac{x}{2}\sqrt{x^2+36} + 18\ln\left|x+\sqrt{x^2+36}\right|+C$ **11.** $\frac{1}{8}\ln\left|\frac{x-4}{x+4}\right|+C$ **13.** $-\frac{1}{41}\ln\left|\frac{x+8}{5x-1}\right|+C = \frac{1}{41}\ln\left|\frac{5x-1}{x+8}\right|+C$ **15.** $\frac{7x^6}{6}\left(\ln x - \frac{1}{6}\right)+C$

17. $\frac{x}{8} + \frac{1}{5.6}\ln\left|8e^{-0.7x}-5\right|+C$ **19.** $-2\ln\left|\frac{x}{3x-1}\right|+C = 2\ln\left|\frac{3x-1}{x}\right|+C$ **21.** $\frac{14}{9}(6x-5)^{\frac{3}{2}}+C$

23. $-8\ln\left|\frac{2x+5}{x}\right|+C = 8\ln\left|\frac{x}{2x+5}\right|+C$ **25.** $\frac{x^3e^{1.5x}}{1.5} - \frac{3x^2e^{1.5x}}{(1.5)^2} + \frac{6xe^{1.5x}}{(1.5)^3} - \frac{6e^{1.5x}}{(1.5)^4}+C$ **27.** $\frac{1}{37}\left(\frac{2}{5}\ln|5x+2| + 7\ln|x-7|\right)+C$

29. $\ln\left|x+\sqrt{x^2-12}\right|+C$

Exercises 7.4, pages 586 - 587

1. 0 **3.** $+\infty$ **5.** 0 **7.** 2 **9.** 0 **11.** $\frac{1}{2}$ **13.** Divergent **15.** $3e^{-20} \approx 6.2\text{E}^{-9}$ **17.** $3e^{-\frac{2}{3}} \approx 1.54$ **19.** Divergent **21.** $\frac{1}{3}$
23. Divergent **25.** Divergent **27.** $\frac{1}{3}$ **29.** $\frac{1}{2}$ **31.** $\frac{1}{2(\ln 2)^2} \approx 1.04$ **33.** 1 **35.** 2 **37.** Area unbounded, integral divergent
39. Converges for case **d**.

Exercises 7.5, pages 595 - 597

1. $\int_0^4 \frac{3}{16}\sqrt{x}\,dx = \frac{3}{16}\left(\frac{2}{3}\right)x^{\frac{3}{2}}\Big]_0^4 = 1$ **3.** $\int_{-1}^{2} \frac{4}{15}\left(\frac{1}{2}x+1\right)dx = \frac{4}{15}\left(\frac{x^2}{4}+x\right)\Big]_{-1}^{2} = 1$ **5.** $\int_1^9 \frac{3}{68}\left(x-\sqrt{x}\right)dx = \frac{3}{68}\left(\frac{x^2}{2}-\frac{2}{3}x^{\frac{3}{2}}\right)\Big]_1^9 = 1$
7. $\int_1^{e^2} \frac{1}{2x}dx = \frac{1}{2}\ln|x|\Big]_1^{e^2} = 1$ **9.** $\int_{-2}^{2} \frac{3}{32}\left(4-x^2\right)dx = \frac{3}{32}\left(4x-\frac{x^3}{3}\right)\Big]_{-2}^{2} = 1$
11. $\int_0^{+\infty} 2e^{-2x}dx = 2\lim_{b\to+\infty}\int_0^b e^{-2x}dx = 2\lim_{b\to+\infty}\left(\frac{e^{-2x}}{-2}\right)\Big]_0^b = -\lim_{b\to+\infty}\left(e^{-2b}-1\right) = -(-1) = 1$ **13. a.** $\frac{2}{9}$ **b.** 1 **15. a.** $\frac{1}{2}$ **b.** $\frac{7}{3}$ **17. a.** $\frac{1}{4}$ **b.** 4
19. a. $\frac{32}{15}$ **b.** $\frac{8}{5}$ **21.** $\frac{13}{27}$ **23.** 0.018 **25.** $\frac{2}{5}$; 10 minutes **27.** $\frac{1}{64}$; 1.8 seconds **29.** 0.26

Exercises 7.6, page 602

1. $\frac{8\pi}{3}$ **3.** 32π **5.** $\frac{\pi}{2}\left(e^4-e^{-2}\right) \approx 27.2\pi$ **7.** $\frac{16\pi}{15}$ **9.** $\frac{128\pi}{3}$ **11.** $\frac{32\pi}{3}$ **13.** $\pi\ln 6 \approx 1.79\pi$ **15.** 53.3π

Chapter 7 Test, pages 607 - 608

1. $\frac{5}{8}xe^{1.6x} - \frac{25}{64}e^{1.6x} + C$ **2.** $\frac{1}{3}xe^{3x} + \frac{14}{9}e^{3x} + C = \frac{1}{9}e^{3x}(3x+14)+C$
3. $\frac{2}{9}x(3x+1)^{\frac{3}{2}} - \frac{4}{135}(3x+1)^{\frac{5}{2}} + C = \frac{2}{135}(3x+1)^{\frac{3}{2}}(9x-2)+C$ **4.** $-\frac{1}{4}x(2x+1)^{-2} - \frac{1}{8}(2x+1)^{-1} + C = -\frac{1}{8}(2x+1)^{-2}(4x+1)+C$
5. $\frac{1}{5}x(x+6)^5 - \frac{1}{30}(x+6)^6 - \frac{3}{5}(x+6)^5 + C = \frac{1}{30}(x+6)^5(5x-24)+C$ **6.** $\frac{3}{4}x^{\frac{4}{3}}\ln|6x| - \frac{9}{16}x^{\frac{4}{3}} + C = \frac{3}{16}x^{\frac{4}{3}}\left(4\ln|6x|-3\right)+C$
7. $-\frac{1}{2}x^{-2}\ln|x| - \frac{1}{4}x^{-2} + C = -\frac{1}{4}x^{-2}\left(2\ln|x|+1\right)+C$ **8.** $\frac{1}{2}x^2\ln|7x| - \frac{1}{4}x^2 + 2x\ln|7x| - 2x + C$ **9.** $8e^{-1} - 12e^{-2} \approx 1.32$
10. $-2(4)e^{-1.5} + 2(1) \approx 0.215$ **11.** $\frac{5508}{135} = 40.8$ **12.** $\frac{4^4(-19)-2^4(-27)}{20} = -221.6$ **13.** $12\ln 8 - \frac{135}{16} \approx 16.52$
14. $2\ln 6 - \frac{1}{2} \approx 3.08$ **15.** $\frac{x}{2}\sqrt{x^2+121} + \frac{121}{2}\ln\left|x+\sqrt{x^2+121}\right| + C$ **16.** $\frac{1}{16}\left(4x - 3\ln|4x+3|\right)+C$ **17.** $\frac{1}{240}\ln\left|\frac{x}{40-x}\right| + C$
18. $\frac{x}{2} - \frac{5}{2}\ln\left|2+5e^{0.2x}\right| + C$ **19.** Divergent **20.** -2 **21.** $\frac{5}{4}$ **22. a.** $\int_0^3 \frac{2}{45}\left(2x+\frac{3}{2}x^2\right)dx = \frac{2}{45}\left(x^2+\frac{1}{2}x^3\right)\Big]_0^3 = 1$
b. $\int_1^4 \frac{10}{49}\left(x^{\frac{3}{2}}-x\right)dx = \frac{10}{49}\left(\frac{2}{5}x^{\frac{5}{2}}-\frac{1}{2}x^2\right)\Big]_1^4 = 1$ **23.** $a = 0.25$ **24.** $\frac{198}{7}\pi$ **25.** $\frac{393}{5}\pi$ **26.** $\left(\frac{11}{2}-2e^{-3}-\frac{1}{2}e^{-6}\right)\pi \approx 5.40\pi$ **27.** 8π
28. \$32,178.07 **29.** \$477,293.05 **30.** \$32,411.16 **31.** \$115,577.27 **32.** 0.105 **33.** 2.37 minutes **34.** 39.1 minutes
35. 36.85 months

Chapter 7 Cumulative Review, pages 609 - 610

1. a. 10 **b.** $f(x)=x^2$ and $f'(5)=10$ **2. a.** 0.25 **b.** $f(x)=\sqrt{x}$ and $f'(4)=0.25$ **3. a.** 7.389 **b.** $f(x)=e^x$ and $f'(2)=7.389$ **4.** $R(x)=15+0.15x$ and $R'(x)=0.15$ **5.** $C(x)=6.00+0.02x$ and $C'(x)=0.02$ **6.** $P'(x)=0.13$ **7.** $f'(x)=0.5x^{-0.5}+3x^{-2}+3x^{-0.75}$ **8.** $f'(x)=10(2x+1)^4$ **9.** $f'(x)=3x(3x+1)^{-\frac{1}{2}}+2(3x+1)^{\frac{1}{2}}$ **10.** $f'(x)=e^x(10x+8)$

11. $f'(x)=\dfrac{(4x-1)\left(3+\dfrac{1}{x}\right)}{(4x-1)^2}-\dfrac{4(3x+\ln x)}{(4x-1)^2}$ **12.** $f'(x)=2^x\ln 2+\dfrac{1}{x\ln 10}$ **13.** $P'(x)=\dfrac{6e^{-x}}{\left(1+2e^{-x}\right)^2}$ **14.** $(x, y) = (0.69, 1.5)$.
In August of 2000 population growth reached a maximum. Although population continues to increase, growth rate declines. **15.** $y = 3$ **16.** Increasing on $(-\infty, 1)$, decreasing on $(1, \infty)$

17. Increasing on $(-\infty, -6)$ and $(0, \infty)$, decreasing on $(-6, 0)$ **18.** 152.25 **19.** 3.53

20. a. $\int_0^7\left[(22-x)-(1+2x)\right]dx$ **b.** 73.5 **21.** $f(x)=2x(2x+1)^{\frac{3}{2}}-\frac{2}{5}(2x+1)^{\frac{5}{2}}+C$ **22.** $f(y)=\frac{1}{2}ye^{6y}-\frac{1}{12}e^{6y}+C$
23. $f(x)=\frac{1}{3}(2x+1)^{\frac{3}{2}}\ln(2x+1)-\frac{2}{9}(2x+1)^{\frac{3}{2}}+C$ **24.** $f(x)=\frac{1}{4}(\ln x)^4+C$ **25.** \$26,808

26. \$222,763.36 **27.** $f(x)=\ln\left|x+\sqrt{x^2+9}\right|+C$ **28.** $\frac{1}{6}\ln\left|\frac{x-3}{x+3}\right|+C$ **29.** $f(x)=\ln|x-3|-\ln|x-2|+C$

30. 1.4715 **31.** 0.2194 **32.** 0.75 **33.** One suitable sequence is e^{e^x} for $x = 1, 2, 3, \ldots$ **34.** $k=\frac{2}{85}$ **35.** $\bar{y}=8.5$
36. $p=\frac{26}{25}-\frac{78}{25e^2}\approx 0.62$ **37.** 278.4 **38.** 140.96 **39.** 56.55 **40.** 26.28

Chapter 8

Exercises 8.1, pages 620 - 623

1. a. 6 **b.** −12 **3. a.** 11 **b.** 9 **5. a.** $\frac{11}{4}$ **b.** undefined **7. a.** undefined on $\mathbb{R}$ **b.** $96\sqrt{2}$ **9. a.** $6e^3\approx 120.51$ **b.** $-3e^{-1}\approx -1.10$
11. a. 1 **b.** $2e^2+2\approx 16.78$ **13. a.** 1377.13 **b.** 674.93 **15. a.** 1134 **b.** 758

17.

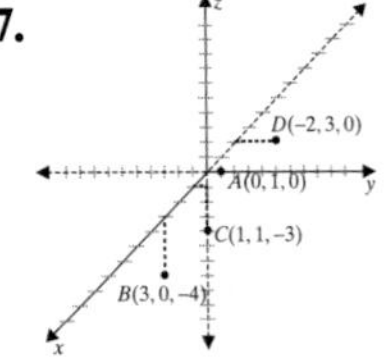

19.

21.

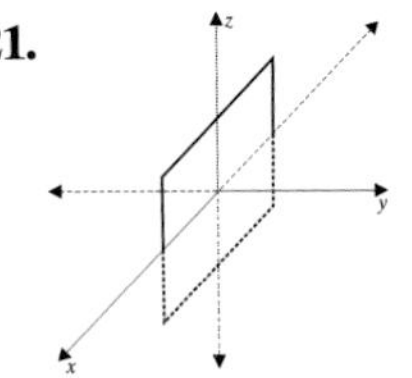

23.

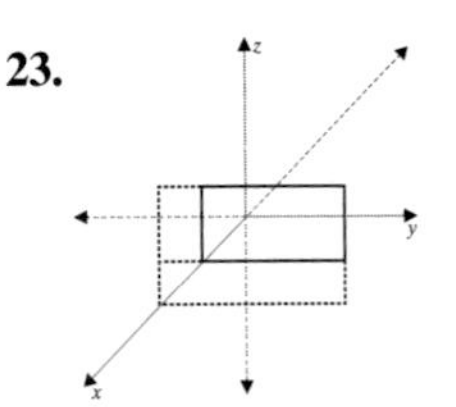

25.

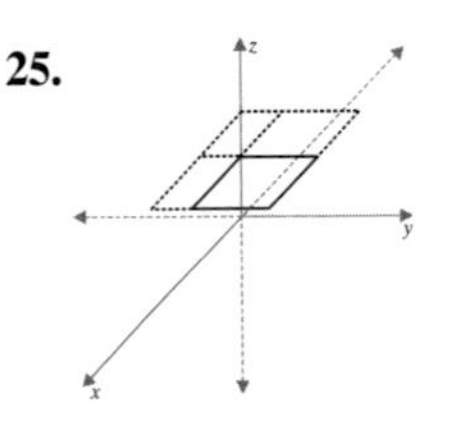

27. 118 **29. a.** \$10,400 **b.** \$11,580 **31. a.** $C(x, y) = 1360 + 160x + 220y$ **b.** \$6400
33. a. $R(x, y)=-20x^2-22y^2+64x+52y+34xy$ **b.** \$336.50 **35. a.** $V(x,y,z) = xyz$ **b.** $S(x,y,z) = xy + 2xz + 3yz$

37.

Number of Years (t)		
3	5	10
1195.62	1346.86	1814.02
1268.24	1485.95	2208.04
1344.89	1638.62	2685.06

Exercises 8.2, pages 632 - 635

1. $\frac{\partial f}{\partial x}=4,\ \frac{\partial f}{\partial y}=7$ **3.** $\frac{\partial f}{\partial x}=4x,\ \frac{\partial f}{\partial y}=10y$ **5.** $\frac{\partial f}{\partial x}=2xy+4y^3,\ \frac{\partial f}{\partial y}=x^2+12xy^2$ **7.** $\frac{\partial f}{\partial x}=\frac{xy}{\sqrt{25+x^2}},\ \frac{\partial f}{\partial y}=\sqrt{25+x^2}$

9. $\frac{\partial f}{\partial x}=-\frac{x}{\sqrt{49-x^2-y^2}},\ \frac{\partial f}{\partial y}=-\frac{y}{\sqrt{49-x^2-y^2}}$ **11.** $\frac{\partial f}{\partial x}=4e^{x-y},\ \frac{\partial f}{\partial y}=-4e^{x-y}$ **13.** $\frac{\partial f}{\partial x}=\ln y,\ \frac{\partial f}{\partial y}=\frac{x}{y}$

15. $\frac{\partial f}{\partial x}=\frac{2x+3y}{x^2+3xy},\ \frac{\partial f}{\partial y}=\frac{3x}{x^2+3xy}$ **17.** $\frac{\partial f}{\partial x}=\frac{4x}{y^2+1},\ \frac{\partial f}{\partial y}=-\frac{4x^2y}{(y^2+1)^2}$ **19.** $\frac{\partial f}{\partial x}=\frac{x^2y+6x}{(xy+3)^2},\ \frac{\partial f}{\partial y}=-\frac{x^3}{(xy+3)^2}$

21. $\frac{\partial f}{\partial x}=3x^2e^y+2xye^{x^2},\ \frac{\partial f}{\partial y}=x^3e^y+e^{x^2}$ **23.** $\frac{\partial f}{\partial x}=\frac{3xy+4}{2\sqrt{xy+2}},\ \frac{\partial f}{\partial y}=\frac{x^2}{2\sqrt{xy+2}}$ **25.** $\frac{\partial f}{\partial x}=x^3e^{xy}(xy+4),\ \frac{\partial f}{\partial y}=x^5e^{xy}$

27. $\frac{\partial f}{\partial x}=\frac{2xy^5}{x^2-5y^2},\ \frac{\partial f}{\partial y}=\frac{-10y^6}{x^2-5y^2}+5y^4\ln(x^2-5y^2)$ **29.** $f_x(x,y,z)=y+2z,\ f_y(x,y,z)=x+9z,\ f_z(x,y,z)=2x+9y$

31. $f_x(x,y,z)=32x(8x^2+5y^2-2z^2),\ f_y(x,y,z)=20y(8x^2+5y^2-2z^2),\ f_z(x,y,z)=-8z(8x^2+5y^2-2z^2)$

33. $f_{xx}(x,y)=2y^3,\ f_{xy}(x,y)=3+6xy^2,\ f_{yx}(x,y)=3+6xy^2,\ f_{yy}(x,y)=6x^2y$

35. $f_{xx}(x,y)=12x^2y^{\frac{2}{3}},\ f_{xy}(x,y)=\frac{8}{3}x^3y^{-\frac{1}{3}},\ f_{yx}(x,y)=\frac{8}{3}x^3y^{-\frac{1}{3}},\ f_{yy}(x,y)=-\frac{2}{9}x^4y^{-\frac{4}{3}}$

37. $f_{xx}(x,y)=0,\ f_{xy}(x,y)=2e^{2y},\ f_{yx}(x,y)=2e^{2y},\ f_{yy}(x,y)=4xe^{2y}$

39. $f_{xx}(x,y)=\frac{160}{9}(4x-3y)^{-\frac{1}{3}},\ f_{xy}(x,y)=-\frac{40}{3}(4x-3y)^{-\frac{1}{3}},\ f_{yx}(x,y)=-\frac{40}{3}(4x-3y)^{-\frac{1}{3}},\ f_{yy}(x,y)=10(4x-3y)^{-\frac{1}{3}}$

41. $f_{xx}(x,y)=0,\ f_{xy}(x,y)=-\frac{15}{(5y+3)^2},\ f_{yx}(x,y)=-\frac{15}{(5y+3)^2},\ f_{yy}(x,y)=\frac{50(3x+1)}{(5y+3)^3}$

43. $f_{xx}(x,y)=\frac{4y^2}{(x-y)^3},\ f_{xy}(x,y)=-\frac{4xy}{(x-y)^3},\ f_{yx}(x,y)=-\frac{4xy}{(x-y)^3},\ f_{yy}(x,y)=\frac{4x^2}{(x-y)^3}$

45. $\frac{\partial S}{\partial m}=90m+22b-320,\ \frac{\partial S}{\partial b}=22m+6b-80$ **47.** $\frac{\partial S}{\partial m}=562m+33b-3494,\ \frac{\partial S}{\partial b}=66m+8b-412$

49. $\frac{\partial F}{\partial x}=8+15y+\lambda,\ \frac{\partial F}{\partial y}=15x-4y+\lambda,\ \frac{\partial F}{\partial \lambda}=x+y-60$

51. $\frac{\partial F}{\partial x}=10x+3y+14\lambda,\ \frac{\partial F}{\partial y}=3x-20y+17\lambda,\ \frac{\partial F}{\partial \lambda}=14x+17y-49\lambda$

53. $f_L(27,64)=\frac{640}{9},\ f_K(27,64)=15$ **55.** $\frac{\partial C}{\partial x}=7+0.06x-y,\frac{\partial C}{\partial y}=5.8-x+0.04y$

57. a. $P(x,y)=16x+12y-x^2+2xy-2y^2-8$ **b.** $P_x(20, 14) = 4$ $P_y(20, 14) = -4$. The marginal profit with respect to model A is $4 and the marginal profit with respect to model B is –$4 when 20 units of model A and 14 units of model B are being produced. **59. a.** $P_x(x, y) = 5000 + 10{,}000y - 1000x$ **b.** $P_y(x, y) = 10{,}000x$

Exercises 8.3, pages 641 - 643

1. $f(3,-1)=-14$ is a local minimum. **3.** $(5,1,-15)$ is a saddle point. **5.** $f\left(\frac{5}{2},4\right)=\frac{133}{4}$ is a local maximum.
7. $(1,-2,-3)$ is a saddle point. **9.** $f(-2,-1)=-18$ is a local minimum. **11.** $f(2,2)=-3$ is a local minimum.
13. $f(1,-2)=13$ is a local maximum. **15.** $f(1,3)=0$ is a local minimum; $(-1,3,4)$ is a saddle point.
17. $(2,5,7)$ is a saddle point; $f(2,1)=-25$ is a local minimum. **19.** $(2,1,8)$ is a saddle point; $(-2,-1,-8)$ is a saddle point.
21. $f(0,0)=0$ is a local minimum; $(2,2,4)$ is a saddle point; $(-2,2,4)$ is a saddle point. **23.** $f(1,2)=6$ is a local minimum.
25. 30 of model X, 25 of model Y. **27.** 200 of the 50-dollar racks, 240 of the 45-dollar racks.
29. \$16 for the quartz clock, \$13 for the keywind clock. **31.** 9 inches by 6 inches by 3 inches

Exercises 8.4, pages 649 - 651

1. $f(2,2)=8$ **3.** $f(1,-1)=7$ **5.** $f(9,6)=612$ **7.** $f(4,4)=128$ **9.** $f(5,2)=30$ **11.** $f(4,5)=112$ **13.** $f(5,2)=69$
15. $f(8,15)\approx 175$ **17.** $f(6,3)=f(-6,-3)=72$ is the maximum; $f(6,-3)=f(-6,3)=-72$ is the minimum
19. $f(12,-6)=864$ is a maximum; $f(4,2)=96$ is a minimum. **21.** 3 units in LA, 37 units in OKC.
23. approximately 3605 units **25.** 10 days in each area **27.** 36 feet by 69 feet with the interior lanes 36 feet long

Exercises 8.5, pages 658- 659

1. a. $y=0.96x+4.1$ **b.**

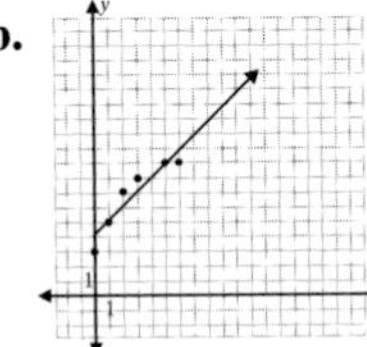

3. a. $y=-1.18x+10.96$ **b.**

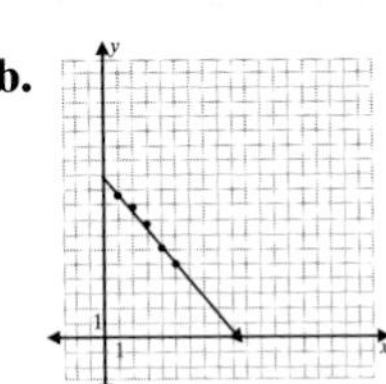

5. $y=-0.12x+8.19$

7. $y=14.7x+221.9$ **9.** $y=1.1x+4.02$ **11. a.** $y=3.77x+21.91$ **b.** 37,000 units **13. a.** $y=11.17x+4.07$ **b.** \$82,260
15. a. $y=-0.663x+76.04$ **b.** 71.40 cents per pound

Exercises 8.6, pages 667 - 669

1. $\frac{3}{2}$ **3.** $\frac{117}{2}$ **5.** $5\ln 2\approx 3.47$ **7.** $\frac{320}{3}$ **9.** $e^3-e^2-e+1\approx 10.98$ **11.** $\frac{4}{3}$ **13.** 56 **15.** $\frac{1}{2}\left(e^{11}-e^8-e^3+1\right)\approx 28437.05$
17. $\frac{288}{5}$ **19.** $\frac{508}{21}$ **21.** $\frac{26}{3}$ **23.** $\frac{98}{3}$ **25.** $\frac{1}{6}$ **27.** $\frac{31}{60}$ **29.** $\frac{e^6}{3}-e^2+\frac{2}{3}\approx 127.75$ **31.** $\frac{3}{20}$ **33.** $\frac{5}{3}$ **35.** 34 **37.** $\frac{32}{3}$ **39.** $\frac{1}{3}$

Chapter 8 Test, pages 674 - 676

1. –26, 96 **2.** Undefined, $-88+2\ln 2\approx -86.61$ **3.** $2, 16\sqrt{6}\approx 39.19$ **4.** $6{,}150e^{-2}\approx 20.30$ **5.** $\frac{\partial f}{\partial x}=10x-\frac{4}{y}$, $\frac{\partial f}{\partial y}=\frac{4x}{y^2}+4y$

6. $\frac{\partial f}{\partial x}=\frac{8e^{\frac{x}{y}}}{y}$, $\frac{\partial f}{\partial y}=\frac{-8xe^{\frac{x}{y}}}{y^2}$ **7.** $\frac{\partial f}{\partial x}=3x\ln\left(y^2+1\right)$, $\frac{\partial f}{\partial y}=\frac{3x^2y}{y^2+1}$ **8.** $\frac{\partial f}{\partial x}=-\frac{15y}{(x-5y)^2}$, $\frac{\partial f}{\partial y}=\frac{15x}{(x-5y)^2}$

9. $\frac{\partial f}{\partial x}=-\frac{xy^2}{\sqrt{6-x^2y^2}}$, $\frac{\partial f}{\partial y}=-\frac{x^2y}{\sqrt{6-x^2y^2}}$ **10.** $\frac{\partial f}{\partial x}=xy^2e^{xy^2}+e^{xy^2}=e^{xy^2}\left(xy^2+1\right)$, $\frac{\partial f}{\partial y}=2x^2ye^{xy^2}$

11. $\frac{\partial f}{\partial x}=3x^2y+9\lambda$, $\frac{\partial f}{\partial y}=x^3+1-2\lambda$, $\frac{\partial f}{\partial \lambda}=9x-2y-6$ **12.** $\frac{\partial f}{\partial x}=7y+2x\lambda$, $\frac{\partial f}{\partial y}=7x+2y\lambda$, $\frac{\partial f}{\partial \lambda}=x^2+y^2-16$

13. $f_{xx}(x,y)=20x^3$, $f_{xy}(x,y)=0$, $f_{yx}(x,y)=0$, $f_{yy}(x,y)=-4+4y$

14. $f_{xx}(x,y)=-\frac{y}{x^2}$, $f_{xy}(x,y)=2y+\frac{1}{x}$, $f_{yx}(x,y)=2y+\frac{1}{x}$, $f_{yy}(x,y)=2x-4e^{2y}$ **15.** $\left(-1,\frac{1}{2},\frac{61}{2}\right)$ is a saddle point.

16. (–4, 1, –18) is a local minimum. **17.** $\left(-4,-12,\frac{56}{3}\right)$ is a local maximum; $\left(1,3,-\frac{13}{6}\right)$ is a saddle point.

18. (0, 0, 11) is a saddle point; $\left(-2,2,\frac{29}{3}\right)$ is a local minimum. **19.** (5, 1, –140.5) is a local minimum; (–2, 1, 31) is a saddle point.

20. (4, 2, 44) is a saddle point. **21.** $f(9, 15) = 252$ is a minimum. **22.** $f(270, 250) = 393{,}800$ is a maximum.

23. $f(10, 24) = 1208.70$ is a maximum. **24.** $f(4, 8) = f(-4, -8) = 224$ is a maximum; $f(4, -8) = f(-4, 8) = -224$ is a minimum.

25. $y=-0.8x+15$ **26.** $y=0.57x+2.95$ **27.** –3 **28.** 3 **29.** $\frac{122}{3}$ **30.** –48.9 **31.** $\frac{1}{2}\left(e^2-e\right)\approx 2.34$ **32.** $\frac{63}{2}=31.5$ **33.** 14 **34.** $\frac{1}{3}$

35. $\frac{1}{3}\left(e^6-2e^3+1\right)\approx 121.42$ **36.** $\frac{709}{72}\approx 9.85$ **37.** $\frac{81}{4}=20.25$ **38.** $\frac{81}{2}=40.5$

39. $V(x,y,z)=xyz$, $S(x,y,z)=xy+4yz+2xz$ **40.** $f_L(40,30)=26.93, f_K(40,30)=53.85$. For each additional unit of labor, production will increase about 27 units. For each additional unit of capital, production will increase about 54 units.

41. 150 units on line X and 105 units on line Y. **42. a.** $y=0.05x+1.98$ **b.** 2.28 million units **43.** $\frac{25}{6}=4.1\overline{6}$

Chapter 8 Cumulative Review, pages 677 - 679

1. a. –3 ft./sec **b.** –2 ft./sec^2 **2.** $\frac{3}{2}e^{2x}-5\ln|x|+3x^2-x^{-3}+C$ **3.** $2\ln|x|+6x-\frac{5}{4}x^4+C$ **4.** \$78.40 **5.** \$4500 **6.** $5.\overline{51}$

7. \$24,187.50 **8.** $2\sum_{i=1}^{10}(0.5i)e^{-0.5i}$ **9.** $1.\overline{3}$ or $\frac{4}{3}$ **10.** $y=e^{3x^2}$ **11.**

(0.736,10)

12. a. $(x,y)=(5,27)$ **b.** CS = 37.5 **c.** PS = 83.33

13. $y=2e^{x^3}$ **14.** $f(x)=10\ln\left|\frac{x-6}{x+6}\right|+C$

15. 4 **16.** 75.40 or 24π **17.** $k=0.333745$ **18.** $p=\frac{4}{9}$ **19.** $1.1\overline{6}=\frac{7}{6}$ **20.** 61.276 **21.** 2 **22.** 9 **23.** $e-1$ **24.** $2\ln 8-\ln 4-2$ **25.** 1

26. $\frac{1}{2}\left(e^2-1\right)^2$ **27.** $f_x=2,\ f_y=3,\ f_{xx}=f_{yy}=f_{xy}=f_{yx}=0$ **28.** $f_x=2+6x,\ f_y=-3-44y^3,\ f_{xx}=6,\ f_{yy}=-132y^2,$

$f_{xy}=f_{yx}=0$ **29.** $f_x=f_y=f_{xx}=f_{yy}=f_{xy}=f_{yx}=e^{x+y}$

30. $f_x=2xe^{x^2+y^2},\ f_y=2ye^{x^2+y^2},\ f_{xy}=f_{yx}=4xye^{x^2+y^2}\ f_{xx}=2e^{x^2+y^2}+4x^2e^{x^2+y^2}\ ,f_{yy}=2e^{x^2+y^2}+4y^2e^{x^2+y^2}$

31. $f_x = f_y = \frac{1}{3+x+y}$, $f_{xy} = f_{yx} = f_{xx} = f_{yy} = \frac{-1}{(3+x+y)^2}$ **32.** $f_x = 6xy^2$, $f_y = 6x^2y$, $f_{xx} = 6y^2$, $f_{yy} = 6x^2$, $f_{xy} = f_{yx} = 12xy$

33. $f_x = \frac{-2y}{(x-y)^2}$, $f_y = \frac{2x}{(x-y)^2}$, $f_{xx} = \frac{4y}{(x-y)^3}$, $f_{yy} = \frac{4x}{(x-y)^3}$, $f_{xy} = f_{yx} = \frac{-2x-2y}{(x-y)^3}$

34. $f_x = e^{-y}$, $f_y = -xe^{-y}$, $f_{xx} = 0$, $f_{yy} = xe^{-y}$ **35.** local minimum at $(6, 3, -17)$ **36.** local minimum at $(-1, 8.67, -183.33)$

37. local min at $\left(1, \sqrt{2}, 14.686\right)$, local max at $\left(-1, -\sqrt{2}, 45.314\right)$, and saddle points at $\left(-1, \sqrt{2}, 22.69\right)$ and $\left(1, -\sqrt{2}, 37.314\right)$

38. $x = 8.772$, $y = z = 5.848$

Chapter 9

Exercises 9.1, pages 691 - 692

1. $\frac{1}{2}$ **3.** $\frac{1}{\sqrt{2}}$ **5.** $\frac{1}{\sqrt{3}}$ **7.** 0 **9.** -2 **11.** $-\frac{1}{\sqrt{2}}$ **13.** 2.0778 **15.** -0.9143 **17.** -0.6737 **19.** 0.2482

21.

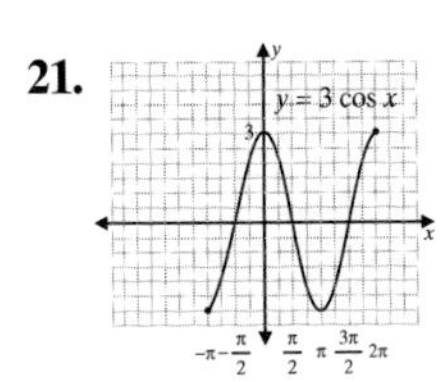

23.

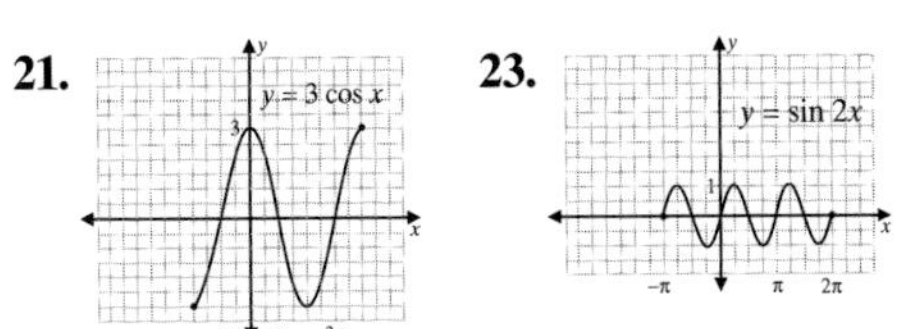

25. Answers may vary **27.** Answers may vary

29. $\sin\theta = \frac{8}{17}$, $\csc\theta = \frac{17}{8}$, $\cos\theta = \frac{15}{17}$, $\sec\theta = \frac{17}{15}$, $\tan\theta = \frac{8}{15}$, $\cot\theta = \frac{15}{8}$

31. 2.233 miles $\approx$ 11,808 ft.

33. a. $N(1) = 40, N(4) = 16, N(7) = 40, N(10) = 16, N(13) = 40$ **b.** 22

Exercises 9.2, pages 699 - 700

1. $-24\sin 4x$ **3.** $-9\sin^2 x\cos x$ **5.** $\frac{1}{x} + 2\sec^2 2x$ **7.** $-8x^2\sin 8x + 2x\cos 8x$ **9.** $\cos^2 x - \sin^2 x$ **11.** $\frac{x\cos x - \sin x}{x^2}$

13. $3e^{3x}\sec^2 e^{3x}$ **15.** $-\frac{5\sin 5x}{2\sqrt{\cos 5x}}$ **17.** $4\sec 4x\tan 4x$ **19.** $2\tan x\sec^2 x$ **21.** 0 **23.** $-3e^{\cos x}\sin x$ **25.** $\frac{\cos x}{\sin x} = \cot x$ **27.** $\frac{1}{1+\cos x}$

29. $\frac{\partial f}{\partial x} = 4 + y\cos xy$, $\frac{\partial f}{\partial y} = x\cos xy$ **31.** $\frac{\partial f}{\partial x} = ye^{xy} + y\cos x$, $\frac{\partial f}{\partial y} = xe^{xy} + \sin x$

33. $\frac{\partial f}{\partial x} = -15x^3\sin(3x+y) + 15x^2\cos(3x+y)$, $\frac{\partial f}{\partial y} = -5x^3\sin(3x+y)$ **35.** Answers will vary **37.** Answers will vary

39. $y = -\frac{\pi}{3}x + \frac{2\pi}{3} + \sqrt{3}$ **41. a.** \$9800 **b.** 0 **c.** The monthly revenue is not changing for September.

Exercises 9.3, pages 707 - 708

1. $\frac{1}{4}\sin 4x + C$ **3.** $\frac{1}{4}\tan 4x + C$ **5.** $-\frac{1}{7}\ln\left|\cos(7x+1)\right| + C$ **7.** $-\frac{1}{2}\cos x^2 + C$ **9.** $\frac{3}{4}$ **11.** 2 **13.** $\frac{1}{8}$ **15.** $\sin e^x + C$

17. $-e^{\cos x} + C$ **19.** $-2\ln\left|\cos\sqrt{x}\right| + C$ **21.** $-\cos(\ln x) + C$ **23.** $\ln 1.5 \approx 0.4055$ **25.** $-\frac{1}{6}$ **27.** $\ln\left|\sin x\right| + C$

29. $\frac{1}{2}x - \frac{1}{4}\sin 2x + C$ **31.** $\frac{1}{2}x\sin 2x + \frac{1}{4}\cos 2x + C$ **33.** $x\tan x + \ln\left|\cos x\right| + C$ **35.** 6 **37.** $\sqrt{2} - \frac{1}{2} \approx 0.9142$

39. $P(t) = 1700 - 800\cos\left[\frac{\pi}{12}(t+4)\right]$ **41.** 19.74 cubic units

Exercises 9.4, pages 715 - 717

1. 60° **3.** −60° **5.** $\frac{3\pi}{4}$ **7.** $\frac{\pi}{6}$ **9.** −0.63 **11.** 0.62 **13.** −0.72 **15.** $\frac{-2x}{\sqrt{1-x^4}}$ **17.** $\frac{1}{2\sqrt{x-x^2}}$ **19.** $\frac{e^x}{1+e^{2x}}$ **21.** $\frac{-2}{\sqrt{1-(2x+1)^2}}$ **23.** $\frac{-3}{\sqrt{1-(1-3x)^2}}$ **25.** $\frac{1}{2}\tan^{-1}2x+C$ **27.** $\tan^{-1}e^x+C$ **29.** $2\sin^{-1}\sqrt{x}+C$ **31.** $\frac{\pi}{12}$ **33.** $\frac{\pi}{2}$ **35. a.** −2.05 minutes per item **b.** The time required to produce each item is decreasing at a rate of 2.05 minutes per item. **37.** 58.5%

Chapter 9 Test, pages 721 - 723

1. $-\frac{1}{2}$ **2.** $-\frac{1}{\sqrt{3}}$ **3.** 1 **4.** $-\sqrt{2}$ **5.** 0.7524 **6.** 0.5867 **7.** −1.4843 **8.** −2.7328 **9.** **10.**

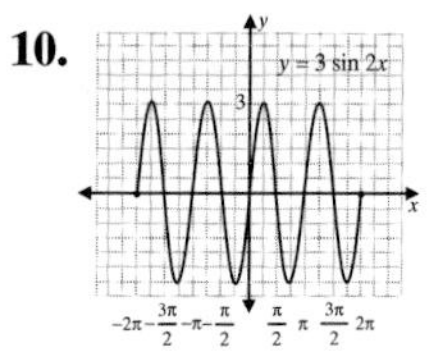

11. Answers will vary **12.** Answers will vary

13. $\sin\theta=\frac{9}{16}$, $\cos\theta=\frac{5\sqrt{7}}{16}$, $\tan\theta=\frac{9}{5\sqrt{7}}$, $\cot\theta=\frac{5\sqrt{7}}{9}$, $\sec\theta=\frac{16}{5\sqrt{7}}$, $\csc\theta=\frac{16}{9}$

14. 15.3 ft. **15. a.** 690, 756, 780, 600 **b.** 3764, 3610, 3400, 2980 **16.** $4\pi\cos\pi x$ **17.** $2\sec^2 xe^{\tan x}$ **18.** $15x^2\cos 4x-20x^3\sin 4x$ **19.** $-6\csc^2 3x\cot 3x$ **20.** $-\frac{\sin x}{\cos x}=-\tan x$ **21.** $-\frac{\sin 2x}{\sqrt{1+\cos 2x}}$ **22.** $\frac{1}{1+\cos x}$ **23.** $\frac{\partial f}{\partial x}=y\cos xy+e^y$, $\frac{\partial f}{\partial y}=x\cos xy+xe^y$

24. $\frac{\partial f}{\partial x}=\tan y+\frac{y}{x}$, $\frac{\partial f}{\partial y}=x\sec^2 y+1+\ln xy$ **25.** $y=-\frac{\pi}{12}x+\frac{\pi}{3}+\frac{\sqrt{3}}{2}$ **26. a.** 13.4 hours **b.** 0.0001 **c.** The number of hours of daylight is not changing significantly on day 274 **27.** $-\frac{1}{2}\cos(x^2-1)+C$ **28.** $-\frac{2}{(2+\sin x)^{\frac{1}{2}}}+C$ **29.** $\frac{1}{3}(e-1)\approx 0.5728$ **30.** 1

31. $-\frac{2}{3}(\cos x+1)^{\frac{3}{2}}+C$ **32.** $\ln|1+\tan x|+C$ **33.** $-8x\cos\frac{x}{2}+16\sin\frac{x}{2}+C$ **34.** $\frac{1}{4}x\tan 4x+\frac{1}{16}\ln|\cos 4x|+C$

35. $\frac{\pi^2}{18}+\frac{1}{2}\approx 1.0483$ **36.** 624 deer **37.** −45° **38.** 0° **39.** $-\frac{\pi}{4}$ **40.** $\frac{\pi}{6}$ **41.** 1.06 **42.** −0.85 **43.** $\frac{4}{\sqrt{1-16x^2}}$

44. $\frac{1}{x\left[1+(\ln|x|)^2\right]}$ **45.** $\frac{\cos x}{1+\sin^2 x}$ **46.** $\frac{-4e^{4x}}{\sqrt{1-e^{8x}}}$ **47.** $\sin^{-1}x^2+C$ **48.** $\frac{1}{2}\sin^{-1}e^{2x}+C$ **49.** $-\tan^{-1}(\cos x)+C$

50. $5\tan^{-1}2\sqrt{x}+C$

Chapter 9 Cumulative Review, pages 723 - 725

1. $f'=2x\cos x+2\sin x$ **2.** $f'=\frac{(5+2x)(-\sin x)}{(5+2x)^2}-\frac{(3+\cos x)(2)}{(5+2x)^2}$ **3.** $\frac{1}{2}(\cos x)(5+\sin x)^{-\frac{1}{2}}$ **4.** $f'=8(200+8x-3x^2)^7(8-6x)$

5. $f'=-(1-2x)(3+x-x^2)^{-2}=\frac{2x-1}{(3+x-x^2)^2}$ **6.** $f'=e^x(\sin 2x+2\cos 2x)$ **7.** $e^3\cos(1)-1$ **8.** $e-\frac{3}{2}$

9. $(x,y)=(3.925, 1.367)$ **10.** $T=\pi$ **11.** $(x_0,y_0)=\left(\frac{\pi}{4},2\right)$ **12.** $(x_1,y_1)=\left(\frac{\pi}{2},1\right)$

13. 1.285 or $\dfrac{\pi+2}{4}$ **14. a.** $x_0=\dfrac{-b}{3a}$ **b.** If $\dfrac{-b}{3a}$ is negative, y'' changes from negative to positive at $x=\dfrac{-b}{3a}$ and y changes from concave down to concave up. If $\dfrac{-b}{3a}$ is positive, then y changes from concave up to down. In either case, $x=\dfrac{-b}{3a}$ gives an inflection point. **15.** $C(x)=2x^2+\dfrac{x^3}{300}+10$ and $C(6)=82.72$ **16.** $\sqrt{82}=9+\dfrac{1}{18}=\dfrac{163}{18}$ **17.** 2.377 hrs.

18. a. 2 sec. **b.** 104 ft. **19. a.** $x=6,\ y=18$ **b.** $CS=\$36$ **c.** $PS=\$72$ **20.** 5.622 **21.** $k=\dfrac{1}{72}$ **22.** $\bar{x}=\dfrac{26}{9}$

23. $p=0.1\overline{6}=\dfrac{1}{6}$ **24.** $\dfrac{2}{\pi}$ **25.** $\dfrac{\sin^4 x}{4}+C$ **26.** $\dfrac{-\cos^5 x}{5}+C$ **27.** $\dfrac{1}{2}x+\dfrac{1}{4}\sin 2x+C$ **28.** $\dfrac{1}{2}x-\dfrac{1}{4}\sin 2x+C$

29. $f_x=e^x\sin 2xy+2ye^x\cos 2xy$ and $f_y=2xe^x\cos 2xy$ **30.** $\dfrac{1}{2}\arctan\left(\dfrac{3x}{4}\right)+C$

Chapter 10

Exercises 10.1, pages 739 - 741

1. $\left\{-1,\dfrac{1}{2},-\dfrac{1}{3},\dfrac{1}{4},-\dfrac{1}{5}\right\}$ **3.** $\left\{\sqrt{3},\sqrt{5},\sqrt{7},\sqrt{9},\sqrt{11}\right\}$ **5.** $\{0,1,2,3,4\}$ **7.** $f(n)=(-1)^{n+1}(2n+1)$ **9.** $f(n)=5n$ **11.** $f(n)=n^2+1$

13. 165 **15.** 25 **17.** 3 **19.** $\sqrt{2}+2$ **21.** $-\dfrac{2}{3}$ **23.** Diverges **25.** $3.\overline{63}$ **27. a.** $\sum_{j=0}^{\infty} jx^{j-1}$ **b.** $G'(x)=\dfrac{1}{(1-x)^2}$ **c.** Both the expansions $f'(x)$ and $G'(x)$ are same. $f'(0.25)=G'(0.25)=1.\overline{7}$; $f'(-0.25)=G'(-0.25)=0.64$; $f'(-0.5)=G'(-0.5)=0.\overline{4}$; $f'(0.5)=G'(0.5)=4$. **29. a.** 1, 4, 9, 16, 25, 36, 49, 64, 81, 100 **b.** $\sum_{n=1}^{100}(2n-1)=100^2=10{,}000$

31. a. $\dfrac{43}{99}$ **b.** **33.** 14 milligrams

Exercises 10.2, pages 748 - 749

1. $\cos x\approx P_6(x)=1-\dfrac{x^2}{2!}+\dfrac{x^4}{4!}-\dfrac{x^6}{6!}$

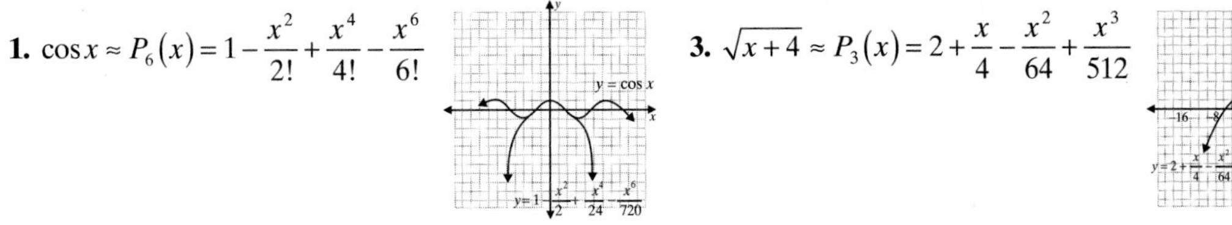

3. $\sqrt{x+4}\approx P_3(x)=2+\dfrac{x}{4}-\dfrac{x^2}{64}+\dfrac{x^3}{512}$

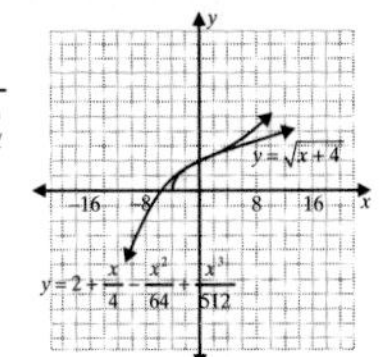

5. $\ln(1+x)\approx P_3(x)=x-\dfrac{x^2}{2}+\dfrac{x^3}{3}$

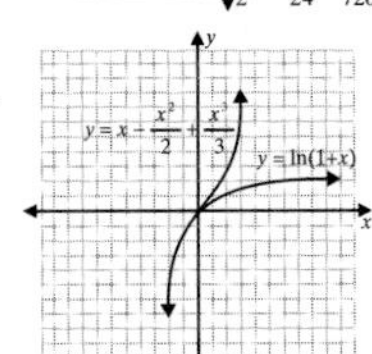

7. $e^{x+1}\approx P_3(x)=e\left(1+x+\dfrac{x^2}{2!}+\dfrac{x^3}{3!}\right)$

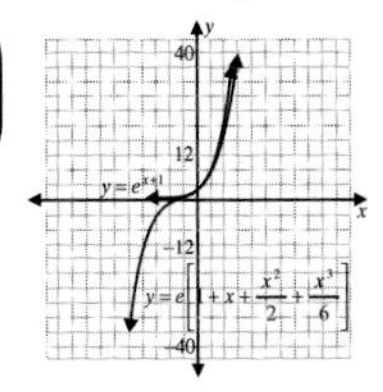

9. $e^x\approx P_4(x)=e\left(1+(x-1)+\dfrac{(x-1)^2}{2!}+\dfrac{(x-1)^3}{3!}+\dfrac{(x-1)^4}{4!}\right)$ **11.** $\cos x\approx P_6(x)=-1+\dfrac{(x-\pi)^2}{2!}-\dfrac{(x-\pi)^4}{4!}+\dfrac{(x-\pi)^6}{6!}$

13. $\sqrt{x}\approx P_4(x)=1+\dfrac{1}{2}(x-1)-\dfrac{1}{4}\dfrac{(x-1)^2}{2!}+\dfrac{3}{8}\dfrac{(x-1)^3}{3!}-\dfrac{15}{16}\dfrac{(x-1)^4}{4!}$

15. $\frac{1}{1-x} \approx P_4(x) = 2 + 4(x-0.5) + 8(x-0.5)^2 + 16(x-0.5)^3 + 32(x-0.5)^4$ **17.** 2.21 **19.** $\sqrt{2} \approx 1.4375$

21. $\sqrt[4]{0.5} \approx 0.8447$ **23.** $\sqrt[3]{27.5} \approx 3.0184$ **25.** answers will vary **27.** $\frac{d}{dt}(\sin t) = \cos t \approx 1 - \frac{x^2}{2!} + \frac{x^4}{4!} - \frac{x^6}{6!}$. The derivative of the Taylor polynomial is the same as the Taylor polynomial of the derivative.

29. $\cos^2 t \approx P_3(x) = 1 - x^2 + \frac{x^4}{3} - \frac{2x^6}{45}$

Exercises 10.3, pages 760 - 763

1. $y = \sum_{n=0}^{\infty} \frac{1}{n!} x^{n+2}$ **3. a.** $y = \sum_{n=0}^{\infty} (-1)^n t^n$ **b.** $y = \sum_{n=0}^{\infty} (-1)^n t^{2n}$ **c.** $y = \sum_{n=0}^{\infty} (-1)^n t^{n+1}$ **d.** $y = \sum_{n=0}^{\infty} (-1)^n t^{2n+1}$ **5. a.** 0.6931

b. $-\ln\frac{1}{2} \approx 0.6931$ **c.** $0.5 + 0.125 + 0.0417 + 0.0156 + 0.00625 + \ldots \approx 0.6931$ **d.** $\sum_{n=1}^{\infty} \frac{t^n}{n}\Big]_0^{\frac{1}{2}} \approx 0.6931$

7. a. $y = 1 - \frac{x}{1!} + \frac{x^2}{2!} - \frac{x^3}{3!} + \ldots + \frac{(-1)^n x^n}{n!} + \ldots$ **b.** $y = 1 - \frac{x^2}{1!} + \frac{x^4}{2!} - \frac{x^6}{3!} + \ldots + \frac{(-1)^n x^{2n}}{n!} + \ldots$

c. $y = 1 + \frac{7x}{1!} + \frac{(7x)^2}{2!} + \frac{(7x)^3}{3!} + \ldots + \frac{(7x)^n}{n!} + \ldots$ **d.** $y = x - \frac{x^2}{1!} + \frac{x^3}{2!} - \frac{x^4}{3!} + \ldots \frac{(-1)^n x^{n+1}}{n!} + \ldots$ **9. a.** $\cosh(t) = 1 + \frac{t^2}{2!} + \frac{t^4}{4!}$

b. $\cosh(t) = 1 + \frac{t^2}{2!} + \frac{t^4}{4!} + \frac{t^6}{6!} + \ldots$

11. a. $g'(x) = \frac{2x}{(1-x^2)^2}, g''(x) = \frac{2(1+3x^2)}{(1-x^2)^3}, g'''(x) = \frac{24x(1+x^2)}{(1-x^2)^4}, g^{(4)}(x) = \frac{24(1+10x^2+5x^4)}{(1-x^2)^5}$ **b.** $P_4(x) = 1 + x^2 + x^4$

c. $g(x) = \frac{1}{1-x^2} = 1 + x^2 + x^4 + \ldots$ **13.** All even coefficients are zero. **15. a.** $\frac{1}{(1-x)^2}$ **b.** $f'(x) = 1 + 2x + 3x^2 + 4x^3 + \ldots$

c. $\frac{1}{(1-x)^2} = \sum_{n=0}^{\infty} (n+1)x^n$ **17.** $\lim_{t\to 0}\left(\frac{\sin t}{t}\right) = 1$ **19.** Answers will vary **21. a.** $f(x) = \frac{x^3}{1-x}$ **b.** $f(x) = \frac{1}{(1-x)^2}$

c. $f(x) = \frac{1}{1-x} - x^3$ **d.** $f(x) = (1+x)^6$ **23.** $f(x) \approx P_4(x) = 1 + 2x + x^2 + \frac{1}{3}x^3 + \frac{1}{12}x^4$

25. $f(x) \approx P_5(x) = 1 + x - \frac{1}{6}x^3 + \frac{1}{120}x^5$

Chapter 10 Test, page 766

1. $a_1 = \frac{2}{3}, a_2 = \frac{4}{3}, a_3 = 2, a_4 = \frac{8}{3}, a_5 = \frac{10}{3}$ **2.** $a_1 = 0, a_2 = \frac{1}{2}, a_3 = \frac{2}{3}, a_4 = \frac{3}{4}, a_5 = \frac{4}{5}$ **3.** $a_1 = 1, a_2 = 3, a_3 = 7, a_4 = 15, a_5 = 31$

4. $a_1 = 1, a_2 = 0, a_3 = -1, a_4 = 0, a_5 = 1$ **5.** 1 **6.** no limit **7.** $\frac{17}{18}$ **8.** 2 **9.** $\frac{7}{6}$ **10.** $\frac{10}{3}$ **11.** $\frac{11}{4}$ **12.** $\sqrt{2}$ **13.** $\frac{17}{99}$

14. 90 milligrams **15.** $y = (e+1)x - 1$ **16.** $T_3(x) = 1 + x + x^2 + x^3$ **17.** $f(83) \approx 9 + \frac{1}{9} - \frac{1}{1458}$ **18.** $T(x) = 1 - \frac{1}{3!}x^2 + \frac{1}{5!}x^4$

19. $P(x) = \sqrt{2} + \frac{\sqrt{2}}{4}x - \frac{\sqrt{2}}{32}x^2 + \frac{\sqrt{2}}{128}x^3$ **20.** $1 + x^3 + x^6 + x^9 + x^{12} + \ldots + x^{3k} + \ldots$ **21.** $\cos\left(\frac{2\pi k}{7}\right) + i\sin\left(\frac{2\pi k}{7}\right)$ for $k = 0, 1, \ldots, 6$

22. $\frac{3-2x}{(1-x)^2}$ **23.** $A_{2n} = \frac{1}{2^n \cdot n!}, A_{2n+1} = 0$, or $f(x) = \sum_{k=0}^{\infty} \frac{x^{2k}}{2^k k!}$

Chapter 10 Cumulative Review, pages 767 - 768

1. $f' = 2\pi\cos\left(2\pi x - \frac{\pi}{4}\right)$ **2.** $f' = e^{2x}(2\sin 3x + 3\cos 3x)$ **3.** $f' = \sum_{k=1}^{\infty} \frac{k(-1)^k x^{k-1}}{2^k}$ **4.** $\frac{(1+\sin x)\sec^2 x - \sin x}{(1+\sin x)^2}$

5. $\ln 3 \approx 1.0986$ **6.** $P_5(x) = (x-1) + \frac{(x-1)^2}{2} - \frac{(x-1)^3}{6} + \frac{(x-1)^4}{12} - \frac{(x-1)^5}{20}$ **7.** -0.25 **8.** $-\frac{1}{2}\cos(x^2-2)+C$

9. $y' = \frac{4}{\sqrt{1-16x^2}}$ **10.** $y' = 2\tan x \sec^2 x$ **11.** $\frac{1}{2}\ln(x^2+1)+C$ **12.** $\frac{1}{6}(x^2+1)^3+C$ **13.** $\arctan x + C$ **14.** $\frac{-1}{4(x^2+1)^2}+C$

15. 0.83 **16.** $e-1$ **17.** $\frac{e^{-\pi}+1}{2} \approx 0.52$ **18.** Answers may vary. **19.** 1.6 degrees per hour **20.** $P = \frac{6000}{1+19e^{-0.225t}}$ **21.** $2+0.5\pi^2$

22. $x = \frac{\pi}{2}$ and $y = 60$, at the first inflection point. **23.** $x = 62.8$; on the 63rd day the date was Sept. 1. **24.** $y = -\pi x + \pi^2$

25. yes **26.** $\frac{1}{e^{\pi}+1} = \frac{1}{24.14}$ **27.** $V = 2\pi$

Index

A

B

C

D

E

F

M

N

O

P

Q

R

S

T

U

V

X

Y

Z